Pollution Control in Chemical and Allied Industries

With Focus on Air and Water Pollution

Pollution Control in Chemical and Allied Industries

With Focus on Air and Water Pollution

N Hanley

CBS Publishers & Distributors Pvt Ltd

New Delhi • Bengaluru • Chennai • Kochi • Kolkata • Mumbai
Bhopal • Bhubaneswar • Hyderabad • Jharkhand • Nagpur • Patna • Pune • Uttarakhand • Dhaka (Bangladesh)

Disclaimer

Science and technology are constantly changing fields. New research and experience broaden the scope of information and knowledge. The author has tried his best in giving information available to him while preparing the material for this book. Although, all efforts have been made to ensure optimum accuracy of the material, yet it is quite possible some errors might have been left uncorrected. The publisher, the printer, and the author will not be held responsible for any inadvertent errors, omissions or inaccuracies.

Pollution Control in Chemical and Allied Industries

ISBN: 978-81-239-2878-4

Copyright © Publisher

First Edition: 2016
Reprint: 2019

All rights reserved. No part of this book may be reproduced or transmitted in any form or by any means, electronic or mechanical, including photocopying, recording, or any information storage and retrieval system without permission, in writing, from the author and the publisher.

Published by Satish Kumar Jain and produced by Varun Jain for
CBS Publishers & Distributors Pvt Ltd
4819/XI Prahlad Street, 24 Ansari Road, Daryaganj, New Delhi 110 002, India.
Ph: 23289259, 23266861, 23266867   Website: www.cbspd.com
Fax: 011-23243014   e-mail: delhi@cbspd.com; cbspubs@airtelmail.in.
Corporate Office: 204 FIE, Industrial Area, Patparganj, Delhi 110 092
Ph: 4934 4934   Fax: 4934 4935   e-mail: publishing@cbspd.com; publicity@cbspd.com

Branches

- **Bengaluru:** Seema House 2975, 17th Cross, K.R. Road, Banasankari 2nd Stage, Bengaluru 560 070, Karnataka, India.
  Ph: +91-80-26771678/79   Fax: +91-80-26771680   e-mail: bangalore@cbspd.com
- **Chennai:** 7, Subbaraya Street, Shenoy Nagar, Chennai 600 030, Tamil Nadu, India.
  Ph: +91-44-26680620, 26681266   Fax: +91-44-42032115   e-mail: chennai@cbspd.com
- **Kochi:** 42/1325, 1326, Power House Road, Opp. KSEB Power House, Ernakulam 682 018, Kochi, Kerala, India.
  Ph: +91-484-4059061-65   Fax: +91-484-4059065   e-mail: kochi@cbspd.com
- **Kolkata:** 6/B, Ground Floor, Rameswar Shaw Road, Kolkata-700 014, West Bengal, India.
  Ph: +91-33-22891126, 22891127, 22891128   e-mail: kolkata@cbspd.com
- **Mumbai:** 83-C, Dr E Moses Road, Worli, Mumbai-400018, Maharashtra, India.
  Ph: +91-22-24902340/41   Fax: +91-22-24902342   e-mail: mumbai@cbspd.com

Representatives

• Bhopal	0-8319310552	• Bhubaneswar	0-9911037372	• Hyderabad	0-9885175004	• Jharkhand	0-9811541605
• Nagpur	0-9021734563	• Patna	0-9334159340	• Pune	0-9623451994	• Uttarakhand	0-9716462459
• Dhaka (Bangladesh)	01912-003485						

Printed at India Binding House, Noida, UP

Preface

Chemical processes provide a diverse array of valuable products and materials used in applications ranging from health care to transportation and food processing. Yet these same chemical processes that provide products and materials essential to modern economies also generate substantial quantities of wastes and emissions. Managing these wastes costs billions of dollars each year, and as emission and treatment standards continue to become more stringent, these costs will continue to escalate. In the face of rising costs and increasingly stringent performance standards, traditional end-of-pipe approaches to waste management have become less attractive and a strategy variously know as *environmentally conscious manufacturing*, *eco-efficient production*, or *pollution prevention* has been gaining prominence. The basic premise of this strategy is that avoiding the generation of wastes or pollutants can often be more cost effective and better for the environment than controlling or disposing of pollutants once they are formed.

The intent of this reference textbook is to describe environmentally preferable or 'green' approaches to the design and development of processes and products.

Section I is devoted to general concepts and engineering considerations. Chapter 1 deals with introduction to environmental issues which have increasingly important implications for enterprises and other organisations. Chapter 2 explains risk concepts, risk assessment is a systematic analytical method used to determine the probability of adverse effects. A common application of risk assessment methods is to evaluate human health and ecological impacts of chemical releases to the environment. Chapter 3 focuses on role and responsibilities of chemical engineers. This chapter discusses that chemical engineers should be aware of their responsibilities to general public, colleagues, employers, environment and also to their profession. This can be achieved by inducing approaches to design safe chemical processes. Chapter 4 is devoted to evaluating exposures and introduces readers to some methods of predicting potential exposure, in particular, occupational exposure and community exposure. Chapter 5 focuses on unit operations and pollution prevention. Chapter 6 deals with green chemistry for sustainable future. Green chemistry is the utilisation of a set of principles that reduces or eliminates the use or generation of hazardous substances in the design, manufacture and application of chemical products. Chapter 7 explains the life-cycle concepts and green engineering. Life-cycle techniques have been adopted in industry to serve a variety of purposes, including product comparison, strategic planning, environmental labelling, and product design and improvement. Chapter 8 is devoted to industrial ecology. The phrase 'Industrial Ecology' evokes powerful images and strong reactions, both positive and negative. To some, the phrase conjures images of industrial systems that mimic the mass conservation properties of natural ecosystems. Chapter 9 concentrates on environmental care in manufacture of chemicals. Chapter 10 deals with air pollution control and is defined as the reducing of sulphur dioxide, nitrogen oxide and particulate matter. Chapter 11 is devoted to water for safe use in chemical and allied industries. Chapter 12 focuses on physical, thermal and chemical methods of waste management. The effect of chemical waste on the environment are reflected by the effects on organisms and effects on the overall ecosystem. Various methods of treatment of these wastes are discussed in detail. Chapter 13 focuses on zero waste for sustainable development. The concept of zero waste seeks to reduce quantities and toxicities of materials used and wastes generated. The zero waste approach can only be sustainable in the long run, since it reduces resources consumption, increases efficiencies, eliminate toxic products and ensure that products are produced to be safely used, reused, recovered and recycled back or disposed into nature in an environment-friendly manner.

Section II is devoted to pollution control in chemical industries. Chemical process industries over the years have developed several methodologies aimed at early detection of hazards and assessment of consequences of an untoward accidental release of hazardous materials in order to effectively attenuate residual risks. The comprehensive criteria developed for the acceptability of risks provides a satisfactory means of evaluating the technology at various stages of inplementation. Some of the polluting chemical industries are discussed in this section. Chapter 14 deals with petroleum refinery and petrochemicals. Various methods of treatments of waste-water and refinery effluents are discussed in detail. Chapter 15 explains the processes used in tanneries, the characteristics and effects of waste generated and methods of pollution control. Chapter 16 focuses on pulp and paper which contributes lot towards the pollution of our water environment. Disposal of wastes by irrigation and land treatment methods are highlighted in this chapter. Chapter 17 is devoted to soap and detergents which are formulated products designed to meet various cost and performance standards. Waste-water control, treatment and recycling operations are discussed. Chapter 18 deals with pesticides which has been identified as one of the highly polluting industries needing pollution control on top priority. The possible sources of pollutants are raw materials used in pesticide synthesis in excess of their stoichiometric requirements and solvents used as an extractive medium. Methods of recycling and reuse of waste-water within the industry has been discussed which help minimise fresh water requirements. Chapter 19 concentrates on fertiliser which discharge variety of wastes as water pollutants in the form of processing chemical such as sulphuric acid, phosphoric acid, ammonium sulphate. Various methods of waste-water treatment are discussed. Chapter 20 is devoted to paint and dyes which uses varied raw material such as resins, solvents, drying oils, pigments and extenders. Characteristics of waste-water and effluents control methods are discussed. Chapter 21 focuses on drugs and pharmaceuticals which discharge highly alkaline and toxic wastes. Air emissions are also discharge from reactors, distillation, crystallisers, dryers and during storage operations. Treatment methods of emissions and waste-water utilisation are discussed. Chapter 22 focuses on sugar, distillery and fermentation products. The waste generated includes water used in splashes to extract the maximum amount of juice and thus waste-water has high BOD due to presence of sugar and oil from machineries. Chapter 23 is devoted to miscellaneous process industries such as caustic soda, sodium carbonate, nitric acid, phosphoric acid and ammonium compounds. Various waste disposal and treatment techniques are discussed. Chapter 24 deals with textile which represents a range of industries with operations and processes as diverse as its products. Various physical, chemical and biological treatment methods of waste-water are discussed.

Section III focuses on pollution control in food processing industries. Effluents from food and allied industries are considered to be close to domestic sewage, since they mainly contain biodegradable material and suspended solids. The large volumes of effluents, seasonal variation, high BOD, high suspended solids, presence of fats, oil and grease, solid wastes are typical of food industry wastes. Microbial processes are used to tackle food wastes.

Chapter 25 is devoted to food processing industry: a review. Chapter 26 and 27 deals with dairy and bakery. Various methods of treatment of wastes and utilisation of bioproducts are discussed. Chapter 28 concentrates on seafood processing and meat products. Economical methods of treatment of seafood processing wastes are discussed.

Section IV deals with special topics. Chapter 29 explains biotechnology for pollution prevention in chemical industry. For removing specific types of pollutants selectively, new types of proteins, peptides and biomimetic adsorbers are being developed using biotechnology. For proper understanding of our environment, we must have a clear idea of the identities and quantities of pollutants and other chemical species in air, water, soil and biological samples, etc. Keeping this in mind chapter 30 focuses on instrumental techniques for environmental pollution monitoring. Various instrumental techniques such as spectrophotometric methods, atomic emission, gas chromatography, etc. are discussed in detail. Chapter 31 deals with environmental management in chemical process industries. The objective of environmental management is to integrate environmental considerations into decision making at all levels. To achieve this, steps should be taken to prevent pollution at source, encourage, develop and apply the best available technical solutions, ensure that the polluter pays for pollution, and involve the public in decision making.

Diagrams, figures, tables and index supplement the text. All the topics have been covered in a cogent and lucid style to help the reader grasp the information quickly and easily.

It may not be wrong to hold that the present reference textbook of *Pollution Control in Chemical and Allied Industries* is essential reading for all students and teachers of engineering, environment and life sciences. The researchers and consultants in environmental field with special emphasis on chemical industries, chemists and industrialists will also find it highly useful and informative.

It has been prepared with meticulous care, aiming at making the book error-free. Constructive suggestions are always welcome from users of this book.

N Hanley

Contents at a Glance

Preface v

SECTION I

General Concepts and Engineering Considerations 1–252

1. Introduction to Environmental Issues — 3–19
2. Risk Concepts — 20–35
3. Roles and Responsibilities of Chemical Engineers — 36–42
4. Evaluating Exposures — 43–59
5. Unit Operations and Pollution Prevention — 60–95
6. Green Chemistry for a Sustainable Future — 96–124
7. Life-cycle Concepts and Green Engineering — 125–151
8. Industrial Ecology — 152–160
9. Environmental Care in Manufacture of Chemicals — 161–175
10. Air Pollution Control — 176–188
11. Water for Safe Use in Chemical and Allied Industries — 189–208
12. Physical, Thermal and Chemical Methods of Waste Management — 209–233
13. Zero Waste for Sustainable Development — 234–252

SECTION II

Pollution Control in Chemical Industries 253–540

14. Petroleum Refinery and Petrochemicals — 255–296
15. Leather and Tannery — 297–339
16. Pulp and Paper — 340–367
17. Soap and Detergent — 368–386
18. Pesticide — 387–408
19. Fertiliser — 409–435
20. Paints and Dyes — 436–455
21. Drugs and Pharmaceuticals — 456–480
22. Sugar, Distillery and Fermentation Products — 481–502
23. Miscellaneous Process Industries — 503–528
24. Textile — 529–540

SECTION III

Pollution Control in Food Processing Industries — 541–576

25. Food Processing Industry: A Review — 543–546
26. Dairy — 547–554
27. Bakery — 555–561
28. Seafood Processing and Meat Products — 562–576

SECTION IV

Special Topics — 577–619

29. Biotechnology for Pollution Prevention in Chemical Industry — 579–585
30. Instrumental Techniques for Environmental Pollution Monitoring — 586–604
31. Environmental Management in the Chemical Process Industries — 605–619

References — 621
Index — 623–630

Contents

Preface   v
Contents at a Glance   ix

SECTION I

General Concepts and Engineering Considerations   1–252

1. **Introduction to Environmental Issues**   3–19

   Introduction   3
   Role of Chemical Processes and Chemical Products   4
       Formulation of an Industrial Cleaner   4
       Formulation of a Paint Solvent   5
       Choice of Refrigerant for a Low-temperature Condenser   5
   Overview of Major Environmental Issues   6
   Global Environmental Issues   6
       Global Energy Issues   6
       Global Warming   7
       Photochemical Smog   8
       Greenhouse Effect   8
       Ozone Depletion in the Stratosphere   11
   Air Quality Issues   12
       Criteria Air Pollutants   13
       Air Toxics   17
   Water Quality Issues   17
   Ecology   18
   Natural Resources   18

2. **Risk Concepts**   20–35

   Introduction   20
   Description of Risk   20
       Voluntary Risk   20
       Natural Disasters   21
       Involuntary Risk   21
   Value or Risk Assessment in the Engineering Profession   22
   General Overview of Risk Assessment Concepts   23
       Hazard Assessment   23
       Dose-Response   23
       Exposure Assessment   23
       Risk Characterisation   23

	Hazard Assessment	24
	Cancer and Other Toxic Effects	24
	Hazard Assessment for Cancer	25
	Hazard Assessment for Non-cancer Endpoints	25
	Structure Activity Relationships (SAR)	26
	Dose-Response Curve	27
	Exposure Assessment	28
	Risk Characterisation	29
	Risk Characterisation of Cancer Endpoints	29
	Risk Characterisation of Non-Cancer Endpoints	30
	Adding Risks	30
	Sources of Ignition in Hazardous Area and How to Control Them	30
	Type of Hazards	31
	Hazard Operability	31
	Sources of Ignition	33
	Controlling Ignition Sources	33
	Check List for Saftey Assessor	34

3. Roles and Responsibilities of Chemical Engineers   36–42

Introduction	36
Responsibilities for Chemical Process Safety	37
Major Air Pollution Episodes	37
Responsibilities of Chemical Engineers for Environmental Protection	40
Engineering Ethics	42

4. Evaluating Exposures   43–59

Introduction	43
Occupational Exposures: Recognition, Evaluation and Control	44
PEL (Permissible Exposure Limits)	44
Characterisation of the Workplace	45
Exposure Pathways	46
Monitoring Worker Exposure	47
Modelling Inhalation Exposures	49
Dispersion Models	51
Assessing Dermal Exposures	51
Exposure Assessment for Chemicals in the Ambient Environment	53
Exposure to Toxic Air Pollutants	53
Dermal Exposure to Chemicals in the Ambient Environment	54
Effect of Chemical Releases to Surface Waters on Aquatic Biota	54
Ground Water Contamination	55
Designing Safer Chemicals	57
Reducing Dose	57
Reducing Toxicity	58

5. Unit Operations and Pollution Prevention   60–95

Introduction	60
Pollution Prevention in Material Selection for Unit Operations	61

Pollution Prevention for Chemical Reactors	63
Material Use and Selection for Reactors	65
Reaction Type and Reactor Choice	68
Reactor Operation	72
Pollution Prevention for Separation Devices	77
Choice of Mass Separating Agent	78
Process Design and Operation Heuristics for Separation Technologies	78
Pollution Prevention Examples for Separations	82
Separators with Reactors for Pollution Prevention	84
Pollution Prevention Applications for Separative Reactors	84
Pollution Prevention in Storage Tanks and Fugitive Sources	87
Storage Tank Pollution Prevention	87
Reducing Emissions from Fugitive Sources	88
Pollution Prevention Assessment Integrated with HAZ-OP Analysis	91
Integrating Risk Assessment with Process Design — A Case Study	93
Acrylonitrile Reactor Risk-based Input-Output Analysis of a Reactor	93

6. Green Chemistry for a Sustainable Future   96–124

Introduction	96
Need for Green Chemistry	96
Emergence of Green Chemistry	97
Breakthroughs and Success Stories of Green Chemistry in Indian Industries	97
Case Studies	97
Cleaner Ranitidine Process	97
Recycle of Mother Liquor	98
Future	98
Important Issues in Green Chemistry and Examples of Green Processes	98
Principles	99
Supramolecular Chemistry	99
Supercritical Water	99
Nanocatalysts in Green Chemistry	100
Clean Hydrogen Peroxide Synthesis via a Nanocatalyst Process	101
Case Studies	101
Case Study I	101
Background	101
Conventional Synthetic Chemistry Based Process	102
Green Chemistry Based Process	102
Principles of Green Chemistry Followed	103
Case Study II	104
Background	104
Conventional Synthetic Chemistry Based Process	104
Green Chemistry Based Process	104
Principles of Green Chemistry	106
Principles of Green Chemistry	106
Prevention	106
Atom Economy	106
Less Hazardous Chemical Syntheses	107

Designing Safer Chemicals	107
Safer Solvents and Auxiliaries	107
Design for Energy Efficiency	107
Use of Renewable Feedstocks	107
Reduce Derivatives	107
Catalysis	107
Design for Degradation	107
Real-time Analysis for Pollution Prevention	107
Inherently Safer Chemistry for Accident Prevention	107
Widening the Scope of Green Technologies through Chemical Engineering	108
Sources of Pollution	108
Selectivity Engineering Principle	109
Green Chemistry to Produce Amines	118
Catalyst Does the Trick	119
Atom-efficient Processes	119
Greener Routes to Propylene Oxide — An Overview	119
Current Commercial Routes for PO	119
Newer Routes for PO Manufacture for a Greener Environment	121

7. Life-cycle Concepts and Green Engineering   125–151

Introduction	125
Life-cycle Assessment	127
Definitions and Methodology	127
Life-cycle Inventories	130
Life-cycle Impact Assessments	136
Classification	137
Characterisation	137
Valuation	139
Interpretation of Life-cycle Data and Practical Limits to Life-cycle Assessments	140
Streamlined Life-cycle Assessments	141
Streamlined Data Gathering for Inventories and Characterisation	141
Qualitative Techniques for Inventories and Characterisation	142
Pitfalls, Advantages, and Guidance	145
Uses of Life-cycle Studies	145
Product Comparison	145
Strategic Planning	145
Public Sector Uses	146
Product Design and Improvement	146
Process Design	148
Comprehensive Life-cycle Management Strategy Needed to Ensure Sustainable Future for Indian Chemical Industry	150
Life-cycle Costing Concept	150
Reducing the Carbon Footprint	150
Boosting Exports	151

Contents xv

8. Industrial Ecology — 152–160

- Introduction — 152
- Environmental Performance of Chemical Processes — 152
- Material Flows in Chemical Manufacturing — 156
- Eco-Industrial Parks — 158
- Assessing Opportunities for Waste Exchanges and By-product Synergies — 160

9. Environmental Care in Manufacture of Chemicals — 161–175

- Introduction — 161
- Material Selection — 161
- Process Conditions — 162
- Zero Discharge Technologies — 162
- Parellel/Series Operations — 162
- Plant Capacity — 162
- Process Patterns — 163
- Utilities — 164
- Catalysts — 165
- Prevent Rather Than Cure — 165
- Pollution Control and Environmental Protection in Chemical Industry — 166
  - Dust Removal — 166
  - Phenol Removal — 166
  - Solvent Vapour Removal — 167
  - BOD Reduction — 167
  - Removal of Chlorine/Hydrochloric Acid Vapours — 167
  - Removal of Hydrogen Sulphide — 167
  - Incinerators — 167
  - Chemical Treatment/Filtration — 167
- Optimising Energy Use in Chemical Process Industry — 168
  - Methodology of Optimising Energy Use — 169
  - Areas for Energy Optimisation in Chemical Process Industries (CPI) — 169
- Integrating Environment, Energy, Technology and Business Processes — 172
  - A Case Study — 172

10. Air Pollution Control — 176–188

- Introduction — 176
- Volatile Organic Compounds (VOCs) — 176
  - Odours — 176
  - Technology for Removal of VOCs and Odour — 177
  - Biological Oxidation or Biofiltration — 177
  - Catalytic Oxidation — 179
- Advanced Scrubber Technology Solutions for Pollution Control — 181
  - Requirements for Acid Gas Removal — 182
  - Mass Transfer — 182
  - Reagent Kinetics — 183
  - Reagents Use-Economics — 185
  - Effluent Characteristics — 185

	Fundamental Principles for Particulate Removal	186
	Distinction of SO_3 Removal Mechanism	187
	Achieving Effluent Requirement Norms	188

11. Water for Safe Use in Chemical and Allied Industries   189–208

Introduction		189
Methods for Detection of Ions and Microbes		190
Chemical Methods		190
Microbiological Method		192
Purified Water Requirement		193
Technologies for Recycling of Waste-water		193
Pollution Prevention in Industry		193
Total Water Management Approach to Recycle of Waste-water		193
Recycle Technologies		195
Effluent Treatment		195
Tertiary Treatment		196
Advanced Treatment		196
Technologies for Recycling of Effluents		196
Automation and Control in Water and Waste-water Treatment Plants		202
Role of Automation in Water and Waste-water Treatment Industry		202
New Generation Automatic Demineralisation System		203
New Generation Reverse Osmosis-Electro-Deionisation (RO/EDI) System		203
New Generation Waste-water Recycling Technology		204
Water Treatment Process Analysers		204
High Purity Water Systems		204
Case Study Waste-water Reclamation		205
Cooling Water Blowdown Recovery at Madras Fertilisers Ltd.		206
Cost Economics		208
Other Benefits		208

12. Physical, Thermal and Chemical Methods of Waste Management   209–233

Introduction		209
Physical Methods		210
Phase Separation		210
Phase Transfer		212
Phase Transition		214
Phase Conversion		215
Membrane Separation		216
Thermal Methods		218
Incineration		219
Thermal Desorption		222
Pyrolysis		223
Solidification and Stabilisation		223
Chemical Treatment of Wastes		224
Categorisation		224
Parameters		225
Process		225

Operations	227
Disinfection	228
Ion Exchange	228
Bioremediation	228
Solid Phase Bioremediation	232
Soil Heaping	233
Composting	233
In Situ Bioremediation	233

13. Zero Waste for Sustainable Development — 234–252

Introduction	234
Sources of Waste	234
Cleaner Technologies	235
Waste Reduction at Source	235
Cleaner Suitable Environmental Technologies	236
Action Plan	237
Towards a Sustainable Chemical Industry	238
Drivers for a Sustainable Industry	238
Principles of Green Chemistry and Green Engineering	239
Need for the Correct Framework	239
Illustrations of Implemented Sustainable Chemistry	240
Waste Reduction	243
Green Chemistry and the Biorefinery	245
Biorefinery Concept	246
Biofuels from Biomass: A Chemical Industry's Perspective	249
Psychrometric Evaporation: A Novel Zero Discharge System	250
Operating Principle	250
Psychrometric Evaporation System	250
Waste Heat Source for Psychrometric Evaporation System	250
Eco-friendliness of Psychrometric Evaporator	252
Potential of Energy, Resource and Water Saving	252
Advantages	252

SECTION II

Pollution Control in Chemical Industries — 253–540

14. Petroleum Refinery and Petrochemicals — 255–296

Introduction	255
Refining of Petroleum	255
Waste Generation in Petroleum Refinery	256
Crude Oil Producing and Handling	258
Refining	258
Emulsions	260
Condensate Waters	261
Acid Wastes	261
Waste Caustics	263

Alkaline Water	264
Special Chemicals	265
Cooling Water	265
Waste-water from Refinery	266
Solvents and Effluents	266
Waste-water Treatment	267
Physical Treatment	267
Chemical Methods	269
Biological Methods	270
Control of Air Emissions in Refinery	271
Air Pollution Control Measures	271
Vapour Control Systems	271
Flares	271
Incinerators	272
Refinery Fuel Gas	273
Hydrocarbon Reuse	273
Covering a Liquid Space	273
Inspection and Maintenance	274
In-Line Sampler	274
Rupture Disks	274
New Valve Technology	274
Mechanical Seals	274
Quick Change Blind/Manifold Design	275
New Gaskets, Bolts, and Welding	275
Steam Stripping	275
Carbon Monoxide Boiler	275
Cyclone	275
Electrostatic Precipitators	275
Scrubbers	276
Combustion Controls	276
Internal-Combustion Engine	276
Combustion Turbines	276
Heaters and Boilers	276
Petrochemical and Allied Products	276
Water Use	277
Cooling Water	277
Steam Generation	277
Process Water	277
Recycling and Reuse	279
Sources and Characteristics of Waste-water	279
Waste-water Treatment Practices	279
Water Pollution Control in Petrochemical Industries	281
Solid Wastes Pollution in Petrochemical Industries	286
Basic Factors to be Considered	287
Disposal/Treatment	288
Treatment of Polymers	293
Control of Air Emissions in Petrochemicals	293
Phthalic Anhydride	293
Air Pollution Control Measures	295
Minimal National Standards (MINAS) for Oil Refineries	296

15. Leather and Tannery — 297–339

- Introduction — 297
- Tanning Process — 297
  - Pretanning Operation — 297
  - Types of Tanning Process and Their Unit Operations — 298
- Sources of Wastes — 300
  - Liquid Wastes — 301
  - Impact of Liquid Wastes on Environment — 304
- Control of Water Pollution — 306
  - In-plant Control Measures — 306
  - Chemical Treatment — 307
  - Secondary Biological Treatment — 308
  - Aerobic Systems — 309
  - Anaerobic Systems — 309
  - Combined Process — 309
  - Treatment through Non-conventional Methods — 310
- Treatment of Waste-water — 310
  - Primary Treatment — 310
  - Secondary Biological Treatment — 310
- Waste Control Measures — 314
  - Good Housekeeping and Water Conservation — 314
  - Recovery and Reuse — 314
  - Process Changes for Reducing Pollution Load — 314
- Guidelines for Quantum Limits of Pollutants in the Treated Waste-water — 315
- Clean Technology in the Leather Industry — 316
  - Characteristics of Waste-water — 317
  - Need for Chrome Recovery and Reuse — 317
  - Choice of Chromium Recovery and Reuse System — 317
  - Pilot Plant for Chromium Recovery — 318
- Recycle, Reuse and Management of Waste-water in Tanneries — 320
- Waste-water in Leather Industry — 320
- Pretreatment and Tanning Operations—Their Contribution to Environmental Problems — 322
  - Approaches to Better Water Management — 324
  - Washings — 329
- Gaseous Pollution — 330
  - Impact of Gaseous Pollutants on Environment — 330
  - Control of Gaseous Pollutants — 331
- Use of Computers in Leather Processing — 331
  - Use of Computers in Technology — 331
  - Artificial Intelligence — 332
  - Automation of Tannery Procedures — 332
  - Present Trends in Tannery Automation — 332
  - Electronic Information System on Leather Processing Chemicals — 333
  - Classification of Leather Auxiliaries — 333
  - About TFEISLPC — 334
  - Salient Features — 334

	Minimal National Standards (MINAS)	338
	Comparative Analysis of Standards	339

16. Pulp and Paper — 340–367

Introduction	340
Grades of Paper and Board and Demand Projection	340
Fibre Sources	341
Forests	341
Agricultural Residues	341
Waste Paper	341
Mill Capacity	341
Basic Process Steps in Paper and Board Manufacture	342
In Wood and Agro based Mills	342
Chemical Recovery Operation	342
Improvement of Paper Quality	342
Effluent Bench Marking of Indian Mills	343
Technology Trends in Pulp Production	343
Water Usage and Policy	344
Waste-water Generation	345
Sources of Waste-water	345
Water Pollution	345
Big Paper Mills	345
Characteristics of Effluents	349
Treatment for the Waste-water	349
For Mills Using Agricultural Residues as Raw Materials	349
Solid Wastes	353
Liquid Wastes	353
For Mills Using Waste Paper and Purchased Pulp as Raw Material	353
Black Liquor Segregation and Disposal	354
Case Study	355
Star Paper Mills Ltd., Saharanpur	355
Characteristics of Waste-water Management in Small Paper Mills	356
Waste Minimisation/Management Options	358
Segregation of Black Liquor Generating Stream	361
Potentiality of Water Hyacinth	362
Enzymatic Treatment	364
Gaseous Emissions and Air Pollution	365
Air Pollution from Big Paper Mills	365
Air Pollution from Small Paper Mills	365
Minimal National Standards (MINAS)	366

17. Soap and Detergent — 368–386

Introduction	368
Classification of Surfactants	368
Sources of Detergents in Waters and Waste-waters	369
Impacts on Waste-water Treatment Processes	369

Industrial Operation and Waste-Water	370
Manufacture and Formulation	370
Soap Manufacture and Processing	370
Neat Soap Manufacture and Waste Streams	370
Fatty Acid Neutralisation	371
Glycerine Recovery Process	371
Production of Finished Soaps and Process Wastes	372
Detergent Manufacture and Waste Streams	372
Surfactant Manufacture and Waste Streams	373
Detergent Formulation and Process Wastes	373
Waste-water Characteristics	374
Waste-water Control and Treatment	375
In-Plant Control and Recycle	375
Waste-water Treatment Methods	376
Biological Treatment	379
Air Emissions	379
Slurry Preparation	379
Spray Drying	380
Granule Handling	385

18. Pesticide   387–408

Introduction	387
Classification of Pesticides	388
Pesticides/Fertilisers Causing Pollution	388
Fate of Pesticides in the Environment	390
Effects of Pesticides in the Environment	390
Possible Health Hazards of Pesticides Residues	394
Effect of Fertilisers on the Environment	394
Pollution Prevention and Control	395
Waste-water Generation in Pesticides Industry	395
Treatment Technologies	398
Pretreatment Technologies	398
Secondary Treatment	398
Tertiary Treatment Technologies	399
Separation Technologies for Removal of Organic and Pesticidal Chemicals from Waste-water	401
Toxicity of Chemical Pesticides	401
Management	402
Environmentally Safe Layout Code for Manufacturing Units	402
Minimal National Standards	406

19. Fertiliser   409–435

Introduction	409
Manufacturing Process	409
Steps in Manufacture	409
Waste-water from Fertiliser Plants	410
Effects of Wastes on Receiving Streams	410
Treatment of Fertiliser Waste-water	411

Pollution Control in Single Superphosphate (SSP) Plant	411
Straight Nitrogenous Fertilisers	412
Phosphate Fertiliser	415
Industrial Operation and Waste-water	415
Waste-water Characteristics and Sources	417
Phosphate Fertiliser Manufacture	420
Waste-water Control and Treatment	420
Waste-water Treatment Methods	422
Ammonium Phosphate Fertiliser and Phosphoric Acid Plant	424
Furnace Wastes from Phosphorus Manufacture	424
Phosphoric Acid and N-P-K Fertiliser Plant	426
Environmentally Balanced Industrial Complexes	427
Fluoride and Phosphorus Removal from a Fertiliser Complex Waste-water	429
Air Emissions	429
Control of Air Emissions	430

20. Paints and Dyes   436–455

Introduction	436
Manufacturing Process	436
Paint Manufacture	436
Varnish Manufacture	436
Lacquer Manufacture	437
Distemper Manufacture	437
Manufacture of Resins and Emulsions	437
Bituminous Paint	437
Air Pollution	438
Emission—Fuel Burning	438
Emission—Process	438
Waste-water Generation	438
Waste-water Quality	438
Waste-water Quantity	439
Waste-water Characterisation	439
Waste-water Treatment	440
Reduction of Waste-water	441
Status of Waste-water Treatment in India	443
Treatability Aspects of Waste-water	443
Waste-water Treatment	443
Standards to Satisfy Environmental Requirements	445
Dye and Dye Intermediates	446
Basic Operations in Dyeing	446
Waste Generation	447
Characterisation of Waste Effluents	448
Reduction of Waste-water	451
Environmental Impact of Dye and its Intermediates	452
Minimal National Standards (MINAS)	454

21. Drugs and Pharmaceuticals — 456–480

Introduction — 456
Production Process — 456
Physical Methods — 456
    Antibiotics, Vitamins and Enzymes — 457
Waste Generation — 457
    Air Emissions — 457
    Effluent Generation — 458
    Solid Waste Generation — 458
    Human and Animal Contribution — 458
    Physical Processes — 458
    Chemical Methods — 459
    Fermentation — 460
Waste-Water — 461
    Physical Treatment — 462
    Primary Treatment/Chemical Treatment — 465
    Secondary Treatment — 466
    Anaerobic Filter and Lagooning — 466
Performance of Existing Treatment Systems — 467
    Cost of Waste-water Treatment — 468
Air Emissions Characterisation — 468
    Emission Sources — 468
    Types of Emissions — 470
Air Pollution Control Measures — 471
    Emission Control Considerations — 472
Minimal National Standards (MINAS) — 479
    Objective — 479

22. Sugar, Distillery and Fermentation Products — 481–502

Introduction — 481
Sugar — 481
    Bagasse Utilisation — 481
    Molasses — The Key Raw Material for Distillery — 483
    Sugar Mill Wastes — 483
    Manufacturing Process — 483
    Sources of Waste-water and the Characteristics of the Wastes — 484
    Effects of the Waste on Receiving Water — 485
    Treatment of the Wastes — 485
Distillery Industry — 486
    Process — 486
    Treatment of Distilling Industry Effluents — 487
    Breweries — 489
    Cleaner Technology for Distillery Industry — 491
Fermentation Products — 496
    Industrial Fermentations — 497
    Auxiliary Operations — 499
    Beverage Alcohol — 500

xxiv Pollution Control in Chemical and Allied Industries

Other Industries	501
Industrial Alcohol	501
Antibiotics	501
Treatment of Non-recoverable Waste	502

23. Miscellaneous Process Industries   503–528

Introduction	503
Inorganic Chemical Industry	503
Raw Materials	504
Waste Characteristics	505
Waste Disposal Techniques	506
Chemical Techniques	508
Physical Techniques	509
Caustic Soda Industry	511
Air Emissions Characterisation	518
Air Pollution Control Measures	519
Disposal of Wastes from Specific Industries	521
Air Emissions Characterisation	525
Air Pollution Control Measures	525
Recent and Future Industry Trends	527
Minimal National Standard (MINAS)	527

24. Textile   529–540

Introduction	529
Water Pollution from Boilers	529
Water Pollution from Water-treatment Plants	529
Operations Involved in Finishing	530
Cotton	530
Wool	531
Synthetic Fibres	531
Dyes	531
Characteristics of Textile Waste-waters	532
Treatment of Textile Waste-waters	533
Pollution Control	533
In-Plant Control	533
Treatment to Reuse	534
Direct Discharge to POTWs	535
On-site Treatment of Textile Waste-waters	535
Pretreatment of Textile Waste-water	536
Physical/Chemical Treatment	536
Coagulation/Clarification	536
Multimedia Filtration	536
Granular Carbon	537
Dissolved Air Flotation	537
Ozonation	538

Biological Treatment	538
Land Treatment of Textile Waste-waters	539
Air Emissions Characterisation	539
Air Pollution Control Measures	539

SECTION III

Pollution Control in Food Processing Industries — 541–576

25. Food Processing Industry: A Review — 543–546

Introduction	543
Water Use	543
Sources and Characteristics of Waste-water	543
Waste-water Treatment Practices	544
Air Pollution	546

26. Dairy — 547–554

Introduction	547
Sources of Waste	547
BOD of Whole Milk and By-products	548
Waste Prevention	548
Disposal of Spoiled Products	548
Utilisation of By-products	548
Methods of Treatment	549

27. Bakery — 555–561

Introduction	555
Waste-water	555
Bakery Waste Treatment	556
Pre-treatment Systems	556
Flow Equalisation and Neutralisation	557
Screening	557
FOG Separation	558
Acidification	558
Coagulation-Flocculation	558
Sedimentation	559
Dissolved Air Flotation (DAF)	559
Biological Treatment	559
Aerobic Treatment	559
Activated Sludge Process	559

28. Seafood Processing and Meat Products — 562–576

Introduction	562
Treatment of Seafood Processing Wastes	562
Seafood-processing Waste-water Characterisation	562

    Primary Treatment 563
    Biological Treatment 564
    Physico-chemical Treatments 566
  Economic Considerations of Seafood-processing and Meat Processing Waste-water Treatment 568
  Treatment of Meat Wastes 568
    Processing Facilities and Wastes Generated 568
    Waste-water Minimisation 570
    Waste-water Treatment Processes 571

SECTION IV

Special Topics 577–619

29. Biotechnology for Pollution Prevention in Chemical Industry 579–585

Introduction 579
Industrial Pollution 580
Biotechnology for Air Pollution Abatement and Odour Control 580
  The Problem 581
  Deodourisation Processes 581
Application of Biotechnology to Pollution Problems of Various Industries 582
  Food and Allied Industries 582
  Paper Industry 583
  Chemical Industry 583
  Tanning Industry 584
  Textile Industry 584
  Pesticide Industry 585

30. Instrumental Techniques for Environmental Pollution Monitoring 586–604

Introduction 586
Chemical Analysis 587
Techniques for Analysis 587
  Spectrophotometric Methods 587
  Gas Chromatography 591
  Mass Spectrometry 594
  Atmospheric Monitoring 595
  Methods of Analysis 597
  Analysis of Oxidants 600
  Analysis of Carbon Monoxide 600
  Analysis of Hydrocarbons 601
  Analysis of Particulate Matter 602
  Direct Spectrophotometric Analysis of Gaseous Air Pollutants 604

31. Environmental Management in the Chemical Process Industries 605–619

Introduction 605
Status of Environmental Management 605

Current Developments	606
Global Environmental Challenges	607
Stress on Environmental Health	607
Source Reduction	608
Promote Integrated Complexes	608
Review Existing Control Limits for Pollutants	608
Call for Beyond Compliance of Statutory Stipulations	609
Harness Environmental Biotechnology	609
Municipal Solid Waste	609
Reduce Emission of Green House Gases	609
Ensure Water Availability	610
Develop Pollution Inventory Database	610
Changing Role of the Regulator	610
Hazardous Waste Disposal	611
Environmental Health Education	611
Emergency Planning for Disaster Mitigation	611
Key Issues in Environmental Management	612
Addressing Challenges: National Environment Policy	612
Role of Top Management	612
Environmental Policy Statement	612
Environment, Health, Safety (EHS) Vision Statement	612
Environmental Targets	613
Control Strategies	613
Risk Management	613
Staff Training	613
Monitoring	613
Public Information	614
Annual Reports	614
Cost Management in the Chemical Process Industry	614
Tackling Costs-Shifting from Products to Processes	615
Combining Environmental Goals with Cost Optimisation	615
Cost Control Through Waste Minimisation: Tools and Techniques	616
Cost Management	617
Methodology for Cost Control	618

References   621

Index   623–630

SECTION I

General Concepts and Engineering Considerations

1. **Introduction to Environmental Issues** — 3
2. **Risk Concepts** — 20
3. **Roles and Responsibilities of Chemical Engineers** — 36
4. **Evaluating Exposures** — 43
5. **Unit Operations and Pollution Prevention** — 60
6. **Green Chemistry for a Sustainable Future** — 96
7. **Life-Cycle Concepts and Green Engineering** — 125
8. **Industrial Ecology** — 152
9. **Environmental Care in Manufacture of Chemicals** — 161
10. **Air Pollution Control** — 176
11. **Water for Safe Use in Chemical and Allied Industries** — 189
12. **Physical, Thermal and Chemical Methods of Waste Management** — 209
13. **Zero Waste for Sustainable Development** — 234

Chapter 1
Introduction to Environmental Issues

INTRODUCTION

In this chapter a wide array of environmental issues are introduced, and their impacts are related to chemical production and use. The pertinent chemicals and the environmental reactions of those chemicals are discussed. For many environmental problems, the chemicals causing the adverse environmental or health impacts are not the same chemical originally emitted from the production process or from the use of a chemical. Thus, the environment is a complex system with a large number of transport and transformation processes occurring simultaneously. Fortunately for the chemical engineer, it is not necessary to understand these processes in great detail in order to gain the insights needed to design chemical processes to be more efficient and less polluting. A focal point for improving process designs is to understand that the properties of chemicals can have an important influence on their ultimate fate in the environment and on their potential impact on the environment and human health. With a basic understanding of environmental issues, the chemical engineer will be able to spot environmental problems earlier and will contribute to the solution of those problems by improving the environmental performance of chemical processes and products.

Recently environmental issues have gained increasing prominence all over the world. Global population growth has led to increasing pressure on worldwide natural resources including air and water, arable land, and raw materials, and modern societies have generated an increasing demand for the use of industrial chemicals. The use of these chemicals has resulted in great benefits in raising the standard of living, prolonging human life and improving the environment. But as new chemicals are introduced into the marketplace and existing chemicals continue to be used, the environmental and human health impacts of these chemicals has become a concern. Today, there is a much better understanding of the mechanisms that determine how chemicals are transported and transformed in the environment and what their environmental and human health impacts are, and it is now possible to incorporate environmental objectives into the design of chemical processes and products.

The challenge for future generations of chemical engineers is to develop and master the technical tools and approaches that will integrate environmental objectives into design decisions. The purpose of this chapter is to present a brief introduction to the major environmental problems that are caused by the production and use of chemicals in modern industrial societies. With each environmental problem introduced, the chemicals or classes of chemicals implicated in that problem are identified. Whenever possible, the chemical reactions or other mechanisms responsible for the chemical's impact are explained. Trends in the production, use, or release of those chemicals are shown. Finally, a brief summary of adverse health effects is presented. This chapter's intent is to present the broad range of environmental issues which may be encountered by chemical engineers and a review of selected environmental regulations that may affect chemical engineers. It is hoped that this information will elevate the environmental awareness of chemical engineers and will lead to more informed decisions regarding the design, production, and use of chemicals.

ROLE OF CHEMICAL PROCESSES AND CHEMICAL PRODUCTS

In this text, a number of design methodologies are covered for preventing pollution and reducing risks associated with chemical production. Figure 1.1 shows conceptually how chemical processes convert raw materials into useful products with the use of energy. Wastes generated in chemical manufacturing, processing, or use are released to the environment through discharges to streams or rivers, exhausting into the air, or disposal in a landfill. Often, the waste streams are treated prior to discharge.

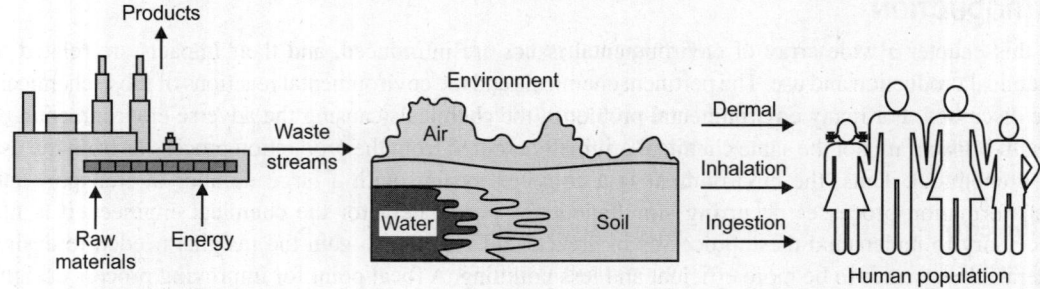

Fig. 1.1. Generalised scenario for exposure by humans to environmental pollutants released from chemical processes.

We may be exposed to waste stream components by three routes: dermal (skin contact), inhalation, and ingestion. The route and magnitude of exposure is influenced by the physical, chemical, and reactivity properties of the waste stream components. In addition, waste components may affect the water quality of streams and rivers, the breathability of ambient air, and the well-being of terrestrial flora and fauna. What information will a chemical engineer need to make informed pollution prevention and risk reduction decisions? A few generalised examples will aid in answering such a question.

Formulation of an Industrial Cleaner

Company A plans to formulate a concentrated, industrial cleaner, and needs to incorporate a solvent within the product to meet customer performance criteria. A number of solvents are identified that will meet cost and performance specification. Further, company 'A' knows that the cleaning product (with the solvent) will be discharged to water and is concerned about the aquatic toxicity of the solvent. The company conducts a review of the pertinent data to aid in making the choice. In aquatic environments, a chemical will have low risk potential if it has the following characteristics:

1. High Henry's Law constant (substance will volatilise into the air rather than stay in the water).
2. High biodegradation rate (it will dissipate before exerting adverse health effects)
3. Low fish toxicity parametre (a high value of the concentration lethal to a majority of test organisms or LC_{50}).
4. Low Bioconcentration Factor, BCF (low tendency for chemicals to partition into the fatty tissue of fish, leading to exposure and adverse health effects upon consumption by humans).

Company 'A' assembles the data and chooses a solvent with the least adverse environmental consequences. Methods are presented in this text to provide estimates of environmental properties. In addition, measured data for some of these properties are tabulated.

Formulation of a Paint Solvent

Company 'B' is formulating a paint for an automobile refinishing. The formulation must contain fast-drying solvents to ensure uniform coating during application. These fast-drying solvents volatilise when the paint is sprayed and are exhausted by a fan. Workers in the booths may be exposed to the solvents during application of the paint and nearby residents may inhale air contaminated by the exhausted solvents.

The company is concerned about the air releases and problems that arise with worker exposure to toxic agents and impact to air quality. A number of solvents having acceptable cost and coating performance characteristics have been identified. A chemical will have low risk potential in the air if it has the following characteristics:
1. Low toxicity properties (high Reference Dose [RfD] for inhalation toxicity to humans or a low cancer potency).
2. Low reactivity for smog formation (ground level ozone production).

Candidate solvents may be screened for these properties to identify the environmentally optimal candidate.

Choice of Refrigerant for a Low-temperature Condenser

A chemical engineer is in charge of redesigning a chemical process for expanded capacity. One part of the process involves a vapour stream heat exchanger and a refrigeration cycle. In the redesign, the company decides to use a refrigerant having low potential for stratospheric ozone depletion. In addition, the engineer must also ensure that the refrigerant possesses acceptable performance characteristics such as thermodynamic properties, materials compatibility, and thermal stability. From the list of refrigerants that meet acceptable process performance criteria, the engineer estimates or finds tabulated data for:
1. Atmospheric reaction-rate constant.
2. Global warming potential.
3. Ozone depletion potential.

From an environmental perspective, an ideal refrigerant would have low ozone depletion and global warming potentials while not persisting in the atmosphere.

These three examples illustrate the role the chemical engineer plays by assessing the potential environmental impacts of product and process changes. One important impact the chemical engineer must be aware of is human exposure, which can occur by a number of routes. The magnitude of exposure can be affected by any number of reactive processes occurring in the air, water, and soil compartments in the environment. The severity of the toxic response in humans is determined by the toxicology properties of the emitted chemicals. The chemical engineer must also be aware of the life cycle of a chemical. What if the chemical volatilises but is an air toxicant? What if the biodegradation products (as, for example, with DDT) are the real concern? For example, terpenes, a class of chemical compounds, were touted as a replacement for chlorinated solvents to avoid stratospheric ozone depletion, but terpenes are highly reactive and volatile and can contribute to photochemical smog formation.

The next sections present a wide range of environmental problems caused by human activities. Trends in the magnitude of these problems are shown in tabular form, and contributions by industrial sources are mentioned whenever possible.

OVERVIEW OF MAJOR ENVIRONMENTAL ISSUES

The next several sections present an overview of major environmental issues. These issues are not only of concern to the general public, but are challenging problems for the chemical industry and for chemical engineers. The goal of the following sections is to provide an appreciation of the impacts that human activities can have on the environment. Also, the importance of healthy ecosystems are illustrated as they affect human welfare, the availability of natural resources, and economic sustainability.

When considering the potential impact of any human activity on the environment, it is useful to regard the environment as a system containing interrelated sub-processes. The environment functions as a sink for the wastes released as a result of human activities. The various subsystems of the environment act upon these wastes, generally rendering them less harmful by converting them into chemical forms that can be assimilated into natural systems. It is essential to understand these natural waste conversion processes so that the capacity of these natural systems is not exceeded by the rate of waste generation and release.

The impact of waste releases on the environment can be global, regional, or local in scope. On a global scale, man-made (anthropogenic) greenhouse gases, such as methane and carbon dioxide, are implicated in global warming and climate change. Hydrocarbons released into the air, in combination with nitrogen oxides originating from combustion processes, can lead to air quality degradation over urban areas and extend for hundreds of kilometres. Chemicals disposed of in the soil can leach into undergound water and reach groundwater sources, having their primary impact locally, near to the point of release. The timing of pollution releases and rates of natural environmental degradation can affect the degree of impact that these substances have. For example, the build-up of greenhouse gases has occurred over several decades. Consequently, it will require several decades to reverse or stall the build-up that has already occurred. Other releases, such as those that impact urban air quality, can have their primary impact over a period of hours or days.

The environment is also a source of raw materials, energy, food, clean air, water, and soil for useful human purposes. Maintenance of healthy ecosystems is therefore essential if a sustainable flow of these materials is to continue. Depletion of natural resources due to population pressures and/or unwise resource management threatens the availability of these materials for future use.

The following sections provide a short review of environmental issues, including global energy consumption patterns, environmental impacts, ecosystem health, and natural resource utilisation.

GLOBAL ENVIRONMENTAL ISSUES

Global Energy Issues

The availability of adequate energy resources is necessary for most economic activity and makes possible the high standard of living that developed societies enjoy. Although energy resources are widely available, some such as oil and coal are non-renewable, and others, such as solar, although inexhaustible, are not currently cost effective for most applications. An understanding of global energy usage patterns, energy conservation, and the environmental impacts associated with the production and use of energy are therefore very important.

Often, primary energy sources such as fossil fuels must be converted into another form such as heat or electricity. As the second law of thermodynamics dictates, such conversions will be less than 100 per cent efficient. An inefficient user of primary energy is the typical automobile, which converts into motion about 10 per cent of the energy available in crude oil.

The global use of energy has steadily risen since the dawn of the industrial revolution. Currently, fossil fuels make up roughly 85 per cent of the world's energy consumption, while renewable sources such as hydroelectric, solar, and wind power account for only about 8 per cent of the power usage. Nuclear power provides roughly 6 per cent of the world energy demand, and its contribution varies from country to country. The United States meets about 20 per cent of its electricity demand, Japan 28 per cent, and Sweden almost 50 per cent from nuclear power

The disparity in global energy use is illustrated by the fact that 65–70 per cent of the energy is used by about 25 per cent of the world's population. Energy consumption per capita is greatest in industrialised regions such as North America, Europe, and Japan.

Another interesting aspect of energy consumption by industrialised countries and the developing world is the trend in energy efficiency, the energy consumed per unit of economic output. Future chemical engineers will need to recognise the importance of energy efficiency in process design.

Many environmental effects are associated with energy consumption. Fossil fuel combustion releases large quantities of carbon dioxide into the atmosphere. During its long residence time in the atmosphere, CO_2 readily absorbs infrared radiation contributing to global warming. Further, combustion processes release oxides of nitrogen and sulphur oxide into the air where photochemical and/or chemical reactions can convert them into ground level ozone and acid rain. Hydropower energy generation requires widespread land inundation, habitat destruction, alteration in surface and groundwater flows, and decreases the acreage of land available for agricultural use. Nuclear power has environmental problems linked to uranium mining and spent nuclear rod disposal. 'Renewable fuels' are not benign either. Traditional energy usage (wood) has caused widespread deforestation in localised regions of developing countries. Solar power panels require energy-intensive use of heavy metals and creation of metal wastes. Satisfying future energy demands must occur with a full understanding of competing environmental and energy needs.

Global Warming

The atmosphere allows solar radiation from the sun to pass through without significant absorption of energy. Some of the solar radiation reaching the surface of the earth is absorbed, heating the land and water. Infrared radiation is emitted from the earth's surface, but certain gases in the atmosphere absorb this infrared radiation, and re-direct a portion back to the surface, thus warming the planet and making life, as we know it, possible. This process is often referred to as the greenhouse effect. The surface temperature of the earth will rise until a radiative equilibrium is achieved between the rate of solar radiation absorption and the rate of infrared radiation emission. Human activities, such as fossil fuel combustion, deforestation, agriculture and large-scale chemical production, have measurably altered the composition of gases in the atmosphere. Some believe that these alterations will lead to a warming of the earth-atmosphere system by enhancement of the greenhouse effect. Figure 1.2 summarises the major links in the chain of environmental cause and effect for the emission of greenhouse gases.

Table 1.1 is a list of the most important greenhouse gases along with their anthropogenic (man-made) sources, emission rates, concentrations, residence times in the atmosphere, relative radiative forcing efficienies, and estimated contribution to global warming. The primary greenhouse gases are water vapour, carbon dioxide, methane, nitrous oxide, chlorofluorocarbons, and tropospheric ozone. Water vapour is the most abundant greenhouse gas, but is omitted because it is generally not from anthropogenic sources. Carbon dioxide contributes significantly to global warming due to its high emission rate and concentration.

8 Pollution Control in Chemical and Allied Industries

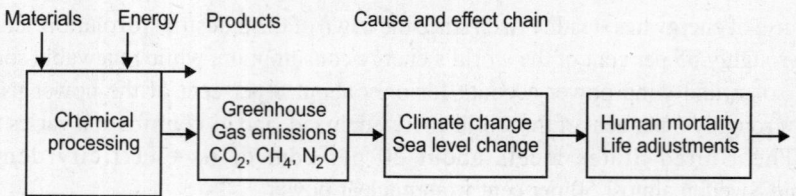

Fig. 1.2. Greenhouse emission from chemical processes and the major cause and environmental effect chain.

The major factors contributing to global warming potential of a chemical are infrared absorptive capacity and residence time in the atmosphere. Gases with very high absorptive capacities and long residence times can cause significant global warming even though their concentrations are extremely low. A good example of this phenomenon is the chlorofluorocarbons, which are, on a pound-for-pound basis, more than 1000 times more effective as greenhouse gases than carbon dioxide.

Photochemical Smog

Photochemical smog is initiated by the photochemical dissociation of NO_2 and the consequent secondary reactions involving unsaturated hydrocarbons, other organic compounds and free radicals, leading to the formation of organic peroxides and ozone. This phenomenon takes place during sunny days with low winds and low level inversion. Photochemical smog and the consequent formation of aerosols reduce visibility, cause irritation to eyes and damage plants and rubber goods.

The oxidation of SO_2 can also take place by interaction with the free radical $HO\cdot$ present in photochemical smog

$$SO_2 + HO\cdot \longrightarrow HOSO_2\cdot$$
$$HOSO_2\cdot + O_2 \longrightarrow HOSO_2 O_2\cdot$$
$$HOSO_2 O_2\cdot \longrightarrow HOSO_2 O\cdot + NO_2\cdot$$
$$\text{(sulphate)}$$

Chemical oxidation of SO_2 may also take place in water droplets, present in aerosols. This reaction is accelerated in the presence of NH_3 and catalysts, e.g. oxides of Mn, Fe, Cu, Ni.

Solid particles, such as soot, bring about catalytic oxidation of SO_2 by providing a heterogeneous phase for contact. Soot is formed during combustion of solid and liquid fuels in domestic and industrial operations, and automobile emissions.

Sulphur dioxide is a pollutant responsible for smog formation, acid rains and corrosion of metals and alloys.

Oxidation of organic compounds

Organic compounds such as hydrocarbons, aldehydes and ketones absorb solar radiation and undergo various photochemical and chemical reactions involving free radicals. Some of these reactions are catalysed by particulate matter such as soot and metal oxides. Some of the resultant intermediates and final products contribute to photochemical smog formation.

Greenhouse Effect

The earth is heated by sunlight and some of the heat that is absorbed by the earth is radiated back into space.

Table 1.1. Greenhouse gases and global warming contribution (m stands for million)

Gas	Source (natural and anthropogenic)	Estimated anthropogenic emission rate	Pre-industrial global concentration	Approximate current concentration	Estimated residence time in the atmosphere	Radiative forcing efficiency (absorptivity capacity) ($CO_2 = 1$)	Estimated contribution to global warming
Carbon dioxide (CO_2)	Fossil fuel combustion; deforestation	6000 million tons/year	280 ppm	355 ppm	50–200 years	1	50%
Methane (CH_4)	Anaerobic decay (wetlands, landfills, rice paddies), ruminants, termites, natural gas, coal mining, biomass burning	300–400 million tons/year	0.8 ppm	1.7 ppm	10 years	58	12–19%
Nitrous oxide (N_2O)	Estuaries and tropical forests; agricultural practices, deforestation, land clearing, low-temperature fuel combustion	4–6 million tons/year	0.285 ppm	0.31 ppm	140–190 years	206	4–6%
Chlorofluoro-carbons (CFC-11 and CFC-12)	Refrigerants, air conditions, foam-blowing agents, aerosol cans, solvents	1 million tons/year	0	0.004–0.001 ppm	65–110 years	4860	17–21%
Tropospheric ozone (O_3)	Photochemical reactions involving VOCs and NOx from transportation and industrial sources	Not emitted directly	NA	0.022 ppm	hours-days	2000	8%

However, some of the gases in the lower atmosphere, acting like glass in a greenhouse, allow solar radiations (in the range 300 to 2,500 nm, i.e. near UV, visible and near infrared region, while filtering the dangerous UV radiations, i.e. < 300 nm) but do not allow the earth to re-radiate the heat into space. In other words, these gases in the atmosphere are transparent to the sunlight coming in, but they strongly absorb infrared radiation, which the earth sends back as heat. A part of the heat so trapped in these atmospheric gases is re-emitted to the earth's surface. The net result is the heating of the earth's surface by this phenomenon, called the 'greenhouse effect'. The gases that are responsible for this greenhouse effect are CO_2, water vapour, CH_4 and man-made chlorofluorocarbons (CFCs). Water vapour strongly absorbs infrared radiations in the range 4,000 to 8,000 nm and CO_2 in the range 12,000 to 16,300 nm. The radiations in the range 8,000 to 12,000 nm escape unabsorbed and this is known as the region of atmospheric window.

Carbon dioxide is released by volcanoes, oceans, decaying plants as well as human activities, such as deforestation and combustion of fossil fuels. Automobile exhausts account for 30 per cent of CO_2 emissions in developed countries.

Methane is released from coal mines, decomposition of organic matter in swamps, rice paddy cultivation, guts of termites in forest debris and stomachs of ruminants.

Chlorofluorocarbons (CFCs) are used as coolants in refrigerators, propellants in aerosol sprays, plastic foam materials like 'thermocoles' or 'styrofoam' and in automobile air-conditioners.

In fact, the 'greenhouse gases' (particularly CO_2 and water vapour) are responsible for keeping our planet warm and thus sustaining life on the earth. If the greenhouse gases were very less or totally absent then the average temperature on the earth would have been at sub-zero levels. But, however, if the concentration of greenhouse gases increases, they may trap too much of heat, which may threaten the very existence of life on earth. For instance, the CO_2 present in the atmosphere of the planet *Venus*, is about 60,000 times more than that on earth. Hence, the average temperature of *Venus* is about 425°C, making the existence of life impossible there.

Oceans and bio-mass are the major sinks for atmospheric CO_2. Oceans convert CO_2 into soluble bicarbonates. The photosynthetic activity in the green plants increases with the increase in CO_2 level in the atmosphere. Forests are the places where lot of photosynthetic activity occurs. They also act as vast reservoirs of fixed but readily oxidisable carbon in the form of vegetation, wood and humus. Hence, forests maintain a balance in the atmospheric CO_2 level, and deforestation upsets this balance and increases the atmospheric CO_2 level.

It is estimated that the atmospheric CO_2 content has increased by 25 per cent during the last two centuries. This is mostly attributed to the industrial revolution and is one of the reasons for the slight increase in the global temperature (about 0.5°C). Since the concentrations of greenhouse gases have been continuously increasing because of deforestation, industrialisation, increased burning of fossil fuels, mining, exhausts from increasing number of automobiles and other anthropogenic activities, there is an increasing concern about the possible 'global warming'. Some scientists fear that if proper precautions are not taken, the concentration of greenhouse gases in the atmosphere may double within the next 50–100 years. If this happens, the average global temperature may increase by 4°–5°C. This will increase the evaporation of surface waters, which may influence climatic changes depending upon the pattern of cloud formation. For instance, low-level dense clouds may exert cooling effect whereas high-level thin cloud formation may exert heating effect due to increased greenhouse effect.

The projections from computer modelling regarding the climatic changes that could be triggered off due to 'global warming' reveal alarming scenarios. Even a 1.5°C rise in surface temperature can adversely affect food production in the world. Thus, the wheat growing zones in the northern latitude may be shifted from the USSR and Canada to the polar regions, i.e. from fertile soils to poor soils near the North Pole. The biological productivity of the ocean would also decrease due to warming of the earth's surface layer, which in turn, may reduce the transport of nutrients from deeper layers to the surface by vertical circulation. Computer modelling also indicates the following effects due to 'global warming': melting of the polar ice caps; dry areas becoming drier; humid areas like the Amazon suffering more intense tropical storms; drastic drop in food production, particularly in lands within 35 degrees north and south of the Equator; increased breeding of pests and diseases due to more humid conditions; shorter, wetter and warmer winters and longer, hotter and drier summers, particularly in mid-continental areas. Global warming may also trigger increased thermal expansion of oceans and melting of glaciers, which may result in an increase in the sea-level by 20 cm to 1.5 metres by the latter part of the 21st century. Thus, cities like Mumbai, Miami, London, Venice, Bangkok and Leningrad may become extremely vulnerable. Defences against the rising sea-levels and expanding oceans are very difficult and expensive, which many nations cannot afford. Further, a global temperature rise is likely to cause more floods, hurricanes, and tornadoes.

There are differences of opinion among experts regarding the dynamics and effects of 'global warming' due to the complexity of natural phenomena that might be operating simultaneously. More accurate future climatic projections will be possible with better super-computer models, based on greater understanding of the complex natural climatic forces involved. But until that time, the possible devastating effect due to 'global warming' by the 'greenhouse effect' cannot be underestimated. Some of the steps suggested to minimise the 'greenhouse effect' include reduction in the use of fossil fuels, encouraging the use of alternative sources of energy (e.g. solar, geothermal, wind, bio-gas, etc.), conservation of forests, extensive afforestation, encouraging community forestry, reduction in the use of automobiles, research in the development of more efficient automobile engines, ban on CFCs and nuclear explosions, development of environmentally compatible technologies with the help of intensive inter-disciplinary research, effective check on the growth of population and imparting of non-formal and formal environmental education.

Ozone Depletion in the Stratosphere

There is a distinction between 'good' and 'bad' ozone (O_3) in the atmosphere. Tropospheric ozone, created by photochemical reactions involving nitrogen oxides and hydrocarbons at the earth's surface, is an important component of smog. A potent oxidant, ozone irritates the breathing passages and can lead to serious lung damage. Ozone is also harmful to crops and trees.

Stratospheric ozone, found in the upper atmosphere, performs a vital and beneficial function for all life on earth by absorbing harmful ultraviolet radiation. The potential destruction of this stratospheric ozone layer is therefore of great concern.

The stratospheric ozone layer is a region in the atmosphere between 12 and 30 miles (20–50 km) above ground level in which the ozone concentration is elevated compared to all other regions of the atmosphere. In this low-pressure region, the concentration of O_3, can be as high as 10 ppm (about 1 out of every 1,00,000 molecules).

Ozone is formed at altitudes between 25 and 35 km in the tropical regions near the equator where solar radiation is consistently strong throughout the year. Because of atmospheric motion, ozone migrates to the polar regions and its highest concentration is found there at about 15 km in altitude. Stratospheric ozone concentrations have steadily declined over the past 20 years.

Ozone equilibrates in the stratosphere as a result of a series of natural formation and destruction reactions that are initiated by solar energy. The natural cycle of stratospheric ozone creation and destruction has been altered by the introduction of man-made chemicals. CFCs are highly stable chemical structures composed of carbon, chlorine, and fluorine. One important example is trichlorofluoromethane, CCl_3F, or CFC-11.

CFCs reach the stratosphere due to their chemical properties; high volatility, low water solubility, and persistence (non-reactivity) in the lower atmosphere. In the stratosphere, they are photo-dissociated to produce chlorine atoms, which then catalyse the destruction of ozone:

$$\begin{aligned} Cl + O_3 &\longrightarrow ClO + O_2 \\ ClO + O &\longrightarrow O_2 + Cl \\ \hline O_3 + O &\longrightarrow O_2 + O_2 \end{aligned}$$

The chlorine atom is not destroyed in the reaction and can cause the destruction of up to 10000 molecules of ozone before forming HCl by reacting with hydrocarbons. The HCl eventually precipitates from the atmosphere. A similar mechanism as outlined above for chlorine also applies to bromine, except that bromine is an even more potent ozone destroying compound. Interestingly, fluorine does not appear to be reactive with ozone. Figure 1.3 summarises the major steps in the environmental cause and effect chain for ozone-depleting substances.

CFC's were first introduced in the 1930's for use as refrigerants and solvents. By the 1950's significant quantities were released into the atmosphere. Releases reached a peak in the mid-eighties (CFC-11 and CFC-12 combined were about 700 million kg). Releases have been decreasing since about 1990. The Montreal Protocol, which instituted a phase-out of ozone-depleting chemicals, is the primary reason for the declining trend. The growth in accumulation of CFCs in the environment has been halted as a result of the Montreal protocol.

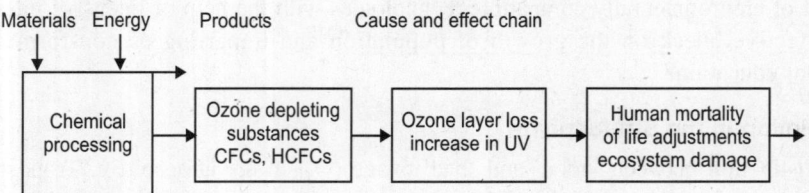

Fig. 1.3. Ozone-depleting chemical emissions and the major steps in the environmental cause and effect chain.

AIR QUALITY ISSUES

Air pollution arises from a number of sources, including stationary, mobile, and area sources. Stationary sources include factories and other manufacturing processes. Mobile sources are automobiles, other transportation vehicles, and recreational vehicles such as snowmobiles and watercraft. Area sources are emissions associated with human activities that are not considered mobile or stationary. Examples of area sources include emissions from lawn and garden equipment, and residential heating. Pollutants can be classified as primary, those emitted directly to the atmosphere, or secondary, those formed in the atmosphere after emission of precursor compounds. Photochemical smog (the term originated as a contraction of smoke and fog) is an example of secondary pollution that is formed from the emission of volatile organic compounds (VOCs) and nitrogen oxides (NO_x), the primary pollutants. Air quality problems are closely associated with combustion processes occurring in the industrial and transportation sectors of the economy. Smog formation and acid rain are also closely tied to these processes. In addition, hazardous air pollutants, including chlorinated organic compounds and heavy metals, are emitted in sufficient quantities to be of concern. Figure 1.4 shows the primary environmental cause and effect chain leading to the formation of smog.

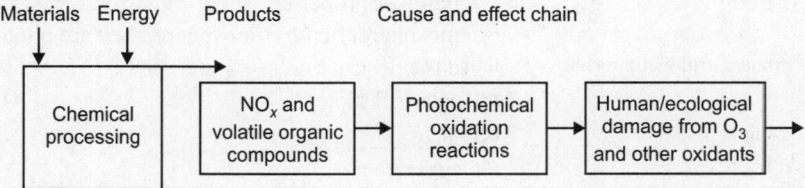

Fig. 1.4. Environmental cause and effect chain for photochemical smog formation.

Criteria Air Pollutants

Clean Air Act which charged the Environmental Protection Agency (EPA) with identifying those air pollutants which are most deleterious to public health and welfare, and empowered EPA to set maximum allowable ambient air concentrations for these criteria air pollutant EPA identified six substances as criteria air pollutants. Table 1.2 and promulgated primary and secondary standards that make up to the National Ambient Air Quality Standards (NAAQS). Primary standards are intended to protect the public health with an adequate margin of safety. Secondary standards are meant to protect public welfare, such as damage to crops, vegetation, and ecosystems or reductions in visibility.

Table 1.2. Criteria pollutants and the national ambient air quality standards.

Pollutant	Primary standard (Human health related)		Secondary (welfare related)	
	Type of average	Concentration[a]	Type of average	Concentration
CO				
[–38%][h]	8 hour[b]	9 ppm (10 mg/m^3)	No secondary standard	
{–25%}	1 hour[b]	35 ppm (40 mg/m^3)	No secondary standard	
Pb				
[–67%]	Maximum Quarterly	1.5 mg/m^3	Same as primary standard	
{–44%}	Average			
NO$_2$				
[–14%]	Annual Arithmetic Mean	0.053 ppm	Same as primary standard	
{–1%}		(100 µg/m^3)		
O$_3$				
[–19%]	1 hour[c]	0.12 ppm	Same as primary standard	
		(235 µg/m^3)		
	8 hour[d]	0.08 ppm	Same as primary standard	
		(157 µg/m^3)		
PM$_{10}$				
[–26%]	Annual Arithmetic Mean	50 µg/m^3	Same as primary standard	

(Contd...)

Pollutant	Primary standard (Human health related)		Secondary (welfare related)	
	Type of average	Concentration[a]	Type of average	Concentration
{−12%}	24 hour[e]	150 μg/m^3	Same as primary standard	
PM$_{2.5}$	Annual arithmetic mean[f]	15 μg/m^3	Same as primary standard	
	24 hour[g]	65 mg/m^3	Same as primary standard	
SO$_2$				
[−39%]	Annual arithmetic mean	0.03 ppm	3 hours[b]	0.50 ppm
{−12%}		(80 μg/m^3)		(1300 (μg/m^3))
	24 hour[b]	0.14 ppm		
		(365 μg/m^3)		

[a] Parenthetical value is an equivalent mass concentration.
[b] Not to be exceeded more than once per year.
[c] Not to be exceeded more than once per year on average.
[d] 3-year average of annual 4th highest concentration.
[e] The pre-existing form is exceedance-based. The revised form is the 99th percentile.
[f] Spatially averaged over designated monitors.
[g] The form is the 98th percentile.
[h] Air quality concentration. % change 2004–2005.
[i] Emissions. % change 2004–2005.

Criteria pollutants are a set of individual chemical species that are considered to have potential for serious adverse health impacts, especially in susceptible populations. These pollutants have established health-based standards and were among the first airborne pollutants to be regulated.

NO$_x$, hydrocarbons and VOCs — ground-level ozone

Ground-level ozone is one of the most pervasive and intractable air pollution problems in the United States and other developing countries. We should again differentiate between this 'bad' ozone created at or near ground level (tropospheric) from the 'good' or stratospheric ozone that protects us from UV radiation.

Ground-level ozone, a component of photochemical smog, is actually a secondary pollutant in that certain precursor contaminants are required to create it. The precursor contaminants are nitrogen oxides (NO$_x$, primarily NO and NO$_2$) and hydrocarbons. The oxides of nitrogen along with sunlight cause ozone formation, but the role of hydrocarbons is to accelerate and enhance the accumulation of ozone.

Oxides of nitrogen (NO$_x$) are formed in high-temperature industrial and transportation combustion processes. Health effects associated with short-term exposure to NO$_2$ (less than three hours at high concentrations) are increases in respiratory illness in children and impaired respiratory function in individuals with pre-existing respiratory problems. Industry makes a significant contribution to the 'fuel combustion' category from the energy requirements of industrial processes. Major sources of hydrocarbon emissions are the chemical and oil refining industies, and motor vehicles. Solvents comprise 66 per cent of the industrial emissions and 34 per cent of total VOC emissions. It should be noted that there are natural (biogenic) sources of HCs/VOCs, such as isoprene and monoterpenes that can contribute significantly to regional hydrocarbon emissions and low-level ozone levels. Ground-level ozone concentrations are exacerbated by certain physical and atmospheric factors. High-intensity solar radiation, low prevailing wind speed (dilution), atmospheric inversions, and proximity to mountain ranges or coastlines (stagnant air masses) all contribute to photochemical smog formation.

Human exposure to ozone can result in both acute (short-term) and chronic (long-term) health effects. The high reactivity of ozone makes it a strong lung irritant, even at low concentrations. Formaldehyde, peroxyacetylnitrate (PAN), and other smog-related oxygenated organics are eye irritants. Ground-level ozone also affects crops and vegetation adversely when it enters the stomata of leaves and destroys chlorophyll, thus disrupting photosynthesis. Finally, since ozone is an oxidant, it causes materials with which it reacts to deteriorate, such as rubber and latex painted surfaces.

Carbon monoxide (CO)

CO is a colourless, odourless gas formed primarily as a by-product of incomplete combustion. The major health hazard posed by CO is its capacity to bind with haemoglobin in the blood stream and thereby reduce the oxygen-carrying ability of the blood. Transportation sources account for the bulk (76.6 per cent) of total national CO emissions. Areas with high traffic congestion generally will have high ambient CO concentrations. High localised and indoor CO levels can come from cigarettes (second-hand smoke), wood-burning fireplaces, and kerosene space heaters)

Lead

Lead in the atmosphere is primarily found in fine particulates, up to 10 microns in diametre, which can remain suspended in the atmosphere for significant periods of time. Tetraethyl lead $(CH_3CH_2)_4$–Pb) was used as an octane booster and antiknock compound for many years before its full toxicological effects were understood. The Clean Air Act of 1970 banned all lead additives and the dramatic decline in lead concentrations and emissions has been one of the most important yet unheralded environmental improvements of the past twenty-five years. In 1997, industrial processes accounted for 74.2 per cent of remaining lead emissions, with 13.3 per cent resulting from transportation, and 12.6 per cent from non transportation fuel combustion.

Lead also enters waterways in urban runoff and industrial effluents, and adheres to sediment particles in the receiving water body. Uptake by aquatic species can result in malformations, death, and aquatic ecosystem instability. There is a further concern that increased levels of lead can occur locally due to acid precipitation that increases lead's solubility in water and thus its bioavailability. Lead persists in the environment and is accumulated by aquatic organisms.

Lead enters the body by inhalation and ingestion of food (contaminated fish), water, soil, and airborne dust. It subsequently deposits in target organs and tissue, especially the brain. The primary human health effect of lead in the environment is its effect on brain development, especially in children. There is a direct correlation between elevated levels of lead in the blood, especially in the urban areas of developing countries that have yet to ban lead as a gasoline additive.

Particulate matter

Particulate matter (PM) is the general term for microscopic solid or liquid phase (aerosol) particles suspended in air. PM exists in a variety of sizes ranging from a few Angstroms to several hundred micrometres. Particles are either emitted directly from primary sources or are formed in the atmosphere by gas-phase reactions (secondary aerosols).

Since particle size determines how deep into the lung a particle is inhaled, there are two NAAQS for PM, $PM_{2.5}$, and PM_{10}. Particles smaller than 2.5 µm are called 'fine'; are composed largely of inorganic salts (primarily ammonium sulphate and nitrate), organic species, and trace metals. Fine PM can deposit deep in the lung where removal is difficult. Particles larger than 2.5 µm are called 'coarse' particles, and are composed largely of suspended dust. Coarse PM tends to deposit in the upper respiratory tract, where removal is more easily accomplished.

Coarse particle inhalation frequently causes or exacerbates upper respiratory difficulties, including asthma. Fine particle inhalation can decrease lung functions and cause chronic bronchitis. Inhalation of specific toxic substances such as asbestos, coal mine dust, or textile fibres are now known to cause specific associated cancers (asbestosis, black lung cancer, and brown lung cancer, respectively).

An environmental effect of PM is limited visibility in many parts of the United States including some National Parks. In addition, nitrogen and sulphur containing particles deposited on land increase soil acidity and alter nutrient balances. When deposited in water bodies, the acidic particles alter the pH of the water and lead to death of aquatic organisms. PM deposition also causes soiling and corrosion of cultural monuments and buildings, especially those that are made of limestone.

SO_2, NO_x, and acid deposition

Sulphur dioxide (SO_2) is the most commonly encountered of the sulphur oxide (SO_x) gases, and is formed upon combustion of sulphur-containing solid and liquid fuels (primarily coal and oil). SO_x are generated by electric utilities, metal smelting, and other industrial processes. Nitrogen oxides (NO_x) are also produced in combustion reactions; however, the origin of most NO_x is the oxidation of nitrogen in the combustion air. After being emitted, SO_x and NO_x can be transported over long distances and are transformed in the atmosphere by gas phase and aqueous phase reactions to acid components (H_2SO_4 and HNO_3). The gas phase reactions produce microscopic aerosols of acid-containing components, while aqueous phase reactions occur inside existing particles. The acid is deposited to the earth's surface as either dry deposition of aerosols during periods of no precipitation or wet deposition of acid-containing rain or other precipitation. There are also natural emission sources for both sulphur and nitrogen-containing compounds that contribute to acid deposition. Water in equilibrium with CO_2 in the atmosphere at a concentration of 330 ppm has a pH of 5.6. When natural sources of sulphur and nitrogen acid rain precursors are considered, the 'natural' background pH of rain is expected to be about 5.0. As a result of these considerations, 'acid rain' is defined as having a pH less than 5.0. Figure 1.5 shows the major environmental cause and effect steps for acidification of surface water by acid rain.

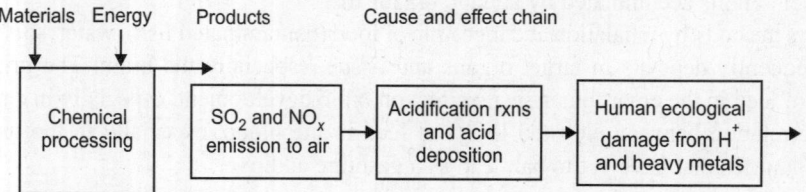

Fig. 1.5. Environmental cause and effect chain for acid rain.

Major sources of SO_2 emissions are non-transportation fuel combustion (84.7 per cent), industrial processes (8.4 per cent), transportation (6.8 per cent), and miscellaneous (0.1 per cent). The goal of this programme is to decrease acid deposition significantly by controlling SO_2 and other emissions from utilities, smelters, and sulphuric acid manufacturing plants, and by reducing the average sulphur content of fuels for industrial, commercial, and residential boilers.

There are a number of health and environmental effects of SO_2, NO_x, and acid deposition. SO_2 is absorbed readily into the moist tissue lining the upper respiratory system, leading to irritation and swelling of this tissue and airway constriction. Long-term exposure to high concentrations can lead to lung disease and aggravate cardiovascular disease. Acid deposition causes acidification of surface water, especially in regions of high SO_2 concentrations and low buffering and ion exchange capacity of soil and surface water. Acidification of water can harm fish populations, by exposure to heavy metals, such as aluminum which is leached from soil. Excessive exposure of plants to SO_2 decreases plant growth and yield and has been shown to decrease the number and variety of plant species in a region.

Air Toxics

Hazardous air pollutants (HAPs), or air toxics, are airborne pollutants that are known to have adverse human health effects, such as cancer. Currently, there are over 180 chemicals identified on the Clean Air Act list of HAPs (US EPA 1998). Examples of air toxics include the heavy metals mercury and chromium, and organic chemicals such as benzene, hexane, perchloroethylene (perc), 1,3-butadiene, dioxins, and polycyclic aromatic hydrocarbons (PAHs).

The Clean Air Act defined a major source of HAPs as a stationary source that has the potential to emit 10 tons per year of anyone HAP on the list or 25 tons per year of any combination of HAPs. Examples of major sources include chemical complexes and oil refineries. The Clean Air Act prescribes a very high level of pollution control technology for HAPs called maximum achievable control technology (MACT). Small area sources, such as dry cleaners, emit lower HAP tonnages but taken together are a significant source of HAPs. Emission reductions can be achieved by changes in work practices such as material substitution and other pollution prevention strategies.

HAPs affect human health via the typical inhalation or ingestion routes. HAPs can accumulate in the tissue of fish, and the concentration of the contaminant increases up the food chain to humans. Many of these persistent and bioaccumulative chemicals are known or suspected carcinogens.

WATER QUALITY ISSUES

The availability of freshwater in sufficient quantity and purity is vitally important in meeting human domestic and industrial needs. Though 70 per cent of the earth's surface is covered with water, the vast majority exists in oceans and is too saline to meet the needs of domestic, agricultural, or other uses. Of the total 1.36 billion cubic kilometres of water on earth, 97 per cent is ocean water, 2 per cent is locked in glaciers, 0.31 per cent is stored in deep groundwater reserves, and 0.32 per cent is readily accessible freshwater (4.2 million cubic kilometres). Freshwater is continually replenished by the action of the hydrologic cycle. Ocean water evaporates to form clouds, precipitation returns water to the earth's surface, recharging the groundwater by infiltration through the soil, and rivers return water to the ocean to complete the cycle. In the United States, freshwater use is divided among several sectors; agricultural irrigation 42 per cent. electricity generation 38 per cent, public supply 11 per cent, industry 7 per cent, and rural uses 2 per cent. Groundwater resources meet about 20 per cent of US water requirements, with the remainder coming from surface water sources.

Contamination of surface and groundwater originates from two categories of pollution sources. Point sources are entities that release relatively large quantities of waste-water at a specific location, such as industrial discharges and sewer outfalls. Non-point sources include all remaining discharges, such as agricultural and urban runoff, septic tank leachate, and mine drainage. Another contributor to water pollution is leaking underground storage tanks. Leaks result in the release of pollution into the subsurface where dissolution in groundwater can lead to the extensive destruction of drinking water resources.

Besides the industrial and municipal sources we typically think of in regard to water pollution, other significant sources of surface and groundwater contamination include agriculture and forestry. Contaminants originating from agricultural activities include pesticides, inorganic nutrients such as ammonium, nitrate, and phosphate, and leachate from animal waste. Forestry practices involve widespread disruption of the soil surface from road building and the movement of heavy machinery on the forest floor. This activity increases erosion of topsoil, especially on steep forest slopes. The resulting additional suspended sediment in streams and rivers can lead to light blockage, reduced primary production in streams, destruction of spawning grounds, and habitat disruption of fisheries.

Transportation sources also contribute to water pollution, especially in coastal regions where shipping is most active.

ECOLOGY

Ecology is the study of material flows and energy utilisation patterns in communities of living organisms in the environment, termed ecosystems. This area of science is very important in pollution prevention because of the possibility that pollutants entering sensitive ecosystems might disrupt the cycling of essential nutrients and elements for life, with potentially unforeseen negative consequences. Ecosystems, whether aquatic or terrestrial, share a common set of characteristics. They extract energy from the sun and store this energy in the form of reduced carbon-based compounds (biomass) in a process termed *photosynthesis*. Another very important function of ecosystems is to cycle elements and molecules through the environment, alternating between organic and inorganic forms of carbon, nitrogen, phosphorus, and sulphur.

Organisms that capture solar energy are primary producers which inhabit the first trophic level of the food chain in ecosystems. Examples of primary producers are plants in terrestrial ecosystems. For aquatic systems, members include aquatic plants, algae, and phytoplankton. The second trophic level is inhabited by the primary consumers, such as grazing animals on land and zooplankton and insects in aquatic environments, which prey upon the primary producers. The third trophic level is occupied by the secondary consumers, which prey upon the primary consumers. Examples are birds of prey, mammalian carnivores, fish, and many others. Additional trophic levels are possible depending upon the particular ecosystem.

Carnivores at the highest trophic levels in ecosystem food chains can encounter increased exposure to certain classes of anthropogenic pollutants. Chemicals that are hydrophobic (water-hating, non-polar organic compounds of high molecular weight), persistent (do not biodegrade or react biologically in ecosystems), and toxic are of particular concern because these chemicals bioaccumulate in animal fat tissue and are transferred from lower to higher trophic levels in the food chain. High levels of polychlorinated biphenyls (PCBs), certain pesticides, and mercury compounds have been detected in fish of the Great Lakes. The use of the pesticide DDT in the 1950s and 1960s caused dramatic reductions in birth rates of certain birds of prey that were consuming contaminated fish and other contaminated animals. Such examples demonstrate the need to understand the workings of ecosystems so that one can mitigate the harm that chemicals released into the environment can cause to ecosystems.

NATURAL RESOURCES

The production of industrial materials and products begins with the extraction of natural resources from the environment. The availability of these resources is vital for the sustained functioning of both industrialised and developing societies. Examples of natural resources include water, minerals, energy resources like fossil fuels, solar radiation, wind, and lumber. Renewable resources have the capacity to be replenished, while non-renewable resources are only available in finite quantities. The management of natural resources is intended to assure an adequate supply of these materials for anticipated future uses, also known as sustainable use. Non-renewable resources are of particular importance because of their inherently finite supply. For example, most energy requirements of today and of the foreseeable future will be met using non-renewable fossil fuels, such as oil, coal, and natural gas. As the availability of resources is diminished, the costs and energy consumption for producing these materials are likely to increase. Resource management techniques like conservation, recycling of materials, and improved technologies can be used to ensure the availability of these materials for the future. In some cases, materials already in use can be continuously recycled into new products (for instance, lead from batteries, steel from scrap cars, aluminium from beverage cans).

Chapter 2
Risk Concepts

INTRODUCTION

Risk is a quantitative assessment of the probability of an adverse outcome. Risk may result from voluntary exposure to hazardous conditions in one's occupation, involuntary exposure to radiation, chemicals, pathogens, or the reckless behaviour of others, or natural disasters.

There are four components of risk assessment; hazardous assessment; doseresponse; exposure assessment; and risk characterisation. The engineer should work with chemists, toxicologists, and others when a risk assessment is needed. Although there may be uncertainties in performing risk assessments, it can assist in choosing between process options. Risk is a concept used in the chemical industry and by practicing chemical engineers. The term risk is multifaceted and is used in many disciplines such as: finance (rate of return for a new plant or capital project, process improvement, etc.), raw materials supply (single source, back integration), plant design and process change (new design, impact on bottom line), and site selection (foreign, political stability). Though the term risk used in these disciplines can be discussed either qualitatively or quantitatively, it should be obvious that these qualitative or quantitative analyses are not the same in all fields (financial risk process change risk). This chapter will focus on the basic concept of environmental risk and risk assessment as applied to a chemical's manufacturing, processing, or use, and the impact of exposure to these chemicals on human health or the environment.

Risk assessment is a systematic, analytical method used to determine the probability of adverse effects. A common application of risk assessment methods is to evaluate human health and ecological impacts of chemical releases to the environment. Information collected from environmental monitoring or modelling is incorporated into models of human or worker activity and exposure, and conclusions on the likelihood of adverse effects are formulated. As such, risk assessment is an important tool for making decisions with environmental consequences. Almost always, when the results from environmental risk assessment are used, they are incorporated into the decision-making process along with economic, societal, technological, and political consequences of a proposed action.

DESCRIPTION OF RISK

Risks can be grouped into three general categories.

Voluntary Risk

A result of actions taken by choice or out of necessity. Examples include firefighting, driving, bungee cord jumping, and lifestyle choices such as diet and smoking.

Natural Disasters

These include floods, hurricanes, earthquakes, and other disasters that are beyond human control. However, the risk to natural disasters can be exacerbated by such voluntary actions as living in a known flood plain or on an active earthquake fault.

Involuntary Risk

Risk resulting from uncontrollable actions of others. Examples include pesticide residues or pathogens in food, occupational exposure to industrial chemicals. These risks tend to have more uncertainty and are not as well known.

Quantitatively, in the above categories, risk in the first two groupings (voluntary and natural) is frequently determined by actuarial-based statistics (e.g. fatalities are correlated with activity, location, and other parameters). In involuntary exposure, such as those to chemicals, risk, for the most part, is based on inferred data (animal tests, analogs, extrapolation). People are more familiar with expressions of risk associated with various activities than they are with risks associated with chemical exposure. Table 2.1 lists one assessor's evaluation of various risk factors.

Table 2.1. Risks from various environmental pollutants.

Risk factor
Cancer risks associated with environmental pollutants
Indoor radon
Worker chemical exposure
Pesticide residues in food
Indoor air pollution
Consumer products use
Stratospheric ozone depletion
Inactive hazardous waste sites
Carcinogens in air pollution
Drinking water contaminants
Noncancer risks associated with environmental pollutants
Lead
Carbon monoxide
Sulphur dioxide
Radon
Air pollutants (e.g. carbon tetrachloride, chlorine)
Drinking water contaminants (e.g. lead, pathogens, nitrates, chlorine disinfectants)
Industrial discharge into surface water
Sewage treatment plant sludge
Mining wastes

Risks from toxic chemicals, depending on the context, may be defined, described, and calculated in different ways. Risk is normally defined as the probability for an individual to suffer an adverse effect from an event. What is the probability that certain types of cancer will develop in people exposed to aflatoxin in peanut products or benzene from gasoline? What is the likelihood that workers exposed to lead will develop nervous system disorders? In the context of this text, a chemical release is an example of an event. As with any relationship expressing or using probability, there is no defined way of expressing (mathematically or with scientific rigor) a single deterministic value of a phenomenon that is probabilistic. A fairly simple conceptual way of expressing chemical risk is shown below.

$$Risk = f (Hazard, Exposure)$$

Hazard is the potential for a substance or situation to cause harm or to create adverse impacts on persons or the environment. The magnitude of the hazard reflects the potential adverse consequences, including mortality, shortened life-span, impairment of bodily function, sensitisation to chemicals in the environment, or diminished ability to reproduce. Exposure denotes the magnitude and the length of time the organism is in contact with an environmental contaminant, including chemical, radiation, or biological contaminants. When risk is in terms of probability, it is expressed as a fraction, without units. It has values from 0.0 (absolute certainty that there is no risk) to 1.0 (absolute certainty that an adverse outcome will occur).

For chemicals the term hazard is typically associated with the toxic properties of a chemical specific to the type of exposure. Similar chemicals would have similar innate hazards. However, one must examine the exposure to that hazard to determine the risk. For example, let us say you have three pumps that are all transporting the same chemical (same hazard), but one pump has a seal leak. Which pump poses the greatest risk to the worker? The pump with the seal leak has the greatest potential for exposure, while the hazards are equal (same chemical), so the seal leak pump poses the greatest risk. To expand, let's say we have three pumps that are transporting different chemicals; which one poses the greatest risk to the worker? In this case the engineer would need to examine the hazard—or innate inherent toxicity—of each of the chemicals, as well as the operation of the pumps to determine which poses the greatest risk.

If a chemical is known to present dermal hazard, the exposure would be expressed as surface area of potentially exposed skin multiplied by the mass of the chemical per unit of surface area of skin that it contacts. In this text, the exposure term in the above equation, unless otherwise stated, will be for human exposure (ingestion, inhalation, and dermal).

The concept of exposure and hazard equating to risk may be applied in different ways, depending on the information available. In addition, risks may be described across pathways or routes, or as a comparison between, say, using one chemical versus another. In the future, research is likely to reveal completely new sources of chemical risks which were previously unknown. For example, stratospheric ozone depletion and endocrine disrupters were emerging concepts. Today, the level of risk can be much more accurately characterised. The risk assessment framework presented in this chapter is sufficiently flexible to apply even to new sources of risk from chemical releases as they are recognised.

VALUE OR RISK ASSESSMENT IN THE ENGINEERING PROFESSION

Risk assessment may be conceptualised as simply a means of organising and analysing all available scientific information that addresses the question, what are the risks associated with a chemical manufacturing process or use of a chemical product? If an engineer is asked to conduct a comprehensive assessment, such as developing an environmental impact statement for a proposed new facility, a major study of this magnitude would necessitate the formation of a team of appropriate professionals (engineer, toxicologist, ecologist, chemist, industrial hygienist, medical and legal staff, etc.). It is critical that the resulting assessment focus not only on the quantitative aspects of risk but also the qualitative character of risk. This need for qualitative assessment is often driven by serious data gaps in health and ecotoxic effects, which preclude precise quantification of all impacts.

From an engineering perspective, it may be useful to think of risk as safety issues extrapolated from the present to the long term. That is, safety may be thought of as the likelihood of immediate adverse consequences, and risk as the likelihood of long-term adverse consequences. Engineers can elevate risk concerns from chronic exposures to toxic chemicals to the same level of concern as safety issues. As with safety issues, the potential for chemical risks is only one of many factors that influences decisions. Financial considerations will always be paramount in economic, social, and environmental costs and benefits of the use of any pesticide. In other words, economic and other factors may or may not be combined with risk issues as regulations are developed. These details become important if the engineer is required to understand and follow the regulations or even requested to comment regarding proposed regulations.

GENERAL OVERVIEW OF RISK ASSESSMENT CONCEPTS

Risk assessment consists of four major components: hazard assessment, dose-response, exposure assessment, and risk characterisation.

Hazard Assessment

What are the adverse health effects of the chemical(s) in question? Under what conditions? For example, does it cause a certain kind of cancer? Toxicologists usually perform this analysis. Since this information is pertinent to use of a chemical, sometimes hazard information can be obtained from reference data.

Dose-Response

How much of the chemical causes a particular adverse effect? There may be multiple adverse health effects, or responses, for the same chemical at different concentrations. Each adverse effect has a unique dose-response curve. The dose-response curve is non-linear because some members of the population are more sensitive than others.

For our purposes, dose is defined as the quantity of a chemical that crosses a boundary to get into a human body or organ system. The term applies regardless of whether the substance is inhaled, ingested, or absorbed through the skin. Dose-response, then, is a mathematical relationship between the magnitude of a dose and the extent of a certain negative response in the exposed population.

Exposure Assessment

Who is exposed to this chemical? How much of the chemical reaches the boundary of a person, and how much enters the person's body? Exposure may be measured, estimated from models, or even back-calculated from measurements called biomarkers taken from exposed people.

Risk Characterisation

How great is the potential for adverse impact from this chemical? What are the uncertainties in the analyses? How conclusive are the results of these analyses?

This general risk assessment framework has been tailored to human health risk assessment from exposure to chemicals. A risk assessment team may decide that specific aspects of the eco-assessment require attention. This level of activity is critical for new plant siting (grass-roots), which must include a thorough examination of the ecosystems in-place as well as unique areas (wetlands, forests, endangered

species habitat). The risk assessment process can be iterative. That is, if a cursory or screening risk assessment identifies concerns, a more rigorous process may be called for. This process may in turn illustrate that there are important data gaps that need to be filled to render the process sufficiently conclusive for risk management. The data gaps may be filled with recommendations for special studies with varying cost and time requirements, such as:
1. Proceeding with testing for health effects.
2. Evaluating the effectiveness of engineering controls and personnel protective equipment (PPE) to limit exposures.
3. Defining the kinetics and decomposition products of a waste stream and the impact of the chemical waste and its degradation products on local flora and fauna.

If it is reasonably clear from the risk assessment that a risk exists, the next step is risk management.

Risk management is the process of identifying, evaluating, selecting, and implementing actions to reduce risk to human health and to ecosystems. The goal of risk management is scientifically sound, cost effective, integrated actions that reduce or prevent risks while taking into account social, cultural, ethical, political, and legal considerations.

Risk managers must clearly answer many questions, some of which are:
1. What level of exposure to a chemical risk agent is an unacceptable risk?
2. How great are the uncertainties and are there any mitigating circumstances?
3. Are there any trade-offs between risk reduction, benefits, and additional cost?
4. What are the chances of risk shifting, that is, transferring risk to other populations?
5. Are some of the risks worse than others?

The answers to these questions often depend on the culture and values of the organisation that commissioned the risk assessment. Minimising risk through improved engineering design and proactive process development should be core values of the engineer.

HAZARD ASSESSMENT

In the context of this text, a hazard is an adverse health effect related to chemical exposure. This section begins with a discussion of hazard assessment. It continues with a discussion about structural activity relationships, which are tools used to screen hazards in the absence of chemical-specific laboratory data, based on the known hazards of materials with similar chemical structures.

As noted above, a chemical exposure hazard assessment answers the question: What are the adverse effects of a chemical? The most common adverse effects, or endpoints, studied are various kinds of cancer, but other types of adverse health effects such as endocrine disruption or reproductive toxicity are also currently being studied. Effects immediately dangerous to life or health may result from a high but brief or acute exposure, while long-term effects may result from chronic exposures to low levels of a toxin that are insufficient to cause any acute effects. Health effect studies are usually performed on rodents; these studies are called subchronic effects studies, and provide the basis for estimating a particular hazard or hazards.

Cancer and Other Toxic Effects

Cancers of various organs or systems are among the most thoroughly studied toxic effects. Other examples of toxic effects which are known to be caused by chemical substances are decreased pulmonary capacity caused by inhalation of asbestos, and damage to the nervous system and internal organs resulting from ingestion of lead, mercury, and other metals. Chemical exposures may also induce neurotoxicity, reproductive toxicity, or developmental toxicity. A relatively new discipline, developmental toxicity, refers to birth defects and other toxic effects which become apparent after birth, and which may be rooted in the prenatal period.

From the perspective of hazard assessment, cancer can be caused by two different types of chemical substances — genotoxic carcinogens and nongenotoxic carcinogens. Genotoxic chemicals are believed to have no threshold amount below which they will not cause cancer. Theoretically, one molecule of a genotoxic carcinogen could alter DNA and cause a mutation. In most cases, such an exposure would not cause cancer because of natural mechanisms which can repair internal damage caused by exposures at this level. Unfortunately, only (expensive) mechanistic studies can distinguish whether a carcinogen is genotoxic or not. In the absence of these studies, genotoxicity is generally assumed.

In contrast, nongenotoxic carcinogens are believed to have a safe threshold quantity. This becomes clearer from dose-response assessment. For the purpose of risk assessment, nongenotoxic substances are analysed much like chemicals with endpoints other than cancer by using the concept of a Hazard Quotient, which is also discussed below. There are other concepts for addressing quantitative estimates of risk, but they are beyond the scope of this chapter.

Hazard Assessment for Cancer

US EPA has developed guidelines for hazard assessment of chemical carcinogens. There are three types of information used to make the hazard determination: human data, animal data, and supporting data. The data are first evaluated as to their conclusiveness. Then, the substance is classified, usually as part of one of the following groups:

Group A: Carcinogenic to humans.
Group B1:   Probably carcinogenic to humans based on limited human evidence of carcinogenicity.
Group B2:   Probably carcinogenic to humans based on sufficient animal evidence, but inadequate human evidence.
Group C: Possibly carcinogenic to humans.
Group D: Not classifiable for human carcinogenicity.
Group E: Evidence of noncarcinogenicity for humans.

Hazard Assessment for Non-Cancer Endpoints

Adverse effects other than cancer and gene mutations are generally assumed to have a dose or exposure threshold. As a result, a different approach is used to evaluate potential risk for these non-cancer effects, which include liver toxicity, neurotoxicity, and kidney toxicity. The first step in this approach requires the identification of a critical effect for which the magnitude of the response can be assessed. The Reference Dose (RfD) or Reference Concentration (RfC) approach is used to evaluate such chronic effects. The Rf+D is defined as 'an estimate (with uncertainty spanning perhaps an order of magnitude) of a daily exposure to the human population that is likely to be without appreciable risk of deleterious effects during a lifetime' and is expressed as a mg pollutant/kg body weight/day. The RfC is expressed as a concentration, or mg/m^3. Roughly speaking, it is the baseline 'safe' dose or concentration to which a real exposure may be compared.

The RfD or RfC is usually based on the most sensitive known effect, i.e. the effect that occurs at the lowest dose. The basic approach for deriving an RfD or RfC involves determining a 'no-observed-adverse-effect level (NOAEL)' or 'lowest-observed-adverse-effect level (LOAEL)' from an appropriate animal study or human epidemiologic study, and applying various uncertainty and modifying factors to arrive at the RfD/RfC. Each factor represents a specific area of uncertainty. The uncertainty surrounding these terms spans about an order of magnitude. For example, an RfD based on an NOAEL from a long-term animal study might incorporate a factor of 10 to account for the uncertainty in extrapolating from the test species to humans, and another factor of 10 to account for the variation in sensitivity within the human population. Another common uncertainty factor may be used to extrapolate from subchronic test exposures to potentially chronic human exposures. An RfD based on an LOAEL typically contains an additional factor of 10 to account for the extrapolation from LOAEL to NOAEL. Finally, another modifying factor (between 1 and 10) is sometimes applied to account for uncertainties in data quality.

The combination of these uncertainty factors can result in highly conservative interpretations. One can conclude from an RfD that a chemical is quite toxic when the reality is that little is known about the chemical's toxicity. The engineer must be sure to use appropriate caveats when presenting data, and should understand the reason or reasons for a specific RfD value.

The NOAEL described above is based on a single study, or data point from a more complete data set. However, it is well known that when drawing conclusions, it is preferable to use all of the available data rather than a single point. To this end, the US EPA is moving to a method that entails developing the RfD or RfC from a dose-response relationship derived from all the data. This new approach, the Benchmark Dose concept, has a goal of improving the quality of the RfD and RfC estimates, and reducing the number of uncertainty factors used.

Structure Activity Relationships (SAR)

Structure Activity Relationships (SAR) are an effective technique for estimating the hazard, as well as other properties, of a chemical. The US EPA often use SARs when estimating hazard and other elements of risk. SARs estimate hazards by drawing analogies with chemically similar substances whose hazard has been studied. The similar substance is called a structural analog. This technique requires the expert judgment of toxicologists. Although an engineer should not be expected to perform this kind of analysis (for assessing hazard), an engineer might request that such an analysis be conducted. Therefore, a brief explanation is provided here.

The definition of structural activity, as it applies here, is the relationship between the structural property of a molecule and its biological activity. Health effects which can be evaluated are many, and include absorption into the body; metabolism by the body, oncogenicity (capability to produce tumors); mutagenicity (capability to induce DNA mutations); and acute, chronic, and subchronic toxicity, neurotoxicity, developmental, and reproductive effects (adverse effects on fertility). Some examples of chemical classes of concern that are amenable to SAR review are: acrylamides, vinyl sulphones, dianilines, sulphoniums, epoxides, benzothiazoliums, hindered amines, acrylates, and dichlorobenzene pigments.

The basis for choosing an appropriate structural analog may be structure, substructure, or physical/chemical properties. For example, an unsaturated ketone may be a good analog for an unsaturated ester. In some instances, the toxicologist predicts metabolites (biotransformation products) of the chemical of interest and assesses the hazards of the metabolites.

Some information about environmental fate is required to complete an SAR assessment. Environmental fate is determined by physical-chemical properties of the chemical. Examples are octanol/water partition coefficient and water solubility. The intrinsic problem with using Structural Activity Relationships to estimate toxicity is the uncertainty associated with extrapolating information from one chemical to another. This uncertainty limits the accuracy of toxicity estimates made using SAR techniques, although they can be helpful when no other data are available. However, direct, accurate data on hazard should be used rather than SAR estimates whenever the data are available.

DOSE-RESPONSE CURVE

A dose-response curve (Fig. 2.1) is a graph of the quantitative relationship between exposure and to estimate a 'safe' dose. Actual dose is compared to safe dose. Actual dose is compared to safe dose to estimate risk. Dose-response answers the question: How large a dose causes what magnitude of effect? Larger dose cause greater and more serious effects. For a given chemical, there is a separate curve for each adverse health effect.

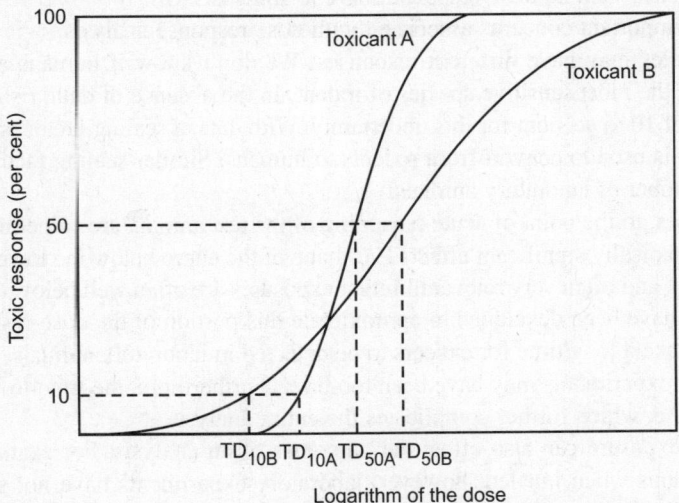

Fig. 2.1. Dose-response curves for two compounds that have different relative threshold limit values.

The basic shape of the dose-response curve is determined by the biological mechanism of action. On a subtler level, the curve illustrates the sensitivity of different members of the population. It is a plot of dose in mg chemical per kg of body weight, versus percent of the population affected by that dose. For example, an LD50, or lethal dose 50 per cent is a statistic frequently tabulated for some chemicals. It is the dose, in mg/kg, at which 50 per cent of the rats or other tested species die. This statistic emerges from a dose-response assessment. Rats, mice and rabbits are frequently tested species. They are like humans in that they are mammals, but they are also small, and breed and mature quickly, which can make the testing process more manageable. Nonetheless, these species may react very differently from humans to exposure to a particular chemical. Significant research efforts have been under way for some time to find reliable substitutes for animal testing of chemical hazards.

The curvature of the dose-response curve illustrates the varying sensitivity of different members of the exposed population. That is if sensitivity to the chemical were constant, dose-response would be a straight line. The curvature illustrates that some people (or, more likely, rodents) are especially vulnerable, while others are more resistant. Among humans, common examples of sensitive subpopulations are children, the elderly, and the immunosuppressed. Developing the data to support a dose-response curve is expensive, time consuming, and rigorous. It is generally not performed until some screening has suggested that it could be useful. When this testing is performed, it often begins with a rangefinder study. The purpose of this preliminary study is to determine what order of magnitude of dose generates adverse effects. This improves the quality of the dose-response testing.

The outcome of the overall dose-response effort helps tell the assessor what the toxicological endpoint of concern is. Are we concerned about neurotoxicity in young children, whose nervous system is still developing? Are we studying cancer in a particular organ? The dose-response study also provides the NOAEL and the benchmark dose (BMD). These quantities can provide a basis for risk assessment.

Since dose-response testing is so resource-intensive, risk assessors sometimes use structural-activity relationships to estimate a NOAEL or BMD, generally incorporating a coefficient to account for uncertainty. That is, we find a chemical whose NOAEL or BMD is known and has similar (chemical) functional groups to the substance of interest. The structural analog is then used to estimate a NOAEL or BMD for the substance with no dose-response curve available.

There are several important concerns associated with dose-response analysis:
1. Different species may have different responses. We don't know if humans are more or less sensitive than the most sensitive species of rodent. In the absence of data, risk assessors use a safety factor of 10 to account for this uncertainty. With data, a scaling factor of body weight to the $3/4$ power is used to convert from rodents to humans. Similar scaling factors are available for a large number of laboratory animals.
2. Very high doses, to the point of acute poisoning of the test animal, are sometimes necessary to generate a statistically significant effect. The shape of the curve below the lowest dose tested is truly unknown, and often very relevant. Actual exposures are often well below the lowest tested dose. Models have been developed to approximate this portion of the dose-response curve.
3. Since it may take a long time for cancers to be detected in laboratory animals, some otherwise well-designed experiments may have been too brief. Furthermore, the time-to-tumor may be a function of dose, which further complicates the entire analysis.
4. The route of exposure can also effect the outcomes of an analysis. For example, Chromium (VI) is hazardous when inhaled; however, laboratory experiments have not shown evidence that exposure through ingestion causes any adverse effects. Therefore, it is extremely important to be cognizant of the route of exposure when assessing risk.

EXPOSURE ASSESSMENT

The amount of a substance that comes into contact with the external boundaries of a person is called exposure. The quantity that crosses the external boundary is defined as dose and the amount absorbed is the internal dose. The ratio of the internal dose to exposure is called the bioavailability of the substance. While some organisations and older sources of information use slightly different definitions, these have been adopted by the US EPA.

Two common routes of exposure to chemicals are through the skin (dermal) and the lungs (inhalation). Because exposure to chemical in the workplace can occur through inhalation and skin absorption, the engineer must be aware of potential pathways into the body. The exposure pathway model highlights potential pathways leading from the process to the worker and provides a framework for evaluating pathways for exposure to chemical in the workplace. Most dermal exposures result from hand contact and may occur while performing common worker activities such as sampling, drumming, filter changing, and maintenance. Since skin is a protective barrier for many kinds of chemicals, the bioavailability of these substances is often low, perhaps 5 per cent. Inhalation exposure may be in the form of vapours, aerosols, or solid particulates. Exposure to vapours, for example, may occur due to vapours generated during activities such as drumming and sampling, or from fugitive emissions from small process leaks. Unlike dermal exposure, the bioavailability of these inhaled vapours can be quite high, often close to 100 per cent.

A third route of exposure is ingestion, through either eating or drinking. Exposure through ingestion is not usually of interest in scenarios likely to be assessed in occupational settings. However, the engineer should note that ingestion can be a major source of exposure to workers who may eat, drink, or smoke on the job without adequate time or facilities for washing up, or where clean rooms are not available in situations where surfaces are contaminated with chemicals and eating, drinking, or smoking is allowed. A fourth route of exposure is percutaneous exposure, or injection through the skin. However, this type of exposure is rarely seen in the workplace. A notable example is the potential for needlesticks in healthcare settings.

The preferred approach for assessing exposure is to use personal monitoring data for the chemical of interest at the site. If not available, monitoring data for the chemical at sites with similar operations is the next choice. If there are no data available on the chemical of interest, exposure can be assessed using data for a surrogate chemical. A surrogate chemical is one whose physical and chemical properties are as similar as possible, and is used in similar operations. Finally, in the absence of any relevant data, exposure can be assessed using models. For example, a mass balance model can be used to estimate inhalation exposure to vapours.

A different approach to addressing exposure is to measure some appropriate biomarker. This applies to people who have already been exposed. A biomarker is a measurable substance whose presence in the body is a direct result of exposure to a specific chemical. Exposure may be estimated from models and based upon the biomarker measurements. Unfortunately, there are few substances that pose an exposure concern for which a biomarker has already been identified and measured. Some substances have metabolites which can be detected in blood or urine; these are common testing approaches for biomarkers.

As the engineer proceeds down this hierarchy of methods for assessing exposure, the degree of uncertainty increases. Information about this uncertainty must be communicated before risk management decisions are made. On the other hand, a high degree of uncertainty may be acceptable for some decisions. In addition to workplace exposure, exposures in the ambient environment resulting from plant emissions may also be part of the exposure assessment. Using emissions as a starting point, exposure can be estimated from a variety of models which consider environmental fate and transport.

RISK CHARACTERISATION

Risk characterisation is the amalgamation of available hazard and exposure information, i.e. risk, as well as all major issues developed during the assessments, including the uncertainty of all aspects of the analysis. It embodies the effects of potential concern, the route and the magnitude of the expected exposure and the numbers of the populations estimated to be exposed. As stated above, the primary human health concern is carcinogenicity. Generally, the potential for carcinogenicity is assessed using pharmaco-kinetics, chronic toxicity data from analogs, and mechanistic information (when these data are available).

Risk Characterisation of Cancer Endpoints

The classical treatment of cancer risk defines risk as the probability of developing cancer from a particular chemical if a sub-population is exposed to that chemical over a lifetime. A person can contract cancer from many sources besides exposure to a particular chemical. This concept is called the background cancer level, and must be separated from the probability of developing cancer from a particular chemical exposure.

Thus, risk is defined in this particular context as the cancer probability in excess of the background cancer level. Our basic equation of risk is:

$$\text{Risk} = f(\text{Hazard, Exposure})$$

The basis for cancer risk assessment is the dose-response curve (risk of incidence of cancer vs. dose of an agent). Unfortunately, for cancer, bioassays are usually run at only two doses to describe the carcinogenic response of the test species. The relationship is typically non-linear. Since it is assumed that carcinogens do not have thresholds, the 'cancer' model generates a non-linear curve. There is never enough data to provide a complete dose-response curve. To deal with this reality, the risk assessor is left with the option of applying one of a number of mathematical models to the limited data set so as to describe the relationship. For a new chemical, with limited dose-response data, one methodology is to use the slope of the dose-response curve or (per cent response per mg pollutant per kg of body weight per day) as a measure of hazard. Exposure is the quantity that arrives at the surface of a person's body, in mg of pollutant per kg body weight per day. This simple application of the basic risk equation often provides the risk manager with sufficient information to make risk management decisions.

Risk Characterisation of Non-cancer Endpoints

Non-cancer risk also has a dose-response curve. The model relationship in this case is linear. Therefore, simplifying assumptions allow us to characterise the risk of adverse health effects as a simple ratio or Hazard Quotient. The hazard quotient is the ratio of the estimated chronic dose or exposure level to the RID or RfC. Hazard quotient values below unity imply that adverse effects are very unlikely. The more the hazard quotient exceeds unity, the greater the level of concern. However, the hazard quotient is not a probabilistic statement of risk.

Adding Risks

The discussion above presumes risk occurs from one chemical at one source. In fact, there are multiple chemicals, multiple pathways, and multiple exposure routes. It is necessary either to estimate what the most important risks are, or to calculate all sources and pathways. Aggregate and cumulative risk are fairly recent terms in the lexicon. Aggregate means adding risks together from multiple exposure routes: dermal, inhalation and ingestion.

The use of the term endpoint becomes important in the emerging area of cumulative risk assessment. Sometimes, the risks from one chemical may be too low to generate concern. However, several different chemicals may have the same toxicological endpoint.

That is, they affect an organ or system adversely in the same way. Exposures from these chemicals need to be combined to determine whether the adverse effect may occur as a result of a combination of chemical exposures.

SOURCES OF IGNITION IN HAZARDOUS AREA AND HOW TO CONTROL THEM

Basic rules of hazardous area classification are included in this section. Ignition sources are identified along with their effective control measures. The selection of equipment for particular zones since the required protection for respective hazardous area are discussed. The roles of safety assessor and engineer during design of area are also presented.

The rapid industrialisation in India during the last few years involved establishment of numerous projects in petrochemicals fertilisers and refineries. These sectors need additional safety measures in view of the hazardous environment in which equipment are required to work in.

The hazardous nature of the gases, vapours and dust present in the atmosphere in such locations make it necessary to go in for special type of enclosures and design features of equipment. The selection of equipment depends upon the designated areas wherein they are installed.

Hazardous areas are defined by three main criteria:
1. Type of hazards.
2. Likelihood of the hazard being present in flammable concentrations.
3. Ignition temperature of the hazardous material.

Type of Hazards

Hazards could be in the form of a gas, vapour, dust or fibre.

Gases and vapours

Gases and vapours are categorised in terms of their ignition energy or the maximum experimental safe gap (in respect of flameproof protection). This categorisation leads to the gas groups (Table 2.2).

Table 2.2. Gas group of hazards.

Mining		Surface industry	
Group I		Group II	
Methane	Group IIA	Group IIB	Group IIC
	Propane	Ethylene	Hydrogen

Note: The gases noted in the table are typical gases for each group.

Gases belonging to group IIC are the most severe and are highly inflammable.

Dust and fibres

Dust and fibres are also defined in terms of their ignition properties including dust cloud ignition properties.

Hazard Operability

The likelihood of the hazard being present in flammable concentrations will vary from place to place. Three zones are defined.

Gases and vapours

Three zones for gases and vapours are:
1. Zone 0: An area in which an explosive gas atmosphere is present continuously or for long periods.
2. Zone 1: An area in which an explosive gas atmosphere is likely to occur in normal operation;
3. Zone 2: An area in which an explosive gas atmosphere is not likely to occur in normal operation and, if it occurs, will only exist for a short time.

Dust

There are three zones for dust:
1. Zone 20: Dust cloud likely to be present continuously or for long periods.
2. Zone 21: Dust cloud likely to be present occasionally in normal operation.

3. Zone 22: Dust cloud unlikely to occur in normal operation, but if it does, will only exist for a short period. Ignition temperature of the hazardous material (temperature classes) is given in Table 2.3.

Table 2.3. Ignition temperature of the hazardous material (temperature classes).

Temperature class	Ignition temperature °C		Limiting temperature °C
	Above	Up to and including	
T6	85	100	85
T5	100	135	100
T4	135	200	135
T3	200	300	200
T2	300	450	300
T1	450	–	450

Selection of type of protection for electrical equipment located in different zones of hazardous areas: Electrical apparatus for use in hazardous areas need to be designed and constructed in such a way that it will not provide a source of ignition. There are ten recognised types of protection for electrical apparatus in hazardous areas. The different types of protection and the zones for which they are suitable are given in Table 2.4.

Table 2.4. Different types of protection.

Zone of hazardous area	Type of equipment/ protection	Remarks
Zone-0	i-Intrinsically safe	No electrical equipment should be allowed. When this is not practicable, intrinsically safe electrical equipment shall be employed
Zone-1	d-Flameproof equipment	Depends on sub classification based on the grouping of gases/vapours (Group-I, Group-IIA/IIB/IIC)
	p-Pressurised enclosure	
	q-Powder/sand filled equipment	Installation to be mutually agreed upon between purchaser and seller
	o-Oil immersed equipment	
	s-Special equipment	
Zone-2	d-Flameproof equipment	
	p-Pressurised enclosure	
	q-Powder/sand filled equipment	
	o-Oil immersed equipment	
	s-Special equipment	
	e-Increased safety equipment	Depends on sub-groups T1, T2, T3, T4, T5 depending upon auto ignition temperature of hazardous gases present
	n-Non-sparking equipment	Same as type e but starting of motor is not considered as a normal operation (starting is considered as abnormal)

Sources of Ignition

The sources of ignition include:
1. Sparks in electrical equipment.
2. Lightning strikes.
3. Hot surfaces.
4. Heated process vessels such as dryers and furnaces.
5. Hot process vessels.
6. Cutting and welding flames.
7. Space heating equipment.
8. Mechanical machinery.
9. Spontaneous heating.
10. Friction heating or sparks.
11. Impact sparks.
12. Stray currents from electrical equipment.
13. Electrostatic discharge sparks.
14. Vehicles, unless specially designed or modified are likely to contain a range of potential ignition sources.

Controlling Ignition Sources

Sources of ignition should be effectively controlled in all hazardous areas by a combination of design measures, and systems of work:
1. Classification of hazardous areas of the plant by preparing hazardous area layout.
2. HAZOP recommendations shall be in totality.
3. Using electrical equipment and instrumentation classified for the zone in which it is located (Table 2.4).
4. Effective earthing of all metallic equipment as per applicable standards.
5. Elimination of surfaces above auto-ignition temperatures of flammable materials being handled/stored, if possible by adopting any other method of construction.
6. Provision of lightning protection as per applicable standards.
7. Vehicles working in hazardous areas should ensure that fumes from exhaust pipe are controlled or arrested.
8. Correct selection of equipment to avoid high intensity electromagnetic radiation sources, e.g. limitations on the power input to fibre optic systems, avoidance of high intensity lasers or sources of infrared radiation.
9. Control over activities that could create intermittent hazardous areas, e.g. tanker loading/unloading.
10. Control of maintenance activities that may cause sparks/hot surfaces/naked flames through a hot 'permit to work' system.
11. Prohibition of smoking/use of matches/lighters.
12. For direct fired heaters, hot oil systems and processes operating above auto-ignition temperatures-safety should be achieved by a combination of a high standard of integrity of fuel and process, pipelines, together with the means of rapid detection and isolation of any pipe that fails. The consequences of the failure of a pipe carrying process materials within the furnace should be considered in any HAZOP study.

13. Dust explosions: Where toxic dusts are processed, releases into the general atmosphere should be prevented, and the extent of any zone, 21 or 22 outside the containment system should be minimal or non-existent. The interiors of different parts of the plant may need to be zoned as 20, 21 or 22, depending on the conditions at particular locations.

Classification of dust relating to auto-ignition and minimum ignition current is undertaken similarly for gases/vapours, but involves additional complications.

The explosibility of dust is dependent upon a number of factors:
1. Chemical composition.
2. Particle size.
3. Oxygen concentration.

Where toxic dusts are handled, in most cases occupiers will need to carry out testing of the product for its explosion properties.

The most likely hazard would arise in drying processes, if substantial quantities were held for extended periods, hot enough to start self heating or smouldering combustion.

Check List for Saftey Assessor

1. Have all ignition sources been considered?
2. Is a full set of plans identifying hazardous areas available? For a large site they need not all be provided in the report, but those examples relevant to the representative set of major accidents must be included.
3. Have all flammable substances present been considered during area classification, including raw materials, intermediates and by-products, final product and effluents? Commonly these will be grouped for the purposes of any area classification study.
4. Locations where a large release is possible and where the extent of hazardous areas has been minimised by the use of mechanical ventilation should be identified, e.g. gas turbine power generation units, compressor houses. Some reference to design codes, and commissioning checks to ensure that the ventilation satisfies the design code, should be assessed. The consequences of a loss of power to the system should be included in any section.
4. Have appropriate standards been used for selection of equipment in hazardous areas? Does the report identify old electrical equipment still in service in a hazardous area, and what assessment has been made to ensure it remains safe for use?
5. Is there a reference to the impact upon extent and classification of hazardous areas in the section describing plant modification?
6. What control measures over ignition sources are adopted in hazardous areas during maintenance; where ignition sources must be introduced and whether typical precautions include the use of supplementary ventilation and portable gas detectors.
7. If there are any large areas of zone 1 on the drawings; is there evidence that by design and operation, controls, the sources of release and consequently the location and extent of hazardous areas have been minimised?
8. Do any zone 2 areas extend to places where the occupier has inadequate control over activities that could create an ignition source, or is there any suggestion that the zone boundaries have been arbitrarily adjusted to avoid this?
9. Have ignition protected electrical equipment been installed and maintained by suitably trained staff.

Thus, hazardous can arise from the operation of a process, from process equipment and from many other sources including electrical equipment. While attention most of the time is on critical process equipment and on areas and substances known to be hazardous, due attention, many a times is not paid to electrical equipment which could pose hazard.

As sources of ignition cannot be avoided in any industry, the only way to control them in hazardous areas and avoid any flashover or major damage to equipment are: to identify the sources of ignition and its auto-ignition temperature; prepare a hazardous area layout; identify the various hazardous area of the plant and classify them as zone 1 or zone 2; select the proper equipment to suit respective area.

Chapter 3
Roles and Responsibilities of Chemical Engineers

INTRODUCTION

Many chemical engineers design and operate large-scale and complex chemical production facilities supplying diverse chemical products to society. In performing these functions, a chemical engineer will likely assume a number of roles during a career. The engineer may become involved in raw materials extraction, intermediate materials processing, or production of pure chemical substances; in each activity, the minimisation and management of waste streams will have important economic and environmental consequences.

Chemical engineers are involved in the production of bulk and speciality chemicals, petrochemicals, integrated circuits, pulp and paper, consumer products, minerals and pharmaceuticals. Chemical engineers also find employment in research, consulting organisations and educational institutions. The engineer may perform functions such as process and production engineering, process design, process control, technical sales and marketing, community relations and management.

As engineers assume such diverse roles, it is increasingly important that they be aware of their responsibilities to the general public, colleagues and employers, the environment, and also to their profession. One of the central roles of chemical engineers is to design and operate chemical processes yielding chemical products that meet customer specifications and that are profitable. Another important role is to maintain safe conditions for operating personnel and for residents in the immediate vicinity of a production facility.

Finally, chemical process designs need to be protective of the environment and of human health. Environmental issues must be considered not only within the context of chemical production but also during other stages of a chemical's life cycle, such as transportation, use by customers, recycling activities and ultimate disposal.

This chapter introduces approaches to designing safe chemical processes. The point of briefly introducing this important topic is to demonstrate that the evolution of the methods used to design safe processes mirrors the evolution of methods described in this text, which are used to design processes that minimise environmental impacts. Another section of this chapter reviews, in slightly more detail, the types of procedures that will be used in designing processes that minimise environmental impacts, and the responsibilities of chemical engineers to reduce pollution generation within chemical processes, and briefly notes some of the other professional responsibilities of chemical engineers, i.e. issues dealing with engineering ethics.

RESPONSIBILITIES FOR CHEMICAL PROCESS SAFETY

A major objective for chemical process design is the inclusion of safeguards that minimise the number and severity of accidental releases of toxic chemicals and the incidence of fires and explosions. A number of chemical plant accidents have occurred in the relatively recent past illustrating the importance of integrating safety into process designs. These accidents resulted in loss of life, permanent disability, and the destruction of chemical plant, process equipment and neighbouring residences. The most famous accidents occurred in Flixborough, England (1974) and Bhopal, India (1984).

Major Air Pollution Episodes

Flixborough

The Flixborough Works of Nypro Limited was designed to produce 70,000 tons per year of caprolactam, a raw material for the production of nylon. The process used cyclohexane as a raw material and oxidised it to cyclohexanol in the presence of air within a series of six catalytic reactors. Under process conditions, cyclohexane vapourises immediately upon depressurisation, forming a cloud of flammable cyclohexane vapour mixed with air. Reactor 5 was found to have a small crack in the stainless steel structure and was removed. The number 4 reactor was connected to the last reactor in the series using a 20" pipe, even though the reactors are normally connected using 28" pipe. The temporary section of piping was not properly supported and it ruptured upon pressurisation, releasing an estimated 30 tonnes of cyclohexane in a large cloud. An unknown ignition source caused the cloud to explode, levelling the entire plant facility. A total of 28 people died, another 36 were injured, and damage extended to nearby homes, shops and factories. The resulting fire in the plant burned for over 10 days. The accident could have been prevented by following proper safety design and operating procedures, including reducing the inventory of flammable liquids on site.

Meuse Valley, Belgium (1930)

A strong atmospheric inversion got settled over Meuse Valley on December 1, 1930 and remained until December 5. Effluents from several factories in the valley, chiefly oxides of sulphur, various inorganic acids, metallic oxides and soot were then trapped in the stable atmosphere. Sixty three people (generally the old and infirm) died and several hundred others deemed ill. Although many suspected sulphur oxides and hydrofluoric acid, the actual lethal substance could not be proved.

Donora, Pennsylvania, 1948

Donora, Pennsylvania is an industrial town on the banks of the Monongahela River about 30 miles south of the heart of Pittsburgh. The major industrial installations were steel and wire mill, a zinc smelter and a sulphuric acid plant. During a particularly calm and meteorologically stable period from October 27 to 31, 1948, air pollutants accumulated because of this many people were hospitalised and 20 died. Illness of several thousand persons was blamed on the episode and over 130 separate lawsuits were filed. The causative agent of the deaths and illness could never be determined incontrovertibly but sulphur compounds were present in the air in abnormally high quantities.

London smog, 1952

Historically, the longest record of intermittent air pollution problems belongs to the city of London, England. The notorious pea-soup fogs become especially offensive when mixed with coal smoke. The word smog (smoke and fog) was coined to describe this foul condition. In 1661, John Evelyn got published his well-known pamphlet, 'Fumifugium or the inconvenience of the air and smoke of London dissipated'. His major recommendation had been the removal of all smoke-producing plants from London. But London did little about it until the famous London smog of December 1952, truly a major air pollution disaster. The smog lasted 5 days from 5th to 9th December and caused 4000 deaths (principally among the old, the infirm and those with respiratory diseases). The onset of fog was followed by acute respiratory symptoms. Almost exactly ten years later, December 3 to 7, 1962 London experienced another black fog, with 340 excess deaths. The improvement over the 1952 episode was laid to smoke reduction brought about by the Clean Air Act and public awareness of the harmful effects of smog, which restrained many respiratory cripples from going outdoors.

Bhopal

Bhopal is located in a central state of India and on December 3, 1984, an accidental release of methyl isocyanate (MIC) occurred, killing 2000 nearby residents and injuring over 20000. The plant, which was partially owned by Union Carbide and partially owned by local investors, manufactured pesticides. One of the intermediates was MIC. MIC is a liquid at ambient conditions, it boils at 39.1°C, its vapour is heavier than air, and it is very toxic even at low concentrations. The maximum allowable exposure concentration of MIC for workers during an eight-hour period is only 0.02 parts per million (ppm). Death at large dose is due to respiratory damage. MIC reacts with water exothermically, but slowly, and the heat released can cause MIC to boil if cooling is not provided. On the day of the accident, the unit using MIC was not operating due to a labour dispute. The storage tank holding the MIC was contaminated with water from an unknown source. A reaction between MIC and water occurred in the tank causing the temperature to rise above the boiling point of MIC. The vapours generated escaped the pressure relief valve on the tank and were diverted into a scrubber and flare system designed to control MIC releases. Unfortunately, the release control system was not operating on this day and an estimated 25 tons of MIC vapour was released into the surrounding community with catastrophic effects. The accident could have been prevented by any number of steps, including the use of proper safety review procedures, by redesigning the process to accommodate a lower inventory of MIC, or by using alternative reaction chemistries that eliminate MIC.

In incidents such as this, loss of life and injuries are tragic, and economic consequences are severe. Engineers have a special role to play in preventing such incidents. Part of an engineer's professional responsibility is to design processes and products that are as safe as possible. Traditionally, this has meant identifying hazards, evaluating their severity and then applying several layers of protection as a means of mitigating the risk of an accident. Figure 3.1 shows the layer of protection concept and includes examples of layers that might be found in a typical chemical plant. This approach can be very effective and has resulted in significant improvement of the safety performance of chemical processes. However, the layer of protection approach has disadvantages that place limitations on its effectiveness: (i) the layers are expensive to build and maintain, and (ii) the hazard remains and there is always a finite risk that an accident will happen despite the layers of protection.

Inherently safer design is a fundamentally different approach to chemical process safety. Instead of working with existing hazards in a chemical process and adding layers of protection, the engineer is challenged to reconsider the design and eliminate or reduce the source of the hazard within the process. Approaches to the design of inherently safer processes have been grouped into the four categories listed below.

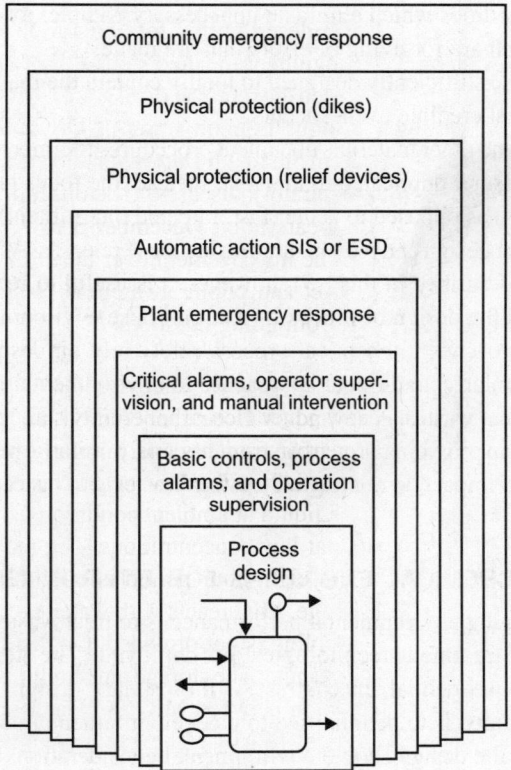

Fig. 3.1. Typical layers of protection for a chemical plant. SIS is safety interlock system and ESD is emergency shutdown.

This list contains a short checklist of questions related to inherently safer processes.
1. Minimise: Use smaller quantities of hazardous substances.
   (a) Have all in-process inventories of hazardous materials in storage tanks been minimised?
   (b) Are all of the proposed in-process storage tanks really needed?
   (c) Can other types of unit operations or equipment reduce material inventories (for example, continuous in-line mixers in place of mixing vessels)?
2. Substitute: Use a less hazardous material in place of a more hazardous substance.
   (a) Is it possible to completely eliminate hazardous raw materials, process intermediates, or by-products by using an alternative process or chemistry?
   (b) Is it possible to substitute less hazardous raw materials or to substitute noncombustible for flammable solvents?
3. Moderate: Use less hazardous conditions or facilities which minimise the impacts of a release of a hazardous material or energy.
   (a) Can the supply pressure of raw materials be limited to less than the working pressure of the vessels they are delivered to?
   (b) Can reaction conditions (temperature, pressure) be made less severe by using a catalyst, or by using a better catalyst?

4. Simplify: Design facilities which eliminate unnecessary complexity and make operating errors less likely, and which are forgiving of errors that are made.
   (a) Can equipment be sufficiently designed to totally contain the maximum pressure generated, even if the 'worst credible event' occurs?

Textbooks, case studies, and other materials document procedures for improving the safety of chemical processes and that material is not duplicated in this text. Instead, the focus in this text is the prevention of chronic (slow, continuous) as opposed to acute (fast, rare and intermittent) releases, and the role that chemical process and product design can play in minimising these releases. As these tools for minimising environmental impacts are described in this text, however, it is useful to recognise analogies between chemical process safety and the design of processes that minimise environmental impacts. As noted in this section, traditional approaches to chemical process safety rely on designing layers of protection around process hazards. Similarly, traditional approaches to environmental management have focused on designing processes to treat wastes. A new generation of inherently safer processes relies on designs that reduce hazards, rather than providing protection from hazards. Similarly, new generations of processes that minimise environmental impact do not rely on treating wastes, but instead are designed so that they do not generate wastes.

RESPONSIBILITIES OF CHEMICAL ENGINEERS FOR ENVIRONMENTAL PROTECTION

When the method for managing environmental performance is to treat wastes, the process is designed, wastes are generated, and treatment technologies are deployed. The design method for meeting environmental objectives is sequential. In contrast, if the primary design, rather than the design of peripheral waste treatment units, is to be modified to meet environmental objectives, a key question to answer is 'At what stage in the design should environmental considerations be considered?'

Designs for new processes and retrofitting of existing procedures are multistep procedures. The first step is the definition of a primitive problem, such as identifying the chemical to be produced and the annual quantity. This is followed by a process creation step that includes choosing reaction chemistry, the use of design heuristics to identify process equipment and operating conditions, development of a base case flowsheet, and process simulation.

The third step is a more detailed process synthesis of separation trains and a heat/power integration analysis. What follows is a detailed design and simulation of the flowsheet, profitability analysis, and optimisation. The final steps include a plantwide control- lability assessment, startup assessment, and reliability and safety analysis.

As part of their professional responsibilities, engineers should, through their designs, continuously improve the environmental performance of chemical processes. Recently the chemical manufacturers association (CMA, now American chemistry council) has adopted the pollution prevention code of management practice, which outlines tangible steps along a path to continuous reductions in the amounts of all contaminants released to air, water, and soil. Table 3.1 shows the set of management practices and specific chapters in this textbook that will aid engineers and other decision makers in achieving pollution prevention objectives. These practices demonstrate a clear commitment by senior management, a path to quantify waste generation and prioritise waste reduction, a preference for source reduction and reuse/recycle rather than pollution control, and a plan to measure and report on progress in achieving reduction goals.

Table 3.1. CMA pollution prevention code of management practices (Now the American Chemistry Council).

This Code is designed to achieve ongoing reductions in the amount of all contaminants and pollutants released to the air, water, and land from member company facilities, The Code is also designed to achieve ongoing reductions in the amount of wastes generated at facilities. These reductions are intended to help relieve the burden on industry and society of managing such wastes in future years

Management practices

Each member company shall have a pollution prevention programme that shall include:

1. A clear commitment by senior management through policy, communications, and resources, to ongoing reductions at each of the company's facilities, in releases to the air, water, and land and in the generation of wastes
2. A quantitative inventory at each facility of wastes generated and releases to the air, water, and land, measured or estimated at the point of generation or release
3. Evaluation, sufficient to assist in establishing reduction priorities of the potential impact of releases on the environment and the health and safety of employees and the public
4. Education of, and dialogue with, employees and members of the public about the inventory, impact evaluation, and risks to the community
5. Establishment of priorities, goals and plans for waste and release reduction, taking into account both community concerns and the potential health, safety, and environmental impacts as determined under items 3 and 4
6. Ongoing reduction of wastes and releases, giving preference first to source reduction, second to recycle/reuse, and third to treatment. These techniques may be used separately or in combination with one another
7. Measurement of progress at each facility in reducing the generation of wastes and in reducing releases to the air, water, and land, by updating the quantitative inventory at least annually
8. Ongoing dialogue with employees and members of the public regarding waste and release information, progress in achieving reductions, and future plans. This dialogue should be at a personal, face-to-face level, where possible, and should emphasise listening to others and discussing their concerns and ideas
9. Inclusion of waste and release prevention objectives in research and in design of new or modified facilities, processes, and products
10. An ongoing programme for promotion and support of waste and release reduction by others, which may, for example, include:
    (a) Sharing of technical information and experience with customers and suppliers
    (b) Support of efforts to develop improved waste and release reduction techniques
    (c) Assisting in establishment of regional air monitoring networks
    (d) Participation in efforts to develop consensus approaches to the evaluation of environmental, health, and safety impacts of releases
    (e) Providing educational workshops and training materials
    (f) Assisting local governments and others in establishment of waste reduction programmes benefiting the general public
11. Periodic evaluation of waste management practices associated with operations and equipment at each member company facility, taking into account community concerns and health, safety, and environmental impacts and implementation of ongoing improvements
12. Implementation of a process for selecting, retaining, and reviewing contractors and toll manufacturers taking into account sound waste management practices that protect the environment and the health and safety of employees and the public

(Contd...)

13. Implementation of engineering and operating controls at each member company facility to improve prevention of and early detection of releases that may contaminate groundwater
14. Implementation of an ongoing programme for addressing past operating and waste management practices and for working with others to resolve identified problems at each active or inactive facility owned by a member company taking into account community concerns and health, safety, and environmental impacts

ENGINEERING ETHICS

Process safety and environmental protection are not the only responsibilities of professional engineers. Engineers also have responsibilities to clients, to colleagues, and to the profession. The American institute of chemical engineers has assembled a Code of Ethics that highlights the main issues in the area of professional conduct.

This code can be found at AIChE website (http://www.aiche.org/membership/ethics.htm). Some of the responses dealt with putting health, safety, and environmental issues ahead of profits; placing self-respect as professionals above loyalty to companies; working within organisations versus whistleblowing to promote ethical behaviour; and taking career risks in order to get a company to do the right thing.

Chapter 4
Evaluating Exposures

INTRODUCTION

The human health risk associated with a chemical is dependent on the rate at which the chemical is released, the fate of the chemical in the environment, human exposure to the chemical, and human health response resulting from exposure to the chemical. In simpler terms, risk is a function of hazard (or toxicity) and exposure. This chapter, addresses the exposure component of the risk equation. Ideally, exposure is quantified by monitoring the work area or environmental setting where a chemical will be used or released; however, when monitoring data are not available to measure exposures, exposures can be estimated using methods described in this chapter.

The methods for estimating exposure will be separated into two sections—occupational and community. Occupational exposure occurs in the workplace. Workers in chemical production facilities may be exposed to toxins used or produced in the chemical process. Exposure to chemicals may occur from the inhalation of workplace air, ingestion of dust or contaminated food, or from contact of the chemical substance with the skin or eyes. In addition, chemical engineers must be aware of community exposures resulting from releases into the air and water, and from solid and hazardous waste disposal. Chemical releases to rivers, lakes, and streams may accumulate in fish and other marine life, which are subsequently used as a source of food, or may be ingested by persons using the downstream reaches of rivers as a supply of potable water. Persons living downwind of a chemical manufacturing facility may be exposed to fugitive and point source releases of chemical toxins to the atmosphere. Disposal of solid and hazardous wastes on the land, either in repositories such as landfills or into subterranean strata by injection into wells may result in contamination of potable groundwater if the waste is not isolated from the water supplies. The intent of this chapter is to introduce students to some methods for predicting potential exposure, in particular, occupational exposure and community exposure. During process design, it may be useful to predict potential exposures to workers from chemical emissions (i.e. 'occupational exposure'), or potential exposures to nearby residents from chemical emissions or releases from the plant (i.e. 'community or general population exposure'). There are other exposure areas, such as consumer exposure, which are not discussed in this chapter. The chemical engineer, in addition to selecting chemicals with low toxicity, also needs to select solvent chemicals and design unit operations to minimise potential exposure as well.

There are many good references on exposure assessment. Interested students are encouraged to consult references on other types of exposure not covered in this chapter. EPA has a website specifically for exposure which contains computerised tools for all exposure areas (http://www.epa.gov/oppt/exposure). This information can be useful in selecting and designing unit operations.

OCCUPATIONAL EXPOSURES: RECOGNITION, EVALUATION AND CONTROL

The basic components of assessing occupational exposure are to recognise all sources of exposure to chemicals, evaluate the exposure, determine if the exposure is within permissible limits, and at the minimum, control those exposures that exceed permissible limits.

Recognising exposures involves developing a list of all sources of chemical exposure in the work environment. Workers may be exposed to chemical substances during the performance of tasks making or utilising chemicals, in sampling reaction vessels, or in transfer of chemicals from the reactor to storage or transportation containers. As mentioned before, contact with the chemicals may occur through inhalation of vapours or by dermal contact as the chemicals are sampled or transferred. Although the highest exposures usually result from tasks performed directly by the worker, significant exposures may occur from nearby tasks performed by other workers or from incidental contact with background contamination in the workplace.

To evaluate the significance of an occupational exposure to a chemical substance, both the level and the duration of exposure must be known. Exposures to chemicals that have no cumulative or persistent effects may be tolerated at low levels in the workplace over long periods of time. However, short-term exposures to higher concentrations may result in acute toxicity to the worker. For other chemicals, exposures to low levels of the chemicals over long periods of time may result in chronic effects even though no acute effects are seen in short-term exposures.

Limitations on occupational exposures to chemicals are set by the occupational safety and health administration (OSHA), a division of the US Department of Labour.

PEL (Permissible Exposure Limits)

Listed in Table 4.1 are limitations for air contaminants set by OSHA for representative chemical substances. The relative toxicity of a chemical substance can be gauged by comparing the OSHA PEL for a chemical substance with that for a known poison, hydrogen cyanide, or an irritating but generally nontoxic gas, ammonia.

Table 4.1. OSHA permissible exposure limits for air contaminants.

Chemical Substance	CAS no.*	ppm by volume**	mg/m^3***
Acetic acid	64-19-7	10	25
Acetone	67-64-1	1000	2400
Acrolein	107-02-8	0.1	0.25
Ammonia	7664-41-7	50	35
Bromine	7726-95-6	0.1	0.7
2-Butanone(Methyl ethyl ketone)	78-93-3	200	590
Ethyl benzene	100-41-4	100	435
Hydrogen cyanide	74-90-8	10	11
Nitric oxide	10102-43-9	25	30
Dichlorodifluoro-methane (CFC 12)	75-71-8	1000	4950

*The CAS number is a unique number assigned as a means of identification to distinct chemical substances by the Chemical Abstracts Service.
**Parts of vapour or gas per million parts of contaminated air by volume at 25°C and 760 torr.
***Milligrams of chemical substance per cubic metre of air.

All limitations given in Table 4.1 are expressed as time-weighed averages for the chemical substance in any 8-hour work shift of a forty-hour work week. For PELs the action level is not the actual PEL but one-half the PEL, meaning action must be taken at this level to reduce the emissions. An overexposure is observed when monitoring demonstrates an average concentration of a chemical in the workplace greater than the occupational exposure limit over the appropriate time period.

Control and elimination of unacceptable exposures require information on the source, pathway, and worker exposed to the chemical substance. Control measures can be applied at any step, e.g. process changes can reduce the amount of emissions from various sources. Adjustments in ventilation systems can intercept chemical contaminants and eliminate the pathway for exposure. Finally, personal protective equipment can provide additional protection when other measures are inadequate.

Characterisation of the Workplace

The first step in an occupational exposure assessment is to characterise the workplace. Description of the workplace begins with a schematic or written description of the chemical manufacturing process and identification of unit operations where exposure to chemicals may occur. The schematic diagram is used to highlight unit operations and activities where exposure to chemicals may occur, provide a description of production activities and process chemistry and identify ventilation and other mechanisms that reduce worker exposures. Written descriptions should also include releases and exposures that do not take place in the chemical manufacturing facility, such as transportation and disposal of empty shipping containers.

From an occupational exposure viewpoint, the key elements of a process flow diagram are the sources of potential exposure. A source of potential exposure is a unit operation or worker task that brings the worker into potential contact with the chemical substance. For this reason, sampling points and transfer operations must be highlighted in the schematic diagram or description. Likewise, transfers of materials entering or leaving the process should be described (bagging, drumming, tank truck filling, etc.) since this highlights handling problems that could result in exposure to chemical substances. Waste streams leaving the process should be identified to indicate possible sources of exposure and to provide a resource for environmental studies. The completed flow diagram should highlight possible sources of exposure and minimise the possibility that potential hazards will be overlooked.

The written description should explain the activities occurring in the work area and should emphasise locations where potential exposure to chemicals may occur. It should also include important details such as component stream concentrations, operating temperatures, and pressures. Other factors that affect the potential for exposure (ventilation systems, open-top or closed vessels, use of protective equipment) should be noted in the description. If respiratory or dermal protection is used to limit exposures, the appropriate protection factor provided by the protective equipment should be listed.

Knowledge of the process and its component operations is needed to assess the likelihood and magnitude of exposure of workers to chemical substances. The frequency and duration of sampling events, the duration of batch processes, the type and frequency of transfer operations, and the number of workers involved in each operation are needed to make quantitative estimates of exposures to chemical substances. For convenience, the workers may be separated into groups performing similar operations and thus having similar exposures. A detailed description of the time engaged in each work task (sampling, monitoring unit operations, transferring raw materials and products) and in each work area should be developed where the potential for significant exposure exists.

The schematic and written descriptions can be used to prepare a relatively complete inventory of the chemicals that may be encountered in the work environment and the rates of use or generation of each chemical. For each chemical of concern, the engineer can assemble physical property data (boiling point, vapour pressure, particle size distribution, etc.) which will be of assistance in assessing the potential for exposure. For solid particles, knowledge of the particle size distribution will enable the engineer to evaluate the fraction of airborne particles that are potentially respirable. An excellent source of information on the health effects (nuisance, irritant, toxicity, carcinogenicity, potential for birth defects, etc.) of chemical substances is the Material Safety Data Sheet (MSDS) prepared for each chemical substance by the manufacturer. The MSDS will also provide occupational exposure guidelines established by regulatory or consensus organisations. These include the OSHA PELs and Threshold Limit Values (TLVs).

Exposure Pathways

Because exposure to chemicals in the work environment can occur through inhalation, skin absorption, or ingestion, the engineer must be aware of these potential pathways into the body. The exposure pathway model in Fig. 4.1 highlights potential pathways leading from process to worker and provides a framework for evaluating pathways for exposure to chemicals in the workplace. Used in conjunction with the schematic diagram, process description, and physical properties of the chemical substance, the important exposure pathways and controls to minimise exposure can be identified.

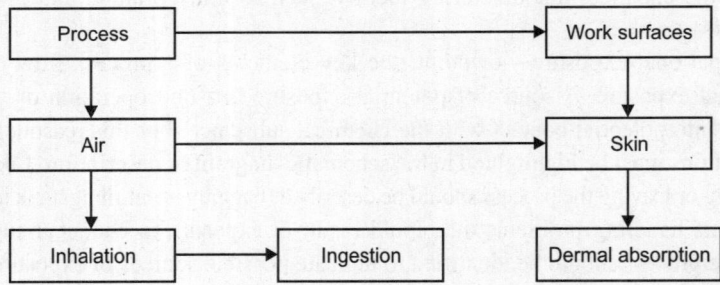

Fig. 4.1. Exposure pathway model.

Inhalation exposure is often the most significant route of workplace exposure. Chemicals can volatilise from the process or evaporate from work surfaces where they are deposited. Exposure to a high vapour pressure solvent can be evaluated solely from the rate of vapourisation and the effectiveness of ventilation controls unless there is also significant skin contact with liquid or vapour. With lower vapour pressure chemicals, longer term volatilisation of spills from work surfaces may be important. Dusty environments created by the generation of fine particles, like carbon dust, or in the cleaning of a manufacturing line can also contribute to inhalation exposure.

Figure 4.2 presents the framework for calculating exposure to chemical substances by the inhalation route. Exposure with units of mass is the product of the severity (mass/time) of exposure and the duration (time) of exposure. Severity is, in turn, the product of the environmental concentration (mass/volume) and the breathing rate (volume/time).

Similarly, duration is the product of the frequency (number of exposures) and period (time/exposure) of exposure. A separate estimate of the rate of absorption of inhaled materials is necessary to calculate the intake of a chemical into the body.

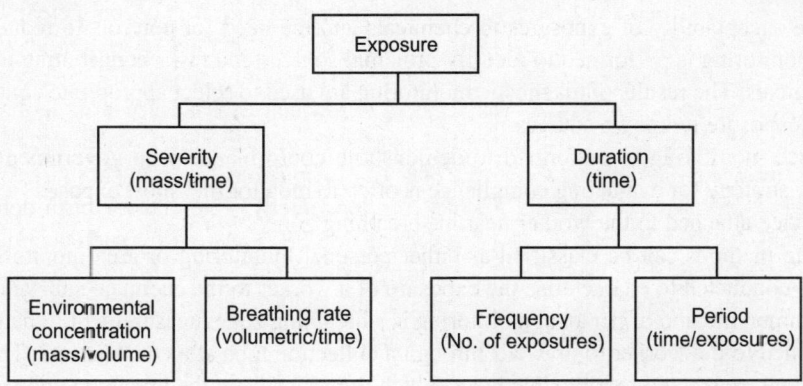

Fig. 4.2. Inhalation exposure framework.

Dermal contact can also represent an important route of exposure for some chemicals, particularly those that readily absorb through the skin in immediately toxic amounts and those that pass through the skin and accumulate in the body. This exposure pathway usually results from direct contact of chemicals with the skin. Exposure may also result from contact with work surfaces that are contaminated.

Figure 4.3 presents the framework for calculating exposure to chemical substances by the dermal route. Exposure (mass) is the product of severity (mass adsorbed per incident) and frequency (number of incidents). Severity of dermal exposure is, in turn, the product of the surface area of exposed skin and the mass adsorption per incident. Dermal intake into the body requires a separate estimate of the rate of uptake of the chemical from the exposed skin surface.

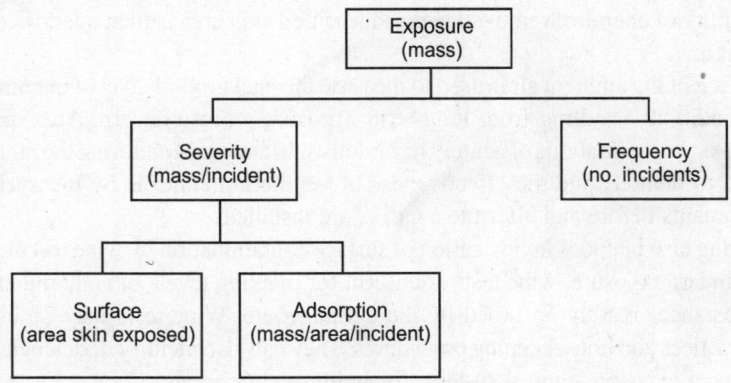

Fig. 4.3. Dermal exposure framework.

Oral ingestion of chemicals is usually a relatively minor route of exposure in the workplace, particularly when dining areas are separate from work areas and employees practice a reasonable level of personal hygiene. However, this may be an important route of exposure for chemicals that accumulate in the body over long periods of exposure.

Monitoring Worker Exposure

Monitoring objectives can be grouped into three categories: baseline, diagnostic, and compliance. Baseline monitoring is performed to evaluate the range of worker exposures. The baseline data are used to determine the acceptability of exposures to chemicals and the need for controls to reduce exposures. Diagnostic monitoring is performed to identify principal sources and tasks contributing to exposure to specific chemicals. The results of diagnostic monitoring are used to select appropriate control strategies for reducing exposure to known sources.

Compliance monitoring is performed to demonstrate conformance with government regulations. The sampling strategy for evaluating compliance is often to monitor the 'most exposed' worker using a collection device attached to the worker near his breathing zone.

Monitoring methods can be classified as either personal monitoring or area monitoring. Personal monitoring is conducted to characterise the exposure of a worker to the chemical substance of interest. The most common method of personal monitoring is a breathing zone measurement. A battery-powered pump is attached to the worker to draw air through a collection tube at a constant rate. The inlet to the collection tube is connected to a flexible hose, which draws air from the breathing zone of the worker The sample is collected for a designated period and the monitoring result is reported as a time-weighed average over the designated period. Two common sample averaging times are 8 hours, a normal work shift, and 15 minutes, a common short-term exposure time limit. These durations correspond to the averaging times of regulatory limits. Eight-hour sampling has the disadvantage that peak exposure information is usually lost. A mixture of full-shift and short-term sampling is usually the best technique for evaluating worker exposures.

Personal monitoring of skin absorption is often difficult. Patch testing is conducted by affixing a patch of absorbent material to an exposed skin surface of a worker for a known period of time. At the end of a specified time period, the patch is removed and the chemical of concern is extracted from the patch and quantified.

Skin washes are used to remove the chemical of concern from the skin surface using a suitable solvent. The quantity of chemical removed from a measured skin area is then quantified and reported as exposure per unit area.

Area monitoring of the ambient air is used to measure the background level of chemical contaminants when chronic conditions resulting from long-term exposure are of concern. Area monitoring is also used to warn of toxic concentrations of acutely hazardous substances. Monitoring the ambient atmosphere can also be used to demonstrate the effectiveness of ventilation controls by measuring the levels of chemical contaminants before and after the controls are installed.

Area monitoring also includes investigation of surface contamination by wipe test methods. Although not a direct method of exposure, wipe tests are useful for tracking levels of contamination, particularly for chemical substances readily absorbed by the dermal route. Wipe tests may be used to document trends in work practices and housekeeping procedures. They can also identify deficiencies in maintenance or operation of local exposure control systems, including safety hoods.

The number of samples collected in a monitoring programme is determined by regulatory requirements or professional judgment. The cost and difficulty of sample collection and analysis will limit the number of samples collected. Conversely, a greater number of samples will decrease the likelihood of significant errors in the sample means and decrease the variance about the mean. The standard error of the mean decreases rapidly as the first few replicates are collected but the likely sample error decreases only slightly with each additional sample after 6 to 10 samples are collected. The variance about the sample mean is inversely proportional to the square root of the number of samples analysed. Similarly, the variance about the sample mean decreases significantly over the initial samples and less so after 10 or more samples.

Modelling Inhalation Exposures

It is not always convenient or possible, in the case of a new or proposed process, to undertake a monitoring programme to determine airborne concentrations of chemicals. In some instances, a more rapid estimate of potential worker exposures to chemical substances is needed. In this situation, the engineer may utilise models which simulate worker exposures.

Mass balance model

A simple model often used to estimate the concentration of airborne contaminants in the workplace is the mass balance model also known as the box model. The work area is modelled as a box in which the contaminant is uniformly distributed. In this case, a mass balance can be written for the contaminant concentration within the work area.

$$V\frac{dC}{dt} = G - kQ(C - C_o) \qquad \ldots (4.1)$$

where,
- C is the concentration of airborne contaminant in the work area (mass/length^3).
- V is the volume of the work area (length^3).
- t is the time during which the contaminant has been emitted.
- G is the emission rate of the contaminant to the air (mass/time).
- Q is the ventilation rate in the work area (length^3/time).
- k is a mixing factor to account for incomplete mixing in the work area (unitless)
- C_o is the concentration of the airborne contaminant entering the work area (mass/length^3).

If the emission rate and ventilation rate are constant, the concentration will reach a steady state and eq. 4.1 becomes:

$$C = C_o + \frac{G}{kQ} \qquad \ldots (4.2)$$

At times, emissions are episodic. Consider a work area that initially contains contaminant at concentration C_o. At some time, $t = 0$, an emission source, releasing contaminant at rate G, is placed in the work area. In this case, the box model can be used to estimate the rise in concentration of the contaminant in the workplace. Again, assuming that the ventilation rate is constant, eq. 4.1 can be integrated to yield:

$$C = C_o + \frac{G}{kQ}[1 - \exp(-kQt/V)] \qquad \ldots (4.3)$$

The mixing factor (k) typically ranges from 0.3 to 0.7 in small rooms without fans. Others have used mixing factors of 0.5 for work areas with average ventilation and 0.1 for poorly ventilated work areas.

The determination of G may be simple or complex, depending on the nature of the emission source. As an example, assume that the source is a pool of liquid that is evaporating at a constant rate. Estimating this emission rate requires input of the vapour pressure of the contaminant, the surface area of the evaporating liquid, and the relationship between the velocity of the air over the liquid surface and mass transfer from the liquid into the flowing air stream.

The penetration model provides acceptable estimates of evaporation rates at low air speeds characteristic of indoor work areas:

$$G = 8.79 \times 10^{-5} \frac{(MW^{0.883})(VP)[1/MW + 1/29]^{0.25}](v^{0.25})}{(T^{0.05})(\Delta x^{0.5})(P^{0.5})} \quad \ldots (4.4)$$

where,
- G is the evaporation rate (g/sec.cm^2).
- MW is the molecular weight of the evaporating species (g/mole).
- VP is the vapour pressure of the evaporating contaminant (atm).
- v is the air velocity parallel to the surface of the evaporating liquid (cm/sec).
- T is the surface temperature of the evaporating liquid (°K).
- Δx is the length of the evaporating pool in the direction of airflow (cm).
- P is the ambient pressure (atm).

The simple mass-balance model does not account for all of the phenomena that influence the exposure to chemicals released into the workplace atmosphere. Exposure may be mitigated by adsorption of the chemical to walls and other surfaces in the work room. In this case, the mass balance on the airborne concentration is given by:

$$V \frac{dC}{dt} = G - kQ(C - C_o) - rC \quad \ldots (4.5)$$

where r is the nonventilatory removal coefficient of airborne contaminant (volume/time). If the ventilation and emission rates are constant, the box model predicts a steady state concentration of:

$$C = \frac{kQC_o + G}{(kQ + r)} \quad \ldots (4.6)$$

Solvents are often volatile and significant accumulations of their vapours may occur in the workplace air. In this instance, the concentration of solvent may exert a significant back pressure retarding the evaporation of additional solvent. Jayjock has published the solution of the mass balance equation when back pressure is significant.

The approach used in development of the mass-balance model, adjusting the ventilation rate to account for imperfect mixing and unventilated areas, has been criticised because the model is still used to describe an imperfectly mixed room. In addition, the mixing factor is an empirical adjustment that must be developed by experimental measurements. An alternative model divides the work area into two perfectly-mixed zones, one near the source of an airborne contaminant and the other removed from the source. Mixing in the work area occurs as a result of ventilation between the two zones. The steady-state or upper bound on concentration in the zone of the work area nearest the source is given by:

$$C = \frac{G(B + Q)}{(BQ)} \quad \ldots (4.7)$$

where,
- C is the concentration of the contaminant in the work area near the source (mass/length^3).
- G is the rate of vapourisation of the contaminant (mass/time).
- B is the rate of exchange of air between the zones located near and removed from the source (length^3/time).
- Q is the ventilation rate of the zone removed from the source (length^3/time).

Although this model does not require estimation of an empirical mixing factor, the air exchange rate, B, must be determined from the physical dimensions of the zones or other criteria.

Dispersion Models

Diffusion of contaminants in workplace air results in the net movement of the contaminants from regions of higher concentration to regions of lower concentration. The spread of the contaminant is aided by the convective mass transfer driven by the ventilation system. The combination of these influences results in movement of contaminants away from their source into the surrounding room.

The mass-balance model described above presumes a uniform concentration of the chemical contaminant in the work area. Dispersion models have a notable advantage; they describe the variation of contaminant concentration with distance from the source. The concentration gradient is described by the following equation when convection occurs in the x-direction only and dispersion occurs equally in all directions:

$$u \frac{dC}{dx} = \frac{D}{r^2} \frac{d}{dr} \left[r \frac{dc}{dr} \right] \qquad \ldots (4.8)$$

where,
- u is the wind velocity in the x direction (length/time).
- C is the concentration of airborne contaminant (mass/length^3).
- D is the diffusion coefficient (length^2/time).
- x is the distance downwind from the source (length).
- r is the distance from the source to the sampling point (length).

This equation has been solved for concentrations resulting from emissions into an infinite space:

$$C = \frac{G}{4\pi Dr} \exp[(-u/2D)(r-x)] \qquad \ldots (4.9)$$

where G is the contaminant emission rate from the source (mass/time).

The diffusion coefficient (D) can be derived from measurements at the sampling site or estimated from values available in the literature. Measurements of the diffusion coefficient in indoor industrial environments have ranged from 0.05 to 11.5 m^2/minute, with 0.2 m^2/min being a typical value.

Molecular diffusion theory strictly applies to vapours and gases; however, particulate matter with aerodynamic diameters less than 10 μm are distributed in workplace air in a similar manner. Using the dispersion models to describe the distribution of dusts and fumes is reasonable for small particles.

Assessing Dermal Exposures

Dermal hazards refer to chemicals that can cause dermatitis or otherwise damage the skin as well as to chemicals that can enter the body through the skin and cause toxic effects in other organs. Dermatitis refers to inflammation or damage to the skin which is localised and does not spread to other areas of the body. Acids, alkalis, and other irritating or corrosive chemicals damage skin which they contact. Repeated contact with epoxy resins may result in skin sensitisation and dermatitis. The National Institute of Occupational Safety and Health has recognised allergic and irritant dermatitis as the second most common occupational disease (after hearing loss), accounting for 15 to 20 per cent of all reported occupational diseases. Because of the often readily apparent reaction to chemicals causing dermatitis, exposures are usually quickly eliminated or protective clothing is used to preclude skin contact with toxic chemicals.

In contrast to contact dermatitis, toxic chemicals may be absorbed through the skin, mucous membranes, or eyes either by direct skin contact with the chemical or deposition of aerosols. This absorption can contribute to toxic effects on other organs. Some substances, such as amines and nitriles, pass through the skin so rapidly that the rate at which they enter the body is similar to rates of inhalation or ingestion.

The three mechanisms of dermal exposure are: (i) direct contact between the worker's skin and a liquid or solid chemical as from splashing or immersion, (ii) transfer of a chemical from a contaminated surface to the skin following direct contact, and (iii) deposition or impaction on the skin as a vapour or aerosol. Aerosols are created when chemicals are applied by spraying or when fluids contact moving surfaces, e.g. metal working fluids interacting with machinery. Aerosols tend to settle rapidly, making an increased separation between the worker and operations using the chemical of concern a feasible means of controlling dermal exposures.

The amount of a chemical remaining on the skin depends on the processes of contamination, removal, and penetration through the skin. Possible removal processes are evaporation, incidental transfer to other surfaces, or intentional decontamination. Dermal exposure is often highly variable between workers over time and between different anatomical locations of the body. Since dermal penetration varies across the anatomical locations of the body, an overall average value for skin exposure is often insufficient.

Direct methods for measurement of skin exposure include collection of chemical contaminants on absorbent pads or clothing and wipe sampling of contaminated surfaces. The absorbent pad technique utilises gauze pads, treated cloth, or alpha-cellulose pads which are attached to various sites on the worker's skin or outer clothing to capture chemicals that would have been deposited on the skin or clothing. Collection pads are exposed for a representative period of time to the work environment and are subsequently removed and analysed for the chemical of concern. An estimate of the potential dermal exposure can be obtained by multiplying the amount of contaminant deposited on a unit area of the absorbent pad by the surface area of the body region that the pad is positioned to represent. This technique is generally used for sampling nonvolatile contaminants or compounds with low vapour pressure; charcoal impregnated cloth has been used for sampling of volatile compounds.

Wipe samples are collected by washing the skin or clothing with water, surfactants, alcohol, acetone, or other solvents. Chemicals remaining on the skin or clothing are collected but those that have penetrated the skin are not collected by this technique. Wipe samples have been used to determine routes of dermal exposure to aromatic amines used as anti-oxidants, intermediates, and curatives in epoxy resins and urethanes. Wipe samples taken from the inside of protective gloves indicated that methylene dianiline used in aircraft composites had penetrated the gloves of workers engaged in the hand lay-up operations in aircraft and aerospace industries. The wipe samples revealed that chemical breakthrough of the protective clothing was the cause of elevated levels of the aromatic amine detected by biological monitoring. Wipe tests have also been used to identify significant exposure to toxic chemicals from handling contaminated tools and improper removal of contaminated clothing.

Computerised image analysis techniques can be used together with fluorescent whitening agents to indirectly quantify exposure of the total body surface. Visual observation of fluorescent tracer deposition on skin has been used to characterise exposure in a variety of pesticide applications. The behaviour of the fluorescent tracer in the application process must be similar to that of the chemical of concern. This technique provides a means of assessing exposures without use of toxic chemicals.

Methods to control dermal exposure to chemicals can take many forms. Substitution of a less toxic chemical is almost always a good option, unless the alternative chemical has a much higher vapour pressure and is likely to cause an inhalation hazard. Consideration should also be given to redesigning the work process to avoid splashes or immersion. Where that is not feasible, personal protection in the form of chemical protective gloves, an apron, or clothing may be selected. Performance characteristics of glove materials must be matched to the hazard to be avoided, i.e. cuts, abrasions, and dermal contact with toxic chemicals. Glove manufacturers can provide information on the ability of a variety of glove materials (natural rubber, polyvinyl chloride, neoprene, nitrile, butyl rubber, polyvinyl alcohol, viton, or norfoil) to preclude penetration of toxic chemicals.

EXPOSURE ASSESSMENT FOR CHEMICALS IN THE AMBIENT ENVIRONMENT

Exposure to chemicals in the ambient environment can occur through inhalation, ingestion, or dermal contact. Typically, exposure by ingestion is not as important as dermal or inhalation exposure. However, ingestion may be a significant route of exposure to chemical substances when animals used for food, such as fish or shellfish, accumulate and concentrate chemical contaminants. Ingestion may occur when particles are trapped and swallowed following respiration or when small children eat dust or soil. Ordinarily, exposure by inhalation or dermal absorption will accompany ingestion and result in more significant uptake of the chemicals of concern. In this section, only inhalation and dermal exposure will be considered.

Assessment begins with identification of all wastes and releases containing a chemical of concern and an estimate of the quantity of waste disposed from each source. Next, the concentration of the chemical of concern in the waste or release is measured or estimated and the characteristics of the waste matrix, such as whether it is a liquid, gas, or solid, are identified. The treatment and disposal practices associated with each waste are identified and the quantity of the chemical of concern released to the air, surface waters, groundwater, and land by the treatment and disposal practices are estimated. Finally, the transport and transformation of the chemical of concern through the air, surface waters, and ground water is modelled, along with the uptake through inhalation, ingestion, or dermal contact. In this section, exposure assessment is used to determine the amount of a chemical of concern potentially contacting a member of the general population.

Exposure to Toxic Air Pollutants

Exposure assessment for toxic air pollutants is a four-step process. The first step is to identify pollutants likely to be in the ambient air. Many chemicals found in factories, consumer goods, and waste treatment plants can be released to the air as toxic air pollutants. Some commonly released chemicals include perchloroethylene from dry cleaners, methylene chloride from industrial cleaning and consumer products such as paint strippers, and chromium from metal plating operations.

The toxic release inventory, available at the EPA website (www.epa.gov) provides an extensive source of data on toxic chemical releases.

The second step in exposure assessment for toxic air pollutants is to estimate the quantities of pollutants released by point, area, and mobile sources. Point sources are sites with a specific, usually fixed, location. Point sources include chemical plants, steel mills, oil refineries, and hazardous waste incinerators. Pollutants can be released when equipment leaks, when chemicals are transferred from one area to another, or when pollutants are emitted from stacks. Area sources of toxic air pollutants are comprised of many small sources releasing pollutants to the outdoor air in a defined area. Examples include dry cleaners, small metal plating operations, and gas stations. Mobile sources include automobiles, trucks, buses, etc. which are important contributors of oxides of nitrogen and sulphur, hydrocarbons, carbon dioxide, and particulates in the air.

Routine releases, such as those from industry, cars, landfills, or incinerators, may follow regular patterns and occur continuously over time. Other releases may be routine but intermittent, such as when production is done in batches. Accidental releases can occur during an explosion, equipment failure, or a transportation accident; the timing and, often, the amount released during accidental releases are difficult to predict.

To estimate the amount of a routine or intermittent release, engineers will often sample the effluent from a facility as it is released. The sample is taken to a laboratory and analysed to quantify the amount released during the collection period. The amount collected during the test is used to predict the amount released each operating day. For example, if 0.1 kilogram of sulphur dioxide is collected in one hour by a collector which samples 1 per cent of the airflow from a stack, a plant operating 24 hrs per day would be expected to emit 240 kg of sulphur dioxide per day.

Alternatively, engineers can use an emission factor to estimate the amount of toxic air pollutant released by a particular facility. Emission factors are averages of emission measurements from a few representative facilities that relate the quantity of a pollutant released to the level of production associated with the release of that pollutant.

The third step is to estimate the concentration of the toxic pollutant at the location where exposure occurs. The concentration of a pollutant decreases as it disperses from the point of release. The decrease in concentration or dispersion of the toxic air pollutant is a function of the wind direction and speed and the terrain over which the air flows, whether flat or hilly, whether flowing over a mountain or through a valley. The location of the release, whether from a tall smokestack or a leak at ground level, will affect the distribution of the pollutant near the facility; a toxic air contaminant released from high stacks is dispersed and diluted while descending to ground level. Other factors that affect the concentration include the temperature and speed of the gas exiting the smoke stack and the location of the release within the facility.

The last step in an exposure assessment is to estimate the number of persons exposed to a toxic air pollutant. Demographers can estimate the number of persons living in areas surrounding a source using census data. Combining the concentration estimates and the census data, engineers can estimate the numbers of people exposed to the pollutant at varied concentrations. To aid decision makers, these results can be compared to a selected benchmark such as an air quality standard or a level with a known health effect. Data on population densities in regions surrounding point sources are available at the Envirofacts section of the EPA website (http://www.epa.gov/enviro/index_java.html).

Dermal Exposure to Chemicals in the Ambient Environment

Swimming in rivers, lakes, and streams is generally the only activity considered to cause significant dermal exposure. Although other activities, e.g. water skiing, fishing, standing in the rain—could lead to human dermal exposure, the frequency, duration, and the amount of skin surface available for exposure are small; therefore, for general and long-term assessments, these activities are considered negligible. Because swimming is an episodic activity, it is necessary to consider both frequency and duration of exposure. In addition, the surface area exposed is an important factor in dermal exposure calculation. These activity-related parameters, when coupled with data on the aquatic ambient concentration of a chemical toxin, yield an estimate of dermal exposure.

Effect of Chemical Releases to Surface Waters on Aquatic Biota

Waste-water generated in the manufacture, processing, or use of a chemical may contain a fraction of the chemical produced and the raw materials used in the manufacturing process. This loss may occur during reaction to produce the chemical, purification, blending, or cleaning of the reactors, piping, and equipment used to process the chemical substance. The waste-water must be either treated by facilities at the plant site or, more often, commingled with the wastes of others and treated at a publicly owned treatment works (POTW). Using physical-chemical property data and estimates of biodegradability, the effectiveness of the treatment can be estimated, so that the amount actually entering the receiving water body can be predicted. The receiving water body will dilute the discharge from the plant site or POTW so that the concentration in the receiving stream can be calculated if the flow in the stream is known. Stream in this context means the receiving body of water and, in this sense, can include creeks, rivers, lakes, bays, or estuaries.

Removal of chemicals during waste-water treatment is controlled by the physical and biological processes employed in the treatment works. The following processes are commonly used to remove chemicals during waste-water treatment:
1. Adsorption to suspended solids in the primary clarifier, aeration basin, and secondary clarifier.
2. Volatilisation through surface vapourisation in the primary and secondary clarifiers and through air-stripping in the aeration basin.
3. Biodegradation by aerobic micro-organisms, most commonly in an activated sludge aeration basin.

Under optimal conditions, a POTW will remove a large percentage, i.e. 70 to 99+ per cent, of many organic pollutants from the waste-water, but treatment efficiency varies with the chemical and physical properties of the pollutant. The POTW is typically less efficient in removal of inorganic pollutants and many of these pass through the POTW unchanged.

An important issue for surface water is the effect that a chemical may have on aquatic organisms including algae, freshwater crustaceans, and fish. A healthy stream with a wider variety of organisms will have a better ability to assimilate chemical releases than a stream whose quality is already compromised. If any link in the food chain in a stream is impacted, the effect can be deleterious to other organisms as well as the health of the stream. Organisms lower on the food chain, such as algae, have shorter lives; for these organisms short-term exposures to high concentrations of chemicals are critical. Consequently, the concentration of a chemical in the receiving body of water when the dilution is least is used to assess the impact of chemical releases on the aquatic biota. For this purpose, the historical stream flow representing the seven consecutive days of lowest flow over a ten-year period is often used to generate estimates of chronic concentrations of chemicals of concern for aquatic life.

The following formula can be used to calculate surface water concentrations of the chemical of concern in free-flowing rivers and streams:

$$SWC = [Release \times (1 - WWT/100)/Stream\ flow] \quad \ldots (4.10)$$

where,
    SWC is the surface water concentration (mass/volume).
    Release is the quantity of chemical released in waste-water (mass/time)
    WWT is the per cent removal in waste-water treatment (dimensionless)
    Stream flow is the measured or estimated flow of the receiving stream (volume/time)

Ground Water Contamination

Industrial solid wastes are often sent to land disposal in municipal, industrial, or hazardous waste landfills. Although less common, surface impoundments and land treatment may be used to contain and treat industrial wastes. Chemicals may leach from the wastes, either in free liquid contained in the waste or in rainwater percolating through the waste, and be carried into the underlying soils. Chemicals entering the soil solubilise in water contained in the pore space between the soil particles. This interstitial water, called groundwater, may subsequently percolate downward into the water table, carrying the chemical contaminants with it.

Groundwater contamination is most common beneath urban areas, agricultural areas, and industrial complexes. Frequently, groundwater contamination is not discovered until long after the actions leading to the contamination have occurred. One reason for this is the slow movement of groundwater through soils and underlying rock strata; in fine-grained soils and low permeability rock strata, groundwater movement is often less than one foot per day. Contaminants in groundwater do not mix or spread quickly, but remain concentrated in slow-moving, localised plumes that may persist for many years. This often results in a delay in the detection of groundwater contamination. In some cases, groundwater contamination discovered today is the result of agricultural, industrial, and municipal practices several decades ago. This also means that the land disposal practices of today may have effects on groundwater quality many years from now.

Groundwater is a vital natural resource. It is used for public and domestic water supply, for irrigation of crops, and for industrial, commercial, mining, and thermoelectric power production purposes. In 2004, the United States Geological Survey reported that groundwater supplied 51 per cent of the nation's total population with drinking water. Unfortunately, groundwater is vulnerable to contamination and, once contaminated, is difficult to remediate. Table 4.2 lists National Primary Drinking Water Standards prescribed by the US Environmental Protection Agency which must be met by all drinking water supplies after treatment, if any.

Table 4.2. National primary drinking water standards for maximum contaminant limits (MCL).

Contaminant	MCL (mg/l)	Potential health effects from ingestion of water	Sources of contamination in drinking water
Benzene	0.005	Cancer	Gasoline, paint, plastics industry
Carbon tetrachloride	0.005	Cancer	Solvents and degradation products
Chlorobenzene	0.1	Nervous system, liver	Metal degreasing processes
o-Dichlorobenzene	0.6	Liver, kidney damage	Paints, dyes, chemical wastes
p-Dichlorobenzene	0.075	Cancer	Room and water deodourants
1,2-Dichloroethane	0.005	Cancer	Leaded gas, fumigants, paints
1,1-Dichloroethylene	0.007	Cancer, liver, kidney damage	Plastics, dyes, perfumes, paints
1,2-Dichloroethylene	0.07	Liver, kidney, nervous system	Waste industrial extraction solvents
Diethylhexyl phthalate	0.006	Cancer	Polyvinyl chloride, other plastics
Ethylbenzene	0.7	Liver, kidney, nervous system	Gasoline, chemical manufacturing
Pentachlorophenol	0.001	Cancer, liver, kidney damage	Wood preservative
PCBs	0.0005	Cancer	Transformer oils, plasticisers
Styrene	0.1	Liver, nervous system damage	Plastics, rubber, landfill leachate
Tetrachloroethylene	0.005	Cancer	Dry cleaning solvent, other solvents
Toluene	1.0	Liver, kidney, nervous system	Gasoline, chemical solvent
1,1,1-Trichloroethane	0.2	Liver, nervous system damage	Adhesives, paints, metal degreasing
1,1,2- Trichloroethane	0.005	Kidney, liver, nervous system	Rubber processing, chemical mfg.
Vinyl chloride	0.002	Cancer	PVC pipe, solvent degradation
Xylenes	10	Liver, kidney, nervous system	Gasoline refining, paints, inks

The transport of a chemical in the subsurface depends on physical-chemical properties of the chemical and the characteristics of the subsurface environment. Some of the more important properties influencing the spread of chemical contaminants in the subsurface include water solubility, soil organic carbon partition coefficient, and vapour pressure. A chemical that is readily soluble in water will be carried deeper into the subsurface by rainwater and once it reaches the water table it will mix intimately with the groundwater. A chemical that has an affinity for organic solvents is likely to be adsorbed onto soil organic matter which constitutes a range of less than 1 to 20 per cent of topsoils with the concentration generally decreasing with increasing depth; adsorption from the groundwater retards the movement of dissolved chemicals. In addition, cationic species can be expected to attach to soil particles that are negatively charged. Chemicals with significant vapour pressure may vapourise to the atmosphere from shallow pore water before precipitation carries the chemical downward to the saturated zone.

Even the simplest descriptions of contaminant migration in groundwater often rely on numerical solutions of the equations governing flow, physical equilibrium, and chemical reaction. Analytical solutions are available for a variety of conditions when only a single spatial dimension is considered. Hydrodynamic dispersion is due to mixing of the groundwater and molecular diffusion of the dissolved species.

DESIGNING SAFER CHEMICALS

A challenge for chemical engineers is to use the general principles outlined in this chapter in designing chemicals that will reduce toxicity. The remainder of this chapter presents semi-quantitative principles and guidelines that can be used in designing safer chemicals.

In designing safer chemicals, it is useful to think about modifying properties so that:
1. Persistence and dispersion in the environment are minimised, reducing exposures.
2. Uptake by the body is minimised, reducing dose.
3. Toxicity is minimised.

This section will consider property modifications that can lead to reduced exposure, dose, and toxicity. Consider first the issue of dose.

Reducing Dose

Converting an exposure (e.g. inhaling a chemical) into a dose (e.g. absorption by the blood through the lung membrane) generally involves the transport of a chemical across a membrane. The three primary membranes of interest are the lung, which controls uptake of chemicals that are inhaled; the skin, which controls the uptake of compounds from dermal exposures; and the gastrointestinal tract, which controls the uptake of chemicals that are ingested. Some of the characteristics of these membranes are listed in Table 4.3.

Table 4.3. Characteristics of membranes that control chemical uptake by the body.

Membrane	Surface area (m^2)	Thickness of absorption barrier (μm)	Blood flow (l/min)
Skin	1.8	100–1000	0.5
Gastrointestinal tract	200	8–12	1.4
Lung	140	0.2–0.4	5.8

From Table 4.3, it is apparent that the gastrointestinal tract has one of the greatest surface areas available for uptake of chemicals by the body. The uptake of chemicals across this membrane is controlled by lipid solubility, water solubility, dissociation constant, and molecular size.

High water solubility enhances uptake through the gastrointestinal tract because water soluble materials are more easily mobilised in the large and small intestine and the materials are more easily mobilised in the large and small intestine and the materials, therefore, experience less mass transfer resistance in migrating to the intestine wall. In contrast, high lipid solubility enhances uptake and transport across the membrane. Thus, the compounds that are likely to be transported from the gastrointestinal tract into the blood streams are compounds with moderate water solubility and moderate lipid solubility. Highly water soluble (lipid insoluble) and highly lipid soluble (log $K_{ow} > 5$, water insoluble) compounds are less likely to be taken up through the gastrointestinal tract.

Molecular weight also plays a role in determining uptake through the gastrointestinal tract. A general guideline is that molecules with molecular weights less than 300 that are both lipid and water soluble are well absorbed, and those with molecular weights in excess of 1000 are only sparingly absorbed.

The lung also provides a relatively large surface area for uptake of chemicals. The lung is a relatively thin membrane and because the membrane is so thin, lipid solubility plays less of a role in chemical uptake than for the gastrointestinal tract. High water solubility will promote uptake through the lung, as will the delivery of the compound on fine particles (less than 1 micron in diameter). Small particles can be inhaled deeply and will deposit deep in the lung, allowing the chemicals adsorbed on or dissolved in the particles to reside in the lung for very long periods.

The skin presents a formidable barrier to chemicals transport. For a chemical to be taken up through the skin, it must pass through multiple layers. As with the gastrointestinal tract, moderate lipophilicity (log $K_{ow} < 5$) promotes absorption through the skin because transport must occur through both largely lipid and largely aqueous layers. Finally, note that once a compound is absorbed into the blood stream, it must still reach a target organ. Many organs have their own barriers to uptake that may influence dose (e.g. the blood-brain barrier is more easily crossed by lipophilic materials). In addition, chemicals may be removed by the body through urine and feces before the target organ is reached (water solubility enhances elimination via this mechanism).

Reducing Toxicity

Designing safer chemicals by reducing toxicity requires a knowledge of the mechanisms by which compounds exert a toxic effect. While these mechanisms are not known in many cases, there are a few general mechanisms for toxicity that can be examined, leading to safer chemical designs.

One group of mechanisms associated with toxic effects are the reactions of electrophilic species with nucleophilic substituents of cellular macromolecules such as DNA, RNA, enzymes and proteins. Table 4.4 presents the possible effects of a number of common electrophiles.

Table 4.4. Examples of electrophilic substituents and the reactions they undergo with biological nucleophiles, and the resulting toxicity*.

Electrophile	General structure	Nucleophilic reaction	Toxic effect
Alkyl halides	R–X where X = Cl,Br,I,F	Substitution	Various; e.g. cancer
α,β unsaturated carbonyl and related groups	C=C–C=O C≡C–C=O C=C–C≡N	Michael addition	Various; e.g. cancer, mutations, hepatoxicity, nephrotoxicity, neurotoxicity, hematoxicity
γ Diketones	R_1–C(=O)–CH_2–CH_2–C(=O)–R_2	Schiff base formation	Neurotoxicity

(Contd...)

Electrophile	General structure	Nucleophilic reaction	Toxic effect
Terminal epoxides	–CH—CH$_2$ \ O / –O–CH$_2$–CH–CH$_2$ \ O /	Addition	Mutagenicity, testicular lesions
Isocyanates	–N=C=O –N=C=S	Addition	Cancer, mutagenicity, immunotoxicity

*The presence of these substituents in a substance does not automatically mean that the substance is or will be toxic. Other factors, such as bioavailability, and the presence of other substituents that may reduce the reactivity of these electrophiles can influence toxicity as well.

Ideally, the use of these groups would be avoided, however, in many cases the electrophilic groups are necessary to produce a desired property. For example, for the case of the unsaturated carbonyls, the Michael addition reaction that causes the toxic effect may be the desired commercial property. Nevertheless, the toxic effects can sometimes be reduced by introducing selected substituents. For example, the addition of a methyl substituent to ethyl acrylate reduces potential health effects:

$$CH_2=CH-C(=O)-O-CH_2-CH_3$$
Ethyl acrylate (carcinogenic)

$$CH_2=C(CH_3)-C(=O)-O-CH_3$$
(Methyl methacrylate, noncarcinogenic)

Isocyanates present another example. In this case, the electrophilic nature of the isocyanate can be masked in some applications by converting the material to a ketoxime derivative.

$$R-NH-C(=O)-O-N=C(CH_3)-CH_2-CH_2-CH_3$$

The ketoxime derivative is then removed, *in situ*, during the use of the compound. This reduces potential exposures and the resultant toxicity.

Clearly, the identification of such structural modifications requires a detailed knowledge of the mechanism of the potential toxicity and the structural sensitivity of that mechanism.

Chapter 5

Unit Operations and Pollution Prevention

INTRODUCTION

In developing a flowsheet for the production of a chemical, it is desirable to consider the environmental ramifications of each unit operation in the process rather than postponing this consideration until the flowsheet is finished. This 'front end' environmental assessment is more likely to result in a chemical process that has less potential to cause environmental harm. In many instances, this environmentally benign design will also be more profitable, because the improved design will require lower waste treatment and environmental compliance costs and will convert a higher percentage of raw materials into salable product.

In considering pollution prevention for unit operations in the design of chemical processes, the following considerations are important.

1. Material selection: Many of the environmental concerns can be addressed by reviewing material properties and making the correct choice of unit operation and operating conditions. The materials used in each unit operation should be carefully considered so as to minimise the human health impact and environmental damage of any releases that might occur.
2. Waste generation mechanisms: Often, a careful evaluation of the mechanisms of in-process waste generation can direct the process designer toward environmentally sound material choices and other pollution prevention options.
3. Operating conditions: The operating conditions of each unit should be optimised in order to achieve maximum reactor conversion and separation efficiencies.
4. Material storage and transfer: The best material storage and transfer technologies should be considered in order to minimise releases of materials to the environment.
5. Energy consumption: Energy consumption in each unit should be carefully reviewed so as to reasonably minimise its use and the associated release of utility-related emissions.
6. Process safety: The safety ramifications of pollution prevention measures need to be reviewed in order to maintain safe working conditions.

In the following sections, we apply this framework for preventing pollution in unit operations by considering choices in materials, technology selection, energy consumption, and safety ramifications. The material choices that are generic to most chemical processes, like process water and fuel type, are analysed with respect to in-process waste generation and emission release. Other process materials that are more specific to various unit operations and chemical reactors are also discussed in the subsequent section. The environmental issues related to the use of reactants, diluents, solvents, and catalysts are discussed first. Then the effects of reaction type and order on product yield and selectivity are covered.

The effects of reaction conditions (temperature and mixing intensity) on selectivity and yield are illustrated. Finally, the benefits of additional reactor modifications for pollution prevention are tabulated. In subsequent section, the most important topics include the choice of material (mass separating agent) to be used in separations, design heuristics, and examples of the use of separation technologies for recovery of valuable components from waste streams, leading eventually to their reuse in the process. Separative reactors are the hybrid unit operations, have special characteristics to help achieve higher conversions and yields in chemical reactors compared to conventional reactor configurations. Various methods for reducing emissions from storage tanks and fugitive sources and the safety aspects of pollution prevention and unit operations are also discussed. It is shown that many pollution prevention efforts tend to make chemical processes more complex, necessitating a higher level of safety awareness.

In making pollution prevention decisions that include choices of materials, unit operations technologies, operating conditions, and energy consumption, it is very important to consider health and environmental risk factors. It is also important to incorporate cost factors and to be aware of safety ramifications. In the end of chapter, we review a method for minimising the potential environmental impact of unit operations by considering the optimum reactor operating conditions as an example application.

Finally, it is also important to introduce the concept of 'risk shifting'. Pollution prevention decisions that are targeted to reduce one kind of risk may increase the level of risk in other areas. For example, a common method of conserving water resources at chemical manufacturing facilities is to employ cooling towers. Process water used for cooling purposes can be recycled and reused many times. However, there is an increased risk for workers who may be exposed to the biocides used to control microbial growth in the cooling water circuit. Also, in some cooling water processes, hazardous waste is created by the accumulation of solids—for example, from the use of hexavalent chromium (a cancer causing agent) as a corrosion inhibitor.

Another example of shifting risk from the environment and the general population to workers involves fugitive sources (valves, pumps, pipe connectors, etc.). One strategy for decreasing fugitive emissions is to reduce the number of these units by eliminating backup units and redundancy. This strategy will decrease routine air releases but will increase the probability of a catastrophic release or other safety incidents. Simply put, the objective of pollution prevention is to reduce the overall level of risk in all areas and not to shift risk from one type to another.

POLLUTION PREVENTION IN MATERIAL SELECTION FOR UNIT OPERATIONS

One very important element of designing and modifying process units for pollution prevention is the choice of materials that are used in chemical processes. These materials are used as feedstocks, solvents, reactants, mass separating agents, diluents, and fuels. In considering their suitability as process components, it is not sufficient to consider only material properties that are directly related to processing; it is increasingly important to consider the environmental and safety properties as well. Use of materials that are known to be persistent, bioaccumulative, or toxic should be avoided as they are under increasing regulatory scrutiny and many manufacturers are moving away from their use. Questions regarding material selection include:
1. What are the environmental, toxicological, and safety properties of the material?
2. How do these properties compare to alternative choices?
3. To what extent does the material contribute to waste generation or emission release in the process?
4. Are there alternative choices that generate less waste or emit less while maintaining or enhancing the overall yield of the desired product?

If processing materials can be found which generate less waste and if the hazardous characteristics of those wastes are less problematic, then significant progress may be made in preventing pollution from the chemical process.

Materials are involved in a wide range of processing functions in chemical manufacturing, and depending on the specific application, their environmental impacts vary greatly. For example, reactants for producing a particular chemical can vary significantly with respect to toxicity and inherent environmental impact potential (global warming, ozone depletion, etc.), and they can exhibit various degrees of selectivity and yield toward the desired product. In addition, the properties of reaction by-products can vary widely, similarly to reactants. Some catalysts are composed of hazardous materials or they may react to form hazardous substances. For example, catalysts used for hydrogenation of carbon monoxide can form volatile metal carbonyl compounds, such as nickel carbonyl, that are highly toxic. Many catalysts contain heavy metals, and environmentally safe disposal has become an increasing concern and expense. After the deposition of inhibitory substances, the regeneration of certain heterogeneous catalysts releases significant amounts of SO_x, NO_x, and particulate matter. For example, the regenerator for a fluid catalytic cracking unit (FCCU) is a major source of air pollutants at refineries. Note that the removal of the FCCU would result in very low yields and consequently unacceptable waste generation at facility level. The choice of a mass separating agent for solids leaching or liquid extraction applications can affect the environmental impacts of those unit operations. Agents that are matched well to the desired separation will consume less energy and release less energy-related pollutants than those that are not well suited for the application. Typically, agents that have lower toxicity will require less stringent clean up levels for any waste streams that are generated in the process.

The choice of fuel for combustion in industrial boilers will determine the degree of air pollution abatement needed to meet environmental regulations for those waste streams. As an illustration, using fuel types having lower sulphur, nitrogen, and trace metals levels will yield a flue gas with lower concentrations of acid rain precursors (SO_x and NO_x) and particulate matter.

Less toxic materials, such as water and air, can still have important environmental implications due to the waste streams that are generated in their use. Air is often used in chemical reactions either as a diluent or as a source of oxygen. For certain high temperature reactions, the nitrogen and oxygen molecules in the air react, forming oxides of nitrogen. Upon release, NO_x will participate in photochemical smog reactions in the lower atmosphere. Therefore, it is important to consider alternative sources of oxidants, such as enriched air or pure oxygen, and diluents, such as carbon dioxide or other inert reaction by-products. Water is used for many purposes in chemical processes; as boiler feed, a cooling medium, reactant, or a mass separating agent. The following example illustrates that the quality of the feed water can have a profound influence on the generation of hazardous waste in a refinery.

Example 5.1: Figure 5.1 shows the many uses of process water in a refinery. Water is brought into contact with crude oil in order to remove salts and other solid contaminants that could disrupt the operation of downstream equipment. The spent water from this operation is sent to the waste-water treatment facility to recover residual oil and to remove toxic constituents. Water that is used as feed to the boilers is softened in an ion exchanger. Steam generated in the boilers is used for process heating and a fraction is returned to the boiler as condensate.

Solution: Solids accumulate in the boiler and excessive levels of suspended solids lead to fouling of heat transfer surfaces in the process, a decrease in heat transfer efficiency, and requires periodic shutdown and cleaning of these surfaces to restore normal operation.

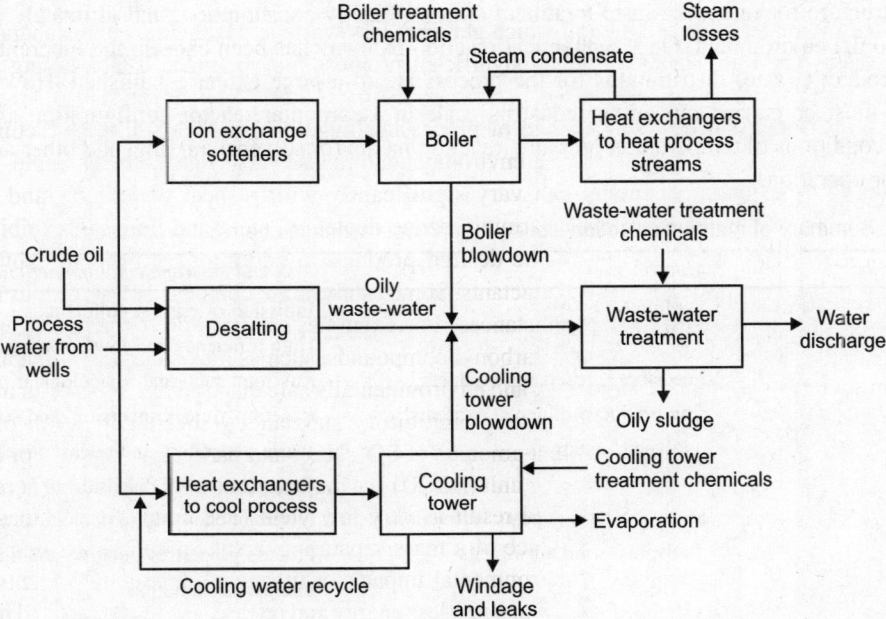

Fig. 5.1. Conceptual diagram of process water at a refinery.

To control solids accumulation, the contents of the boiler are sent to waste-water treatment when the dissolved solids content is above a cut-off level, in a step termed 'blowdown'. Similarly, in the cooling system of the refinery, dissolved solids accumulate because the mechanism of cooling in the cooling tower is evaporation, which effectively retains solids in the process. When the high calcium solids from the cooling tower blowdown meet the alkaline boiler blowdown, precipitation occurs. This precipitate can clog waste-water treatment equipment and can form oily sludges upon being blended with the waste-water from the desalter unit. It has been shown that every kg of solids precipitate in oily waste-water creates about 10 kgs of oily sludge. The oily sludge is a RCRA hazardous waste and is costly because of expensive disposal fees and the oil lost from the process that could have been made into products.

Pollution prevention solution: At a southwest United States petroleum refinery, the solution to this waste disposal problem involved the pretreatment of all process water using reverse osmosis to separate dissolved solids from the feed water, thereby eliminating the source of solids for the oily sludge. This solution proved to be cost effective because the savings in disposal costs alone was enough to pay for the pretreatment equipment and its operation. Additional savings were realised because fewer boiler and cooling tower treatment chemicals were needed (90 per cent reduction in use). Also, maintenance costs were lower since scale build-up on the cooling waste heat exchanger surfaces was reduced.

Additional examples of material use and pollution prevention in unit operations will be provided in the sections of this chapter on chemical reactors and separation equipment. A summary to relating unit operations, materials selection, and risk factors is presented in Table 5.1.

POLLUTION PREVENTION FOR CHEMICAL REACTORS

From an environmental perspective, reactors are the most important unit operation in a chemical process. The degree of conversion of feed to desired products influences all subsequent separation processes, recycle structure for reactors, waste treatment options, energy consumption, and ultimately pollutant releases to the environment. Once a chemical reaction pathway has been chosen, the inherent product and by-product (waste) distributions for the process are to a large extent established. However, the synthesis must be carried out on an industrial scale in a particular reactor configuration and under specified conditions of temperature, pressure, reaction media (or solvent), mixing, and other aspects of the reactor operation.

Table 5.1. Summary of material selection issues for unit operations in chemical processes.

Unit operation	Materials	Risk and environmental impact issues
Boilers	Fuel type	Emission of criteria pollutants
		High efficiency and low emissions boilers
Reactors	Feedstocks, reactants, products, by-products, diluents, oxidants, solvents, catalysts	Environmental and toxicological properties
		Reaction yield, conversion, and selectivity
		Waste generation and release mechanisms
		Catalyst reuse or disposal
Separators	Mass separating agents, extraction solvents, solid adsorbents	Environmental and toxicological properties
		Process properties (relative volatility, etc.)
		Energy consumption
		Regeneration of solid adsorbents
Storage tanks	Feedstocks, products, solvents	Environmental and toxicological properties
		Air emissions
		Vapour pressure of liquids
Fugitive sources	Feedstocks, products, solvents	Same as storage tanks
Cooling towers	Water, biocides	Environmental/toxicological properties of biocides
		Waste generation by dissolved solids
Heat exchangers	Heat transfer fluids	Environmental and toxicological properties

In designing chemical reactors for pollution prevention, there are many important considerations. The raw materials, products, and by-products should have a relatively low environmental and health impact potential. This means that the environmental and toxicological properties of the chemicals involved should indicate that they are relatively benign. In addition, the conversion of reactants to desired products should be high and their conversion toward by-products should be low. In other words, the reaction yield and selectivity for the desired product should be as high as possible. Finally, energy consumption for the reaction should be low. Another consideration is that the life cycle impacts of reactants, products, and by-products should be relatively low. For example, cumulative emissions and impacts of raw materials should be relatively low, environmental impacts during subsequent use by consumers should be small, and if possible the reaction products should be recyclable. Engineers must balance all of these considerations. For the discussion in this chapter, these reactor considerations will be classified as:

1. Material use and selection.
2. Reaction type and reactor choice.
3. Reactor operation.

The following discussion proceeds from the most general to the more specific topics on preventing pollution for chemical reactors.

Material Use and Selection for Reactors

Issues involving the use of materials in a chemical reactor include the choice of feed entering the reactor, the catalyst if one is needed, and solvents or diluents. However, it is important to mention material selection here in light of their influence on the environmental impacts of reactors in chemical processes.

Raw materials and feedstocks

Raw materials used in chemical reactions can be highly toxic or can cause undesirable by-products to form. Although some of these raw materials may be converted to relatively benign chemicals through chemical reactions in the process, their presence may be a concern because of the potential for uncontrolled release and exposure to humans in the workplace and also in the environment. An important strategy for environmental risk reduction for chemical processes is to eliminate as many of these toxic raw materials, intermediates, and products as possible.

The elimination of a raw material or the use of a more benign substitution may necessitate the adoption of new process chemistry. For example, phosgene is used in large volumes all over the world in the manufacture of polycarbonates and urethanes. Phosgene $-COCl_2$ is highly toxic and may pose risks for workers at manufacturing facilities and to the surrounding population if large releases occur. In the phosgene process for producing polycarbonates, polycarbonate is produced from bisphenol-A monomer and phosgene in the presence of two solvents, methylene chloride and water. A new process for polycarbonate synthesis has been demonstrated using solid-state polymerisation in the absence of both phosgene and methylene chloride (also toxic), by including diphenyl carbonate (DPC) and phenol instead. Similarly, alternative phosgene-free routes to urethanes have recently been developed. The 'Tier 1' environmental performance tools are very useful for evaluating these and other alternative reactions chemistries.

In the production of fuels for transportation, petroleum refineries are required to remove sulphur from their products. If not removed, SO_2 is released to the atmosphere upon combustion of the fuels in automobiles, trucks, or stationary combustion sources. Exposure to SO_2-contaminated air causes lung irritation and other more serious health effects, and SO_2 emissions contribute to acidification of surface water and ecosystem damage. Choosing a crude oil raw material with lower sulphur content (sweet crude) reduces the amount of sulphur that needs to be removed and reduces operating costs, but is considerably more expensive to purchase. Another option to consider would be to use, and therefore incur the associated costs with, a hydrodesulphurisation or a hydrotreating unit to remove the sulphur. The sulphur can then become a salable product.

In partial oxidation reactions of hydrocarbons to form alcohols or other oxygenated organics, air has traditionally been the source of oxygen in the reaction, and the nitrogen in the air has acted as a heat sink agent (diluent) to help control temperature rise for the exothermic reaction. Some CO_2 and H_2O are produced, and due to the presence of N_2, some NO_x is formed. NO_x is a precursor in the formation of photochemical smog in urban atmospheres and its emission from industrial facilities is regulated under the Clean Air Act. One method to reduce or eliminate the formation of NO_x in partial oxidation reactions is to use pure oxygen or enriched air as the oxidising agent, thus preventing NO_x formation. Carbon dioxide that is recovered and recycled from the reactor effluent or water vapour could be used as the heat sink instead of nitrogen. Another method is to install NO_x control equipment on the original process. An important issue in this case is whether the costs associated with purchasing and operating the CO_2 recovery equipment are lower than operating NO_x control equipment. Another consideration is whether the additional pollutant releases associated with NO_x prevention equipment are lower than the releases in the original process.

Solvents

Another important class of raw materials used in chemical reactors are solvents. This is especially true in 'solution' and 'emulsion' polymerisation reactions in which the reaction of monomers to create high molecular weight polymers occurs either within the solvent phase or within dispersed droplets of monomer in the solvent phase. In some polymerisations, addition of solvents can enhance precipitation of polymer in solid form, co-solubilise monomer and initiator, and act as a diluting medium to modulate the rate of reaction and rate of heat removal. The most commonly used polymers are low-density polyethylene (LDPE), high-density polyethylene (HDPE), polyvinyl chloride (PVC), polypropylene, and polystyrene. Solvents used in the production of some of these high volume polymers include xylene, methanol, lubricating oil, hexane, heptane, and water. Solvents are concern due to their high volatility and potential to cause low-level ozone during smog formation reactions in the atmosphere. They may also be a health concern for workers and the general population in the vicinity of the facility. In addition, there are several on-line resources for evaluating substitute solvents. The solubility, toxicological, cost, and environmental properties of the candidate solvents can be compared with each other and with the original solvent. Supercritical carbon dioxide is being studied as a substitute solvent in many reaction systems. Applications include both homogeneous and dispersed phase polymerisation reactions in which the supercritical CO_2 replaces volatile organic compounds and chlorofluorocarbons as traditional solvents in the reaction mixture.

Catalysts

A catalyst is a substance that is added to a chemical reaction mixture in order to accelerate the rate of reaction. Catalysts are either homogeneous, being dissolved in the reaction mixture, or heterogeneous, typically existing as a solid within a reacting fluid mixture. The choice of catalyst has a large impact on the efficiency of the chemical reactor and ultimately upon the environmental impacts of the entire chemical process. Advances in catalysts can improve the environmental performance of a chemical reactor in several ways. Catalysts can allow the use of more environmentally benign chemicals as raw materials, can increase selectivity toward the desired product and away from unwanted by-products (wastes), can convert waste chemicals to raw materials, and can create more environmentally acceptable products directly from the reactions.

The production of reformulated gasoline (RFG) and diesel fuels from crude oil is a clear example of how improved catalysts can create chemicals that are better for the environment. Because of recent trends in the petroleum refining industry, improved catalysts are being used in several reaction processes within modern refineries. These trends include:
1. Increased processing of crude oils with lower quality (higher percentages of sulphur, nitrogen, metals, and carbon residues).
2. More demand for lighter fuels and less for heavy oils.
3. Environmental regulations that limit the percentages of sulphur, heavy metals, aromatics, and volatile organic compounds in transportation fuels.

In particular, the inclusion of RFG in the clean air act (CAA) has prompted many changes in catalyst formulation and reactor configurations.

Table 5.2 is a summary of conventional and improved catalytic reaction processes for RFG and diesel production. As seen in the Table, the major emphasis is on catalyst improvements for sulphur and nitrogen removal from heavier crude fractions, reduced aromatics content, and increased production of branched C5–C7 alkanes for octane enhancement.

Table 5.2. Summary of conventional and improved catalysts for reformulated gasoline and diesel production.

Process	Objective	Conventional catalyst	Improved catalyst	Benefits of improved catalyst
Reformulated gasoline				
FCC	Conversion of heavy oils to gasoline	Zeolites ReY zeolites	USY zeolites USY + ZSM-5 USY/matrix GSR	Increased gasoline yield Reduced coking Increased light olefins/ selectivity Gasoline sulphur reduction
Reforming	Gasoline octane enhancement	Pt/Al_2O_3	$Pt/Ir/Al_2O_3$ $Pt-Re/Al_2O_3$ $Pt-Re/Al_2O_3$ + zeolite $Pt-Sn/Al_2O_3$	Low-pressure operation Reduced coking Increased octane Improved catalyst stability
Alkylation	Production of branched alkanes for gasoline octane enhancement	H_2SO_4 HF	Supported BF_3 Modified SbF_3 Supported liquid-acid catalysts	Less corrosive Safe handling Fewer environmental problems
Isomerisation	Conversion of C5/C6 alkanes into high-octane branched isomers	Pt/Al_2O_3 $Pt/SiO_2-Al_2O_3$ Mordenite (zeolite)	Solid super acid catalysts (e.g. sulphonated zirconia)	Low temperatures Increased conversion Less cracking
Diesel production				
Middle distillate hydrotreating	Diesel desulphurisation	$Co-Mo/Al_2O_3$	High metal $Co-Mo/Al_2O_3$ with modified support pore structure	Sulphur removal to < 500 ppm
Middle distillate aromatics hydrdogenation	Production of low aromatics diesel	$Ni-Mo/Al_2O_3$	$Ni-Mo/Al_2O_3$ Noble metal-zeolite combination in two-stage process	Aromatics hydrogenation to acceptable low aromatics levels in diesel
VGO hydrotreating	FCC feed pretreatment to reduce sulphur and nitrogen levels	$Co-Mo/Al_2O_3$	$Co-Mo/Al_2O_3$ with improved formulation and pore structure Ni-Mo/Zeolite + amorphous $SiO_2-Al_2O_3$	Increased activity for N & S removal Improved cycle length Increased throughput Mild hydrocracking Increased middle distillate selectivity

(Contd...)

Process	Objective	Conventional catalyst	Improved catalyst	Benefits of improved catalyst
Gas oil hydrocracking	Conversion of heavy gas oils to lighter products (gasoline and diesel)	Ni-Mo/Al$_2$O$_3$ Ni-W/Al$_2$O$_3$	Ni-W/modified Al$_2$O$_3$ Ni-W/SiO$_2$-Al$_2$O$_3$ Ni-W/Zeolite + amorphous SiO$_2$-Al$_2$O$_3$	Increased middle distillate selectivity Superior quality middle distillates Increased catalyst life

FCC: Fluidised catalytic cracker
GSR: Gasoline sulphur reduction
RE: Rare earth
US: Ultra stable
VGO: Vacuum gas oil
Y & ZSM-5: Crystalline forms y zeolite catalyst

In the above section, we presented examples of how material selection in chemical reactors can impact the environment and reviewed cases where waste generation or toxicity were reduced. In the next section, we investigate the effects of reaction type and reactor choice on waste generation in chemical reactors.

Reaction Type and Reactor Choice

The details of any chemical reaction mechanism, including the reaction order, whether it has series or parallel reaction pathways, and whether the reaction is reversible or irreversible, influences pollution prevention opportunities and strategies for chemical reactors. These details will determine the optimum reactor temperature, residence time, and mixing. In addition, reactor operation influences the degree of reactant conversion, selectivity, and yield for the desired product, by-product formation, and waste generation. As a general rule, in designing chemical reactors for pollution prevention; one would like a reaction with a very high conversion of the reactants, high selectivity toward the desired product, and low selectivity toward any by-products. A typical reactor efficiency measure pertaining to reactant conversion is the reaction yield, defined as the ratio of the exiting concentration of product to inlet reactant ($[P]/[R]_0$). Reaction selectivity is defined as the ratio of exiting product concentration to the undesired by-product concentration. We define a modified selectivity as the ratio of exiting product concentration to the sum of product and by-product (waste) concentrations ($[P]/([P]+[W]) = [P]/[$Reactant consumed$]$). This allows us to display both yield and selectivity on the same scale, from 0 to 1. Yields and selectivity values that are very close to unity indicate an efficient reaction, with little waste generation or reactant to separate in downstream unit operations.

Parallel reaction pathways are very common in the chemical industry. An example of an industrial parallel reaction is the partial oxidation of ethylene to ethylene oxide, whereas the parallel reaction converts ethylene to by-products, carbon dioxide, and water.

We will begin our discussion of reaction types and their implications for pollution generation with the simple irreversible first-order parallel reaction mechanism shown below:

$$R \xrightarrow{k_p} P$$

$$R \xrightarrow{k_w} W \qquad \ldots (5.1)$$

R is the reactant, P the product, W a waste by-product, and k_p and k_w are the first-order reaction rate constants for product formation and waste generation (time^{-1}), respectively. The relative concentrations of products and waste components are significantly affected by the ratio of the first order reaction rate constants, k_p/k_w. Figure 5.2 illustrates the dependence of the reactor effluent concentrations of products and waste as a function of reactor residence time for several values of these rate constant ratios. In order to achieve maximum reactor yields, the residence time must be about 5 times the reaction time constant $(k_p + k_w)^{-1}$. The reaction selectivity is constant and independent of reactor residence time for first-order irreversible, isothermal parallel reactions. As shown next, in a series reaction, selectivity is affected by reactor residence time, and therefore this parameter must also be considered for pollution prevention in chemical reactors.

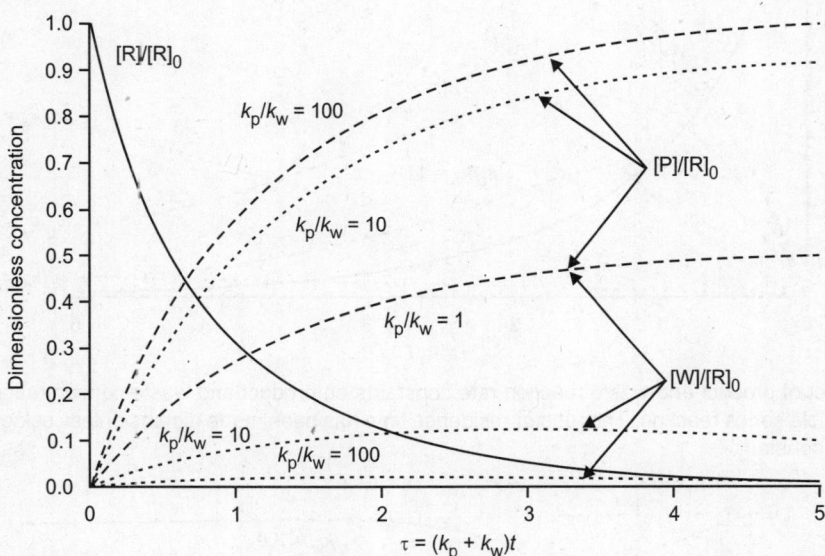

Fig. 5.2. Effect of k_p/k_w on the reactor outlet concentrations of products, reactants, and by-products (waste) for a simple irreversible first-order parallel reaction mechanism.

In a series reaction, the rate of by-product (waste) generation depends on the rate of product formation, as shown by the first-order irreversible series reaction below:

$$R \xrightarrow{k_p} P \xrightarrow{k_w} W \qquad \ldots (5.2)$$

Longer reactor residence times lead to not only more product formation but also more by-product generation. The amount of waste generation for a series reaction depends on the ratio of product formation rate constant (k_p) to the by-product generation rate constant (k_w) and also on the residence time in the reactor. Figure 5.3 illustrates the effect of reactor residence time on reactant, product, and by-product concentrations for several reaction rate constant ratios (k_p/k_w). For each ratio, there is an optimum reactor residence time that maximises the product concentration. Figure 5.4 shows the product yield $([P]/[R]_0)$ and modified selectivity $([P]/([P]+[W]))$ over a range of reactor residence time for several reaction rate ratios. For irreversible series reactions, the modified selectivity continues to decrease with time. At longer residence times, the rate of waste generation is greater than the rate of product formation. To minimise waste generation in series reactions, it is important to operate the reactor so that the ratio k_p/k_w is as large as possible and to control the reaction residence time. Also, if there is a way to remove the reaction product as it is being formed and before its concentration builds up in the reactor, then by-product generation can be minimised. We discuss this point more when the topic of separative reactors is covered later in this chapter.

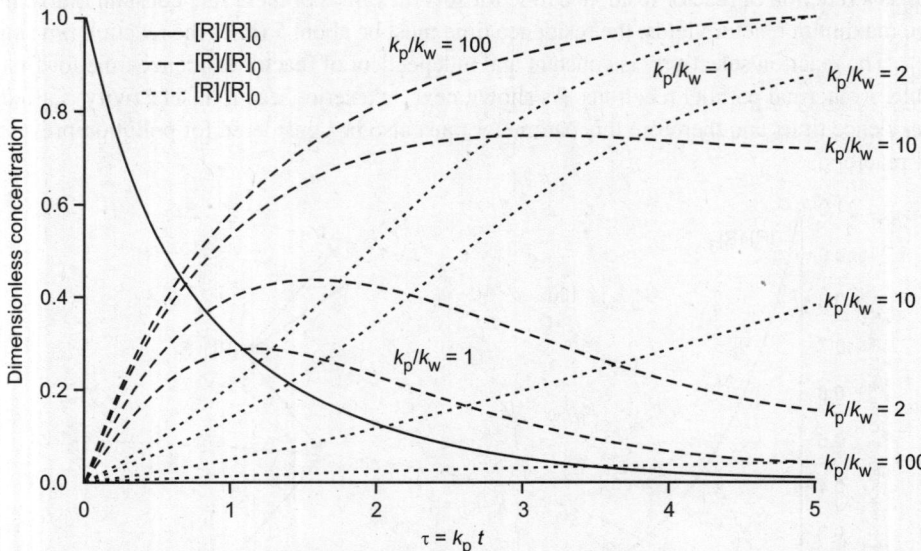

Fig. 5.3. Effect of product and waste reaction rate constants on product and waste concentrations in a first-order irreversible series reaction. The reactor residence time has been made dimensionless using the product reaction rate constant.

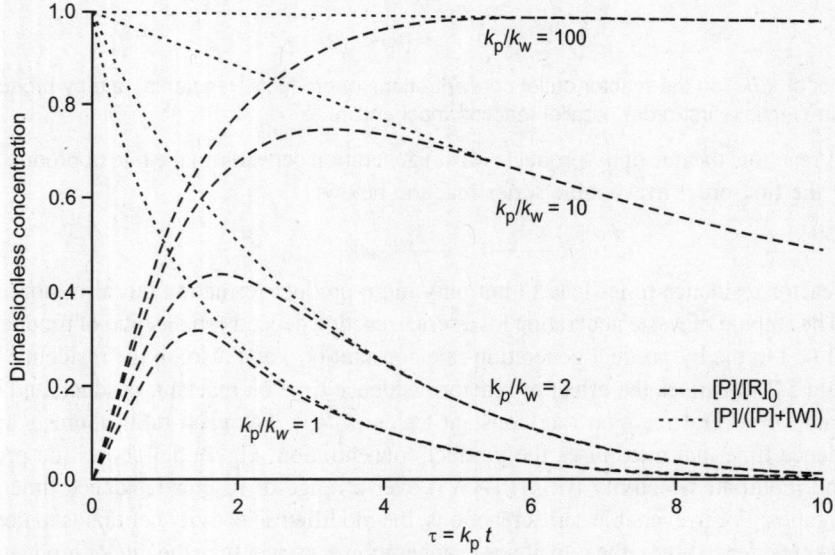

Fig. 5.4. Effect of product and waste reaction rate constants on product yield ($[P]/[R]_0$) and product modified selectivity ($[PV([P]+[W])$) for a first-order irreversible series reaction. The reactor residence time has been made dimensionless using the product reaction rate constant.

Reversible reactions are another important category of chemical reactions. Figure 5.5 shows the reactant, product, and by-product concentrations profiles in parallel and series reversible reactions for a wide range of reaction rate constants. It is evident that reversible reactions inhibit full conversion of reactants to products. Also, the reactor residence time is a key operating parameter for reversible reactions. Selectivity improvements for reversible reactions, operated at equilibrium, can be achieved by utilising the concept of recycle to extinction. As an example of this concept, consider the steam reforming of methane to form synthesis gas ($CO + H_2$) for methanol production.

$$CH_4 + H_2O \longleftrightarrow + CO + 3H_2$$

$$CO + H_2O \longleftrightarrow + CO_2 + H_2$$

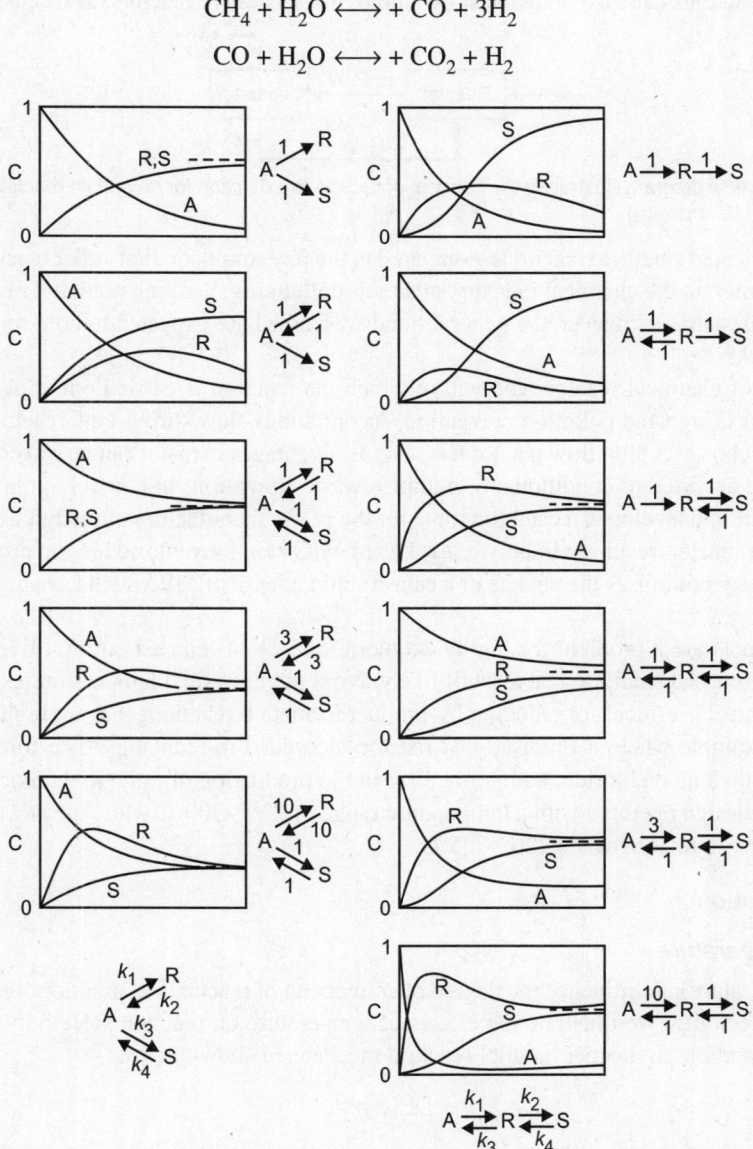

Fig. 5.5. Product and by-product profiles for reversible parallel and series reactions.

Both reactions are reversible and at equilibrium. When CO_2 is recovered and recycled back to the reactor, it decomposes in the reactor as fast as it forms, and no net conversion of methane to CO_2 occurs. This requires additional operating costs, but there is no selectivity loss of reactant, the process is cleaner, and it may be the lowest cost option overall.

Figure 5.6 shows a process flow diagram for a reactor combined with a separator that recycles reactants and by-products back to the reactor. This configuration can be operated such that all reactants fed to the reactor are converted to product with no net waste generation from the process. Selectivity improvements for reversible reactions can also be realised by employing separative reactors, as discussed later.

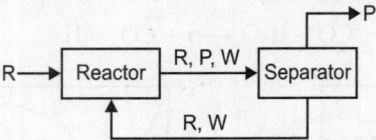

Fig. 5.6. Process flow diagram illustrating the concept of recycle to extinction for reversible reactions (R = reactant, P = product, and W = waste).

More complicated chemical reactions, compared to the few simplistic first-order reactions mentioned above, are common in the chemical industry, and their pollution-generating potential must be evaluated on a case-by-case basis. However, the general trends discussed are expected to hold for more complex reaction networks.

The choice of chemical reactor type within which the reaction is carried out is also an important issue for process design and pollution prevention. A continuous-flow stirred-tank reactor (CSTR) is not always the best choice. A plug flow reactor has several advantages in that it can be staged and each stage can be operated at different conditions to minimise waste formation. In a novel application of a plug flow reactor, DuPont developed a catalytic route for the *in situ* manufacture of methyl isocyanate (MIC) using a pipeline reactor, resulting in only a few kgs of MIC being inventoried in the process at any one time. This strategy minimises the chance of a catastrophic release of MIC, such as happened at Bhopal, India, in 1984.

When hot spots are a problem for highly exothermic reactions carried out in a fixed-bed catalytic reactor, a fluidised-bed catalytic reactor will likely avoid the unwanted temperature excursions. Good temperature control is critical for reducing by-product formation reactions that are highly temperature-sensitive. An example where a fluidised-bed reactor succeeded in reducing waste formation is in the production of ethylene dichloride, an intermediate in the production of polyvinyl chloride (PVC). The prior fixed-bed design operated with a temperature range of 230°–300°C while the newer fluidised-bed design was able to run at between 220°–235°C.

Reactor Operation

Reaction temperature

Reaction temperature can influence the degree of conversion of reactants to products, the product yield, and product selectivity. We illustrate the effects of temperature on reaction selectivity by considering the simple irreversible first-order parallel reaction mechanism shown below:

$$R \xrightarrow{k_p} P$$

$$R \xrightarrow{k_w} W \qquad \ldots (5.3)$$

where R is the reactant, P the product, W a waste by-product, and k_p and k_w are the first-order reaction rate constants for product formation and waste generation (time^{-1}), respectively. The ratio of the reaction rates for product formation to by-product generation is an important indicator of reaction selectivity.

$$\frac{k_p[R]}{k_w[R]} = \frac{k_p}{k_w} = \frac{A_p e^{-(E_p/RT)}}{A_w e^{-(E_w/RT)}} \quad \ldots (5.4)$$

where A_p and A_w are the frequency factors (time^{-1}) and E_p and E_w are the activation energies (kcal/mole) for product and waste respectively, R is the gas constant [1.987×10^{-3} kcal/(mole·K)], and T is absolute temperature. Because the reaction rate constants, k_p and k_w, are functions of temperature, their ratio is also a function of temperature. For the purpose of illustration, we can calculate the change in this ratio [$\Delta(k_p/k_w)$] as the temperature is changed to a new value (T_1) above or below a given initial temperature (T_0).

$$\Delta \frac{k_p}{k_w} = \frac{e^{-(E_p/RT_1)} / e^{-(E_w/RT_1)}}{e^{-(E_p/RT_0)} / e^{-(E_w/RT_0)}} = \frac{e^{-(E_p-E_w)/RT_1}}{e^{-(E_p-E_w)/RT_0}} \quad \ldots (5.5)$$

Figure 5.7 shows the expected change in the ratio of product/by-product rate constants when temperature is changed (ΔT) above and below T_0. When $E_p > E_w$, the ratio increases with increasing temperature and decreases with decreasing temperature. Therefore, pollution can be prevented in parallel (and also series) reactions by increasing reactor temperature when $E_p > E_w$. The opposite holds true for when $E_p < E_w$. Also, as the difference between E_p and E_w increases, temperature has a more pronounced influence on the change in the rate constant ratios.

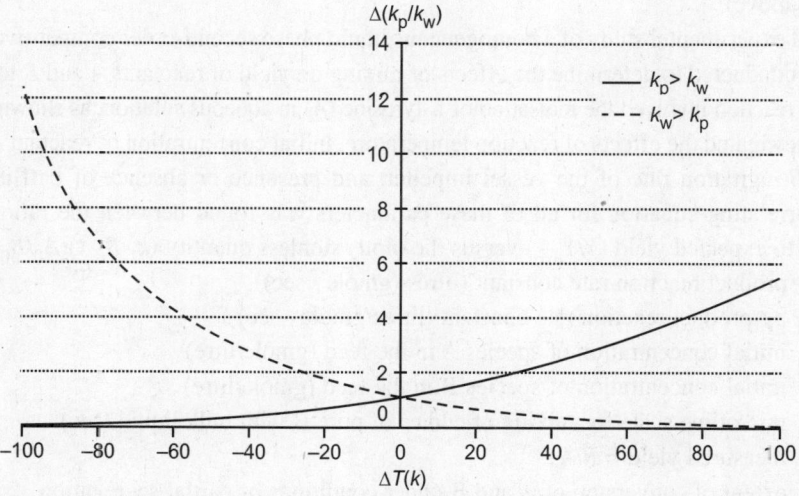

Fig. 5.7. Effect of reaction temperature on the ratio of rate constants for a first-order parallel and irreversible reaction. For $E_p > E_w$, E_p was set to 20 kcal/mole and E_w to 10 kcal/mole. For $E_p < E_w$, E_p was set to 10 kcal/mole and E_w to 20 kcal/mole.

Mixing

When a reactant in one inlet stream is added to another reactant that already exists in a well-stirred reactor, the course of complex multiple reactions can be affected by the intensity of mixing in the vessel. For irreversible reactions, the reaction yield and selectivity may be altered compared to the case where the reactants are mixed instantaneously to a molecular level. This may lead to a greater amount of waste by-product generation. In addition, the rate of reaction can be reduced because of diffusional limitations between segregated elements of the reaction mixture. The complications that arise from imperfect mixing are particularly evident for rapidly reacting systems. In these situations, reactants are significantly converted to products and by-products before mixing is complete.

To illustrate the effects of mixing, it is illustrative to examine the competitive-consecutive reaction carried out in a constant stirred tank reactor (CSTR), using the reaction mechanisms shown below:

$$A + B \xrightarrow{k_1} R$$

$$R + B \xrightarrow{k_2} S \qquad \ldots (5.6)$$

This reaction is also sometimes referred to as the series-parallel reaction. This reaction type is a good kinetic representation of the nitration and halogenation of hydrocarbons and saponification of polyesters, among its many industrially-relevant examples. Reactant A is initially charged in the reactor and B is added as a solution through a feed pipe in a continuous manner until a stoichiometric amount of B is added. Species R is the desired product and S is a by-product. If the reactions are first order, mixing will not affect selectivity.

However, if the reactions are second order, the presence of local excess B concentrations can cause overreaction of R to S via the second reaction. This effect of mixing occurs for both homogeneous and heterogeneous reaction systems and for batch or semi-batch reactors (B added to an initial charge of A as described above).

A detailed experimental study of a homogeneous liquid phase second-order competitive-consecutive reaction was conducted to determine the effects of mixing on yield of reactants A and B to product R in a CSTR. The reaction involved the iodisation of L-tyrosine (A) in aqueous solution, as shown in Fig. 5.8.

Smith investigated the effects of reaction temperature, initial concentration of reactant A (A_0), rate of addition of B, agitation rate of the vessel impeller, and presence or absence of baffling within the reactor. A correlating equation for all of these parameters was found between the ratio of measured reactor yield to expected yield (Y/Y_{exp}) versus the dimensionless quantity $(k_1 B_0 \tau)(A_0/B_0)$, where,

k_1 = product reaction rate constant (litres/(gmole · sec))
k_2 = by-product reaction rate constant (litres/gmole · sec))
A_0 = initial concentration of species A in the feed (gmole/litre)
B_0 = initial concentration of species B in the feed (gmole/litre)
τ = microtime scale for mixing of eddies of pure B with bulk liquid (sec)
Y = measured yield = R/A_0
$(k_1 B_0 \tau)$ = extent of conversion of A and B under conditions of partial segregation

$$Y_{exp} = \text{'expected yield' (perfect mixing)} = \frac{R}{A_0} = \frac{1}{(k_2/k_1 - 1)} \left[\frac{A}{A_0} - \left(\frac{A}{A_0}\right)^{k_2/k_1} \right]$$

A/A_0 = fraction of reactant A remaining at the end of the reaction.

$$\ldots (5.7)$$

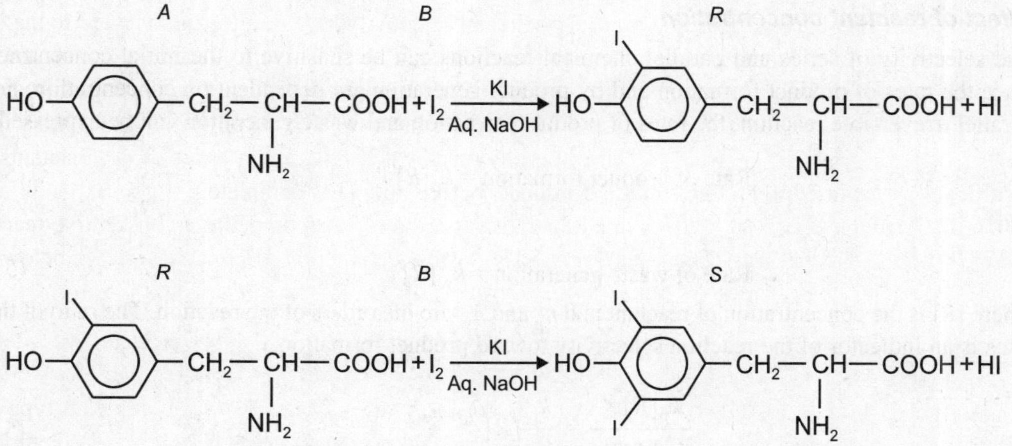

Fig. 5.8. Iodisation of L-tyrosine in aqueous solution.

To give an idea of the range of observed yields in the experiments, values of Y/Y_{exp} were measured from 0.66 to 0.98, depending upon mixing intensity and other parameters. A correlation fitting the data is presented in Fig. 5.9. It was found that when the quantity $(k_1 B_0 \tau)(A_0/B_0)$ was less than or equal to 10^{-5}, $Y = Y_{exp}$. This criterion allows us to 'set' the mixing intensity for any second-order competitive-consecutive reaction.

$$10^{-5} = (k_1 B_0 \tau)(A_0/B_0) = (k_1 \tau A_0) \qquad \ldots (5.8)$$

Rearranging for τ from the above equation, and incorporating the Kolomogoroff universal equilibrium theory for turbulent motion, we get:

$$\tau = \frac{10^{-5}}{A_0 k_1} = \frac{0.882 v^{3/4} L_f^{3/4}}{(u')^{7/4}} \qquad \ldots (5.9)$$

where,
L_f = a characteristic length scale of the vessel (ft)
u' = fluctuating turbulent velocity (ft/sec.)
v = kinematic viscosity (ft^2/sec.)

We can rearrange the equation above for u' and incorporate a correlation for turbulent fluctuation velocity in an agitated CSTR for feed entering at the impeller ($u' = 0.45 \pi D N$). We arrive at:

$$u' = \left[\frac{0.882 v^{3/4} L_f^{3/4}}{\left(\frac{10^{-5}}{A_0 k_1}\right)} \right]^{4/7} = 0.45 \pi D N \qquad \ldots (5.10)$$

Thus, with this equation, we can establish the necessary impeller agitation speed (N, revolutions per second) to ensure that mixing will not adversely affect the yield, given k_1, v, L_f, and D, the impeller diameter (ft).

Effect of reactant concentration

The selectivity of series and parallel chemical reactions can be sensitive to the initial concentration, since the rates of product formation and by-product generation are dependent on concentration. For a parallel irreversible reaction, the rates of product formation and waste generation can be expressed as

$$\text{Rate of product formation} = k_p[R]^{n_p} \quad \ldots(5.11)$$

and

$$\text{Rate of waste generation} = k_w[R]^{n_w} \quad \ldots(5.12)$$

where $[R]$ is the concentration of reactant and n_p and n_w are the orders of the reaction. The ratio of these rates is an indicator of the reaction selectivity toward product formation.

$$\frac{k_p[R]^{n_p}}{k_w[R]^{n_w}} = \frac{k_p}{k_w}[R]^{(n_p - n_w)} \quad \ldots(5.13)$$

If $n_p > n_w$, then increasing the concentration of reactant will increase the reaction selectivity toward the product and away from the waste by-product. Conversely, if $n_p < n_w$, then increasing reactant concentration will decrease selectivity toward the desired product.

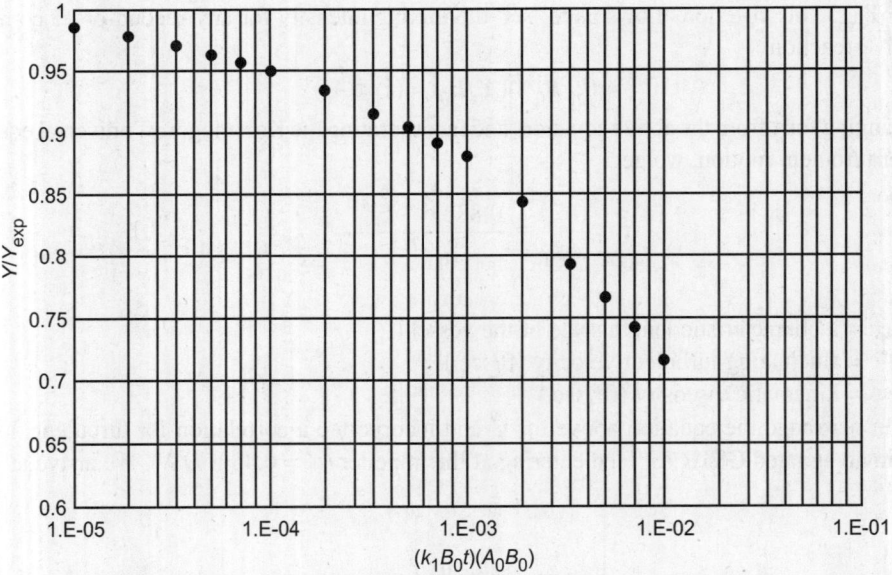

Fig. 5.9. Correlation of reaction yield efficiency with the mixing parameters of Paul and Treybal for an irreversible consecutive-competitive second-order reaction.

Summary of other methods

There are numerous operational modifications for improving the environmental and economic performance of reactors. A summary of many important modifications is shown in Table 5.3, along with a short description of the nature of the problem addressed, the modification, and the observed benefit.

Table 5.3. Additional reactor operation modifications leading to pollution prevention.

Improve reactant addition

    Problem: Non-optimal reactant addition can lead to segregation and excessive by-product formation

    Solution: Premix liquid reactants and solid catalysts before their introduction into a reactor using static in-line mixers

    Benefits: This will result in more efficient mixing of reactants and reduced waste generation by side-reactions for 2nd order or higher competitive-consecutive reactions

    Solution: Improve dip tube and sparger designs for tank reactors. Do not add low-density material above the liquid surface of a batch reactor. Control residence time of gases added to liquid reaction mixtures

    Benefits: Improved bottom-nozzle dip-tube design and improved gas residence time; control strategy reduced hazardous waste generation by 88 per cent

Catalysts

    Problem: Homogeneous catalysts can lead to heavy metal contamination of water and solid-waste streams

    Solution: Consider using a heterogeneous catalyst where the metals are immobilised on a solid support

    Problem: Old catalyst designs emphasised conversion of reactants over selectivity to the desired product

    Solution: Consider a new catalyst technology that features higher selectivity, and better physical characteristics (size, shape, porosity, etc.)

    Benefits: Lower downstream separation and waste treatment costs for by-products—for example. a new catalyst for making phosgene ($COCl_2$) minimised for formation of carbon tetrachloride and methyl chloride, and eliminating an end-of-pipe treatment device

Distribute flows in fixed-bed reactors

    Problem: Reactants entering a fixed-bed reactor are poorly distributed. The flow preferentially travels down the centre of the reactor. The residence time of the fluid in the centre is too short and at the reactor walls is too long. Yield and selectivity suffer

    Solution: Install a flow distributor at the reactor entrance to ensure uniform flow across the reactor cross-section.

Control reactor heating/cooling

    Problem: Conventional heat exchange design is not optimum for controlling reactor temperature

    Solution: For highly exothermic reactions, use cocurrent flow of cooling fluid on the external surface of tubular reactors at the inlet where reaction rates and heat generation rates are highest. Use countercurrent flow of cooling fluid near reactor exit where reaction rates and heat generation rates are smallest

    Problem: Diluents added to gas phase reactions, often nitrogen or air, help to dissipate heats of reaction but can result in the generation of wastes, such as oxides of nitrogen in partial oxidation reactions

    Solution: Use a non-reactive substitute diluent, such as carbon dioxide in partial oxidation reactions or even water vapour. Carbon dioxide will need to be efficiently separated from product streams, cooled, and recycled back to the reactor. If water vapour is used, it can be condensed but might result in a waste-water stream for certain reactions

Additional reactor operation issues

    Improve measurement and control of reactor parameters to achieve optimum state

    Provide a separate reactor for recycle streams

    Routinely calibrate instrumentation

    Consider using a continuous rather than a batch reactor to avoid cleaning wastes

POLLUTION PREVENTION FOR SEPARATION DEVICES

Separation technologies are some of the most common and most important unit operations found in chemical processes. Because feedstocks are often complex mixtures and chemical reactions are not

100 per cent efficient, there is always a need to separate chemical components from one another prior to subsequent processing steps. Separation unit operations generate waste because the separation steps themselves are not 100 per cent efficient, and require additional energy input or waste treatment to deal with off-spec products. In this section, we discuss the use of separation devices with respect to pollution prevention in chemical processes. First, the importance of the choice of material (mass separating agent) to be used in the separation step is presented. Next, design heuristics regarding the use of separation technologies in chemical processes are covered. Finally, we present examples of the use of separation technologies for recovery of valuable components from waste streams, leading eventually to their reuse in the process.

Choice of Mass Separating Agent

The correct choice of mass separating agent to employ in a separation technology is an important issue for pollution prevention. A poor choice may result in exposure to toxic substances for not only facility workers but also consumers who use the end product. This is especially important in food products, where exposure to residual agents is by direct ingestion into the body with the food. For example, decaffeinated coffee beans and instant coffee used to be extracted with a chlorinated solvent. While the solvent was extremely effective in extracting caffeine from the bean, residuals in the final product posed an unacceptable health risk to consumers. Caffeine is now extracted using supercritical carbon dioxide (among other benign agents), whose residuals pose no health risk. Edible oils are extracted from plant material using volatile solvents. The oil is recovered from the solvent using distillation while the solvent is recycled back to the process. Residuals can be present in the final product and therefore, it is important to use the lowest toxicity mass separating agent in these applications. In addition to these toxicological issues, a poor choice of mass separating agent may lead to excessive energy consumption and the associated health impacts of the emitted criteria pollutants (CO, CO_2, NO_x, SO_x, particulate matter).

Choice of a mass separating agent in an adsorption application can be illustrated using a simple example. Adsorption is a technology whereby a chemical dissolved in a liquid or a gas phase will preferentially become immobilised on the surface of a solid matrix (adsorbent) packed within a column. Separation and recovery of toxic metal ions from aqueous streams is one very important application of adsorption. Granular activated carbon (GAC) is a very common type of adsorbent, but for the recovery of metals, it has been found that typical strong cation exchange resins have approximately a 20-fold higher capacity to adsorb Cu^{2+} than GAC. The metal must be recovered from the regenerated adsorbent using a strong acid. In this case, the use of GAC would require more energy consumption and would generate more acid waste than the cation exchange resin.

Process Design and Operation Heuristics for Separation Technologies

A typical chemical process might be depicted as shown in Fig. 5.10, where a reactor converts feed materials into products and by-products that must be separated from each other by the additional input of energy. There are waste streams leaving the process and entering the air, the water, and the soil compartments of the environment. While it may be difficult or impossible to eliminate all waste streams, there is every reason to believe that wastes can be minimised by the judicious choice of mass separating agent, by the correct choice and sequencing of separation technologies, and the careful control of system parameters during operation.

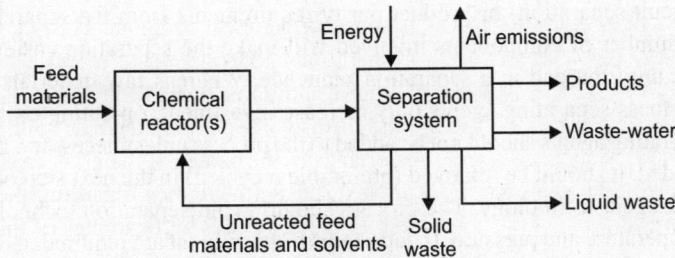

Fig. 5.10. Typical chemical process.

The first step in minimising wastes generated from separation units in chemical processes is to choose the correct technology for the separation task. Making the correct choice, based on the physical and chemical properties of the molecules to be separated, will lead to processes that generate less waste and use less energy per unit of product. The separation task may pertain to a process stream or a waste stream. Table 5.4 shows unit operation choices and property differences between the components to be separated.

Table 5.4. Unit operations for separations and property differences.

Unit operation	Property difference
Adsorption	Surface sorption
Chromatography	Depends on stationary phase
Crystallisation	Melting point or solubility
Dialysis	Diffusivity
Distillation	Vapour pressure
Electrodialysis	Electric charge and ionic mobility
Electrophoresis	Electric charge and ionic mobility
Gel filtration	Molecular size and shape
Ion exchange	Chemical reaction equilibrium
Liquid-liquid extraction	Distribution between immiscible liquid phases
Liquid membranes	Diffusivity and reaction equilibrium
Membrane gas separation	Diffusivity and solubility
Reverse osmosis	Molecular size
Micro- and ultrafiltration	Molecular size

After selecting the best separation technology, it is worth considering several pollution prevention heuristics to guide the design of the flowsheet and operation of the units. Table 5.5 shows several design and operation heuristics for separation processes. Streams of similar enough composition can be combined in order to reduce the number of unit operations and their associated capital costs and emission sources. Corrosive materials should never be added unless necessary and if added or generated in the process, should be separated immediately. Their removal can minimise investment and the generation of trace metals. Unstable materials should be removed, preferably at low temperatures, to reduce the formation of undesirable waste products, such as tars. Removing the highest volume components first in a process will minimise downstream equipment investment, associated energy-related costs, and the addition of materials required for processing that could become another source of waste. If component properties are very close (difficult separation) or product purity requirements from the separation are extremely high, reducing the number of components involved will make the separation easier. Therefore, these cases should be left until the end in a separation sequence. Whereas raw materials and products add value to the design, mass separating agents only increase investment, operating cost, and waste loads. Therefore, mass separating agents should not be added to the process unless necessary. If a mass separating agent needs to be added, it should be removed (preferable recycled) in the next step of the process using an energy separating agent technology. The process should avoid separation technologies that operate far from ambient temperature and pressure. If departures from ambient are required, it is more economical to operate above rather than below ambient.

Table 5.5. Separation Heuristics to prevent pollution.

Combine similar streams to minimise the number of separation units
Remove corrosive and unstable materials early
Separate highest-volume components first
Do the most difficult separations last
Do high-purity recovery fraction separations last
Use a sequence resulting in the smallest number of products
Avoid adding new components to the separation sequence
If a mass separating agent is used, recover it in the next step
Do not use a second mass recovery agent to recover the first
Avoid extreme operating conditions

Distillation accounts for over 90 per cent of the separation applications in chemical processing in the United States and other developing countries. Because of its prominence, we will present a number of pollution prevention techniques that are specific to distillation. Distillation columns contribute to process waste in four major ways:

1. By allowing impurities to remain in a product.
2. By forming waste within the column itself.
3. By inadequate condensing of overhead product.
4. By excessive energy use.

Product impurities above allowable levels must eventually be removed, leading to additional waste streams and energy consumption. Waste is formed within distillation columns in the reboiler where excessive temperatures and unstable materials combine to form high molecular weight tars or polymers on the heat transfer surfaces. The condenser vent must be open to the external environment to relieve non-condensable gases that build up in the column. If the condenser duty is insufficient for the internal vapour load in the column, excess vapour will exit the vent as a waste stream. Energy use leads to the direct release of criteria pollutants (CO, NO_x, SO_x, particulate matter, volatile organic compounds) and global warming gases (primarily CO_2).

The most common way to increase product purity in distillation is to increase the reflux ratio. However, this increases the pressure drop across the column, raises the reboiler temperature, and increases the reboiler duty. But for stable materials, this may be the easiest way to decrease waste generation due to inadequate product purity. If a column is operating close to flooding (increasing reflux ratio is not an option), then adding a section to the column leads to higher-purity products. Replacing existing column internals (trays or packing) with high-efficiency packing results in greater separation for an existing column, and results in both lower pressure drop and reboiler temperature. Changing the feed location to the optimum may increase product purity without changing any other system parameters. In one documented case, relocating the feed to the optimum position reduced the loss of product to waste from 30 to 1 lb./hr, increased column capacity by 20 per cent, and decreased the refrigeration cooling load by 10 per cent. Additional ways to increase column separation efficiency are to insulate the column and reduce heat losses; improve feed, reflux, and liquid distribution; and pre-heat the column feed, employing cross-exchange with other process streams. Finally, if the overheads product contains a light impurity, it may be possible to withdraw the product from a side stream near the top of the column. A bleed stream from the overhead condenser can be recycled back to the process to rid the column of the light component.

Example 5.2: Energy savings in ethanol-water distillation: Side stream case.

When a product from a distillation is needed with a composition between the distillate (x_D) and bottoms (x_B) products, a side stream with this composition collected from the column will always save energy compared to combining the top and bottom product streams. Consider the distillation column with a side stream of composition x_S and flow rate of S (moles/hr), as shown in Fig. 5.11. Using a McCabe-Thiele analysis, demonstrate energy savings can be achieved using a side stream. Mole fractions are ethanol.

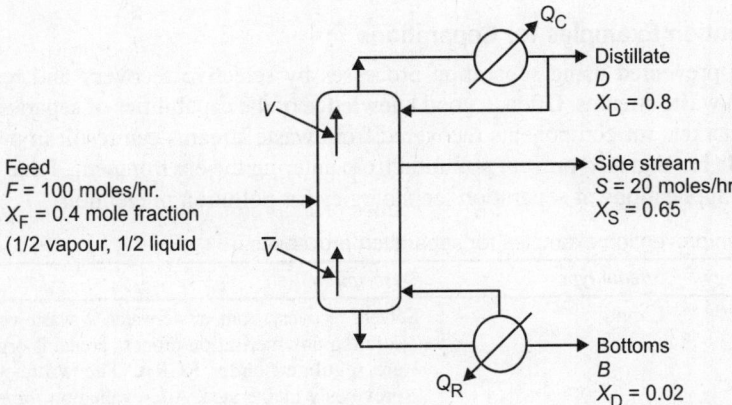

Fig. 5.11. Schematic diagram of distillation column with side stream.

Solution: It can be easily shown, using graphical methods presented in any standard textbook, that the required separation for this column, feed conditions, and separation requirements can be achieved using a column with 12 equilibrium stages operating with a reflux ratio of L/D = 2.5. The side stream is taken from the fifth stage from the top of the column. Similarly, for a 12-stage column without a side stream, the required separation for this feed can be accomplished using a reflux ratio of only 2.0. Nonetheless, the side stream has clear energy savings as shown in the Table below.

Column design	L/D	D	V	Q_R (cal/hr)
Side stream	2.5	32.56	63.96	63.96×10^4
No side stream	2.0	48.72	96.16	96.16×10^4

Using a side stream design, the energy savings are (96.16 − 63.96)/96.16 × 100 = 33.5 per cent. The energy savings will translate to reduced impacts for global warming and acid rain/deposition.

Note: $$D = F\frac{x_F - x_B}{x_D - x_B}; \quad Q_R = \bar{V}\lambda_W; \quad \bar{V} = D\left(\frac{L}{D} + 1\right) - F(1-q)$$

In this example, the feed quality (q) is taken to be 1/2 (1/2 vapour and 1/2 liquid) and λ_w is the latent heat of vapourisation of water (in the reboiler; 10^4 cal/mole).

There are several ways to decrease the generation of tars in the reboiler of the column. One way is to reduce the column pressure, resulting in lower reboiler temperatures. Caution must be taken, as this affects the condenser temperature and efficiency. The above step should be coupled with steps to reduce the steam temperature. Reboiler temperature can be reduced by desuperheating the steam, by using a lower pressure steam, by installing a thermocompressor, or by using an intermediate heat transfer fluid. The existing reboiler may be retrofitted with high flux tubes allowing for the use of lower pressure steam.

If the overheads condenser is undersized relative to the vapour loading in the column, it can be re-tubed or replaced with a larger capacity unit. This step will reduce the likelihood that fluctuations of column operation will result in hot vapour being expelled from the column vent. An additional way to reduce distillation column emissions and waste generation is by improving the process control technology. This step will assure that product purity specifications will be met and reduce the possibility that off-spec product will be created.

Pollution Prevention Examples for Separations

Pollution can be prevented using separation processes by selective recovery and reuse of valuable components from waste streams. Often, a good knowledge of the capabilities of separation technologies combined with markets for components recovered from waste streams can result in processes that are not only profitable but that also prevent pollution from entering the environment. Table 5.6 summarises many successful applications of separation technologies for pollution prevention.

Table 5.6. Pollution prevention examples for separation processes.

Separation technology	Stream type	Description
Distillation	Liquid	Solvent recovery from waste-water. A waste-water stream from a solution polymerisation process contains organic solvents that are regulated under RCRA. The waste-water stream was previously incinerated. A re-evaluation found that distillation-followed by extraction could be used to recover more than 10 million lb/yr. of solvent and reduce incineration loads by 4 million lb/yr., and had a payback period of only 2 years
Distillation	Liquid	Ink and solvent recycle. Waste ink from newspaper printing contains organic solvent (20 per cent), water (15 per cent) and ink (65 per cent). A flash distillation is used to separate the high-boiling ink from the solvent/water solution and binary distillation is used to separate the solvent from the water. Solvent and ink are reused in the process
Distillation	Liquid	Batch distillation of used antifreeze. Pure ethylene glycol is recovered and blended with water plus other additives to make new antifreeze
Distillation	Liquid	Acid recovery from spent acid streams. In the electroplating industry, spent acids from etching tanks, cleaning tanks, and pickling tanks can be processed using distillation to recover pure acids (HCl, HNO_3, etc.)

(Contd...)

Separation technology	Stream type	Description
Distillation	Liquid	Solvent recovery and reuse in automobile paint operations. A closed-loop solvent utilisation system has been established for cleaning out paint lines between colour changes. The collected paint/solvent mixture is transported to a central reprocessing facility, the solids are separated, and pure solvent is recovered by distillation. The solvent is reused in automobile painting
Extraction	Liquid	Extraction of a batch process residue. A batch process has difficulties using distillation, resulting in about 1/3 of the production run being incinerated. A low-boiling-point material recovered from the batch residue was found to be an effective extraction solvent to recover more product from the residue
Extraction	Liquid and sludges	Hydrocarbon recovery from refinery waste-water and sludge. Triethylamine is used as a solvent to recover hydrocarbons from refinery waste-water and sludges. The hydrocarbons are recycled back to the process
Reverse osmosis	Liquid	Closed-loop rinsewater for process electroplating. Reverse osmosis is able to return pure water and a concentrated metals-containing stream to the plating bath. There are over 200 documented industrial applications
Reverse osmosis	Liquid	Recovery of homogeneous metal catalysts.
Ultrafiltration	Liquid	Polymer recovery from waste-water. Cleaning of polymerisation reactors generates a stream from which polymers such as latex can be recovered. Also, polyvinyl alcohol can be recovered in the manufacture of synthetic yarn
Adsorption	Gas	Natural gas dehydration. Molecular sieve adsorbents are being used to dehydrate natural gas, thereby eliminating the use of a solvent (triethyleneglycol)
Adsorption	Liquid	Replacement of azeotropic distillation. Azeotropic solvents, such as benzene and cyclohexane, can be eliminated by contacting azeotropes (ethanol/water or isopropanol/water) with molecular sieve adsorbents
Membrane	Gas	Recovery and recycle of high-value volatile organic compounds. Examples include recovery of olefin monomer from polyolefin processes, gasoline vapour recovery from storage facilities, vinyl chloride recovery from PVC reactor vents, and chlorofluorocarbons (CFCs) recovery from process vents and transfer operations. Emerging applications included also
Membrane	Liquid	Recovery of organic compounds from waste-water streams. Pervapouration is a membrane process used to recover organics from low flow (10–100 gal/min.) and moderate concentration (0.02 to 5 per cent by wt.) waste-water
Membranes (RO, NF, UF, MF, ED)*	Liquid	Metal ion recovery from aqueous waste streams

*RO–reverse osmosis, NF–nanofiltration, UF–ultrafiltration, MF–microfiltration, ED–electrodialysis.

Separators with Reactors for Pollution Prevention

Separators can be combined with reactors to reduce by-product generation from reactors and increase reactant conversion to products. These combinations of separators and reactors can involve either distinct units or integrated units.

POLLUTION PREVENTION APPLICATIONS FOR SEPARATIVE REACTORS

An exciting new reactor type that has a very high potential for reducing waste generation is the separative reactor. These hybrid systems combine chemical reaction and product separation in a single process unit. When chemical reaction and separation occur in concert, the requirements for downstream processing units are reduced, leading to lower capital costs. The key feature allowing for the prevention of waste generation and maximising product yield is the ability to control the addition of reactant and the removal of product more precisely than in traditional designs. Unfavourable chemical equilibrium can be shifted to maximise reactant conversion and product yield. Unwanted by-product generation can be minimised in series reactions by the removal of the desired product within the reaction zone and before significant secondary reactions can occur. Separation units that have been integrated with reaction include distillation, membrane separation, and adsorption. A recent review has been written on emerging uses of separative reactors for pollution prevention employing membranes and solid adsorbents.

A good demonstration of reaction coupled with adsorption is oxidative coupling of methane (OCM). Methane reacts with oxygen in the presence of metal oxide catalysts at a temperature of about 1000K to form ethane and ethylene.

$$2CH_4 + 1/2 O_2 \longrightarrow C_2H_6 + H_2O$$
$$2CH_4 + O_2 \longrightarrow C_3H_4 + 2H_2O$$

A parallel path in which methane is completely oxidised is shown below:

$$CH_4 + 2O_2 \longrightarrow CO_2 + 2H_2O$$

There is also a concern that the ethylene product can be oxidised to carbon dioxide, thereby reducing product yield and increasing waste generation. Successful application of OCM in industry would allow the use of methane, a high-production chemical that is difficult to transport, as a feedstock for ethylene, an important intermediate in polymer production. The difficulty with traditional OCM in a fixed-bed or fluidised-bed reactor is that the feed ratio of CH_4/O_2 must be kept around 50 or more to limit complete oxidation reactions from occurring. This results in relatively high selectivity (80–90 per cent) but limits the yield of C_2S to less than 20 per cent. A separative reactor composed of a series of reactor/adsorber sections substantially improved the yield of C_2 (50–65 per cent). Each section contained a fixed-bed catalytic reactor operating at high temperature (1000K), followed immediately by a cooler adsorption bed. Figure 5.12 shows the arrangement of a four-section simulated countercurrent moving-bed chromatographic reactor (SCMCR) in which the smaller columns are the catalytic fixed-bed reactors and the larger columns are for adsorption. In section 1, the carrier gas (N_2) sweeps unreacted adsorbed CH_4 into the next section (feed section 2).

A small make-up feed stream comprised of CH_4/O_2 in stoichiometric amounts (to make up for consumption in reaction) is combined with a carrier gas stream from section 1, which then enters the reactor in the feed section. Reaction products (C_2S) and unreacted CH_4 are adsorbed in the large column of section 2, with the C_2 products being retained in the upper portion and the more mobile CH_4 in the bottom. Section 3 is isolated from flow, yet contains unreacted CH_4 in the adsorption column. Section 4 is the product removal section in which the C_2 products are swept off the adsorption bed and into a side stream roughly midway in the column. After maintaining the SCMCR in this configuration for a prescribed time interval, the flow configurations are advanced one section to the left so that section 1 is the product removal section, section 2 is the carrier section, section 3 is the feed section, and section 4 is the isolated section.

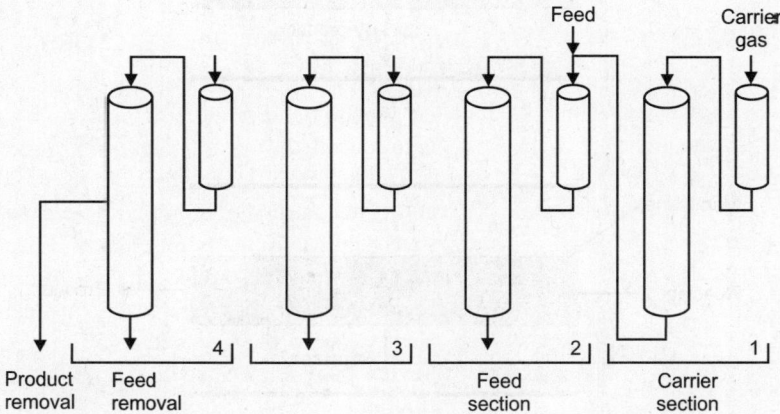

Fig. 5.12. Schematic of a four-section SCMCR for the oxidative coupling of methane.

Another application of the coupled catalytic reactor with adsorbent beds is the partial oxidation of methane to methanol. Methanol is in demand as a fuel oxygenate, is a feedstock for other oxygenates in reformulated gasoline, and is being investigated as an alternative fuel for gasoline and diesel engines. Reformulated gasoline reduces emissions of CO, NO_x, volatile organic compounds, and benzene from automobiles, as required by the Clean Air Act of 1990. The current process for methanol production is a costly two-step process, consisting of steam reforming of methane to produce CO and H_2, followed by methanol formation by passing CO and H_2 over a metal oxide catalyst. The overall reaction of CO and H_2 to form methanol is endothermic by 125 kJ/mole, requiring significant energy input. In contrast, the partial oxidation reaction of methane has the overall reaction

$$CH_4 + 1/2O_2 \longrightarrow CH_3OH$$

and is exothermic by 126 kJ/mole. However, in order to minimise over-oxidation of methanol to CO_2 in a series reaction, the feed CH_4/O_2 ratio must be kept high, leading to disappointingly low per-pass methanol yields of less than 10 per cent. In recent experiments, methanol yield increased from 3–4 per cent in a tubular reactor to 17 per cent when the reactor was interfaced with adsorber beds and operated in the SCMCR mode. This demonstrated that the SCMCR mode of reaction is useful for increasing performance of low conversion per pass reactions.

Reaction coupled with membrane separation is another often-studied configuration for increasing the efficiency of chemical reactions. Much like adsorption-based separative reactors, the equivalent membrane-based unit can be used to selectively remove either products or by-products from the reaction zone, thereby overcoming low conversions in equilibrium-limited reactions and reduce waste generation in series reactions. However, membrane-based separative reactors can also be used to selectively permeate reactant into the reaction zone in order to control excessive by-product formation (e.g. permeation of O_2 in partial oxidation or oxidative coupling reactions). Both of these modes of operation are shown in Fig. 5.13. Membrane materials can be organic, porous inorganic, or nonporous (dense) inorganic, and either can be constructed of inert (non-reactive) material or can contain catalysts in various configurations.

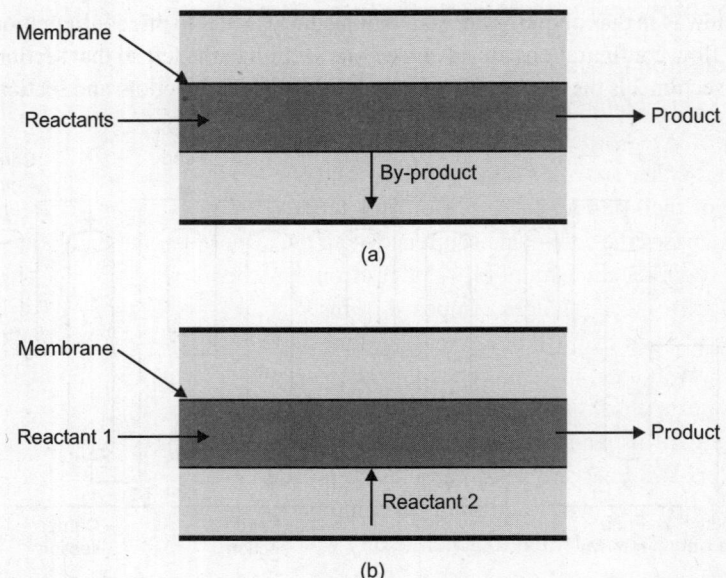

Fig. 5.13. Main modes of operation for membranes separative reactors: (a) selective removal of products/by-products, and (b) selective permeation of reactants.

Applicable reaction types that can be improved by membrane separative reactors include:
1. Thermodynamically-limited reactions (e.g. $C_6H_{12} \longleftrightarrow C_6H_6 + 3H_2$).
2. Parallel reactions in which product formation has a lower reaction order than by-product generation.
3. Series reactions such as selective dehydrogenations and partial oxidations.
4. Series-parallel reactions.

Applications of membrane separative reactors for partial oxidation have shown encouraging results. The test reaction of ethane oxidative dehydrogenation to ethylene showed that per pass yields increased from 12 to 52 per cent. Positive results were also demonstrated for membrane separative reactors for dehydrogenation of ethylbenzene to produce styrene, where conversions as high as 70 per cent were observed, approximately 15 per cent higher than conventional methods, and an increase from 2–5 per cent in styrene selectivity.

The potential for membrane separative reactors or other means to improve the environmental and economic performance of some of the top 50 commodity chemicals in the US chemical industry formed through partial oxidation or by dehydrogenation reactions was recently studied. The maximum energy saving was estimated to be 0.25 quadrillion BTU if every commodity chemical formed through either selective oxidation or dehydrogenation achieved maximum efficiency.

Additional challenges remain before commercial application of membrane separative reactors can be realised. These include:
1. Economical manufacture of thin, defect-free selective membrane layers over large surface areas.
2. Leak-free reaction systems with high temperature seals.
3. Elimination or reduction of sweep gases which dilute product streams.
4. Enhanced membrane and catalyst performance, including resistance to fouling and deactivation.

POLLUTION PREVENTION IN STORAGE TANKS AND FUGITIVE SOURCES

Storage Tank Pollution Prevention

Storage tanks are very common unit operations in several industrial sectors, including petroleum production and refining, petrochemical and chemical manufacturing, storage and transportation, and other industries that either use or produce organic liquid chemicals. Tanks are used for many purposes, including storage of fuels and for feedstock or final product buffer capacity. The main environmental impact of storage tanks is the continual occurrence of air emissions of volatile organic compounds from roof vents and the periodic removal of oily sludges from tank bottoms.

Tank bottoms are solids or sludges composed of rusts, soil particles, heavy feedstock constituents, and other dense materials that are likely to settle out of the liquid being stored. There are various methods of dealing with these materials once they are present. They may be periodically removed and either treated via land application or disposed of as hazardous waste. As long as the bottoms components are compatible with downstream processes, they may be prevented from settling to the tank bottom by the action of mixers that keep the solid particles suspended in the liquid. Another method is to use emulsifying agents that keep water and solids in solution and out of the tank bottoms. A concern with the use of this method is the potential to generate oily waste downstream in the refinery processes from the presence of the emusifiers.

Air emissions of volatile organic compounds from storage tanks are a major source of airborne pollution from petroleum and chemical processing facilities. These emissions stem from the normal operation of these units in response to the changes in liquid level within the tank and the action of ambient changes in temperature and pressure. These loss mechanisms are termed working losses and standing losses, respectively. The emissions are dependent upon the vapour pressure of the stored liquids, tank characteristics such as tank type, paint colour and condition, and also the geographic location of the tank. There are six major types of storage tanks. A listing of these tank types, short descriptions, a summary of emission mechanisms, and pollution reduction measures are listed in Table 5.7.

Table 5.7. Storage tank types and pollution reduction strategies.

Storage tank type	Description	Loss mechanisms	Pollution reduction
Fixed roof	Cylindrical shell with permanent roof (flat, cone, or dome), freely vented or with pressure/vacuum vent	Working losses – VOCs in headspace above liquid are expelled when tank is filled. Standing losses – headspace gas expands/contracts by ambient T and P	Pressure/vacuum vents reduce standing losses, heating the tanks reduces standing losses, pollution control equipment on vent (adsorption, absorption, cooling) reduce emissions 90–98%. Vapour balancing approach
External floating roof	Cylindrical shell without a fixed roof, a deck floats on the liquid surface and rises and falls with liquid level, deck has flexible seals on shell inner wall to scrape liquid off shell wall	Working losses – evaporation from wetted shell wall or columns as liquid is withdrawn. Standing losses– small annular space between deck system and shell wall is source of these losses	Little reduction can be accomplished to control or prevent the wind-driven emissions from the shell wall. Emissions actually greater than fixed-roof tanks

(Contd ...)

Storage tank type	Description	Loss mechanisms	Pollution reduction
Internal floating roof	Same as external floating roof with a permanent fixed roof above. Roof is either column or self-supported	Same as external floating roof tank. Permanent roof blocks wind and reduces working losses	60–99% emission reduction compared to a fixed-roof tank
Domed external floating roof	Similar to an internal floating roof tank but has a self-supported domed roof	Similar to self-supported permanent roof	60–99% emission reduction compared to a fixed-roof tank
Variable vapour space	Roof telescopes to receive expelled vapours. Diaphragm used to accept expelled vapours	Working losses occur when liquid level is raised. Standing losses are eliminated	No data available on emissions reduction
Pressure tanks	Low pressure (2–15 psig) and high pressure (> 15 psig)	No losses from high pressure tanks. Working losses from low pressure tanks during filling operations. No standing losses	No data available on emissions reductions

Vapour balancing involves routing the expelled vapours during tank filling to another tank that is supplying the liquid.
T are daily changes in ambient temperature.
P are changes in barometric pressure.

Reducing Emissions from Fugitive Sources

Fugitive emission sources in chemical processes include valves, pumps, piping connectors, pressure relief valves, sampling connections, compressor seals, and open-ended lines. There may be thousands of these components in a typical synthetic organic chemical manufacturing industry (SOCMI) facility and tens to hundreds of thousands in a large petroleum refinery. These emission sources are significant contributors to air pollution from SOCMI facilities, as estimates have shown that as much as one third of air emissions occur from fugitive sources.

Within individual components, leaks are localised near seals, valve packings, and gaskets. Components in good working order rarely leak quantities of process fluids that are of significant concern. When leaks occur due to a seal, packing, or gasket failure, the exact timing, location, and rate of release is difficult to predict. These leaks are of two types-either low-level leaks that may persist for long periods of time until detected, or sudden episodic failures resulting in a large release. However, the leaks can be prevented or repaired, and leakless technologies are available for situations where even small rates of release cannot be tolerated.

In this section, we identify which of the components from fugitive sources listed above have the greatest potential for emission reductions as a result of pollution prevention efforts. Next, established methods for reducing or preventing fugitive emissions are presented. Finally, a study summarising the emissions reductions that are possible in chemical manufacturing facilities is presented.

Fugitive emission profiles

The average rate of emission of volatile organic compounds (VOCs) from fugitive components of different types can vary significantly within a given facility. To demonstrate this, we estimate the emission rate from all fugitive sources within two processing units at a refinery, a cracking unit and a hydrogen plant. The average emission factors presented in Table 5.8 will be used along with a knowledge of the numbers of sources within a given component type and the mass fraction of VOC in the stream serviced by the component. The equation used to calculate the emission rate for each component is:

$$E = m_{VOC} f_{av}$$

where E is the emission rate (kg/hr/source), m_{VOC} is the mass fraction of VOC in the stream, and f_{av} is the average emission factor. The numbers of fugitive sources and their contributions to the emissions from the two processing units at a refinery are given in Table 5.8. Valves in all service are by far the largest source for emissions from these process units, comprising 55.3 and 63.4 per cent of the total for the cracker unit and H_2 plant, respectively. These emissions are disproportionately large for the relative number of valves in the processes—22.5 and 22.8 per cent for the cracker unit and H_2 plant, respectively. The component present in the largest number is connectors in all service, being 74.4 and 75.1 per cent of the total for the cracker unit and H_2 plant, respectively. Relief valves appear to be significant emission sources, as do seals on pumps and compressors.

Table 5.8. Distribution of fugitive components and emission rates from a cracking unit and a Hydrogen plant at a refinery.

		Cracker				Hydrogen plant[b]		
Component	Service[a]	Equipment count	m_{VOC}	Emissions, kg/hr	(%)	Equipment count	Emissions, kg/hr.	(%)
Pump seals	LL	6	0.75	0.51	4.0	2	0.22	1.9
	HL	9	0.55	0.1	0.81	2	0.042	0.36
Compressor seals	HC gas	4	1.0	2.60	20	0	0	0
	H_2 gas	0		0	0	6	0.30	2.6
Valves	HC gas	200	1.0	5.3	42	70	1.9	16
	H_2 gas	0		0	0	80	0.66	5.7
	LL	196	0.75	1.6	13	427	4.7	41
	HL	294	0.55	0.037	0.29	427	0.85	0.73
Connectors	All	2277	0.75	0.42	3.3	3313	0.83	7.2
Relief valves	Gas	11	1.0	1.8	14	15	2.4	21
	Liquid	15	0.63	0.066	0.52	2	0.014	0.12
Open-ended lines	All	32	0.75	0.054	0.43	42	0.084	0.72
Sampling taps	All	17	0.75	0.19	1.5	24	0.36	3.1
Total	–	–	–	13	100	–	12	100

[a] HL: heavy liquid, LL: light liquid, HC: hydrocarbon
[b] m_{VOC} = 1.0 for all.

Methods to reduce fugitive emissions

There are two methods for reducing or preventing emissions and leaks from fugitive sources. They are:
1. Leak detection and repair (LDAR) of leaking equipment.
2. Equipment modification or replacement with emissionless technologies.

Both methods can be effective in reducing low-level as well as large episodic leaks of process fluid.

In an LDAR programme, equipment such as pumps and valves are monitored periodically using an organic vapour analyser (OVA). The wand of the OVA is directed towards the suspected source of leakage on each piece of equipment, i.e. at a packing nut on a valve, at a shaft seal on a pump, or a gasket or weld on a flange or connector. Guidance documents are available from the US EPA on detailed procedures to monitor for leaks and estimate emissions for fugitive sources (US EPA 1993b). If the source registers an OVA reading over a threshold value (>10,000 ppm), the equipment is said to be leaking and repair is required. Progress towards achieving desired fugitive emission reduction targets can be measured by using the OVA screening values and US EPA emission correlations for fugitive sources, as shown in Table 5.9. The nature of the repairs depends upon the piece of equipment, but may involve something as simple as tightening a packing nut on a valve or it may require replacement of a seal on a pump or a gasket in a connector. Repairs may require shutdown of the process, and would be conducted during regularly-scheduled shutdown times in order to minimise the number of upsets in process operation and reduce repair-related emissions.

Table 5.9. Correlations for estimating fugitive emissions and their default values.

Equipment	Service	Leak rate from correlation, kg/hr/source[a]		Default emissions,[b] kg/hr./source
		SOCMI	Refinery	
Valves	Gas	$1.87 \times 10^{-6} \, C^{0.873}$	$2.18 \times 10^{-7} \, C^{1.23}$	6.56×10^{-7}
	Light liquid	$6.41 \times 10^{-6} \, C^{0.797}$	$1.44 \times 10^{-5} C^{0.80}$	4.85×10^{-7}
Pump seals	Light liquid	$1.9 \times 10^{-5} \, C^{0.824,c}$	$8.27 \times 10^{-5} \, C^{0.83,d}$	$7.49 \times 10^{-6,c}$
	Heavy liquid		$8.79 \times 10^{-6} \, C^{1.04}$	
Compressor seals	Gas	–	$8.27 \times 10^{-5} \, C^{0.83}$	
Pressure-relief valves	Gas		$8.27 \times 10^{-5} \, C^{0.83}$	
Flanges/other connectors	All	$3.05 \times 10^{-6} \, C^{0.885}$	$5.78 \times 10^{-6} \, C^{0.88}$	6.12×10^{-7}

[a] C: screening value in ppm.
[b] These values are applicable to all source categories.
[c] This correlation/default-zero value can be applied to compressor seals, pressure-relief valves, agitator seals, and heavy liquid pumps.
[d] This correlation can be applied to agitator seals.

Industrial LDAR programmes vary greatly in their frequency of monitoring and in their effectiveness. Constant monitoring of emissions using area monitors is possible when contaminants are detectable in very low concentrations. For cases where constant monitoring is either technically impossible or too expensive, periodic monitoring on a monthly, quarterly, or annual basis using an OVA is the preferred approach. Monitoring on a more frequent basis may be more costly, but has been shown to be more effective in reducing emissions. For example, monitoring and repairing valves in light liquid service at monthly intervals is a third more effective in reducing emissions compared to monitoring quarterly, and is three times as effective as monitoring every six months (US EPA 1982).

Equipment modification to reduce fugitive emissions might involve redesigning a process so that it has fewer pieces of equipment and connections, replacing leaking equipment with new conventional equipment, or the inclusion of new emissions-reducing technology, and sealant injection. In this discussion, we focus only on equipment replacement for the major fugitive emission sources, valves, connectors/flanges, compressors, and pumps. We discuss the types of equipment, where the leaks are likely to occur, and what equipment changes can be made to reduce or eliminate releases.

Connectors are the most numerous pieces of equipment and are used to connect pipe to other pipes or to equipment and vessels. Connectors on smaller piping (<2 inches in diameter) are either threaded pipes and couplers or nut-and-ferrule connectors. Flanges are connectors with a flexible seal junction that are used for pipes greater than 2 inches in diameter. Another connector type is welded pipe. Leaks may occur from connectors due to thermal deformation on correct assemblies or may result from cross-threading and incorrect assembly of nut-and-ferrule types. Monitoring of these pieces of equipment would occur near threaded junctions, gaskets, and welds.

Seals around moving parts are common locations of leaks for fugitive sources such as valves, pumps, and compressors. For valves, the moving part is the stem that connects the internal components of the valve with the outside. The packing is subject to degradation or the stem may have surface defects, both of which may promote leaks. Valve packing technologies that use rings to keep the packing from extruding and springs to maintain the packing under constant pressure and contact with the stem have been developed. These systems can reduce leak rates and reduce maintenance requirements for 10–50 times as long as conventional packing. There are two main types of 'leakless' valves that have no emissions through the stem. They are bellows valves; which are expensive and are mostly used in the nuclear power industry, and diaphragm valves, in which a physical barrier (diaphragm) exists between the process fluid and the valve stem.

Seals around pumps generally occur where a rotating shaft meets the stationary casing. Two main types of seals are used; packed seals and mechanical seals. Mechanical seals that are well-maintained are superior to packed seals, but because they are expensive and time-consuming to repair, a second packed or mechanical seal is commonly used. Sealless designs for pumps include canned motor pumps, where the bearings are in the process fluid, and diaphragm pumps, where a moving diaphragm pumps the fluid. Magnetic pumps are also available in which the impeller is driven by magnets. Mechanical seals for pumps have improved greatly over the last 10–20 years, making them a viable alternative for leakless pumps for many applications.

Compressors are similar to pumps in that they move fluid, but in this case the fluid is a gas. Both packed seals and mechanical seals are used, but packed seals are found only on reciprocating devices. Mechanical seals are not necessarily of the contact design used for pumps, but rather include carbon rings, labyrinth-type, and oil film seals.

The emissions control effectiveness of various emission reduction measures is shown in Table 5.10 as a percentage reduction. Leakless technologies are 100 per cent effective in eliminating emissions for properly functioning equipment, but are expensive to purchase and maintain. For example, pumps with dual mechanical seals are estimated to have an amortised annual cost roughly 10 times quarterly or monthly LDAR. Compared to facilities that do not have emissions reduction programmes, fugitive emissions from a moderately-sized petroleum refinery can be reduced by approximately 70 per cent by using the most effective reduction techniques shown in Table 5.10. Similarly, for SOCMI facilities, fugitive emission reductions are expected to be around 60–70 per cent.

POLLUTION PREVENTION ASSESSMENT INTEGRATED WITH HAZ-OP ANALYSIS

The hazard and operability study (HAZ-OP) is a formal procedure that can be applied to individual chemical process units to identify potential hazards. Hazards are not only identified, but their possible causes are also investigated, the consequences of those hazards are defined, and actions to mitigate the hazard are summarised. It can be applied to a new process design before construction begins or can be applied to existing process units to improve safety performance. The study is typically conducted by a team of engineering and operations personnel who are familiar with the process. During the study, team members identify potential hazards through a structured examination of the design. One of the members of the team is assigned the task of recording the hazards and their suggested solutions. In this text, we demonstrate a limited use of HAZ-OP for the purpose of identifying hazards when process changes are made for pollution prevention. We do not identify actions to mitigate those hazards.

Table 5.10. Effectiveness of various fugitive emission reduction techniques.

Equipment	Control technique	Control effectiveness (%) SOCMI	Petroleum refinery
Pumps, light liquid service	Dual mechanical seals	100	100
	Monthly leak detection and repair	60	80
	Quarterly leak detection and repair	30	70
Valves, gas/light liquid service	Monthly leak detection and repair	60	70
	Quarterly leak detection and repair	50	60
Pressure-relief devices	Tie to flare; rupture disk	100	100
	Monthly leak detection and repair	50	50
	Quarterly leak detection and repair	40	40
Open-ended lines	Caps, plugs, blinds	100	100
Compressors	Mechanical seals, vented to degassing reservoirs	100	100
Sampling connections	Closed purge sampling systems	100	100

The methodology for a HAZ-OP study is to apply a series of guide words to the process design intention. The design intention relates to what the certain steps or units in the process are intended to do. Examples of process intentions are: (i) cooling water flow through a reactor or distillation condenser, (ii) inerting system for a reactor, separator, or storage tank, and (iii) air supply for pneumatically-driven process control valves. The guide words associated with process intentions are shown in Table 5.11. For example, the use of the guideword 'NO' in the process intention 'cooling water flow through a reactor' would indicate a loss of cooling water.

Table 5.11. HAZ-OP procedure guide words.

Guide words	Meaning
NO or NOT	The complete negation of the intention
MORE	Quantitative increases
LESS	Quantitative decreases
AS WELL AS	A qualitative increase
PART OF	A qualitative decrease
REVERSE	The logical opposite of the intention
OTHER THAN	Complete substitution

The consequences of this might be an overheated reactor, a runaway reaction, or, potentially, an explosion. If we apply each guideword to a process intention, we arrive at a set of possible consequences. The HAZ-OP approach is applied to each unit and to each pipeline into or out of each unit, and continues until every unit in the processes has been analysed.

For large and complex processes, this approach can be very time-consuming and often tedious. Nonetheless, this procedure is finding increasing use in the chemical industry.

The additional complexity of the internal floating-roof tank introduced more possible modes of failure. Although the safety hazards associated with any process modification for pollution prevention would have to be evaluated on a case-by-case basis, many would result in a more complicated process. The rule of thumb in avoiding safety hazards in processes is to 'keep it simple'. The additional complexity of many pollution prevention applications makes safety assessment an important component of waste and risk reduction efforts in chemical process designs.

INTEGRATING RISK ASSESSMENT WITH PROCESS DESIGN — A CASE STUDY

Thus far in this chapter, we have incorporated environmental, health, and safety concerns into the design and operation of unit operations. We have used quantitative assessment measures only to a limited extent. It is very useful to provide a more quantitative risk assessment capability for the evaluation and optimisation of unit operations.

One important application could be screening of by-products generated in a chemical reactor. Decisions regarding optimum reactor operation can then be made based on the risks posed by the individual by-products generated rather than on just the mass rate of generation for each component. The case study deals with choosing residence time in a fluidised bed reactor for the production of acrylonitrile.

Acrylonitrile Reactor Risk-based Input-Output Analysis of a Reactor

Acrylonitrile is produced in a fluidised-bed reactor containing a catalyst (Bi-Mo-O). The main reaction for acrylonitrile is ammonoxidation represented by:

$$CH_2=CH-CH_3 + NH_3 + \tfrac{3}{2} O_2 \rightarrow CH_2=CH-CN + 3H_2O$$
$$\text{propylene} \quad \text{ammonia} \quad \text{oxygen} \quad \text{acrylonitrile} \quad \text{water}$$

In addition there are five other possible side reactions including:

$$CH_2=CH-CH_3 + O_2 \rightarrow CH_2=CH-CHO + H_2O$$
$$\text{acrolein}$$

$$CH_2=CH-CH_3 + NH_3 + \tfrac{9}{4} O_2 \rightarrow CH_3-CN + \tfrac{1}{2} CO_2 + \tfrac{1}{2} CO + 3H_2O$$
$$\text{acetonitrile}$$

$$CH_2=CH-CHO + NH_3 + \tfrac{1}{2} O_2 \rightarrow CH_2=CH-CN + 2H_2O$$

$$CH_2=CH-CN + 2 O_2 \rightarrow CO_2 + CO + HCN + 3H_2O$$
$$\text{hydrogen cyanide}$$

$$CH_3-CN + \tfrac{3}{2} O_2 \rightarrow CO_2 + CO + HCN + H_2O$$

Hopper and coworkers constructed a set of reactor models for the above set of chemical reactions assuming first-order reaction kinetics with respect to the reactant, product, and by-product species. The model also included mole balance and energy balance equations for the reactor. The model was used to predict the effects of reaction temperature, residence time, and reactor type — continuous flow stirred tank reactor (CSTR), plug flow reactor (PFR), and fluidised bed reactor (FBR) — on the generation of reaction by-products in the acrylonitrile reaction.

Here we illustrate the use of the FBR model predictions in determining the optimum residence time for minimum waste generation and acceptable economic performance. The evaluation is based on both mass generation as well as risk generation approaches.

The predicted concentrations of feed, product, and by-product species from the reactor as a function of reactor residence time are shown in Fig. 5.14. These results show that acrylonitrile concentration increases with residence time up to about 10 seconds. Thereafter, the increase in acrylonitrile concentration is slower and after 15 seconds, there is no further increase. Reactants (propylene and ammonia) continue to decline with increasing reactor residence time due to conversion of the reactant to product and by-product species. By-products, hydrogen cyanide (HCN) and acetonitrile, exhibit complex profiles with respect to residence time. HCN is generated in significant amounts only above about 5 seconds residence time. HCN is the dominant reaction by-product on a mass basis at higher residence times. Acetonitrile is generated in higher amounts than HCN at low residence times, but tends to remain at a constant concentration as residence time increases to 20 seconds. Based on these results, Smith recommended operating the reactor at a temperature of 400°–480°C, with a reactor residence time of 2–10 seconds, and to use a fluidised-bed reactor.

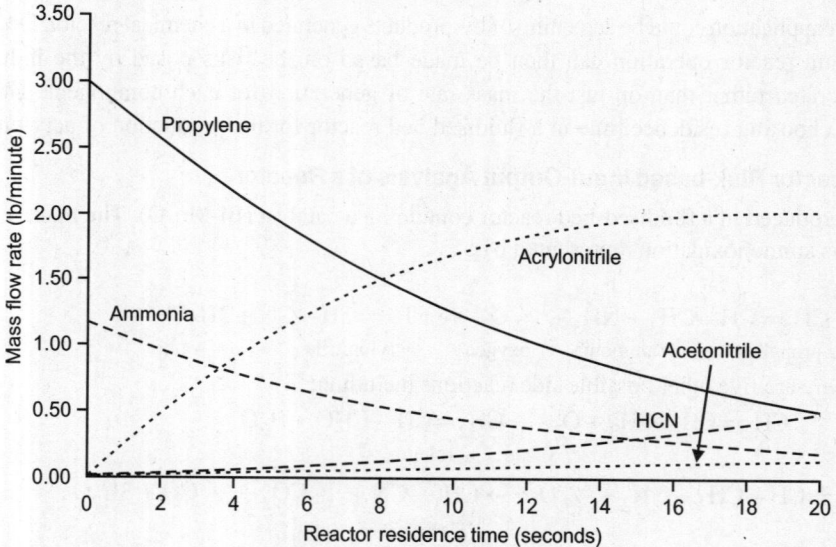

Fig. 5.14. Effect of reactor residence time on the conversion of propylene and ammonia to product (acrylonitrile) and by-products (hydrogen cyanide (HCN), and acetonitrile). The model is of a fluidised-bed reactor at 400°C. By-product generation is shown on a mass basis.

A presentation of the same reactor results on a risk basis is shown in Fig. 5.15. In generating this figure, stoichiometric coefficients at each residence time were calculated using the data shown in Fig. 5.14 by taking the ratio of the mass flow rates of reactants and by-products with respect to acrylonitrile. The environmental index of the product acrylonitrile dominates the other environmental indexes shown in Fig. 5.15, regardless of the reactor residence time. The total environmental index for this reactor is shown in greater detail in Fig. 5.16. There is a minimum in the total index at a residence time of 10 seconds. This case study demonstrates that the 'Tier 1' environmental assessment with a screening economic analysis provides valuable insights into the overall performance of the reactor design (residence time). There is little economic benefit in operating at residence times greater than 10 seconds, and the total environmental risk index is a minimum at 10 seconds. Therefore, a residence time of 10 seconds is a logical operating point for this reactor and reaction system.

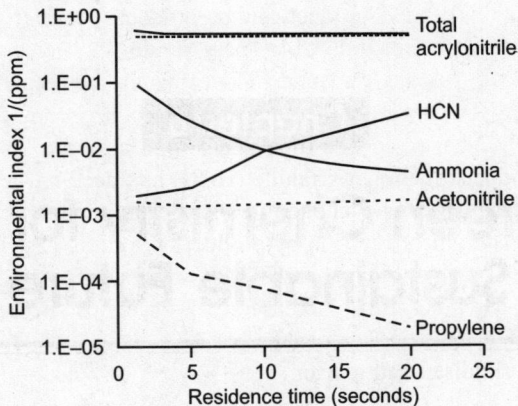

Fig. 5.15. Effect of reactor residence time on the conversion of propylene to product (acrylonitrile) and by-products (hydrogen cyanide (HCN), and acetonitrile). The model is of a fluidised-bed reactor at 400°C. By-product generation is shown on a risk basis.

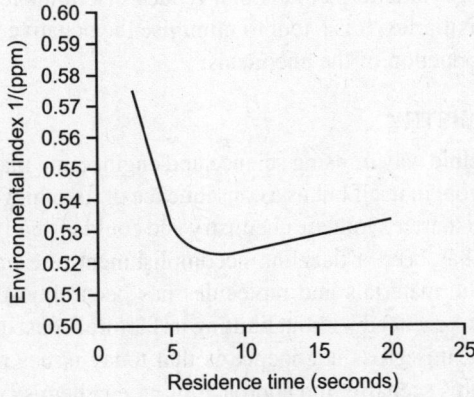

Fig. 5.16. The total environmental index as a function of reactor residence time for acrylonitrile production via the ammonoxidation pathway.

Chapter 6
Green Chemistry for a Sustainable Future

INTRODUCTION

Green Chemistry is the utilisation of a set of principles that reduces or eliminates the use or generation of hazardous substances in the design, manufacture and application of chemical products. The goal of Green Chemistry is to design synthetic methods that reduce or eliminate the use of toxic substances, wastes, solvents and other auxiliaries; it is a tool to minimise the negative impacts of the chemicals and processes involved in the production of the chemicals.

NEED FOR GREEN CHEMISTRY

Green Chemistry is a responsible way of using science and engineering that strive to improve the public image of chemistry, not as a goal in itself but as a consequence of its achievements. During the twentieth century, chemists were able to master synthetic chemistry and could virtually make all possible chemicals found in nature. While this has been a dazzling accomplishment, the information and education on toxicity and ecotoxicity of the materials and molecules has been almost completely neglected. As a result, chemistry has long been practiced without limiting its harmful consequences upon the environment and society. The net result of this gross negligence is that today issues related to sustainability have become quite grim. Against this scenario, the approach of green chemistry comes as a major respite.

The constant and at times even bitter fight between industry and environmental organisations need not be contentious anymore because green chemistry enables the environmental movements and the industrialists to work cooperatively. Green chemistry gives equal respect to the cause of environment protection and industrial success.

A survey conducted by industrial toxicology research centre (ITRC) Lucknow, India, revealed that more than 70 per cent of colours used in eatables were azodyes which are not permitted according to prevention of food adulteration rules (PFAR) in India. WHO Expert Committee on Food Additives has placed these non-permitted colours in the category CII, meaning that the data for long-term effects are not available/ inadequate for these dyes. Similarly, there are a number of chemicals used in daily life such as preservatives in soft drink preparations, and kitchen preparations, etc. which are not safe for human health.

Various pesticides, insecticides and colours are drained out from agricultural lands, chemical/ pharmaceutical and other industries, and tanneries, etc. which ultimately contribute to all three: water, air and soil pollution. These and many more such examples are enough reasons for incorporation of green chemistry methodologies, techniques and practices.

Green chemistry not only includes the shift of the use of harmful chemicals but also directs to set the usage of systems where the use of chemicals could be minimised with maximum atom economy—a concept introduced by green chemistry. Thus, this science involves the design and redesign of chemical synthesis and processes to prevent pollution, rendering the products cost effective as well as environment-friendly.

EMERGENCE OF GREEN CHEMISTRY

Green chemistry, is very appropriately known as sustainable chemistry. The reason for its rapid adoption around the world is the realisation of it being a pathway to ensure economic and environmental prosperity. Green chemistry starts at the molecular level and ultimately delivers more environmentally benign products and processes.

The term 'green chemistry' was coined by Professor Paul Anastas, considered as the father of green chemistry, who worked with the environmental protection Agency, USA (EPA) 'Twelve principles of green chemistry' and has been translated to more than five languages.

BREAKTHROUGHS AND SUCCESS STORIES OF GREEN CHEMISTRY IN INDIAN INDUSTRIES

Green chemistry and engineering is vigorously changing the way we invent, manufacture and use chemical substances. Outstanding and environmentally benign reaction conditions and safer chemicals have been developed for manufacturing synthetic chemicals using the 12 principles of green chemistry. Today, the growing number of green chemistry methods developed by academic and industrial researchers enables companies to build strategies for industrial implementation of green chemistry.

CASE STUDIES

In order to recognise and promote the efforts of industries in the design and implementation of the green chemistry technology. Two Indian companies—SMS Pharmaceuticals and Newreka Green Synth Technologies Pvt. Ltd. were awarded with the Indo-US green chemistry network centre (GCNC) Award for the successful commercialisation of green chemistry-based technologies at the Indo-US Science and Technology Forum Workshop on Green Chemistry held in January, 2008.

Cleaner Ranitidine Process

SMS Pharmaceuticals Ltd. Hyderabad has developed a catalytic green method for the conversion of foul smelling toxic gases like methyl mercaptan generated in the manufacture of ranitidine base (daftmen base), and similar emissions from other drugs/intermediates to useful by-products recyclable in the same process.

Ranitidine HCl is a commonly used antiulcer drug. Six or seven industries produce Ranitidine HCl in India with the total production amounting to a little over 50 per cent of the global requirement. SMS Pharmaceuticals Ltd. located at Hyderabad in India is the largest manufacturer of Ranitidine HCl. Importantly, in the process of production of this drug, a large quantity of methyl mercaptan is generated in the plant. Methyl mercaptan is a low boiling liquid that rapidly turns into gas at ambient temperature and pressure, having obnoxious odour of rotten cabbage.

The low vapour density of the gas facilitates its easy diffusion into and rapid mixing with the atmospheric air thereby rendering the air stinky. Moreover, methyl mercaptan is a health hazard because it causes dizziness, headache, nausea, respiratory arrest, and even coma and unconsciousness. A little longer exposure to high concentration of the gas can be fatal. The sustained contact with liquid or the gas may cause frostbite as well. Therefore, it was very important to control the formation of methyl mercaptan. Current treatment methods to control methyl mercaptan through incineration, scrubbing, adsorption, all have attendant problems.

In a significant development, SMS pharmaceuticals Ltd. and Prof Mihir K. Chaudhuri of IIT Guwahati have come up with an improved process rendering the manufacturing of Ranitidine HCl clean and environmentally benign. In this process, the metyl mercaptan has been first converted to dimethyl sulphide that is catalytically converted to the odourless dimethyl sulphoxide (DMSO) (raw material used in manufacturing process of Ranitidine HCl) in a quantitative manner. H_2O_2 is the terminal oxidant. New catalysts are based on titanium, vanadium, aluminium and a solid acid in one case. The catalytic process has been assigned to RCHEM Pvt. Ltd. at Humnabad for green production of Ranitidine HCl. The DMSO produced by the catalytic reaction is used in the process itself thereby reducing the total cost of production by nearly 40 per cent. This is a real world example of 'green chemistry and green technology' practiced in India.

Recycle of Mother Liquor

Similarly, Newreka Green Synth Technologies Pvt. Ltd. has bagged the award at the GCNC symposium based on case studies of its work in two specific areas—the recycle of mother liquor from chemical process streams and the use of its proprietary catalysts to replace harsh conditions. Mother liquor typically contains raw materials, finished product and both organic and inorganic impurities. Depending on the reaction medium, it can be acidic, neutral and basic.

Newreka Recycle Solution (NRS), which covers all of these areas, has several important advantages in addition to the green chemistry principles it is based on, that is high selectivity leading to high quality, no need for purification, the ability to customise catalyst formulation, close to theoretical yield and recovery, low capital investments, savings on cost and legislative compliance, thus freeing up room for expansion, and better access to markets in the developed world.

FUTURE

There is no doubt that the emerging area of green chemistry has identified scientific principles, approaches, and methodologies that have demonstrated the most positive aspects of chemistry in industry. While the success of green chemistry thus far seems quite large in terms of quantitative benefit to human health and the environment, they are merely the tip of the iceberg when compared to the potential. To reach this full potential, greater awareness, adoption, and development of green chemistry practices are necessary.

Industries have also realised that when their professional chemists are knowledgeable about pollution prevention concepts, they are able to identify and implement effective pollution prevention technologies.

IMPORTANT ISSUES IN GREEN CHEMISTRY AND EXAMPLES OF GREEN PROCESSES

Attempts are being made not only to quantify the greenness of a chemical process but also to factor in other variables such as chemical yield, the price of reaction components, safety in handling chemicals, hardware demands, energy profile and ease of product workup and purification.

In 2005 Ryoji Noyori identified three key developments in green chemistry: use of supercritical carbon dioxide as green solvent, aqueous hydrogen peroxide for clean oxidations and the use of hydrogen in asymmetric synthesis. Examples of applied green chemistry are supercritical water oxidation, on water reactions and dry media reactions.

Bioengineering is also seen as a promising technique for achieving green chemistry goals. A number of important process chemicals can be synthesised in engineered organisms, such as shikimate, a Tamiflu precursor which is fermented by Roche in bacteria.

Principles

Paul Anastas, then of the United States Environmental Protection Agency, and John C. Warner developed 12 principles of green chemistry, which help to explain what the definition means in practice. The principles cover such concepts as:

1. The design of processes to maximise the amount of raw material that ends up in the product.
2. The use of safe, environment-benign substances, including solvents, wherever possible.
3. The design of energy efficient processes.
4. The best form of waste disposal: do not create it in the first place.

Supramolecular Chemistry

Research is currently ongoing in the area of supramolecular chemistry to develop reactions which can proceed in the solid state without the use of solvents. The cycloaddition of *trans*-1,2-*bis*(4-pyridyl)ethylene is directed by resorcinol in the solid state. This solid-state reaction proceeds in the presence of UV light in 100 per cent yield (Fig.6.1).

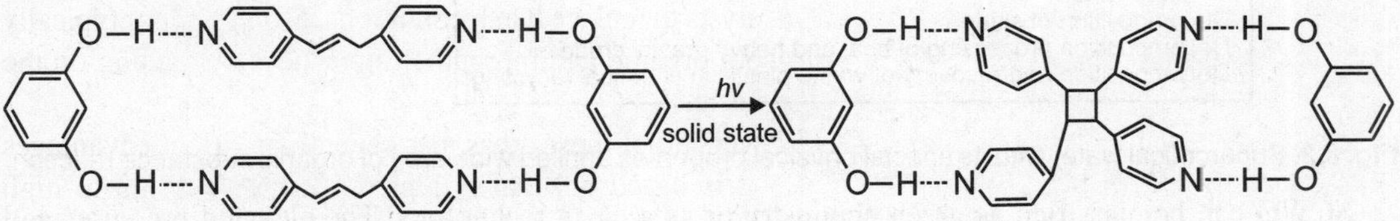

Fig. 6.1. Solid-state reaction proceeds in the presence of UV light.

Supercritical Water

A supercritical fluid possesses the characteristics of both fluid and gaseous substances: the fluid behaviour of dissolving soluble materials, and the gaseous behaviour of excellent diffusibility. In the case of supercritical water particularly, the effect as a reaction-causing solvent is great, and because continuous control over the physical properties of fluid is allowed by regulation of the pressure and/or the temperature, various uses in the field of reacting systems are being considered.

Synthesis of artificial crystal (which uses the difference between the solubility in an intense solvent action under conditions of both high temperature and high pressure), has been in use for some time as an industrial application of supercritical water as a reaction solvent; more recently, supercritical water with its special physical properties has been applied in the wide area of organic substance reactions (Fig. 6.2).

Supercritical water oxidation or SCWO is a process that occurs in water at temperatures and pressures above a mixture's thermodynamic critical point. Under these conditions water becomes a fluid with unique properties that can be used to advantage in the destruction of hazardous wastes such a PCBs. The fluid has a density between that of water vapour and liquid at standard conditions, and exhibits high gas-like diffusion rates along with high liquid-like collision rates. In addition, solubility behaviour is reversed so that chlorinated hydrocarbons become soluble in the water, allowing single-phase reaction of aqueous waste with a dissolved oxidiser. The reversed solubility also causes salts to precipitate out of solution, meaning they can be treated using conventional methods for solid-waste residuals. Efficient oxidation reactions occur at low temperature (400°–650°C) with reduced NO_x production.

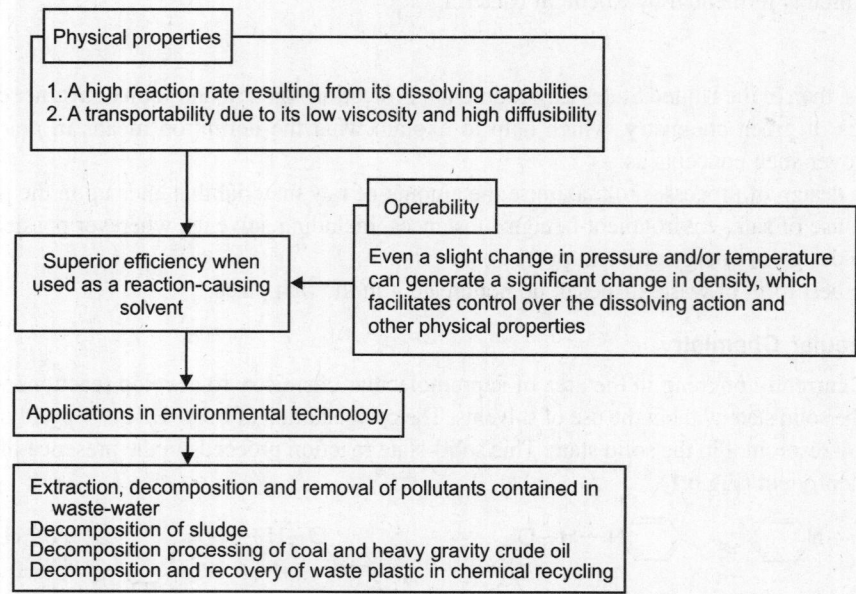

Fig. 6.2. Supercritical water with its special physical properties applied wide area of organic substance reaction.

SCWO can be classified as green chemistry or as a clean technology. The elevated pressures and temperatures required for SCWO are routinely encountered in industrial applications such as petroleum refining and chemical synthesis. Partial oxidation in supercritical water has also found some success in the thermal depolymerisation of non-hazardous agricultural wastes, recycling them into heating oil and other petroleum-like products. In Japan a number of commercial SCWO applications exist, among them one unit for treatment of halogenated waste built by Organo. In Korea two commercial size units have been built by Hanwha. In Europe, Chematur Engineering AB of sweden has commercialised the SCWO technology for treatment of spent chemical catalysts to recover the precious metal, the AquaCat process. The unit has been built for Johnson Matthey in the UK. It is the only commercial SCWO unit in Europe and with its capacity of 3 m^3/hour, it is the largest SCWO unit in the world.

Nanocatalysts in Green Chemistry

The catalysts on which more than 20 per cent of world industrial production is based—including the expensive platinum employed to scrub clean the exhausts of millions of vehicles and the molecules pharmaceutical giants use to manufacture drugs—soon could be replaced in large part by more effective nanotechnology upgrades. Reducing catalytic substances to nanometers in size greatly increases the surface area available per gram, which in turn boosts the level of catalytic activity.

In the near term, 'the chemical industry will be more environmentally friendly because nanotechnology can make catalysts 100 per cent selective to the desired product. 'One of the chemical industry's biggest problems in the last century were reactions that were not completely selective to the desired product and producing too much waste and environmentally harmful materials'.

Clean Hydrogen Peroxide Synthesis via a Nanocatalyst Process

Zhou and colleagues have developed a nanocatalyst to generate hydrogen peroxide, the chemical often used as a bleach that finds use in manufacturing materials such as polyurethane. 'Whenever you make hydrogen peroxide with conventional methods, you use a really complicated process with starting materials that are toxic and the process cost is high,' he explained. 'Using a nanocatalyst, you can just combine hydrogen and oxygen—two very clean and simple molecules to procure—to produce hydrogen peroxide, and that's with 90 per cent selectivity in large-scale tests, much higher than any other catalysts used to make hydrogen peroxide'.

A nanotechnology platform developed at headwaters technology innovation (HTI) has many applications. H_2O_2 is a clean, environment-friendly oxidant that can substitute for environmentally harmful chlorinated oxidants in many manufacturing operations. However, the current H_2O_2 manufacturing process is complex, high cost, and energy intensive. The process uses an anthra-quinone working solution containing several toxic and hazardous chemicals, which causes environmental pollution when disposed. HTI has developed a nanocatalyst technology that enables the synthesis of H_2O_2 from hydrogen and oxygen. The breakthrough technology eliminates all the hazardous chemicals of the existing process, along with its undesirable by-products. It produces H_2O_2 more efficiently, cutting both energy use and costs, and generates no toxic waste. The technology enables a simple, environment-friendly, commercially viable H_2O_2 manufacturing process. In partnership with Evonik (a major H_2O_2 manufacturer), HTI successfully demonstrated its technology in a pilot plant. Tests in a commercial demo plant are being conducted. Construction of a full-scale commercial plant is planned to start in 2011. The project received 2007 presidential green chemistry challenge award from EPA.

At the nanoscale, substances can demonstrate catalytic activity where they once did not. This increased specificity can lead to safer drugs also. 'Often in making pharmaceuticals, you make a lot of side products also, some of which are really toxic and can cause health problems,' Zhou said. 'You can design a nanocatalyst at the molecular level that is more selective and will only deliver the drug molecule you want.'

'Gold is famously inactive', Zhou said, 'but when we make gold less than 6 nanometers, it becomes an active catalyst, helping oxygen combine with carbon monoxide to make carbon dioxide'.

The ability of nanotechnology to enhance catalytic activity opens the potential to replace expensive catalysts with cheaper nanocatalysts.

CASE STUDIES

We present here two case studies where we replaced conventional 'synthetic chemistry based processes' with 'green chemistry based processes' successfully at commercial scales and the impact it had on the economics as well as on the environment. It's worth mentioning that these technologies won the prestigious Indo-US science and technology forum green chemistry award in January, 2008.

CASE STUDY I

Green chemistry based recycle solution for neutral mother liquor in unit process 'reduction' (an alternative to catalytic hydrogenation with hydrogen gas and raney nickel).

Background

This process was developed for one of the pharmaceutical Industry. For manufacturing one of their pharmaceutical intermediates, the customer was using conventional synthetic chemistry based process to convert a nitro compound into an amino compound. The conventional technology involved using hydrogen gas along with pyrophoric catalyst like raney nickel at high pressure.

Besides the hazardous nature of the process, the customer's other concern was economics. Continuously increasing prices of raney nickel (prices of raney nickel have increased five folds in the last 5 years) and solvents (prices of solvents have almost doubled) were putting lot of pressure on their costing.

Conventional Synthetic Chemistry Based Process

The conventional technology involved reduction of nitro group into the corresponding amine using hydrogen gas at high pressure along with pyrophoric raney nickel as the catalyst. The reaction was being carried out in a high pressure autoclave. Initially, solvent 1 was taken in the 'High Pressure Autoclave' (Fig. 6.3) along with nitro compound (RNO_2) and raney nickel. Then, the autoclave was pressurised with hydrogen gas at the desired temperature. Then, hydrogen gas was purged till the reaction was complete. After completion of reaction, spent raney nickel was filtered and the solution of amino compound in solvent 1 was taken for distillation. After solvent 1 was distilled out, the mass in the distillation vessel was cooled and chilled to isolate 1st crop of amine (RNH_2). This was usually 99%+ in purity but light brown in colour.

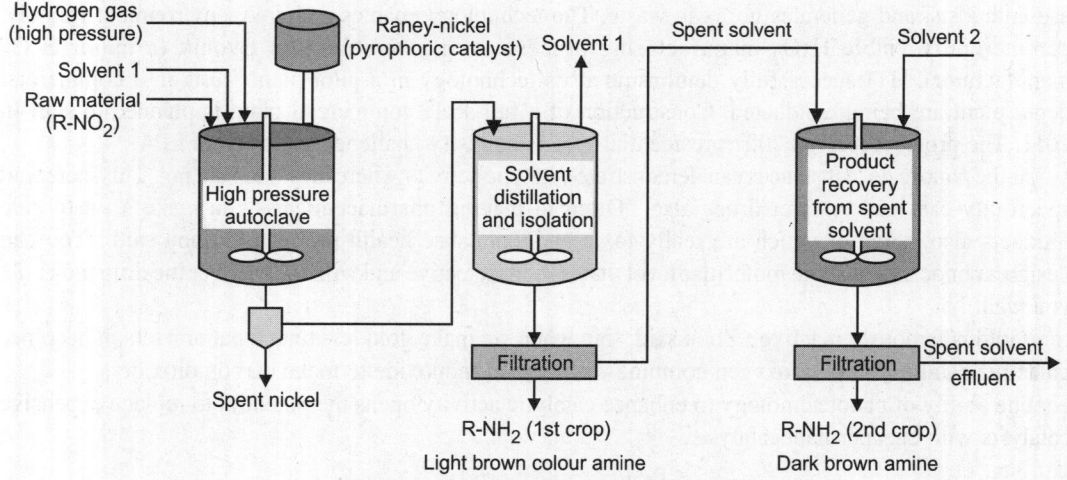

Fig. 6.3. Conventional reduction technology.

The spent solvent obtained after centrifuging first crop of amine still contains substantial quantity of amine and hence was taken to another vessel for recovering the product. Here, solvent 2 was introduced to suppress the solubility of amine in solvent 1 and isolate amine. The second crop of the product is recovered after centrifuging. This was usually dark brown in colour and required additional purification to get quality similar to first crop. The overall yield in the process was around 85 per cent of theoretical.

After recovery of second crop of amine, the mother liquor (containing solvent 1, solvent 2, some quantity of amine, and other organic impurities) obtained from the centrifuge is a waste and had to be incinerated.

Green Chemistry Based Process

The green chemistry based process developed by our R & D team involved reduction of nitro group into the corresponding amine using water as the reaction and extraction medium. The reaction is carried out in a simple stainless steel (SS) vessel at atmospheric pressure, temperature of 95°–100°C and pH in the range of 5 to 7. Initially, water is taken (for the first batch, subsequently the neutral mother liquor is recycled) in the reduction vessel and heated up to 95°–100°C. Some quantity of Newreka's proprietary, customised catalytic formulation, newreka reduction catalyst (NRC), is added to initiate the reaction. Then, the nitro compound (RNO_2) is added along with balance quantity of NRC over a period of 4–5 hours. The reaction is completed within 6 hours (Fig. 6.4).

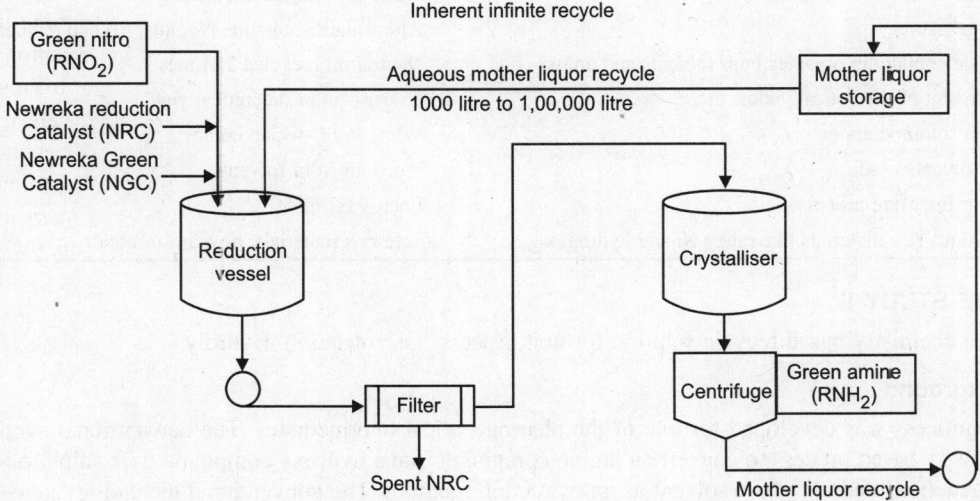

Fig. 6.4. Newreka recycle technologies (NRT).

Then, Newreka's second proprietary customised catalytic formulation, newreka green catalyst (NGC), is added. NGC ensures that all ppm level nitro present is completely converted into amine and also removes most of the organic and inorganic impurities present in the reaction mass. This ensures that no part of the catalyst is carried forward with the product and also ensures that the mother liquor is clean enough to be recycled back for the next batch. After NGC addition, the reaction mass is filtered and the solution containing water and amine is taken into a crystalliser where the amine isolates on cooling and chilling. The spent NRC obtained is a by-product for another industry.

The amine is centrifuged and the neutral mother liquor obtained is taken for recycling back in the next reduction batch. Amine obtained is off-white in colour and 99%+ purity (even after 30 recycles). The yield is around 95 per cent of theoretical. The surprising result for our customer was that the yields in the downstream processes increased by 2.5 per cent due to improved quality of amine.

Principles of Green Chemistry Followed

1. Prevent waste: Design chemical syntheses to prevent waste, leaving no waste to treat or clean up.
2. Design safer chemicals and products: Design chemical products to be fully effective, yet have little or no toxicity.
3. Design less hazardous chemical syntheses: Design syntheses to use and generate substances with little or no toxicity to humans and the environment.
4. Use safer solvents and reaction conditions: Avoid using solvents, separation agents, or other auxiliary chemicals. If these chemicals are necessary, use innocuous chemicals.

5. Increase energy efficiency: Run chemical reactions at ambient temperature and pressure whenever possible.

Comparison of processes and benefits achieved by our customer are given in Table 6.1.

Table 6.1. Case study—neutral mother liquor.

A comparison of conventional technology and NRS	
Solvent as 'reaction medium'	Water as 'reaction medium'
High pressure	Atmospheric pressure, Nominal pH and temperature
Effluent containing organics both solvents and amine	Neutral ml recycled 25 times
85 per cent of theoretical yield	95 per cent or theoretical yield
Brown coloured amine	White to off-white amine
Two solvents used	Freedom from solvents
Energy Intensive process	Energy efficient
Hazardous raw materials like raney Ni and hydrogen gas	Safe raw materials, no acids or alkali

CASE STUDY II

Green chemistry based recycle solution for unit process 'Diazotisation–Hydrolysis'.

Background

This process was developed for one of the pharmaceutical intermediates. The conventional synthetic chemistry based process to convert an amino compound into a hydroxy compound uses sulphuric acid as a reaction medium and a solvent as an extraction medium. The conventional technology generated huge quantity of acidic liquid effluents and also the yield in the process was low.

Conventional Synthetic Chemistry Based Process

In the conventional technology, initially dilute sulphuric acid medium was taken and amino compound was dissolved in it. Then, at the desired temperature, sodium nitrite was added slowly for diazotisation. After diazotisation, the temperature of reaction mass was raised to hydrolyse the mass. Then the reaction mass was taken in another vessel where it was treated with copper sulphate, neutralised with sodium hydroxide and the product, i.e. the hydroxy compound was extracted using ethyl acetate as a solvent.

After extraction, the organic layer was separated and taken for solvent distillation and isolation while the aqueous layer (which was sulphuric acid containing highly acidic liquid) was removed as mother liquor which had to be treated and then incinerated. After ethyl acetate was distilled, the product was isolated by adding acetic acid followed by cooling, chilling and centrifuging. The product (crude hydroxy compound) obtained was brown in colour with approximately 70 per cent of theoretical yield. The mother liquor obtained from centrifuge was acidic mother liquor containing some quantity of ethyl acetate. This also had to be neutralised and then incinerated (Fig. 6.5).

Green Chemistry Based Process

The green chemistry based process developed by MIIT R & D team involved questioning the current reaction conditions and methodology and altering them such that side reactions and hence organic impurities are minimised. Secondly, they developed a proprietary customised recycle catalytic formulation (RCat) for selectively removing organic and inorganic impurities from acidic mother liquor such that it can be recycled back. Unwanted raw materials like copper sulphate and acetic acid were eliminated from the process (Fig. 6.6).

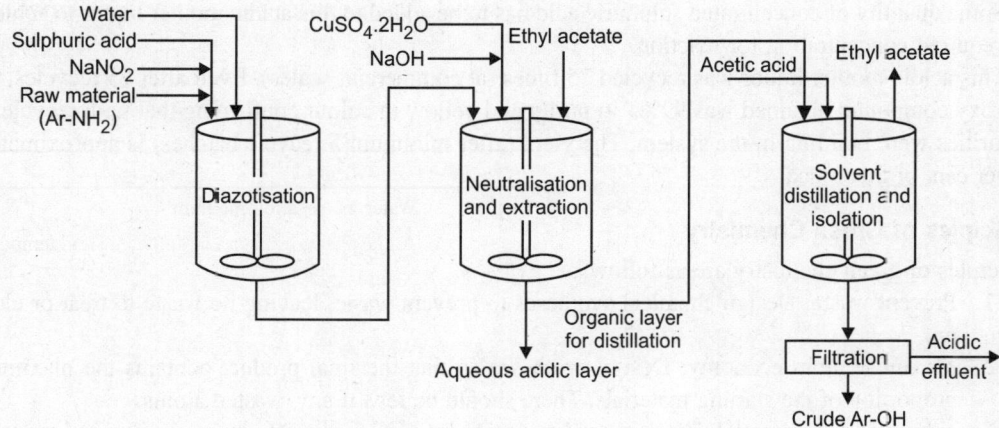

Fig. 6.5. Conventional diazotisation technology.

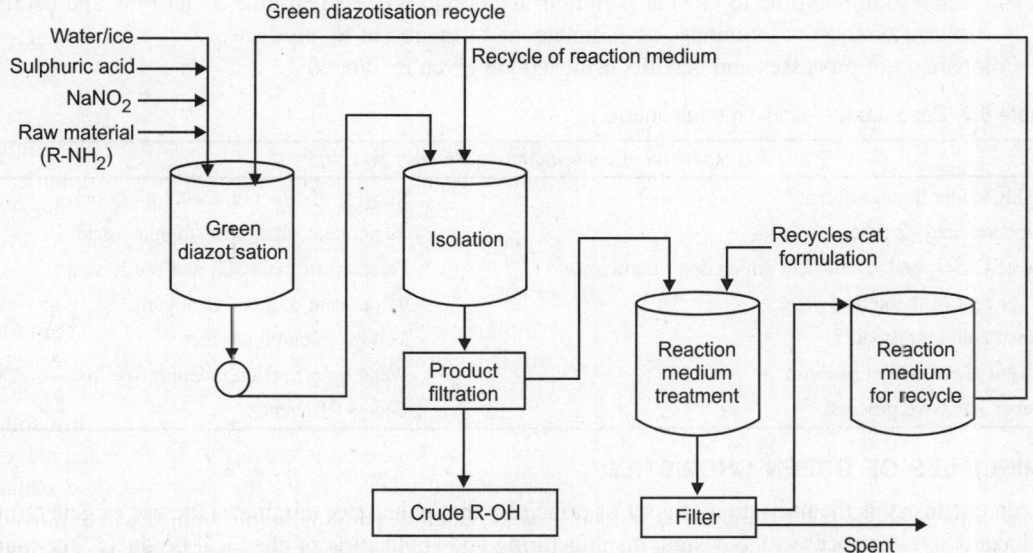

Fig. 6.6. Newreka diazotization technology (NDT).

Initially, dilute sulphuric acid medium is taken (for the first batch, subsequently the acidic mother liquor is recycled) and amino compound is dissolved in it. Then, at the altered conditions, diazotisation is carried out with sodium nitrite. After this, the reaction mass is hydrolysed at higher temperature. After the reaction is complete, the product is isolated from water (use of ethyl acetate as extraction medium was eliminated). The hydroxy compound obtained after centrifuging is yellow in colour. The acidic mother liquor contains sulphuric acid, hydroxy compound, some organic impurities and some inorganic impurities. This acidic mother liquor is then reacted with Newreka's proprietary, Customised, Catalytic Formulation called 'Recycle Cat (RCat)' to selectively remove unwanted organic and inorganic impurities (without neutralising sulphuric acid). The mass is then filtered to remove Spent RCat and the acidic mother liquor is ready to be recycled back in diazotisation and hydrolysis step as reaction medium.

Some quantity of concentrated sulphuric acid has to be added to this acidic mother liquor to achieve the required concentration for reaction.

This acidic mother liquor was recycled 75 times (at commercial scales). Even after 75 recycles, the hydroxy compound obtained was 99%+ in purity and yellow in colour confirming that with recycle no impurities were building in the system. The yield (after minimum 5 recycle batches) is approximately 95 per cent of theoretical.

Principles of Green Chemistry

Principles of green chemistry are as follows:
1. Prevent waste: Design chemical syntheses to prevent waste, leaving no waste to treat or clean up.
2. Maximise atom economy: Design syntheses so that the final product contains the maximum proportion of the starting materials. There should be few, if any, wasted atoms.
3. Use safer solvents and reaction conditions: Avoid using solvents, separation agents, or other auxiliary chemicals. If these chemicals are necessary, use innocuous chemicals.
4. Analyse in real time to prevent pollution: Include in-process real-time monitoring and control during syntheses to minimise or eliminate the formation of by-products.

Comparison of processes and benefits achieved are given in Table 6.2.

Table 6.2. Case study—acidic mother liquor.

\multicolumn{2}{c}{A comparison of conventional technology and NRS}	
Highly acidic liquid effluent	Recycle of 'reaction medium' 75 times
Sulphuric acid –2 times	50 per cent savings in sulphuric acid
Use of $CuSO_4$ and acetic acid which don't participate	Freedom from $CuSO_4$ and acetic acid
70 per cent of theoretical yield	95 per cent of theoretical yield
Brown coloured product	Yellow coloured product
Solvent as extraction medium	Water as extraction medium free from solvent
Energy intensive process	Energy efficient

PRINCIPLES OF GREEN CHEMISTRY

Green chemistry is the utilisation of a set of principles that reduces or eliminates the use or generation of hazardous substances in the design, manufacturing and application of chemical products. The main principles of Green chemistry may be expressed as follows:

Prevention

'Prevention is better than cure'. It is better to prevent waste than to treat or clean up waste after it has been created. Wealth is converted into waste in this way and wealth is needed to create the waste. Treatability costs are raising steadily, pollution control are becoming stringent making waste treatment almost an unviable approach. However, regulatory requirements necessitate us to do the same by draining the profits.

Atom Economy

Synthetic methods should be designed to maximise the incorporation of all materials used in the process into the final product.

Less Hazardous Chemical Syntheses

Wherever practicable, synthetic methods should be designed to use and substances generate that possess little or no toxicity to human health and the environment.

Designing Safer Chemicals

Chemical products should be designed to effect their desired function while minimising their toxicity.

Safer Solvents and Auxiliaries

The use of auxiliary substances (e.g. solvents, separation agents, etc.) should be avoided wherever possible and innocuous when used. The discovery of next-generation solvents such as supercritical fluids and ionic liquids or the development of solventless systems provide a future alternative to the ubiquitous use of organic and halogenated solvents that are of great concern to human health and the environment.

Design for Energy Efficiency

Energy requirements of chemical processes should be recognised for their environmental and economical impact and should be minimised. If possible, synthetic methods should be conducted at ambient temperature and pressure. The making and breaking of chemical bonds is intrinsically an energetic process and the material transformation and the basis for energy efficiency must be designed at the molecular level. New methods for improved fundamental design of reaction conditions and pathways are needed to reduce energy intensiveness.

Use of Renewable Feedstocks

Raw materials or feedstocks should be renewable, rather than depleting whenever technically and economically practicable.

Reduce Derivatives

Unnecessary derivatisation (use of blocking groups, protection/deprotection, and temporary modification of physical/chemical processes) should be minimised or avoided if possible, because such steps require additional reagents and can generate waste.

Catalysis

Catalytic reagents (as selective as possible) are superior to stoichiometric reagents.

Design for Degradation

Chemical products should be designed so that at the end of their function they break down into innocuous degradation products and do not persist in the environment.

Real-time Analysis for Pollution Prevention

Analytical methodologies need to be further developed to allow for real-time, in-process monitoring and control prior to the formation of hazardous substances.

Inherently Safer Chemistry for Accident Prevention

Substances and the form of substance used in a chemical process should be chosen to minimise the potential for chemical accidents, including releases, explosions, and fires.

WIDENING THE SCOPE OF GREEN TECHNOLOGIES THROUGH CHEMICAL ENGINEERING

Green chemistry has become the darling of practitioners and learners alike. Chemical engineers will be at the forefront of developing green technologies. Heterogeneous catalysis, solventless synthesis, use of microwaves, ultrasound, etc. have been attempted by many researchers all over the world to suggest how atom economical processes are also the most cost effective. Sustainable development will then be a reality. The principles of selectivity engineering can be applied to a variety of important reactions in pharmaceutical and fine chemical industries with reference to environmental issues and a few examples, covering phase transfer catalysis and solid acid catalysis, are cited to illustrate these aspects. The process engineering and reactor engineering aspects are also briefly outlined.

In pharmaceutical and fine chemical industries as also speciality, dyestuff, agrochemical and fragrance industries, multistep and multiphase batch reactions are frequently encountered, which involve several cumbersome and expensive separation/purification stages that result into poor atom economy. Pharmaceutical and drug industry typically produce 80–100 kg of by-product or waste per kg of desired product. A challenging aspect of multistage synthesis, particularly of bioactive compounds, is that the complexity of the process goes on increasing with each additional stage. This affects the desired selectivity because the number of functional groups and the number of chemically reactive centres in the molecule increase, steric hindrances influence, and thermal stability of compound decrease. Furthermore, recovery and reuse of solvents and/or reactants add to costs.

Sources of Pollution

The basic building blocks in multi-step synthesis are mostly aromatic compounds with incorporation of such groups as halo, $-NH_2$, $-OH$, $-NO_2$, $-SO_3H$, alkyl, acryl, aryl, etc. which change the reactivity of the cyclic compound and sometimes the colour and odour. Multi-step synthesis is typically done via routes which do not take into account the atom economy or the amount and nature of co-products or by-products. Catalysis, if any, is limited to homogeneous catalysis involving highly corrosive substances posing disposal problems. Use of solvents is also very common. Low boiling, halogen containing hazardous inflammable solvents are used.

Some of the important precursors used in these industries are: aniline, acetanilide, benzyl chloride, benzoic acid, monochlorobenzene, chloronitrobenzenes, cresols, cresylic acid, cumene, cyclohexane, cyclohexanone, dichlorobenzenes, dihydroxybenzenes, dimethyl sulphate, ethylbenzene, bisphenol-A, alpha-methylstyrene, nitrobenzene, nonyl-phenol, xylenes, phenol, phthalic anhydride, pyridine, picolines, salicylic acid, etc.

Most of these are hazardous to handle and wasted during separation or washings. Irrespective of the best solutions to reduce waste at source, it may not be totally avoided and in such cases, it should provide an impetus to convert the waste into an asset. Plant engineers should look for integrated facility to convert liabilities into assets. The environmental challenges faced by these industries during the 21st century will be:

1. Development of theoretically zero waste (or minimum waste) processes.
2. Minimisation of hazardous products and greenhouse gases.
3. Replacement of corrosive and nonreusable catalysts.
4. Evolution of sustainable systems; zero energy input processes.
5. Development of single-pot or cascade engineered processes.
6. Reduction in number of steps.
7. Minimum use of solvents, environmentally benign solvents or solvent-free synthesis.

8. Design of electrically engineered systems (catalysts, different forms of energies, ultrasound; microwave radiation).
9. Replacement of petroleum feedstock by renewable resources.

Selectivity Engineering Principle

From the process economics and environmental viewpoints, the intensification of reaction rates and enhancement of selectivities can be achieved through selectivity engineering principle which adopts the following strategies:

1. In a biphasic reaction, addition of a third solid phase (porous or non-porous and/or catalytic or noncatalytic) with particle size smaller than diffusion film thickness. The triphasic system can lead to an enhancement in rate by orders of magnitude with reference to the original gas–liquid, and liquid–liquid systems (microphase catalysis).
2. Converting a liquid–liquid reaction into liquid–liquid–liquid or liquid–liquid–solid phase transfer catalysis.
3. Heterogenising a homogeneous catalysis.
4. Cascade engineering or single pot synthesis.
5. Electrically engineered composite catalysts.
6. Deliberate incorporation of mass transfer and/heat transfers resistance.
7. On a larger scale, integration of different plants on the same site to achieve environmental goals.

Use of third phase: solid particles

The desired selectivity of the product can be improved by using an inert solvent or an adsorbent or a membrane leading to the generation of third phase which can remove the product or coproduct to improve the conversion. In cefoxitin process, in the acrylation step with acetyl chloride, HCl is selectively removed by using powdered molecule sieves (3Å/4Å). In the alkylation of hydroquinone, phenol and aniline with MTBE or *tert*-butanol, yields were improved by using these adsorbents.

Creation of a third liquid phase: L-L-L PTC

Phase transfer catalysis (PTC) is now a mature technique for conducting multi-phase reactions. In liquid-liquid-liquid phase transfer catalysis (L-L-L PTC), the third liquid phase is the main reaction phase. The advantages of L-L-L PTC over normal PTC are:

1. Increase in reaction rates by orders of magnitude.
2. Easier catalyst recovery and reuse.
3. The catalyst need not be bound to a solid support.
4. Better selectivity, hence the attendant difficulties of reduced activity and mechanical strength associated with liquid-liquid-solid (L-L-S) PTC can be avoided.

The disadvantages of L-L-L PTC are:

1. More amount of catalyst is required, which is expensive.
2. The method is not applicable for systems where a very high temperature is required to carry out the reaction.

As the temperature increases, the stability of third liquid phase decreases. However, if the catalyst is stable, then by lowering the temperature at the end of the reaction it could be easily separated into a third phase for recovery and reuse. Synthesis of *p*-nitroanisole (PNA), *p*-nitro-phenetole (PNP) and *p*-butoxynitrobenzene from *p*-chloronitrobenzene (PCNB), sodium hydroxide and methanol, ethanol, *n*-butanol, respectively was found to be very selective and efficient with tetrabutylammonium bromide (TBAB) as catalyst under the L-L-L PTC systems in comparison with the bi-liquid PTC.

The nature and concentration of catalyst and the amount of sodium hydroxide are important factors that influence the formation of the third liquid phase (catalyst-rich middle phase) and distribution of catalyst. It was observed that the third liquid phase was the main reaction phase, but at a certain critical concentration of the catalyst, sodium hydroxide was formed as the third phase. The reuse of the third phase is important from economical point of view and it is found that the third phase can be reused effectively.

Cascade engineered phase transfer catalysts (CEPTC)

Another related term is the so-called cascade engineering which embraces the single pot (reactor) wherein more than one step, involving complex reactions, are combined elegantly to produce the desired product, minimising both by-product formation and separation stages. It could be used to produce different products as well, starting from the same precursor but under different mode of operation.

Cascade engineering inherently or tacitly assumes the role of catalysts. In the realm of PTC, the technique can be used under L-L, L-S (reactant), L-L-S (bound catalyst) and L-L-L phase transfer catalysis for certain molecules such as chloronitrobenzenes or benzyl chloride to produce a variety of chemicals. For instance, chloronitrobenzenes are amenable to a variety of chlorine substitution and nitro reduction reactions.

In order to study cascade engineering with phase transfer catalysis, chlorobenzenes could be used as the starting materials. Both the chloro and nitro groups could be manipulated to obtain a variety of products, all having commercial values (Fig. 6.7).

Reduction of chloronitrobenzenes could lead to chloroanilines, aminophenols, aminothiophenols or substituted diphenyl sulphides or polysulphides or mixture thereof in the presence of Na_2S under alkaline condition using a suitable PTC. The PTC could be used under L-L, S-L, L-L-L and L-L-S (catalyst) and depending on the type of the system used, the product selectivity could be varied.

Benzyl chloride is used in a number of PTC reactions. Of these, the reaction with sodium cyanide leading to phenylacetonitrile has a lot of commercial significance. Since sodium cyanide and potassium cyanide are hazardous chemicals, it is safer to take them as limiting reactants, whereby the final reaction mixture would contain only the unreacted benzyl chloride (b.p. 179°C), phenylacetonitrile (b.p. 233.5°C) and the dissolved catalyst.

These could be separated by distillation under vacuum. In order to avoid the handling of unreacted metal cyanide, if benzyl chloride is taken in excess, then it can be subjected to further reaction under PTC conditions without separation of the catalyst. For instance, benzyl chloride can be transferred to benzyl alcohol under alkaline conditions as well as phenylacetonitrile hydrolysed to phenylacetic acid. In these reactions under alkaline conditions, instead of phenylacetic acid and benzyl alcohol, C-alkylations of phenylacetonitrile might take place leading to different products which are even otherwise useful in pharmaceutical industry for the preparation of pentapiperdine, idoxifen (nonsteroidal antiestrogen), oxelodine, phenoperidine and *d*-cylonine.

Even the higher acetonitriles, namely, benzyl phenylacetonitrile and dibenzyl phenylacetonitrile can also be used for this purpose. Furthermore, the formation of benzyl ether will be there because benzyl chloride hydrolysis leads to benzyl ether as a by-product. In this reaction, phenylacetic acid, used in the synthesis of Penicillin G, could be extracted *in situ* as the sodium salt.

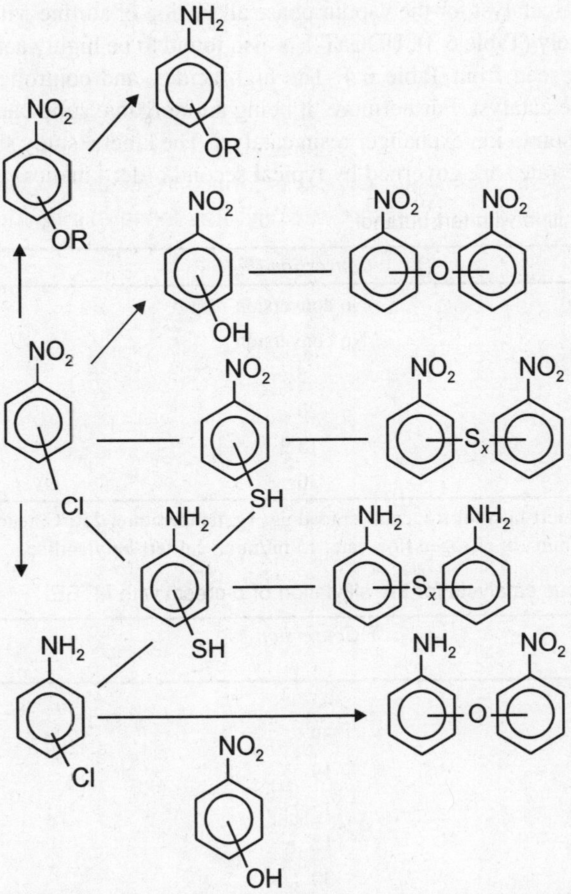

Fig. 6.7. Cascade engineered PTC: chloronitrobenzene based products.

Process hazards

Industrial accidents are reported to have occurred mostly in batch processes involving hazardous reactants, solvents and exothermic reactions. Nitrations, oxidations, alkylations, polymerisations, halogenations and hydrogenations are important unit processes, the risk being in that order approximately. Reagents such as phosgene, dimethyl sulphate, formaldehyde/hydrogen chloride (for chloromethylation), sodium azide, hydrogen fluoride, hydrogen cyanide, methyl isocyanate, bromine and for that matter chlorine are hazardous to handle and their use calls for special precautions. Some of these reagents can be replaced by catalytic options such as carbonylation instead of phosgenation, solid superacids for HF, etc.

Engineered catalysts

In certain cases, it is possible to combine two different functionalities synergistically to get better activity and selectivity. A synergistic combination of sulphated zirconia and mesoporous molecular sieve catalyst (UDCaT-1), with high surface area and acidity is prepared by *in situ* deposition of zirconium hydroxide in a mesoporous molecular sieve (HMS) and then promoting its acidity by sulphating agents. UDCaT-1 was found to be an active catalyst for the vapour phase alkylation of aniline with *tert*-butanol to give 4-*tert*-butylaniline selectively (Table 6.3). UDCaT-1 is also found to be highly active for the alkylation of *p*-cresol with MTBE as seen from Table 6.4. The high acidity and controlled activity of UDCaT-1 makes it a truly desirable catalyst. Furthermore, it being an inorganic solid can be employed at higher temperatures in comparison to ion exchanger resin catalyst. The kinetic study shows that all species are weakly adsorbed and the rates are governed by typical second order kinetics.

Table 6.3. Alkylation of aniline with tert-butanol.

Catalyst	Conversion (%)	Selectivity (%)[c]
HMS	No conversion	–
ZrO_2	No conversion	–
Sulphated-ZrO_2	5	98
UDCaT-1 (i)[a]	20	99
UDCaT-1 (ii)[a]	13.9	99
UDCaT-1 (i)[b]	30	99

Reaction conditions: aniline: tert-butanol 1:4, catalyst loading: 1g, temperature: 250°C, reactant feed rate: 6.5 ml/hour; a: nitrogen flow rate: 30 ml/min., b: nitrogen flow rate: 15 ml/min., c:4-tert-butylaniline.

Table 6.4. Activity of various catalysts for the alkylation of *p*-cresol with MTBE.

Catalyst	% Conversion	% Selectivity of 2-tert-butyl-p-cresol
UDCaT-1	45	97
Indian 130	39	92
Filtrol-24	19	96
Sul. zirconia	15	91
K-10	12	96
HPA/K-10	30	96

Reaction conditions: MTBE (220 mmol), *p*-cresol (220 mmol), catalyst loading; 3.5% of reaction mixture, temperature: 100°C, speed: 700 rpm, reaction time: 3 hours; by-products formed are: isobutylene, diisobutylene and methanol.

Another such catalyst is based on the sulphated zirconia and carbon molecular sieves. Table 6.5 shows the effect of different catalysts on conversion and *p:o* selectivity in nitration of chlorobenzene and toluene. S-ZrO_2 yielded high selectivity for *para*-isomer in the nitration of chlorobenzene and no *meta*-isomer or dinitrated by-products were observed. Modification of S-ZrO_2 by coating it with carbon molecular sieve (CMS) to yield UDCaT-2 further enhanced the selectivity for the *para*-isomer though there was no considerable change in conversion.

This can be attributed to the fact that the CMS around S-ZrO_2 acts as a barrier for the bulkier *ortho*-isomer and hence favouring the formation of kinetically smaller *para*-isomer. By the use of eclectically engineered catalyst, UDCaT-2, in nitration of chlorobenzene and toluene, the use of mixed acids was completely eliminated which also facilitated easy separation of the products. UDCaT-2 was found to be highly selective catalyst for nitration of aromatic compounds, chlorobenzene, in particular. The *p:o* ratio of nitro-substituted chlorobenzene and toluene was found to be dependent on the pore size of the catalyst and was found to increase with decrease in the pore size of the catalyst. Also, UDCaT-2 was found to be highly reusable with very small decrease in activity after every use.

Table 6.5. Nitration of substituted benzenes.

Catalyst	Conversion of acetyl nitrate (%)		P:O ratio	
	Chlorobenzene	Toluene	Chlorobenzene	Toluene
No catalyst	No conversion	60	–	0.55:1
S-ZrO$_2$	47	98	10.6:1	0.75:1
20% DTP/K-10	–	73	–	0.80:1
UDCaT-2	45	98	13.2:1	0.86:1
5%DTP/S-ZrO$_2$	41	92	8.6:1	0.74:1
5%DTP/S-ZrO$_2$/CMS	21	95	10.2:1	0.80:1

Reaction conditions: chlorobenzene: nitric acid 2:1, temperature: 30°C; catalyst loading: 0.029 g/cm^3; toluene: nitric acid 2:1, temperature: 20°C, catalyst loading: 0.012 g/cm^3; speed 2,500 rpm, flow rate of nitric acid 1.85×10^{-5} mol/s, reaction time 90 minutes.

The synergism between heteropolyacids (HPA) and K-10 clay was found to be an excellent catalyst for condensation, etherification, alkylation, etc. Due to the problems associated with unavailability, transportation and handling or *iso*-butylene, particularly for usage in low-tonnage fine chemical and speciality manufacture, it is advantageous to generate pure isobutylene *in situ*. Two attractive sources of isobutylene are *tert*-butanol and methyl-*tert*-butyl ether (MTBE). *Tert*-butanol is available as a by-product in the Arco process for propylene oxide. MTBE is a good source for the generation of pure isobutylene and the by-product, methanol, is also a very important raw material in chemical industry. On the contrary, the dehydration of *tert*-butanol *in situ* leads to water as a coproduct in the alkylation reaction and thus different yields of the alkylated product are expected *vis a vis* MTBE as the alkylation agent.

Synthesis of MTBE from *tert*-butanol and methanol has been studied in this laboratory by using a variety of solid acids. Heteropoly acids (HPA) supported on clays have shown superior activity as catalyst in comparison to others in the alkylation and etherification reactions (Tables 6.6 to 6.10).

Table 6.6. Efficacies of different catalysts in alkylation of aniline with tert-butanol.

Catalyst	% Conversion of aniline	% Product selectivity	
		2-TBA	4-TBA
DTP/K-10	34	50	50
DTP/Filtrol-24	31	64	36
K-10	19	58	42
Filtron-24	22	50	50
DTP	31	61	39
DTS	28	57	43
Al pillared clay	21	66	34
KSF	21	58	42
S-ZrO$_2$	14	51	49
Na montmorillonite	10	55	45

Reaction conditions: aniline: tert-butanol 1:4 mol, temperature 150°C, catalyst loading 0.05 g/c^3, speed 1,000 rpm, time: 4 hours, autogeneous pressure: 200 psi, autoclave: 100 ml.

Table 6.7. Efficacies of different catalysts in alkylation of aniline with MTBE.

Catalyst	% Conversion of aniline	% Product selectivity		
		2-TBA	4-TBA	2,4-DTBA
DTP/K-10	70	53	31	16
DTP/Filtrol-24	51	54	28	18
DTP/KSF	56	56	32	12
DTP/SWy2	47	53	36	11
Filtrol-24	35	43	41	16
K-10	40	44	37	19
Al pillared clay	37	67	26	7
DTP	52	63	29	8
DTS/K-10	56	58	34	8
DTS	48	60	31	9
$AlCl_3/FeCl_3$/K-10	41	54	36	10
$S-ZrO_2$	26	53	36	11
KSF	24	45	45	10
Na montmorillonite	18	45	42	13

Reaction conditions: aniline: MTBE 1:4 mol, catalyst loading 0.05g/cm^3, temperature 175°C, speed 1,000 rpm, time: 4 hours, autogeneous pressure: 350 psi, autoclave: 100 ml.

Table 6.8. Effect of different catalyst on alkylation of phenol with MTBE.

Catalyst	% Conversion of phenol	Per cent selectivity of butylated phenol				
		2-TBP	4-TBP	2,4-DTBP	2,6-DTBP	2-/4-ratio
DTP (20 %)/K-10	68	38	38	22	2	1.0
DTP	55	4	5	86	5	0.8
K 10	52	33	39	25	3	0.84
$S-ZrO_2$	30	40	48	10	2	0.83
$ZnCl_2$ (20%)/K-10	32	38	42	19	1	0.90
Na montmorillonite	23	53	47	–	–	1.12
Al exchanged K-10	20	48	45	7	–	1.06
Zr exchanged K-10	14	52	45	3	–	1.15
Cr exchanged K-10	11	59	41	–	–	1.43

Reaction conditions: phenol: MTBE 1:2 mol, catalyst loading 0.04 g/cm^3, temperature: 150°C, speed 1,000 rpm, time: 4 hours.

Table 6.9. Effect of alkylating agent on alkylation of phenol.

Alkylating agent	% Conversion of phenol	% Product distribution				
		2-TBP	4-TBP	2,4-DTBP	2,6-DTBP	2-:4-ratio
MTBE	68	38	38	22	2	1.0
Tert-butanol	55	50	43	7	–	1.16

Reaction conditions: phenol: MTBE or tert-butanol 1:2 mol, catalyst: 20% DTP/K-10, catalyst loading: 0.04 g/cm^3, temperature: 150°C, speed 1,000 rpm, time: 4 hours; DTP: dodecatugstrophosphoric acid; TBP: tert-butylphenol; DTBP: di-*tert*-butylphenol.

Table 6.10. Alkylation of hydroquinone with MTBE.

Catalyst	% Conversion	% Selectivity of	
		2-TBHQ	2,5-DTBHQ
DTP (20%)/K-10	46	72	28
TSA (20%)/K-10	42	42	58
K-10 montmorillonite	29	100	–
$ZnCl_2$ (20%)/K-10	25	100	–
S-ZrO_2	26	100	–
DTP/$ZnCl_2$/K-10	24	89	11

Reaction conditions: hydroquinone: MTBE 1:3 mol, catalyst loading: 0.014 g/cm^3, temperature: 150°C, speed: 1,000 rpm, solvent: 1,4-dioxane, reaction time: 4 hours; autogeneous pressure: 150 psi, DTP: dodecatungsto phosphoric acid; TSA: tungsto silicic acid; TBHQ: tert-butylhydroquinone; DTBHQ: di-tert-butylhydroquinone.

MTBE was found to be very effective *tert*-butylating agent *vis a vis* tert-butanol 20 per cent DTP/K-10 was found to be the best catalyst for the alkylation of aniline. The alkylation of hydroqunione was studied with MTBE and *tert*-butanol as alkylating agents with a variety of catalysts amongst which 20 per cent w/w dodecatungstrophosphoric acid on K-10 (DTP/K-10) was found to be the best catalyst. The rate of alkylation of hydroquinone with MTBE was much faster then that with *tert*-butanol. The reaction mechanism involves weak adsorption of MTBE on the catalyst followed by surface reaction with hydroquinone leading to typical second order kinetics at a fixed catalyst loading.

Reactors and process engineering

Innovation in processes can be achieved through reactor design and making the processes continuous. In contrast to the traditional stirred tank, several other combinations appear promising in pharmaceutical and fine chemical industries. The range of options available include:

1. Fixed bed of catalysts with film flow or flooded, loop reactors.
2. Fluidised beds.
3. Moving beds.
4. Multi-stage suspended beds.
5. Pipeline reactors with static mixers.
6. Distillation column reactors for equilibrium limited reactions.
7. Catalysts in tea-bags.
8. Catalyst-coated structured packings.
9. Honey-comb catalysts.

The conventional stirred tank reactor can be modified in several ways to improve mass transfer and heat transfer rates, via new designs of energy saving impellers, leading to drastic reduction in batch times, favourable mixing environments for microbes and enzymes without affecting their activity and selectivity.

Spargers have been efficiently designed to aid mass transfer rates and reactors and separators have been combined (combo-reactors); for instance, reactive distillation, membrane reactors for overcoming thermodynamic barriers on conversions and yields.

The role of mixing and micro-mixing are not properly understood in several reactions which are typically conducted in laboratory in round bottom flasks without proper suspension of catalyst or dispersion of immiscible liquids in multi-phase reactions. The data so collected are invariably inadequate for scale-up and lead to lower yields or different product distribution altogether on plant scale. In fact, a multipronged approach is essential in process development with a proper flow-sheet and optimisation of a number of variables simultaneously starting from a microreactor, bench-scale prototype, pilot plant, and commercial reactor. Such an optimisation must take into account reaction chemistry, kinetics, molecular modelling, hydrodynamics, safety, hazard and process control. The role of computers for control and optimisation and computational fluid dynamics will be very critical.

For making pharmaceutical and speciality chemicals in a multi-product plant, retro-synthesis, chemical tree building and batch scheduling will also be the key elements, where experience gathered in handling hazardous but atom efficient routes will come handy.

Eco-friendly solvents

The use of solvents has to be properly understood and a variety of eco-friendly solvents including water and supercritical carbon dioxide should be considered.

The role of microemulsions with water is also a beneficial way of conducting reactions. Various ecofreindly solvents are aqueous blends, oxygenates, terpenes, N-methyl-2-pyrrolidone, propylene glycol, methyl isopropyl ketone and methyl amyl ketone, ethyl acetate, n-butyl, isopropyl and n-propyl acetate, hybrid aqueous-non aqueous solvents, γ-butyrolactone, terpene d-limonene (citrus-peel derived), tropical derived esters, aliphatic and cycloaliphatic hydrocarbons, isopropyl alcohol, butyl acrylate, 3-ethoxy propionate, butyl propionate, pentyl propionate, dimethyl glutarate, dimethyl succinate, dimethyl adipate and methylal.

Continuous processes

Continuous processes are inherently safer not only due to low inventory of reactants, but also due to improvements in selectivity. Stirred reactors and tubular reactors of small capacity provide high heat transfer rates, withstand high pressures, can be stopped easily and in some cases can be operated adiabatically without leading to run-away situations by introduction of pre-cooled reactants.

In the manufacture of intermediates for vitamin A, a nozzle reactor has been used adiabatically.

Limitations of thermodynamic equilibrium have been overcome in multistage sparged reactors, e.g. in the manufacture of phthalimide from molten phthalic anhydride and NH_3. The production of hazardous peroxy esters from the corresponding acid chloride and hydroperoxide is much safer in a continuous mode as also the production of 4,4'-bipyridyl from pyridine.

Miniature technology has been exploited by using microreactors, which are 3-D structures imbedded like microchip circuits. The principle advantages of microchannel reactors are enhanced heat and mass transfer rates, suitability for highly exothermic and hazardous reactions with mass transfer limitations leading to unstable products.

Zero discharge processes

Replacement of a large number of liquid acids by solid acid catalyst in alkylation, acylation, isomerisation, esterification, nitration, hydration has been now widely studied leading to zero discharge processes. Reduction of nitro compounds with hydrogen rather than iron/acid, and epoxidations with hydrogen peroxide or hydroperoxides, rather than using chlorohydrin or peracetic acid, lead to practically zero discharge.

m-Phenoxybenzaldehyde can be made using a safer and cleaner process based on m-phenoxytoluene, which in turn is obtained by vapour phase dehydration of m-cresol and phenol, instead of the polluting process based on bromination of benzaldehyde followed by Ullmann reaction with phenol in the presence of copper catalyst.

2,4-Dihyroxyacetophenone can be synthesised from resorcinol and acetic acid in diisopropyl ether over Amberlyst-15. The synthesis of paracetamol is typically done starting from p-chloronitrobenzene or phenol in 3–4 stages involving high raw material cost and effluent costs. Instead, the oximation of p-hydroxyacetophenone followed by rearrangement by the Celanese process looks attractive.

The Friedel-Crafts acrylation of aromatics is the method of choice in today's organic chemistry for synthesising aromatic ketones as reactive intermediates for the production of fine chemicals. The conventional method of preparation of these aromatic ketones is the homogeneous Friedel-Crafts acrylation of aromatic hydrocarbons with carboxylic acid derivatives using Lewis acids ($AlCl_3$, $FeCl_3$, BF_3, $ZnCl_2$, $TiCl_4$, $ZrCl_4$) or Brönsted acid (polyphosphoric acid, HF). For this purpose, stoichiometric-to-excess amounts of the catalyst are required for the reaction to proceed.

On the industrial scale, the use of metal halide type of acids, which are preferred catalyst, creates work-up and effluent problems. Indeed, during the work-up of acrylation mixtures, catalysts are destroyed which produce relatively large amounts of hydrochloric acid in the off-gas or in the effluent. This hydrochloric acid, which is to be disposed off, originates both from the catalyst and from acryl chloride employed for the acrylation. In addition, to this disposal as a considerable environmental problem, the corrosion problem, due to the hydrochloric acid, must be solved. Pressure from legislative and environmental bodies together with a growing awareness within the chemical industry led to a search for new eco-freindly processes to replace unacceptable outdated reactions. Therefore, a process that could be environmentally friendly and inexpensive, with respect to the disadvantage indicated, is clearly desirable. Solid acids that give the desired level of activity, but which can be easily removed from the reaction mixture with no residual inorganic contamination of the organic products, offer behaviour advantage over existing methods.

A 100 per cent atom economical process for the preparation of cyclohexyl esters from the carboxylic acids and cyclohexene over ion exchange resin catalysts where all atoms of the reactants are utilised without any by-product formation is given in Table 6.11.

Table 6.11. Effect of different substrates.

Acid	% Conversion
Phenylacetic acid	74
Benzoic acid	39
Butyric acid	11
Iso-butyric acid	10
Anthranilic acid	15
Acetic acid	30
Propionic acid	21
Heptanoic acid	11

Reaction conditions: acid: cyclohexene, 1:3 mol, time: 6 hours, catalyst: Amberlyst-15, solvent: cyclohexane, catalyst loading: 0.015 g/cm^3, temperature 100°C speed: 1,000 rpm.

The preparation of perfumery esters such as cyclohexyl-phenyl acetate, cyclohexyl acetate, cyclohexyl anthranilate, cyclohexyl benzoate, cyclohexyl butyrate, cyclohexyl isobutyrate, cyclohexyl heptoate, and cyclohexyl propionate is covered. A complete theoretical and experimental analysis is presented for the model studies with phenylacetic acid and cyclohexene.

Esterification

For the esterification of phthalic anhydride with various alcohols such as isoamyl alcohol, *n*-butanol and 2-ethylhexanol, a large number of catalysts have been reported; for instance, H_2SO_4, *p*-toluenesulphonic acid (PTSA), methanesulphonic acid (MSA), hydrochloric acid and phosphoric acid, which are all liquid-phase catalysts.

Dialkyl phthalates have been prepared from the reaction of phthalic anhydride with alkanols by using various solid acids. Esters of salicylic acid such as methyl salicylate and isoamyl salicylate are known to be very good perfumery products used in industry. Methyl salicylate can be prepared from methanol and salicylic acid by using ion-exchange resins as the catalyst.

Ultrasound-promoted acylation of yara-yara in the presence of solid-acid catalysts

Acylation of 2-methoxynaphthalene with acetic anhydride can be carried out by using different solid acid catalysts such as zeolites, acid activated clays, ion-exchange resins, sulphated zirconia and triflic acid supported zeolites under irradiation by sonic waves. The product obtained is 2-acetyl-1-methoxy naphthalene. Acrylation of anisole gives products, which are extremely useful in fine chemical industry.

Acrylation of anisole with acetic anhydride is carried out by using triflic acid supported on mesoporous molecular sieve (HMS), under irradiation by sonic waves. The final product obtained is 4-acetyl-1-methoxy benzene and the selectivity towards *para*-product is more than 95 per cent.

Thus green chemistry has demonstrated its effectiveness in achieving its mission of meeting environmental and economical goals simultaneously. In order to realise the potential of green chemistry science, technology and commerce, every chemist has to develop products, processes and services in a sustainable manner to improve the quality of life, the natural environment and industry competitiveness. green chemistry issues are here to stay. The most successful chemical companies of the future will be those who exploit its opportunities to their competitive advantages, and the most successful chemists of the future will be those who use green chemistry concepts in R & D, innovation and education.

Thus, many unit processes involve use of sulphuric acid as a reaction medium, like nitration, diazotisation, hydrolysis, methylation etc. These processes generate huge quantities of sulphuric acid containing mother liquor. It also contains unreacted raw materials, finished product, organic and inorganic impurities. It is possible to recycle such mother liquor back in the process and at the same time increase process efficiency further. To have a sustainable planet, chemists, chemical engineers, scientists, academicians, government, policy makers, and common man, together with entire industry and research organisations need to take on a new conscience, a 'green conscience'.

GREEN CHEMISTRY TO PRODUCE AMINES

Chemists at UC riverside have discovered an inexpensive, clean and quick way to prepare amines—nitrogen-containing organic compounds derived from ammonia that have wide industrial applications such as solvents, additives, anti-foam agents, corrosion inhibitors, detergents, dyes and bactericides. Currently, industries produce amines in a costly two-step process that results in massive amounts of by-products as waste. Although there are several methods to prepare amines on laboratory scales, most of them are not suitable for commodity chemical production not only because of the formation of waste materials, but also because the cost of the starting substances used to prepare amines is high,' said Guy Bertrand, a distinguished professor of chemistry, whose laboratory made the discovery.

Catalyst Does the Trick

Betrand explained that, currently, companies use hydrochloric acid, a highly corrosive solution, to produce amines. To generate one ton of amines, manufacturers must discard three tons of by-products, adding to the overall cost of production. However, 'green chemistry' method, produces no waste, which makes it inexpensive, Bertrand said. Moreover, the reaction is a quick one-step reaction, and you need a tiny amount of a catalyst to do the trick.

The catalyst in question—a gold atom linked to a cyclic alkyl amino carbene (CAAC)—is a ligand that Bertrand's laboratory discovered in 2005. The gold compound readily catalyses the addition of ammonia—a colourless, pungent gas composed of nitrogen and hydrogen—to a number of organic compounds. One such chemical reaction involves ammonia combining with acetylene to produce an amine derivative; a carbon-nitrogen bond is created in this reaction. One of the greatest challenges in chemistry is to develop atom-efficient processes for the combination of ammonia with single organic molecules to create carbon-nitrogen bonds, Bertrand said.

Atom-efficient Processes

In atom-efficient processes, the amount of starting materials equals the amount of all products generated, with no atoms wasted. More than 100 MT of ammonia are produced annually in the world, and the production of amines similarly is huge. Essential to life as constituents of amino acids, amines occur in drugs and vitamins, and are used also to manufacture cosmetics, cleaning and crop protection agents, plastics, and coating resins.

The study paves the way for finding catalysts that mediate the addition of ammonia to simple alkenes, which are organic compounds containing a carbon-carbon double bond, Bertrand said. This process is widely considered to be one of the ten greatest challenges for catalytic chemistry. UCR's Office of Technology Commercialisation has filed a patent application on the new catalyst developed in Bertrand's laboratory, and is seeking commercial partners to develop it.

GREENER ROUTES TO PROPYLENE OXIDE—AN OVERVIEW

Propylene oxide (PO), derived from propylene, is a key global petrochemical building block for polyurethanes and for glycols. Major global producers include Dow Chemical, Lyondell, Huntsman, BASF, Repsol and Shell Chemical. Global propylene oxide (PO) demand is growing at 20 per cent annually. Hence the drive to develop and commercialise process without co-product is on anvil.

This overview looks at the prevailing commercial routes for the manufacture of propylene oxide, the latest trends in developing new routes, with an emphasis on a greener route.

Current Commercial Routes for PO

Following are the current commercial routes prevailing for the manufacture of PO:
1. Chlorohydrin with integrated chloralkali.
2. Ethylbenzene hydroperoxidation/styrene coproduct.
3. t-Butyl hydroperoxidation/t-butyl alcohol coproduct.
4. Sumitomo process.

Chlorohydrin process

This process involves reacting propylene with hypochlorous acid formed by action of chlorine with water (Fig. 6.8).

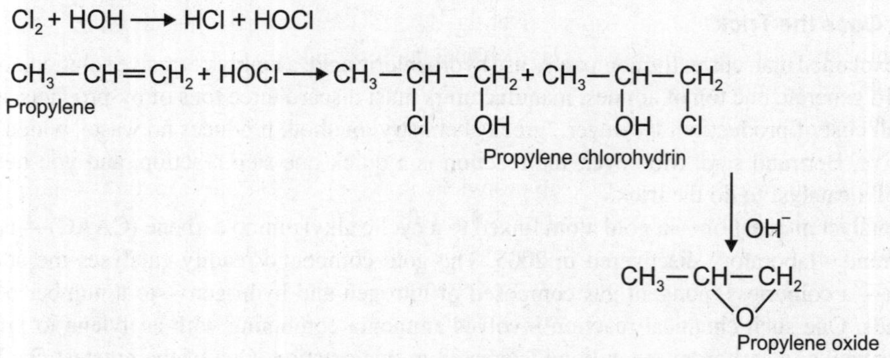

$$Cl_2 + HOH \longrightarrow HCl + HOCl$$

$$\underset{\text{Propylene}}{CH_3-CH=CH_2} + HOCl \longrightarrow \underset{\underset{\text{Propylene chlorohydrin}}{Cl \quad\;\; OH}}{CH_3-\underset{|}{CH}-\underset{|}{CH_2}} + \underset{OH \quad\;\; Cl}{CH_3-\underset{|}{CH}-\underset{|}{CH_2}}$$

$$\downarrow OH^-$$

$$\underset{\text{Propylene oxide}}{CH_3-CH-CH_2 \atop \diagdown O \diagup}$$

Fig. 6.8. Reaction of propylene with hypochlorous acid.

The resulting propylene chlorohydrin is hydrolysed using milk of lime or dilute sodium hydroxide, giving PO and corresponding chloride of the hydroxyl compound.

Ethyl benzene hydroperoxidation process (SMPO)

The process involves transfer of O-atom from ethylbenzene hydroperoxide to propylene in the presence of a proprietary catalyst to give PO and styrene monomer.

t-Butyl hydro peroxidation process

This is same as the above process and *t*-butyl hydro peroxide is used instead of ethylbenzene hydroperoxide to give PO and *t*-butyl alcohol.

Although significant PO capacity is based on chloro hydrin process, this route suffers from environmental liabilities and large capital investment requirements. The hydroperoxidation routes to PO that coproduce styrene monomer and *t*-butyl alcohol are responsible for majority of current global production, but require large capital investments and pose difficulties in balancing the markets for PO and coproducts leading to volatile economic performance of operations. Their operational performance is being incrementally improved (Fig. 6.9).

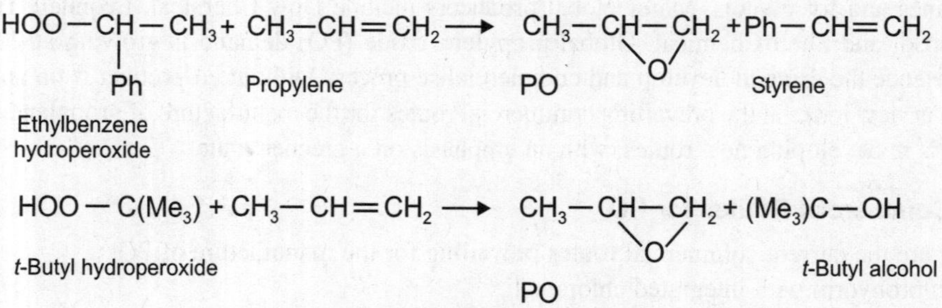

$$\underset{\substack{\text{Ethylbenzene}\\ \text{hydroperoxide}}}{HOO-\underset{\underset{Ph}{|}}{CH}-CH_3} + \underset{\text{Propylene}}{CH_3-CH=CH_2} \longrightarrow \underset{PO}{CH_3-CH-CH_2 \atop \diagdown O \diagup} + \underset{\text{Styrene}}{Ph-CH=CH_2}$$

$$\underset{\text{t-Butyl hydroperoxide}}{HOO-C(Me_3)} + CH_3-CH=CH_2 \longrightarrow \underset{PO}{CH_3-CH-CH_2 \atop \diagdown O \diagup} + \underset{\text{t-Butyl alcohol}}{(Me_3)C-OH}$$

Fig. 6.9. *t*-Butyl hydroperoxide process.

Sumitomo process

Sumitomo has patented the use of cumene hydroperoxide (CHP) for epoxidation of propylene. The CHP is obtained by liquid phase oxidation of cumene with air by free radical mechanism. Selectivity of cumene to CHP is 95–98 per cent.

The use of CHP has potential advantages over the other hydroperoxidation such as:
1. Stability of CHP, as compared to ethylbenzene hydroperoxide.
2. Smaller and simpler equipments.
3. Reduction in capital cost.

In the epoxidation, a molar ratio of propylene to CHP 10:1 is used at a reaction temperature of 60°C and sufficient pressure is maintained to keep propylene in liquid phase. The reaction is catalysed by a proprietary silylated Ti-containing SiO catalyst. Selectivity to PO based on CHP is 95 per cent whereas selectivity to PO based on propylene is 99 per cent.

CHP is converted to cumyl alcohol, which is dehydrated to alpha-methyl styrene, which in turn is hydrogenated over multiple beds of Cu-Cr oxide catalyst to give back cumene. The chemistry is shown below.

Newer Routes for PO Manufacture for a Greener Environment

Direct oxidation of propylene with oxygen

PO should be attainable by direct oxidation process analogous to that of ethylene oxide. The challenge in the case of propylene oxidation has been to find a catalyst that gives sufficiently high selectivity to PO by suppressing the competing combustion reaction.

$$CH_3-CH=CH_2 + \tfrac{1}{2}O_2 \rightarrow CH_3-CH-CH_2 \text{ (PO, epoxide)}$$

Recent ARCO patent, sites the use of silver-based catalyst supported on alkaline earth metal support in presence of tungsten and potassium. The best results were 8.8 per cent propylene conversion with 53 per cent selectivity of propylene to PO.

US patent 2,985,668 (May 23, 1961) Mr. Haruo Shingu involves bringing a vapour mixture of propylene and an elemental oxygen containing gas, in contact with finely divided CuO/Ag/AgO suspended in organic solvent (having boiling point greater than the products) at 150–230°C. The PO is recovered as effluent vapour mixture from liquid phase.

Plant design for direct oxidation of propylene would likely be based on pure oxygen feed to gain yield advantage and lower battery limit capital cost.

Hydro-oxidation of propylene

The process involves one mole of each propylene, hydrogen and oxygen reacting to give one mole of PO and water.

$$CH_3 - CH = CH_2 + O_2 + H_2 \longrightarrow PO + water$$

Competing side reactions include propylene hydrogenation to propane and propylene combustion to CO_2 and water. Additional hydrogen and oxygen consumption occurs to form water.

US patent 6,710,194 B1 to ARCO describes the following:
1. Reaction medium: methanol.
2. Catalyst: 0.4 per cent Pd coated over $x\text{TiO}_2(1-x)\text{SiO}_2$, where, x is 0.01 to 0.125.
3. Feed: 20 per cent v/v propylene, 20 per cent v/v hydrogen and 60 per cent v/v propane (ballast gas).

4. Reaction condition: 60°C and 12.8 bar pressure.
5. Conversion rate: 53 per cent and 90 per cent selectivity.

Bayer inventors (US patent 6,504,039 B2) claim high selectivity of PO from propylene by hydro-oxygenation process as follows:
1. Reactor: Fixed bed tube reactor.
2. Temperature: 46°C.
3. Feed: $O_2/H_2/C_3H_6$ 0.1/1.3/0.4 l/hr.
4. Catalyst: Ti oxide with 0.2–1.0 wt per cent sulphate content coated with <6 nm gold particles (0.1–1.0 wt per cent).
5. Yield: 3.4–5.8 per cent per pass and >97 per cent selectivity.

Similar detailed study has been done using Ag/TS-1 catalyst system by Xinwen Guo.

Epoxidation with in situ hydrogen peroxide

Invention of Titanium silicate (TS-1) catalyst by Taramasso (US patent 4,410,501) opened a new route for synthesis of PO according to Enichem technology using hydrogen peroxide as oxidant.

The approach makes use of conventional anthraquinone (AHQ) process for production of hydrogen peroxide by feeding propylene and oxygen to the oxidation stage of AHQ. This leads to the formation of hydrogen peroxide, which is then consumed in oxidation of propylene to PO catalysed by TS-1 (Fig. 6.10).

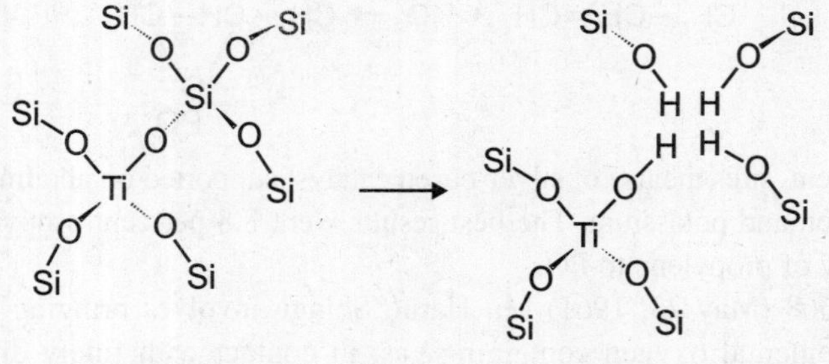

Fig. 6.10. Schematic view of a fully tetrahedrally bonded Ti ion substituted for a Si ion at one of the lattice position of TS-1 (a) and the Same Ti site located near a silicon vacancy terminated with hydrogen atoms forming a silanol nest (b).

Epoxidation with purchased hydrogen peroxide

The direct use of hydrogen peroxide as reagent is desirable for its mild reaction conditions and fewer co-products except water as co-product.

$$CH_3-CH=CH_2 + H_2O_2 \longrightarrow PO + water$$

A hydroxylic solvent such as water or alcohol is generally needed, which may result in initial loss of PO due to epoxide ring opening reaction (Fig. 6.11).

Another problem associated with the process is the cost and shipping hazards of standard grades of hydrogen peroxide. Even shipping cost of diluted hydrogen peroxide is prohibitive. Cost and logistics problem can be mitigated by *in situ* hydrogen peroxide production.

Alternative approach is a large colocated hydrogen peroxide plant feeding the crude hydrogen peroxide which has not incurred concentration, purification and shipping cost. This allows raw material storage and inventory cost minimisation. This approach has been disclosed by Dow, BASF and Solvay. Another plant had been planned in S. Korea using UHDE PO technology and Degussa Headwaters hydrogen peroxide technology. This process generally involves the use of TS-1 type catalyst.

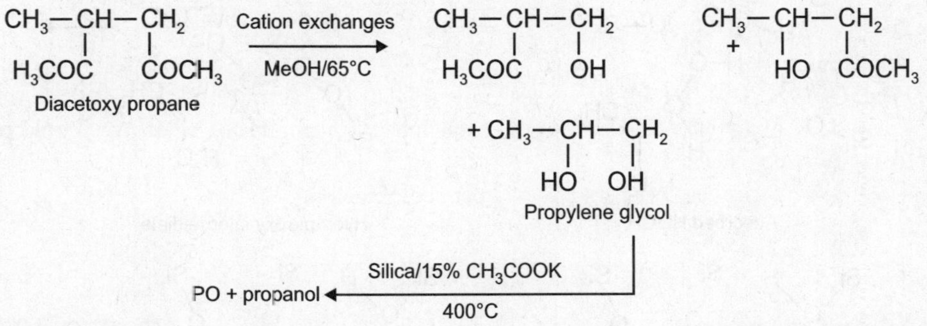

$$\underset{\text{Diacetoxy propane}}{\underset{|\;\;\;\;\;\;\;\;\;\;\;|}{CH_3-CH-CH_2}} \xrightarrow[\text{MeOH/65°C}]{\text{Cation exchanges}} \underset{H_3COC\;\;\;\;OH}{\underset{|\;\;\;\;\;\;\;\;\;|}{CH_3-CH-CH_2}} + \underset{HO\;\;\;\;COCH_3}{\underset{|\;\;\;\;\;\;\;\;\;|}{CH_3-CH-CH_2}}$$

$$+ \underset{\text{Propylene glycol}}{\underset{HO\;\;\;\;OH}{\underset{|\;\;\;\;\;|}{CH_3-CH-CH_2}}}$$

$$\text{PO + propanol} \xleftarrow[400°C]{\text{Silica/15% } CH_3COOK}$$

Fig. 6.11. Reaction of diacetoxy propane products formed are propylene glycol, silica, propylene oxide and propanol.

The Ti-OOH species formed by interaction between framework Ti-atom and hydrogen peroxide molecule transfer O-atom to propylene to give PO.

TS-1 develops acid centres in presence of hydrogen peroxide in alcoholic and aqueous solutions, which catalyse epoxide ring opening.

So use of anhydrous hydrogen peroxide may help to avoid the side reaction with urea-H_2O_2 adduct (UHP).

The major advantage of UHP lies in controlled release of H_2O_2 into solution. This was demonstrated by epoxidation of cyclohexene over TS-1 catalyst in acetonitrile solvent using UHP and found satisfactory. This may be extended to propylene epoxidation also.

Kendall T. Thomson of Purdue University has found that Ti-sites (in TS-1 catalyst) located adjacent to Si-vacancies also catalyse the reaction besides fully co-ordinated Ti-sites using density functional theory (DFT).

The epoxidation near a Si-vacancy occurs through a sequential pathway where hydrogen peroxide first forms a hydroperoxy intermediate Ti-OOH and then reacts with propylene by proximal oxygen abstraction.

The transition state for this step exhibits a six-fold oxygen co-ordination on Ti, which is due to less constraint environment of Ti adjacent to vacancy site (Fig. 6.12).

Other methods cited in Patents

US patent 4,226,780 to BASF involves hydrolysis of diacetoxy propane to acetoxy propane and 1,2-propanediol with cation exchanger in methanol in stirred reactor/distillation unit at 65°C.

The product mixture is then fed in tubular reactor/vapouriser with silica (2–3 mm particle size) and potassium acetate and heated to 400°C. The distillate contains PO and propanal.

US patent 6,403,840 B1 to GRT. Inc (Texas USA), involves reaction of propylene bromohydrin and /or dibromopropane with metal oxide to form PO. The metal bromide is converted to form original metal oxide and bromine, both of which are recycled.

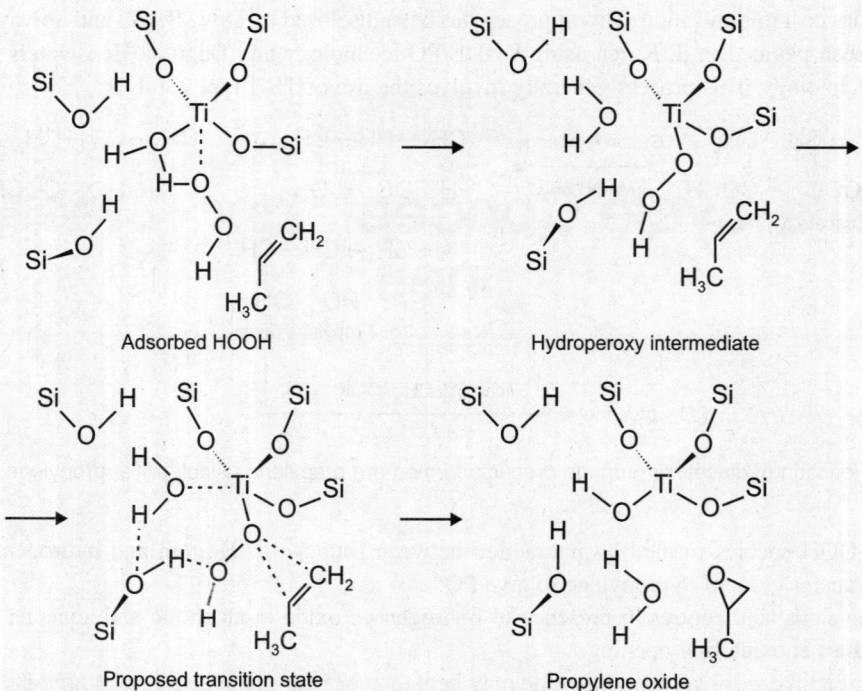

Fig. 6.12. Mechanism for HOOH/propylene epoxidation to PO on a defective Ti site in TS-1.

Chapter 7

Life-cycle Concepts and Green Engineering

INTRODUCTION

Life-cycle studies are a uniquely useful tool for assessing the impact of human activities. These impacts can only be fully understood by assessing them over a life cycle, from raw material acquisition to manufacture, use, and final disposal. Life-cycle techniques have been adopted in industry and the public sector to serve a variety of purposes, including product comparison, strategic planning, environmental labelling, and product design and improvement.

Life-cycle assessments have four steps. The first is scoping, where boundaries are determined and strategies for data collection are chosen. The second step is an inventory of the inputs and outputs of each life-cycle stage. Next is an impact assessment, where the effects of the inputs and outputs are evaluated. The final step is an improvement analysis. Even for simple products, comprehensive life-cycle studies require a great deal of time and effort. Also, no matter how much care is taken in preparing a study, the results obtained have uncertainty.

Products, services, and processes all have a life cycle. For products, the life cycle begins when raw materials are extracted or harvested. Raw materials then go through a number of manufacturing steps until the product is delivered to a customer. The product is used, then disposed of or recycled. These product life-cycle stages are illustrated in Fig. 7.1, along the horizontal axis. As shown in the figure, energy is consumed and wastes and emissions are generated in all of these life-cycle stages.

Processes also have a life-cycle. The life-cycle begins with planning, research and development. The products and processes are then designed and constructed. A process will have an active lifetime, then will be decommissioned and if necessary, remediation and restoration may occur. Figure 7.1, along its vertical axis, illustrates the main elements of this process life cycle. Again, energy consumption, wastes, and emissions are associated with each step in the life-cycle.

Traditionally, product and process designers have been concerned primarily with product life-cycle stages from raw material extraction up to manufacturing. That focus is changing. Increasingly, chemical product designers must consider how their products will be recycled. They must consider how their customers use their products.

Process designers must avoid contamination of the sites at which their processes are located. Simply stated, design engineers must become stewards for their products and processes throughout their life cycles. These increased responsibilities for products and processes throughout their life cycles have been recognised by a number of professional organisations. Table 7.1 describes a code of product stewardship developed by the chemical manufacturers association (now named the american chemistry council).

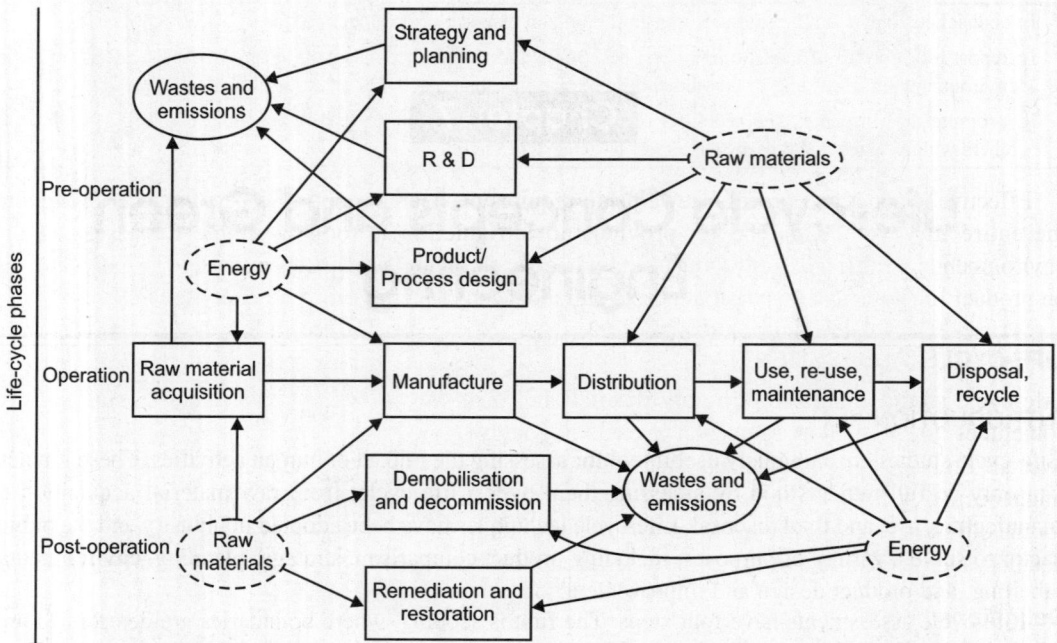

Fig. 7.1. Product life-cycles include raw material extraction, material processing, use, and disposal steps, and are illustrated along the horizontal axis. Process life-cycles include planning, research, design, operation, and decommissioning steps and are shown along the vertical axis. In both product and process life-cycles, energy and materials are used at each stage of the life-cycle and emission and wastes are created.

Table 7.1. The chemical manufacturers association (American Chemistry Council) product stewardship code.

The purpose of the Product Stewardship Code of Management Practices is to make health, safety and environmental protection an integral part of designing, manufacturing, marketing, distributing, using, recycling and disposing of our products. The code provides guidance as well as a means to measure continuous improvement in the practice of product stewardship

The scope of the code covers all stages of a product's life. Successful implementation is a shared responsibility. Everyone involved with the product has responsibilities to address society's interest in a healthy environment and in products that can be used safely. All employers are responsible for providing a safe workplace, and all who use and handle products must follow safe and environmentally sound practices

The code recognises that each company must exercise independent judgment and discretion to successfully apply the code to its products, customers and business

Relationship to guiding principles

Implementation of the code promotes achievement of several of the Responsible Care Guiding Principles:

    To make health, safety and environmental considerations a priority in our planning for all existing and new products and processes

    To develop and produce chemicals that can be manufactured, transported, used and disposed of safely

    To extend knowledge by conducting or supporting research on the health, safety and environmental effects of our products, processes and waste materials

(Contd...)

- To counsel customers on the safe use, transportation and disposal of chemical products
- To report promptly to officials, employees, customers and the public, information on chemical-related health or environmental hazards and to recommend protective measures
- To promote the principles and practices of responsible care by sharing experiences and offering assistance to others who produce, handle, use, transport or dispose of chemicals

Effective product and process stewardship requires designs that optimise performance throughout the entire life-cycle. This chapter provides an introduction to tools available for assessing the environmental performance of products and processes throughout their life-cycle. The primary focus is on product life-cycles, but similar concepts and tools could be applied to process life-cycles.

LIFE-CYCLE ASSESSMENT

Life-cycle studies range from highly detailed and quantitative assessments that characterise, and sometimes assess, the environmental impacts of energy use, raw material use, wastes, and emissions over all life stages, to assessments that qualitatively identify and prioritise the types of impacts that might occur over a life-cycle. As shown in this chapter, different levels of detail and effort are appropriate for the different ways the life-cycle information is used. In this section, the steps involved in conducting detailed, highly quantitative life-cycle assessments are described.

Definitions and Methodology

There is some variability in life-cycle assessment terminology, but the most widely accepted terminology has been codified by international groups convened by the society for environmental toxicology and chemistry (SETAC). Familiarity with the terminology of life-cycle assessment makes communication of results easier and aids in understanding the concepts presented later in this chapter. To begin, a life-cycle assessment (LCA) is the most complete and detailed form of a life-cycle study. A life-cycle assessment consists of four major steps.

Step 1: The first step in an LCA is to determine the scope and boundaries of the assessment. In this step, the reasons for conducting the LCA are identified; the product, process, or service to be studied is defined; a functional unit for that product is chosen; and choices regarding system boundaries, including temporal and spatial boundaries, are made. But what is a functional unit, and what do we mean by system boundaries? Let's look first at the system boundaries.

The system boundaries are simply the limits placed on data collection for the study. The importance of system boundaries can be illustrated by a simple example. Consider the problem of choosing between incandescent light bulbs and fluorescent lamps in lighting a room. During the 1990s the US EPA began its green lights programme, which promoted replacing incandescent bulbs with fluorescent lamps. The motivation was the energy savings provided by fluorescent bulbs.

Like any product, however, a fluorescent bulb is not completely environmentally benign, and a concern arose during the green lights programme about the use of mercury in fluorescent bulbs. Fluorescent bulbs provide light by causing mercury, in glass tubes, to fluoresce. When the bulbs reach the end of their useful life, the mercury in the tubes might be released to the environment. This environmental concern (mercury release during product disposal) is far less significant for incandescent bulbs. Or is it? What if we changed our system boundary? Instead of just looking at product disposal, as shown in the first part of Fig. 7.2, what if the entire product life cycle were considered, as shown in the second part of Fig. 7.2.

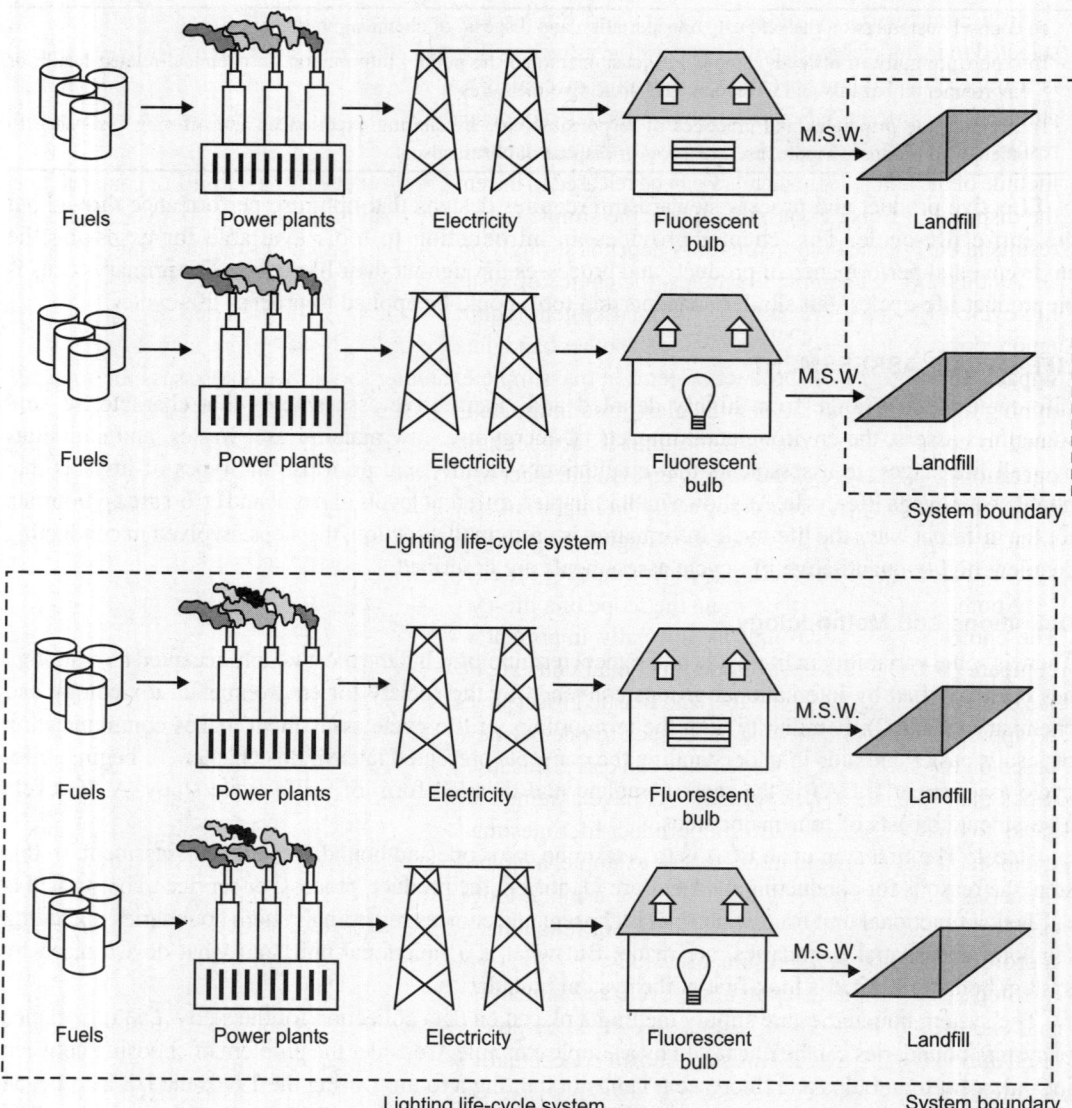

Fig. 7.2. The importance of system boundaries in life-cycle assessment is illustrated by the case of lighting systems. As noted in the text, fluorescent bulbs contain mercury and if these bulbs are sent directly to municipal solid waste landfills, mercury might be released into the environment. Use of incandescent bulbs would result in a smaller amount of mercury in the municipal solid waste stream. Thus, an analysis focusing on just municipal solid waste disposal would conclude that fluorescent bulbs release more mercury to the environment than incandescent bulbs. If a larger system is considered, however, the conclusion changes. Mercury is a trace contaminant in coal, and when coal is burned to generate electricity, some mercury is released to the atmosphere. Since an incandescent bulb requires more energy to operate, the use of an incandescent bulb can result in the release of more mercury to the atmosphere than the use of a fluorescent bulb. Over the lifetime of the bulbs, more mercury can be released to the environment due to energy use than due to disposal of fluorescent bulbs.

In a comparison of the incandescent and fluorescent lighting systems, if the system boundary is selected to include electric power generation as well as disposal, the analysis changes. Although mercury is a trace contaminant in coal, the burning of coal is the greatest contributor of Hg releases to the atmosphere. Since an incandescent bulb requires more energy to operate, the use of an incandescent bulb results in the release of more mercury to the atmosphere than the use of a fluorescent bulb. Over the lifetime of the bulbs, more mercury can be released to the environment due to the burning of coal than due to the disposal of fluorescent bulbs. Thus, the simple issue of determining which bulb, over its life-cycle results in the release of more mercury depends strongly on how the boundaries of the system are chosen.

As this simple example illustrates, the choice of system boundaries can influence the outcome of a life-cycle assessment. A narrowly defined system requires less data collection and analysis, but may ignore critical features of a system. On the other hand, in a practical sense it is impossible to quantify all impacts for a process or product system. In our simple example, should we also assess the impacts of mining the metals, and making the glass used in the bulbs we are analysing? In general, we would not need to consider these issues if the impacts are negligible, compared to the impacts associated with operations over the life of the equipment. On the other hand, for specific issues, such as mercury release, some of these ancillary processes could be important contributors. What is included in the system and what is left out is generally based on engineering judgement and a desire to capture any parts of the system that may account for 1 per cent or more of the energy use, raw material use, wastes or emissions.

Another critical part of defining the scope of a life-cycle assessment is to specify the functional unit. The choice of functional unit is especially important when life-cycle assessments are conducted to compare products. This is because functional units are necessary for determining equivalence between the choices. For example, if paper and plastic grocery sacks are to be compared in an LCA, it would not be appropriate to compare one paper sack to one plastic sack. Instead the products should be compared based on the volume of groceries they can carry. Because fewer groceries are generally placed in plastic sacks than in paper sacks, some LCAs have assumed a functional equivalence of two plastic grocery sacks to one paper sack. Differing product lifetimes must also be evaluated carefully when using life-cycle studies to compare products. For example, a cloth grocery sack may be able to hold only as many groceries as a plastic sack, but will have a much longer use lifetime that must be accounted for in performing the LCA. However, the choice of functional unit is not always straightforward and can have a profound impact on the results of a study.

Step 2: The second step in a life-cycle assessment is to inventory the inputs, such as raw materials and energy, and the outputs, such as products, by-products, wastes, and emissions, that occur and are used during the life cycle. This step, shown conceptually in Fig. 7.3, is called a life-cycle inventory, and is often the most time consuming and data intensive portion of a life-cycle assessment. Examples of life-cycle inventories and more detail concerning the structure of a life-cycle inventory are provided in the next section.

Step 3: The output from a life-cycle inventory is an extensive compilation of specific materials used and emitted. Converting these inventory elements into an assessment of environmental performance requires that the emissions and material use be transformed into estimates of environmental impacts. Thus, the third step in a life-cycle assessment is to assess the environmental impacts of the inputs and outputs compiled in the inventory. This step is called a life-cycle impact assessment.

Step 4: The fourth step in a life-cycle assessment is to interpret the results of the impact assessment, suggesting improvements whenever possible. When life-cycle assessments are conducted to compare products, for example, this step might consist of recommending the most environmentally desirable product. Alternatively, if a single product were analysed, specific design modifications that could improve environmental performance might be suggested. This step is called an improvement analysis or an interpretation step.

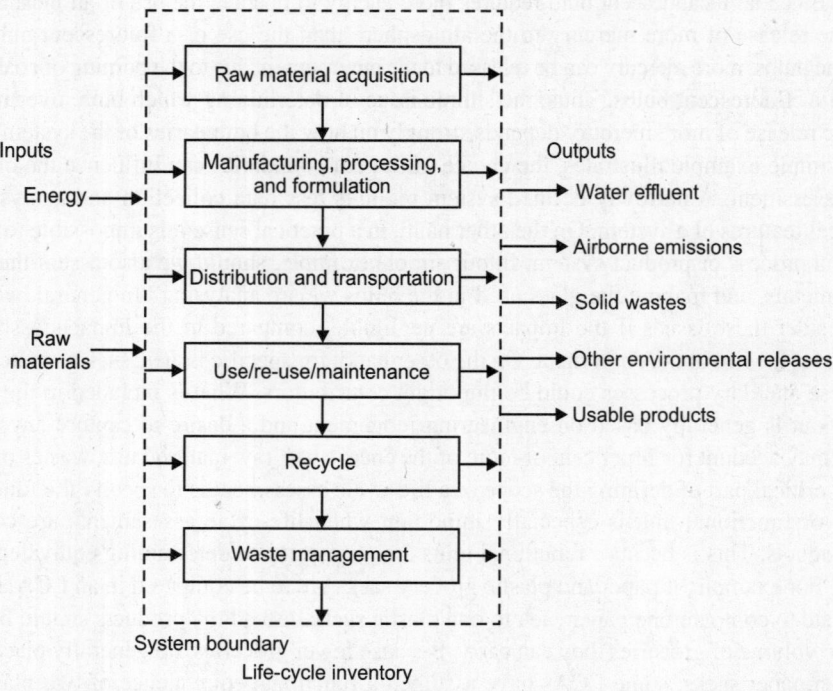

Fig. 7.3. Life-cycle inventories account for material use, energy use, wastes, emissions, and coproducts over all of the stages of a product's life cycle.

Life-cycle Inventories

A life-cycle inventory is a set of data and material and energy flow calculations that quantifies the inputs and outputs of a product life cycle. Some of the values that are sought during the inventory process are objective quantities derived using tools such as material and energy balances. As shown later in this section, other values are more subjective and depend on choices and assumptions made during the assessment.

Before describing in detail the data elements associated with a life-cycle inventory, take a moment to review the stages of a product life-cycle. The first stage in a product life-cycle, as shown along the horizontal axis of Fig. 7.1 is raw material acquisition. Examples of raw material acquisition are timber harvesting, crude oil extraction, and mining of iron ore. After raw material acquisition is the material manufacture stage, where raw materials are processed into the basic materials of product manufacture. Felled trees are processed into lumber and paper, for example. Crude oil is processed into fuels, solvents, and the building blocks of plastics. These materials move to the product manufacture stage where they are used to make the final product. In this stage, paper and plastic may be made into cups, steel turned into car bodies, or solvents and pigments turned into paints. The next stage of the life-cycle stage is use. Some products, such as automobiles, generate significant emissions and wastes during use, while other products, such as grocery sacks, have negligible material and energy flows associated with the use of the product. The final life-cycle stage consists of disposal or recycling.

Recycling can occur in several ways. A product might be reused, which is what happens when a ceramic cup is washed and reused instead of being thrown away. The product could be re-manufactured, where the materials it contains are used to make another product. A newspaper, for example, might be made into another newspaper or might be shredded and used for animal bedding. Finally, products might be recycled to more basic materials, through processes such as plastics depolymerisation or automobile disassembly which yield commodity materials such as monomers and steel.

Tracking material flows, over all of the stages of a life cycle, is required for a comprehensive life-cycle inventory. Even for a simple product made from a single raw material in one or two manufacturing steps, the data collection effort can be substantial. Table 7.2 shows a summary of an inventory of the inputs and outputs associated with the production of one kilogram of a relatively simple product: ethylene. Consider each element in the table.

Table 7.2. Life-cycle inventory data for the production of 1 kg of ethylene.

Category	Input or output	Unit average
Energy content fuels, MJ	Coal	0.94
	Oil	1.8
	Gas	6.1
	Hydroelectric	0.12
	Nuclear	0.32
	Other	<0.01
	Total	9.2
Feedstock, MJ	Coal	<0.01
	Oil	31
	Gas	29
	Total	60
Total fuel + feedstock	Iron ore	200
Raw materials, mg	Limestone	100
	Water	1,900,000
	Bauxite	300
	Sodium chloride	5400
	Clay	20
	Ferromanganese	<1
Air emissions, mg	Dust	1000
	Carbon monoxide	600
	Carbon dioxide	5,30,000
	Sulphur oxides	4000
	Nitrogen oxides	6000
	Hydrogen sulphide	10

(Contd...)

Category	Input or output	Unit average
Water emissions, mg	Hydrogen chloride	20
	Hydrocarbons	7000
	Other organics	1
	Metals	1
	Chemical oxygen demand	200
	Biological oxygen demand	40
	Acid, as H^+	60
	Metals	300
	Chloride ions	50
	Dissolved organics	20
	Suspended solids	200
	Oil	200
	Phenol	1
	Dissolved solids	500
	Other nitrogen	10
Solid waste, mg	Industrial waste	1400
	Mineral waste	8000
	Slags and ash	3000
	Nontoxic chemicals	400
	Toxic chemicals	1

The first set of data in the table are energy requirements. These are the hydrocarbon fuels and electric power sources used in extracting the raw materials for ethylene production and for running the ethylene manufacturing process (an energy-intensive operation). The next set of data elements are referred to as feedstock energy. The main raw materials of ethylene production (oil and gas) are also fuels. The energy content of this feedstock for ethylene production is reported in units of energy rather than mass (a common practice among life-cycle study practitioners) so that it can be combined with the energy that was required in the production process.

A second set of entries in the table describes non-fuel raw material use. These include iron ore, limestone, water, bauxite, sodium chloride, clay, and ferromanganese. As shown in this table, these data are often aggregated over the life-cycle and reported as aggregate quantities. Thus, water use would include water used in oil field production as well as steam used in the ethylene cracker. Some of the entries may seem obscure, but only serve to point out the complex nature of product life-cycles. For example, the limestone use is due in part to acid gas scrubbing in various parts of the product life-cycle.

A final set of inventory elements are the wastes and emissions. Some subjectivity is introduced here in deciding which materials to report. For example, some life-cycle inventories do not report the release of carbon dioxide, a global warming gas, or the use of water. Neglecting these inventory elements implies that they are not important. More subtle subjectivity can arise in defining exactly what is and what is not a waste. Consider the example of a paper plant that debarks wood. The wood that is not used in the pulp making operation is commonly burned for energy recovery within the pulping operation. Some life-cycle practitioners may count this material as a waste that is subsequently used as a fuel.

Other life-cycle practitioners might regard the material as an internal process stream. The environment does not recognise a difference between these two material accounting methods, but a life-cycle inventory applying one type of material accounting would appear to predict larger quantities of solid waste that a life-cycle inventory that employed different material accounting practices.

Take a moment to review the entries in Table 7.2 in order to obtain an idea of the level of effort necessary to inventory the inputs and outputs.

Table 7.2 provides life-cycle inventory data for a single material: ethylene. A complex product such as a computer would have a very complicated life-cycle framework. Computers are made up of many products (semiconductors, casing, display, etc.) that are themselves made from diverse materials, some of which require sophisticated manufacturing technologies. Life-cycle inputs and outputs for each of these sub-products would need to be inventoried in a life-cycle assessment of a computer.

Other complexities in life-cycle inventories arise when processes have coproducts. To illustrate the concept of co-product allocation, consider the allocation of inputs and outputs for the processes shown in Fig. 7.4. The left-hand side of this figure shows a process where one input results in two products and one type of emission. If a life-cycle inventory is being performed on one of the two products, then the input and emissions must be allocated between the two products. Part of the life-cycle of ethylene can be used as an example. Ethylene is made, in part, from a petroleum liquid referred to as naphtha. Naphtha is produced in petroleum refineries, which have crude oil as their primary input.

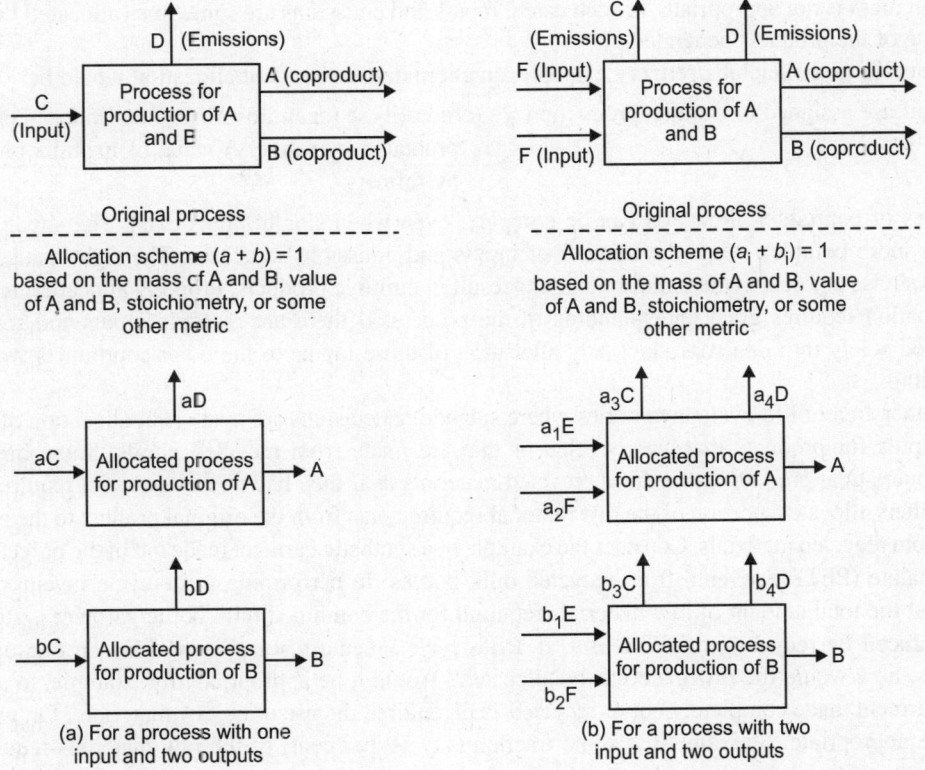

Fig. 7.4. Allocating material use, energy use, and emissions among multiple products that are manufactured in the same processes can be difficult.

The refinery produces a variety of products, including gases, gasoline, other fuels, asphalt, and the naphtha used to make ethylene. Data on emissions and crude oil usage are generally available for the refinery as a whole, and the fraction of the refinery's crude oil use and emissions due to naphtha production must generally be assigned using an allocation procedure. One commonly used allocation procedure is based on mass of products. As shown in Fig. 7.4, the input and emissions attributed to each of the products can be allocated based on the mass of the coproducts. In the naphtha/refinery example, the crude oil usage and emissions might be assigned in the following way:

Crude oil use assigned to naphtha production = (crude oil use for entire refinery/total mass of products produced by refinery) mass of naphtha produced by refinery

In most life-cycle inventories, allocation of material use, energy use, and emissions among co-products is based on mass. Sometimes, however, the coproduct is a by-product that would not be produced solely for its own merit, and allocation based on value might be more appropriate. As a graphic example, consider raising cattle, which produces beef and manure product streams. Clearly, the cattle rancher is in business to produce beef, not manure. Yet, if inputs and emissions to the cattle ranching were allocated based on the mass of the two products, most of the inputs and emissions would be assigned to the manure (other, less graphic examples could have been chosen, such as a pharmaceutical production process that creates a recyclable solvent by-product). Clearly, in some situations, allocation based on mass is not appropriate. In such cases, inputs and emissions are sometimes allocated based on the value of the products generated.

Returning to the naphtha/refinery example, an alternative co-product allocation would be:

Crude oil use assigned to naphtha production = (crude oil use for entire refinery/total value of products produced by refinery) value of naphtha produced by refinery

Issues of coproduct allocation can be complex, even when single inputs exist. The situation can become more complex when the number of inputs and emissions increases. The right-hand side of Fig. 7.4 shows a process where multiple inputs result in multiple products. Properly allocating inputs in this situation requires great understanding of the process. If there are multiple inputs and some are converted solely into one coproduct, any allocation of those inputs to the other coproducts would be misleading.

Another area of life-cycle inventories where subjective decisions are made is in allocation of inputs and outputs for products that are recycled or that are made from recycled goods. Some life-cycle practitioners treat products made from recycled materials as if they had no raw material requirements, while others allocate a portion of the raw material requirements from the original product to the product made from recycled materials. Consider the example of a synthetic garment made out of the polyethylene terephthalate (PET) recovered from recycled milk bottles. In performing a life-cycle inventory, it is clear that the total amount of raw materials required for the combined milk bottle/garment system has been reduced by recycling the PET. But, if a life-cycle inventory were to be done on one of these products, how would the raw materials be allocated? Would it be appropriate, for example, to assume that a garment made completely out of recycled PET required the use of no raw materials? Or, would it be more appropriate to assume that some fraction (say 50 per cent) of the raw materials required to produce the milk bottles should be assigned to the life-cycle inventory of the garment? There are no correct answers to these questions, and different life-cycle practitioners make different assumptions.

Sometimes these assumptions, which do not appear explicitly in tables of results such as Table 7.3, can have a significant effect on the inventory data.

Perhaps the most important uncertainty in life-cycle inventories, however, is due to the quality of data available on the processes being inventoried and the level of aggregation of the data. Overall data quality issues, such as whether data are direct measured values or are based on engineering estimation methods, are fairly straightforward to identify and deal with. Data aggregation issues can be more subtle. Consider two examples of cases where data aggregation has impacted the findings of a life-cycle inventory. A first example is provided by the comparison of electric vehicles to gasoline-powered vehicles in the Los Angeles area. Table 7.3 and Fig. 7.5 provide data summaries from two different life-cycle inventories that compared electric vehicles to gasoline-powered vehicles. The data shown in Table 7.3 indicate that driving an electric vehicle results in far less emission of reactive organic gases, nitrogen oxides, carbon monoxide and particulate matter, than driving an ultra-low emission vehicle (ULEV) an equivalent distance. In contrast, the data from Fig. 7.5 indicates that an electric vehicle emits more of certain pollutants (such as NO_2 and sulphur oxides) than a gasoline-powered vehicle (based on EPA emission data). Why is there such a dramatic difference? The answer is related to data aggregation. The data reported in Fig. 7.5 were based on a life-cycle inventory that used nationally averaged emissions data for electric power generation. In contrast, the data reported in Table 7.3 were based on a life-cycle inventory that used emissions data for electric power generation in Southern California. Since emissions from power generation in Southern California are much lower than the average for the rest of the United States, and because a large fraction of the total emissions associated with an electric vehicle are due to the power generation required to fuel the vehicle, the studies lead to very different results. It can be presumed that both studies are technically correct, but they present a different picture of the relative benefits of electric vehicles.

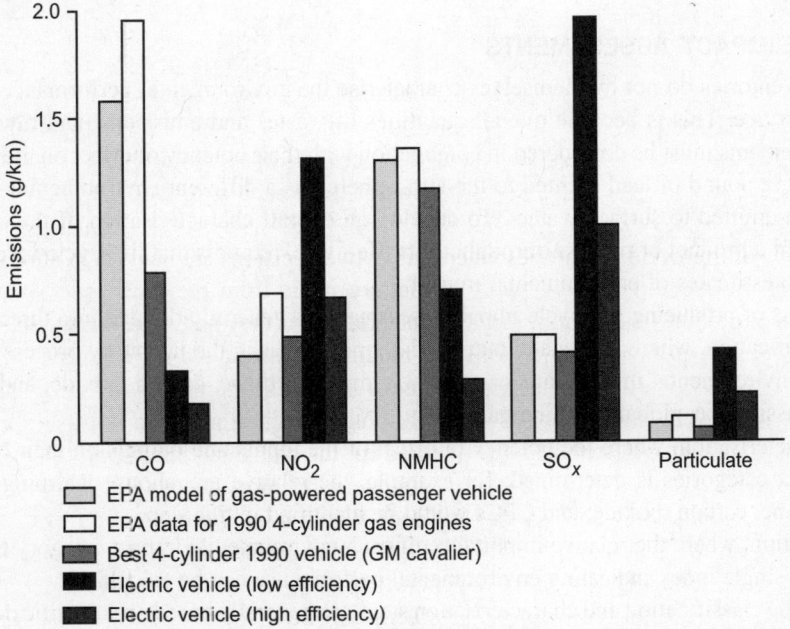

Fig. 7.5. Comparison of air pollutant emissions for gasoline- and electricity-powered motor vehicles.

A second example of the importance of the level of data aggregation emphasises the difference between well-operated and poorly operated facilities. Epstein tabulated the emissions and transfers of wastes associated with 166 refineries in the United States. It was found that the total emissions and waste transfers reported by refineries, through the Toxic Release Inventory, averaged 3.5 kg per barrel of refinery capacity. The 10 per cent of the refineries with the highest reported emissions and waste transfers released more than 9 kg per barrel, while the 10 per cent of the refineries with the lowest reported emissions and transfers released less than 0.15 kg pounds per barrel. Thus, a life-cycle inventory that used emissions data from a refinery with low emissions might lead to very different conclusions than a life cycle inventory based on data from a refinery that has high emissions.

Table 7.3. Comparison of electric vehicle and gasoline powered vehicle emissions based on electricity generation within Southern California.

Emission type	Emissions of gasoline-powered ultra low emission vehicles (ULEVs) in california (g/mi)	Emissions of electric vehicles based on an average of power plants within air quality districts in california (g/mi)
Reactive organic gases	0.191	0.003
Nitrogen oxides	0.319	0.011
Carbon monoxide	1.089	0.024
Particulate matter	0.018	0.004

In summary, this section has described the basic elements of a life-cycle inventory. In performing the inventory, a number of assumptions are made concerning functional units, system boundaries, coproduct allocation, data aggregation methods, and other parameters. It is prudent to define and explore these assumptions and uncertainties before arriving at conclusions based on life-cycle inventory data.

LIFE-CYCLE IMPACT ASSESSMENTS

Life-cycle inventories do not by themselves characterise the environmental performance of a product, process, or service. This is because overall quantities of wastes and emissions, and raw material and energy requirements must be considered in conjunction with their potency of effect on the environment. Simply stated, a pound of lead emitted to the atmosphere has a different environmental impact than a pound of iron emitted to surface waters. To develop an overall characterisation of the environmental performance of a product or process, throughout its life-cycle, requires that life-cycle inventory data be converted into estimates of environmental impact.

The process of producing life-cycle impact assessments is generally divided into three major steps.
1. Classification, where inputs and outputs determined during the inventory process are classified into environmental impact categories; for example, methane, carbon dioxide, and CFCs would be classified as global warming gases.
2. Characterisation, where the potency of effect of the inputs and outputs on their environmental impact categories is determined; for example, the relative greenhouse warming potentials of methane, carbon dioxide, and CFCs would be identified in this step.
3. Valuation, where the relative importance of each environmental impact category is assessed, so that a single index indicating environmental performance can be calculated.

Note that the classification and characterisation steps are generally based on scientific data or models. The data may be incomplete or uncertain, but the process of classification and characterisation is generally objective. In contrast, the valuation step is inherently subjective, and depends on the value society places on various environmental impact categories. Each of the three steps is discussed in more detail below.

Classification

As a first step in life-cycle impact assessment, inputs and outputs that were the subject of the inventory are classified into environmental impact categories. Examples of environmental impact categories are given in Table 7.4. Note that some impact categories might apply to very local phenomena (for example, aquatic toxicity to organisms found only in certain ecosystems), while other impact categories are global (for example, stratospheric ozone depletion and global warming).

Table 7.4. Examples of environmental impact categories.

Global warming
Stratospheric ozone depletion
Photochemical smog formation
Human carcinogenicity
Atmospheric acidification
Aquatic toxicity
Terrestrial toxicity
Habitat destruction
Depletion of nonrenewable resources
Eutrophication

As an example of classification, consider the list of air emissions inventoried for a study that examined polyethylene, shown in Table 7.5. Nitrogen oxides emissions would be classified as photochemical smog precursors, global warming gases, and acid precipitation and acid deposition precursors. Carbon monoxide emissions, on the other hand, would be classified as a smog precursor.

Table 7.5. Selected air emissions from the production of one kilogram of polyethylene

Description	Kg emissions per kg of polyethylene
Nitrogen oxides	0.0012
Sulphur dioxide	0.009
Carbon monoxide	0.0009

Characterisation

The second step of impact assessment, characterisation, quantifies impact for each inventory item by integrating the inventory amount with the potential to cause an impact; i.e. potency factor. For example, if the impact category is global warming, then relative global warming potentials can be used to weight the relative impact of emissions of different global warming gases. Other weighting factors were presented for smog formation potential, atmospheric acidification potential, and other categories. Once these potency factors are established, the inventory values for inputs and outputs are combined with the potency factors to arrive at impact scores.

In this chapter, the emphasis is on the new issues that arise when applying these impact scoring methods to life-cycle data, and on the range of variation in impact scoring systems that have been employed around the world.

Consider first the new issues that arise when impact scoring systems are applied to life-cycle data. When the impact scoring systems are applied to processes, the location of the emissions can be specified and the time at which the emissions occur can be specified. In contrast, for life-cycle assessment data, the spatial location of the emissions may not be known and the temporal distribution of the emissions may be uncertain. For example, in a life-cycle assessment for an automobile, emissions, energy use, and material use may be distributed all over the world since automotive components are manufactured all over the world and users may operate vehicles all over the world. The energy use, material use, and emissions would also be distributed over a product lifetime that may last for more than a decade. In general, life-cycle impact assessments do not account for this spatial and temporal distribution of energy use, material use, and emissions. Energy use, material use, and emissions are summed over the life cycle and the weighting or potency factors are then applied to these summed inventory elements. Does it make sense to sum the emissions of (for example) carbon dioxide from activities all over the world over a period of more than a decade, as would be done in a life-cycle impact assessment of an automobile? The answer to that question, of course, depends on project boundaries, and the spatial scales and time scales over which the impact occurs. It may be appropriate, for example, to sum worldwide emissions of global warming gases in a life-cycle study. It may be inappropriate to do the same summation for a type of impact that depends strongly on local conditions. Compounds that contribute to acid rain, for example, may not be an environmental concern in areas where the soil is well-buffered and acid rain is not a problem. Similarly, the release of nitrates in one area might cause eutrification, while the release of phosphates might be the cause of eutroification in another area.

A summary of the concerns associated with spatial and temporal averaging of emissions is given in Table 7.6. Some recent life-cycle impact assessment methods have attempted to account for spatial and temporal variability of potency factors, but this remains a relatively underdeveloped area of life-cycle impact assessment. Most life-cycle impact assessments continue to assume that inventory data can be summed over the entire life cycle without accounting for spatial and temporal distributions.

Table 7.6. Impact Categories frequently considered in life-cycle Assessments. Listed in Table 7.4, these range from local to global in their spatial extent and operate over time scales ranging from hours to decades. These spatial and temporal characteristics of impacts should be compared to the spatial and temporal resolution of data collected in life-cycle studies.

Impact categories	Spatial scale	Temporal scale
Global warming	Global	Decades/centuries
Stratospheric ozone depletion	Global	Decades
Photochemical smog formation	Regional/local	Hours/day
Human carcinogenicity	Local	Hours (acute)-decades (chronic)
Atmospheric acidification	Continental/regional	Years
Aquatic toxicity	Regional	Years
Terrestrial toxicity	Local	Hours (acute)-decades (chronic)
Habitat destruction	Regional/local	Years/decades
Depletion of nonrenewable resources	Global	Decades/centuries
Eutrophication	Regional/local	Years

Another issue that arises in impact assessment is the choice of potency factors. In practice, there are numerous impact scoring system available. Many of these characterisation schemes have been developed by life-cycle researchers. Also, a number of schemes for weighting releases to the environment have been developed for reasons other than life-cycle assessment, and these can be adopted for life-cycle characterisation. At times, different life-cycle impact systems will lead to different results.

Why would different potency schemes lead to different results? The answer is simple. The methods are often based on different criteria. Some commonly used potency factors are based on data from environmental regulations. In these systems, each emission is characterised based on the volume of air or water that would be required to dilute the emission to its legally acceptable limit. For example, if air quality regulations allowed 1 part per billion by volume of a compound in ambient air, then one billion moles of air (22.4 billion litres of air at standard temperature and pressure) would be required to dilute one mole of the compound to the allowable standard. This volume per unit mass or mole of emission is called the critical dilution volume and can vary across political boundaries. Other potency factor systems are based on relative risk, but establishing relative risks requires assumptions about the type of environment that the emissions are released to. These assumptions may differ in the various impact assessment schemes.

Valuation

The final step in life-cycle impact assessment, valuation, consists of weighting the results of the characterisation step so that the environmental impact categories of highest importance receive more attention than the impact categories of least concern. There is no generally accepted method for aggregating values obtained from the evaluations of different impact categories to obtain a single environmental impact score. Some of the approaches that have been employed are listed in Table 7.7. Some methods assign valuations of high, medium, or low to the impact categories based on the extent and irreversibility of effect, so that stratospheric ozone depletion might receive a high rating and water usage might receive a low rating.

Table 7.7. Strategies for valuing life-cycle impacts.

Life-cycle impact assessment approach	Description
Critical volumes	Emissions are weighted based on legal limits and are aggregated within each environmental medium (air, water, soil)
Environmental priority system	Characterisation and valuation steps combined using a single weighting factor for each inventory element. Valuation based on willingness-to-pay surveys
Ecological scarcities	Characterisation and valuation steps combined using a single weighting factor for each inventory element. Valuation based on flows of emissions and resources relative to the ability of the environment to assimilate the flows or the extent of resources available
Distance to target method	Valuation based on target values for emission flows set in the Dutch national environmental plan

Valuation schemes based on the 'footprint' of the inputs and outputs have been suggested. In these schemes, characterisation would be conducted so that the air, water, land, and other resources required to absorb the inputs and outputs are quantified. These quantities could then be normalised according to the amount of each resource available, on either a local or global basis, and added within resource category. The resource with the highest combined normalised value is the one that is being most adversely impacted. In fact, it would be possible to arrive at a single value that represented the total fraction of the earth's resources required to buffer the inputs and outputs over the life-cycle being studied.

Data on the public's willingness to pay for various environmental health categories have also been used in developing valuation schemes. However, there is very little data of actual scenarios where people paid a premium based solely on environmental preferability, and most willingness-to-pay information is based on surveys.

The following example illustrates the use of the environmental priority strategy (EPS) system, developed in Sweden, which combines characterisation and valuation into single values. Impact categories for this system include biodiversity, human health, ecological health, resources, and aesthetics. Environmental indices are assigned to compounds by considering six factors:

1. Scope: the general impression of the environmental impact.
2. Distribution: the extent of the affected area.
3. Frequency or intensity: the regularity and intensity of the problem in the affected area.
4. Durability: the permanence of the effect.
5. Contribution: the significance of one kilogram of the emission of that substance in relation to the total effect.
6. Remediability: the relative cost to reduce the emission by one kilogram.

Data from willingness-to-pay studies were used in developing the indices. Note that with this system, impacts are aggregated, and environmental value judgements and priorities are built into the indices.

Valuation occurs implicitly in every life-cycle study, because the attributes chosen for inventorying, such as air emissions and energy usage, reflect the values of the practitioners and the organisation funding the study. Also, the choice of impact categories to be evaluated in the classification and characterisation steps implicitly includes valuation. For example, odour is not typically included as an impact category, implicitly suggesting that it is of minimal importance relative to impacts such as ecotoxicity and human toxicity.

While there is no widely accepted procedure for aggregating impact scores across different impact categories, aggregation within impact categories takes place widely. It would be impractical, for example, to have a separate impact category for every biological species. Some impact assessment schemes have separate impact factors for aquatic and terrestrial life but within those broad categories, the response of different species to the same dose of a compound is very different.

Because valuation is subjective, many practitioners stop at the characterisation step. If a life-cycle study was conducted to compare two products and the impact scores for each impact category were higher for one product than the other, valuation is not needed to determine which product is environmentally superior. This rarely happens, however; typically products being compared and design alternatives for a single product have some positive features and some less desirable features. Each alternative has an environmental footprint with unique characteristics, meaning that any design choice typically means tradeoffs between categories of impacts.

Interpretation of Life-cycle Data and Practical Limits to Life-Cycle Assessments

While the process of a life-cycle assessment might seem simple enough in principle, in practice it is subject to a number of practical limitations. In performing the inventory, system boundaries must be chosen so that completion of the inventory is possible, given the resources that are available. Even if sufficient resources are available, the time required to perform a comprehensive life-cycle inventory may be limiting. Then, even if the necessary time and resources are available, life-cycle data are subject to uncertainty for the reasons cited earlier in this chapter.

The limitations of life-cycle inventories are then carried forward into the impact assessment stage of life-cycle studies, and the impact assessment methodologies add their own uncertainties. For example, potency factors are not available for all compounds in all impact categories. Issues of temporal and spatial aggregation, as described in this section arise. Finally, valuation adds an element of subjectivity into the analyses.

This is not to say that life-cycle assessments are without value. Rather, despite the uncertainties involved, these assessments provide invaluable information for decision-making and product stewardship. They allow environmental issues to be evaluated strategically, throughout the entire product life cycle. The challenge is to take advantage of these valuable features of life-cycle assessments while bearing in mind the difficulties and uncertainties.

The next section describes methods for managing the uncertainties and effort required for life-cycle assessments. Once these tools are described, application and interpretation of life-cycle information will be examined.

STREAMLINED LIFE-CYCLE ASSESSMENTS

The use of life-cycle studies falls along a spectrum from a complete spatial and temporal assessment of all the inputs and outputs due to the entire life cycle (which may never be accomplished in practice, both because of a lack of information and because it would require a tremendous amount of effort and expense) to an informal consideration of the environmental stresses that occur over a product or process life-cycle. This spectrum is illustrated in Fig. 7.6. The further a study falls to the right on the spectrum, the more expensive and time-consuming the study will be. In this chapter, an analysis that includes an inventory of all inputs and outputs and all life-cycle stages (including an assessment of which ones are significant enough to be included in the inventory), an impact assessment, and an improvement analysis will be called a life-cycle assessment and a study that falls to the left in the spectrum of complexity will be said to involve the use of life-cycle concepts. Studies in between the two extremes will be called streamlined life-cycle assessments. Streamlined life-cycle assessments are conducted in order to find the most important life-cycle stages or type of inputs and outputs for more detailed study. Also, they can be used to identify where the most significant environmental issues occur.

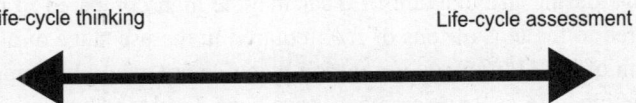

Fig. 7.6. Life-cycle studies fall along a spectrum of difficulty and complexity, beginning with the use of life-cycle concepts and ending with complete life-cycle assessments.

Streamlined Data Gathering for Inventories and Characterisation

The importance of product stewardship and growing awareness of the importance of product and process life cycles, coupled with a growing frustration with the complexity and data intensity of traditional life-cycle assessments has led to a new type of product life-cycle evaluation, often referred to as a streamlined life-cycle assessment. There are many ways that a life-cycle assessment can be streamlined. A study might build extensively on previously completed life-cycle assessments. A life-cycle assessment for polyethylene, for example, might rely on the data presented in the previous section on ethylene and focus on extending the supply chain through the polymerisation of ethylene into polyethylene. Similarly, data collected in previous studies may indicate that certain impact categories or life-cycle inventory categories can be safely neglected without a meaningful effect on the results of the study.

Other approaches for making life-cycle studies easier to accomplish include omission of product components or materials. The omission can be based on whether the components or materials contribute significantly to the product's overall environmental impacts. Some practitioners routinely exclude any component that accounts for less than 1 per cent of the total product weight. This could result in inadequate study results, because some small components, such as semiconductor devices in computers, can have large environmental impacts relative to their weight. There are other ways to decide whether a component or material should be included or omitted in a life-cycle study, such as its economic value, which in turn reflects resource scarcity and ease of manufacturing and is at least loosely tied to environmental importance. Energy use (which is sometimes relatively simple to find data for) or toxicity might also be considered.

Environmental impact categories are sometimes neglected in streamlined life-cycle studies. Similarly, a selected set of inputs or outputs might be chosen for inventorying. Some products are known to have heavy impacts due to process wastes but require little energy, making an inventory of energy requirements less necessary than an inventory of gaseous, liquid, and solid residues.

Another possible shortcut to completing a life-cycle study would be to leave out life-cycle stages. Short-lived products, such as single-use packaging, usually have environmental impacts that are dominated by raw material acquisition and materials manufacture and disposal. In contrast, the use phase dominates for long-lived products that require resources during use.

Qualitative Techniques for Inventories and Characterisation

One of the more common techniques employed in streamlined life-cycle studies involves conducting qualitative rather than quantitative analyses. For example, instead of quantifying the number of units of energy required to produce a product, the energy usage could be characterised as high, medium, or low. Qualitative evaluations can be enormously helpful in reducing the time and resources necessary for providing life-cycle information, because detailed inventory information is not necessary. However, there is a risk of failing to assess different life-cycle stages and products in a comparable manner. For example, energy usage during manufacture of a car may be high compared to manufacture of many products, but compared to the tens of tons of fuel required in the use stage of a typical car during its lifetime, an evaluation of high for energy use during manufacture would be inappropriate. Qualitative approaches for streamlined life-cycle assessments have been developed by a number of researchers.

Some life-cycle practitioners evaluate both the quantity of the inputs and outputs and their impacts in a single qualitiatve process. An example of such a system is the Environmentally Responsible Product Assessment, which relies on the use of expert evaluations of extensive checklists, surveys, and other information. Scores from 0 to 4 (with 4 indicating environmental preferability) are assigned to the life stages and inventory categories listed in Table 7.8. The table shows that there are five life-cycle stages and five inventory categories to assess, for a total of 25 assessments per product. After all the scores are assigned, they are added together to arrive at an overall score. The maximum value for this overall score, therefore, is 4×25, or 100.

As an example, the scoring guidelines and protocols for just one of the 25 elements in the environmentally responsible product assessment matrix is given in Table 7.9. These guidelines and protocols are for the materials choice category of the premanufacture life-cycle stage.

Table 7.8. Life stages and inventory categories evaluated in the environmentally responsible product assessment matrix.

Inventory category/ life stage	Premanufacture	Product manufacture	Product delivery	Product use	Refurbishment, recycling, disposal
Materials choice					
Energy use					
Solid residues					
Liquid residues					
Gaseous residues					

Table 7.9. Environmentally responsible product matrix scoring guidelines for the materials choice inventory category during the premanufacture life-cycle stage.

Score	Condition
0	For the case where supplier components/subsystems are used: No/little information is known about the chemical content in supplied products and components.
	For the case where materials are acquired from suppliers: a scarce material is used where a reasonable alternative is available. (Scarce materials are defined as antimony, beryllium, boron, cobalt, chromium, gold, mercury, platinum, iridium, osmium, palladium, rhodium, rubidium, silver, thorium, and uranium.)
4	No virgin material is used in incoming components or materials.
1, 2, or 3	Is the product designed to minimise the use of scarce materials (as defined above)?
	Is the product designed to utilise recycled materials or components wherever possible?

The results of the environmentally responsible product assessment can be plotted on a chart such as the one shown in Fig. 7.7. A product that is relatively environmentally benign would have all the points of the chart clustered around the centre, and an environmentally damaging product would have points that fell towards the outside circumference.

In these qualitative, streamlined life-cycle analyses, functional units and allocation methods are not explicitly considered. However, use of virgin materials is penalised, so credit is given for using recycled materials. Scores developed by different individuals tend to fall within 15 per cent of each other, which is an indication of the uncertainty in the results. Evaluations tend to be based on comparisons to a standard and focus on whether or not best practices are being followed. Therefore, this scheme might be useful in improving already-designed products rather than a product that is in the early design phases; however, it is not as useful for comparing completely different means to fulfilling a need. For example, they may be helpful in identifying whether aqueous or chlorinated organic solvents are environmentally preferable, but not for comparing a process change that makes cleaning unnecessary compared to the use of aqueous or chlorinated solvents.

Streamlined life-cycle assessment methods could be devised to produce results with an absolute basis. For example, inputs and outputs could be assigned qualitative inventory scores that correspond to a specific functional unit and an absolute value. Product inventory matrices with types of inputs and outputs for rows and life-cycle stages for columns would be filled out with evaluations of high, medium, low, and none. For illustrative purposes, consider the life-cycle inputs and outputs of one kilogram of glass. One might assign a score of low to compounds whose air emissions are believed to be greater than zero but less than 0.001 kilogram, a score of medium if air emissions of a compound are believed to be between 0.001 kilogram and 0.1 kilogram, and a score of high if air emissions of a compound are believed to be greater than 0.1 kilogram. While it might be difficult (or even impossible) to inventory the inputs and outputs of a life cycle to within one or two significant digits, less effort is required to arrive at estimates accurate to within an order of magnitude.

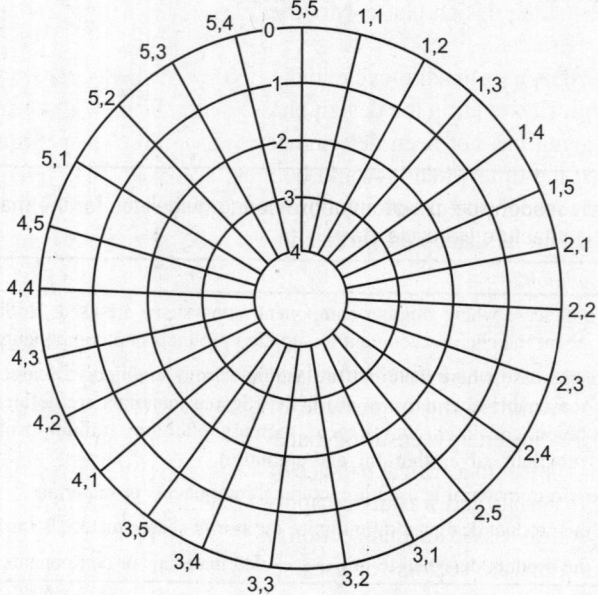

Fig. 7.7. The target plot for the environmentally responsible product assessment matrix. Each radial axis represents one of the 25 life-cycle stage/inventory category combinations from Table 7.8.

Potency factor matrices with evaluations of high, medium, low, and none for each of the inputs and outputs from the product inventory matrix could also be devised. The rows of these matrices would be types of inputs and outputs and the columns would be environmental impact categories. Numerical scores could then be assigned to the qualitative evaluations. For example, low could be assigned a score of one; medium, a score of two; and high, a score of three. The inventory scores could then be added to the potency scores to arrive at impact assessment scores. This streamlined life-cycle assessment technique would require a large number of relatively simple evaluations. Complex products made from many separate components would be more amenable to this type of streamlined assessment because scores with an absolute basis could be weighted and summed over the components involved. Also, the potency factor scores would need to be developed only once for each type of input and output.

Many such streamlined assessments could be devised and it is beyond the scope of this chapter to review all of the methods that have appeared. Nevertheless, it is useful to keep in mind the features of a well-designed, streamlined study. A good streamlined life-cycle assessment has a goal, an inventory, an impact assessment, and an improvement analysis, just as a comprehensive life-cycle assessment does. All of the relevant life stages are evaluated, if only to say they are being omitted and why, and all of the relevant inventory categories are evaluated (again, perhaps only to say why they are being omitted). In a streamlined life-cycle assessment, evaluations for inventory categories and impact assessments may be qualitative instead of quantitative.

Pitfalls, Advantages, and Guidance

Streamlined life-cycle assessments and life-cycle concepts play a particularly important role in green engineering—more so than comprehensive life-cycle assessments. This is because of the nature of the design cycle of a product or process. There is a rule of thumb that 80 per cent of the environmental costs of a product are determined at the design phase. Modifications made to the product at later stages can therefore have only modest effects.

Thus, it is in the early design phase that life-cycle studies for improving the environmental performance of a product are most useful. However, in the design phase, materials have not been selected, facilities have not been built, packaging has not been determined, so a comprehensive, quantitative life-cycle assessment is impossible at the time when it would be most useful. Instead, preferable materials and processes can be identified through the use of an abbreviated life-cycle study early in the design cycle where it is most effective. This is discussed further in the later sections on design of products and processes.

USES OF LIFE-CYCLE STUDIES

According to a survey of organisations actively involved in life-cycle studies, the most important goal of life-cycle studies is to minimise the magnitude of pollution. Other goals include conserving nonrenewable resources, including energy; ensuring that every effort is being made to conserve ecological systems, especially in areas subject to a critical balance of supplies; developing alternatives to maximise the recycling and reuse of materials and waste; and applying the most appropriate pollution prevention or abatement techniques. As discussed in this section, life-cycle studies have been applied in many ways in both the public and private sectors for uses such as developing, improving, and comparing products.

Product Comparison

The most widely publicised life-cycle studies are those that have been conducted for the purpose of comparing products. A life-cycle study comparing cloth and disposable diapering systems, another study comparing plastic and paper cups, and another one comparing polystyrene clamshells and paper wrappings for sandwiches are examples of studies that received a great deal of attention from the press. Product comparison studies are often sponsored by organisations that have a vested interest in the results, and because of the open-ended nature of life-cycle studies, there is always room for criticism of the assumptions that were made and the data that were gathered in the course of the study. Because the results of these high-profile product comparison life-cycle studies have generated a great deal of controversy and debate, they have created skepticism over the value of life-cycle studies. This has diverted attention away from some of the less controversial applications, such as studies conducted in order to improve products.

Strategic Planning

One of the most important uses for manufacturers of life-cycle studies is to provide guidance in long-term strategic planning concerning trends in product design and materials. By their nature, life-cycle studies include environmental impacts whose costs are external to business (e.g. habitat destruction) as well as internal (e.g. the cost of waste generation). Assessing these external costs is key to strategic environmental planning, as regulations tend to internalise what are currently external costs of doing business.

Public Sector Uses

Life-cycle studies are also used in the public sector. Policymakers report that the most important uses of life-cycle studies are: (i) helping to develop long-term policy regarding overall material use, resource conservation, and reduction of environmental impacts and risks posed by materials and processes throughout the product life-cycle; (ii) evaluating resource effects associated with source reduction and alternative waste management techniques; and (iii) providing information to the public about the resource characteristics of products or materials.

Some of the most visible of the applications of life-cycle studies are environmental or ecolabelling initiatives. Besides environmental labelling programmes, public sector uses of life-cycle studies include making procurement decisions and developing regulations.

Product Design and Improvement

Product comparisons have received the most attention from the press, but in a survey, manufacturers state that the most important uses of life-cycle studies are: (i) to identify processes, ingredients, and systems that are major contributors to environmental impacts; and (ii) to compare different options within a particular process with the objective of minimising environmental impacts.

Manufacturers have more potential for influencing the environmental impacts of products than any other 'owners' of life-cycle stages. This is because they can exert some influence over the environmental characteristics of the supplies they use, because manufacturing processes account for a large portion of the wastes generated in the United States and other developed countries, and because manufacturers determine to some extent the use and disposal impacts of the products they make.

Choosing suppliers

As stated before, manufacturers have some potential to influence the environmental characteristics of the companies from whom they purchase supplies. This is illustrated by the efforts of Scott Paper Company of USA in their procurement of pulp for paper products. Scott decided to use a life-cycle approach to environmental impacts as opposed to its traditional focus on environmental concerns only at plants that it owns when it found that the issues of major concern were not in the life-cycle stages directly controlled by the company.

Scott's first step was to require its pulp suppliers in Europe to provide detailed information about their emissions, energy use, manufacturing processes, and forestry practices. In their impact assessment, Scott ranked the environmental impact categories by consulting with opinion leaders. They found that there was considerable variation in performance among suppliers, and that the suppliers that were ranked worst in one environmental impact category tended to be worst in other categories as well. The poorest performing suppliers were shown the potential for improvement, and if they did not choose to proceed, Scott no longer used them as a supplier. As a result of this programme, Scott changed about 10 per cent of its pulp supply base. Scott publicised their efforts and its products were seen as environmentally preferable by consumers and environmental advocacy groups.

Improving existing products

The results of a life-cycle study conducted for the purpose of product improvement are shown in Table 7.10. This table shows the results of an inventory of the energy required to produce one kilogram of polyethylene. The majority of fuel required to make polyethylene is in the organic matter that instead of being burned for energy is converted to polyethylene. In fact, the values in the column titled 'Feedstock Energy' are about 75 per cent of the total energy requirements. This inventory showed that the focus of efforts to reduce the life-cycle energy requirements of polyethylene are best spent on reducing the mass of polyethylene in products (i.e. to make them as light as possible).

Table 7.10. Average gross energy required to produce one kilogram of polyethylene.

Fuel type	Production and fuel energy, MJ	Delivered energy, MJ	Feedstock energy, MJ	Total energy, MJ
Electricity	5.31	2.58	0.00	7.89
Oil fuels	0.53	2.05	32.76	35.34
Other	0.47	8.54	33.59	42.60
Total	6.31	13.17	66.35	85.83

Another life-cycle study conducted for the purpose of product improvement showed that the energy requirements of the use stage of a polyester blouse are 82 per cent of the total energy requirements over the life-cycle. Furthermore, the greatest potential for reducing the energy requirements over the life cycle consisted of switching to a cold water wash and line dry instead of a warm water wash and drying in a clothes dryer. Such a switch would reduce the energy requirements of the use stage by 90 per cent. Thus, one of the greatest environmental improvements that could be made in a garment is to make it cold water washable.

In another life-cycle study for product improvement of clothing, it was shown that the means of transportation used in delivering a garment to a customer can have a profound impact on the garment's life-cycle energy requirements. This study showed that in the case where next-day shipping is used, transportation and distribution energy requirements can be 28 per cent of manufacturing life-cycle energy requirements. Transportation and distribution of products generally contribute negligibly to the energy requirements of a product and are frequently neglected in life-cycle studies. Prior to this study, the garment manufacturer was unaware that their choice of delivery mode could contribute significantly to the energy required over the life cycle of their products.

In yet another life-cycle study conducted for product improvement, the components of a computer workstation were assessed to reveal which were responsible for the majority of raw material usage, wastes, emissions, and energy consumption

The components studied included semiconductors, semiconductor packaging, printed wiring boards and computer assemblies, and display monitors. One of the findings of the study was that the majority of energy usage over a workstation's life-cycle occurs during the use phase of the display monitor. Therefore, to reduce the overall energy usage of a computer workstation, efforts are best directed at reducing the energy consumed by the monitor. Semiconductor manufacture was found to dominate hazardous waste generation and was also found to be a significant source of raw material usage. This is inspite of the fact that by weight, semiconductors are a very small portion of a workstation.

Another life-cycle study was conducted by a light switch maker in Europe as a result of a competitor gaining market share by claiming to manufacture a cadmium-free switch. The life-cycle inventory showed that the cadmium contained in the contactor of the switch for both manufacturers was negligible compared to the cadmium used in plating operations during manufacture. In effect, only one of the manufacturers made switches that contained cadmium, but neither switch was truly 'cadmium-free'. Also, the life-cycle study revealed that the biggest environmental gains could be had by reducing the electricity consumed by the switch over its ten-year lifetime. This result is surprising because the electricity consumed is small per event and only becomes important when totalled over the life cycle.

Using life-cycle concepts in early product design phases

Traditionally, performance, cost, cultural requirements, and legal requirements have set the boundaries for the design of products. Increasingly, environmental aspects are included with this core group of design criteria and life-cycle studies can be used to assess environmental performance. Optimising environmental performance from the beginning of the design process has the possibility of the largest gains, but it is a moving target as markets, technologies, and scientific understanding of impacts change. However, as stated earlier, roughly 80 per cent of the environmental costs of a product are determined at the design phase, and modifications made to the product at later stages may have only modest effects. Thus, it is in the early design phase that life-cycle studies for improving the environmental performance of a product are most useful.

Motorola has developed a matrix for streamlined life-cycle assessment that is intended in part to specifically address early design. The matrix is shown in Table 7.11. There are five life-cycle stages and three impact assessment categories (one of which is divided into two sub-categories) in the matrix. Motorola intends to use this matrix in three succeedingly quantitative phases: the initial design concept phase, the detailed drawings phase, and the final product specifications phase. In the initial design phase, the matrix elements can be filled out by asking a series of yes and no questions. An overall score is computed by adding the yes answers, and changes in that score show progress in the product's environmental characteristics. This example is typical of emerging trends in product design.

Table 7.11. Motorola's life-cycle matrix.

Impact		Part sourcing	Manufacturing	Transportation	Use	End of life
Sustainability	Resource use					
	Energy use					
Human health						
Eco health						

Process Design

Industrial process changes should be given strategic thought because they are generally in place for decades and retrofits tend to be expensive and difficult. While the life-cycle stages of a process are different from those of a product (as shown in Fig. 7.1), the types of inputs and outputs and impact categories are the same. Generally, process choices (including choices of feed materials for the process) are likely to have more impact over the life-cycle of the process than production of the equipment itself.

Jacobs Engineering has developed a life-cycle matrix tool that has been applied to processes (as opposed to products). This matrix is shown in Table 7.12. This tool identifies five inventory categories and seven environmental impact categories at two spatial scales (shop level and global). A base case process is determined and elements in the matrix are assigned +1, 0, or −1, depending on whether the alternative is an improvement, equivalent, or worse than the base case. Note that not all life-cycle stages are explicitly identified in this scheme.

Table 7.12. Jacobs engineering impact analysis matrix for evaluating alternative processes.

	Impacting parameters									
	Shop level					Global level				
Risk area	Material inputs	Energy inputs	Atmospheric emissions	Aqueous wastes	Solid wastes	Total	Material inputs	Energy inputs	Atmospheric emissions	Aqueous wastes	Solid wastes	Total
Global warming											
Ozone-depleting resource utilisation											
Non-renewable resource utilisation											
Air quality											
Water quality											
Land disposal											
Transportation effects											
Total											

COMPREHENSIVE LIFE-CYCLE MANAGEMENT STRATEGY NEEDED TO ENSURE SUSTAINABLE FUTURE FOR INDIAN CHEMICAL INDUSTRY

The chemical industry must evolve a comprehensive life-cycle management strategy if it is to ensure a sustainable future for itself. According to Dr. Raghavan (Chairman, Research Advisory Committee, defence research and development organisation (DRDO)) stressed the need for industry to reassess material and energy efficiency of its products and services; recognise parallels between environmental and economic impacts; identify tradeoffs between manufacture, use, and end-of-life requirements; and integrate sustainable practices into its core operations. Environmental issues do not occur in isolation and must be considered in decision making at all levels.

Life-cycle Costing Concept

Dr. Raghavan stressed the need for industry to take into account the 'life-cycle costing' concept, to assess the environmental impact of its products. This evolving methodology involves factoring in the cost of R & D, production, use and end-of-life cost while calculating the cost of a chemical product. 'It covers all costs associated with system for a defined life cycle including material and energy flows, labour costs, knowledge costs, transaction costs and marketing expenses. The industry should focus on the three pillars of environmental, economic and social sustainability, in order to ensure a future for it. Environmental sustainability, will involve reducing the ecological footprint of the industry through innovations for clean processing; design and manufacture of 'green' products and minimising greenhouse gas and ozone effects. Economic sustainability, on the other hand will involve optimal utilisation of resources, and social sustainability can only be ensured by raising living standards through a combination of pro-active labour policies, pragmatic public-private initiatives and risk communication. All of them require industry to take the path of innovation and market competitiveness.

Reducing the Carbon Footprint

In order to reduce the carbon footprint of the industry, the waste sludge transport and drying have significant impact on the carbon footprint of process treatment technologies. It is advisable to minimise these impacts by reducing transport distance and minimising water content through flocculation.

Yet another approach — although of limited nature — was to use carbon dioxide as a chemical building block to manufacture agrichemicals such as urea and phosgene; to make polymers such as polycarbonate and polyureas; or to produce industrial chemicals such as oxalic acid (by electrolytic reduction and coupling of two carbon dioxide molecules) and carbon monoxide (by electrolytic reduction). Although it is technically feasible to use CO_2 as feedstock, the process is energy intensive and highly expensive. 'Carbon dioxide can be coaxed to participate in, chemical reactions such as reduction, addition and condensation.'

'Polymer synthesis is one of best options to produce new CO_2 based structurals in the 21st century, but R & D has to be intensified in this field.'

The countries should also intensify efforts for use of natural gas as energy and chemical feedstock, especially as gas provides the largest heat of combustion per unit of CO_2 produced; the ease of sulphur removal; and the fact that it is the cheapest source of hydrogen.

'Major gas processing complexes are in the offing, which use methane as feedstock to produce basic building blocks, feedstock, solvents and bulk, intermediates, specialities and performance chemicals will be produced in satellite units.'

Boosting Exports

In order to boost chemical exports from countries, Dr. Raghavan suggested a twinpronged strategy that included a renewed focus on eco-friendly products and highend specialities. Stressing the role of product and process innovation in the chemical industry, he referred to a McKinsey study that showed that globally 60 per cent of companies failed to generate positive returns on R & D, while 20 per cent earned marginal positive returns and a mere 20 per cent achieved substantial returns. Highlighting the importance of managing innovation correctly in an organisation so as to reap its reward, he stressed the need to be competitive by effectively integrating creative abilities with organisation processes and marketing plans. In his view, management must focus on new knowledge creation; incremental ideas for better products/processes; quick conversion into working prototypes; and new models for manufacture and distribution. 'Incremental technological breakthroughs may be adequate to start with; but they need to coalesce for bigger gains,' he pointed out.

Chapter 8

Industrial Ecology

INTRODUCTION

The phrase 'Industrial Ecology' evokes powerful images and strong reactions, both positive and negative. To some, the phrase conjures images of industrial systems that mimic the mass conservation properties of natural ecosystems. Powerful analogies can be drawn between the evolution of natural ecosystems and the potential evolution of industrial systems.

Billions of years ago, the Earth's life forms consumed the planet's stocks of materials and changed the composition of the atmosphere. Our natural ecosystems evolved slowly to the intricately balanced, mass-conserving networks that exist today. Can our industrial systems evolve in the same way, but much more quickly? These are interesting visions and thought-provoking concepts. But is industrial ecology merely a metaphor for these concepts? Is there any engineering substance to the emerging field of industrial ecology?

As demonstrated in this chapter, industrial ecology is much more than a metaphor and it is a field where engineers can make significant contributions. At the heart of industrial ecology is the knowledge of how to reuse or chemically modify and recycle wastes — making wastes into raw materials. Chemical engineers have practiced this art for decades.

The history of the chemical manufacturing industries provides numerous examples of waste streams finding productive uses. Nonetheless, even though the chemical manufacturing industries now provide excellent case studies of industrial ecology in practice — tightly networked and mass-efficient processes — there is much left to be done. While the chemical manufacturing industries are internally integrated, there is relatively little integration between chemical manufacturing and other industrial sectors and between chemical manufacturers and their customers. Engineers could take on design tasks such as managing the heat integration between a power plant and an oil refinery or integrating water use between semiconductor and commodity material manufacturing. The goal is to create even more intricately networked and efficient industrial processes — an industrial ecology. Not all of the tools needed to accomplish these goals are available yet, but this chapter begins to describe the basic concepts and suggests the types of tools that the next generation of process engineers will require.

ENVIRONMENTAL PERFORMANCE OF CHEMICAL PROCESSES

The environmental performance of chemical processes is governed not only by the design of the process, but also by how the process integrates with other processes and material flows. Consider a classic example — the manufacture of vinyl chloride.

Billions of tons of vinyl chloride are produced annually. Approximately half of this production occurs through the direct chlorination of ethylene. Ethylene reacts with molecular chlorine to produce ethylene dichloride (EDC). The EDC is then pyrolysed, producing vinyl chloride and hydrochloric acid.

$$Cl_2 + H_2C=CH_2 \longrightarrow Cl\,H_2C-CH_2\,Cl$$

$$Cl\,H_2C-CH_2\,Cl \longrightarrow H_2C=CH\,Cl + HCl$$

In this synthesis route, one mole of hydrochloric acid is produced for every mole of vinyl chloride. Considered in isolation, this process might be considered wasteful. Half of the original chlorine winds up, not in the desired product, but in a waste acid. But the process is not operated in isolation. The waste hydrochloric acid from the direct chlorination of ethylene can be used as a raw material in the oxychlorination of ethylene. In this process, hydrochloric acid, ethylene, and oxygen are used to manufacture vinyl chloride.

$$HCl + H_2C=CH_2 + 0.5\,O_2 \longrightarrow H_2C=CHCl + H_2O$$

By operating both the oxychlorination pathway and the direct chlorination pathway, as shown in Fig. 8.1, the waste hydrochloric acid can be used as a raw material and essentially all of the molecular chlorine originally reacted with ethylene is incorporated into vinyl chloride. The two processes operate synergistically and an efficient design for the manufacture of vinyl chloride involves both processes.

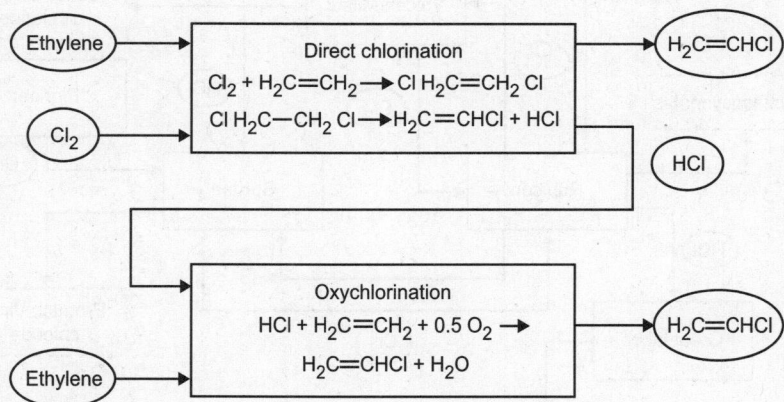

Fig. 8.1. By-product hydrochloric acid from the direct chlorination of ethylene is used as a raw material in the oxychlorination process; by operating the two processes in tandem, chlorine is used efficiently.

Additional efficiencies in the use of chlorine can be obtained by expanding the number of processes included in the network. In the network involving direct chlorination and oxychlorination processes, both processes incorporate chlorine into the final product. Recently, more extensive chlorine networks have emerged linking several isocyanate producers into vinyl chloride manufacturing networks. In isocyanate manufacturing, molecular chlorine is reacted with carbon monoxide to produce phosgene:

$$CO + Cl_2 \longrightarrow COCl_2$$

The phosgene is then reacted with an amine to produce an isocyanate and by-product hydrochloric acid:

$$RNH_2 + COCl_2 \longrightarrow RNCO + 2HCl$$

The isocyanate is subsequently used in urethane production, and the hydrochloric acid is recycled. The key feature of the isocyanate process chemistry is that chlorine does not appear in the final product.

Thus, chlorine can be processed through the system without being consumed. It may be transformed from molecular chlorine to hydrochloric acid, but the chlorine is still available for incorporation into final products, such as vinyl chloride, that contain chlorine. A chlorine-hydrogen chloride network incorporating both isocyanate and vinyl chloride has developed in the Gulf Coast of the United States. The network is shown in Fig. 8.2. Molecular chlorine is manufactured by Pioneer and Vulcan Mitsui. The molecular chlorine is sent to both direct chlorination processes and to isocyanate manufacturing. The by-product hydrochloric acid is sent to oxychlorination processes or calcium chloride manufacturing. The network has redundancy in chlorine flows, such that most processes could rely on either molecular chlorine or hydrogen chloride.

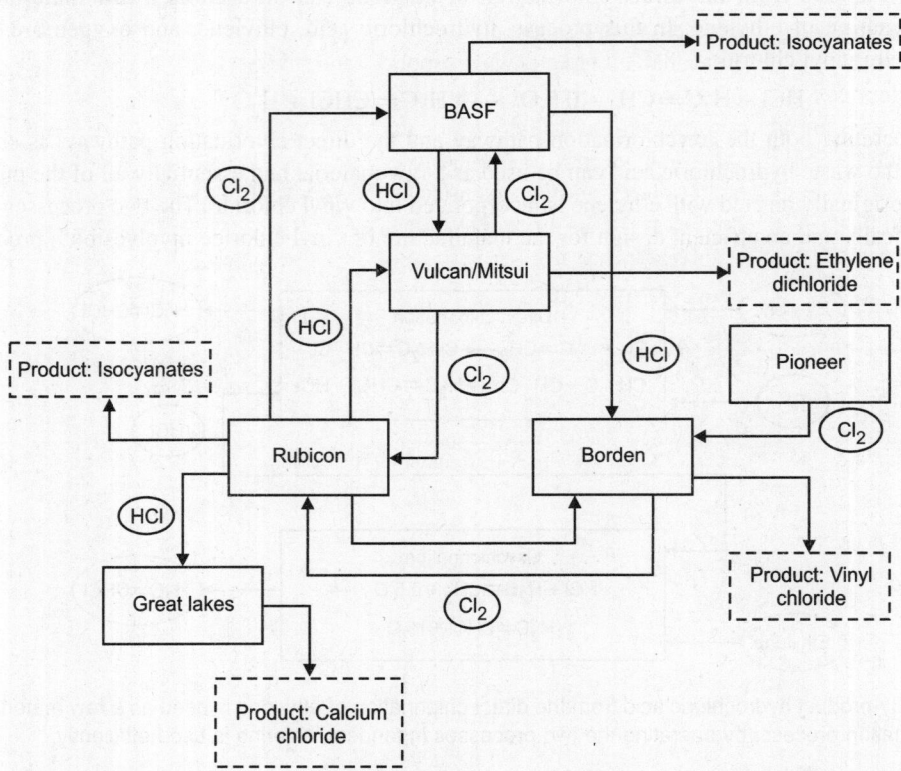

Fig. 8.2. Chlorine flows in combined vinyl chloride and isocyanate manufacturing.

Consider the advantages of this network to the various companies:
1. Vulcan/Mitsui effectively rents chlorine to BASF and Rubicon for their isocyanate manufacturing; the chlorine is then returned in the form of hydrochloric acid for ethylene dichloride/vinyl chloride manufacturing.
2. BASF and Rubicon have guaranteed supplies of chlorine and guaranteed markets for their by-product HCl.

Even more complex networks could, in principle be constructed. As shown in Table 8.1, chlorine is used in manufacturing a number of non-chlorinated products. Table 8.1 lists, for selected reaction pathways, the pounds of chlorinated intermediates used along the supply chain, per pound of finished product. This ranking provides one indication of the potential for networking these processes with processes for manufacturing chlorinated products.

An examination of individual processes, such as those listed in Table 8.1, can be useful in building process networks, but the individual process data do not reveal whether efficient use of chlorine is a major or a minor issue in chemical manufacturing. To determine the overall importance of these flows, it is useful to consider an overall chlorine balance for the chemical industry. The overall flows of chlorine into products and wastes, as well as the recycling of chlorine in the chemical manufacturing sector, is shown in Fig. 8.3. The data indicate that roughly a third of the total chlorine eventually winds up in wastes. By employing the types of networks shown in Figs. 8.1 and 8.2, the total consumption of chlorine could be reduced.

Identifying which processes could be most efficiently integrated is not simple and the design of the ideal network depends on available markets, what suppliers and markets for materials are nearby, and other factors. What is clear, however, is that the chemical process designers must understand not only their process, but also processes that could supply materials, and processes that could use their by-products. And the analysis should not be limited to chemical manufacturing. Continuing with our example of waste hydrochloric acid and the manufacture of vinyl chloride, by-product hydrochloric acid could be used in steel making or by-product hydrochloric acid from semiconductor manufacturing might be used in manufacturing chemicals.

Table 8.1. Partial listing of nonchlorinated chemical products that utilise chlorine in their manufacturing processes.

Product	Synthesis pathway	Kg of chlorinated intermediates per kg of product
Glycerine	Hydrolysis of epichlorohydrin	2.2
Epoxy resin	Epichlorohydrin via chlorohydrination of allyl chloride, followed by reaction of epichlorohydrin with bisphenol-A	1.7
Toluene diisocyanate	Phosgene reaction with toluenediamine	1.6
Aniline	Chlorobenzene via chlorination of benzene, followed by reaction of chlorobenzene with ammonia	1.6
Phenol	Chlorobenzene via chlorination of benzene, followed by dehydrochlorination of chlorobenzene	1.5
Methylene diphenylene diisocyanate	Phosgene reaction with aniline (also produced with chlorinated intermediates)	0.75
Propylene oxide	Chlorohydration of propylene	0.73

Finding productive uses for by-products is a principle that has been used for decades in chemical manufacturing. What is relatively new, however, is the search for chemical by-product uses in industries that extend far beyond chemical manufacturing. This chapter will examine both of these topics—the overall flows of raw materials, products and by-products in chemical manufacturing industries—as well as the potential for combining material and energy flows in chemical manufacturing with material and energy flows in other industrial sectors. Variously called by-product synergy, zero waste systems, or even industrial ecology, the goal of this design activity is to create industrial systems that are as mass-efficient as possible.

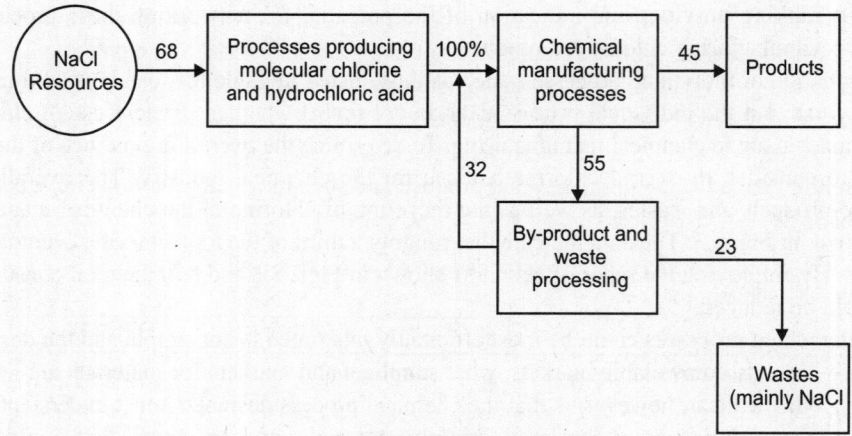

Fig. 8.3. A summary of flows of chlorine in the European chemical manufacturing industry.

Next section provides an overview of material flows in chemical manufacturing and describes analysis methods that can be used to optimise flows of materials. Another section examines case studies of exchanges of materials and energy across industrial sectors and the emerging concept of ecoindustrial parks and briefly attempts to assess the potential benefits of by-product synergies.

MATERIAL FLOWS IN CHEMICAL MANUFACTURING

The chemical manufacturing industries are a complex network of interrelated processes. An individual process typically relies on other chemical manufacturing processes for raw materials and as markets for its products. Take the manufacture of styrene as an example. Styrene manufacturing relies on ethylene and benzene, manufactured in other processes, for raw materials. The styrene is not sold as a consumer product; rather, it is used as a raw material for polystyrene manufacturing. Additional complexity arises from the fact that most sequences of chemical manufacturing process are not unique. There are generally a variety of pathways available for manufacturing products.

As a relatively simple example of the multiple pathways available in chemical synthesis, again consider styrene. Styrene is produced from ethylene and benzene, but the source of the ethylene might be naptha, or refinery gases. Benzene might be produced by dehydrogenation of cyclohexane, dealkylation of toluene or separation from crude oil. These options provide multiple pathways from raw materials to styrene. Each route has raw material requirements, energy requirements, water usage, and rates of emissions and waste generation.

Selecting the most environmentally benign and most economical route is a difficult proposition. It is made even more difficult when the entire chemical supply chain is considered. For example, in methanol production, the methanol is produced using carbon monoxide. The carbon monoxide in turn may be produced through a partial oxidation of a material that is currently wasted by another process. On the other hand, to convert the carbon monoxide into methanol requires hydrogen, which is an energy-intensive material. Evaluations of the environmental features of producing a chemical product should examine the entire chemical raw material supply chain, but to realistically examine these supply chains requires comprehensive, integrated models of material flows in the chemical process industries. Fortunately, such models have been developed. Rudd and coworkers have developed basic material and energy flow models of over 400 chemical processes associated with the production of more than 200 chemical products, describing a complex web of chemical manufacturing technologies.

An understanding of material flows in these networks can be used at a variety of levels. First, the material flow networks can be used simply to identify potential users and suppliers of materials, and to identify networks of processes that are strategically related. For example, for the types of networks, it would be useful to have lists of processes that produce and consume hydrochloric acid. A partial list is given in Table 8.2.

Table 8.2. Partial list of processes that produce or consume hydrochloric acid. Such lists are useful in identifying potential material exchange networks.

Processes that consume hydrochloric acid	Processes that produce hydrochloric acid
Chlorobenzene via oxychlorination of benzene	Adiponitrile via chlorination of butadiene
Chloroprene via dimerisation of acetylene	Benzoic acid via chlorination of toluene
Ethyl chloride via hydrochlorination of ethanol	Carbon tetrachloride via chlorination of methane
Glycerine via hydrolysis of epichlorohydrin	Chloroform via chlorination of methyl chloride
Methyl chloride via hydrochlorination of methanol	Ethyl chloride via chlorination of ethanol
Perchloroethylene via oxychlorination of ethylene dichloride	
	Methyl chloride via chlorination of methane
Trichloroethylene via oxychlorination of ethylene dichloride	
	Perchloroethylene via chlorination of ethylene dichloride
	Phenol via dehydrochlorination of chlorobenzene
	Trichloroethylene via chlorination of ethylene dichloride

Once consumers and producers of targeted chemicals are identified, material and energy flow models can be used to construct networks. The network that makes the most sense depends on the features that are to be optimised. Analyses have been performed to identify networks that minimise energy consumption, the use of toxic intermediates, and chlorine use. Other analyses have considered the response of networks to perturbations in energy supplies and restrictions on the use of toxic substances. Regardless of the application, however, the material flow model of the chemical manufacturing web provides the basic information necessary to identify and optimise networks of processes.

Yet another use of comprehensive material flow models is in the evaluation of new technologies. Consider once again the case of chlorine use in chemical manufacturing. Rather than generating complex networks involving HCl and molecular chlorine, it might be preferable to use a chemistry that converts waste HCl into molecular chlorine. Several processes have been proposed and are listed in Table 8.3.

Table 8.3. Processes for reducing chlorine use in chemical manufacturing.

Process description
Chlorine via electrolysis of hydrogen chloride (Ker-Chlor process)
Chlorine via oxidation of hydrogen chloride ($CuCl_2$ catalyst)
Chlorine via oxidation of hydrogen chloride (HNO_3 catalyst)

These processes will only be successful if they can compete with the reuse of by-product HCl. Data on material and energy flows in the chemical manufacturing web can again be used to assess the competitiveness of new chemical pathways, such as the technologies listed in Table 8.3.

ECO-INDUSTRIAL PARKS

The examples of process networking described in the beginning of this chapter dealt exclusively with chemical manufacturing. Yet, the types of material and energy flows found in chemical manufacturing (solvents, acids, water, energy, salts) are used in a wide variety of industrial sectors. It would, therefore, seem reasonable to consider designing industrial networks that involve a variety of industries.

One of the classic examples of this type of network is a group of facilities located at Kalundborg in Denmark. At Kalundborg, an oil refinery, a sulphuric acid plant, a pharmaceutical manufacturer, a coal burning power plant, a fish farm, and a gypsum board manufacturer form an industrial network, exchanging flows of energy and mass. As shown conceptually in Fig. 8.4, the power plant and the refinery exchange steam, gas, and cooling water. Waste heat from the power plant is used in district residential heating and to warm greenhouses and a fish farm. Ash from coal combustion at the power plant is shipped to cement manufacturers. Calcium sulphate from the scrubbers at the power plant is sent to the gypsum board manufacturer. Treated process sludges from the pharmaceutical plant are sent to local farmers for use as fertiliser, and the refinery sends hot liquid sulphur from the desulphurisation of crude oil to a sulphuric acid manufacturer.

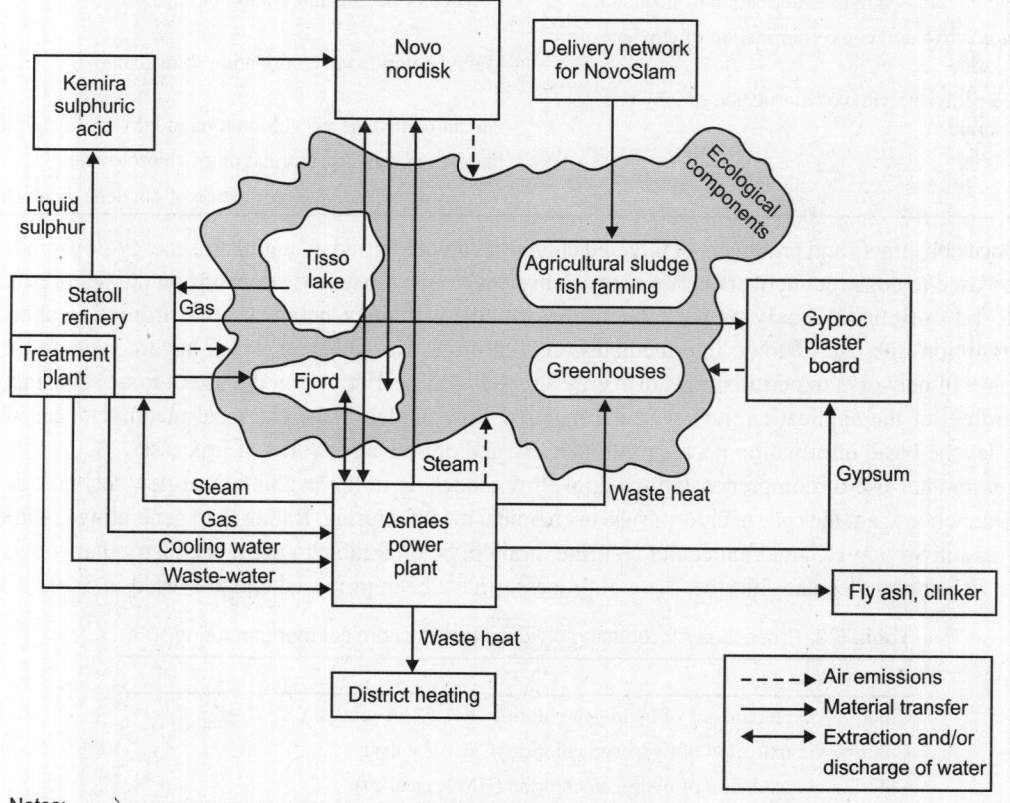

Notes:
(1) This figure is not drawn to scale, nor is it an accurate geographic depiction.
(2) Unused residuals resulting from all activities in the industrial ecopark are eventually released into the biosphere.

Fig. 8.4. The industrial network at Kalundborg, Denmark.

A more detailed examination of the exchanges of material and energy at Kalundborg reveals a number of interesting features:
1. The ecopark developed over a period of more than 30 years. Some material and energy exchanges have occurred for decades and the exchanges continue to grow in extent.
2. The exchanges have the potential to be remarkably efficient. For example, the power plant is able to use some of the waste heat and steam produced through power generation by sending it to the refinery, greenhouses, the fish farm, and the district heating system. If markets were found for all of the waste steam, up to 90 per cent of the heat from the plant's combustion of coal could be utilised. The only losses would be energy escaping with the stack gases. By contrast, typical coal-burning power plants in the United States use heat from combustion solely to generate electricity, at an efficiency of about 40 per cent.
3. Material and energy exchanges provide economic benefits to the participants. In some cases, such as the power plant's sale of calcium sulphate to the gypsum board manufacturer, the direct economic benefits do not fully cover the recovery costs. In these cases, the exchanges are driven by regulations, such as those requiring the scrubbing of power plant stack gases to remove SO_2. The exchanges simply lower the cost of compliance by making it unnecessary to landfill or otherwise dispose of the waste generated by the scrubbers. In other cases, such as the use of power plant waste heat in the refinery, the exchanges are self-supporting.

The central facilities in the Kalundborg Ecopark are the power plant and the oil refinery. Many of the exchanges either originate from or go to the power plant or the refinery. While using a power plant or a refinery as a central facility is a concept that could be successful in other locations, many other approaches are possible. Consider, for example, an eco-industrial park in North Texas where the central facility is a steel mill. This facility, shown conceptually in Fig. 8.5, utilises scrap cars as the primary feed material. The steel from the vehicles goes to an electric arc furnace, producing a variety of steel products. The furnace also produces a significant quantity of electric arc furnace (EAF) dust, which contains significant quantities of zinc, lead, and other metals. In the North Texas facility, the EAF dust is sent to a cement kiln where the trace metals (copper, sulphur, manganese, chromium, nickel, zinc, lead, and others) have value. Automobile Shredder Residue can be burned for energy recovery, or some of the plastics in the residue can be separated.

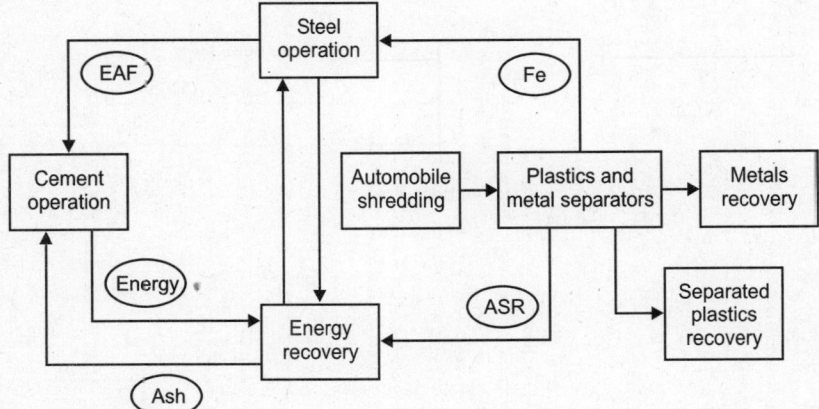

Fig. 8.5. Material flows in an eco-industrial park in North Texas.

Another alternative for electric arc furnace dust, currently being explored in Europe, is as a feed for zinc and lead recovery operations. The recovered zinc can then be used in producing galvanised steel products and batteries can be used as an alternative source of zinc.

These two case studies illustrate the basic principles of ecoparks—integrating flows of energy and materials in diverse industrial operations, increasing mass and energy efficiency. The two cases examined in this section involved exchanges between facilities that are located adjacent to each other; however, co-location of facilities is not always necessary.

ASSESSING OPPORTUNITIES FOR WASTE EXCHANGES AND BY-PRODUCT SYNERGIES

In the previous sections we have discussed anecdotally that productive uses can be found for selected waste streams. Are these anomalies, or are there large quantities of waste materials that can be productively used? This question is difficult to answer with certainty, but a few simple examples may illustrate the potential for finding new uses for waste.

One estimate of the potential for industrial exchanges of materials and energy can be drawn from a simple examination of energy flows in the United States. Approximately a third of the 80–100 quadrillion BTU of energy consumed annually in the United States is used for electric power generation. Of the energy used in electricity generation, roughly $\frac{2}{3}$ is lost as waste heat. This means that roughly a quarter of all energy demand in the United States could be met through the utilisation of lost heat. Combined heat and power systems are emerging throughout the country to take advantage of such opportunities, but much remains to be done.

A second example of the potential for conservation through material exchanges involves another ubiquitous material—water. Water is used in virtually all industrial processes and major opportunities exist for reuse since, in general, only a small amount of water is consumed; most water in industrial applications is used for cooling, heating, or processing of materials, not as a reactant. Further, different industrial processes and industrial sectors have widely varying demands for water quality. For example, waste-water from a semiconductor manufacturing facility that requires ultrapure water may be suitable for a variety of other industrial applications. Thus, water exchanges and reuse provide a significant opportunity.

Chapter 9
Environmental Care in Manufacture of Chemicals

INTRODUCTION

Two approaches are possible for any chemical manufacturing or chemical utilising units in order that its wastes be taken care of:
1. Adaptation of otherwise traditional forms of controlling and managing wastes to new, more stringent ecological requirements.
2. Striving to overhaul the processes and plant equipment in such a manner that it will result in cutting down, doing away or doing without wastes.

Making unavoidable production less detrimental is another consideration and is also an acceptable one. From the viewpoint of environment control, a perfect or near perfect achievement of this goal is possible only when the process steps have been studied with respect to many factors:
1. Possible production of detrimental effluents at any stage of the process.
2. Likely sources that pave the way for wastage of materials in any form.
3. Applicable remedial measures for the above aspects including recovery of any loss of energy or materials at an early stage of its production or detection in the system; or utilisation of these produce as secondary materials at some stage or the other of the process on hand, etc.

Although it is easier said than done, it is not to be brushed aside as impossible. Nonstop experimentation, study, and deeper introspection are required. A major portion of such a study is required to be completed in the design stage itself as once the plant is on stream, any alterations comes with production loss and related hurdles.

Few routes to achieve this are:
1. Design of tonnage plants on mathematical methods.
2. Techniques for process optimisation.
3. Syntheses of closed material systems.

MATERIAL SELECTION

Proper selection of material of construction of equipment in the designing of a process has a long lasting contribution in the manufacture of a commodity. It saves energy and labour costs for not only fabrication, but also operation.

Efficiency in the usage of utilities will rise to optimum levels and the corrosion characteristics of the material will induce longevity.

PROCESS CONDITIONS

Low pressure systems, medium pressure systems and high-pressure systems applicable to a given process step must be studied carefully and the right one selected for implementation. The adoption of the right pressures, with or without catalysing or accelerating agents, boosts the performance of the total set-up to a desired efficiency and avoids the otherwise possible environmental deterioration, in the near or distant future. With application of various pressure systems, there is total possibility of using different types of equipment and different forms of the materials in deriving a given product. The selection of the pressures or for that fact, any other parameters of operation, must be viewed with respect to all these connected factors too. In case of exothermic reactions, it is generally easier to obtain a high degree of conversion by keeping down the temperature and maintaining isothermal conditions. As it is highly difficult to maintain a temperature profile that is isothermal, at least in fixed-bed reactors, multisection reactors with intermediate cooling arrangements are being used in modern designs.

It is a generally adopted rule to have one of the reactants in excess of the stoichiometric proportion, although it leads to marked increase in capital costs and the size of the plant. Instead, reactions can be carried out in two stages, continually withdrawing one of the product streams. Wherever the reaction is followed by a drop in volume of the feed components, raising the pressure in the system can be a useful alternative to get rid of certain operational hassles.

ZERO DISCHARGE TECHNOLOGIES

Zero discharge technologies are possible in many process systems, thanks to modern science and technology. These involve utility cycles that are specially designed. Closed configurations are capable of permitting all-round utilisation of total materials and utilities, and ensure total improvement in the process efficiency, and cost reduction via minimised usage of energy and utilities.

PARELLEL/SERIES OPERATIONS

Production of chemicals is generally categorised as a selection or mixture of operations that are either parallel or series in nature. This is done to shorten the line of production to the minimum possible. Combinations of parallel/series operations or flow patterns are nothing but individual plants that are of self-contained nature. The hidden aim of such designs is to envisage minimal in-process losses, of either materials or utilities that have something to do with that stage of process. Long lines of production tend to give material loss as well as possible problems of maintenance of equipment, besides increasing the probability of pollution to an uncalled-for magnitude.

PLANT CAPACITY

With a decision to employ the best possible technology, it is always better to go in for a plant of higher capacity. This brings in savings via usage of lesser manpower, reduced maintenance of utilities, etc. and easy procedures for abatement of pollution. Higher capacity plants have been proven to use considerably fewer quantities of utilities, compared to their smaller counterparts. Increase in plant capacity need not be necessarily via scaling up the physical dimensions of the equipment. This approach generally brings with it a number of hurdles, viz. elaborate designs, huge requirement of space, utility consumption and disposal problems associated with the utility generators, increase in cost for provision of storage facilities for raw materials, piping, increase in production costs, etc. The other methods that need to be studied with respect to the type and characteristics of the product are:

1. Increase in feedstock concentration.

2. Usage of higher process pressures.
3. Usage of higher process temperatures.
4. Usage of better catalysts.
5. Usage of better reactants.
6. Better flow sheets and processes.

Explorers in quest of small equipment for a given process have ultimately come out with techniques of combining several unit operations in an equipment, although such ideas have posed stricter norms for fabrication and erection.

It has to be added here that these developments and challenges have ended up with success and have resulted in innovative outcomes like efficient recovery systems for outlet streams, improving ecological and safety aspects.

PROCESS PATTERNS

Possibility of recovery of energy released within the process, as input at some stage or other of the same process, is more in large capacity plants. Energy output from the stages of a process is fed to systems designed for generation of utilities for manufacture of many chemicals. Cooling by adsorption is considered beneficial in some designs, compared to indirect heat transfer systems, as they depend on low potential heat requirements.

Application of inert gas atmosphere in reaction systems has been proved to be efficient, safe and economical in many designs of chemical manufacture. Catalysts and accelerators are of importance for speedy processing. Methods of recovery of catalysts must be designed well. Stepping up the reactivity of catalysts is a development of interest in this direction. Removal of by-product streams may facilitate high conversion rates in some cases. Selective sorption techniques may call for intentional introduction of a foreign stream into a system. Here, the recovery of the foreign stream must be thought of as an economic measure, if not compelled by the end product quality demands. Closed circuits can help in the following respects:

1. Recovery of un-reacted source materials.
2. Saving in usage of heat transfer media.
3. Re-use of heat or other forms of energy used in the system.
4. Maintain effect of process on environment to be under confinable and acceptable limits.

In order to reduce the hassles posed by fine dust or mist in the process, one or more of the following equipment can be made use of:

1. Emulsifiers.
2. Filter mist eliminators.
3. Fluidised beds.
4. Froth apparatus.
5. Packed towers.
6. Suspended packing units.
7. Venturi scrubbers.
8. Wetted wall spray towers.
9. Condensers (coagulation).
10. Alkaline absorption units.
11. Sorption equipment that envisage the sorption based on the usage of ion-exchange filters.
12. Usage of activated charcoal and/or alumina and/or clay and/or silica gel.

The reduction in intake quantities result in reduction in quantity to be treated, as well as quantity of untreated coolant. It has been practically demonstrated that plants of higher production capacity require comparatively lesser quantities of energy (electrical or otherwise). Pollution of emitted air too is considerably minimised.

Plant integration helps develop multiple products under one facility at the same time, using a given set of equipment. This helps in avoiding repeated shutdowns/start-ups of utility mechanisms, as well as other crucial equipment. Closed circuit processes often demand the design and efficient usage of some or all of the following systems, in general.

1. Better reactors.
2. Centrifugal compressors.
3. High-speed steam turbines.
4. Sensible heat utilisation in reactors.
5. Usage of cheaper heat transfer agents, etc.

UTILITIES

Cleaning of waste gases is mandatory, for the performance of the equipment, as well as reuse of emitted gases elsewhere in the process or even for ensuring a better environment. Irrecoverable loss of low potential energy from various cooling systems is an unavoidable stream in some process flows. Generally, these are harmless in their environment-related aspects. However, these must not be neglected, as they tend to get themselves involved in some fashion or other into larger requirement of energy for the specific needs of the processes. The ultimate attainment of such a flow is addition to loss of heat in systems that produce energy.

Methods of improving combustion will pay off. The organic components of tail gases can be eliminated by burning them in a flare, if not by a catalytic digestion. The amount of escaping gases can be kept under permissible control levels by adopting timely inspection procedures and prompt and immediate attention of any fault. Usage of gas streams as fuels during start-ups is worth considering. Reduction in number of start-ups and shutdowns are important.

Using gas-fired furnaces for production of power steam relieves the burden of electric power stations, but adds to emissions of gaseous materials that demand careful attention and treatment or re-usage before being let off. Irrespective of the extent to which this emission is contaminated with toxic constituents, at least volume-wise, these need proper treatment. It is an undeniable fact that in many industrial processes, the exit gases generally carry a good amount of heat that can be used somewhere in the process itself, after a thorough study of compatibility and a bit more of additional investment. The investment is worth considering as it will always be cheaper than investment in fresh fuels (Table 9.1).

Table 9.1. Gas emissions from small/large closed circuit units.

Constituent	Small tonnage units	Closed circuit large units
Inorganics	3.00	Nil
Nitrogen oxides	1.55	1.22
Sulphur oxides	5.90	0.10

Bringing down the combustion temperature is another step that can help. Catalytic cleaning units can come in handy in doing away with emission of toxic gaseous streams. These units are designed to operate with the introduction of a reducing gas such as hydrogen or natural gas. Oxygen concentration must be kept low at least during the initial stages of the combustion zone. Here, optimal and efficient utilisation of clean air is emphasised.

Reactions in the high temperature zones of a reactor must be free from any type of resistance whatsoever to the smooth continuation to completion of the reaction in as short a span of time as possible and feasible. Waste gases leaving an absorber must be subjected to a secondary cleaning process, before discharge into the atmosphere.

Combustible gases such as natural gas, hydrogen, carbon monoxide and ammonia can be used for catalytic reduction of the valuable product. Cleaning plants of modern age make use of palladium on a substrate (to reduce the consumption of palladium).

Wherever possible, the present day technologies involve use of enriched gas, instead of the otherwise commonly used natural gas as reducer. This has resulted in avoiding waste heat recovery boiler, etc. and other connected expenses. High temperature catalytic cleaning of exhaust gases will be the basis of designs to come in future. However, the efficiency of operation of such a process depends on the reliability of its components and the ease of start-up/shut-down, in addition to the cost of power. Zeolites make for a great option as absorbing agents.

CATALYSTS

Present day technologies demand the use of catalysts. Wherever found to be of use for smooth and speedy reactions, catalysts are required to possess some or all of the following properties:
1. High activity at low temperatures.
2. Low pressure drop across the bed.
3. Facilitate low energy consumption.
4. Low sensitivity towards dust.
5. High mechanical strength.
6. Stability at elevated temperatures.

PREVENT RATHER THAN CURE

One promising way to prevent water bodies from getting polluted is to do away with effluents themselves. It may seem, at first sight that this situation is not achievable in practice, but it will be necessary to forge ahead. The few possible roads to this destination are:
1. Recycle all the water used for any purpose at any stage of the process.
2. Utilise the effluents, either unpurified or partly purified, within the process.
3. Collect all slurry at one point, neutralise them, allow the neutralised one to settle, allow the clear water to cool and reuse the water while filtering the slurry and remove the cake to a dump pile.
4. Localised recycle loops help reduce quantity of effluents.
5. Use process liquids and not extra water or fluids to wash or absorb gaseous streams or for make-ups.
6. Reduce the contents of impurities in the effluents. Ion exchange techniques come in handy here.
7. Avoid leakage from any container that is in use at any stage of operation, storage, transportation or during breakdowns.
8. Use submersible pumps for handling liquefied gases.
9. Vacuum type transfer facilities help handle vapours in a better fashion.
10. Avoid human element at possible and crucial control points.

Always be conscious about the fact that wash water is not a single entity, but a name given to the collection of varying quantities of starting materials, products, by-products, wash waters, overflows, wastes from the equipment and process stages, leakage, precipitations, oil fractions and carry-overs.

Defects in plant designs, improper running of plants and disorganised manufacture contribute heavily to emissions to environment.

A typical example of what a proper design can do to minimise dangers to atmosphere is given in Table 9.2. The two sets of results enumerated here are those of the analyses of waste-water emanating from a dichloromethane plant before and after treatment by available and appropriate methods, respectively.

Table 9.2. A glimpse of effect of treatment.

Component	Before treatment	After treatment
Dichloroethane	2.55	0.30
Chlorine	0.84	Nil
HCl	3.77	Nil
Vinyl chloride	0.65	Nil
Methyl/ethyl chlorides	0.12	Nil
NaOH	Nil	Traces

Extending the duration between shut-downs can help in minimising the after effects of purging, a process that results in discharge of gases and vapours that remain in the system and flushing, the process that leads to hazardous chemicals being dumped in the sewage treatment area. This is possible only by making use of equipment that is highly reliable for a given performance.

POLLUTION CONTROL AND ENVIRONMENTAL PROTECTION IN CHEMICAL INDUSTRY

Chemical industry occupies a major position in any developing nation. Be it plastics, fibres or refineries there is some chemical process involved which generates pollutants. It is the duty of designers, operators or owners to run chemical industry so that pollutants are treated, disposed and possibly made use to generate useful products or energy and keep environment clean for public at large. Here some practical ways are enumerated to keep pollution under control meeting statutory regulations. Number of consultancy companies undertake detailed engineering and design work for large chemical projects. Number of small sector projects would benefit from these practical suggestions.

Dust Removal

In cement or calcined bauxite units, a venturi scrubber is installed to scrub dust. A powerful blower creates enough vacuum to suck air laden with dust. This air is made to go through venturi scrubber. Dust gets separated and pure air is vented to atmosphere. This method can be employed for meeting pollution control requirement for a calcined bauxite unit.

Phenol Removal

Phenol is harmful at a very low level in effluent and its reduction to few ppm is achieved by special algae which uses phenol for its growth. The resultant effluent is clarified or filtered to meet the statutory requirement.

Solvent Vapour Removal

Activated carbon filters are available through which effluent gases are passed. Solvents get adsorbed on activated carbon. These solvents can be removed and carbon can be regenerated for reuse by passing steam through bed of activated carbon. Gases can be further scrubbed in water tower. This method can be used for designing a grease plant where solvent vapours are generated in the reactor.

BOD Reduction

Surface aerators are put to induce fresh oxygen from air into effluent, to achieve desired biological oxygen demand (BOD). Floating surface aerators are available with very low power requirement. The purpose of reducing BOD is to minimise the damage done by effluent stream going into river or sea as fish live on dissolved oxygen in water and any effluent which takes away dissolved oxygen from water is harmful to fish and other marine life.

Removal of Chlorine/Hydrochloric Acid Vapours

Water scrubbers are installed to absorb the hydrochloric acid vapours and make weak acid. This is used by various dyes, lube additives and bulk drug companies.

Removal of Hydrogen Sulphide

Hydrogen sulphide is a potent poisonous and dangerous gas. Exposure to small quantity of hydrogen sulphide gas can kill plant operator. It is best handled by sodium hypochlorite scrubber followed by burning in a flare unit or burning in incinerator. This method is used in organo-phosphorus based plants.

Incinerators

Certain waste cannot be handled except by burning at very high temperature of 1400° to 1800°C in incinerator. This incinerators can be made profitable by generating steam or providing source of heat. Intelligent entrepreneur can keep incinerators not only for their plant waste but offer services to neighbouring plant and earn enough money to pay for incinerator's cost in two years. Such cases are there in Andhra Pradesh in pesticides/bulk drug industry.

Chemical Treatment/Filtration

To reduce certain toxic elements, they are treated with other chemicals like alum or caustic soda or sulphuric acid or any other strong reactant to produce harmless chemicals, which can be separated by filtration. This can be used as a source of income also, like calcium chloride can be manufactured by treating lime and hydrochloric acid waste.

Special additives

Fuel additives have been developed which reduce pollution and increase efficiency of engine and offer better mileage.

Catalytic converters

Palladium converters are available to reduce emission of harmful vapours in fuel exhaust.

Membranes

Special membranes are available to selectively, remove oxygen, nitrogen or carbon dioxide gases and purify the air.

Ultraviolet rays

Sterile atmosphere is necessary for injectable area in pharmaceutical industry by use of ultraviolet rays. Air is properly dehumidified, filtered and cleaned before exposing to ultraviolet rays.

Environment in plant

Chemical plant area has to be made comfortable by proper air changes. The fresh air is to be sent. The vapours generated by handling of solvent drums should be sucked by downcomer pipes leading to scrubbing system. The dust is also contained by keeping crusher, pulveriser, storage silos under nitrogen or carbon dioxide. Later is preferred for phosphorus or sulphur pulverisers as it is heavier than air and descends when added from top of the system displacing air. Explosions have occurred in handling phosphorus pentasulphide, when they were not properly covered by inert gas.

Safety and environment protection

Both aspects are complimentary. A good clean plant will attain better safety standard as operators will be healthy and alert and exposed to minimum risk.

Spread of data

A number of publications are available giving dangerous property of every chemical, its TLV limit, and its effect on health when handled. Reputed manufacturers are benefitting from use of this data while handling chemicals, but common public is mostly unaware of dangerous properties of chemicals, particularly carried by tankers/trucks on road.

Government/Media — expectation

It is expected that Government educate public at large through TV, Radio and Press about the dangers posed by various chemicals, particularly transported via tanker/trucks. Even Indian Chemcial Manufacturing Association (ICMA); Federation of Indian Chambers of Commerce and Industry (FICCI) or Tanker Owner Association should spend some money on TV advertisements or programmes to bring about awareness. Mobile foam fighters on highway should be kept for emergency fire.

Some budget must be kept by every industrialist for advertisement in educating the need of handling dangerous chemicals his factory produces. Government can take care through Directorate of Public Health, Safety and Factory Inspectorate. Public should not suffer for the mistake of tank lorry drivers or chemical goods carriers. A comprehensive insurance policy against such disaster should be made compulsory so that public exposed to chemical disaster is awarded suitable compensation.

Thus, for the safety of mankind, our future generation and our own happy life, it is mandatory that we give attention to our rights and obligations to society and expect the same from manufacturers of chemicals. It is better to design a good plant and do preventive work to reduce pollution and safeguard environments. It is very costly to protect environment when factories are not designed properly. The initial investment for pollution control means will go a long way in running the industry without fear or favour of government or public at large.

OPTIMISING ENERGY USE IN CHEMICAL PROCESS INDUSTRY

During the last four decades the induction of energy efficient technologies has lead to dramatic reduction in energy usage in chemical process industries. Due to compulsions from global competition to be highly cost competitive and the awareness thereof, companies are on a drive to reduce costs. Energy consumption in Chemical Process Industries (CPI) is dependent on the products manufactured and process employed. Energy cost in caustic chlorine plant is around 60 per cent of the manufacturing cost. But on an average the energy cost in CPI lies between 10–20 per cent of the total manufacturing cost. Therefore, energy cost reduction can play a significant role in increasing the profitability of any CPI. On an average the net profit of a company is around 3 per cent of the revenue. For companies having utility cost as 10 per cent of revenue, a saving of 20 per cent in utility consumption can result in an increase of profit by 66 per cent.

Energy conservation is many times understood as a cut in energy consumption but actually it is a cut in the misuse/waste of energy. Successful firms concentrate on efficiency first, products second, and then on marketing and sales.

Revenue expansion based on inefficient operations results in severe operating losses. Successful companies reduce cost to watch existing revenue levels. Unsuccessful companies attempt to increase revenue to cover existing costs.

Methodology of Optimising Energy Use

1. Measure and benchmark consumption. Compare with globally accepted norms.
2. Carryout energy audit and energy balance.
3. Examine availability of more energy efficient processes and equipment with higher efficiencies. Implement new technologies bringing in a reduction in energy and raw material consumptions.
4. Reduce cycle time by eliminating non-value adding activities.
5. Identify areas of losses and plan methods to reduce losses.
6. Reuse waste, harness waste streams.
7. Replace higher form of energy use by low grade/low cost/renewable energy.
8. Minimise transmission losses.
9. Measure and control.

Areas for Energy Optimisation in Chemical Process Industries (CPI)

Boilers and steam usage

1. For Solid fuel fired boilers: Convert stoker fired boilers to FBC
2. Optimise excess air. Provide continuous monitoring with auto adjustment of oxygen trim in large boilers and periodical checking in smaller boilers.
3. Preheat combustion air with waste heat
4. Install variable frequency drives (VFD) on large boiler combustion air fans having variable loads.
5. Burn waste stream if permitted, use bio waste like coconut kernel, rice husk, instead of conventional fuels.
6. Recycle condensate.
7. Recover flash steam from higher pressure condensate.
8. Pass steam through back pressure steam turbine rather than through pressure reducing station for low pressure steam.
9. Attend to steam leakages and repair damaged insulation.
10. Examine possibility of installation of cogeneration systems (combined electricity and steam generation)/trigeneration system (combined electricity, steam and refrigeration generation).

Pumps

1. Select the right pump to match head and flow requirements.
2. Make maximum use of gravity flow. Avoid intermediate storages to avoid pumping. For circulation system use siphon effect; avoid free fall (gravity) return.
3. Avoid throttling/bypass; to control flow, prefer speed controls or sequenced operation of pumps.
4. In pumping to systems having a number of non-continuous users, auto on-off valves/control valves need to be provided on users and VFD on pumps.
5. Check if pumps are operating near best efficiency points (BEP).
6. Segregate high head and low head loads and install separate pumps.
7. Operate booster pumps for small loads requiring higher heads, in place of operating complete system at higher head.
8. Operate cooling/chilling system with higher fluid differential temperature to decrease flow and hence save pumping energy.
9. Replace old pumps by high efficiency pumps.
10. Trim impellers wherever pumps are over designed.
11. Valve throttling indicates pump over design; replace pump with correct size pump or install lower size impeller.
12. Coat hydraulic passages of pumps with resins having better surface finish to reduce internal friction and increase efficiency.
13. Minimise pressure drop in piping by rerouting of pipeline, removing valves, which never need to be operated, and resizing of pipeline.

Process

Process is defined as a systematic sequence of chemical, physical or biological activities for conversion of raw materials into products.

1. Eliminate non-value adding activities.
2. Attempt to reduce cycle time.
3. Increase batch size to get higher throughput with minimal increase in utility consumption.
4. Reduce reflux in distillation by redesign of column height/replacement by more efficient packings.
5. Reexamine the usage of utilities. Attempt switching over from chilled brine to chilled water and chilled water to cooling water.
6. Examine the possibility of intermittent stirring. Control operation of stirrers by interlocking, providing timers.
7. Examine possibility of speed reduction.
8. Examine switching over from batch to continuous process.
9. Examine correction stages if any. Attempt getting it right the first time so that corrections are avoided. This will reduce utility consumption.
10. Remove redundant piping/equipment from the plant to make all leakages visible and easy to attend to.
11. Find alternatives to electrical heating systems.

Compressed air

1. Clean suction filter regularly.
2. Check/clean intercooler/aftercooler regularly.

3. Attend to suction and discharge valves of reciprocating compressors.
4. Reduce air compressor discharge pressure to lowest allowable setting.
5. Replace compressed air usage by air blowers wherever possible. Use hydraulic system rather than pneumatic cylinders. Use electronic controls in place of pneumatic controls.
6. Check for leakages regularly and attend to them immediately.
7. Install VFD for single compressor operation or control system for coordinating multiple compressor operation.
8. For small compressors (<25 HP) convert from load/unload operation to on/off operation.
9. For compressed air dryers use highest allowable dew point settings. Use refrigerated air dryer if within the operating range or else if desiccant dryer is required use heat of compression (HOC) dryer.
10. Check pressure drops in piping systems, and optimise pipe size/provide ring main system.
11. Check actual free air delivered (FAD) regularly, calculate specific power consumption and rectify/replace inefficient compressors.

Cooling towers

1. Control CT fans based on cold well temperature; use two speed or VFD if fans are few and on-off stage control if cells are many.
2. Select CT with low pressure drop, high efficiency PVC cellular fills in place of splash bars.
3. Periodically clean, water distribution nozzles. Ensure that no channelling of water flow is taking place. Uniform flow distribution will improve performance of cooling tower.
4. Optimise cooling water chemical treatment.
5. Replace aluminum fans by aerodynamic FRP fans.

Refrigeration Systems

1. Challenge the need of refrigeration system, particularly, for old batch processes. Optimise the temperature requirement.
2. Examine the possibility of vapour absorption system operating with waste heat streams in place of vapour compression systems.
3. Check regularly for correct refrigerant charge levels.
4. Check for damaged insulation/sweating.
5. Select multistage compressors with intercooling for low temperature applications.
6. Operate chillers with lowest possible condensing temperature and highest possible chiller (evaporator) temperature. For vapour compression chillers use the lowest possible cooling water temperature.
7. Carryout regular cleaning of condenser to ensure proper heat transfer.

Lighting

1. Select high efficiency lighting luminaries having highest lumens/watt output, e.g. compact fluorescent lamp (CFL), low pressure sodium vapour lamp.
2. Provide lighting transformer to reduce the voltage of lighting loads.
3. Make use of task lighting.
4. Make most use of daylighting by providing skylight.
5. Paint walls and ceiling with light colours.
6. Lower height of light fixtures.
7. Control lighting with clock timers, occupancy sensors, photocells and master switch.

8. Select ballast with high efficiency and high power factors.
9. Use LED lamps for indicating purpose.

Fans and blowers

1. Select fans with aerofoil fan blades, replace old in efficient fans by modern high efficiency fans/blowers.
2. Ensure that design of fans/blowers are matching with operating conditions if not replace with correct size fan/blower.
3. Replace throttle/bypass control by speed control.
4. Minimise speed to minimum possible.
5. Reduce pressure drops in system by proper design/sizing of ducting. Minimise bends in ductings.
6. Eliminate leakages.
7. Clean screens, filters, fan blades regularly.
8. Avoid idle running of fans by interlocking with main equipments.

Motors

1. Properly size the motor for optimum efficiency.
2. Use energy efficient motors for continuous operating loads.
3. Balance three phase loads. An imbalanced voltage can reduce efficiency of motor by 3–5 per cent.
4. Connect motors remaining underloaded (<40 per cent) continuously, in star.
5. Rewound motors should be checked for efficiency.
6. Provide capacitor banks at MCC to correct PF.
7. Use soft starters/VFD instead of fluid coupling for loads having high starting torque or loads prone to jamming.

By following the above proven approach that is simple and easy to implement any organisation can save substantial energy and tap the potential for improving profit margins, which the entire chemical process industry is striving hard to achieve. It is mandatory for all registered companies to include, as a part of the director's report, the energy conservation measures taken and the additional measures being implemented. It is suggested that readers can use these as clues to further work on these areas and attempt for higher energy efficiencies.

INTEGRATING ENVIRONMENT, ENERGY, TECHNOLOGY AND BUSINESS PROCESSES

Chemical industry operations contend with frequent manual material handling; product changeovers, chances of cross-contamination; a complex waste-water stream having widely fluctuating characteristics with varying hydraulic chemical and biological load; spillages during handling; reprocessing of low yield or off-specific batches due to power failures; and the need to handle a number of gaseous pollutants emanating from the same plant. Thus, a total approach to managing the operations and the business process is imperative for survival and growth. Traditionally, the practice often has been to work on issues such as energy conservation, environmental protection, yield improvement and quality assurance separately, although a close look will reveal that all these issues are interlinked and produce an optimum solution when worked with a total approach by a multi-disciplinary team.

A Case Study

Figure 9.1 illustrates the process flow sheet for the manufacture of a high purity fine chemical. In the traditional process technology, highly energy intensive conventional evaporation and distillation techniques were used to concentrate the product solution. The operation consumed a substantial quantum of energy in the form of electricity, steam, chilled water and process water and also generated substantial biological and chemical oxygen demand in the waste-water stream due to significant quantum of raw material in the waste stream because of relatively low yields. Despite the quality of the product being superior, the profit margin was increasingly coming under pressure in the emerging competitive marketplace.

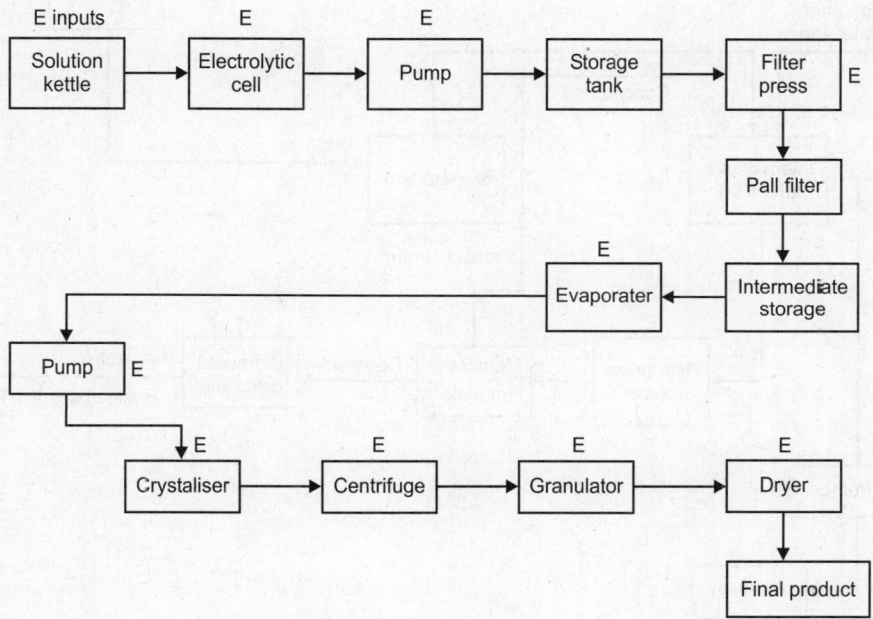

Fig. 9.1. Flow diagram of original process: High purity fine chemical.

In the existing process, the unit operations were fine-tuned. This improved the yields but no quantum leap was feasible until the concentration processes were replaced by an elegant separation process. After detailed investigation by a multi-disciplinary team, it was found feasible to replace the existing concentration processes by a state-of-the-art membrane separation technology. This required a significant amount of development work, but with the promise of substantial savings. Figure 9.2 illustrates the modified process with membrane separation route which finally achieved the objectives.

Energy intensity in the traditional and membrane processes are listed in Table 9.3.

Table 9.3. Energy intensity in the traditional and membrane processes.

Vacuum evaporation process	*Membrane separation process*
Steam	Electricity for solution pump
Chilled water in evaporator	
Electricity for vacuum pump and stirrer	
Chilled water in vacuum pump	
Process water for vacuum pump	
Waste-water from vacuum pump	

174 Pollution Control in Chemical and Allied Industries

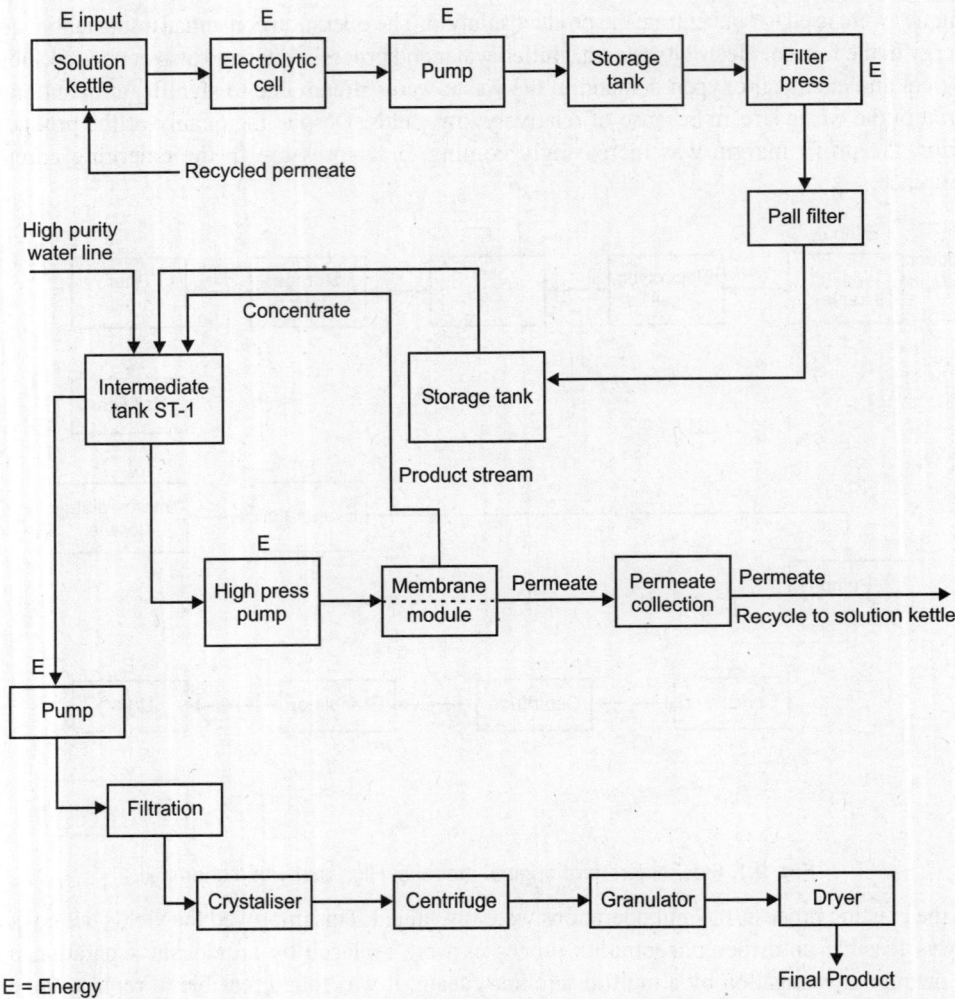

Fig. 9.2. Flow diagram of modified process: Membrane separation route.

To summarise, the benefits in the membrane separation process are:
1. Significant reduction in total energy consumption.
2. Reduced raw material consumption and improved yield due to recycling of the permeate.
3. Reduced effluent load, i.e. significant reduction of COD and BOD in waste-water stream, which, in turn, eliminated the heed for putting up the additional waste-water treatment facility which was being planned.

Figure 9.3 shows a conceptual model for integrating the Product Specification, Process Technology and Business Process in a highly competitive economy. The customer will demand quality product at the least cost from a company which is fully complying with international Quality Assurance and ESH standards. The company, in turn, will be constrained to comply with ISO 9000 and ISO 14000 systems and yet be cost competitive.

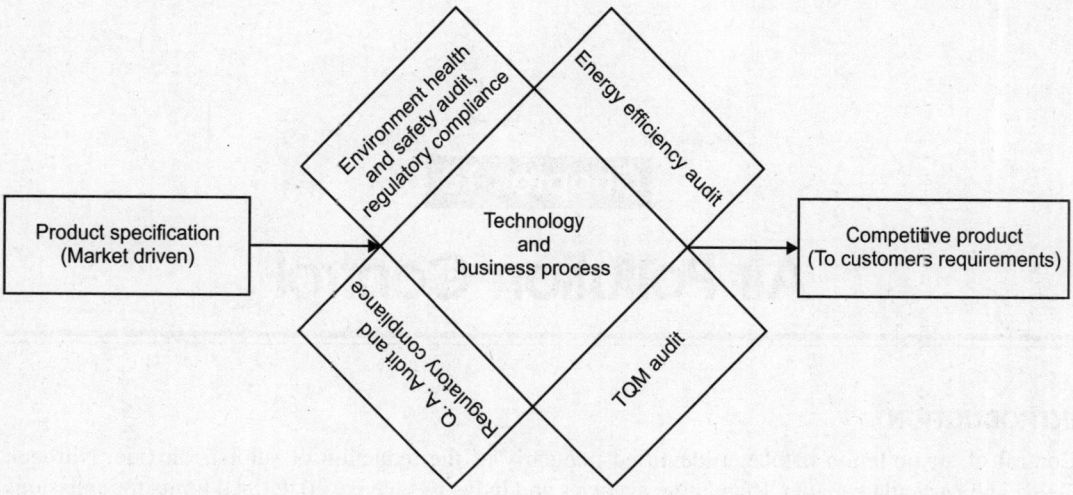

Fig. 9.3. Activity chain for a technology based product.

Chapter 10

Air Pollution Control

INTRODUCTION

Control of air pollution has been identified popularly as the reduction of sulphur dioxide, Nitrogen Oxide and particulate matter. Regulating agencies and industries have well defined limits for emissions and control of these pollutants. However, air pollutants such as volatile organic compounds, organic and inorganic odourous compounds have so far not been regulated by Pollution Boards in India. Attention to the control of these pollutants has been meagre or non-existent by the industry.

VOLATILE ORGANIC COMPOUNDS (VOCs)

Volatile organic matter, particulate matter and nitrogen oxides react in presence of ultra violet light to produce a haze of chemicals and result in an increase of the ground level concentration of ozone. Exposure to ozone is hazardous to mankind and it can cause chest pain, cough, bronchitis, emphysema and asthma. The environmental protection agency (EPA) in the USA imposes an air quality standard of 0.08 ppm of ozone measured over 8 hours. Europe, USA and Japan are introducing regulations and controls to limit all pollutants including VOCs and Ozone. Odours caused mostly by volatile organic compounds such as amines, sulphides and others are also being regulated. For the gasoline industry, European community is limiting VOC emissions to 35 grams TOC (total organic carbon) per m^3 of gasoline loaded. In the USA, this limit is 10 g/m^3 of VOCs. In the absence of limits for VOCs, it is estimated that 2,31,000 tons of gasoline will be emitted by the gasoline distribution industry alone in USA. Similarly 12,000 tpa of organic compounds will be emitted by the US. petrochemical industry in the absence of VOC control. Besides the petroleum and petrochemical industries, volatile organic carbon emissions are to be regulated and controlled in plastics, rubber chemicals, electronics and printing industries.

Odours

In addition to an increase in the overall pollution loads, volatile organic compounds can cause acute discomfort and nuisance due to odours. Industries such as distilleries, paper and pulp, asphalt, paints, printing/film industry, food processing and waste-water treatment plants emit odours. Compounds containing amines, sulphides, aldehydes, mercaptans, ketones and hydrogen sulphide gas cause odours which can be beyond the threshold limits for the surrounding population. Fragrance and flavours of flower and sweet industries may have pleasant odours but they can be a source of discomfort and air pollution for the neighbourhood if they are constantly exposed.

In European countries, several flowers and fruit processing plants have installed odour removal units. The odour threshold limits for concentration of H_2S and some organic compounds in air are given in Table 10.1.

Table 10.1. Odour threshold limits for concentration of H_2S on organic compounds.

Compound	Threshold limit ppm	Type of odour
H_2S	0.00047	Rotten eggs
NH_3	48.8	Sharp, pungent
Cl	0.314	Pungent, suffocating
Dimethyl sulphide	0.001	Decayed vegetables
Ethyl mercaptan	0.00019	Decayed cabbage
Pyridine	0.0037	Irritable, disagreeable
Triethylamine	0.08	Fish

This is only a short list and several other examples can be given for compounds which originate from chemical and food processing industries and cause strong odours. Unfortunately, no legislation or regulation exists in most countries.

Odour removal plants have been put up by a few industries in Europe, USA and Japan only in response to objections by the surrounding population.

Technology for Removal of VOCs and Odour

Both VOCs and odours can be controlled by technologies which are well known and can be selected on the basis of suitability and cost. It is, however, essential to characterise the emission by the flow rate of contaminated air, the concentration and nature of VOCs and odour causing compounds. Some of the technologies for their control are listed below:
1. Biological oxidation.
2. Absorption with chemical reaction.
3. Absorption and desorption.
4. Thermal oxidation-regenerative or recuperative.
5. Catalytic oxidation-regenerative or recuperative.

In this section, we shall briefly discuss biological oxidation and regenerative catalytic oxidation.

Biological Oxidation or Biofiltration

Biological oxidation or biofiltration has been successfully applied in Europe for the control of odours and volatile organic carbon emissions. In this process, the air containing VOCs is passed through a chamber which is humidified by water (Fig. 10.1). It is led to the top of a biofilter which consists of a bed of active material. The biological oxidation is carried out by active organisms which are present on the surface of compost material. Compost or biofilm is supported on a packing material such as rings or saddles made of plastic.

The down flow fixed film reactor can achieve 95–99 per cent removal of VOCs and odorous compounds. Flows of 15,000 m^3/hr. and 3,000–5,000 mg/m^3 of VOC load can be handled at room temperature.

The biological bed requires small amounts of chemical nutrients and has to be kept wet by sprinkling water over it. A typical bed or biofilter has a life of 3 to 5 years.

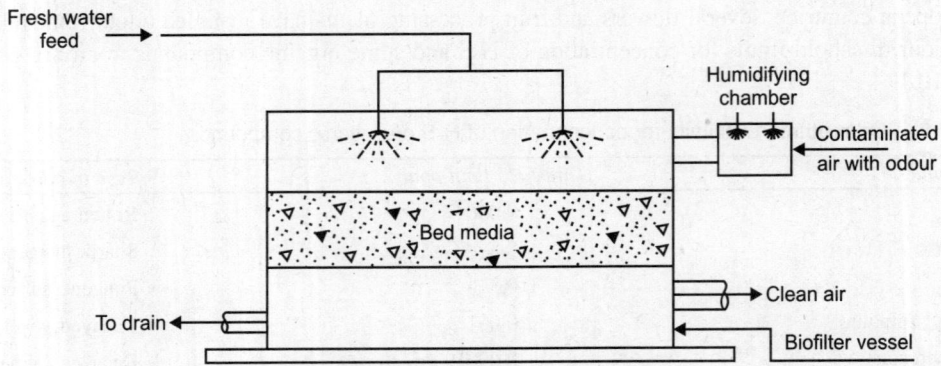

Fig. 10.1. Biofilter for removal of odour.

There are currently more than 500 biofilters operating in USA and Europe for VOC removal. They have the advantage of low operating costs. Products of oxidation are water, carbon dioxide and a small quantity of liquid effluent which contains no hazardous material. Control of moisture in the bed by an automatic control system avoids dryout of bed which can kill the bioactive organisms.

Biofiltration is an attractive technology for economically removing biodegradable organic compounds from air. Relative biodegradability of various VOCs is given in Table 10.2.

Table 10.2. Ease of biodegradibility of VOCs.

Higher biodegradibility	Medium biodegradibility	Lower biodegradibility
Acetaldehyde	Acetonitrile	Dioxane
Butadiene	Benzene	1-1-1-Trichlorethane
Methanol	Carbon disulphide	Perchloroethylene
Cresol	Hexone	
Styrene	Methyl ethyl ketone	
	Toluene	
	Xylene	
	Phenol	

Flow sheet of Monsanto's Dynazyme process including a biofilter is shown in Fig. 10.1. It consists of a bed of proprietary support material durafil for the compost, and moisture control system. Residence time depends on the type of VOCs and is usually in the range of 15–90 seconds. Moisture content in the bed is kept in the range 40–60 per cent and the bed temperature is 20–35°C. pH of the liquid drain is controlled by addition of nutrients.

Biofilters are widely used for removal of odorous compounds from air. In Europe, particularly Holland, they are also used to remove fragrance from exhaust air. Applications are found in food industries such as coffee, sweets and chocolates and in flavour and fragrance industries.

The filter for fragrance removal consists of a polymer or active carbon support on which a film of special bio-seed material with enzyme is grown. The bed has a life of 4–5 years and can be reactivated by fresh inoculation of active material.

Odours can also be removed by chemical reaction of Mercaptans and H_2S with alkaline sodium hypochlorite solution. ICI, UK markets a specially designed odourgard process in which a small quantity

of catalyst is used to enhance the rate of removal of H$_2$S by scrubbing the solution of hypochlorite. There are 30 odourgard installations working in various food processing, chemical and sewage treatment plants.
1. Typical example of a simple Biofilter plant-St. Louis, Lemay Plant, 3500 cfm, inlet 10–200 ppm H$_2$S/mercaptan, outletless than 1 ppm.
2. Typical example of an odourgard plant - Bridlington POTW, 24,000 m^3/hr, inlet 70 ppm organic sulphides, outlet 0.3 ppm.

Catalytic Oxidation

Biological oxidation cannot be applied to VOCs which are not easily bio-degradable and when concentrations of VOCs are high. Absorption and desorption can be applied in special cases depending on the nature of the VOC, suitability of the absorbent material and reclaim value of the VOC material. The most widely applicable method of VOC removal is oxidation which essentially destroys the VOCs whose cost of recovery is prohibitively high.

Two alternatives for oxidation are:

Thermal oxidation

Heat is added to gas stream to incinerate or burn VOCs present. The gases have to be heated to a temperature of about 800°C for 0.5 to 1 second.

Catalytic oxidation

Oxidation of VOCs is carried out in the presence of a catalyst. The activation energy is lower than that for thermal oxidation and the reaction takes place at 200°–500°C with residence times of 0.2 to 1 second.

Several variations of the above processes such as recuperative or regenerative thermal or catalytic oxidation processes have been practiced for removal of VOCs. Regenerative catalytic oxidation has the advantage of recovering the sensible heat of gases leaving the catalyst bed for preheating of inlet gas. They can attain high destruction efficiency of 99 per cent for VOCs and high thermal efficiency of 90–95 per cent. The fuel required to achieve the destruction of VOCs is minimum for regenerative catalytic oxidation.

The catalysts used for oxidation of VOCs are Platinum group metals or base metal oxide catalysts. The Platinum group metals are expensive but are compact and offer lower pressure drops. Base metal oxide catalysts (Fe, Cr, Mn) in pellet form are less expensive and can be designed for a wide range of flow rates and pressure drops.

Ease of oxidation varies from methane which is most difficult to oxidise to oxygenates which are easily oxidised as per Table 10.3.

Table 10.3. Oxidation of VOCs.

Difficult	Easier
Methane ⟶ Saturated hydrocarbons ⟶ Aromatics ⟶ Olefins ⟶ Oxygenates	

Regenerative catalytic oxidation of VOCs

Regenerative catalytic oxidation consists of a bed of base metal oxide catalyst pellets with a layer of ceramic media below and a second ceramic layer above the bed. The ceramic beds act as heat sink and trap the sensible heat available in the gases after the combustion of VOCs and fuel if used. This heat is transferred to cold inlet gases containing VOCs to preheat them to the reaction temperature prior to the oxidation of VOCs in the catalyst bed.

A typical design of Regenerative Catalytic Oxidation system is shown in Fig. 10.2. This is the Dynacycle system offered by Monsanto Enviro-Chem Systems, USA, in which there are two chambers each consisting of a catalyst bed and two ceramic layers. In a typical cycle, air with VOCs enters the bottom ceramic layer which has been heated in a previous cycle. When it reaches the catalyst layer, both the air and catalyst are at the reaction temperature required for oxidation. The temperature of the catalyst bed and that of exit gases increase and the gas temperature falls as the gases flow through the second ceramic layer which has been cooled by loss of sensible heat in the previous cycle. The exit gas temperature reduces and the temperature of the second ceramic layer increases as a result of heat transfer. After a specified interval of time, the entire flow is reversed by opening and closing of valves and the same cycle repeats with the air entering the second ceramic layer in the reverse direction. The concept and plant design of this unsteady state catalysis were first developed by Prof. Yuri Matros at the Institute of catalysis in Novosibirsk. This was commercialised by Monsanto in collaboration with Prof. Matros in USA and has now been applied extensively for the regenerative catalytic oxidation of VOCs. The flow diagram in Fig. 10.2 shows two chambers but the basic principle employed is the same and with two chambers the flow reversal is achieved smoothly without any significant discontinuity. Typical interval for reversal of flow is 5 to 10 minutes.

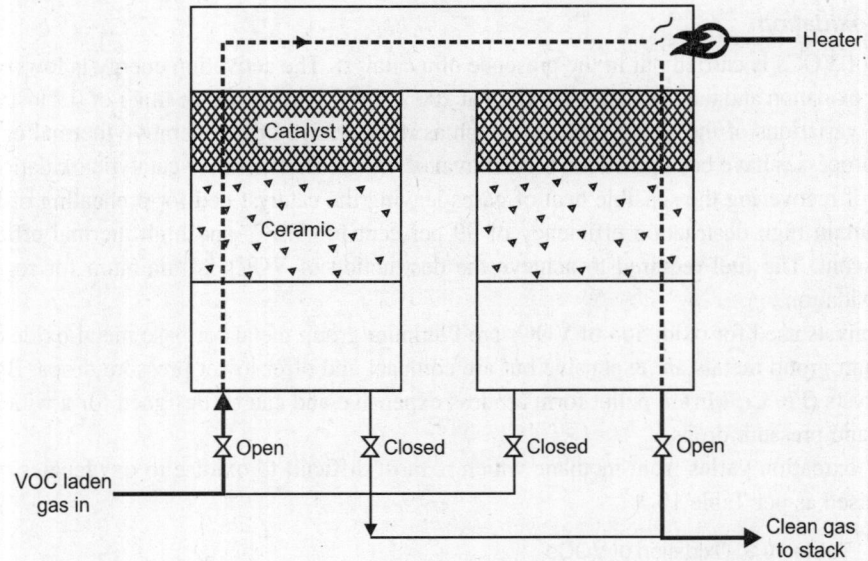

Fig. 10.2. Regenerative catalytic oxidation for VOC removal.

There are several operating installations of regenerative catalytic oxidiser based on the design developed originally by Prof. Matros. These units are known as Dynacycle owing to the cyclic nature and periodic reversal of flow directions. Capacities range from 800 to 80,000 m³/hr, and VOCs such as C_4–C_8 alcohol, aldehyde, toluene, styrene, MEK, mineral oil, etc. are removed effectively. The regenerative catalytic oxidiser handles a flow rate turn down of 10:1 and changing VOC concentrations of 1 ppm to 5000 ppm efficiently. Care must be taken to avoid use of catalytic oxidisers for air streams containing phosphorus, heavy metals and silica which may act as poison for catalysts. The VOC concentration should be lower than that which will generate a continuous flame. VOCs should be at a concentration level that will provide heat for raising air to the catalyst activation temperature. If the concentration is too low, small amounts of fuel may be needed for achieving the reaction temperature of 200°–500°C. Dynacycle units have been successfully used in aluminium rolling mills, electronics and magnetic tape manufacturing units, plastics and plasticiser industries, spray painting/drying shops and phenol/acetone plants.

Thus, the technology for controlling emission of hazardous VOCs and odour causing compounds is now developed and available to industries in India. There is a need to introduce proper legislation, regulations and awareness amongst the industries to apply the most appropriate technology to tackle the problems of VOC and odour.

The implementation of the available proven technologies will improve the environment and benefit the population and industries.

ADVANCED SCRUBBER TECHNOLOGY SOLUTIONS FOR POLLUTION CONTROL

In the wake of stringent pollution control regulations, scrubbers have come a long way in effectively reducing gaseous industrial emissions. Different models of wet gas scrubbers which use circulating liquid to absorb acid vapours, are available for cleaning the flue gas. This section gives an insight on different types of wet gas scrubbers and their mechanism of action together with the factors that impact the performance of the scrubbers.

Throughout the world, more and more stringent air pollution regulations require industrial and utility operations to install pollution control equipment with better efficiencies. In many cases, scrubber technology can, and should, be used. In the past, scrubbers have been used to perform three basic operations:

1. To quench the flue gas down to its adiabatic saturation temperature.
2. To absorb acid gas emissions.
3. To remove particulate material.

In many cases, the scrubber is required to handle not just one, but two, or even three of these functions in one unit. While scrubber technology has been around for many years, changes have been made to the auxiliary equipment surrounding the unit, so that the scrubber is more effective and more reliable. In addition, costs associated with various reagents and disposal of the scrubber effluent, requires that the plant operator investigates all possible options to ensure that the scrubber unit selected meets the overall air emission requirements, while being economical to operate, and be reliable in its performance.

Finally, the plant operator must ensure that solving one environmental issue does not lead to other problems.

There are many types and models of wet gas scrubbers in the marketplace. Basically, all have a section where liquid, (typically water), is contacted with the incoming flue gas. All scrubbers have a method for removing the water droplets from the gas before it leaves the scrubber. How the liquid is introduced, and how droplets are removed vary from one model to another. When water is mixed with a hot gas that is not saturated, adiabatic saturation occurs, and the gas is 'quenched'. Depending upon particle size and pressure drop, a percentage of the particulate matter in the flue gas will collide with the water droplets.

Finally, acid gases are absorbed in the scrubbing liquid, where it can be reacted with a chemical reagent (Fig. 10.3).

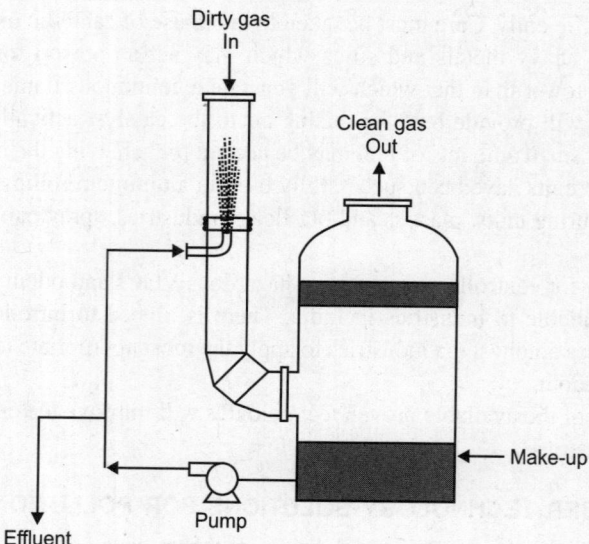

Fig. 10.3. Dyna wave froth scrubber.

Requirements for Acid Gas Removal

One of the most common reasons for use of a wet scrubber is for acid gas removal. Depending upon the upstream process, different acid gases can be present in the flue gas as vapours. Typical acid gases include SO_2, HCl and HF, though there could be many other gases. The amount of the acid vapour that can be removed from the flue gas is a function of:

1. Process parameters of the mass transfer device.
2. Kinetics of the reagent.
3. Characteristics of the effluent.

Mass Transfer

Wet gas scrubbers use circulating liquid to contact the flue gas and absorb the acid vapours into the liquid stream. The amount of circulating liquid is a key process parameter for wet scrubbers. Generally referred to in terms of L/G (liquid to gas ratio; gpm/1000 acfm), the amount of circulating liquid must be enough to fully quench the gas and absorb the appropriate amount of acid vapours from the gas stream.

There are a number of different designs of wet gas scrubbers that are used for acid gas absorption. The type of device may limit which reagent can be used. This will affect the total L/G ratio, which will have a direct impact on operating power. Some examples of the various types of units include packed towers, spray towers and froth scrubbers.

1. Packed tower: Packed tower is an efficient mass transfer device requiring relatively low L/G ratios. However, both the reagent and the resultant salts formed must be soluble, as otherwise, the packed tower is susceptible to pluggage. In addition, packed columns are not effective at removal of particulates from the gas stream since gas side pressure drop is typically very low.
2. Spray towers: Spray towers generally have the highest L/G ratios for a particular application, but can use liquids with either suspended or dissolved solids. Again, spray towers can only operate at low-pressure drop. They are therefore not very effective at removal of particulate from the gas stream.

3. Venturi scrubbers: Venturi scribbers are effective for particulate removal, but are not that good as a mass transfer device.
4. Reverse jet froth scrubbing technology: Reverse jet froth scrubbing technology of MECS Dyna wave can handle a liquid with both dissolved and suspended solids. In addition, L/G ratios are modest, and the gas side pressure drop can be varied. This allows this unit to handle both acid gas emissions and particulate with the same device.

Reagent Kinetics

Simple kinetics of chemical reactions can affect the amount of acid gas vapours that will be absorbed in a scrubbing device. Once the acid gas vapour is absorbed into the liquid, acid is formed, decreasing the pH of the circulating liquid. Lower pH tends to inhibit further acid vapour absorption. To overcome this, reagents are added to the circulating liquid to react with the acid in solution, maintaining an acceptable pH range for further vapour absorption. Figure 10.4 indicates the effects of reagent and L/G selection on SO_2 removal efficiency. There are numerous reagents that can be used, including caustic, limestone and magnesium oxide.

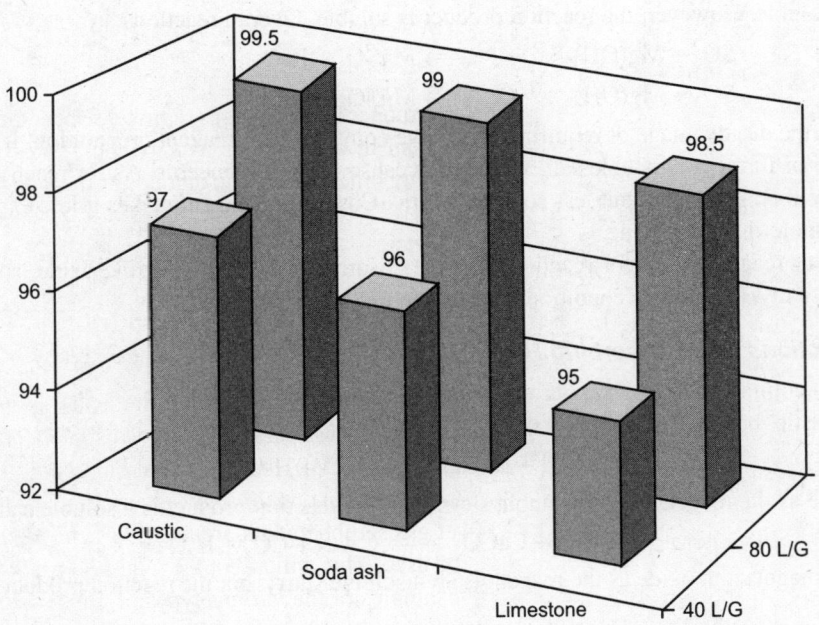

Fig. 10.4. Expected removal efficiency for a Dyan wave froth scrubber.

Typical reactions when absorbing SO_2

Using Caustic as the reagent, it is soluble and the salt products formed are also soluble.

$$SO_2 + NaOH \longrightarrow NaHSO_3$$
$$SO_2 + 2NaOH \longrightarrow Na_2SO_3 + H_2O$$
$$SO_2 + 2NaOH + \tfrac{1}{2} O_2 \longrightarrow Na_2SO_4 + H_2O$$

The absorption of SO_2 with caustic is the easiest method of removing SO_2 from the flue gas. The acid-base reaction is fast and the equipment required is minimal. However, the caustic reagent is the most costly and the resulting by-product, soluble sodium sulphite/bisulphite can be a potential disposal problem. The sulphite/bisulphite composition depends upon operating pH. Higher pH shifts the salt formation to sulphites.

In the case of limestone as the reagent, which is insoluble, the salts formed are also insoluble.

$$SO_2 + CaCO_3 \longrightarrow CaSO_3 + CO_2$$
$$SO_2 + CaCO_3 + \tfrac{1}{2}O_2 \longrightarrow CaSO_4 + CO_2$$
$$SO_2 + CaCO_3 + \tfrac{1}{2}O_2 + 2H_2O \longrightarrow CaSO_4 \cdot 2H_2O + CO_2$$

Limestone is probably the most commonly used reagent for reaction with SO_2. In addition, the salt formed is insoluble which means that it can be filtered from the liquid and then disposed off in solid form. This system does have higher capital cost. Reagent preparation, and vacuum filtration equipment add significantly to the overall project costs. However, for very large flue gas flows, these costs are quickly offset by the lower operating costs. Limestone is also considerably less reactive than most other reagents.

A third possible reagent is magnesium oxide. The reagent slurry used is magnesium hydroxide, which is insoluble. However, the reaction product is soluble. Overall reactions are:

$$SO_2 + Mg(OH)_2 \longrightarrow MgSO_3 + H_2O$$
$$SO_2 + Mg(OH)_2 + \tfrac{1}{2}O_2 \longrightarrow MgSO_4 + HP$$

MgO has the disadvantage of requiring expensive equipment for reagent preparation. It also has the disadvantage of forming a soluble salt, which may cause disposal concerns. As for reactivity, MgO is more reactive than limestone, but less so than caustic. Cost of the reagent makes it less expensive than caustic, but more than limestone.

Some other reagents used for reaction with the absorbed SO_2 are lime, zinc oxide, soda ash, and ammonia. Usually, unique site conditions dictate the use of these reagents.

Typical reactions when absorbing HCl

Another very common acid gas vapour requiring emission controls is HCl.

Using caustic, both the reagent and the reaction products are soluble.

$$HCl + NaOH \longrightarrow NaCl + H_2O$$

Limestone as the reagent is an insoluble slurry, but the reaction product is a soluble salt.

$$2HCl + CaCO_3 \longrightarrow CaCl_2 + CO_2 + H_2O$$

Whereas magnesium oxide as the reagent is an insoluble slurry, but the reaction product is a soluble salt.

$$2HCl + Mg(OH)_2 \longrightarrow MgCl_2 + 2H_2O$$

Typical reactions when absorbing HF

The third acid gas vapour that is encountered frequently is hydrogen fluoride.

Using caustic, both the reagent and the reaction products are soluble.

$$HF + NaOH \longrightarrow NaF + H_2O$$

Using limestone, both the reagent, and the reaction products are insoluble.

$$2HF + CaCO_3 \longrightarrow CaF_2 + H_2O + CO_2$$

Finally, using magnesium oxide, the reagent and the reaction products are insoluble:
$$2HF + Mg(OH)_2 \longrightarrow MgF_2 + 2H_2O$$

Reagents Use-Economics

AS shown above in the listing of some typical reactions, there are numerous options when selecting a reagent. As is usually the case, the economics of the situation dictates which reagent is most attractive for a particular application.

In general, for small gas flow rates where minimal amount of reagent is required, caustic is usually favoured. For large gas flow rates where large amounts of reagent is needed, then limestone is usually favoured.

Caustic is considerably more expensive than limestone. While prices vary for different geographical regions. When significant amounts are going to be used, operating costs favour using limestone. However, limestone has the burden of requiring more expensive material handling equipment upstream of the scrubber. For small applications, the savings in capital cost of not needing this, additional equipment when using caustic, easily offsets the higher operating costs.

Cost for handling the effluent will also be different depending upon which reagent is selected. For each application, the end-user should work closely with his equipment supplier to determine which type of system provides the overall best total cost.

Effluent Characteristics

The reaction products formed can also impact the overall design of the scrubber system, impacting both the capital and operating costs. As is obvious from the data listed above, one of the key differences is whether the salts formed are insoluble or soluble. Insoluble salts mean that the circulated liquid in the scrubber will be slurry with suspended solids. This can be a disadvantage for some of the scrubbers available for use.

Packed columns: Packed columns do not handle suspended solid slurries, and can become plugged very easily. Spray tower can handle limited slurries, provided that the concentration of solids do not become too high. However, because spray towers use numerous spray headers, and hundreds of spray nozzles, they can become prone to erosion if salt concentration becomes high.

Venturi scrubbers: Venturi scrubbers can handle medium amounts of dissolved and suspended solids. They have only limited piping which is essential for slurry piping. However, they do have the drawback of medium pressure drop through the nozzle which limits the total percentage of solids.

Reverse jet froth scrubbers: Reverse jet froth scrubbers with only very few, low pressure nozzles and minimal vessel internals can handle high solids concentrations and minimise the total effluent stream.

Insoluble salt products can minimise disposal costs and water consumption. The insoluble salt products can usually be filtered with vacuum filtration equipment. Much, if not all, of the recovered water can be sent back to the scrubber. The resulting dried, solid products can be disposed off in a landfill.

Soluble salts are easier to handle in the scrubber, since they are in solution, and thus do not tend to either erode, or plug, certain type of scrubber units. However, disposal means that a liquid effluent stream with the dissolved salts must be discharged. In some cases, local sewer systems, or deep well injections can handle this liquid. However, if neither of these options is available, then disposal costs become much higher.

Fundamental Principles for Particulate Removal

Another basic operation that scrubbers are called on to deal with is the removal of particulate from the flue gas. As stated earlier, removal of particulates requires energy, in the form of gas side pressure drop. Two fundamental principles govern the amount of particulate that can be removed:
1. Particle size distribution of the catalyst fines (PSD).
2. Gas side pressure drop utilised by the scrubber.

In order to predict a particulate efficiency for a given application, the particle size distribution of the particulates is required. At a given pressure drop, the removal of particulate at each size range varies. The smaller the particulates, the less is removed. Particulates less than 1 micron in size are especially difficult to remove. Therefore, high removal efficiencies of small PSDs require high pressure drops. Assuming a particle size distribution as given in Table 10.4, a curve (Fig. 10.5) can be created that relates the expected particle removal efficiency with gas side pressure drop.

Table 10.4. Example of particle size distribution.

Particulate size (microns)	Particulate removed (wt%)
<0.1	0.1
<0.3	4
<1.0	35
<1.5	54
<3.0	75
>20	6

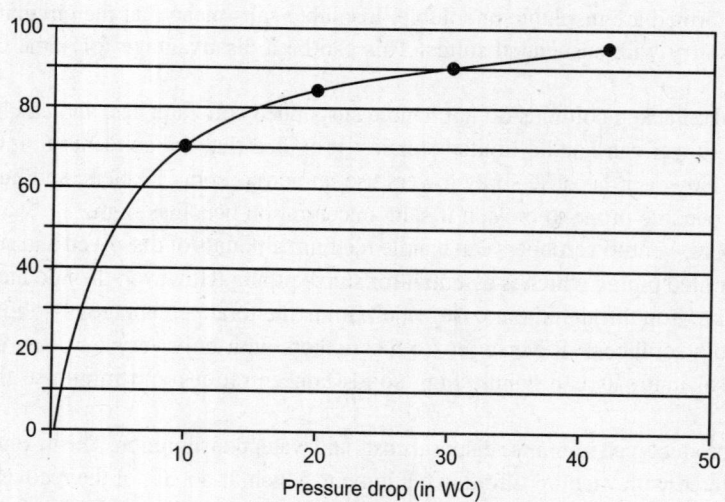

Fig. 10.5. Estimate total removal efficiency.

As can be clearly seen from the examples above, for effective particulate removal, the scrubber must be capable of sufficient gas side pressure drop.

Venturi scrubbers: Venturi scrubbers are very effective in this respect, and if only particulate removal is required, they are very often the type of scrubber chosen.

Packed columns and spray towers: Packed columns and spray towers are very low pressure drop devices. As such, they are not used as a primary means for particulate removal.

Reverse jet froth scrubbers: Reverse jet froth scrubbers can be designed for either low or high, gas side pressure drop. For applications requiring particulate removal and acid gas absorption, they are the ideal choice for a single scrubber that can handle both issues.

Distinction of SO_3 Removal Mechanism

Opacity from a flue gas source is often regulated, and requirements can vary. Opacity is caused by the diffraction of light waves by small particles of solid or mist. Particles having a diameter between 0.3 and 0.6 microns cause the greatest diffraction of light, since their size is closest to the wavelength of visible light. SO_3 hydrolyses to form sulphuric acid mist, H_2SO_4 as it is cooled in the presence of water vapour. The mist formed is a very fine submicron liquid droplet (~0.4 microns) and is thus a major contributor to opacity. For a liquid droplet, the removal mechanism is not an absorption phenomenon, but rather by mechanical means similar to that required for solid particulate matter. This mechanism remains the same for any wet scrubber and makes SO_3 much more difficult to remove than SO_2. In wet gas scrubbers, particulate matter (solid or liquid) requires energy to be imparted on the gas stream to capture the particulates by means of inertial impaction. This energy is usually measured in the form of gas side pressure drop. The smaller the size of the particulate, the more energy (higher pressure drop) must be imparted to capture the particulate. Figure 10.6 reflects approximate removal efficiencies of a scrubbing device using inertial impaction for removal of SO_3 (H_2SO_4 mist) at different gas side pressure drops.

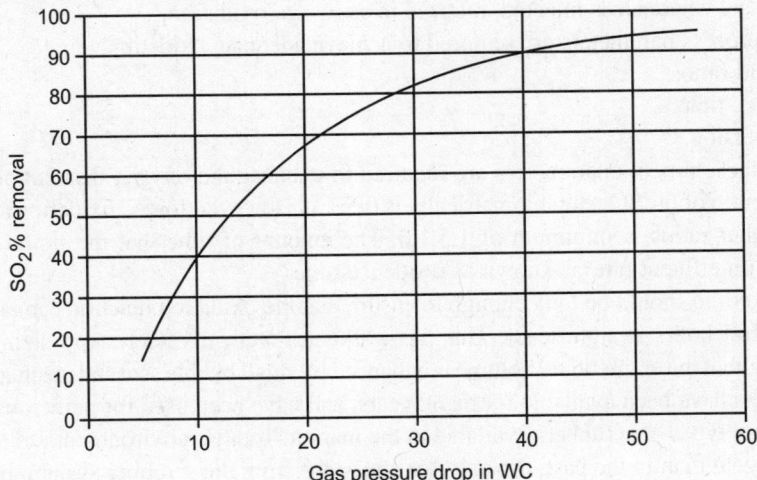

Fig. 10.6. Acid mist removal efficiency vs. gas side pressure drop.

As already mentioned, both venturi scrubbers, and froth scrubbers such as the MECS Dyna Wave, can be designed for the proper gas side pressure drop. Packed columns and spray towers are low pressure drop units, and as such cannot remove significant amounts of SO_3. It should be mentioned that SO_3 and particulates can also be removed by wet electrostatic precipitators (WESPs). These devices are very efficient in removing submicron particulates but can be capital intensive. However, if the acid gas scrubber is a packed column or spray tower, then a device such as a WESP is almost always required if SO_3 is a concern.

Achieving Effluent Requirement Norms

As previously discussed, depending upon the acid gas being removed and the reagent used, either a soluble or insoluble salt is formed. The soluble salts must be discharged from the scrubber to prevent build up. When discharged, they can be sent to either the plant waste-water system, to an evaporation pond, or to deep well injection. Many waste-water treatment facilities are now restricting the amount of salts that they can handle because of limitations set on their discharge. Thus, discharge of soluble salts may require the additional cost of either an evaporation pond, or a deep well injection system. Both can add considerably to the cost of the basic scrubber system. For units handling insoluble salts, it is very common to install a vacuum filtration system. Here, the water is removed, and the salts can be disposed off as a solid waste. Usually, most of the water removed can be recycled back to the scrubber.

In either case, if the primary acid gas is SO_2, it is very common to oxidise the salts generated. In the case of sodium salts, the sodium sulphite and bisulphites formed have a very high chemical oxygen demand (COD). Discharge of these salts to a waste-water treatment system can very quickly overlad the capacity of that system to inject enough oxygen into the liquid. In this case, it is common to oxidise the sulphites in the scrubber sump to the sulphate form.

When limestone is used as the reagent, calcium sulphite is formed in the scrubber. This is an insoluble salt, but due to its structure, is very difficult to filter. Again, the salt can be oxidised to the sulphate form. This salt is also insoluble, but is easily filtered. In both cases, the oxidation of the sulphites to sulphate is required. To handle this, air can be injected into the scrubber sump liquid to oxidise the sulphite to sulphate. This is referred to as *in situ* oxidation. Likewise, the scrubber effluent can be sent to a separate vessel where air is injected, referred to as *ex situ* oxidation.

Three main process parameters are required to achieve adequate oxidation:
1. Air/liquid ratio.
2. Residence time.
3. Air dispersion.

All three of these process parameters are required to enhance the oxygen dissolution rate limiting step. The molar ratio of $O_2:SO_2$ stoichiometrically is 0.5:1.0. However, forced oxidation systems should run at much higher ratios, a minimum of 1.5:1.0. The amount of time that the liquid spends in the scrubber (based on effluent rate) is known as residence time.

The residence time should be long enough to ensure that the oxidation reaction can take place. This duration of several hours is significant. This has a direct affect on vessel sump sizing. Finally, air dispersion throughout the entire liquid volume is enhanced by small bubble size and mechanical agitation.

Thus scrubbers have been available for many years, and have been used for numerous applications. There are different types of scrubbers available in the market. Today's environmental requirements are much more stringent than in the past, not only for emissions from the scrubber system, but also for the resulting liquid effluent discharge from the scrubber. For each application, the scrubber is called on to perform different functions. It is imperative that the end user analyses initial capital cost not only for the scrubber system itself, but also for all the auxiliary equipment required with the scrubber. These capital costs must be balanced against the operating cost of the unit, and the impact the scrubber will have on the operating cost of other downstream equipment, in particular, the waste-water treatment system.

There is now a wealth of experience with the use of scrubber technology. It is particularly important that the end-user works in close collaboration with the technology provider to take advantage of this knowledge, to determine the overall best solution. Looking merely at the initial capital cost of the lowest cost provider may end up being a costly answer.

Chapter 11

Water for Safe Use in Chemical and Allied Industries

INTRODUCTION

Effect of ions, metals, microbes in water range from beneficial to troublesome to dangerously toxic. Some metals, ions are essential, others may adversely affect water used for drinking, water used in analytical process or used in laboratory work. Some metals found in drinking water can cause a variety of health concerns. Some are known or suspected carcinogens. Arsenic, lead, and cadmium fall into this group. But on other hand a number of inorganics are essential to human nutrition in trace amounts, which includes arsenic, selenium, chromium, copper, nickel and zinc.

Inorganic constituents of interest in drinking water are summarised below:
1. Aluminum: It is common in treated drinking water. It shows low acute toxicity. It affects phosphorus absorption that can create weakness, arthralgia and anorexia.
2. Arsenic: In excessive amounts, arsenic causes acute gastrointestinal damage and cardiac damage. It can cause skin and lung cancer. Chronic doses can cause vascular disorders such as Blackfoot disease. It also has teratogenic potential in mammalian species.
3. Asbestos: Inhaled asbestos is clearly carcinogenic. When ingested through water it causes gastrointestinal tract cancer.
4. Barium: Acute effect of barium exposure includes prolonged stimulation of the cardiac, gastrointestinal and neuromuscular system. Chronic exposure may contribute to hypertension.
5. Cadmium: It acts as an emetic at ingested dose of 3–90 mg, becomes toxic at 10–326 mg, and is fatal at 300–3500 mg. Chronic exposure results in renal dysfunction.
6. Chromium: It produces liver and kidney damage, internal hemorrhage and respiratory disorders.
7. Lead: A high blood lead level is associated with interference with heme synthesis, kidney damage, and impaired reproductive function.
8. Silver: Chronic exposure to silver results in argyria, a bluegray discolouration of the skin and organs. This is a permanent aesthetic effect and results from 1 to 5 gm of accumulation.
9. Sodium: Sodium is major constituent in drinking water. Higher concentration of sodium is associated with high blood pressure and heart disease in the 'at risk' population. In addition, certain diseases are aggravated by a high salt intake, including congestive heart failure, cirrhosis of liver and renal diseases.

Hence it becomes essential to detect the ions present in water. Analyses of water include determination of ions in water. It is required for:
1. Determination of purity of drinking water.

2. As a quality control check of ultra pure water that is vital component of most of laboratory procedures.
3. Environmental testing of lakes and rivers.

Direct determination of ions present in water is the main technique used to determine presence of ions in water.

The measurement of specific ionic constituents in water is used in two ways:
1. To provide direct determination of the speciation of various molecular or valence forms of an element.
2. To provide elemental analysis either directly or after chemical conversion into a measurable form.

The major techniques are:
1. Volumetry.
2. Photometry.
3. Potentiometry.
4. Separation of the ions chromatographically.
5. Sequential measurement by non-specific detector, as conductivity.

Further the specific component information is provided by combining separation methods with selective elemental analysis techniques such as atomic absorption spectroscopy (AAS) ICP-MS. Accuracy of all these tests depends on availability of water for preparation of reagents, standards and blanks for sample preparation treatment and rinsing components.

METHODS FOR DETECTION OF IONS AND MICROBES

There are cetain methods for detection of ions and microbe in water used for industrial, laboratory and drinking purposes for ensuring the quality of water. These methods are classified as under.

Chemical Methods

1. Titration.
2. Spectrophotometry.
3. Electrochemical techniques.
4. Ion chromatography.

Microbiological Methods

For coliform group

1. Multiple tube fermentation method.
2. Membrane filter method.

For fecal streptococci

1. Multiple tube dilution.
2. Membrane filtration.

Chemical Methods

Some chemical methods for detection of ions in water are discussed here.

Titration

The well-established titration for alkalinity, hardness and chloride are still widely used. Potentiometric end-point detection with pH or ion selective electrodes has tended to replace indicator colour changes with the greater ease of automation and consistency.

With large volume sample, detection limits can be reduced to ppm levels in some cases. This technique can be applied to chemical addition to water, for i.e. boric acid in the primary water circuits of light water reactors and mixtures of hypochlorite, chloride and chlorate.

Spectrophotometry

Spectrophotometric analysis using colour reactions of specific ions are also still popular.

Advantages

There are some advantages of this method.
1. With automated analysers they can prove quicker and more cost-effective than other techniques like ion chromatography (IC).
2. Most ions can be determined in various types of water at ppm or lower concentrations.
3. In some cases it offers particular advantages over other techniques, e.g. for iron the detection limit is lower than, for example, flame AAS, while for silica the colour reaction is extremely sensitive and also specific to soluble silicates, enabling them to be distinguished from total silicon, as measured by AAS or ICP-OES.

Electrochemical techniques

An electrochemical technique enjoys some advantages like modest cost and their suitability for direct measurement in difficult matrices.

Ionic concentration at or above ppm levels can be detected by voltammetry. In this method, potential applied to a working electrode is varied in small steps and the resulting current recorded as a function of potential. The potential can be varied in steps, pulses or both, giving staircase or normal pulse voltammetry or polarography (with dropping mercury electrode) which are suitable for qualitative analysis and differential pulse or square wave voltammetry for quantitative analysis. For greater sensitivity, stripping steps are used. Stripping analysis involves two stages, electrolysis followed by stripping.

During electrolysis a specific potential is applied to the working electrode for a certain period of time during which the electroactive species in the water sample are deposited on the working electrode. The working electrode is glassy carbon covered with a thin film of gold or mercury, a hanging mercury drop electrode or a solid electrode. Stripping during which deposited species are redissolved into the solution, is performed electrochemically or by chemical oxidation. Elements are identified by their characterised redox potential.

Potentiometric stripping analysis (PSA)

PSA uses chemical oxidation by dissolved oxygen or mercury ions to determine metals such as lead, copper, cadmium, and zinc. The electrode potential is measured against the time during stripping. The rest time at each metal's redox potential is proportional to the amount of the metal in sample.

Constant current stripping analysis (CCSA)

CCSA uses the same measuring principle, but applies a constant current to control stripping. Application includes determination of mercury, nickel, cobalt, and arsenic.

Voltammetric stripping analysis (ASV/CSV)

In this, the electrode potential is gradually changed and resulting current is measured. The stripping current is proportional to the amount of species deposited. Metals which form amalgam such as lead, copper, cadmium, zinc, tin, thallium, manganese, bismuth, gallium can be measured using a mercury film or drop electrode. Voltammetry can differentiate between oxidation states of an element and so can be used for environmentally significant assays. Electrochemical techniques lack the wide elemental range and long linear response of some atomic and mass spectroscopy technologies.

Ion chromatography

It is a major technique for ion determination. Determination of common anions such as bromide, chloride, fluoride, nitrates, phosphate and sulphate is often desirable to characterise water and to assess the need for specific treatment.

Ion chromatography with chemical suppression of eluant conductivity

A water sample is injected into a stream of carbonate, bicarbonate eluant and passed through a series of ion exchangers. The anions of interest are separated on the basis of their relative affinities for a low capacity, strongly basic anion exchanger. The separated anions are directed onto a strongly acidic cation exchanger or through a hollow fibre of cation exchanger membrane or micromembrane suppressor bathed in continuously flowing strongly acid solution. In the suppressor the separated anions are converted to their highly conductive acid forms and the carbonate bicarbonate eluant is converted to weakly conductive carbonic acid. The separated anions in their acid forms are measured by conductivity. They are identified on basis of retention time as compared to standards. In most ion chromatographic methods, analysis is carried out with conductivity detection; amperometric, UV-visible spectrophotometric and fluorescence detectors are available for selective and more sensitive analysis. Only ion chromatography provides a single instrumental technique that may be used for their rapid, sequential measurement. Ion chromatography eliminates the need to use hazardous reagent and it effectively distinguishes among the halides, SO_3^-, SO_4^-, or NO_2^-, NO_3^-. This method is not recommended for the routine determination of fluorine because equivalency studies have indicated positive or negative bias and poor precision in some samples. IC using special techniques such as dilute eluant or gradient elution with fibre suppressor or membrane suppressor can determine fluorine accurately.

Microbiological Method

Following discussion describe procedures for making microbiological examination of water samples to determine water quality for it's use in laboratory investigation and also for domestic purpose.

Tests for detection and enumeration of indicator origanism, rather than of pathogens are used. Coliform group of bacteria as herein defined, is the principle indicator of suitability of water for domestic industrial purpose. Coliform group density is taken as a criterion of degree of pollution and thus of sanitary quality. Significance of tests and interpretation of results are well authenticated and have been used as a basis for standards of bacteriological quality of water supplies. Two standard methods are presented for the detection of the coliform group.

Multiple tube fermentation method

It is customary to report results of coliform test by the multiple tube fermentation procedure as a most probable number (MPN) index. This is an index of the number of coliform bacteria that, more probably than any other number, would give the results shown by the laboratory examination.

Membrane filter method

It is direct plating method. This procedure permits the direct count of coliform colonies. In both procedures coliform density is reported conventionally as the MPN or membrane filter count per 100 ml.

Use of either procedure permits appraising the sanitary quality of water and the effectiveness of treatment processes. Because it is not necessary to provide a quantitative assessment of coliform bacteria for all samples, a qualitative, presence/absence test is included.

Both the multiple tube dilution technique and membrane filter procedures have been modified to incorporate incubation in confirmatory tests at 44.5°C to provide estimate of density of fecal organism.

PURIFIED WATER REQUIREMENT

All the techniques described in this review require purified water for their successful application. For trace IC, and volumetry, the use of ultra-pure water is specified not only for preparation of samples, blanks and standards, but also for eluant and other reagents. The use of water with minimum contaminant levels of all types removes one area of potential variability in the analysis. An impurity in water causes problem like, direct interference from presence of an ion being measured to overlapping peaks, shifts in background, inhibition of electrode or column activity.

TECHNOLOGIES FOR RECYCLING OF WASTE-WATER

Pollution Prevention in Industry

Prevention or minimisation of pollution at source is the best control method. Hence before going into methods of effluent treatment, we should look at the possibilities of preventing or minimising effluent generation. Pollution prevention is defined as the use of materials, processes, or practices that reduce or eliminate the generation of pollutants or wastes at the source. Also known as source reduction, pollution prevention includes practices that reduce the use of hazardous and non-hazardous materials, energy, water, or other natural resources.

Pollution prevention in the manufacturing community can be achieved by changing production processes to reduce or eliminate the generation of waste at the source. As it applies to industry, the environmental management hierarchy stipulates that:
1. Pollution should be reduced at the source whenever feasible
2. Pollution that cannot be reduced should be recycled in an environmentally safe manner whenever feasible
3. Pollution that cannot be reduced or recycled should be treated in an environmentally safe manner whenever feasible and
4. Disposal or other releases into the environment should be used only as a last resort and should be conducted in an environmentally safe manner.

Pollution prevention activities include: product changes, input material process changes, technology 1 process changes, improved operating practices, recycling/reuse activities

Total Water Management Approach to Recycle of Waste-water

Waste-water recycle should take shape at the drawing board stage in contrast to the traditional approach of designing the raw water and waste-water treatment plants (end of pipe solutions) separately (Fig. 11.1). This will enable planning for water recycle at the design stage itself. The benefits are many as shown in Fig. 11.2.

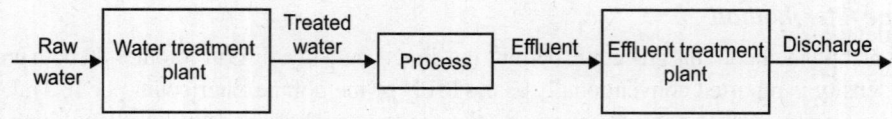

Fig. 11.1. Conventional treatment.

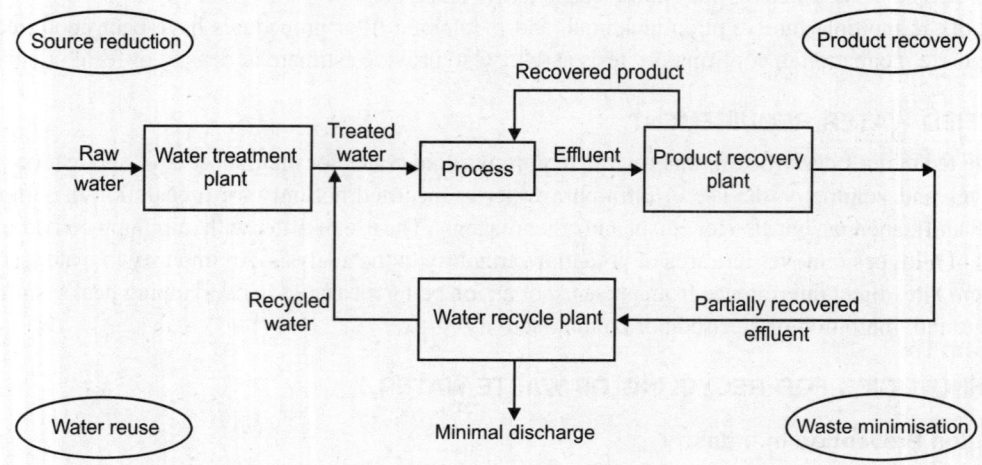

Fig. 11.2. Modern integrated solution.

Firstly, because water is recycled to the process, raw water consumption is reduced. The designer can, therefore, plan for a raw water treatment plant of lower capacity and lower cost. Secondly, effluent treatment is essentially not required and the quantity of waste disposed is less—which again, leads to cost reduction.

Investment is certainly required for the product recovery and water recycle plants; but it gives a good payback and provides value for money spent. Pollution is not just abated but prevented; pollutants are separated not destroyed; energy is saved and the total cost of water and waste-water treatment is reduced. How do we use the experience gained in successful on/off site recycle and integrated solutions for water and waste-water treatment? To achieve the ultimate goal, total water management should be applied right at the design stage. We need to apply these approaches in a complex industry in multiple ways.

Guidelines for selection of recycle schemes

1. Study the manufacturing process thoroughly and identify areas where reduction of water consumption is possible.
2. Identify the process where reduction of pollution load is possible by changing raw material or adopting cleaner manufacturing processes.
3. Identify streams that can be segregated and treated economically. In case of electroplating, the rinse water can be segregated and treated for recovery of plating metal. This not only reduces the overall cost of recycle but also facilitates the recovery of valuable products from the waste-water stream.
4. Identify effluents which are relatively clean and can be treated with simple processes so that they can be recycled internally without letting the water out into an effluent treatment plant.

5. Identify the quality of water required at various manufacturing stages. For steam generation, high quality water may be required. For washing or for cooling water make up, high quality water may not be required. It is always economical to design a recycle system to produce water suitable for lower end uses.
6. Select a technology that is easy to implement, operate and service.
7. Look for the availability of spare parts that may be needed in future.
8. Reliability of performance is important.
9. Low in operating cost.
10. Good service network of the plant supplier

Recycle Technologies

Any waste-water recycling plant requires three stages of treatment:
1. Effluent treatment.
2. Tertiary treatment.
3. Advanced treatment.

Effluent Treatment

A good effluent treatment is a pre-requisite for a good effluent recycle system. Unless we remove the easily removable pollutants with cost-effective methods, it would be difficult to recycle the effluents economically.

Usually effluent treatment plants (ETPs) are designed to meet statutory requirements for disposal. When recycling is considered, the ETP should also be designed considering overall requirements of treatment. For example, in India, disposal standards do not require complete removal of nutrients. But, when we are installing a downstream reverse osmosis system, it is better to remove nutrients in the biological system of the ETP. This will help in reducing fouling of the reverse osmosis system.

There are different technologies available for effluent treatment to remove different pollutants. Table 11.1 lists some generic technologies applied in effluent treatment.

Table 11.1. Effluent treatment technologies.

Pollutant	Treatment technology
Floating matter	Manual bar screens, mechanically cleaned screens, drum screens, etc.
Grit	Manual grit chambers, aerated grit chambers, detritors, etc.
Oil and grease	Oil and grease traps, API oil separators, TPI oil separators, dissolved air flotation systems, ultra filtration, etc.
Acidity/alkalinity	Neutralisation using acid/alkali dosing
Suspended solids	Clarifiers, clariflocculators, high rate solid contact clarifiers, lamella clarifiers, tube settlers, dissolved air flotation, ultra high rate clarifiers, pulsating clarifiers, etc.
BOD/COD	Biological systems such as activated sludge process, trickling filters, sequential batch reactors, membrane bio-reactors, etc. For concentrated effluents, anaerobic systems such as upflow anaerobic sludge blanket (UASB) reactors, Anaerobic contact reactors, upflow anaerobic contact reactors, anaerobic filters, complete stirred anaerobic reactors, etc.
Heavy metals	Precipitation using solid contact clarifiers, ion exchange processes and membrane systems for metal recovery, etc.

(Contd...)

Pollutant	Treatment technology
Toxic substances	Different treatment technologies are adopted based on the nature and concentration of toxic substances. For example, phenols can be removed with biological systems at low concentrations whereas chemical oxidation may be required for higher concentrations.
	Photo-chemical oxidation is also used to remove or break toxic and complex organics such as phenols, benzene, pesticides, etc.

Tertiary Treatment

Treatment beyond disposal norms for re-using effluents for low end usages is called tertiary treatment. It acts as pretreatment to advanced treatment for complete recycle of effluents. Table 11.2 enlists some generic technologies applied in tertiary treatment.

Table 11.2. Tertiary treatment technologies.

Pollutant	Treatment technology
Suspended solids	Clarifiers, clariflocculators, high rate solid contact clarifiers, lamella clarifiers, tube settlers, dissolved air flotation, ultra high rate clarifiers, pulsating clarifiers, etc.
Turbidity	Gravity sand filters, pressure sand filters, dual media filters, multi media filters, continuous sand filters, auto valveless filters, etc.
Bacteria	Chlorination, ozonation, ultraviolet sterilisation, mixed oxidant systems, etc.
Colour	Oxidation, precipitation, adsorption, nanofiltration, etc.
Residual chlorine	Activated carbon filtration, dosing of reducing agents, ultraviolet treatment, etc.

Advanced Treatment

Further treatment is required after tertiary treated effluents for conforming to the requirements of various end usages of treated water.

Technologies for Recycling of Effluents

Biological treatment

Biological treatment is mainly carried out for removal of bio-degradable organic matter in the effluents. Biological treatment can be classified into aerobic and anaerobic treatment. Anaerobic treatment is particularly applied for medium to high concentration effluents (BOD values above 2000 mg/l). At these concentrations, aerobic treatment systems require high power for aeration whereas in anaerobic treatment, it produces power in the form of bio-gas which is used as fuel. Upflow anaerobic sludge blanket (UASB) reactor is one such technology for anaerobic treatment of waste-water.

Activated sludge process (ASP) system is an age old technology for aerobic biological treatment. Sequential batch reactor (SBR) is an improvement over ASP wherein nutrient removal is also carried out in the treatment unit. There are many variants of SBR.

A modified SBR is a cyclic operating activated sludge system with a compact and modular design. This design combines the advantages of the conventional and the sequencing batch reactor technology. Like in the conventional system, the reactor volume and the level in the tanks are always constant. It is a continuous system for both influent feed and effluent discharge. Like the fill-and-draw system, the reactor operates according to a time controlled process cycle that allows for the alternation of all essential processes (aeration, mixing, sedimentation) in a single compartment.

The advanced treatment leads to sustainability (effluent recycling). A biological reactor consists of three units. Each unit is a similar in design, equipment and functional cycle. All units are hydraulically inter-connected and completely redundant.

Main advantages of modified SBR compared to conventional systems:
1. No need for external clarifiers, sludge rakers, recycle pumps/screws/piping.
2. No extra disturbances in clarifier caused by sludge recirculation.
3. Optimal dynamics in substrate gradient and integrated selector effect, favours the formation of well settleable sludge flocks.
4. Control in time enables flexibility by adapting times for nitrification, denitrification, biological phosphorus removal, sedimentation, depending on influent characteristics.
5. Easy and very compact construction, small footprint.
6. Maximum redundancy.
7. Easy to cover.
8. Easy to extend, modular construction.
9. Common walls.

Main advantages of modified SBR compared to conventional sequencing batch reactors:
1. No head loss (volume is always used for 100 per cent).
2. Raw waste-water does not enter the compartment in the phase preceding the sedimentation sequence. This guarantees optimal BOD, COD, N, P removal efficiencies, no short-circuiting.
3. Continuous influent and effluent flow means lower maximal flow rate and therefore smaller piping diameter and less installed capacity of pumps, aerators, weirs.
4. No moving mechanical parts.
5. No varying pressure on walls.
6. Enables use of surface aerators.
7. Enables use of fine bubble aeration.

Clarification

Clarification is a simple process that is used as a polisher after biological treatment for the removal of suspended solids. Chemicals are often dosed to obtain good quality of treated water.

High rate solid contact clarifiers (HRSCC) or ultra high rate clarifiers (UHRC) are very useful as they work on the principle of solid contact which has inherent advantage in terms of reliable performance, low chemical consumption, etc. In many cases, the effluent characteristics may vary with the design values due to some process changes. In such scenarios, HRSCC or UHRC provides flexibility in terms of dosing additional chemicals to keep parameters such as hardness, silica, etc. within the operating limits of subsequent membrane systems. Treatment units such as clariflocculators, tube settlers, parallel plate settlers are not suitable for hardness and silica removal. Parallel plate settlers and tube settlers are ideal for systems where space is a constraint. Ultra high rate systems which use a combination of technologies of solid contact and parallel plate separation are ideal in all cases.

Filtration

Filtration is a basic treatment process in any recycle system. Some systems require only suspended solid removal for recirculation of effluents. These applications include side stream filtration to increase the cycle of concentration in cooling towers, mill scale filtration in steel industry, etc. In many cases, filtration works as a pretreatment to many advanced treatment systems such as membrane filtration or ion exchange process.

Adsorption

Adsorption is a process used for removal of trace organics, TOC, Cl_2 and activated carbon is mainly used for the purpose. Some ion exchange resins are also available for adsorption of colour, organic matter, etc.

Membrane bio-reactor

Membrane bio-reactor (MBR) technology is one of the latest technologies in biological treatment. Submerged membranes are used in place of clarifiers to separate sludge from the waste-water so as to produce high quality permeate. These MBRs can handle very high sludge concentrations in the aeration tank because of which the size of the aeration tank reduces four to five fold. As the membrane acts as a fine filter, it does not require any further treatment using sand filters, activated carbon filters, etc.

Advantages of MBR over conventional treatment processes:

1. The quality of treated water in case of MBR is much superior to conventional biological systems. As the membrane acts as a physical barrier, it does not allow any sludge particles and to a great extent bacteria and viruses to pass through it. Micro-organisms like coliform or cryptosporidium can be easily removed in MBR. This increases the reliability of the system.
2. MBR does not require clarifier tank whereas conventional activated sludge process requires clarifier which further adds to area requirement and cost.
3. Conventional biological systems require further costlier tertiary treatment to match the performance of the MBR system. This may include coagulation, filtration, chlorination, adsorption, UV treatment etc.
4. MBR system has minimum number of treatment units and is very simple to operate. It does not require any regular handling of hazardous chemicals. As the treatment units are less, it is less prone to system breakdowns.
5. MBR requires much less space compared to conventional activated sludge process. Biological reaction in MBR can be carried out under the condition of 4 to 5 times of MLSS compared to conventional activated sludge process. It means that the aeration tank volume is $1/4^{th}$ to $1/5^{th}$ that of a conventional design. Combining this with other features mentioned above results into a very compact design requiring less space than conventional designs.
6. Conventional treatment systems require disinfection with chlorine, which has to be removed completely before applying onto gardens or for green belt development. Otherwise, high amounts of residual chlorine may damage the plants. Also, disinfection with any disinfectant does not remove organisms, it only inactivates them. The effect depends on the amount of disinfectant used, the quality of filtration applied, the retention time available for oxidation and the existence or nonexistence of other competing reaction partners (scavenging). As MBR acts like a physical barrier, it completely removes bacteria and viruses up to a degree of 4–7 log removal (10^4 to 10^7 times reduction), independent of type or life form of organisms. Hence, MBR does not require any other chemical disinfection.
7. As chlorination is not required, MBR does not produce disinfection by-products which are toxic to many living beings.
8. The sludge produced is only $1/5^{th}$ of conventional systems. It is also highly stable and hence easy to dispose of.

9. MBR treatment does not produce any objectionable odours and hence it is very safe to house near or inside basements of housing or commercial complexes.
10. As the MBR system does not require sludge recirculation system, clarifier, filter feed pumps, filter back wash pumps, chemical dosing pumps, UV sterilisation, etc. it results into low energy consumption when compared to other conventional systems.
11. The treated sewage from MBR has low silt density index (SDI) and can be fed directly into RO plant without any further treatment.

Ultra filtration

Ultra filtration (UF) is a very important technology in waste-water treatment. It has many applications in waste-water recycle. Some of the widely used applications are:
1. Pretreatment to RO for producing good quality water.
2. Colloidal silica removal.
3. Caustic recovery.
4. Colour removal.
5. Bacteria and virus removal.
6. Oil emulsion waste treatment.
7. Treatment of whey in dairy industries.
8. Electrocoat paint recovery.
9. Concentration of textile sizing.
10. Concentration of gelatine.

UF is mainly used as a pretreatment to nano filtration (NF) and reverse osmosis (RO) so as to reduce the silt density index (SDI), a parameter important to avoid NF/RO fouling. Advantages of using UF as pretreatment to RO are:
1. Increased flux in RO.
2. Reduced cleaning frequencies of RO, thereby reducing the RO downtime and chemical cleaning costs.
3. Minimising cartridge filter consumption. Cartridge filter is used only as safety barrier after UF to take care of any possible contamination resulting from chemical dosing.
4. Increases the life of RO membranes.

UF also selectively rejects some sparingly soluble dyes such as indigo. But many soluble dyes may pass through the UF system. In some cases, charged UF membranes have been successfully employed to remove soluble dyes.

Nanofiltration (NF)

NF has a pore size much smaller than UF and hence can reject many colour-causing elements. NF can very effectively separate dyes and concentrate them too. This way of concentration and purification reduces the loss for dyes. Also, when dyes are removed from the concentrated salt solution, it can be reused in the process thereby reducing the pollution load and also saving water and salt.

In general, it may be stated that this separation technology which depends both on size separation as well as the charges on the membranes can be economically applied to separate organics and also to separate higher valency cations or anions and associated salts from monovalent salt. Thus, softening of any aqueous stream is possible by separating out Ca^{++} or Mg^{++} or SO_4^{--}, CO_3^{--} from Na^+, Cl^-, etc.

Reverse osmosis (RO)

Reverse osmosis is increasingly used in recycle of waste-water for producing good quality water for reuse in the process. There is good improvement in membrane element design in terms of reducing fouling potential and increasing the life of membrane. Low cost of membranes is making its application a viable option.

RO is a membrane technology used for separation of salts from waste-water so as to make it reusable in the process.

RO is more useful in separating salts and organic compounds from textile effluent which are pretreated for removal of suspended/colloidal matter and certain pollutants which are likely to foul the RO membrane systems. As textile effluents contain high amount of dissolved salts, RO is a suitable technology for separation of these salts and for producing permeate which can be used in the process.

Eletrodialysis (ED)

Eletrodialysis is an electro-membrane process in which the ions are transported through a membrane from one solution to another under the influence of an electrical potential. ED can be utilised to perform several general types of separations such as separation and concentration of salts, acids and bases from aqueous solutions or the separation and concentration of monovalent ions from multiple charged components or the separation of ionic compounds from uncharged molecules. ED membranes are usually made of cross-linked sulphonated polystyrene. Anion membranes can be of cross-linked polystyrene containing quaternary ammonium groups. Usually, ED membranes are fabricated as flat sheets containing about 30–50 per cent water. Membranes are fabricated by applying cation and anion-selective polymers to a fabric material.

The system consists of two kinds of membranes: cation and anion, which are placed in an electric field. The cation-selective membrane permits only the cations, and anion-selective membrane only the anions. The transport of ions across the membranes results in ion depletion in some cells, and ion concentration in alternate ones.

Electrodialysis is used widely for production of potable water from sea or brackish water, production of ultra pure water, recovery of organic acids from salts, deacidification of fruit juices, heavy metal recovery, acid recovery from etching baths, pickling liquors, etc. treatment of plating waste-waters, demineralisation of whey, soya sauce, sugar, fruit juice and organic acid.

ED is an economical process only when used on brackish water, and tends to be most economical at TDS levels of up to 5000 mg/l. But, improved designs in ED makes it suitable for sea water desalination also. There are claims that the power consumption for ED in sea water desalination can be as low as 5 kWh/m^3 which is comparable to RO.

Plate and frame RO system

Plate and frame RO is another technology in reverse osmosis. Although the membrane is same, it offers some advantages because of its construction.

The plate and frame RO technology is designed to treat waste that is higher in dissolved solids content, turbidity, and contaminant levels than waste treated by conventional membrane separation processes. The membrane modules in this design feature larger feed flow channels and a higher feed flow velocity than other membrane separation systems. These characteristics allow the plate and frame RO greater tolerance for dissolved solids and turbidity and a greater resistance to fouling and scaling of the membranes. Suspended particulates are readily flushed away from the membrane during operation.

The high flow velocity, short feed water path across each membrane, and the circuitous flow path create turbulent mixing to reduce boundary layer effects and minimise membrane fouling and scaling. This design also allows easy cleaning and maintenance of the membranes. This technology can use reverse osmosis, ultra filtration, or micro filtration membrane materials. These membranes are more permeable to water than to contaminants or impurities. Water in the feed is forced through these membranes by pressure. The permeate thus consists of a larger fraction of water with a lower concentration of contaminants. The impurities are selectively rejected by the membranes and are thus concentrated in the smaller fraction of the concentrate left behind. The percentage of water that passes through the membranes is a function of the operating pressure, membrane type, and concentration of the contaminants.

The Plate and Frame RO design has following advantages over conventional spiral RO:
1. It can tolerate higher feed pressures (up to 140 bar) and hence can handle higher dissolved salt concentrations in the feed water.
2. It has higher fouling resistance because of wide feed channels and higher flow velocities.
3. It can be easily maintained. If required, the module can be opened for cleaning.
4. It requires less pretreatment units compared to spiral RO design.

Photo chemical oxidation (PCO)

There are several compounds in the effluent streams that are difficult to treat with conventional treatment processes. Furthermore, many of these compounds find their way to ground water sources thereby polluting them. These compounds need to be reduced to levels so as to meet the environmental standards. There are many upcoming technologies that would be effective in the treatment of these complex compounds.

'Advanced oxidation processes' is one such technology that is capable of treating complex organic and refractory compounds by bringing about the oxidation of these compounds. 'Photochemical oxidation' and 'photocatalytic oxidation' are some of the advanced oxidation processes.

In the photochemical oxidation process, photons of appropriate energy levels from a source of light (UV or solar radiation) interact with oxidants such as hydrogen peroxide, ozone, etc. to generate free hydroxyl radicals that are used to oxidise the wastes directly into carbon dioxide and water without the generation of sludge.

In the photocatalytic oxidation process a catalyst, usually a semiconductor (like titanium dioxide) in aqueous solution is used to generate free hydroxyl radicals by interaction of photons of appropriate energy levels from a source of light (UV or solar radiation) with the catalyst.

Photochemical oxidation has been found to degrade a wide number of refractory compounds such as chlorinated hydrocarbons, aromatic compounds, etc. Similarly, photocatalytic oxidation has been able to degrade compounds such as poly-chlorobiphenyls, which cannot be degraded by conventional treatment methods. Major advantages of photochemical and photocatalytic oxidation processes are that they can be carried out at room temperature and atmospheric pressure.

Ion exchange process

Ion exchange is a proven technology in terms of removal of dissolved salts. Ion exchange is also used as pretreatment to RO for removal of hardness, alkalinity, heavy metals, etc. Some special applications for treatment of waste-water using ion exchange process include removal of fluorides, nitrates, iron, arsenic, etc.

Ion exchange process is also used to recover metals from waste-water streams. Many applications are in use for recovery of valuable products from waste-water.

Some of such applications include:
1. Nickel or chromium recovery from electroplating rinse waters.
2. Zinc recovery from waste-water of rayon industry.
3. Copper recovery.
4. Mercury recovery from chlor alkali waste-water.

Thus, waste-water recycle is mandatory for many industries because of water scarcity, legislation, rising water costs, unreliable water supplies, environmental requirements from buyers in case of exporters, etc. Apart from these reasons, industries now identify recycling as their social responsibility for environmental friendly manufacturing of goods.

Many technologies are now available for managing industrial waste-water. It is of utmost importance to involve water management specialists right from the planning stage of the project so that the best optimum solutions can be developed. Priority should always be given to source reduction and product recovery rather than end-of-pipe waste-water treatment. Best technologies should be adopted for recovery and recycle of water from waste-water. Final effluents which cannot be recycled should be treated and disposed of in an environment friendly way.

AUTOMATION AND CONTROL IN WATER AND WASTE-WATER TREATMENT PLANTS

Automation can be defined as the use of scientific techniques to automate the operation and/or control the equipment, process or system; to minimise human intervention to optimise the process efficiency and the use of resources (raw material, power, labour/manpower, etc). Automation can be as simple as a time-based operation of a traffic signal or as complex as a single room control of a large petroleum refinery. In this section, we will restrict ourselves to automation using programmable logic controller (PLC).

Automation has economic impact in the form of costs involved and more prominently in the form of savings. The cost involved depends upon the degree of automation desired (whether simple timer based or a complex SCADA based) and the criticality of the process/equipment. It's important that the organisation is clear about the cost and anticipated returns on its investment (in the form of savings).

Some of the benefits of automation can be summarised as:
1. Improved and/or consistent product quality.
2. Repeatable performance.
3. Greater environmental compliance and values.
4. Increased productivity.
5. Reduction in downtime and maintenance costs.
6. Reduction in operation costs.
7. Use of manpower in other value creating activities.
8. Availability of accurate data/information, enabling timely decision-making and better process control.
9. Greater customer satisfaction and brand leverage.

Role of Automation in Water and Waste-water Treatment Industry

Water being a critical utility in many industries (pharma, electronics, power, etc.), it is important to evaluate the benefits of automation on water and waste-water treatment plants.

Let us consider certain examples to understand the importance of automation in water and waste-water treatment.

Bottled water industry and beverage industry

Conductivity and pH of the treated water are very critical and need to be not monitored and controlled, so as to specifications meet the BIS for packaged/ bottled water.

Pharma industry

This industry requires high purity water for formulations (purified water) and injectables. The quality parameters to be maintained are conductivity TOC (<500 ppb). Apart from this, most high purity water systems today prefer non-chemical treatment processes favouring RO-EDI-UF system. One of the feed limitations to RO is that there should be no free chlorine content. This necessitates the need of an oxidation-reduction potential (ORP) meter.

Waste-water treatment

The key design parameters for efficient waste-water treatment includes monitoring dissolved oxygen (DO) levels during various cycles of operation. In the absence of instrumentation to monitor it, the treated effluent will not be able to consistently meet the discharge norms with respect to BOD/COD/ ammonical nitrogen/ phosphorous levels.

The above examples reflect the necessity for automation of water and waste-water treatment processes, to consistently meet the standards of treatment.

The criticality of the process (softening for washing pipes or for boiler feed water) will determine the degree of automation (timer/PLC/SCADA). A range of automated solutions have been developed, from simple timer based dosing systems to PLC based systems, for monitoring high purity water. A few ultra-modem technologies available are:

New Generation Automatic Demineralisation System

The system comprises of automated twin-bed deionisers incorporating superior counter flow ion exchange technology (previously available only in large custom-built plants). Such systems are required for high purity water generation with resistivity greater than $0.2~\mu S$ cm. The operational cycle of the packaged unit is controlled by sophisticated PLC. The PLC is designed to operate a bank of solenoid valves in the regeneration sequence, which in turn operates the piping valves. Functions of the PLC are:

1. Once the desired volume throughput is achieved, the control system stimulates the rapid-regeneration cycle of 30 minutes. The regeneration of the cation and anion bed is simultaneous, producing a near neutral waste thereby reducing the disposal cost.
2. A message display continuously reads out the system status and provides information of flow rates, throughput, and number of regenerations. The system is fitted with an audible alarm to indicate any defaults and also indicates the 'no-flow' condition.
3. The control system can be scaled up and connected to level sensors in chemical tanks to prevent regeneration when insufficient chemicals are available.

New Generation Reverse Osmosis-Electro-Deionisation (RO/EDI) System

The RO/EDI system generates high purity water at the touch of a single button. It is a compact skid mounted unit which comprises pretreatment (softener/activated carbon), reverse osmosis, electrodeionisation, and post treatment (ultra filtration/ultra-violet).

The system is controlled by a fully integrated PLC and a touchscreen human machine interface (HMI) housed in a SS cabinet. The PLC controls the system operation and provides nine levels of security.

The softener operation and regeneration (including dilution of saturated brine solution) is set in stages and each stage is preset on the HMI.

The system sounds an alarm or shuts down if the set conditions are exceeded. This provides foolproof security to ensure consistent treated water quality. Conductivity and temperature of the treated water are monitored to meet the design specifications.

The hot water cleaning and sanitisation of the RO membranes and EDI is initiated automatically. The temperature and pressure during the cycle are continuously monitored and controlled through the PLC

New Generation Waste-water Recycling Technology

The system is designed to ensure complete and consistent treatment for carbon oxidation, nitrification (N), denitrifcation (DN), and bio-phosporous. The treatment cycle consists of fill-aeration, settling and decantation which is completed in the same tank in four hours.

Plant control is essentially automatic which is a major factor in reducing operating costs. The cycle and its sequence are controlled by a scalable PLC. The automatic basin air distribution system is also regulated by the same PLC. This means that optimum operation and performance are synonymous. Steady state is easily maintained even during high and low load conditions.

The basic feature of the system is its ability to measure the dissolved oxygen content in the aeration basin and its ability to vary the blower speed through a variable speed drive. This results in almost 30–50 per cent power saving (as blowers are the largest consumers of power in the treatment plant).

Water Treatment Process Analysers

The instruments for conductivity, resistivity, redox and pH measurement are available as online process analysers as well as in portable models.

The online process analysers are ultra-modern microprocessor based instruments that measure the quality of water and waste-water. These analysers have been specially developed to meet the need of the discerning user for high-end, value-packaged features at an economical price. The systems are based on application specific integrated circuit (ASIC), with quick access and great flexibility through the menu driven SETUP program. The 0/4–20 mA transmitter output current is galvanically isolated providing accurate and reliable measurements. Moreover, these modern process controllers comply with the electro magnetic compatibility (EMC) standard and are protected against electro magnetic interference.

High Purity Water Systems

Industries like pharma and semiconductor manufacturing, requires high purity water. Quality parameters are to be maintained to meet the purified water requirement are shown in Table 11.3.

Table 11.3. Quality parameters for purified water.

Conductivity	<1µS/cm
pH	5–7
TOC	(<500 ppb)
Total bacterial count	<100 CFU/ml

Further purification is needed to meet the stringent requirement in case of production of water for injection (WFI), which is as follows:

Total Bacterial count	<10 CFU/100 ml.
Endotoxin	<0.25 EU/ml.

To achieve these critical quality parameters, it is very essential to monitor, record and validate the system including the source of water. This is possible only by automation and instrumentation for monitoring critical parameters, maintaining records, reviewing the records, undertaking corrective and preventive methods and continued improvements to face the challenges during audit from many countries to get an approval for exports.

Some of the instruments that are employed for Online Monitoring right from the source are:
1. Conductivity meter to monitor the raw water conductivity.
2. pH meter to monitor the raw water.
3. Chlorine monitor/controller to measure the free chlorine level.
4. ORP monitor/controller to measure oxidation and reduction potential.
5. pH meter to check the feed pH
6. Conductivity meter to monitor feed water condition.
7. Flow meters, either indicators/controllers, at various stages to monitor the flow of the many equipment used in the system design to treat water right from the raw water stage to purified or WFI standards.
8. Pressure indicators/transmitters at various equipments.
9. Temperature indicators/transmitters at various points.
10. Conductivity at final point of production and if necessary at the point of use.
11. Level control devices for storage tanks.
12. Online and off line TOC monitoring.

All these instruments are available in low-medium-high accuracy levels with many features, controls and signals. Many use minimum instruments to meet the requirement. More emphasis is laid on maintaining records and for validation which increases the necessity of sophistication and hook up of these instruments for control, monitoring and recording, to computers through customised SCADA software. The use of instruments has increased recently with continuous upgradation of treated water quality and innovative techniques used to monitor the critical parameters at various sampling points. For example, systems like RO-RO-CDI used highly sophisticated instrumentation and controls, including monitoring of voltage and current which plays a crucial role for the success of CDI.

It is needless to mention that these instruments need to be calibrated to give accurate results and there are authorised institutes that undertake calibration of the instruments.

There are a few parameters that can be measured only in the laboratory either by chemical or by using instrumentation methods.

There is high level of quality consciousness in the pharma industry in general, and the increasing awareness is a trend, which is a welcome one.

Thus, automation in general leads to quality improvement, fewer errors, higher productivity per person, as well as reduction in time required to complete a particular process.

CASE STUDY WASTE-WATER RECLAMATION

This case study shows how Madras Fertilisers Limited, located in the most water-starved city, Chennai, has overcome the water supply problem by adopting waste-water reclamation by installing a cooling water blowdown plant supplied by Ion Exchange (India) Ltd. in the year 2007.

Indiscriminate industrial development and exploitation of limited water resources are compelling every industry to seriously address the problem. Often, industrial units are compelled to close down due to the severe water crisis. Industries have now realised that the limited water supplies can be put to better use by recycling waste-water. Companies have moved on to waste-water reclamation. Waste-water reclamation involves treating or processing waste-water for reuse.

This case study focuses on the waste-water reclamation carried out at Madras Fertilisers Limited (MFL), Chennai. Chennai is one of the worst water-starved cities in India. MFL adopted itself to the situation and set up its first waste-water reclamation plant about 15 years ago, even when the technology was at a nascent stage in India. MFL had the foresight to delink itself from reliance on fresh water supplies and look towards treated sewage from Chennai as an alternate source of industrial water.

After successful implementation of the sewage reclamation plant, MFL decided to also recover water from their other waste streams within the complex. One area identified was the recovery of cooling tower blowdown, which amounted to almost 160 m^3/hour.

Cooling Water Blow down Recovery at Madras Fertilisers Ltd.

This case study focuses on cooling water blowdown recovery as an alternative source of water reclamation adopted by MFL, Chennai. This project has led to a total cost saving of Rs. 36.79/kl compared to raw water cost of Rs. 60/kl and has yielded a return on investment in 14 months. The cooling water blowdown (CWBD) plant is designed to treat cooling tower blowdown that was earlier wasted to drain. The 160 m^3/hr recycle plant is supplied by Ion Exchange (India) Ltd. and incorporates pretreatment followed by ultra filtration which feeds the existing reverse osmosis plant.

Cooling tower blowdown offered the system designer a unique challenge of recovery through a membrane process because of its peculiar characteristics by virtue of the number of cycles of concentration maintained in the cooling tower. One of the prime concerns was the high levels of dissolved silica in the blowdown water (Table 11.4), which hampered the performance of the membrane system by placing a constraint on the recovery possible.

Considering that one of the primary aims of the recycle plant was also to maximise recovery of fresh water, it was essential to bring the dissolved silica levels down to acceptable levels to maximise recovery. Other concerns included trace residual of cooling water treatment chemicals, which had to be taken into account while designing the membrane process.

Table 11.4. Quality of blowdown water.

pH	7–7.5
TH	1300 (as CaCO$_3$)
CaH	850 (as CaCO$_3$)
MgH	450 (as CaCO$_3$)
Cl	1850 (as CaCO$_3$)
SiO$_2$	190 ppm
TPO$_4$	20 ppm
Total Fe	3.5 ppm
Turbidity	20 NTU
SRB	10000 max. (organisms/ml)
THB	100000 max. (colonies/ml)
SDI	Out of range
Oil and grease	Nil (ppm)

Considering all these aspects, a scheme was worked out which incorporated a special high rate solids contact clarifier (HRSCC), where the reaction for effective removal of silica with the aid of dolomite by adsorption on magnesium hydroxide flocs was achieved efficiently (Fig. 11.3).

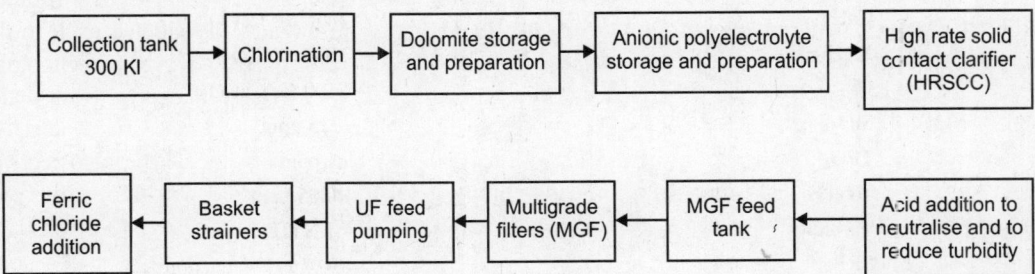

Fig. 11.3. Pre-treatment of water.

The treated water from the overflow of the HRSCC was first passed through a multigrade sand filter to bring down the levels of suspended solids including carryover of any dolomite/magnesium flocs (Table 11.5). The sand filtered water was then fed into a specially designed ultrafiltration plant using hollow fibre modules.

Table 11.5. Quality of water passed through multigrade filter.

pH	7–7.5
TH	1485 (as $CaCO_3$)
CaH	1475 (as $CaCO_3$)
MgH	10 (as $CaCO_3$)
Cl	2100 (as $CaCO_3$)
SiO_2	<25 ppm
TPO_4	<1 ppm
Total Fe	<0.1 ppm
Turbidity	<10 NTU
SRB	<10,000 (organisms/ml)
THB	< 1,00,000 (colonies/ml)
SDI	Out of Range
Oil and grease	Nil (ppm)

The ultrafiltration modules were chosen based on their capability to efficiently reduce the level of SDI in the feed water to the reverse osmosis (RO) plant. The system was designed to ensure peak efficiencies and higher recoveries by operating on a cycle that involved regular flushing and automatic chemically-enhanced back flush.

This ensured that the treated water from the ultrafiltration plant achieved a consistent SDI of less than 2, thus enhancing the performance of the downstream RO plant. The highlights of the treatment by ultrafiltration has been the reduction of turbidity to consistently less than 1 NTU and over 6 log reduction (99.9999 per cent) in SRB (Sulphate Reducing Bacteria) (Table 11.6).

Table 11.6. Quality of water post ultrafiltration

PH	7–7.5
TH	1485 (as $CaCO_3$)
CaH	1475 (as $CaCO_3$)
MgH	10 (as $CaCO_3$)
Cl	2100 (as $CaCO_3$)
SiO_2	<25 ppm.
TPO_4	<1 ppm
Total Fe	<0.05 ppm
Turbidity	<1 NTU
SRB	99.9999% reduction
THB	99.9999% reduction
SDI	<2
Oil and grease	Nil (ppm)

Successful implementation of the cooling water blowdown plant ensured a further recovery of 40 per cent of the effluents generated from the fertiliser complex. 10 per cent of effluents generated are supplied to a neighbouring industry for process use. These measures have achieved a recovery and reuse of almost 80 per cent of effluents and supply of 10 per cent of effluents for beneficial use, thus taking the unit towards 'a near zero effluent discharge' status. Cooling water operating details are given in Table 11.7.

Table 11.7. Cooling water operating details.

Circulation rate	18860 m^3/hr
Hold up volume	6500 cu. m.
Number of cells	9 concrete + 3 wood
Delta T	12°C
COC	4 cycles

Cost Economics

The capital cost of the cooling water blowdown plant was Rs. 4.5 crores. The cost of blowdown water is nil. The operating cost, including operating cost of ultra filtration and reverse osmosis plant is approximately Rs. 23.21/kl. The total cost saving realised is Rs. 36.79/kl with a return on investment in 14 months in the 2007.

Other Benefits

An added advantage of such waste-water recovery projects is that they help in reducing the burden on fresh water supplies and thus increase availability of fresh water for other needs such as drinking water in water starved areas.

Chapter 12

Physical, Thermal and Chemical Methods of Waste Management

INTRODUCTION

The effects of chemical waste on the environment are reflected by the effects on organisms and effects on the overall ecosystem. And, it is also worth remembering that virtually all chemical wastes are poisonous to a degree, some extremely so.

As with all environmental pollutants, such chemical wastes eventually reach a state of physical and chemical stability and equilibrium with the environment, although it may take many centuries to occur. In many cases, the behaviour (or reactivity) and ultimate fate of a chemical waste is a function of its physical properties and surroundings. On the basis that it is not possible to wait for the chemical to reach an equilibrium with its surroundings, methods of remediation are essential.

The first step in considering the appropriate technology to employ to treat a specific waste is to determine whether the waste is hazardous or nonhazardous. And, the definition of a hazardous/nonhazardous waste is determined by the relevant legislation. There is the general tendency to think of a chemical waste as some obscure and ill-defined sludge discarded from a process. However, chemical waste can also be in the form of volatile organic compounds, semi-volatile organic compounds, metals, radioactively contaminated materials, or a mixture of any or all of these types.

The ideal treatment process reduces the quantity of chemical waste to a small fraction of the original amount and converts it to a nonhazardous form if such a conversion is possible. Another consideration in selecting a treatment technology is the location where the wastes are to be treated. For example, wastes may be treated in place (*in situ*), within the confines of the site, or at an off-site facility (*ex situ*).

There are various alternative waste treatment technologies, such as chemical treatment and biological treatment. Physical treatment processes (which also include thermal methods such as incineration) as well as solidification or stabilisation are the subject of this chapter. These processes are used to (i) recycle the waste and reuse waste materials, (ii) reduce the volume and toxicity of the waste stream, and (iii) produce a final residual material that is suitable for disposal.

The effectiveness of the application of each of these technology groups to a specific waste varies depending on (i) the type of waste, (ii) the concentration of the individual components in the waste, (iii) the physical phase of the material, (iv) the desired level of treatment, and (v) the final method or disposal of any remaining residue.

The waste characteristics can include such properties as volatility (gases, volatile solutes in water, gases or volatile liquids held by solids, such as catalysts), liquid phase materials (waste-water, organic solvents), dissolved or soluble materials (water soluble inorganic species, water-soluble organic species, compounds soluble in organic solvents), semisolid materials (sludge, grease), and solid materials (dry solids, including granular solids with a significant water content, such as dewatered sludge, as well as solids suspended in liquids). Waste treatment may occur at three major levels: primary, secondary, and tertiary (polishing).

Primary waste treatment is generally regarded as preparation for further treatment, although it can result in the removal of by-products and reduction of the quantity and hazard of the waste. Secondary waste treatment detoxifies, destroys, and removes hazardous constituents. Polishing usually refers to treatment of a waste product for safe discharge. An example is the treatment of recycled water that is removed from wastes so that it may be safely discharged. In addition, especially in the case of water treatment, the processes employed can be generally categorised as external treatment or internal treatment.

External treatment uses processes such as flotation and clarification to remove material, including suspended or dissolved solids, hardness, and dissolved gases. Following this basic treatment, the water may be divided into different streams, some to be used without further treatment and the rest to be treated for specific applications.

PHYSICAL METHODS

In contrast to the chemical methods of waste treatment, physical processes for the treatment of chemical waste include processes that separate components of a waste stream or change the physical form of the waste without altering the chemical structure of the constituent materials.

These processes are very useful for separating hazardous materials from an otherwise nonhazardous waste stream so that they may be treated in a more concentrated form, separating various chemical components for different treatment processes, and preparing a waste stream for ultimate destruction in a biological or thermal treatment process.

Knowledge of the physical behaviour of wastes has been used to develop various unit operations for waste treatment that are based upon physical properties. These operations include: (i) phase separation (filtration), (ii) phase transfer (extraction, sorption), (iii) phase transition (distillation, evaporation, precipitation), and (iv) membrane separations (reverse osmosis, hyperfiltration and ultrafiltration).

The important aspect of physical methods of remediation is to isolate the contaminant so that it can be recovered/destroyed while being contained within a specific area. In terms of containment, there are several technologies available which usually involve construction of a containment wall (usually concrete) around the contaminated area on the side where leakage into nearby water systems can occur.

One aspect of remediation that is evolving is the application of soil-mix wall technology. This technology consists of mixing soils *in situ* with cement grout using multiple shaft augers to construct overlapped cement columns. The use of augers makes it possible to define the treatment zone clearly and to confine the contaminant(s).

Phase Separation

The most straightforward means of physical treatment involves separation of components of a mixture that are already in two different phases. In many cases the separation must be aided by mechanical means, particularly sedimentation, screening, centrifugation, flotation, and filtration.

Separation is used to divide the wastes into two or more distinct waste streams based upon size, density, or material type. Normally accomplished by either manual or mechanical means, separation allows for a more efficient operation of the subsequent technologies while reducing the quantities of material to be treated.

Sedimentation

Sedimentation is usually accomplished by providing sufficient time and space in special tanks or holding ponds for settling. Chemical coagulating agents are often added to encourage the settling of fine particles.

Screening

Screening is a process for removing particles from waste streams, and it is used to protect downstream pretreatment processes.

Screening is the most common technology employed for separation. Four general categories of screens are available: (i) grizzly screens, sets of parallel bars used for the removal of coarse material, (ii) revolving screens, a cylindrical frame covered with wire cloth, (iii) vibrating screens, normally used when higher capacities are required, and (iv) oscillating screens, used at lower speeds than vibrating, which are used for separating particles by grain size.

Flotation

Flotation is a process for removing solids from liquids and moving the particles to the surface by using tiny air bubbles. Flotation is useful for removing particles too small to be removed by sedimentation.

In the process of dissolved air flotation, air is dissolved in the suspending medium under pressure and when the pressure is released comes out of solution as minute air bubbles attached to suspended particles, which causes the particles to float to the surface.

Flotation can also be used to remove hydrocarbon-type waste from *in situ* locations without the more destructive 'dig-and-move' procedures. For example, a chemical waste might be removed by a relatively simple (and perhaps standard) flushing technique (Fig. 12.1.) or by a somewhat more innovative technique that involves the principles of laminar flow and which can be applied to lighter-than-water liquids and also to heavier-than-water liquids (Fig. 12.2).

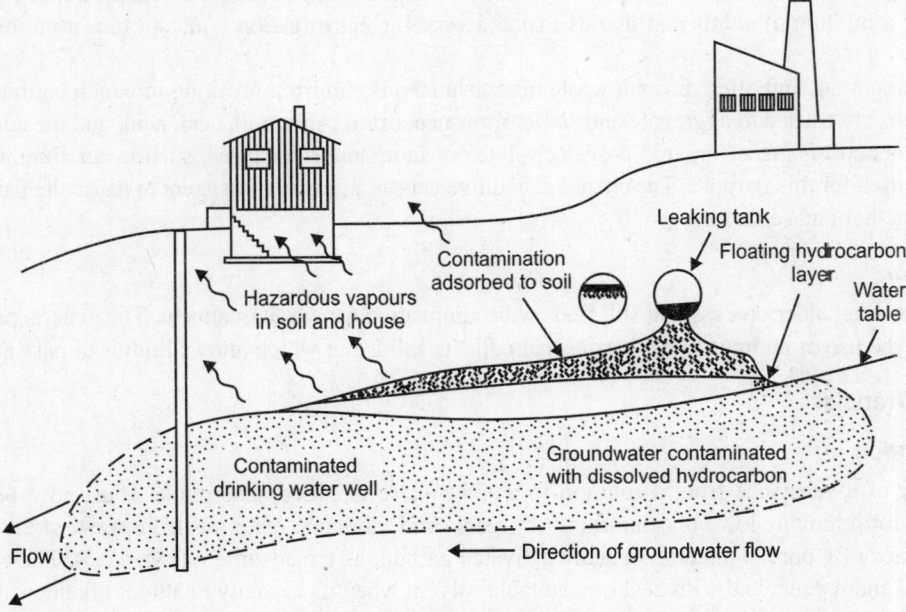

Fig. 12.1. Flow of a contaminant plume in groundwater.

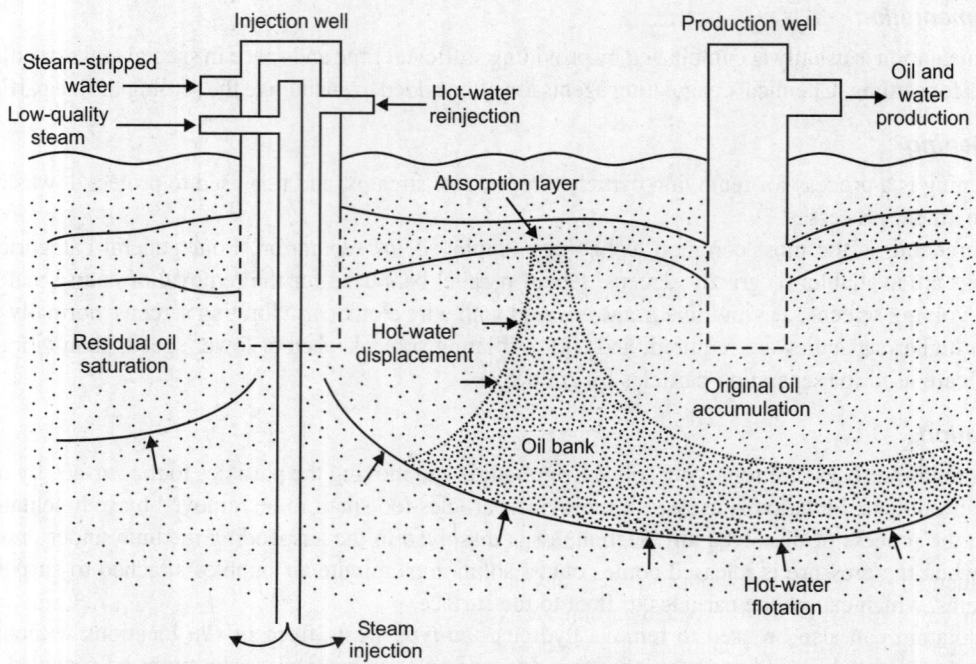

Fig. 12.2. Subsurface cleanup of organic waste by the CROW (contained recovery of oily waste) process.

Centrifugation

Centrifugation is a process for separating solid and liquid components of a waste stream by rapidly rotating a mixture of solids and liquids inside a vessel. Centrifugation is most often used to dewater sludge.

An important and often difficult waste treatment step is emulsion breaking in which colloidal-sized emulsions are caused to aggregate and settle from suspension. Agitation, heat, acid, and the addition of coagulants consisting of organic polyelectrolytes or inorganic substances, such as an aluminum salt, may be used for this purpose. The chemical additive acts as a flocculating agent to cause the particles to stick together and settle out.

Filtration

Filtration is an older process that still finds wide application for waste treatment. The general principles involve the use of an impassable barrier that collects solids but which allows liquids to pass through.

Phase Transfer

Adsorption

Transfer of a substance from a solution to a solid phase is called adsorption. Thus, adsorption is a process for removing low concentrations of organic and inorganic materials from waste streams using the surface of a porous material, usually activated carbon, as the adsorbent (Fig. 12.3). The carbon is replaced and regenerated with heat or a suitable solvent when its capacity to attract organic substances is reduced.

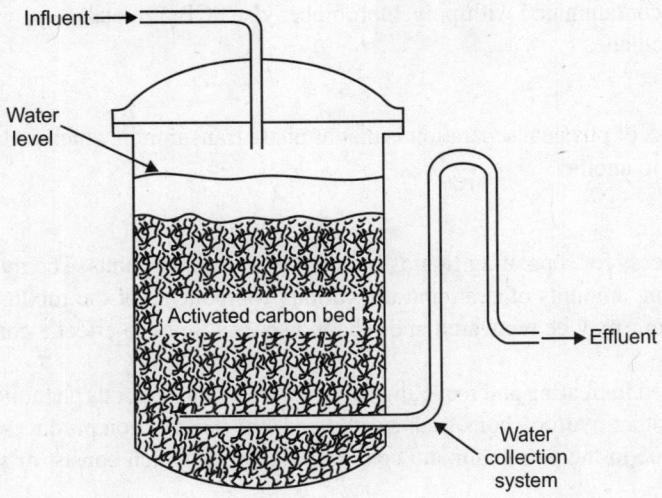

Fig. 12.3. A carbon adsorption system

Effluents from other treatment processes, such as biological treatment of degradable organic solutes in water, can be polished with activated carbon. Activated carbon sorption is most effective for removing from water those chemicals that are poorly water-soluble and that have high molecular masses, such as xylene, naphthalene, cyclohexane, chlorinated hydrocarbons, phenol, aniline, dyes, and surfactants. Activated carbon does not work well for organic compounds that are highly water-soluble or polar.

Solids other than activated carbon can be used for sorption of contaminants from liquid wastes. These include synthetic resins composed of organic polymers and mineral substances. For example, clay and regenerable adsorbents are employed to remove impurities from waste lubricating oils (not yet a RCRA waste) in some oil recycling processes.

Solvent extraction is a process for separating liquids by mixing the stream with a solvent which is immiscible with part of the waste but which will extract certain components of the waste stream. The extracted components are then removed from the immiscible solvent for reuse or disposal.

One of the more promising approaches to solvent extraction and leaching of chemical waste constituents is the use of supercritical fluids, most commonly carbon dioxide, as extractor solvents. A supercritical fluid is one that has characteristics of both liquid and gas and consists of a substance above its critical temperature and pressure (31.4°C/88°F and 72.9 atm 1071 psi respectively, for carbon dioxide). Supercritical fluids exhibit extraordinary solvating power toward various chemical species. Even though carbon dioxide is most frequently used as a supercritical medium due to its low critical temperature, other chemical species are also employed as supercritical fluids.

After a substance has been extracted from a waste into a supercritical fluid at high pressure, the pressure can be released, resulting in separation of the substance extracted. The fluid can then be compressed again and recirculated through the extraction system.

Some possibilities for treatment of chemical wastes by extraction with supercritical carbon dioxide include removal of organic contaminants from waste-water, removal of volatile organic compounds from spent catalysts, extraction of organic halogen pesticides from soil, extraction of oil from emulsions used in aluminum and steel processing, degreasing of mechanical parts, and regeneration of the spent carbon. Waste oils contaminated with polychlorobiphenyls (PCBs), metals, and water can be purified using supercritical ethane.

Phase Transition

A second major class of physical separation is that of phase transition in which a material changes from one physical phase to another.

Distillation

Distillation is a process for separating liquids with different boiling points. The mixed-liquid stream is exposed to increasing amounts of heat, and the various components of the mixture are vaporised and recovered. The vapour may be recovered and reboiled several times to effect a complete separation of components.

Distillation is used in treating and recycling solvents, waste oil, aqueous phenolic wastes, and xylene contaminated with other hydrocarbons such as ethylbenzene. Distillation produces distillation bottoms (still bottoms, residue in the petroleum and coal tar industries), which consist of solids, semisolid tar, and sludge.

Evaporation

Evaporation is a process for concentrating nonvolatile solids in a solution by boiling off the liquid portion of the waste stream. Evaporation units are often operated under some degree of vacuum to lower the thermal energy required to boil the solution. This is due to the fact that the boiling point of a solution is lowered by decreasing the exerted pressure.

Evaporation is usually employed to remove water from an aqueous waste to concentrate it. A special case of this technique is thin-film evaporation in which volatile constituents are removed by heating a thin layer of liquid or sludge waste spread on a heated surface.

Drying

Drying is the removal of solvent or water from a solid or semisolid (sludge) or the removal of solvent from a liquid or suspension, is a very important operation because water is often the major constituent of waste products, such as sludge obtained from emulsion breaking. In freeze drying, the solvent, usually water, is sublimed from a frozen material. Solids and sludge are dried to reduce the quantity of waste, to remove solvent or water that might interfere with subsequent treatment processes, and to remove volatile constituents. Dewatering can often be achieved thermally and can be improved with addition of a filter aid, such as diatomaceous earth, during the filtration step.

Stripping

Stripping is a means of separating volatile components from less volatile components in a liquid mixture by the partitioning of the more volatile materials to a gas phase of air or steam (air stripping or steam stripping) (Fig. 12.4). The gas phase is introduced into the, aqueous solution or suspension containing the waste in a stripping tower that is equipped with trays or packed to provide maximum turbulence and contact between the liquid and gas phases. The two major products are condensed vapours and a stripped bottom residue.

Examples of two volatile components that can be removed from water by air stripping are benzene and dichloromethane. Air stripping can also be used to remove ammonia from water that has been treated with a base to convert ammonium ions to volatile ammonia.

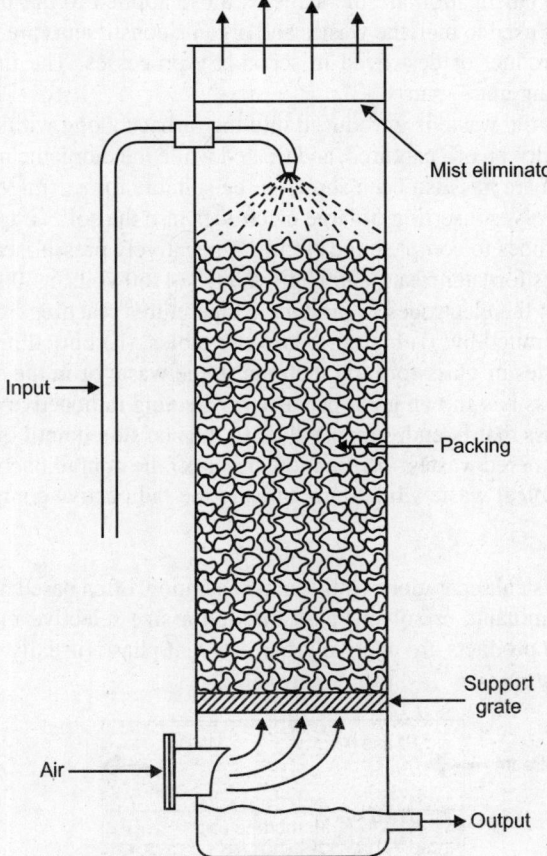

Fig. 12.4. A stripping tower

Precipitation

Precipitation is used here as a term to describe processes in which a solid forms from a solute in solution as a result of a physical change in the solution, as compared to chemical precipitation in which a chemical reaction in solution produces an insoluble material. The major changes that can cause physical precipitation are cooling the solution, evaporation of solvent, or alteration of solvent composition.

The most common type of physical precipitation by alteration of solvent composition occurs when a water-miscible organic solvent is added to an aqueous solution, so that the solubility of a salt is lowered below its concentration in the solution.

Phase Conversion

Vitrification

Vitrification (also referred to as glassification) is a phase conversion process in which a chemical waste (or constituents of the waste) is melted at high temperature to form an impermeable capsule around the remainder of the waste. The process has a benefit over other processes insofar as it can be applied both *in situ* and *ex situ*.

The principles behind vitrification are the same as those applied to the production of glass. High temperature electrodes are used to melt the wastes and organic constituents are transformed by pyrolysis and either collected as product or destroyed in secondary processes. The inorganic components are immobilised in the resulting glass matrix.

In ex situ applications, the waste is introduced into the furnace along with the silica soda, and lime. The organic materials are driven off, captured, and treated while the inorganic materials are incorporated into the glass. The plasma arc has also been shown to be suitable for *ex situ* vitrification.

In situ vitrification involves insertion of large electrodes into the soil. Graphite is spread on the soil surface between the electrodes to complete the circuit. A negatively pressurised hood is placed over the site to collect any off-gases for later treatment. High voltage (4160 volts, a 3000 kW electrical source is required) is applied across the electrodes to produce temperatures reaching 3600°C (6510°F). The use of *in situ* vitrification is limited by: (i) high groundwater tables, (ii) buried metal objects, and (iii) the need for sufficient quantities of glass-forming material in the waste or in the soil.

The vitrification process has shown great promise for treating radioactive and mixed wastes which are immobilized in the glass matrix and, supposedly, can then be stored until the radioactivity decays to a safe level. In the case of mixed wastes, vitrification drives off the nonradioactive components allowing them to be treated as chemical waste while immobilising the radioactive component.

Membrane Separation

Another major class of physical separation is molecular separation, often based upon membrane processes in which dissolved contaminants or solvent pass through a size-selective membrane under pressure (Figs. 12.5 and 12.6). The products are relatively pure solvent phase (usually water) and a concentrate enriched in the solute impurities.

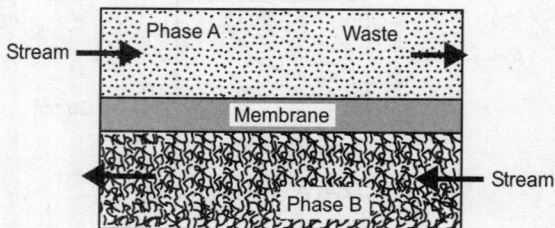

Fig. 12.5. A membrane contactor system.

Water and lower molecular mass solutes under pressure pass through the membrane as a stream of purified permeate, leaving behind a stream of concentrate containing impurities in solution or suspension.

Dialysis

Dialysis is a process for separating components in a liquid stream by using a membrane. Components of a liquid stream will diffuse through the membrane if a stream with a greater concentration of the component is on the other side of the membrane. Dialysis is used to extract pure process solutions from mixed waste streams.

Electrodialysis

Electrodialysis is an extension of dialysis and is used to separate the components of an ionic solution by applying an electric current to the solution, which causes ions to move through the dialysis membrane (Fig. 12.7). It is very effective for extracting acids and metal salts from solutions.

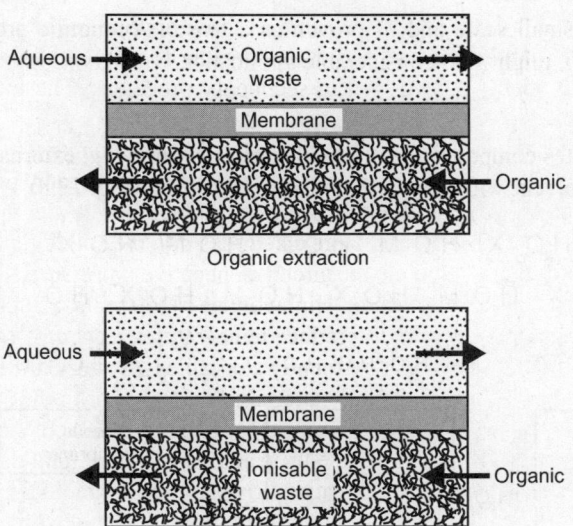

Fig. 12.6. Cleanup of liquid streams using membranes.

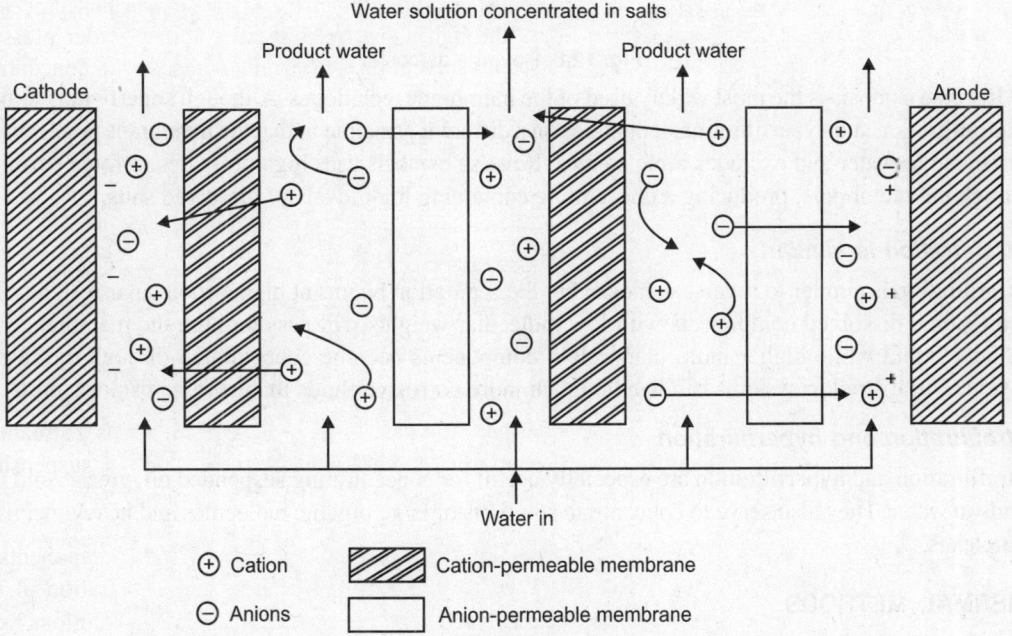

Fig. 12.7. Electrodialysis.

The related process of electrolysis shows promise as a method for the dechlorination of polychlorinated biphenyls. Commercial polychorinated biphenyl mixtures in spiked soil samples have been reduced to biphenyls and hydro-biphenyls. The technique involves mixing the contaminated soil into a microemulsion that is used as the medium to carry out the catalytic electrolysis. The procedure certainly shows promise on the small scale and, if proven as a feasible/economic process on the large scale (especially as an *in situ*), might find considerable application.

Reverse osmosis

Reverse osmosis separates components in a liquid stream by applying, external pressure to one side of the membrane so that solvent will flow in the opposite direction (Fig. 12.8).

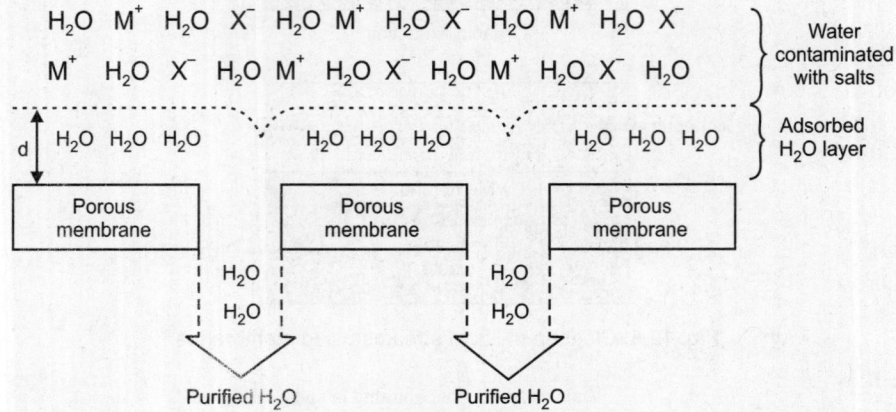

Fig. 12.8. Reverse osmosis.

Reverse osmosis is the most widely used of the membrane techniques. Although superficially similar to ultrafiltration and hyperfiltration, it operates on a different principle in that the membrane is selectively permeable to water and excludes ionic solutes. Reverse osmosis uses high pressures to force permeate through the membrane, producing a concentrate containing high levels of dissolved salts.

Ultrafiltration is similar

Ultrafiltration is similar to reverse osmosis, but the separation begins at higher molecular weights. The result is that dissolved components with low molecular weights will pass through the membrane with the bulk liquid while higher molecular-weight components become concentrated through the loss of solvent. Ultrafiltration systems can handle much more corrosive fluids than reverse-osmosis units.

Ultrafiltration and hyperfiltration

Ultrafiltration and hyperfiltration are especially useful for concentrating suspended oil, grease, and fine solids in water. They also serve to concentrate solutions of large organic molecules and heavy metal ion complexes.

THERMAL METHODS

Outside of the traditional containment methods (land filling and capping), application of some type of thermal process has, until recently, been the most common form of treatment for chemical waste.

Thermal treatments have lost some popularity recently due to the threat of emissions from incomplete combustion. Except for vitrification, thermal technologies are *ex situ* processes requiring the wastes to be transported to the processing unit. However, the use of a thermal destruction technology may be (depending upon the nature of the emissions) superior to wet scrubbing technologies for emissions cleaning.

Incineration

These are designed depending on the quality and quantity of the waste feed, the required process time in consideration with the kind of thermal processing that suits the wastes. Hearth furnaces constitute a common variety used in this process. Wastes are fed through a feed port into the incinerator, which is built with arms to rotate inside to keep the material dynamic. The retention period for the wastes inside the incinerator is selected carefully and maintained, as a result of the design of such arms only. The control over the overall temperature of the incinerator needs to be overemphasised. The resultant gas from the incineration is led through accessories like pre-cooler, particle separator, filter, sub-cooler, fan etc.

In an alternative type, viz. fluidised bed incinerator, the material is retained in a fluidised state with the help of gases blown in. Here, pre-heating of the bed is mandatory. Moreover, excess air is limited to not more than 35–40 per cent above the stochiometric requirement.

The liquid waste incinerators are considered as labour-friendly, as most of the cumbersome requirements of the other types of incinerators are done away with this. The general feed requirement is that the feed wastes act as a liquid, the solids are molten, so that they are pumpable and attract atomisation in the liquid waste incinerator. The waste gas incinerators act in either of the following three principles:
1. Direct flame method.
2. Multiple hearths.
3. Catalytic combustion.

While the multiple hearths have been discussed in the solid waste treatment section, the direct flame method is applicable only when the wastes or their combustion produce do not pose any threat to the environment and demands a very high temperature levels that generally runs in the range of 500°C. The catalytic combustion process uses the most common property that the general gaseous wastes do not possess the fuel value enough to initiate or support or sustain self-ignition. This method is applicable to the waste gases that have hydrocarbon levels lower than lower explosive limits. The process is considered advantageous in the sense that this produces clean hot gas.

Multiple chamber incinerators are most commonly employed to treat wastes of plastic nature. The arrangement of chambers classifies them as either retort or inline type. Molten salt incinerators are composed of a salt mixture, generally containing sodium carbonate and sodium sulphate in 9:1 ratio. Changes in the salt types and mix proportions can be envisaged to result in different temperature ranges. One attractive feature of this process is that eutectic mixtures of reactive salts can be developed which render astonishing additional benefits like containment of toxicity.

As a general rule of nature, every coin has two sides. While this process is attractive in the above said aspect, it also brings in with it the challenge of treating of the salt wastes. They have to be either regenerated or incinerated or used as say, road fills. This also calls for requirement of fuels for start-up and support.

Rotary kilns are designed to keep the material in continuous movement to ensure proper mixing while process is on. Wet air oxidation method is solely applicable for treatment of oxidisables in their aqueous state. The underlying principle here is that organic compounds increase their rate of oxidation with pressure. This will aid in arriving at an incomplete oxidation in liquid phase that will help destroy the organics.

Incineration of solid and liquid chemical wastes, such as polychloro-biphenyl materials, is a common and efficient method of destruction. The destruction of polychlorobiphenyls commences at ca. 800°C (1470°F) and commercial incinerators operate at temperatures in excess of 1000°C (1800°F). Major concerns in this type of incineration involve potential generation and emission of by-products of incomplete combustion. Such by-products, if produced, can often be more harmful to the environment.

Suitable for the purpose are stationary incinerators, cement kilns, incinerator ships, and smaller mobile thermal destruction units. Efforts have also been spent in developing efficient catalysts that allow the use of lower temperature reactions with higher efficiency.

This, of course, is a convenient point at which to note only one of the many ways in which environmental events on the land, in the air, and in the water are connected. And the connection between the air, land, and water systems is not guaranteed to be sequential:

$$\text{Land systems} \longleftrightarrow \text{water systems} \longleftrightarrow \text{atmosphere}$$

Briefly, emissions from the incinerators may be gaseous (SO_x, NO_x, particulate matter), which can pollute the atmosphere or, through the formation of the constituents of acid deposition, the water systems and the land systems, and therefore control is necessary.

Alternatively, these emissions might have a direct effect on the vegetation by deposition of the gases or particulate matter directly on to the land at a point downwind of the incinerator. In addition, solid waste (ash) from an incinerator can pollute the land and, as a result of leaching by rain (acidic or otherwise) pollute the water systems. Therefore, attention must be focused on the disposal of the ash.

The air emission systems from these thermal reactors, as well as from the furnaces that burn solids emanating from waste-water plants, must be designed to prevent chemical and particulate air pollution. These air emission systems include electrostatic precipitators (ESP) and afterburners.

In addition, sulphur-containing fuels emit sulphur dioxide and other sulphur containing gases as a result of combustion and these sulphur-containing gases have the potential for conversion to sulphates (sulphuric acid in the atmosphere), which can be deposited on the land as sulphate.

Nitrogen oxides (also produced during fuel combustion) are converted to nitrates in the atmosphere and the nitrates eventually are deposited on soil. Soil also adsorbs nitric oxide (NO) and nitrogen dioxide (NO_2) readily and these gases are oxidised to nitrate in the soil. Carbon monoxide is converted to carbon dioxide and possibly to biomass by soil bacteria and fungi. Elevated levels of heavy metals (such as lead) from mines and smelters are also found in the soil near such facilities.

Thermal treatment of chemical waste can be used to accomplish most of the common objectives of waste treatment: (i) volume reduction, and (ii) removal of volatile, combustible, mobile organic matter and destruction of toxic and pathogenic materials. Incineration utilises high temperatures, an oxidising atmosphere, and often turbulent combustion conditions to destroy wastes.

The effective incineration of wastes depends upon the combustion conditions, which are: (i) a sufficient supply of oxygen in the combustion zone, (ii) thorough mixing of the waste, the oxidant, and any supplemental fuel, (iii) combustion temperatures above 900°C ($\leftrightarrow$ 1650°F), and (iv) sufficient residence time to allow reactions to occur.

Usually, the heat required for incineration comes from the oxidation of organically bound carbon and hydrogen contained in the waste material or in supplemental fuel:

$$C_{(organic)} + O_2 = CO_2 + \text{heat}$$
$$4H_{(organic)} + O_2 = 2H_2O$$

These reactions destroy organic matter and generate heat required for endothermic reactions, such as the breaking of C-Cℓ bonds in organic chlorine compounds.

Incineration detoxifies chemical waste by destroying organic compounds, reduces the volume of the waste, and converts liquid waste to a solid product by reacting and vaporising any fluids present in the wastes. The primary use of incineration is for the destruction of volatile organic compounds and semi-volatile organic compounds. Incinerators have been extremely capable of destroying organic compounds in waste. Removal efficiencies as high as 99.9999 per cent have routinely been achieved (often referred to as the six-9s treatment level).

If the waste is exposed to high temperatures in an oxygen-poor environment, the process is known as pyrolysis. The products of this process are simpler organic compounds, which may be recovered or incinerated, and a char or ash.

Incineration systems are designed to accept specific types of materials; they vary according to feed mechanisms, operating temperatures, equipment design, and other parameters. The main products from complete incineration include water, carbon dioxide, ash, and certain acids and oxides, depending upon the waste in question. Unlike pyrolysis, incineration is carried out with excess oxygen.

Waste incineration and pyrolysis systems include single-chamber liquid systems, rotary kilns, and fluidised-bed incineration systems. In a single chamber liquid system a brick-lined combustion chamber contains liquids that are burned in suspension. In addition to being the primary parts of an incineration system, these units can be used as after burners for rotary kilns.

A rotary kiln is a versatile large refractory-lined cylinder capable of burning virtually any liquid or solid organic waste; the unit is rotated to improve turbulence in the combustion zone. Fluidised-bed incineration uses a stationary vessel within which solid and liquid wastes are injected into a heated, extremely agitated bed of inert granular material; the process promotes rapid heat exchange and can be designed to scrub off the gases. The ideal wastes for incineration are predominantly organic materials that will burn with a heating value of at least 5,000 Btu/lb and preferably more than 8,000 Btu/lb. Such heating values are readily attained with waste having a high content of organic constituents. In some cases, however, it is desirable to incinerate wastes that will not burn alone and which require supplemental fuel, such as methane and petroleum liquids.

Examples of such wastes are nonflammable organic chlorine wastes, some aqueous wastes, or soil in which the elimination of a particularly troublesome contaminant is worth the expense and trouble of incinerating it. Inorganic matter, water, and organic heteroelement contents of liquid wastes are important in determining their susceptibility to incineration.

Many wastes, including hazardous waste, are burned to produce fuel for energy recovery in furnaces and boilers. This process (coincineration) uses combustible waste material more for energy generation than for waste destruction. In addition to heat recovery from combustible waste, an existing on-site facility, rather than a separate waste incinerator, can be used for waste disposal.

Incinerators can be designed to handle wastes in any physical state and have proven effective in treating solids, liquids, sludge, slurries, and gases. The effectiveness of an incinerator depends on three factors: (i) temperature of the combustion chamber, (ii) residence time in the chamber, and (iii) amount of mixing of the material with air while in the chamber.

Normal combustion temperatures range between 900° and 1500°C (1650°–2280°F) in some instances the temperatures employed are much higher. Many incinerators for hard-to-burn compounds employ two combustion chambers. The first chamber converts the compounds to gas and initiates the combustion process. In the second chamber, combustion of the gases is completed.

There are four major components of chemical waste incineration systems: preparation, combustion, pollutant removal, and ash disposal. Waste preparation for liquid wastes may require filtration settling to remove solid material and water, blending to obtain the most appropriate mixture for incineration, or heating to decrease viscosity. Solids may require shredding and screening, and good examples include tyres. Atomisation is commonly used to feed liquid wastes. Several mechanical devices, such as rams and augers, are used to introduce solids into the incinerator.

The most common kinds of combustion chambers are liquid injection, fixed hearth, rotary kiln, and fluidised bed. Often the most complex part of a waste incineration system is the air pollution control system, which involves several operations. The most common operations in air pollution control from waste incinerators are combustion gas cooling, heat recovery, quenching, particulate matter removal, acid gas removal, and treatment and handling of by-product solids, sludge, and liquids.

The inert portion of the waste remains as ash after incineration. For liquid waste, the amount of ash remaining is generally minimal, whereas for solid waste, the volume of ash can be as much as one-third by weight of the original material. If the ash contains metals or radioactive material, it must be further treated prior to disposal. The most frequently employed method of treating the ash remaining from the incineration process is solidification/stabilisation.

Hot ash is often quenched in water and, prior to disposal, it may require dewatering and chemical stabilisation. A major consideration with waste incinerators and the types of wastes that are incinerated is the disposal problem posed by the ash especially in respect to potential leaching of heavy metals.

Waste incinerators may be divided among the following, based upon type of combustion chamber. (i) rotary kiln incinerators in which the primary combustion chamber is a rotating cylinder lined with refractory materials and an afterburner downstream from the kiln to complete destruction of the wastes; (ii) liquid injection incinerators that burn liquid waste dispersed as small droplets; (iii) fixed-hearth incinerators with single or multiple hearths upon which combustion of liquid or solid wastes occurs; and (iv) fluidisedbed incinerators that have a bed of granular solid (such as sand) maintained in a suspended state by injection of air to remove pollutant acid gas and ash products.

Advanced design incinerators, including plasma incinerators, make use of an extremely hot plasma of ionised air injected through an electrical arc. Medical wastes are often incinerated by these systems. Examples are electric reactors which use resistance-heated incinerator walls at around 2200°C (3990°F) to decompose the waste by radiative heat transfer. There are also infrared systems which generate intense infrared radiation by passing electricity through silicon carbide resistance heating elements. Molten salt combustors use a bed of molten salt, such as sodium carbonate, at 900°C (1650°F) to destroy the waste and retain gaseous emissions through chemical reaction. Finally, there are also molten glass processes which use a molten glass to transfer heat to the waste and to retain products in a non-leachable form.

Incinerators may be mobile, transportable, or stationary (fixed). Mobile reagent incinerators are normally relatively small units that are mounted on a flat bed trailer and transported to the hob site. Transportable incinerators are larger units that can be disassembled into manageable components and move from one site to another by a caravan of trucks. Stationary/fixed incinerators are permanently erected at a site, and the wastes are brought to the site for treatment.

Thermal Desorption

Thermal desorption is the process of heating a waste in a controlled environment thereby volatilising any organic constituents. Thermal Desorption works especially well for volatile organic compounds but can also be employed for semi-volatile organic compounds. Removal efficiencies ranging from 65 to 99 per cent have been achieved depending upon the type of waste.

Prior to entering the thermal desorption unit, the wastes are screened to eliminate coarse pieces. If the wastes have a high moisture content, this excess moisture is also removed. The wastes are then passed to a furnace which operates at temperatures in the range 300°–600°C (570°–1110°F). Volatile organic compounds become gaseous in the process and are either collected on an adsorbent, such as activated carbon, for further treatment or passed through an incinerator connected in-line with the thermal desorption unit.

Pyrolysis

This is a process that is defined, at least theoretically as a zero air indirect heat process. Pyrolysis is an air-starved process, wherein the combustion proceeds with 'less than stochiometric level' air. The aim is to distil or vapourise the compounds to give out combustible gases. Wastes that are considered worthy of treatment via this process are only those that possess a high calorific value, inherent in them. Here the otherwise end product problems that are associated with common incineration, viz. ash and clinker are not derived, as they are also converted to combustibles. The feed to this process can be combined with the otherwise non-treatable wastes of various other treatment methods (a few of which are being discussed), so that they too meet a useful end as getting converted to combustibles.

Pyrolysis is a chemical change brought about by the action of heat. The process differs from incineration, which is the combustive destruction of a material in direct flame in the presence of oxygen. Pyrolysis can also be defined as destructive distillation in the absence of oxygen (or other oxidant) with the simultaneous removal of volatile products.

Pyrolysis converts wastes containing organic material to combustible gas, charcoal, organic liquids, and ash/metal residues. In some instances, the organic liquid fraction produced during pyrolysis has the potential to produce constituents for synthetic crude oil. As a different note, pyrolysis can also be applied to oil shales to produce liquid hydrocarbons for synthetic crude oil.

The effectiveness of waste destruction by pyrolysis depends upon: (i) the residence time within the retort, (ii) the rate of temperature increase, (iii) the final temperature, and (iv) the composition of the feed material.

Pyrolysis units, which operate at temperatures from 500° to 800°C (930°–1470°F) have achieved up to 99.9999 percent destruction/removal efficiencies.

Plasma torch processes apply the principles of pyrolysis at temperatures in the range 5,000° to 15,000°C (9,000°–27,000°F). The wastes are fed into the thermal plasma, where they are dissociated into their basic atomic components which recombine in the reaction chamber to form carbon monoxide, nitrogen, and hydrogen, as well as methane and ethane.

Acid gases which are removed from the emissions by scrubbers and any solid products are either incorporated into the molten bath at the bottom of the chamber or removed from exhaust gases by particulate scrubbers or filtres.

SOLIDIFICATION AND STABILISATION

Solidification and stabilisation are treatment systems which move beyond the older concept of 'dig and move' and are designed to: (i) improve the handling and the physical characteristic of the waste, (ii) decrease the surface area across which transfer or loss of contained pollutants can occur, and (iii) limit the solubility of, or detoxify, any chemical constituents in the waste.

Stabilisation/solidification processes are being used to minimise the potential for groundwater pollution from land disposal of hazardous wastes. Many variations are used, but most rely on pozzolanic reactions to chemically stabilise and physically solidify the waste. Portland cement alone or in combination with fly ash, cement kiln dust lime or other ingredients is the principal solidifying agent used.

Stabilisation and solidification processes are very effective at immobilising most heavy metals present in sludge, contaminated soils and other wastes. They are not as effective at immobilising toxic organic materials. Organically modified clays are now being evaluated as an additive to stabilisation and/or solidification processes in order to adsorb and retain these organic pollutants in the solidified waste form.

The environmental acceptability of stabilisation and solidification processes depends on the long term ability of the waste form to retain contaminants. This will be governed by the chemical binding mechanisms involved and by the durability of the waste for widespread acceptance of stabilisation and solidification processes will be hampered until the longterm durability of the waste form can be demonstrated.

Stabilisation usually involves the addition of materials that ensure that the chemical constituents are maintained in their least soluble or least toxic form. In the solidification process, the results are obtained primarily, but not exclusively, via the production of a monolithic block of treated waste with high structural integrity. Stabilisation techniques limit the solubility of, or detoxify), the waste contaminants; even though the physical characteristics of the waste may not be changed.

Both of these techniques might also have been classified under phase transition or phase conversion. Encapsulation, another perhaps more specific form of waste stabilisation, is a method by which mixed wastes, i.e. waste containing radioactive material and higher level radioactive wastes can be rendered less harmful to the environment. Crosslinked polyethylene, chosen for its durability, has been suggested as a suitable encapsulating reactive polymers such as maleic anhydride grafted polypropylene (pp-g-MA) are also being considered for this purpose.

CHEMICAL TREATMENT OF WASTES

In any industrial organisation, the one way that is important and beneficial both economically and environmentally is to consider excess materials as an additional resource that can be utilised elsewhere either after they get discarded or retreated. It is in the better side of the humanity welfare that this aspect must be given the highest support and importance even without any regulations by any governmental organisations or so.

Environmental protection must be given the top most priority while meeting commercial targets. This is because of the undeniable fact that the costs of protecting the environment must have to be paid, if not today, tomorrow and if not by us, by our offsprings. Incorporation of available techniques that will render only environment-friendly outputs from any set-up has to be thought of and envisaged by industries, humanity and every individual. It may seem costly today in some stray cases but, with inflation, it is never going to be cheaper than this. Is it justified to put our future generations in the altar for saving us a few dimes? Is not justice delayed justice denied? Long-term effects must not be neglected. In order to prevent degradation of the quality of the environment, we should no longer allow the treatment of the wastes remain to be a burning issue in today's world.

Categorisation

At the outset, the wastes can be categorised as of either solid, liquid or gaseous status. Another classification happens to be as enlisted below:

1. Wastes getting discharged directly from the units that manufacture desired products or components. This is by means of normal operations.
2. Materials that get contaminated with storm water, wash fluids, etc. that run off from the processing zones.
3. The outputs of utilities that inevitably result as generated in the blow down, etc. of say, systems that are required to produce the energy or cooling fluids.
4. Wastes created by virtue of handling of products during processing for the desired product.
5. Inadvertent spills too contribute a share to waste production.
6. Sewage from sanitary needs of personnel, due to nature calls or food handling.

Prediction of the effects of such production of wastes becomes mandatory for each and every industrial establishment as a prelude to find an amicable waste disposal or treatment procedure later. This is easier said than done as the resources or the methods selected must be capable of covering a huge variation probability generally due to product demands or processes available for the production of the same product. Each individual unit process must have to be studied in depth for the variety of methods available and the method that embraces the minimum extent of production of wastes is to be adopted. This again is not the ultimate as the process thus selected may not be viable in the existing or expected market scenarios, as cost consideration is of paramount importance for any business to survive, let alone flourish. Another major consideration that should not be given a back seat is the kind of operational safety that is inherited by the selected process.

Wastes produced intentionally must be near zero level, whereas inadvertently must be the least by virtue of selection of the right process for the manufacture. The treatment facilities cannot be reliable unless the wastes have been fully and clearly characterised. Even the performance characteristics of the various treatment facilities that can be thought of or worked out, have to be elaborated with respect to their evaluated treatability studies. It is the best practice to get these derivations proved as far as their eligibility to give the expected outputs via say pilot plant facilities.

Parameters

The study of the capability and foolproof character of the selected treatment operation must be in a position to reveal the effects of the parameters used in the operations. It should always be borne in mind during envisaging any such pilot units or test facility, that the switch over to mass scale is not going to get the results revealed in such tests to a hundred percent level. A few of the operational parameters mentioned above can be enumerated as:

1. Detention time.
2. Hint on any special treatment that may be demanded by the process.
3. Oxygen requirements.
4. Production of sludge.
5. Reliability of the process.
6. Sludge age.
7. Stability of the process.
8. The characteristics of thus produced sludge.
9. Waste removal rates.

Here, it may be worth touching upon the conventional pollution related parameters, viz. acidity, aesthetics, alkalinity, colour, corrosivity, dissolved solids, odour, oxygen demand, particulate, pH, surface activity, suspended solids, taste, temperature, total carbon, toxicity, turbidity, etc.

Process

The ultimate aim of envisaging any treatment facility is to give environment-friendly outputs in the end. It is recommended to adopt a method that operates in a closed system, if possible. Of course, the initial cost of setting up or maintaining a facility to treat pollutants or corrodants calls for high initial investment, but it is worth in the long run.

Reliance on any material that is or that may produce some other material with a high likelihood of posing disposal problems at a later stage, must be curbed, at least by keeping constant attention to arrive at an amicable process to meet this end result—even if not in the near future. It may be of interest to note here that, in several countries, based on the actual disposal costs, different tax slabs have been legislated for items like waste paper, plastics, glass, metals, etc. This step not only offsets the disposal costs but also discourages the manufacturers from adopting makeshift processes to get away from the wastes they may tend to produce. Countries now have commenced caring about minute aspects like proper and efficient methods of collecting the wastes, let alone treating them and curbing their generation (Fig. 12.9).

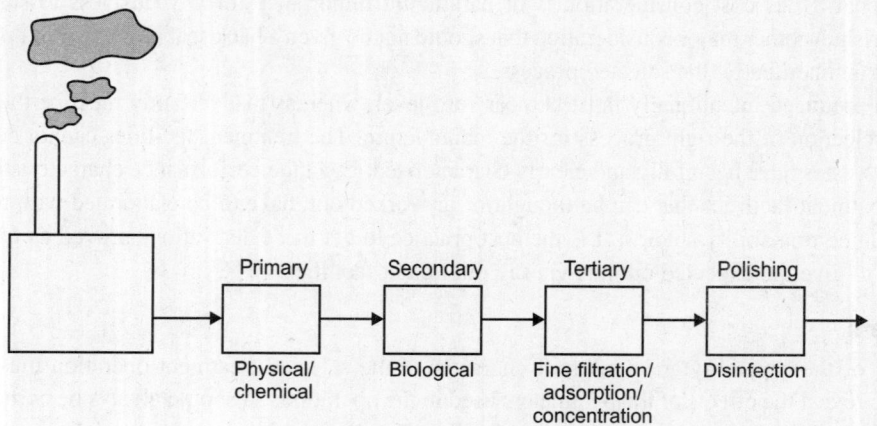

Fig. 12.9. Effluent treatment.

There are seldom cost effective alternative solution available for certain processes in waste treatment technology. Biological treatment processes greatly fall under this category. Treatments of waste by means of a combination of physical and chemical methods has always been welcome as they embrace a lot of advantages, viz.

1. Treatment continues unaffected by toxic substances carried out by various feeds.
2. Only a very small area of work is demanded by this.
3. Reliability on the service expected to be rendered by the facility is on the higher side.
4. Variations in either the quantitative or qualitative compositions do not affect the process much.
5. Completion to a greater degree of processing is possible.
6. No step affects the other step uncontrollably.
7. Delays in one step can be tolerated to moderate levels.
8. Alterations or modifications to any stage of the processing are possible.
9. Higher degree of purity of the end product can be made achievable.
10. Plant automation is made easy.
11. Each method operates selectively.

Wastes, irrespective of their source of generation, can be inferred to be containing the following types of states of matter only: colloidal, dissolved, sub-colloidal, and undissolved.

Special attention has to be given to the treatment and removal of toxic wastes. The problem that they may pose is interference with the treatment by biological methods.

Process modification is the better alternative than pollutant abatement, if only it can be met. With exponential leap of technology in the present day world, it is rather foolish and suicidal to sit thinking

that 'It is not applicable to my process'. A continuous and never ending process of looking out for the better method is made possible via:
1. Selection of a process that produces, by design, minimal wastes.
2. Modification of the process to reduce further the volume or the mass of the wastes produced by virtue of the processing.
3. Exploring the possibility of recycling the so-called wastes to get the maximum benefit out of them before discarding them to the waste treatment area.
3. Tests at least in the Laboratory levels, to derive valuable coproducts from these wastes before the final treatment stage.
4. In case no fruitful method of recycling to end up in a conversion to useful prime product is found, at least they must be turned out to get a face fit for reuse in the process either as an intermediate or reagent.
5. If even the route discussed in the above step is sealed, probabilities of using at as a usable item in some other industry, thus forming a cheaper source to the latter and saving the hassles of waste treatment to the former.
6. In the event of an utter failure with any use for the waste, an ultimate try can be given to see whether it can be used as a fuel substitute at least.
7. Treating them to quench their deleterious effects on the environment.
8. Using energy in waste treatment facilities must have to be considered with an open and vigilant eye on the following factors.
9. The direct cost of the energy used.
10. The effects of the produce generated in the treatment—either directly or indirectly on the environment.
11. The kind of depletion of the nonrenewable resources of energy in treatment of a material destined to be of nil use.
12. The cost factor that may affect the choice of providing the energy source to the treatment facility.

It is rather a tricky fact to say that power is required to abate wastes and production of power also produces wastes that are again to be treated to enrich the environment.

Operations

Generally, the following operations are considered to constitute the waste treatment processing:
1. Chemical clarification.
2. Disinfection.
3. Ion exchange.
4. Pyrolysis.
5. Incineration.

Chemical clarification

This depends on the treatment agents that are added to the streams at stages. The following factors too play a part:
1. Particle charge.
2. The chemical constituents of the material.
3. The physical properties of the material.
4. The rate of agitation.
5. The relative concentration of the additive.

6. The temperature of the reaction contents.
7. The time duration the agitation is effected, etc.

Disinfection

This is generally completed by simple chlorination only. Here, the correct dosage plays a vital role. Insufficient dosage ends up in premature termination of the process and does not produce the desired results, whereas excess chlorination results in further necessity for additional treatment, as toxicity of the treated waste goes high, rendering the total process shoot up in economic non-viability.

The time and duration and the rate of dosage is important here. Chlorine can be dosed in either liquid form or as a derivative, say hypo. Ozone is an alternative disinfectant worth considering, but comparatively expensive. This, however, brings in added advantage like high sanitary quality.

Ion Exchange

These help remove certain salts and tailor their ability to cooperate with the process of waste treatment. They can aid in removing a given salt or metal from a multi component system composition, consisting of complex agents. Sorption is another area that is helped by ion exchanges. The one striking aspect worth considering here is that certain ion exchange resins do not show any sudden reduction in their capability to perform their best even after absorbing quite a few ions from the materials they were allowed to be reacting in the process of ion exchanges. This, when evaluated with the fact that these resins can be reactivated, highlights that they are here to stay.

BIOREMEDIATION

Bioremediation is the use of living organisms (primarily micro-organisms) to degrade pollutants previously introduced into the environment or to prevent pollution through treatment of waste streams before they enter the environment.

Bioremediation is emerging as one of several alternate technologies for removing pollutants from the environment, restoring contaminated sites, and preventing further pollution.

Biodegradation of waste is the conversion of waste materials by biological processes to simple inorganic molecules and to a certain extent, to biological materials. The complete bioconversion of a substance to inorganic species such as carbon dioxide, ammonia, and phosphate is called mineralisation.

Biological waste treatment is a generic term applied to processes that use micro-organisms to decompose organic wastes either into water, carbon dioxide, and simple inorganic substances or into simpler organic substances, such as aldehydes and acids. The purpose of a biological treatment system is to control the environment for micro-organisms so that their growth and activity are enhanced, and to provide a means for maintaining high concentrations of the micro-organisms in contact with the wastes.

Since biological treatment systems do not alter or destroy inorganic substances, and high concentrations of such materials can severely inhibit decomposition activity, chemical or physical treatment may be required to extract inorganic materials from a waste stream prior to biological treatment.

Detoxification refers to the biological conversion of a toxic substance to a less toxic species, which may still be relatively complex or biological conversion to an even more complex material. An example of detoxification is the enzymatic conversion of paraoxon (a highly toxic organophosphate insecticide) to p-nitrophenol, which has only about 1/200 the toxicity of the parent compound.

Biodegradation is usually carried out by the action of micro-organisms, particularly bacteria. However, micro-organisms such as bacteria are not panaceas for chemical wastes. Their function can vary depending upon the specific bacterium and the metabolic processes of the bacteria.

Biotransformation is the conversion of a substance through metabolisation thereby causing an alteration to the substance by biochemical processes in an organism. Metabolism is divided into the two general categories of catabolism, which is the breaking down of more complex molecules, and anabolism, which is the building up of life molecules from simpler materials. The substances subjected to biotransformation may be naturally occurring or anthropogenic (made by human activities). They may consist of xenobiotic molecules that are foreign to living systems.

An important biochemical process that occurs in the biodegradation of many chemical waste materials is co-metabolism. This does not serve a useful purpose to an organism in terms of providing energy or raw material to build biomass, but occurs concurrently with normal metabolic processes.

An example of cometabolism of chemical waste is provided by the white rot fungus which degrades a number of kinds of organic chlorine compounds (including polychlorobiphenyls, cyanides, and chlorodioxins) under the appropriate conditions. The enzyme system responsible for this degradation is one that the fungus uses to break down lignin in plant material under normal conditions.

Biodegradation of chemical waste that can be metabolised takes place whenever the wastes are subjected to conditions conducive to biological processes. The most common type of biodegradation is that of organic compounds in the presence of air, that is, aerobic processes. However, in the absence of air, anaerobic biodegradation may also take place. Furthermore, inorganic species are subject to both aerobic and anaerobic biological processes. Although biological treatment of chemical waste is normally regarded as degradation to simple chemical species such as carbon dioxide, water, sulphates, and phosphates, the possibility must always be considered of forming more complex (in some cases hazardous) chemical species. An example of the latter is the production of volatile, soluble, toxic methylated forms of arsenic and mercury from inorganic species of these elements by bacteria under anaerobic conditions.

Physical, chemical and biological treatment processes are employed for waste-water treatment. In addition, chemicals are introduced for precipitation of nutrients, followed by coagulation and filtration for removing solids remaining after biological treatment. In some cases, granular activated carbon or membrane filtration or a combination of membrane-assisted solvent extraction are used for additional purification of the groundwater streams and waste streams. This higher level of treatment is advisable because of the damage that any visual traces of chemical waste can do to the appearance of the water. In addition, the treatment may combat the potential eutrophic effect that the nutrients phosphorus and nitrogen may have on the receiving water.

For the most part, anthropogenic compounds resist biodegradation much more strongly than do naturally occurring compounds. This is generally due to the absence of enzymes that can bring about an initial attack on the compound. A number of physical and chemical characteristics of a compound are involved in its amenability to biodegradation. Such characteristics include hydrophobicity, solubility, volatility and affinity for lipids. Some organic structural groups impart particular resistance to biodegradation. These structural groups include branched carbon chains ether linkages, meta-substituted benzene rings, chlorine, amines, methoxy groups, sulphonates, and nitro groups.

Acintomycetes are micro-organisms that are morphologically similar to both bacteria and fungi. They are involved in the degradation of a variety of organic compounds, including degradation-resistant alkanes, and lignocellulose. Other compounds attacked include pyridines, non-chlorinated aromatics, and chlorinated aromatics. Fungi are particularly noted for their ability to attack long-chain and complex hydrocarbons and are more successful than bacteria in the initial attack on polychlorobiphenyls.

Phototrophic micro-organisms, which include algae, photosynthetic bacteria, and cyanobacteria (blue-green algae) tend to concentrate organophilic compounds in their lipid stores and induce photochemical degradation of the stored compounds.

Usually the products of biodegradation are molecular forms that tend to occur in nature and that are in greater thermodynamic equilibrium with their surroundings than the starting materials. Detoxification refers to the biological conversion of a toxic substance to a less toxic species.

Microbial bacteria and fungi possessing enzyme systems required for biodegradation of wastes are usually best obtained from populations of indigenous micro-organisms at a chemical waste site where they have developed the ability to degrade particular kinds of molecule. Although it has some shortcomings in the degradation of complex chemical mixtures, biological treatment offers a number of significant advantages and has considerable potential for the degradation of chemical wastes, even *in situ*.

The biodegradability of a compound is influenced by its physical characteristics, such as solubility in water and vapour pressure, and by its chemical properties including molecular mass, molecular structure, and presence of various kinds of functional groups, some of which provide a 'biochemical handle' for the initiation of biodegradation. With the appropriate organisms and under the right conditions, even substances such as phenol that are considered to be biocidal to most micro-organisms can undergo biodegradation.

Properties of chemical wastes and their media can be changed to increase biodegradability. This can be accomplished by adjustment of conditions to optimum temperature, pH (usually in the range of 6–9), stirring, oxygen level, and the amount of the material. Biodegradation can be aided by removal of toxic organic and inorganic substances, such as heavy metal ions.

Biodegradability of polymers has become a focal point in both environmental protection and landfill space management The impact of plastic materials in municipal solid waste (MSW) is even greater when it is considered that while plastics make up only about 9 per cent, by mass, they make up 28 per cent of the volume of MSW. The biodegradability issue of plastic articles has been brought to the public attention via: (i) startling discovery of tragic deaths of baby dolphins due co their mouths getting stuck by discarded plastic loop carriers of 6-pack cans; (ii) finding of undecomposed newspapers, still completely readable, that had been buried over 70 years in landfills; and (iii) an increasing number of deaths of wild birds at national or municipal park sites due to injuries from carelessly discarded snack wrappers. Most commodity polymers of molecular weight of about 1000 or higher are essentially not biodegradable, in their present forms. Biodegradation is often accompanied or preceded by photodegradation by which polymers go through reduction in molecular weights. Efforts have been made to make polymers and polymeric articles biodegradable after serving for intended uses. Some of the successful and noteworthy products include polyhydroxybutyratevalerate (PHBV), ethylene-carbon monoxide (E/CO), copolymer, aliphatic polyester, polycaprolactone and vinyl ketone photodegradable polymers.

Aerobic processes for the treatment of waste utilise aerobic bacteria and fungi that require molecular oxygen. These processes are often favoured by micro-organisms, in part because of the high energy yield obtained when molecular oxygen reacts with organic matter. Aerobic and anaerobic digestion processes are well adapted to the use of an activated sludge process. These treatments can be applied to wastes such as chemical process wastes and landfill leachates. Some systems use powdered activated carbon as an additive to absorb nonbiodegradable organic wastes.

Contaminated soils can be treated in variety of ways to eliminate contaminates. It is possible in principle to treat contaminated soils biologically in place by pumping oxygenated, nutrient-enriched water through the soil in a recirculating system.

Anaerobic processes for the treatment of waste are those processes in which micro-organisms degrade wastes in the absence of oxygen and can be practiced on a variety of organic wastes. Compared to the aerated activated sludge process, anaerobic digestion requires less energy yields, less sludge by-product, generates hydrogen sulphide which precipitates toxic heavy metal ions, and produces methane which can be used as an energy source.

Activated sludge is the biologically active sediment produced by the repeated aeration and settling of sewage and/or organic wastes. The dissolved organic matter acts as food for the growth of aerobic flora. These species produce a biologically active sludge which is usually brown in colour and which destroys the polluting organic matter in the sewage and waste. The process is known as the activated sludge process. Briefly, the activated sludge process (Fig. 12.10) is a versatile and effective waste treatment process. Micro-organisms in the aeration tank convert organic material in waste-water to microbial biomass and carbon dioxide. Organic nitrogen is converted to ammonium ion or nitrate and organic phosphorus is converted to orthophosphate.

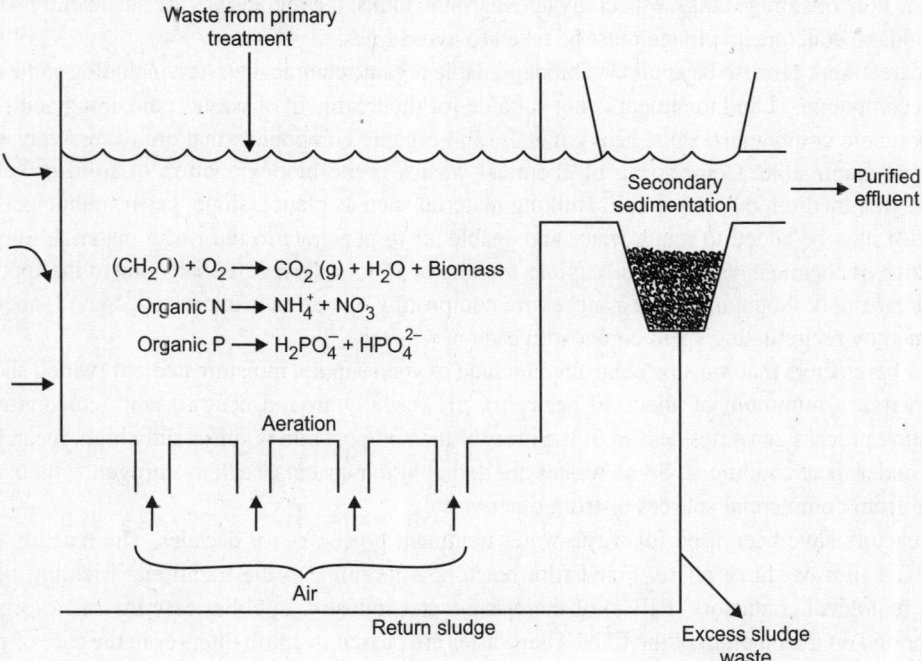

Fig. 12.10. The activated sludge process.

The microbial cell matter formed as part of the waste degradation processes is normally kept in the aeration tank until the micro-organisms are past the log phase of growth, at which point the cells flocculate and separate from the liquid. These solids settle out in a settler and a fraction of them is discarded. Part of the solid, the return sludge, is recycled to the head of the aeration tank and comes into contact with fresh sewage. The combination of a high concentration of micro-organisms in the return sludge and a rich food source in the in-flowing sewage provides optimum conditions for the rapid degradation of organic matter.

However, in terms of pollution by humans, sewage sludge is not the beneficial organic fertiliser that many believe it to be. Sewage sludge spread on land may contaminate water by release of a variety of organic and inorganic contaminants. The spread of disease in ancient and medieval times due to contamination of water supplies will attest to this. Similarly, materials leached from landfills by (acid) rain can also cause serious contamination of land and water systems. In addition, leachates from unlined pits and lagoons containing chemical liquids/sludge may cause a specific pollution (drinking water) problem or a more general pollution problem. Recent studies show that concentrated municipal sludges can be decisively decomposed to final mineralised compounds using supercritical water oxidation process.

Soil is a natural medium for a number of living organisms that may have an effect upon biodegradation of chemical wastes. Of these, the most important are bacteria. Wastes that are amenable to land treatment are biodegradable organic substances. However, in soil contaminated with chemical wastes, bacterial cultures may develop that are effective in degrading normally recalcitrant compounds through acclimation over a long period of time. Land treatment is mostly used for petroleum refining wastes and is applicable to the treatment of fuels and wastes from leaking underground storage tanks.

On the note of storage tanks, especially aboveground tanks, the regulations are sufficiently strict that precautions to counteract spillage must be taken to avoid fines.

Land treatment can also be applied to biodegradable organic chemical wastes, including some organic halogen compounds. Land treatment is not suitable for the treatment of wastes containing acids, bases, toxic inorganic compounds, salts, heavy metals, and organic compounds that are excessively soluble, volatile, or flammable. Composting of chemical wastes is the biodegradation of solid or solidified materials in a medium other than soil. Bulking material such as plant residue, paper, municipal refuse, or sawdust may be added to retain water and enable air to penetrate to the waste material. Successful composting of chemical waste depends upon a number of factors, such as the selection of the appropriate micro-organism or inoculum. Once a successful composting operation is underway; a good inoculum is maintained by recirculating spent compost to each new batch.

Other parameters that must be controlled include oxygen supply, moisture content (which should be maintained at a minimum of about 40 per cent), pH (usually around neutral), and temperature. The composting process generates heat so, if the mass of the compost pile is sufficiently high, it can be self-heating under most conditions. Some wastes are deficient in nutrients, such as nitrogen, which must be supplied from commercial sources or from other wastes.

Bioreactors have been used for waste-water treatment processes for decades. The reactors may be either fixed film or slurry phase. Fixed film reactors are similar to the traditional trickling filters or rotating biological contactors (RBCs) of the waste-water industry. In either case the micro-organisms are supported on the medium of the filter. The wastes are passed over the filter (or in the case of rotating biological contactors the filter is passed over the waste) allowing the micro-organisms to come into contact with the wastes and break down the organic material. Slurry phase reactors are tanks into which the wastes, nutrients, and micro-organisms are placed. The tank is mixed and may be aerated, many instances, contaminated groundwater is used to create the waste slurry. Both fixed film and slurry phase treatments are either batch or continuous mode.

Solid Phase Bioremediation

Solid phase bioremediation, often referred to as land farming, treats wastes using conventional soil management practices to enhance the microbial degradation of the wastes. The wastes are placed directly on the ground or in shallow tanks, if required by RCRA restrictions. Nutrients and micro-organisms are normally added to the wastes which are routinely tilled during the treatment process. This tilling improves aeration and the contact of the organisms with the wastes. While treatment may occur throughout the upper three to five feet of the soil, mostly occurs within the top foot, called the zone of incorporation.

Soil Heaping

Soil heaping is piling wastes in heaps of several feet high on an asphalt or concrete pad. Nutrients, micro-organisms, and air are provided through perforated piping placed throughout the pile. The pile is covered to contain volatile organic compounds, to stabilise the micro-organism's environment, and to control soil erosion. The volatile organic compounds can be further controlled by applying a vacuum to the pile and treating the exhaust.

Composting

Composting is another application of bioremediation. In this process, the wastes are normally mixed with a structurally firm bulking material such as chopped hay and wood chips. As with the other bioremediation technologies, nutrients, air, and micro-organisms must be added. The three major types of composting are open windrow, static windrow, and reactor systems. The differences among the three relate to how aeration is accomplished. In the open windrow system, the compost piles are open to the air whereas in the static windrow system the air is mechanically forced into the compost piles. When reactors are used, the compost is mechanically mixed to ensure aeration.

In Situ Bioremediation

One of the advantages of bioremediation is that it can be effectively applied to treat wastes in place.

The process usually entails introduction of nutrients, micro-organisms and air to the soil/waste through a series of injection wells or infiltration trenches. The term bioventing has also been applied to this technology, although the term could just as easily be applied to composting or to soil heaping. Nevertheless, whatever the name applied to the technology, it has shown some success for hydrocarbon degradation.

If the soil does not have sufficient moisture content, water may also have to be added. *In situ* bioremediation is often applied in conjunction with groundwater pump and treat systems and soil flushing activities.

There is also the concept of gene manipulation as a means degrading polynuclear aromatic hydrocarbons. The concept offers promise for many sites (such as town gas sites where wastes containing polynuclear aromatic hydrocarbons are evident). However, the degradation products from such interactions may require cleanup. But, it is quite possible that the degradation products are easier to clean than the original polynuclear aromatic hydrocarbons. There is also the concept of using biodegradation on such wastes where the waste has been reduced to residual saturation by flushing technologies. A final flushing to remove the biodegraded material will be necessary.

Chapter 13

Zero Waste for Sustainable Development

INTRODUCTION

'Zero waste is an ideal concept that has to be implemented along with a set of conducive policies and practices, and goes beyond conventionally practised waste treatment and recycling/reuse'. 'This concept also goes much beyond statutory requirements, by adopting an ecosystem approach to the vast flow of resources and wastes'. The concept of 'zero waste', seeks to reduce quantities and toxicities of materials used and wastes generated.

The zero waste approach can only be sustainable in the long run, since it reduces resources consumption, increases efficiencies, eliminates toxic products, and ensures that products are produced to be safely used, reused, recovered and recycled back or disposed into nature in an environment-friendly manner.

It is an irony that even after more than 30 years of enactment of Water Act, 25 years of Air Act, and 10 years of Environment Protection Act, we seem to be still bogged down with environmental compliance.

A basic question arises as to when do we move to environmental performance? The regulations focus on pollution control except a mention of 'steps taken, by the applicant wherever feasible, for reduction and prevention in the waste generated or for recycling or reuse' under 'renewal of authorisation' clause in Hazardous Waste (Management and Handling) Rules 2003. Several initiatives are taken by government are noteworthy but not sufficient in bringing leap frog improvements.

It may be appreciated that natural resources are used in the generation of energy, manufacture of goods and consumption of goods. Resources are required for products to be manufactured (in the form of feed-stock, additives, etc.), production processes involved (in the form of utilities), packaging, transportation (in the form of fuel, lubes, etc.) and energy generation (in the form of fuel). From each of these activities, waste gets generated, which can cause air, water and/or land pollution (Fig. 13.1).

SOURCES OF WASTE

Since waste gets produced from manufacturing processes, it should strictly be considered as a by-product arising out of inefficiencies in the manufacturing processes.

Thus, waste should be treated as an undesirable by-product. It could as well be considered as a resource, which is at a wrong place and/or at wrong time. Waste generated at one location may be a useful resource at some other location.

Waste generated at a given time may become a resource at some other time when different technologies get developed and new products/applications get developed.

Zero Waste for Sustainable Development

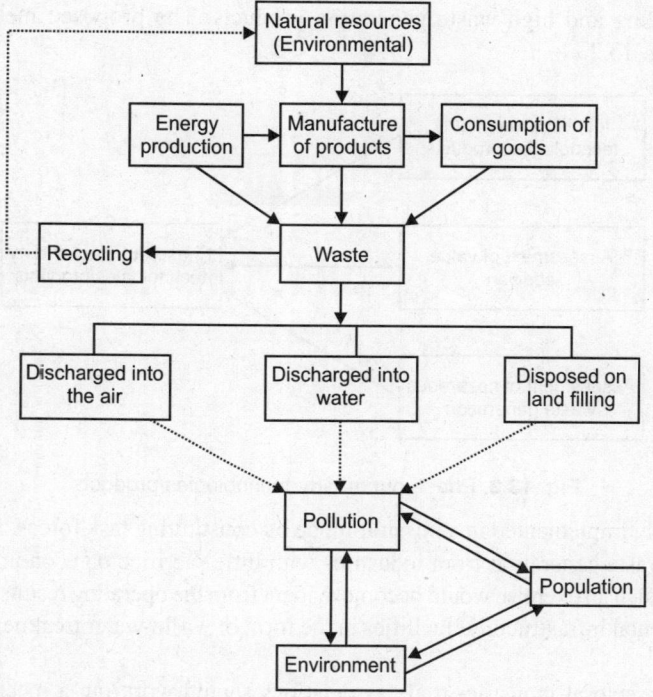

Fig. 13.1. Source and sink.

Considering that source for inputs for manufacture and energy is finite, and sink where wastes get disposed is also finite, how long can the existing practice of suboptimal utilisation of resources sustain?

Cleaner Technologies

The induction of cleaner technologies should start with the identification of dirty technologies/products, followed by a systematic and conscious action plan of phasing out such technologies/products.

The methodology may include a detailed study of different products that are identified by professionals from their knowledge and experience, followed by deliberations and consensus building on the need and plan of phase out. After detailed study of each product and technology, workshops need to be organised. These, among others, should have participation from the manufacturers, technologists, research scientists, concerned industrial/trade associations, financing agencies and industrial development agencies.

Waste Reduction at Source

The potential and extent of waste reduction at source should be assessed first. After a series of such workshops for different technologies/products, a priority list for phasing out technologies/products should be prepared.

One of the simplest bases that could be considered *prima facie* is the quantity of hazardous waste generated per rupee of value added per unit of production of the given product. Based on this yardstick and based on consultations with the stakeholders, action plan needs to be agreed upon for phasing out such low value adding and high waste generating products. The proposed methodology is shown schematically in Fig. 13.2.

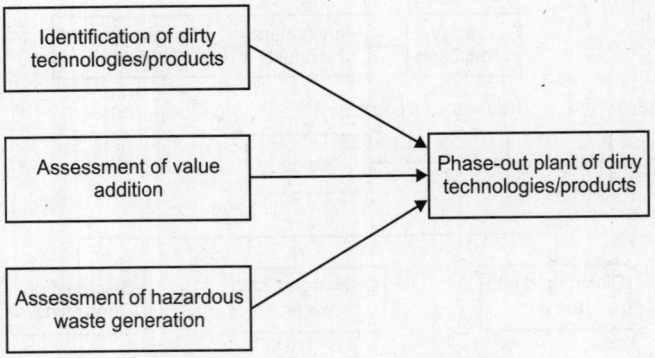

Fig. 13.2. Phase-out of dirty technologies/products.

Such an approach, implemented in a mission-mode by constituting task forces, is expected to reduce overall hazardous waste generation from industries with little sacrifice on economic gains.

The outcome of such an exercise would become evident from the operating results of the corresponding common environmental infrastructural facilities in the form of waste-water treatment plants and secured landfill sites.

In the long run, sectoral industries/trade associations should work out a mechanism of setting up appropriate research and developmental facilities, and networking of existing facilities, focusing on cleaner technologies for the products that the developed countries discontinue to manufacture and plan to source from developing countries from time to time. Government could play a vital role in supporting this endeavor by extending financial incentives in the form of making available technologies/technical services at token charges, which is in line with the WTO requirements

Cleaner Suitable Environmental Technologies

The next stage of induction of cleaner technology that merits consideration is development of suitable cleaner environmental technologies for the utilisation of large quantity of wastes generated from different industrial activities, as shown in Fig. 13.3.

Fig. 13.3. Development of cleaner environmental technologies for waste utilisation.

In a traditional model, an industrial activity consists of individual manufacturing processes that take feed-stock, generate products for sale and waste for disposal.

In an integrated model, industrial system is considered in which consumption of energy and materials is optimised, waste generation is minimised, and wastes of one process serve as feedstock for another process.

The central concept of industrial ecology is evolution of the industrial system from an open system, where resources are consumed and harmful wastes are discharged into the environment, to a more closed system in such a way that no waste would leave the industrial system or have negative impact on natural system.

Such symbiotic industrial arrangements would have positive effects, viz. reduction in the usage of the virgin materials as resource inputs, reduction in pollution, increasing systemic energy efficiency leading to reduce energy use, reduction in volume of waste products requiring disposal, with the added benefit of preventing disposal-related pollution, and increasing the amount and types of process outputs that have market value.

The concept of Sustainable Chemistry is expanding the concept of green chemistry to green production. Suschem's management focus mainly on industrial biotechnology, materials technology (nanotechnology in particular), and reaction and process design (with special emphasis on process intensification).

Beyond greening

Three stages of environmental strategies to be adopted by industries, viz. pollution prevention, product stewardship, and clean technologies.

The Vision 2020 initiative of the US chemical companies has set an ambitious target of reducing energy consumption per unit of chemical manufacture by 30 per cent. In Europe, one initiative has targeted to produce 25 per cent of all organic chemicals from renewable feedstock.

ACTION PLAN

We too may fix some time-frame for our action plan and goals to avoid open-endedness by considering drafting a suitable legislation for green chemistry/cleaner production for which Resource Conservation and Recovery Act of USA could be used as guideline. The action plan may include establishing an Apex body in the form of a Center of Excellence, e.g. on the lines of York University for:

1. Networking disjointed efforts by several agencies on similar objective.
2. Integrating different terminologies being used.
3. Developing manpower by way of regular courses and short-term courses.
4. Involving academic and R & D institutions.
5. Involving venture capital funds.
6. Identifying and prioritisation of areas of work.
7. Improvement of reactions like alkylation, acylation, halogenation, sulphonation, nitration, etc. for aromatic compounds which are precursors for chemicals used in dyestuffs, agrochemicals.
8. Recyclable catalysts.
9. Green/renewable feedstock for biofuels and bioenergy–biorefineries.
10. Biopolymers and biofibres.
11. Green solvents.
12. Green lubricants.
13. Target oriented less toxic agrochemicals by following participatory and mission-mode approach.

A mechanism needs to be put in place whereby inventory of natural resources consumed and waste emissions/discharges is periodically reviewed by each industry and goals are set for reducing natural resources consumption per unit of production by reduction of waste generated at source and by waste utilisation.

Environmental audit and corporate social responsibility reports reflecting these be subjected to third party audit. Technology is still not considered a core driver for growth in India. Waste reduction with an objective of gradually moving towards zero waste, like any other proactive approach, should be a voluntary effort on part of industries to start with rather than forced by regulations right in the beginning.

At the same time a suitable policy frame-work needs to be evolved that encourages waste reduction efforts made by industries by way of incentives, makes waste disposal gradually more expensive, and prohibits secured land filling of waste without some kind of pretreatment, e.g. thermal, biological, etc.

TOWARDS A SUSTAINABLE CHEMICAL INDUSTRY

While there is a lot of discourse on sustainable chemical industry production, many are not very sure that adopting sustainable technologies, techniques and engineering could result in more profitable businesses. Environmental and energy efficiencies in manufacturing are synergetically linked and form the basis for sustainable development. This will result in manifold benefits for industry and business, for society, humankind and for planet Earth. Several pioneering global companies like DuPont, Dow, BASF are already into this. Several processes based on renewable feedstocks are being commercialised. Waste products are being recycled or put to better use adding additional value to business runs.

This section is replete with such examples. The push for sustainable development has opened up absolutely new and exciting opportunities and research in using renewable feedstocks, biomass, bioproducts, biofuels. An integrated biorefinery that can exploit commercially all streams emanating from a biorefinery is becoming a feasibility sooner than what most would anticipate. For a country like India, the emerging basket of biotechnologies in the chemical industry could also kick start our rural and agro economies.

The organisation for economic cooperation and development (OECD), says that Industrial sustainability can be achieved through 'continuous innovation, improvement and use of clean technologies to reduce pollution levels and consumption of resources'. Achieving industrial sustainability means employing technologies and know-how to use less material and energy, maximising renewable resources as inputs, minimising generation of pollutants or harmful waste during product manufacture and use, and producing recyclable or biodegradable products.

The chemical and process industries are well poised to adapt and drive the challenges of industrial sustainability as they enable virtually every other sector from food, shelter, clothing to construction, transportation and health, to provide high quality products and services. The chemical industry creates materials for multiple consumer markets that need to be produced, used and recycled by manufacturing processes that are clean, safe and economical. These processes must use the minimum of resources with the ultimate goal of using only renewable and recyclable feedstock.

Drivers for a Sustainable Industry

The world business council for sustainable development in their document titled 'business case for sustainable development', describes the following business incentives to promote sustainability:
1. Profitability, fiscal changes.
2. Environment conscience, legislation and regulation.
3. Concerns from shareholders, employees and customers.
4. Long term business viability related to public perception and image.

Profitability is, in itself, a prime driver for sustainability. The strategies of sustainable development to create value are given below:
1. Anticipation of costs and risks.
2. Cost reduction in consumption of resources or production of waste.

3. Innovation.
4. Competitive advantage by differentiation of products put on the market.
5. Improvement of the company image.

This section reviews how the global chemical industry is creating new, more sustainable value by implementing 'green chemistry' and engineering concepts, and also reviews some issues and challenges facing the green chemistry and engineering, followed by a review of some select concepts and examples of processes in this area.

Principles of Green Chemistry and Green Engineering

For the chemical industry, innovation is the lifeline ensuring continuous improvement in performance and enabling more sustainable production and use of its products. The challenge is not only to provide products and services to consumer and society, but also to innovate new ways of doing so with markedly lower dependence on materials, energy, labour and waste. If technology was regarded as a cause of environmental degradation, it is now recognised for its key contribution to sustainability by decoupling economic growth from negative environmental impact.

Most chemical companies now have set environmental and societal targets. DuPont has set the following goals for 2010 as part of its sustainability mission:

1. Derive 25 per cent revenue from non-depletable resources.
2. Against 14 per cent in 2002.
3. Reduce global carbon-equivalent greenhouse gases (GHG) emissions by 65 per cent compared to 1990 as base year.
4. Keep energy use at 1990 levels.
5. Source 10 per cent of the company's global energy use from renewable resources.

By the year 2012, as compared to base year 2002, BASF aims to:

1. Cut 10 per cent of GHG emissions per ton of sales.
2. Cut 40 per cent air pollution, 30 per cent heavy metal emissions.
3. Have 80 per cent fewer lost-time accidents, 70 per cent fewer transportation accidents.
4. By 2009 the company will have all relevant information on chemicals handled in volumes of more than 1 ton per year.

Need for the Correct Framework

The principles of green chemistry gave a tangible framework to reduce environmental impact of products. Improving industrial sustainability requires goals not only at the molecular levels, but also at the process and system levels. Green engineering focuses on how to achieve these goals through technology. When designing new materials, products or processes, scientists and engineers are provided with a framework, beyond baseline engineering quality and safety specifications, based on the 12 principles mentioned below, that consider environmental, economic and social factors:

1. Mass and energy in and outputs should be as inherently non-hazardous as possible.
2. Prevention of waste is better than clean-up.
3. Minimise energy in separation or purification processes.
4. Maximise mass, energy, volume and time efficiency in product/process.
5. Output-pulled is preferred to input-pushed.
6. Energy is the main criterion for choice between recycle, reuse or disposal.
7. Durability must be targeted (no external life).

8. Avoid one-size-fits-all, minimise excess.
9. Minimise material diversity in multicomponent products.
10. Integration and interconnectivity are a way to industrial ecology.
11. Design for performance in commercial 'after-life'.
12. Favour mass and energy inputs from renewable resources.

Technical innovation alone will not suffice to ensure that sustainable technology is adopted. Correct framework conditions are required to enable successful investment in green chemistry and engineering. These include:

1. Market related instruments — including incentives to increase research and access to finance for new and potential innovations.
2. Cost effective regulations — that do not inhibit change through over regulation.
3. Appreciation by society — a demand for new products and systems that are real.

Illustrations of Implemented Sustainable Chemistry

The following section will highlight industry innovations *vis-à-vis* current technical challenges which include:

1. Renewable resources.
2. Eco-efficient processes.
3. Energy efficiency and sustainable energy.
4. Waste reduction and waste reuse.
5. Reduction of GHG and other emissions.
6. Inherently safe processes.

Renewable resources biomass

Nature produces a formidable 200 billion tons of biomass per year by photosynthesis, of which only 3–4 per cent is used for food and non-food sectors. Thus, use of biomass can play an important role in sustainable development and handling global warming issues in two major areas of development:

1. Biomass for energy uses, directly or indirectly.
2. Biomass as base resource for chemicals, materials and products.

Three main areas can be identified especially for the chemical industry:

1. Use of renewable raw materials replacing fossil fuel feedstock, where new enzymes and whole cell systems convert biomass into fermentable sugars and then to downstream products.
2. Bioprocesses replacing traditional chemical processes to prepare many organic and other chemicals.
3. Development of new bioproducts such as chemicals, new plastics and high performance polymers.

Use of biomass to replace fossil fuel resources is leading to the development of the most important group of technologies and plays a very important role in reducing CO_2 emissions, particularly the liquid biofuels for sustainable transportation, such as biodiesel (C_{16} to C_{18} fatty acid methyl ester, produced from vegetable oils) and bioethanol and its derivative ETBE (produced from plant sugars). Other liquid biofuels, such as bioethanol and its derivative MTBE, are derived from lignocellulosic material. Europe alone produces about 1.5 m ton per annum (tpa) of biodiesel. Bioethanol is traditionally produced by the fermentation of simple sugars, from starches of potatoes, corn, wheat and other plants. Other feedstocks such as waste products from the beverage, food and forest industries are also used. Research is also being carried out to produce ethanol from agricultural residues such as rice straw, sugarcane baggase, corn, municipal solid waste and energy crops. Bioethanol is used as an automotive fuel, mixed with gasoline to form a fuel called 'gasohol'. Ethanolblended fuels account for 12 per cent of all automotive fuels in the USA.

Oleochemicals

Oleochemicals derived from natural fats and oils from marine or vegetable sources are used in detergents and cosmetics, additives for plastics, rubbers and textiles, in the production of paint and surface coatings and a variety of applications in the food and pharmaceutical industries. The oleochemical industry already uses mainly renewable raw materials such as palm oil, kernel oil and coconut oil.

Biopolymers

In 1997, Cargill and Dow Chemical Company formed a 50:50 joint venture, to develop through a novel synthesis of polylactic acid (PLA), a polymer derived from corn-derived dextrose, NatureWorks PLA, the first synthetic polymer to be produced from a renewable resource. It is fully compostable or recyclable to the monomer. The two key technical advances for large-scale, low-cost production are the synthesis of lactide and efficient vacuum distillation. The Cargill-Dow process has better than 95 per cent yield, requires 30 to 50 per cent fewer fossil resources than conventional plastics and results in a 30 to 60 per cent reduction on GHG emissions. The first plant built in Nebraska (USA) has an annual capacity of 1,40,000 tons. Future plants will be able to use various other biomass feedstocks, such as agricultural waste.

Through metabolic engineering, an Escherichia Coli K 12 micro-organism produces 1,3-propanediol (PDO) in a simple fermentation process developed by Du Pont and Genecor. PDO is used for the production of PTT (poly trimethylene terephthalate), a new polymer which is used for the production of high quality fibres branded as Sorona. Pilot plant yields are reported to reach 135 g/l at a rate of 4 g/l and production is expected to reach 500 ktpa by 2010.

Xanthane, a polysaccharide is obtained by fermentation and used as a viscosifier in food, sauces, paints and cement. Biopol is a natural resource based polyester developed by ICI; the copolymer, poly-3-hydroxy butyrate-3-hydoxyvalerate is produced from wheat carbohydrates by fermentation using *Alcaligenes eutrophius*. Recent developments using modified enzymes at Metabolics are expected to improve both yields and product properties. Researchers have succeeded in manipulating the genetic code of plants and micro-organisms in order to produce polyhydroxybutyric acid.

Dow Corning and Genecor collaborated on an ambitious 35 M USD project to make bio-based organosilicones. This so-called 'Silicon-Biotechnology' is expected to yield biologically mediated organosilicone-based materials for applications in diagnostics, biosensors, controlled delivery of active ingredients and personal care products.

Use of the enzyme 'nitrile hydratase', by Mitsubishi Rayon was among the first applications of biotechnology in the chemicals industry for the manufacture of acrylamide. The conventional process involved hydration of the nitrile with sulphuric acid in the presence of copper catalysts. The new enzymatic transformation in a batch reactor (kept at 5°C to avoid polymerisation) results in more than 99.99 per cent conversion, selectivity and yield to reach a productivity of approximately 2 kg/1/d.

Eco-efficient products

BASF has been using eco-efficiency since 1996 as a way to deliver the right product at the right time. The company compared the ecological footprint of the dyeing process for blue denim using indigo powder from various sources, i.e. biotech-derived indigo granules, synthetic indigo in conventional dyeing and an electrochemical process. The analysis showed the electrochemical process to be clearly favourable in terms of its environmental impact and a new plant was constructed on the basis of these findings. BASF's share of the jeans dyeing market has since risen from 2 to 40 per cent.

Pigments such as cadmium sulphoselenide, lead molybdate and bismuth vanadate, available in the colour range from red to yellow have excellent performance characteristics, but are environmentally undesirable. Hence, Rhodia developed Neolor, nontoxic, inorganic pigments based on cerium sulphides used in automotive coatings, engineering plastics, ceramic colouring and packaging, where high opacity, thermal stability and light-fastness are required.

Many products designed to ecological rules can be cheaper to produce and use, providing higher value for their users. BASF produces a biodegradable compostable polyester called Ecoflex, a copolymer of 1,4-butanediol, adipic acid and terephthalic acid which has good thermoplastic properties and forms semitransparent packaging. Luviset polyurethane copolymers were developed by BASF to reduce VOC emissions in the production and use of hair sprays. Neopor building insulation incorporates infrared absorbing/scattering components that absorb more heat and reduce energy use.

For carpet backing applications, Dow chemical company developed biobalance polymers, polyurethane polymers that are manufactured with a portion of the polyol coming from a renewable resource, the soyabean plant.

These carpet backings can also be manufactured with postindustrial waste and a non-woven outer surface made from 100 per cent recycled PET. Synalox is a polyglycol lubricant having cleaner profiles than fossil fuel based equivalents, and is biodegradable while maintaining and improving performance characteristics.

Deedmac, a fabric softening cationic surfactant developed by Procter and Gamble has a 99 per cent removal in sewage treatment and a much improved eco-toxicological profile.

The following examples illustrate the usefulness of nanotechnologies for sustainability:

1. HDS (Highly dispersible silica) is used by Michelin for its 'green tyre' which significantly improves performance in terms of rolling friction and consequently in the energy consumption of the vehicle and its adherence to wet road surfaces. Silica, in aggregates of a few nanometers, replaces the traditional carbon black. The final properties are controlled at the very first stage of synthesis. The silica is produced in various forms: powders, micropearls, granules, to avoid dusting during its handling.
2. Cerium oxide nanoparticles, pure or doped with iron, found a new application for pollution abatement in diesel engines. Perfectly dispersed in gas oil, the 5 nm cerium oxide particles are trapped in the very heart of the soot particles, which are then oxidised in the exhaust filter. The cerium oxide is totally retained in the catalyst bed. The effectiveness of this nanocatalyst is such that the addition of < 10 ppm in the gas oil allows complete combustion of soot in a few minutes. 2 litres of EOLYS allows the effective depollution of a diesel engine upto 1,50,000 km.

Energy efficiency

More energy efficient processes in chemical industry can be achieved in different ways. Fine control of reaction parameters to obtain better yields and selectivity lead to easier, less energy intensive downstream separation technologies, such as the use of membranes. Process design to minimise multiple drying steps in a process will also significantly reduce energy consumption.

Examples

Polimeri Europa has developed new zeolite-catalysed ethylbenzene and cumene processes, which overcome all the traditional drawbacks of conventional alkylation technologies. Cumene technology is also moving towards zeolite-based processes. BASF jointly with plant builders Durr Systems and Car Company, Daimler-Chrysler, developed a new coating concept for painting the Mercedes A class model, cutting the paint consumption by upto 20 per cent and reducing emissions and energy consumption. The process consists of three consecutive 'wet-on-wet' coating applications; an anti-corrosion dipcoat, whithout heavy metals; a water-based coat and a solvent-free powder-slurry clearcoat deposited by electrostatics. With this new technology, only one final drying is necessary.

Waste Reduction

There is a clear business case for waste reduction as drives for zero targets show. For example, by developing a new technology for manufacturing a key component of Lycra, Du Pont earned an additional 4 million USD while cutting down 2 ktpa of waste. Sumitomo Chemicals and Enichem combined two catalytic processes for a 70 ktpa caprolactum plant. Instead of the classical route with hydroxylamine sulphate, a toxic and corrosive reactant, cyclohexanone is directly ammoximated by ammonia with water over a Ti-MFI type zeolite in the rearrangement of the intermediate oxime to final caprolactam, (a catalyst replaces oleum). Coproduction of ammonium sulphate, with poor fertilising properties, causing economic and environmental concerns to manufacturers, is completely eliminated, NO_x and SO_x emissions are also eliminated.

Waste reuse

New environment-friendly materials can be made using secondary materials and by-products from another process or industry. For instance, Uniqema has developed a process for extraction of squalene from olive oil waste streams (instead of shark livers). Squalene is used in formulations of health, personal care and cosmetics. Uniqema has also developed a process to recover waste frying oil to produce high-grade stearine, which is used in rubber tyres.

High grade carbon fibres are made by Conoco from low value 'bottom-of-the barrel' pitch, thereby transforming a low value, low utility product to a versatile high value material. This carbon fibre business known as 'Cevolution', started in 2002 with a 3,600 tons capacity process. Carbon fibres, a lightweight, yet strong material, opens up opportunities for nontraditional, more durable applications (from car body panels or smart asphalt to high capacity batteries).

Dow Bio Products Ltd. manufactures Woodstalk brand fibreboard, shelving and flooring underlay, at its facility in Elie, Manitoba, Canada. These alternatives to conventional fibreboard and plywood are made from wheatstraw, an annually renewable resource. The wheatstraw fibre used to produce Woodstalk boards would otherwise have been burned after the wheat harvest, adding large quantities of smoke and other products of combustion, such as carbon monoxide into the air. This use of straw residue now provides secondary revenue for farmers in the Winnipeg area.

Emissions reduction

Increased energy efficiency and reuse of renewable energy resources will continue to reduce the chemical industry's CO_2 emissions. It is estimated that 40 per cent of the VOC emissions equivalent to around 15 per cent of global GHGs, originate from organic solvents widely used in the chemicals and related industries, some of which are also highly flammable and toxic. Efforts have been made to replace them by alternative media, such as water, supercritical solvents, ionic liquids and neoteric solvents. The principle of using these alternatives has been demonstrated but only a few industrial processes are commercialised. Supercritical fluids (SFC) are popular in the extraction industry. The Unicarb spray coating process, commercialised by Union carbide in the early 1990's replaces majority of traditional solvents by cosolvent modified SC CO_2. This technique is used in the automotive and furniture industry.

New business models and new value chain for a sustainable

Economy

Examples discussed in the previous sections lead us into the next line of thinking. What is the impact of these developments and what are the emerging technology areas of promise for the future?

Development and implementation of more sustainable technologies have enabled the industry to decouple economic growth from negative environmental impact. For instance, the European chemical industry has increased production by over 35 per cent since 1990 but its energy consumption is roughly at the same level as 1990 while CO_2 emissions have reduced. Furthermore, the industry has improved the sector's competitiveness with increase in productivity by 59 per cent over the period 2001–2002.

The chemical industry is a diverse and complex sector. Nine areas have been clearly identified as having a significant impact towards sustainability of the chemical industry in the next decade:
1. White (industrial) biotechnology.
2. Process intensification and meso technologies.
3. Multiscale process design and scale out.
4. Catalytic processes.
5. Green solvents.
6. New activations.
7. New materials.
8. Thermo-economics.
9. Industrial ecology.

White (industrial) biotechnology

Biobased sustainable industrial chemistry is very high on the international agenda. The Organisation for economic cooperation and development (OECD) publication, 'biotechnology industrial sustainability for clean industrial products and processes: towards industrial sustainability' and many other publications such as the USA Roadmap for biomass technologies note that biotechnology is a powerful tool for the production of commodity chemicals. Much development has already been done on the possible uses of biomass for production of various materials.
1. Direct use of specific materials from plants and trees, for construction materials, paper, textiles, etc.
2. Specific products based on specific constituents of crops (oil, fibre, etc.) by chemical conversion, e.g. for specialty chemicals, biodegradable polymers, etc.
3. Generic conversion of biomass to basic constituents: 'Building blocks' such as CO, H_2, CH_4, and also ethanol, acetone, etc. through fermentation.
4. Potential for production of high value-added products (pharmaceutical products and agrochemicals).

Industrial biochemistry seems to be more acceptable to the general public than agricultural biotechnologies and in the long run, production of chemicals should be increasingly based on biomass as raw material: including non-food resources, restricted residues and wastes.

Chemistry and chemical engineering have essential contributions to make in:
1. Developing the technologies for conversion of biomass into energy carriers and chemicals.
2. Research on which biomass materials are most suited and where and how arable land and water resources can be developed and exploited.
3. Determining the metrics and indicators for the assessment of the effectiveness of the sustainable projects.

There are economic challenges for the development of 'white biotechnology': Feedstocks such as vegetable oil and glucose can be expensive, the enzymes used to convert the material require a high investment in research and long development times. It is also clear that specialty chemicals, with lower volumes of production, are also likely to see the most profound early impact from biotechnology.

Green Chemistry and the Biorefinery

The 20th century saw the emergence and establishment of an organic chemicals manufacturing industry based on petroleum refining. The 21st century will see the development of a new organics industry based on biomass refining. In both the scenarios, the driver is energy. The enormous demand for petroleum as a cheap, single-use fuel gave chemicals manufacturing a large volume, low cost and continuous supply of hydrocarbons from which the petroleum industry was built. Chemistry and chemical engineering technology for cracking, separation, rearranging, polymerising and functionalising facilitated processing of complex mixtures of simple chemicals and transforming them into a multitude of higher value molecules, with a seemingly never-ending range of applications from high volume, low-cost plastics to small volume but highly expensive drugs.

We are at the new beginning of an era where new, renewable sources of energy are sought with increasing vigour; biomass, renewable carbon, is guaranteed a place in the new energy portfolio for the foreseeable future. The growth in the bioenergy (e.g. biomass gasification) and biofuels (e.g. biodiesel) industries will add value to the food industries in the consumption of renewable carbon. Food production generates a lot of waste from the crop residues (e.g. wheat straw) through the processing (where substantial losses occur) to sale and consumption; we waste a dangerously high proportion of food. However, what is waste to food manufacturing can be feed to the chemicals, energy and other industries.

Examples

Wheat straw contains significant quantities of valuable wax compounds (fatty alcohols, alkanes, etc.) and the lignocellulosic fraction which can be used to make paper or ethanol; rice husks from rice farms can be burned to yield the energy needed to drive the farm machinery; the residue is rich in silica that has diverse applications. Used food oils can be recovered and through chemical transformation, turned into biodiesel.

The refining of nature's daily bounty will provide a treasure of chemical potential from pre-treatment to incineration, as well as by devoting some of the raw material to bioprocessing. A biorefinery can be considered as an integral unit that can accept different biological feedstocks and convert them to a range of useful products, including chemicals, energy and materials. Figure 13.4 illustrates the concept of a biorefinery.

Research is underway at the University of York and is directed towards chemical aspects of biomaterials and bioenergy as well as biochemicals.

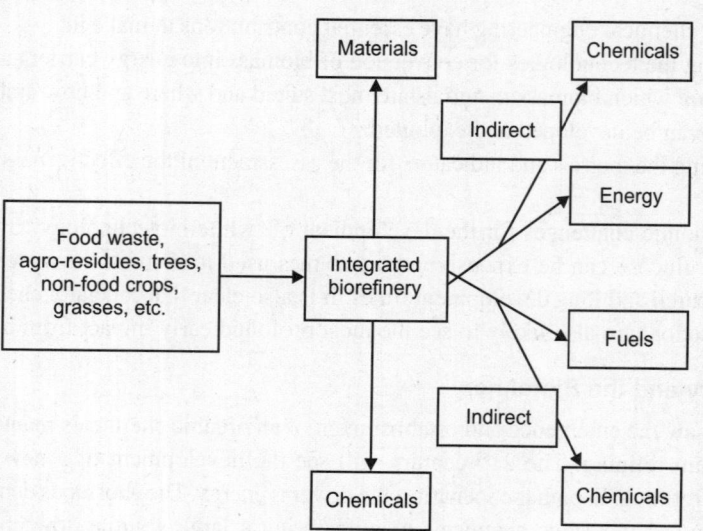

Fig. 13.4. The integrated biorefinery as a mixed feedstock source of chemicals, energy, fuels and materials.

Based on a wide range of renewable feedstocks (including low value plants, such as trees, grass, energy and food crop by-products; marine resource wastes and food wastes), the applications of green chemical technologies (including supercritical fluid extraction; microwave processing, catalytic and other clean synthesis methods) would lead to new, genuinely sustainable, low environmental impact routes for important chemical products, materials and bioenergy. These are well illustrated in Fig. 13.5.

Biorefinery Concept

Some of the key areas being investigated at University of York, USA for the biorefinery concept are:

Supercritical fluid extraction of secondary metabolites from plant surfaces

Wheat straw is an example of low value, high volume agricultural by-product that can provide the basis for a biorefinery. Materials, chemicals and bioenergy/fuels are all accessible from this chemically interesting feedstock (Fig. 13.6).

From trees to value-added chemicals

Trees represent an enormous renewable carbon resource. Finding ways of adding value to such wood resources will help stimulate rural enterprises and maintain the rural economy (Fig. 13.7). This could include the utilisation of residues from the pulp and board industries such as bark. Other sources of wood currently not utilised such as branches could also be used. Some of the examples of new applications include cellulosic fibres for textiles and sawdust resins as adhesives. A quarter of felled wood is converted into timber. The remainder is a rich composite of primary and secondary metabolites that are largely unexploited for novel renewable products.

Trees are also rich in extractable secondary metabolites including terpenes, flavonoids, sterols and resin acids, as well as waxes. Some of these have been shown to be valuable including polyphenolic flavonoids as antioxidants.

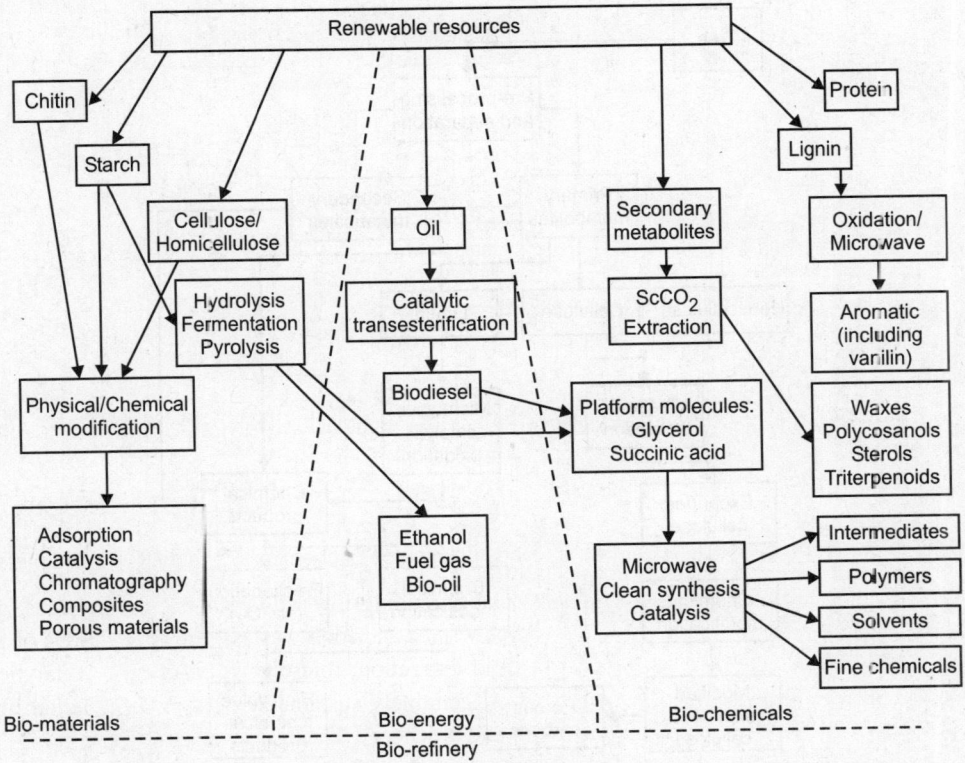

Fig. 13.5. Integrated biorefinery concept using renewable resources and application of Green chemistry.

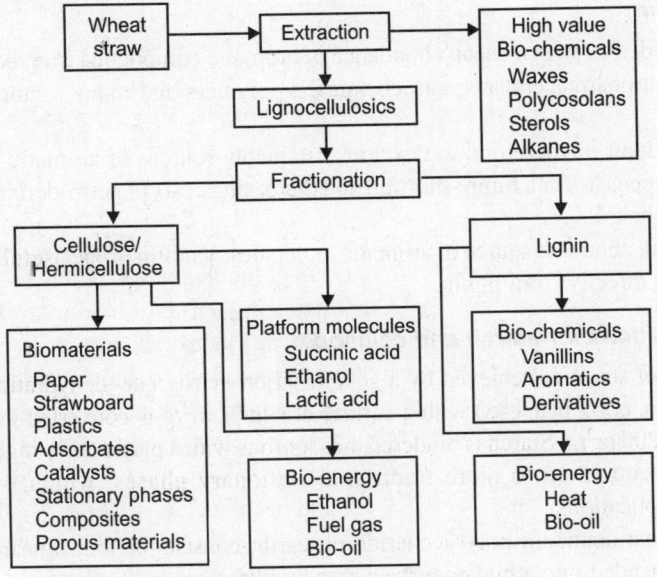

Fig. 13.6. The wheat straw biorefinery concept.

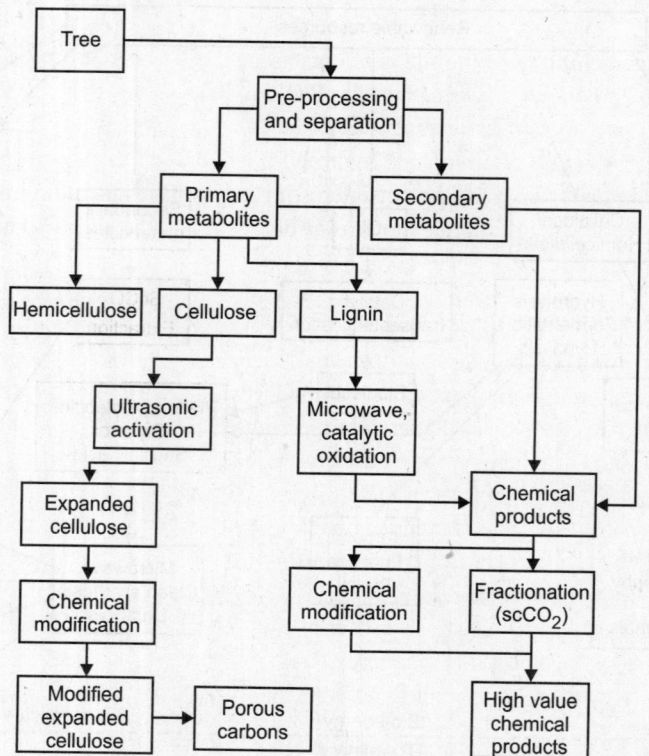

Fig. 13.7. Possible green chemical routes to exploit and add value to tree metabolites.

Vanillin from lignin

Petroleum has provided us with a cheap abundance of aromatic compounds from benzene upwards, and aromatic products in pharmaceuticals, agrochemicals, polymers and many sectors have become well established.

The need of the hour is, therefore, to develop sustainable sources of aromatic compounds that can build up to help compensate for a future shortfall and increasing costs of petro-derived aromatic platform molecules such as phenol.

Lignin is a natural, abundant source of aromatic molecules. Vanillin is one useful aromatic compound that can be obtained directly from lignin.

High value applications for starch and cellulose

Physical expansion of starch is achieved by a gelatinisation–retrograde–dehydration process, whereby natural starch (potato, corn, pea, etc.) with a surface area of 1 m^2/g is converted into a material with a surface area of 100–200 m^2/g. Starch is rendered mesoporous with a particularly high mesopore volume/micropore volume compared to more traditional stationary phases, which is advantageous for chromatographic applications.

Cellulose, the most abundant polysaccharide on earth, consists of long linear chains of glucose. Cellulose can be expanded into a higher surface area by ultrasonic activation from 10 to 70 m^2/g.

Chitosan

Chitosan is deacetylated chitin from the shells of crustaceans, cell walls of fungi and exoskeletons and gut linings of insects, and one of nature's largest volume materials. It is, therefore, truly sustainable, as well as being significantly different to nature's other large volume, renewable materials, cellulose and starch in that it offers a mixture of amino and hydroxyl functions rather than hydroxyl groups alone.

The amino function opens up a number of interesting possibilities including metal binding for the remediation of toxic waste streams; applications as a solid base and the ability to bind metal complexes for catalytic applications.

Biomass platform molecules

Two interesting and promising biomass platform molecules are glycerol and succinic acid. Both are readily obtainable large volume raw materials, although the availability of glycerol has been rapidly accelerated by political drivers such as demand for biodiesel. As a platform molecule, glycerol has good multiple functionality and a number of immediately valuable or promising applications based on the exploitation of green chemistry and green technologies. Succinic acid is currently produced largely from petroleum in a multiple process (butane to maleic anhydride to maleic acid to succinic acid). It has a number of established markets, but these are mostly specialty, due to its high price, and therefore excludes its use as a commodity. It has been shown that bio based succinic acid produced from glucose could be cheaper than petro based substance with easy placement in the commodity chemical price range as well as having assured sustainability of supply (from bioprocessing of agro and food by-products). This would open up large markets, including polymers and solvents.

Biofuels from Biomass: A Chemical Industry's Perspective

Biomass is an interesting starting material for transportation fuels because of its renewable nature. The main constituents of biomass are the carbohydrate polymers, cellulose, hemicellulose, and lignin. Current transportation fuels are mixtures mainly of hydrocarbons derived from crude oil. Biomass derived fuels are preferably used as blends with fossil fuels to avoid the need for adaptations in the existing car fleet. The total blend needs to meet the set specifications for fossil fuels. The conversion of biomass to biofuels, therefore, centres on the removal of oxygen from carbohydrates to obtain hydrocarbons. Oxygen is preferably eliminated in the form of CO_2 or H_2O, because the heat of combustion of these molecules is zero and all energy is concentrated in the remaining products. Based on gross chemical formulae, the conversion options from biomass to biofuels have been proposed by many workers. Conversion routes from carbohydrates to hydrocarbons tend to proceed via C1–intermediates (CO, CH_4, and carbon). Gasification and subsequent Fischer-Tropsch synthesis seems to be the logical processing route to obtain hydrocarbons from biomass via these intermediates. If oxygen-containing products are allowed, conversion can proceed via larger intermediates. For blending with gasoline, ethanol emerges as an interesting component. Longer carbon chain molecules are required for diesel. These can only be obtained by atleast two process steps, involving the combination of carbon chain; (e.g. by etherification or esterification) and reduction of polarity (by hydrogenation or oligomerisation). The lignin present in biomass has a complex, highly cross linked structure and can probably best be converted to synthesis gas by gasification or applied directly as solid fuel.

Thus, the rapidly escalating costs of petroleum and petrochemicals are likely to accelerate the shift towards chemical products derived from renewable, biological feedstocks. In the near and distant future we can expect more biorefinery type facilities where some combination of chemical, energy and food processing are used to add value to biomass feeds. Green chemistry and green engineering together with the biorefinery concept looks to be a promising partnership for a sustainable future.

PSYCHROMETRIC EVAPORATION: A NOVEL ZERO DISCHARGE SYSTEM

Many industrial processes produce abundant quantities of waste-water. This waste-water is to be treated to prevent any injury to the aquatic life in the receiving water. The wisdom of conventional treatment is to convert soluble pollutants into solid pollutants by converting them into bacteria colony by biological treatment or into a chemical sludge by physico-chemical methods or by adsorption, attaching them onto a solid phase. Finally, these solids are removed from the effluent water and treated effluent is discarded. In zero discharge, reverse separation is happening i.e. water is separated from the effluent by evaporation technique so that pollutants are converted directly into a sludge. Moreover, as water is evaporated, there is nothing remaining to be discarded but in conventional treatment after pollutants are removed, the effluent is still required to be discharged. Moreover, zero discharge is almost 100 per cent effective for separation of water (as a vapour) and pollutants (as sludge) and there is no discharge from the plant. Moreover some chemical industries are facing stringent norm of zero discharge and to achieve this objective there are only two systems available in the market:
1. Reverse osmosis (RO) system.
2. Multiple effect evaporators (MEE).

Psychrometric evaporators can be the third alternate. It is highly energy efficient and effective, with very low operating cost and its economy is equivalent to that of a 15-effect steam evaporator.

Operating Principle

In psychrometric evaporation air and water are kept in intimate contact with each other. Evaporation of water takes place by diffusion process at the wet bulb temperature of air, so the heat required for evaporation is very low grade which may be abundantly produced during various plant processes and rejected into atmosphere via 'open' cooling system. In psychrometric evaporation the low grade heat rejection arrangement of plant processes is coupled with the psychrometric evaporator to evaporate waste-water almost free of cost. In contrast, in any other conventional steam evaporator, 90 per cent of its operating cost is because of heating utilities. Moreover, because of lower temperature, at which heat-exchange and evaporation takes place, the tendency of scaling and the rate of corrosion is very low compared to the conventional high temperature steam evaporators as shown in Fig. 13.8.

Psychrometric Evaporation System

The psychrometric evaporation system is very simple system made of two components.
1. Psychro-evaporator: It is venturi shaped equipment in which, fresh air current is created by water jets with jet ejector momentum transfer principle. Because of venturi effect, air is sucked, and effectively mixed with fragmented water particles. This permits evaporative heat transfer to take place by utilising the sensible heat of the warm effluent for diffusion evaporation of its own water content.
2. Effluent pre-heater: It is the suitable type of heat exchanger, employed to transfer waste heat from the hot process vapours, fluids, liquids or hot gases into the effluent to make it warm again.

Waste Heat Source for Psychrometric Evaporation System

Heat with temperature as low as 45–50°C is abundantly available from:
1. The condenser of chillers, solvent recovery systems and reflux condensers of reactors and columns.

2. Surface condensers of the evaporators and crystallisers.
3. Flue gases and hot air from boilers, driers and furnaces.
4. Heat exchangers of engine coolant of DG sets and after-coolers of air compressors. cooled

Psychrometric evaporation systems by virtue of using low temperature heat can be the ultimate sink for such processes. Comparison between steam based evaporator and psychrometric evaporator is shown in Table 13.1.

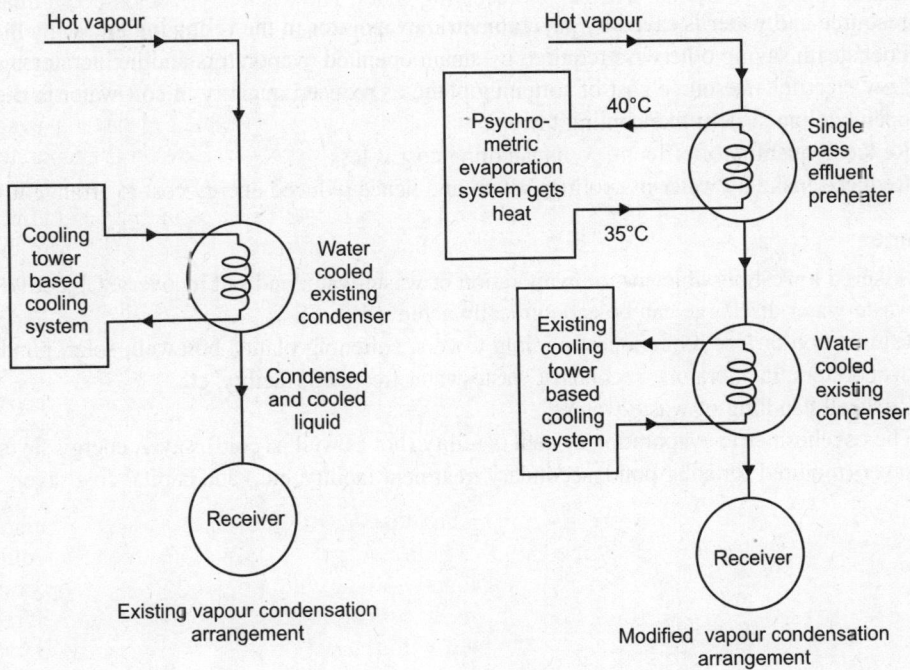

Fig. 13.8. Psychrometric evaporation system.

Table 13.1. Comparison between steam based evaporator and psychrometric evaporator.

Steam based evaporator	Psychrometric evaporator
Water boils in absence of fresh air	Water evaporates in presence of fresh air.
Water boils at more then 60°C (under vacuum) or 100°C (under pressure) temperature	Water, because of presence of fresh air, evaporates at wet bulb temperature of air, which is normally around 30°C
Because boiling temperature is higher, it needs live steam or high grade heat source	To evaporator water in it, the heat source needed shall have temperature just higher than wet bulb temperature of ambient air
It requires high grade heat source, normally purchased and costly one, obtained from hot utility	It requires low grade heat source, which is abundantly produced in various plant processes and needs to be rejected via open, cooling tower like system
80–85% efficient as insulation losses are substantial	Because of near to ambient temperature, insulation losses are negligible and its efficiency is near to 100%
Complicated and costly system	Simple construction and operation

Eco-friendliness of Psychrometric Evaporator

As psychrometric evaporator reduces the thermal load of open cooling systems, it saves precious potable water needed by it as a make-up water. Moreover, it has converted effluent into the evaporated distilled water vapour and given back clean water, instead of waste-water. Again, it has saved energy, fuel and electricity and so reduced their polluting impact on nature.

Potential of Energy, Resource and Water Saving

Energy, resource and water is saved by psychrometric evaporator in the following areas:
1. Fuel/steam saving otherwise required by steam operated evaporators and incinerators.
2. Low electricity/resource cost of softening plant, as reduced quantity of soft water is needed by open cooling system like cooling towers.
3. Reduced quantity of effluent, so its handling cost is less.
4. Reduced make up water in cooling system and hence reduced energy cost to arrange it.

Advantages

1. Assured lowest possible cost of evaporation of waste-water and salt recovery. Objective of zero waste-water discharge can be economically achieved.
2. Elimination or size reduction of cooling towers, softening plants, borewell, solar, pond, steam evaporators, incinerators, secondary waste-water treatment facility, etc.
3. Reduced handling of waste-water.
4. The psychrometric evaporation system is utility (hot as well as cold), saver, energy saver, space saver (required for solar pond, secondary treatment facility, etc.) and capital cost saver.

SECTION II

Pollution Control in Chemical Industries

14. **Petroleum Refinery and Petrochemicals** — 255
15. **Leather and Tannery** — 297
16. **Pulp and Paper** — 340
17. **Soap and Detergent** — 368
18. **Pesticide** — 387
19. **Fertiliser** — 409
20. **Paints and Dyes** — 436
21. **Drugs and Pharmaceuticals** — 456
22. **Sugar, Distillery and Fermentation Products** — 481
23. **Miscellaneous Process Industries** — 503
24. **Textile** — 529

Chapter 14
Petroleum Refinery and Petrochemicals

INTRODUCTION

The term petroleum is used to describe the mixture of hydrocarbons in oil, including the gases above the liquid in oil wells and the gases and solids which are dissolved in the liquid. Petroleum was formed in remote periods of geological time from the remains of living organisms. It is, therefore, a fossil fuel. Weathered rock material eroded from land masses and carried to the sea accumulated in layers over millions of years in subsiding basins and the remains of large quantities of marine plant and animal organisms became incorporated in the sediment. Owing to the great thickness of the sediments, high pressures built up which, probably in conjunction with biochemical activity, led to the formation of petroleum, although the detailed mechanism is obscure.

REFINING OF PETROLEUM

The crude oil pumped out of the well is a deeply coloured, foul-smelling and viscous liquid. It is always found mixed with considerable amount of water and earth particles. In this condition it cannot be used either in engines or as fuel. Neither can it be used as a base for producing other useful components. Chemically it is a complex mixture of hydrocarbons. Just as a sand collector sifts through sieves of finer and finer mesh, one has to use suitable techniques to separate hydrocarbons having different sizes and varying boiling points. The important technique used for this purpose is distillation.

Distillation in an oil refinery is essentially similar to the laboratory process except that the liquid hydrocarbon mixture is allowed to vapourise completely and made to condense into different fractions in the fractionating column.

Petroleum from different oil fields has different compositions. However, we can describe the contents in a general way. Crude oil is not a single chemical compound but a mixture of several unsaturated and saturated hydrocarbons and aromatic compounds.

Unrefined petroleum has constituents with boiling points (b.p.) ranging from 20° to 400°C. The main constituents of the crude oil are:
1. Petroleum gases which are gases at normal temperature.
2. Gasoline, a colourless liquid, boiling between 30° to 200°C.
3. Kerosene, a colourless liquid, boiling between 140° to 300°C.
4. Gas oil, brown liquid, with a boiling range only slightly above kerosene.
5. Diesel oil, a liquid with a boiling range slightly higher than gas oil.
6. Syrupy residue containing fuel oil, lubricating oils, wax and bitumen.

So it becomes necessary to separate these hydrocarbons in order to make each fraction available for appropriate use. All these constituents are skilfully fractionated.

The refining or manufacturing of petroleum products and of petroleum chemicals involves two major branches—physical changes or separation operations and chemical changes or conversion processes. The crude oil is separated into various useful fractions by fractional distillation and finally converted into desired specific products. The process is called 'refining of crude oil' and the plants set up for the purpose, are called the oil refineries. The process of refining involves the following steps.

The crude petroleum is allowed to settle and then centrifuged to remove clay and silica, etc. Refining of petroleum consists of three processes:

1. Fractional distillation: In this process lighter and heavier portions separate out and gas, gasoline, fuel oils, lubricants, wax distillates and asphalts are produced. The primary products obtained are modified by decolourising, deodourising and by separating out hydrocarbon groups and impurities.
2. Cracking: By virtue of this process crude petroleum or their fractions are decomposed by heat to produce products which have lower boiling points.
3. Treating: The process of treating is of three kinds:
   (a) Process by which a product is improved by removing undesired components.
   (b) Process by which the quality of a product can be improved by adding some outside matter which either combines chemically or remains in solution.
   (c) Process through which the chemical structure of hydrocarbons constituting the oil fraction is changed.

Fractional distillation is carried out in a fractionating still. A typical flow sheet for petroleum refining is shown in Fig. 14.1. It consists of a pipe still and bubble tower. In pipe still the crude petroleum is heated and in bubble tower it is fractionated.

Crude oil is piped through a pipe still where it is heated to a temperature 316° to 371°C. All components except the residue is vapourised. All the products obtained by heating oil are introduced in the first section of the bubble tower. The residue is taken out from the bottom. The vapours go up in the bubble tower. As the vapours go up, these become cooler and cooler. Due to this gradual cooling, vapours condense over these plates. Oil condensed at lower plates has higher boiling range than the oil from plates at higher level.

Gasoline vapours go out from the top. Gas oil, kerosene and gasoline are further treated in separate fractionating columns to further purify them and the higher portions from these are returned to the bubble tower. Fractions obtained in this distillation are: (i) gasoline, (ii) kerosene, (iii) gas oil, and (iv) residual oil.

WASTE GENERATION IN PETROLEUM REFINERY

For convenience of discussion, wastes related to the petroleum industry have been grouped according to the following activities:

1. Crude oil producing and handling, including the removal of crude oil from the ground to tankage.
2. Refining.
3. Transportation and marketing, including the movement of crude oil from oil field tankage to the refinery, the movement of refined products to distribution terminals and sales stations and the operation of sales stations.

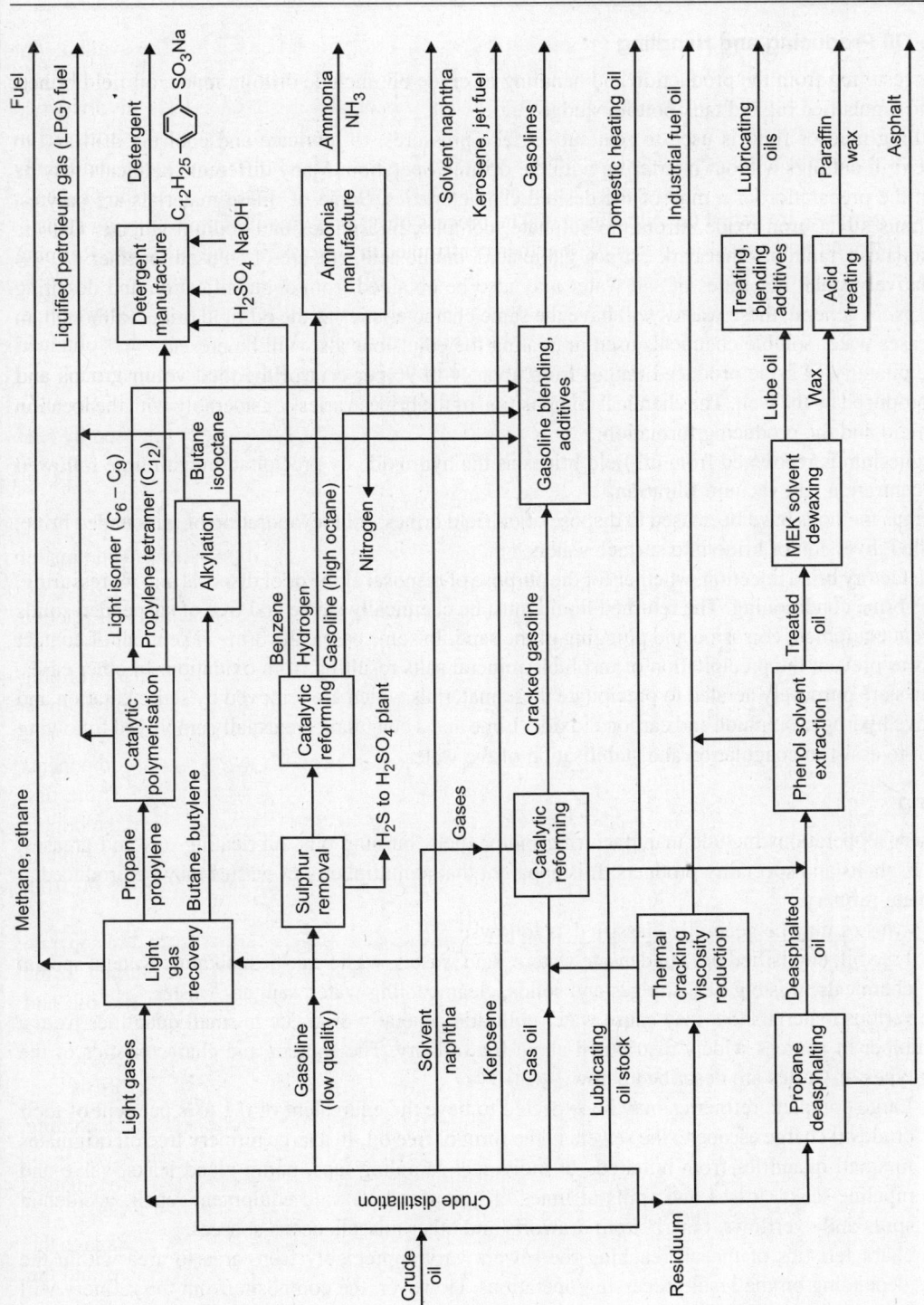

Fig. 14.1. Basic refinery operations.

Crude Oil Producing and Handling

Wastes resulting from the production and handling of crude oil, include drilling muds, oil field brines, free and emulsified oil and tank bottom sludge.

Drilling mud or fluid is used to seal sub-surface pressures, to lubricate and cool the drill and to remove drill cuttings without interfering with the drilling operation. Many different chemicals may be used in the preparation of a mud of the desired characteristics. Some of these materials are barytes, amorphous silica, iron oxide, strontium sulphate, complex phosphates and sodium silicate. Caustic soda, soda ash, tanin or other bark extract, gallic acid, humic acid, etc. are often used together.

Relatively small quantities of salt water may also be obtained from demulsification and desalting processes. In general, these wastes will have the same characteristics as the oil field brines, although, in some cases water-soluble chemicals used in treating the emulsions also will be present.

The quantity of brine produced ranges from about 4 to 96 per cent of the total volume of oil and water produced by the well. The chemical composition of the brines varies considerably with the location of the field and the producing formation.

Magnesium is recovered from oil field brines as the hydroxide by precipitation with lime followed by concentration and vacuum filtration.

Various methods have been used to dispose of oil field brines, solar evaporation of impounded brine, controlled diversion of brine into surface waters.

Satisfactory brine injection, whether for the purpose of disposal alone or of disposal and repressuring, requires brine conditioning. The returned liquid must be chemically stable and free of suspended solids to prevent equipment corrosion and plugging of the sand. In some cases, the brine is kept out of contact with air to prevent the precipitation of insoluble mineral salts resulting from oxidation. In other cases, the brines are purposely aerated to precipitate these materials which are removed by sedimentation and to remove hydrogen sulphide and carbon dioxide. Lime and a coagulant are usually employed following aeration to assist in coagulation and stabilisation of the water.

Refining

A refinery's operations include manufacturing motor fuels, burning oils, lubricating oils and greases, waxes, asphalts and speciality products. It is apparent that a multiplicity of wastes may be produced at a complete refinery.

Such wastes may be generally classified as follows:
> Free oil, emulsified oil, condensate waters, acid wastes, waste caustics, alkaline waters, special chemicals, waste gases, sludges and solids, clean cooling water, sanitary wastes.

The various materials that may cause water pollution generally originate in small quantities from a large number of sources widely distributed about the refinery. The sources and characteristics of the various types of wastes are described below:
> Large complete refineries may be expected to have the equivalent of 0.1 to 2 per cent of their crude oil charge escape to the sewers in the form of free oil. In a large refinery free oil originates in small quantities from hundreds of individual sampling taps, pump gland leaks, valve and pipeline leaks, losses and spills at times of unit shutdown and equipment repair, accidental spills and overflows, tank bottom drawoffs and other miscellaneous sources.

The characteristics of the oil reaching the sewers vary appreciably from area to area within the refinery depending on the local processing operations. However, the composite from the refinery will approach the characteristics of weathered crude oil. The initial boiling point may be as low as 100°F.

Oil exists in the waste-water in two fractions. One fraction is carried in suspension in the form of small droplets, small suspended solids—oil agglomerates or oil in water emulsion. The other fraction floats on the surface of the water, usually, as a water in oil emulsion and is commonly called free oil. The relative concentration of free or floating oil to total oil will vary appreciably. The presence or absence of oil in water emulsions is the largest single variable. Small concentrations of emulsions exist in most waste-waters containing oil although emulsion itself is not drained to the sewers.

In practically all cases gravity differential oil-water separators are provided to recover floating oil and to treat the waste. In the process of separating oil from water, oil rises to the surface, sediment settles to the bottom and relatively small concentrations of oil and suspended solids pass through the separator. Some solid matter rises to the surface with the oil and some oil settles to the bottom with the solids.

The separator skimmings are unfit for return to process units because of their solids and water content. It is generally considered that combined bottom sediment (suspended solids) and water contents in excess of about 1.0 per cent may interfere with efficient processing, may create accident hazards or may shorten 'on stream' running time. Therefore, slop oil treatment is necessary to reduce the solids and water content to 1.0 per cent or less. In most cases these emulsions are rather easily resolved by heating to 190°F, retaining at this temperature for 4 to 6 hours and then settling for 12 to 24 hours. At the end of the settling period three definite layers of material will exist. The top layer will be clean oil, the middle layer will be secondary emulsion and the bottom layer will be water containing soluble components, suspended solids and oil. A small amount of material may settle out as an oily sludge. In some cases it is advantageous or necessary to use caustic (sodium hydroxide) or emulsion breaking organic chemicals for resolving slop oil emulsions. Ordinarily, only enough caustic is necessary to adjust the pH of the slop oil to approximately 8.0. The resolution of slop oils having pH values 6.5 to 7.5 appear to be the most difficult. Both lower and higher pH values give better results. The addition of 2 to 15 barrels of waste caustic per 1000 barrels of slop oil in conjunction with heat treatment has aided in resolving some slop oil emulsions. In such cases the treatment and disposal of this material present a definite problem. Pilot plant experience using an 0.5 square foot Oliver precoat continuous vacuum filter shows that slop oil emulsions may be successfully resolved by filtration through diatomaceous earth. Filtration rates of 2.3 to 6.8 gallons per square foot per hour (the range in numerical values shown includes the central 50 per cent of the data) were obtained when filtering at about 170°F. The solids removed by filtration varied from 0.043 to 0.182 pounds per square foot per hour. The cake obtained averaged about 21 per cent moisture and 28 per cent oil. The fuel value of the cake was 12700 Btu per pound. It should be noted that the pollution load is much less than that from caustic and heat treatment. The fact that filtration with resolve slop oil emulsions indicates that the suspended solids present are probably the stabilising agents. The quantity of these emulsions ordinarily averages about 17 per cent of the charge. The volume is excessively large at localities troubled with biological growths in the sewerage system or separation facilities. These secondary emulsions can be reduced in volume by about 55 to 60 per cent by treatment with organic emulsion breaking compounds.

Gravity differential oil-water separators cannot remove oil in the form of oil-in-water emulsions or in the form of oil-suspended solids agglomerates with specific gravities approximately that of the water. An estimate can be obtained of the probable oil content of gravity-differential separator effluents by determining the susceptibility to separation (STS) number of the waste. The numerical value of the number is representative of that concentration of oil that cannot be separated by normal gravity differential means.

Emulsions

The presence of oil that cannot be separated from waste-waters by conventional gravity differential means is of considerable economic importance to the refiner because of the loss of valuable product and the need for costly facilities to treat the effluent from his oil recovery separators.

An oil-in-water emulsion has turbidity as its chief characteristics and usually has a milky or pearly-gray appearance. This type of emulsion is not removed in the gravity type oil separator and when it is discharged into a large stream or body of water, it usually breaks as a result of dilution and the oil rises to the surface of the water.

Emulsions may be formed in the sewerage system as a result of intimate contact between oil, water and emulsifying agents or may originate directly as process by-products.

The occurrence of coke, clay, sanitary sewage, water treating plant sludges and other flocculent and fibrous solids appears to increase the concentration of nonseparable oil. The presence of tars, asphalts, petroleum sludges, soaps and numerous solvent and treating chemicals also increases the nonseparable oil content of waste-waters. The pumping of waste-water is especially conducive to the formation of emulsions.

The direct formation of emulsions may result from the chemical treatment of lubricating oils, waxes and burning oils, from distillate separators, from barometric condensers, tank drawoffs, desalting operations, pump gland leakage, special chemical manufacturing acid sludge recovery processing, wax deoiling, barrel and truck washing, machine shops and other sources.

Emulsions formed in the water washing of acid treated lubricating and wax stocks following neutralisation are generally known as 'white water' or 'milk water'. These emulsions are very stable. Samples stored at room temperatures have remained stable for several years. They contain 1 to 3 per cent oil by weight. Batch treatment processes will produce about one barrel of milk water per barrel of product. In some cases it is possible to prevent the occurrence of these or similar types of emulsions by the use of special chemicals or special operating techniques.

Sulphonic acids and resinous materials may be recovered from some of these emulsions but generally they can be used only as fuel. This is true because the heterogeneous nature of the recovered material necessitates refining procedures equivalent to those required for the manufacture of similar products of better quality from more reliable sources of raw material.

One refinery utilises another waste material to treat water-soluble petroleum sulphonates produced in the manufacture of white oils. The sulphonate waste is acidified to a pH of 5 with acidic waste-water from a sulphuric acid sludge conversion plant. The sulphonic acid formed are precipitated with hydrated lime at a pH of 7 to 8.4. Reductions of 98 and 93 per cent are obtained in colour and oxygen consumed, respectively. The scum and sludge are pumped to a lagoon. The quantity of emulsions resulting from barometric condensers and jet vacuum pumps has been successfully reduced by the use of a 'dirty water' cooling tower system in a number of cases. In other cases, surface condensers have been substituted for barometric condensers and have been provided on jet vacuum pumps to prevent the formation of this type of emulsion.

Various chemical processes are in use to treat emulsions which must be discharged from the units. In some cases these processes are applied locally to the discharges of individual units, while in other cases these are used on the effluent of plant separators. One refinery presently treats emulsion resulting from steam vacuum jets on crude distillation units with caustic. Oil is the continuous phase in this emulsion which contains about 50 per cent water. Naphthenic acids and probably naphthenic soaps are the emulsifying agents. Approximately 150 to 250 pounds of sodium hydroxide are used per 1000 barrels of emulsion. Although some fresh caustic is used, laboratory studies have shown that spent caustics from the neturalisation of hydrogen sulphides, sulphur dioxide and/or sulphonic acids could be used satisfactorily for this purpose if they contain sufficient alkalinity. The process involves mixing the caustic with the emulsion, heating and settling. The water-caustic solution from the process is drawn to the sewer.

The application to petroleum wastes of the conventional coagulation and sedimentation process as it has been used in the treatment of domestic water supplies, has presented several problems, chiefly in connection with the settling of the floc. The presence of oil in the floc contributes considerable buoyancy to the particles, thus adversely affecting their settling characteristics. As a result, settling tanks must be constructed with larger surface area to settle these light flocs. An overflow rate of 800 to 1000 gallons per day per square foot of surface area is common in the domestic water treating field.

Condensate Waters

Condensate waters, as referred to herein, originate from distillate separators, running tanks and barometric condensers. It has been reported that condensate waters from distillate separators may contain one or more of such compounds as organic and inorganic sulphides, normal or acid sulphites and sulphates, sulphonic acids and their salts, mercaptans, amines, amides, quinalines and pyridines, naphthenic acids, phenols, etc. They may also contain chemicals used for corrosion prevention such as ammonia, caustic soda, calcium hydroxide, etc. Not all these substances will be found in a specific waste-water at the same time. Waste of this type may also contain suspended matter such as coke, iron sulphide, silica, metallic oxides, soaps, emulsions, sulphonic, naphthenic acids and insoluble mercaptides. etc.

It has been possible in some instances to reduce the quantity of condensate wastes through the use of 'dirty water' cooling tower systems for barometric condensers and jet vacuum pumps. In other cases, surface condensers have been substituted for barometric condensers and provided on jet vacuum pumps to control the quantity and strength of this type of waste. Reduction in the quantity of condensate wastes not only lessens the waste disposal problem but also may reduce the loss of valuable water-soluble material.

In case where the water-soluble constituent forms a constant boiling mixture with water, it is sometimes possible to reduce losses by azeotropic distillation. In one case, a phase separation on cooling made it possible to reduce solvent losses from condensate waters by the use of two azeotropic distillations, one for the solvent rich layer and one for the water rich layer.

The installation of a stainless steel mat (called demister) in an asphalt stripper at a propane deasphalting unit has prevented the carry over of asphalt in the tower overhead. As a result, spray condenser water which condenses the stripping steam, can be recycled to a cooling tower. Approximately 500 gpm of asphalt contaminated water has been eliminated from the sewer in this manner.

Most of the hydrogen sulphide in these wastes and a high percentage of the ammonia can be successfully removed by steam stripping.

Pilot plant investigations have indicated that hydrogen sulphide can be successfully stripped from sulphide-bearing wastes with flue gas. Elevated temperatures, 140° to 170°F, appear to appreciably reduce the quantity of gas required.

Acid Wastes

Sulphuric acid is used extensively in the petroleum industry both as a treating agent and a catalyst. Other acids and acid salts also used as catalysts include hydrofluoric acid, phosphoric acid, aluminum chloride and zinc chloride. Acid bearing wastes originate from the acid treatment of gasoline, white oils, lubricating oils and waxes; from the handling of acid sludges and the recovery or manufacture of acid; from the alkylation of motor fuel stocks; from the use of acidic catalysts; and from special chemical manufacturing. The wastes occur as rinse waters, scrubber discharges, spent catalyst, sludges, condensate waters and miscellaneous discharges resulting from sampling procedures, leaks and spills and shutdowns. Acid wastes from pump gland leakage and sampling may be collected locally and returned to the process. Substantial reductions in loss of acid to the plant sewers have been attained by this procedure.

Petroleum products are treated with sulphuric acid to improve colour, odour and stability and to reduce gum content and remove sulphur. Acid requirements and concentrations vary widely with the characteristics of the hydrocarbon and the objective in treating. In general, acid requirements vary from 0 to 60 pounds per barrel using acid concentrations of 85 to 99.9 per cent. One large complete refinery averages about 2.5 pounds of acid per barrel of crude oil charged. Acid treatment of oils results in the formation of sulphur dioxide and an acid sludge. The sulphur dioxide is generally blown out of the oil to the atmosphere. Where this practice creates an atmospheric pollution problem, the sulphur dioxide may be absorbed in water or alkaline solution and discharged to the sewer or may be collected and recovered for acid manufacture or elemental sulphur production.

The acid sludges produced by the sulphuric acid treatment of oils vary considerably depending on the amount of acid used and the characteristics of the treated stock. The consistency of the sludge may vary from that of a low viscosity fluid to that of a highly viscous material that will solidify on standing. The specific gravity of representative acid sludges may vary from about 1.2 to 1.8 and the sulphuric acid content from about 25 to 80 per cent.

The characteristics and quantity of sludge produced afford utilisation opportunities to most refineries. The methods of utilisation are varied and include — burning immediately as fuel, treatment to produce by-products such as oils, tars, asphalts, resins, fatty acids, sulphonic acids, ammonium sulphate, metallic sulphate, coke and other materials, and regeneration for acid recovery.

Although some refineries use acid sludges as fuel alone, after fluxing with fuel oil, after neutralisation with alkaline sludges or alkali or after some combination of these procedures, the probable need of special facilities to minimise corrosion, the high cost of handling and the atmospheric pollution problem due to sulphur dioxide limit the application of this method.

The sludge formed from the acid treatment of light oils is sometimes neutralised with alkaline wastes and discharged to the plant sewers. This practice is not recommended if other alternates are available.

Utilisation of acid sludges to recover sulphuric acid offers the best means of handling the disposal problem at most refineries. It is possible to manufacture 50 per cent or more of a complete refinery's acid requirements from acid sludge. The processes for the recovery of sulphuric acid are basically either hydrolysis or thermal decomposition processes. Sulphur is recovered by a modification of the thermal decomposition process.

Upon hydrolysis, sulphuric acid sludges resolve into two phases — a weak sulphuric acid layer and a tarry acid oil layer. The latter is usually burned as a fuel. About 40 to 65 per cent of the acid used can be recovered by concentrating the weak acid layer. This material is black in colour but contains a relatively small amount of carbonaceous matter. The addition of water in the hydrolysis operation is controlled so that acid concentration of the water layer is not less than 40 per cent. A number of different type concentrators have been used for this material. Concentration is generally carried out by direct contacting of the acid with hot gases at atmospheric pressure or by indirect heat transfer which utilises steam or a similar heat source in a unit which may be operated under vacuum.

In some cases, the weak acid produced by hydrolysis may be used as a source of acid in waste treatment operations such as breaking emulsions, neutralising alkaline wastes, etc. or it may be used to dilute stronger acid to the desired strength instead of using water. In the thermal decomposition process, the sludge is decomposed to sulphur dioxide, water and coke in a continuously operating kiln at temperatures about 500°F. The sulphur dioxide is converted to sulphuric acid by one of the conventional catalytical contact processes or into free elemental sulphur. Experience has been that from 85 to 90 per cent of the sulphur in the sludge can be recovered in this manner.

The sludge conversion-contact acid plant method can produce white acids of 98 to 99.9 per cent concentration or higher. Black acids of the same strength can be produced by feeding weak acids obtained by hydrolysis to the towers in the place of clean water. This type of recovery plant has the advantage that clean acid can be produced, if desired and that auxiliary equipment can be installed for the conversion of hydrogen sulphide to sulphuric acid and facilities can be provided for burning sulphur for acid manufacture. Unfortunately, this type of recovery plant is not in general economically applicable to smaller refineries for production under about 25 tons per day.

The sludge conversion-contact acid plant has its own waste disposal problems. The waste-water streams may contain sulphur dioxide, sulphuric acid, coke and oil-in-water emulsions. In one refinery, a sulphur dioxide bearing waste from a contact cooler in the acid recovery plant is used in the treatment of a petroleum sulphonate waste. Sulphur dioxide may also be stripped from waste-water and utilised. The coke is usually recovered by sedimentation, dried and burned as fuel. Emulsions may be broken using alum or other materials and the oil recovered as fuel.

Sulphuric acid and hydrofluoric acid waste may originate from the alkylation of feed stocks for the manufacture of aviation gasoline. The sulphuric acid alkylation process produces about 50 barrels of approximately 90 per cent acid per 1000 barrels of product. In view of threatened shortages of sulphuric acid, many refineries are re-using and regenerating spent acid from their alkylation operations. In some cases the spent acid is returned to the acid supplier. Applications in the refinery for re-use include acid treatment of naphthas, furnace oil, lubricating oil and waxes. The acid is recovered from the sludges produced as described in the discussion on the disposal of acid sludges.

Waste Caustics

Waste caustics as referred to herein originate from the caustic washing of light oils to remove mercaptans, hydrogen sulphide and other acidic materials that may occur naturally in crude oil or any of its fractions or may be produced by a variety of processing methods. The quantity of waste caustic produced will vary greatly depending on the characteristics of the crude and the methods of processing. The constituents of waste caustics responsible for their pollution characteristics include mercaptans, thiophenol, thiocresols, phenols, cresols, disulphides, alkylsulphides, the sodium salts of any one of a number of saturated mono acids, naphthenic acids or sulphonic acids and other materials.

The satisfactory disposal of waste caustic is a major problem at most refineries. The problem is lessened in some cases by process replacement or improvement to achieve more efficient utilisation of available treating capacity. Regenerative systems have been used to eliminate the discharge of spent caustic from mercaptan and hydrogen sulphide removal operations.

Likewise, the 'down grading' use of spent solutions from one process as a treating agent in another has proven advantageous. For example, discarded solutions, after having been made essentially lead free, may be used as prewashes for other treating operations such as scrubbing of hydrogen sulphide from distillates and gases.

Spent caustics from the treating of motor stock from catalytical cracking processes, commonly called carbolates, have been sold as such for the recovery of phenolic constituents. In some cases, this waste is first acidified to release 'acid oils' which are separated from the waste layer and sold for their phenolic content. Acid oils from carbolate solutions are approximately 90 per cent phenolic and amount to about 50 per cent of the volume of the carbolate. The low acid oil content and the heterogenous nature of the constituents make spent caustic from treating a poor source of recoverable by-products. Studies at one refinery indicate that the acid oil content of this material is 3 to 10 per cent by volume and is 60 per cent phenolic and 40 per cent naphthenic.

Acid oils may be released or 'sprung' from spent caustic solutions by neutralisation with mineral acids and acid gases. Frequently waste acidic material such as spent acids and acid sludges from other refinery operations may be used for this purpose. The sulphides are converted to hydrogen sulphide which may be burned or recovered and the acid oils are burned as fuel. The water layer from this reaction is usually drawn to the sewer, preferably at a low rate since it still contains a relatively high concentration of oxygen consuming materials. Flue gases have been used in some cases for the treatment of waste caustics. The caustic solutions are neutralised by the flue gases at temperatures of 160° to 180°F to about pH 5 to release the acid oils, unoxidised hydrogen sulphide and mercaptans and certain weak organic acids. During this process the easily oxidised mercaptides and sulphides are partially oxidised. The acid oils separated from the solution collected are used for fuel or a source of special chemicals. The hydrogen sulphide and odourous materials released to the gas stream must be conducted to burning facilities for atmospheric pollution control. The remaining water solution contains a mixture of carbonates, bicarbonates, sulphates, sulphites, thiosulphates, phenols, etc. which may require further oxidative treatment prior to discharge to natural waters.

Recently a method for utilising spent pickling acid from the steel industry of the disposal of spent caustics has been developed. The sulphides and many of the organic acids are removed by precipitation and filtration and the remaining organic acids are extracted with naphtha. The final solution is suitable for discharge into the refinery drainage system. The spent pickling acid used in this process contains from 1 to 5 per cent of sulphuric acid and from 20 to 30 per cent of ferrous sulphate. Waste caustics high in sulphides are treated by contact with steam and air in a packed tower at one refinery. The sulphides are converted to thiosulphates. The process operates with approximately 95 per cent conversion.

Spent caustics or the liquors resulting from the neutralisation of spent caustics may be treated by biological oxidation. Investigations using table top aeration experiments and small-scale biological filter experiments and extensive biological treatment experiments all indicate the feasibility of the biological treatment of waste caustics and other refinery wastes. Filter studies showed that oil concentrations of at least 100 ppm had no effect on BOD removal. Sludge from the filter effluent settling tank contained about 31 per cent oil on an oil free, dry basis. Temperatures of 85° to 90°F appeared to be optimum and recycle ratios of 4 to 5 to 1 materially increased reductions. Sulphide concentrations in excess of 10 ppm were found to have a detrimental effect on BOD reduction. Pilot plant investigations using the activated sludge process indicate that the presence of oil does not adversely affect the oxidative capacity of the sludge for materials like phenol but does seriously interfere with the settling characteristics of the sludge and thereby reduces general treatment efficiencies.

Alkaline Water

Alkaline water, as differentiated from alkaline condensate water and waste caustics, may originate from the washings of neutralised acid treated oils, the washings of caustic treated oils, the dehydration of treated light oils, the aqueous tank bottoms of stored caustic treated and washed gasolines, vessel and tower washings at times of shutdowns and miscellaneous sources.

The alkaline water referred to above, originating from continuous treating processes, contribute to the general pollution load but do not create a major problem. The intermittent discharge of aqueous tank bottoms and the washing of towers and vessels at times of shutdown can cause pollution peaks requiring special attention. At one refinery where biological treatment is to be provided for the plant effluent, intermittent flows of alkaline water, such as washings from batch treaters and shutdown drainage is discharged to holding tanks from which they can be drawn to the sewer at a controlled rate.

Special Chemicals

Under the category of special chemicals are included the special solvents and extraction solutions utilised in selective solvent refining, gas purification, light oil treating, etc. and the chemicals, by-products and wastes produced in the manufacture of petrochemicals.

Special chemicals utilised in petroleum processing include phenol, cresols, furfural, salts of isobutyric acid, nitrobenzene, acetone, methyl ethyl ketone, dichlorethyl ether, ethylene dichloride, benzol, tannin, fatty acids, diethanol amine, methanol, sodium hypochlorite, tri-sodium phosphate, lead sulphide, copper chloride, etc. Special chemicals manufactured include various synthetic detergents, methanol, maleic anhydride, phthalic anhydride, benzol, toluol, naphthenic acids, etc.

The special chemicals mentioned above pose special waste control problems. The water soluble organics, for example, can add tremendously to the oxygen demand characteristics of the plant wastes if allowed to discharge into the sewers. Others listed are emulsifying agents and would adversely affect separator operation if allowed to mix with other refinery wastes. In many cases, the value of these materials is sufficient to justify the use of collection and recovery systems. Drainage from leaks, spills, pumps, valves, sampling, routine maintenance activities, etc. is frequently recovered to control losses to a minimum. Distillation facilities for the concentration and/or separation of chemicals from undesirable contaminants are generally provided. Furfural, for example, has been recovered for reuse by the use of two azeotropic distillations.

When recovery of these materials is impractical, they can generally be removed from water by one or a combination of the procedures previously discussed in connection with the treatment of other wastes. Pilot-plant studies on the biological filtration of methyl ethyl ketone wastes indicated that reductions of 78 per cent or more could be expected for concentrations up to 60 ppm at temperatures of 80°F or above.

Cooling Water

Cooling water makes up nearly the entire volume of waste-water from petroleum refining operations. Since these wastes may become oil-contaminated due to equipment failure. Thus it is necessary to provide oil separation facilities as insurance against accidental pollution. Consequently, uncontaminated cooling water is generally turned into a common oil carrying sewerage system. However, when the standards of waste-water quality necessitate costly 'effluent polishing', the separate collection and disposal of cooling waters which are subject to infrequent pollution has proved economical at some refineries. Likewise, the size of effluent treatment facilities has been substantially reduced by decreasing the quantity of waste-water through the use of recirculating cooling systems. At one refinery requiring effluent treatment, the quantity of waste-water has been reduced from 24 mgd to 8 mgd by the installation of cooling towers. The use of dirty water cooling tower systems and the elimination of barometric condensers and jet vacuum pumps, emulsions also serve to keep sizes of treatment facilities at a minimum. Where water is scarce, refineries have used sewage treatment plant effluents as a source of water supply and in other cases have used their own treated effluents.

WASTE-WATER FROM REFINERY

Waste-waters in a refinery originate from the following processes:
1. Storage and transportation.
2. Fractionation by pressure and vacuum distillation.
3. Reforming.
4. Cracking: catalytic and thermal.
5. Hydro-desulphurisation.
6. Solvent treating processes for lube oil manufacture.
7. Utilities function — steam boiler, cooling water, etc.

The water use and waste-water generation in the oil refineries in the country are noted to be greatly influenced by the type of cooling system, once through or recirculation adopted.

It may be noted that in refineries having once through cooling system due to an enormous amount of water used, because of easy availability of sea water, a huge amount of valuable hydrocarbon is lost along with the waste-waters. These refineries have only primary oil separation facilities and taking undue advantage of the large quantity of waste-water, discharge the further recoverable oil along with the effluent. This will result in low concentration of oil in the effluent but the quantum oil of discharged is indeed high. Quantities of pollutants from any process section is a function of type or source of process feed stock and the type of processes involved, e.g. crude oils from different sources, will probably produce different quantities of pollutants, thus cracking processes will produce more phenols. Pollutants in the waste-water depends on the amounts of steam, process water and cooling water used in a refinery. Aqueous wastes will be more if a refinery uses more stripping steam and once-through cooling water than the refinery which uses air cooling.

The approximate quantum of water used by different unit operation in refinery is:
1. 30 l/kg of oil for processing and cooling in once-through basis.
2. 1 l/kg of oil by recycling and to increase the efficiency of the processes.
3. 0.3 l/kg of oil, when air cooling is used in place of water cooling.

Table 14.1 gives the details about the water used by the oil refineries and the waste-water generated.

Correlation of plant effluent analyses should be on the basis of kg of pollutant per unit quantity of plant feedstock, e.g. a particular cracking plant charged with a particular feed stock will produce a fixed amount of water soluble phenols per thousand barrels of feed. The amount of stripping steam and the amount of cooling water discharge will determine the phenol concentration in the cracking plant aqueous effluent.

Solvents and Effluents

Refineries generally use solvents like toluene, methyl ethyl ketone, benzene for removing waxes and furfural for the removal of aromatics during the manufacture of lubricating oil. During the operation with these solvents starting from receiving in the refineries to final operation there is every possibility for the loss of these through spillage, solvent extraction and maintaining the proper ratio. The solvents lost by these processes ultimately accumulates in the effluent water system through surface drains with washout waters. Removal of these solvents contamination from the effluent water is a typical specialised job.

Solvent contaminated water streams from the units in the refineries gets diluted in the total influent water which undergoes the various treatment. The physical treatment removes free oils and suspended solids and chemical treatment removes emulsified oils, sulphides, etc. while the biological treatment removes the dissolved degradable hydrocarbons. So there is every possibility that the solvents remain unremoved in the final effluent water unless special treatment procedures are followed. On the contrary the chemicals like furfural are harmful for the biological treatment through trickling filter and activated sludge because the micro-organisms responsible for biodegradation gets damaged through furfural. Quality of effluent water from petroleum refineries is given in Table 14.1.

Table 14.1. Quality of effluent water from petroleum refineries.

Characteristics	Tolerance limit for effluent water discharge		
	Into Inland surface water IS 2490	MINAS	Into Marine coastal area IS 7968
pH	5.5–9.0	6.0–8.5	5.5–9.0
Phenolic compounds, ppm	1.0	1.0	5.0
Sulphides ppm	2.0	0.5	5.0
Oil and grease ppm	10	1020	–
TSS ppm	100	20	
COD mg O_2/l	250	–	250
BOD @ 20°C for 5 days mg O_2/l	30	15	100
Cadmium (as Cd) ppm	2.0	–	2.0
Chromium (as Cr) ppm	0.10	–	1.0
Copper (as Cu) ppm	3.0	–	3.0
Lead (as Pb) ppm	0.1	–	1.0
Mercury (as Hg) ppm	0.01	–	0.01
Nickel (as Ni) ppm	3.0	–	5.0
Selenium (as Se) ppm	0.05	–	0.05
Zinc (as Zn) ppm	5.0	5.0	

WASTE-WATER TREATMENT

Removal of pollutants from oil refinery can be classified in three types of treatments:

Primary treatment	Gravity separation.
Intermediate treatment	Neutralisation, chemical coagulation and sedimentation, dissolved air flotation.
Final treatment	Biological, physical, chemical. Activated sludge aeration lagoon, trickling filter, cooling tower oxidation.

The classification can also be made on the type of treatment that is: physical, chemical and biological.

Physical Treatment

Waste-water may contain coarse, suspended and floating solids, grease, etc. These need to be removed before waste-water is subjected either to chemical or biological treatment. Common unit operations of physical treatments are bar screens, grinders, grit chambers, grease traps, flocculation, sedimentation, flotation, chemical precipitation, sludge pumping. This treatment basically removes inert material which may hinder the subsequent treatments.

Bar screens

The purpose of bar screen is to remove large floating matter as well as to avoid clogging of filter media and obstructing aeration in activated sludge. These can be hand cleaned or mechanically cleaned. For the hand cleaned screen the velocity of approach should be limited to about 0.456 m/s.

Grit chambers

Grit chambers removes grit, i.e. sand, gravel, etc. which have specific gravities greater than the suspended organic matter. Purpose of providing grit is to protect mechanical equipments and avoid deposition in pipelines. Two types of grit chamber generally provided are horizontal flow grit chamber and aerated grit chamber.

Primary sedimentation

Purpose of this process is to reduce settleable suspended solid contents of waste-water. Here the waste-water is left undisturbed, the particles of higher specific gravity than waste-water will settle and those with lower specific gravity will float. About 50–65 per cent removal of suspended solids and 20–40 per cent of BOD_5 (at 20°C) can be achieved in a properly designed sedimentation tank. Common detention period is 90–150 minutes based on average rate of flow. Surface loading rates for untreated wastes are 20–40 m^3 pd/m^2 with a peak flow of 40 m^3 per day.

Oil separation

Oil can be present as free or emulsified form or both, in the waste-water. It is necessary to remove free oil before breaking the oil water emulsion. Attempts are made to float the free oil in suitably designed tank and then it is skimmed by a rotating arm or disc. The rise velocity of the oil globules is governed by Stoke's law. API separator is generally used for such type of oil separations. These separators are normally operated without using any chemicals or coagulant or aids. Oil and grease separated from these units as oil are recovered and reused in the refinery. Titled plate separators (TPS) or corrugated plate separators are used to separate, the free oil from the waste-water. TPS gives better performance over gravity separators as it reduces the rising distance of the oil droplet before reaching the surface where it coalesce to form oil layer which rises to top from where it is skimmed off.

The gravity separators are essentially rectangular chambers equipped with oil skimming and sludge scrapping device and design of the separators is based on removal of oil globules of 0.015 cm in diameter and above. The efficiency of oil removal is around 98 per cent.

An emulsion is defined as a mixture of two immiscible liquids, one of the liquids being dispersed throughout the other in the shape of very fine droplets. Both the oil-in-water and water-in-oil emulsions are present in a refinery waste-water. The de-emulsifying agents which are normally adsorbed on the surface of the emulsified particles rendering the emulsion stable are soaps, sulphates, sulphonic and naphthenic acids, quartenary ammonium compounds, organic ethers and esters.

To remove emulsified oil, dissolved air flotation (DAF) technique is used. This technique can be used alone or in combination with flocculation. This combination can remove 97 per cent of the oil, 75 per cent of the suspended solids and reduce both biochemical oxygen demand (BOD) and chemical oxygen demand (COD) by 80 per cent. This reduction varies with the characteristics of the waste stream. The basic mechanism by which an oil-in-water emulsion is broken involves neutralisation of the charges carried by the oil droplets by forming a floc that is initially precipitated with a charge opposite of that on the oil. Adsorption of the oil by a flocs which initially has a high adsorptive capacity for the oil and entrapment of the oil as the floc forms and grown around the oil droplets. The above objective can be achieved by adjusting the pH of the waste-water between 5 and 6, subsequently precipitating the emulsion stabilising agents like sulphonic and naphthenic acids as water insoluble calcium salts by the addition of hydrated lime until a pH of 7.5 to 8.5 reached.

Cationic polyelectrolytes are available for de-emulsification of oil. This has remarkable advantage over the conventional methods. Also polypropylene adsorbent can be utilised for de-emulsification. This adsorbent is in a continuous belt form which is oleophillic (attracts oil) and also hydrophobic (water repellent). This when passed in the waste-water adsorbs oil on its surface.

Various coagulating agents like ferrous sulphate, ferric sulphate (chlorinated copper as) calcium chloride, calcium carbonate and hydrated lime have been used to de-emulsify an oil separator effluent. One particular type of coagulant may be effective for the waste-water of a particular refinery but may, however be ineffective for the waste-water of another refinery. The type and behaviour of the emulsifying agents would dictate the application of a particular coagulant best suited for the purpose. The flocs that are formed by such treatment adsorbs most of the oil and depending on the initial oil concentration, an effluent may be produced with an oil content of 10 mg/l or even less.

Chemical Methods

After removal of grit and floating matter, suspended and dissolved organic matter are removed. Important unit operations and processes involved in the chemical treatment of waste-water are:

Chemical coagulation, flocculation and sedimentation

Dissolved air flotation, both with and without flocculation, has been successfully used to obtain a low oil content in a refinery waste-water. Under favourable conditions it is possible to obtain an oil content of as low as 10 mg/l from a flotation cell using a coagulant and coagulant aid. However, dissolved air flotation units require more skilled and regular supervision and power than gravity separators, hence their installations have been restricted to a few refineries in India.

Aluminium sulphate, lime, ferrous sulphate when used in the way they are used as in normal water clarification, on waste-waters, result in producing better effluent than plain sedimentation. With favourable conditions suspended matter can be reduced by about 90 per cent, if the plant is operated skilfully. Interest in chemical treatment of waste-waters has been revived because there is:

1. Decrease in cost of chemicals.
2. Better understanding of mechanisms involved in coagulation.
3. Improvement in sludge handling processes.

Setting velocities of finely divided and colloidal particles are very low and hence require very long detention period. Also special efforts or mechanisms have to be provided so that these small particles aggregate together to settle under gravity at accelerated rates.

Aggregation or building up of flocs is induced by addition of chemical coagulants to decrease the zeta potential which stabilises the colloids. To encourage collisions between destabilised particles, agitation is necessary. Also coagulants have to be used between optimum pH range to achieve optimum efficiencies.

Sulphide precipitation

Sodium hydroxide, when used to scrub cracked hydrocarbons for the removal of sulphur and phenol compounds, forms sodium sulphide and other products.

If the spent caustic is derived from caustic scrubbing cracked gasoline which has been debutanised, then it would contain very little sodium sulphide and the sulphur compounds will be mostly mercaptans and thiophenols. In some refineries which discharge significant amount of spent caustic liquor containing very high sulphide concentration, chemical precipitation of sulphides by iron salts has been found to be effective and more economical than by neutralisation followed by steam or air stripping. Sulphide precipitation is particularly to waste-water containing sulphides and mercaptans as the latter is also precipitated as iron sulphides. By using chlorine with ferrous sulphate, chlorinated copperas is formed which becomes more effective in the precipitation of the sodium sulphide.

Filtration

This removes finely divided suspended material. This may or may not be preceded by chemical coagulation.

Air stripping

This is used for removal of ammonia and hydrogen sulphide.

Ion-exchange

This is used for the removal of phosphates, nitrogen and total dissolved solids.

Reverse osmosis

This is used for removal of organic and inorganic substances.

Carbon adsorption

This is used for reducing organic matter.

Biological Methods

Biological treatment unit is primary meant for removal of pollutants like phenol, residual sulphide and BOD and also the non-recoverable oil present in the secondary effluent. Bacterial seeding and fertilisation of the oily waste with appropriate bacterial species will accelerate biological degradation provided that dissolved oxygen and sufficient time are available.

Approximately 66 per cent of the hydrocarbon oxidising species are *Pseudomonas* species. Next in order of abundance are species of *Mycobacteria*, *Proactinomyces*, *Zctinomyces*, yeast and moulds.

The BOD of mineral oil is relatively high as about 3–4 mg of oxygen are required for the oxidation of 1 mg of various kinds of hydrocarbons. Depending upon their chemical composition, the oxygen demand of various kinds of oils ranges between 3.1 to 3.5 mg/mg of oil. These values may be compared with oxygen demand values of 1.07 for glucose, 1.18 for starch and cellulose, 1.5–1.8 for proteins and 2.5 to 2.9 for vegetable oils. It is found that the bacteria in biofilters and activated sludge to be upto 84 per cent effective in removing oil from refinery waste.

Grease and oils are particularly resistant to anaerobic digestion and when present in sludge they cause excessive scum accumulation in digesters, clog the pores of filters and deter the use of filters and deter the use of sludge fertilisers. When these substances are discharged in waste-waters or treated effluent they often cause surface films and shoreline deposits.

A knowledge of the quantity of grease and oil present in a waste is helpful in overcoming difficulties in plant operation, in determining plant efficiencies and in controlling the subsequent discharge of these materials to receiving stream. A knowledge of the amount of grease present in sludge can aid in the diagnosis of digestion and de-watering problems and indicate the suitability of a particular sludge for use as fertiliser.

Depending on the type of oxidising agent used, the biological methods which employ micro-organisms, is grouped under aerobic (when free molecular oxygen is present) or anaerobic (absence of free oxygen) methods of treatment.

The rate of decomposition by microbes is importantly a function of the oil water interface. The greater the oil water interface the faster the microbial decomposition rate. It is reported that microbial disposal of oil appeared to be feasible with specifically adopted cultures (about 51 to 61 per cent of crude oil was utilised by a species of *Pseudomonas* over a period of 21 days). Now-a-days a concept of engineering the species to utilise a particular waste has been developed. Also mixed cultures can be used to degrade a particular types of waste from oil refinery.

CONTROL OF AIR EMISSIONS IN REFINERY

Air Pollution Control Measures

Control of refinery air emissions can be accomplished by process change, equipment change, procedure change, installation of control equipment, improved housekeeping, increased monitoring, and better equipment maintenance. Specific control devices are discussed below for the refining sector. Some combination of these often proves to be the most effective solution. Several of these controls also result in some form of saving.

Vapour Control Systems

Vapour control systems involve the collection of process or fugitive vapour emissions and their recycling or recovery for hydrocarbon content or fuel value or their destruction. In some instances, where economically feasible, the emission stream itself can be recycled to refinery fuel gas production or the fuel value of the emission stream can be recovered via combustion. Hydrocarbons in vapours can be destroyed via flares or incinerators. The specific configuration of vapour control systems depends on the specific configuration and economics of the refinery.

Flares

Flares are commonly used for the disposal of waste gases during process upsets (e.g. start-up, shutdown) and emergencies. They are basically safety devices that also are used to destroy organic constituents in waste emission streams. Flares can be used for controlling almost any nonhalogenated VOC emission stream. They are designed and operated to handle large fluctuations in flow rate and VOC content. There are several different types of flares, but the prevalent refinery flare, illustrated in Fig. 14.2, is an elevated steam-assisted flare. Flaring generally is considered a control option when the heating value of the emission stream cannot be recovered because of uncertain or intermittent flow, as in process upsets or emergencies. If the waste gas to be flared does not have sufficient heating value to sustain combustion, auxiliary fuel may be required. According to studies conducted by the EPA, 98 per cent destruction efficiency can be achieved by steam-assisted flares when controlling emission streams with heat contents greater than 300 Btu/scf.

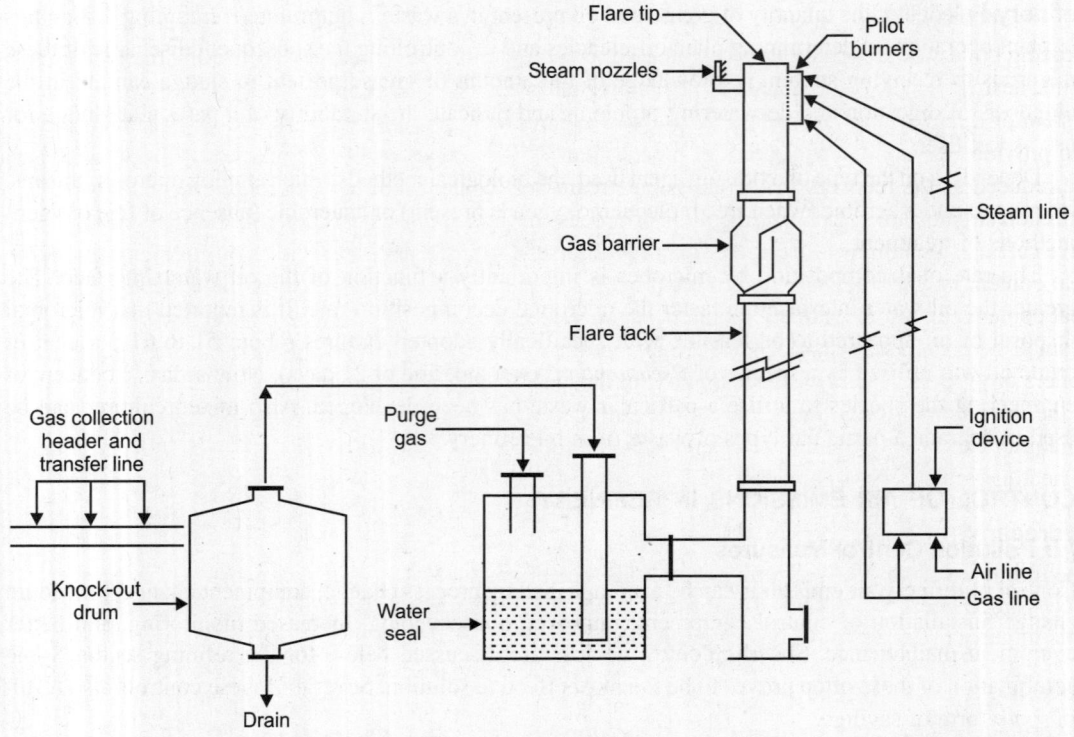

Fig. 14.2. Steam-assisted elevated flare system.

Incinerators

There are two types of incinerators—thermal and catalytic. Thermal incinerators are used to control a wide variety of continuous VOC emission streams. Thermal incineration is preferable to flaring when halogenated or sulphur-bearing compounds are present owing to the corrosive properties of these compounds. Destruction efficiencies up to 98–99 per cent are achievable with thermal incineration. Although they accommodate minor fluctuations in flow, thermal incinerators are not well suited to streams with highly variable flow because of reduced residence time and poor mixing during increased flow conditions. Thermal incineration is typically applied to emission streams that are dilute mixtures of VOCs and air. In such cases, due to safety considerations, the concentration of the VOCs is limited in some instances to 25 per cent of the lower explosive limit (LEL). Thus if the VOC concentration is high, dilution may be required. When emission streams controlled by thermal incineration are dilute (i.e. low heat, content), supplementary fuel is required to maintain the desired combustion temperature. Fuel requirements may be reduced by recovering the energy contained in the hot flue gases from the incinerator.

Catalytic incinerators are similar to thermal incinerators in design and operation, but they employ a catalyst to increase the reaction rate. Since the catalyst allows the reaction to take place at lower temperatures, significant fuel savings are possible. Destruction efficiencies of up to 95 per cent are generally achieved with catalytic incineration. Catalytic incineration is not as broadly applicable as thermal incineration because of catalyst sensitivity to pollutants and process conditions.

Refinery Fuel Gas

Existing boilers or process heaters can be used to control emission streams by recovering the fuel value while destroying the VOCs. This is accomplished by diverting the streams to the refinery fuel gas system or directly into the firebox. When used as emission control devices, boilers or process heaters can provide destruction efficiencies of greater than 98 per cent at a small capital cost. Operating costs are reduced as the recovery of the emission stream heat content reduces fuel requirements. Typically, emission streams are controlled in boilers or process heaters and used as supplemental fuel only if they have sufficient heating value (greater than 150 Btu/scf). In some instances, emission streams with high heat content may be the main fuel to the process heater or boiler. Note that emission streams with low heat content can also be burned in boilers or process heaters when the flow rate of the emission stream is small compared with that of the fuel-air mixture. There are some limitations on the application of boilers or process heaters as emission control devices. Because these combustion devices are essential to the operation of the refinery, only those emission streams that will not reduce their performance or reliability can be controlled. Streams not suitable for control include those with varying flow rate and/or heating value, high-volume/low-heating-value streams, and streams with corrosive compounds.

Hydrocarbon Reuse

The ultimate vapour-recovery process recycles the emission stream to the light ends recovery unit (gas processing), where it is recovered as a salable product, usable feedstock, or gasoline blending component. This requires compressing and separating, by distillation, the stream into its respective components.

Covering a Liquid Space

Covering a liquid surface or the vapour space above a liquid surface suppresses VOC emissions to the atmosphere. Examples of control by covering are floating roof tanks (external and internal) and waste-water systems. The control of waste-water treatment systems requires covering sources where emission generation is greatest, namely, process drains, junction boxes, and oil—water separators. By suppressing emissions through the separator, more hydrocarbon can be recovered as a liquid and recycled back to the refinery. Covering a drain involves either a physical cover at ground level or a liquid seal in the drain pipe. Emission reductions of 40–50 per cent are achievable by drain seals. Junction boxes require venting to prevent siphoning and vapour locks.

Vents on junction boxes should be at least 4 inches in diameter and 3 ft. in length. To minimise VOC emissions, the vent pipe must also have a water seal. Control efficiencies can be assumed to be equal to those of drain seals.

Any VOC emissions from oil–water separators are controlled by installing a fixed or floating roof. A vapour space under a fixed roof may constitute an explosion or fire hazard. In order to eliminate this problem, the vapour space can be blanketed with either plant gas or an inert gas such as nitrogen. Floating roofs eliminate much of the vapour space, thus greatly reducing the potential for volatilisation from the oil layer. Emission reduction estimates based on qualitative information range from 90 to 98 per cent. Data developed by Litchfield indicate that 85 per cent is representative of the emission reduction achievable by a simple fixed or floating roof. The most obvious factor affecting performance is the degree of maintenance.

In addition to covering components in the waste-water system, the system can be completely closed and vented to a combustion device, such as a flare. Emission reductions depend on the efficiency of the control device; for instance, flares would result in 98 per cent reduction.

Inspection and Maintenance

Improved maintenance — including scheduled inspection and monitoring, improved housekeeping, and improved employee training — is a very practical method for reducing hydrocarbon emissions and alleviating odour problems. Moreover, it is often the only control method for some sources, such as valves, relief valves, seals, cooling towers (heat exchanger leaks), and sampling operations. For nearly all sources, especially process drains, waste-water separators, treating units, blind changing, and loading operations, employee awareness of the problem will reduce emissions.

In-Line Sampler

In-line sampling allows the collection of a representative process sample without having to flush any hydrocarbon to the process drain. Typically, small piping/tubing is run parallel to the process piping. A three-way valve is used for sample collection, thus eliminating any 'dead legs'.

Rupture Disks

Rupture disks as a control device are used to protect relief valves from the corrosive process environment. They are typically thin metal disks located on the pressure side of the relief valve. They are designed to burst at the relief valve setting. Owing to their 'one-time' use, rupture disks are applicable for relief valves that are expected to be vented in emergencies only. They also can be used in place of relief valves in certain applications.

New Valve Technology

New valve technology has been developed that can significantly reduce pollutant emissions from the valve stem packing. It includes new types of packing material for valve stems and new seal technologies. New valve designs have been developed for some applications to prevent leaks by eliminating packing material leak sources around valve stems or by encasing potential leak sources in a 'bellows' structure to trap emissions.

The new 'leakless' valve designs do have limitations of applicability. Some designs cannot withstand the high temperatures and pressures of many refinery process streams. Other designs are limited to valve sizes of 6–8 inches in diameter and, therefore, are applicable only to smaller line sizes in refineries. For refineries, the most widely used control technique to minimise leakage from valves is an effective schedule of inspection and preventative maintenance.

Mechanical Seals

Mechanical seals are used to control product emissions from centrifugal pump and compressor glands. A simple mechanical seal consists of two rings with wearing surfaces at right angles to the shaft. One ring is stationary, while the other is attached to the shaft and rotates with it. A spring and the action of the fluid pressure keep the two faces in contact. Lubrication of the wearing faces is effected by a thin film of the material being pumped. The wearing faces are precisely finished to ensure perfectly flat surfaces.

A pressure seal can further reduce emissions. A liquid that is less volatile or dangerous than the product being pumped is introduced between a dual set of seals. Since this liquid is maintained at a higher pressure than the product, some of it passes by the seal and into the product. The pressure differential prevents the product from leaking outward and the sealing liquid provides lubrication. As some of the sealing liquid passes the outer seal (hence the need for low volatility), a means should be provided for its disposal.

Quick Change Blind/Manifold Design

Several devices have been developed to reduce spillage, such as Hamer and Greenwood blinds. These 'line' blinds do not require a complete break of the flange connection as 'slip' blinds do, but rely on a gear mechanism to release the plate. Combinations of these devices in conjunction with gate valves allow changing of the line blind while the line is under pressure from either direction.

In addition, during design stages, the manifold can be designed to minimise the need for blinding, as well as the quantity of liquid that will be spilled. With highly volatile products, water or nitrogen can be used to displace the product, precluding any product spillage during the blind change.

New Gaskets, Bolts, and Welding

Replacement of leaking gaskets on existing pipe and valve flanges and minimising the number of flanges with the use of welded pipe are methods of reducing VOC emissions from flanges.

Steam Stripping

Steam stripping is an effective control for removing VOCs from process streams. Two examples within a refinery include FCC catalyst stripping and sour-water treating. Spent FCC catalyst is steam stripped to remove absorbed hydrocarbons at the reactor exit. Process waste-water is stripped with steam to remove H_2S, ammonia, and light gases.

The process is generally conducted in the sour-water stripper. The sour water and steam are fed into the column. The stripped water and condensed steam are routed to the waste-water system and the undesirable constituents are incinerated.

Carbon Monoxide Boiler

Depending on the refinery, the catalytic cracker may or may not have a CO boiler. Carbon monoxide boilers are used to recover the energy contained in the catalyst regeneration off-gases. The recovered energy is used as process heat for various refinery processes. The fuel used in the CO boiler consists of the process gas from the catalyst regenerator and an auxiliary fuel source. The process gas may contain up to 5–10 per cent CO. Combusting the gas in the boiler produces heat and reduces emissions of CO and VOCs to negligible levels.

Cyclone

Cyclone separators are used for catalyst dust collection in the upper section of both FCC unit reactors and regenerators. The cyclones are employed as a single unit or in multiple two- or three-stage series units. In general, high-efficiency cyclones have dust collection efficiencies of over 90 per cent for particle sizes of more than 15 μm. The efficiency drops off rapidly for particles less than 10 μm.

Electrostatic Precipitators

Electrostatic precipitator (ESP) particle removal is accomplished by charging the particles, collecting the particles, and transporting the collected particles into a hopper. An ESP is very sensitive to the aerosol density and the electrical resistivity of the material, but is less sensitive to particle size. The electrical resistivity of the particles influences the drift velocity or the attraction between the particles and the collecting plate. A high resistivity will cause a low drift velocity, which will decrease the overall collection efficiency. Used to control PM emissions on FCC units, ESPs achieve efficiencies as high as 80–85 per cent.

Scrubbers

Scrubbers are used in refineries to separate and purify gaseous streams containing high concentrations of VOCs, SO_2, and PM. Examples of scrubbing within a refinery include applications on the asphalt blowing airstream and on the FCC regenerator off-gas. A common system on FCC regenerators is a caustic scrubber, where particulate removal and sulphur oxides absorption take place in a venturi scrubber. Particulate is removed by inertial impaction of the scrubbing liquid with the entrained particles. The sulphur oxides absorption reactions that take place are as follows:

$$2NaOH + SO_2 \longrightarrow Na_2SO_3 + H_2O$$

Sulphur dioxide and PM removal efficiencies of 93–98 per cent have been recorded.

Combustion Controls

All the criteria pollutants and several potential hazardous air pollutants (HAPs) are emitted from combustion sources within refineries. Sulphur oxides can be controlled by fuel selection, fuel desulphurisation, or flue gas treatment. Carbon monoxide, hydrocarbons, and PM can be minimised by more efficient combustion. Nitrous oxides can be controlled by combustion modification or by flue gas treatment. Controls specific to each refinery combustion source are described below.

Internal-Combustion Engine

The VOC, CO, and NO_x emissions can be controlled by a three-way catalyst (similar to a catalytic converter in a motor vehicle) on a rich-burn engine (fuel-rich combustion). The NO_x emissions can be controlled by selective catalytic reduction (SCR) on a lean-burn engine. Catalyst systems generally are designed for 80 per cent reduction efficiencies. Precombustion chambers, which control the air-to-fuel ratio in stages, are another NO_x-reduction technique for gas-fired engines. Injection timing retard also reduces NO_x production. The magnitude of the reduction may vary considerably among engine types. In general, reductions of 20–34 per cent, with corresponding 1–4 per cent fuel consumption penalty and slight increases of VOCs and CO, are obtained.

Combustion Turbines

Wet injection using either water or steam is the most prevalent NO_x-reduction technique for turbines. Depending on turbine type and size, NO, emission levels of 25–50 ppm at 15 per cent oxygen are obtainable. Selective catalytic reduction is also applicable to turbines for NO_x reductions of generally 80 per cent. Coupling SCR with wet injection can result in NO_x emission levels as low as 5 ppm.

Heaters and Boilers

In addition to three-way catalysts and SCR, selective noncatalytic reduction processes can be applied to process heater and boiler flue gas for NO_x control. The process using ammonia can achieve 40–60 per cent reduction, and that using urea can achieve 20–80 per cent reduction, depending on temperature. Low-NO_x burners, which stage the mixing of air and fuel to reduce flame temperature, result in NO_x reductions of 30–40 per cent over conventional burners.

PETROCHEMICAL AND ALLIED PRODUCTS

It is difficult to make any general statements on the petrochemical industry owing to the fact that product mix, raw materials and production technology vary from one installation to another. The inputs to the petrochemical industry are almost all petroleum-based products. The primary inputs are ethylene, propylene, butadiene, benzene, toluene, xylenes, ammonia and methanol. These are produced from the following raw materials: petroleum distillates, propane, ethane, natural gas, coal, shale oil and biomass. Thus the existence of these raw materials and a refining industry are important factors in the development of a petrochemical industry within a region. The outputs of the petrochemical industry fall into four basic categories: plastics, synthetic fibres, synthetic rubbers and detergents. The end products of the petrochemical industry are seldom directly available to the consumer, but are intermediate products that serve as input to a limited number of secondary industries.

Water Use

The variation in inputs, processes and products corresponds to a wide range of water uses both in quantity and quality and even more so in a wide range of waste streams. The variety means that such indicators as specific water use depend strongly on the various processes and water-treatment technologies, so that an overall analysis is difficult. However, a general perspective of the alternative technologies and the resulting water use and waste streams will be provided. Possible recycling and reuse of waste-water and effluent treatment practices will also be considered.

The three basic uses for water in the petrochemical industry are: cooling water, which accounts for 80 per cent of intake water; steam generation, which uses 5 per cent of intake water; and process water, which uses 15 per cent of intake water.

Cooling Water

A large amount of heat is generated in the petrochemical industry. This heat is removed from the processes through contact or non-contact cooling with water. The exact amount of water needed is a function of the particular process employed.

In closed (non-contact) systems, the quality of the water is an important factor for the maintenance of the cooling system. In contact cooling the water quality is very important to prevent contamination of the processes outputs.

Steam Generation

There are three major uses of steam generated within a petrochemical plant:
1. Non-contact process heating.
2. Power for a variety of purposes.
3. As a diluent, stripping medium or source of vacuum using steam jet injectors.

Process Water

Although process water accounts for only 15 per cent of total water use, it is vital to the production of petrochemical products. As more of the cooling systems become closed cycles, the percentage of process water used will increase. Petrochemical processes have been categorised by the USEPA into four groups on the basis of water use within the process.

Non-aqueous processes

Contact between water and reactants or products is minimal. Water is not required as a reactant or diluent and is not formed as a reaction product. The only water use is for periodic washes or catalyst hydration.

Processes with process water contact as steam diluent or absorbent

Process water is in the form of dilution steam, direct product quench or absorbent for effluent gases. Reactions are all vapour-phase over solid catalysts. Most processes involve absorption coupled with steam stripping of chemicals for purification and recycling. Steam is also used for the decoking of catalysts.

Aqueous liquid-phase reaction systems

The catalyst is in an aqueous medium containing mineral salts, acids or bases. Continuous regeneration of a catalyst system often requires extensive water use. Substantial removal of spent inorganic salt by-products may also be required. Additional water may be required for final purification or neutralisation of products.

Batch and semi-continuous processes

Processes are carried out in reaction kettles equipped with agitators, scrapers or reflux condensers depending on the nature of the operation. Many reactions are liquid-phase with aqueous catalyst systems. Cleaning of non-continuous production equipment is a major source of waste-water.

Waste-loads from product separation and purification will be at least 10 times those from continuous processes.

A single process may be complicated and embrace more than one of the groups described above. A schematic of the treatment of waste-water from the organic chemicals industry is given in Fig. 14.3.

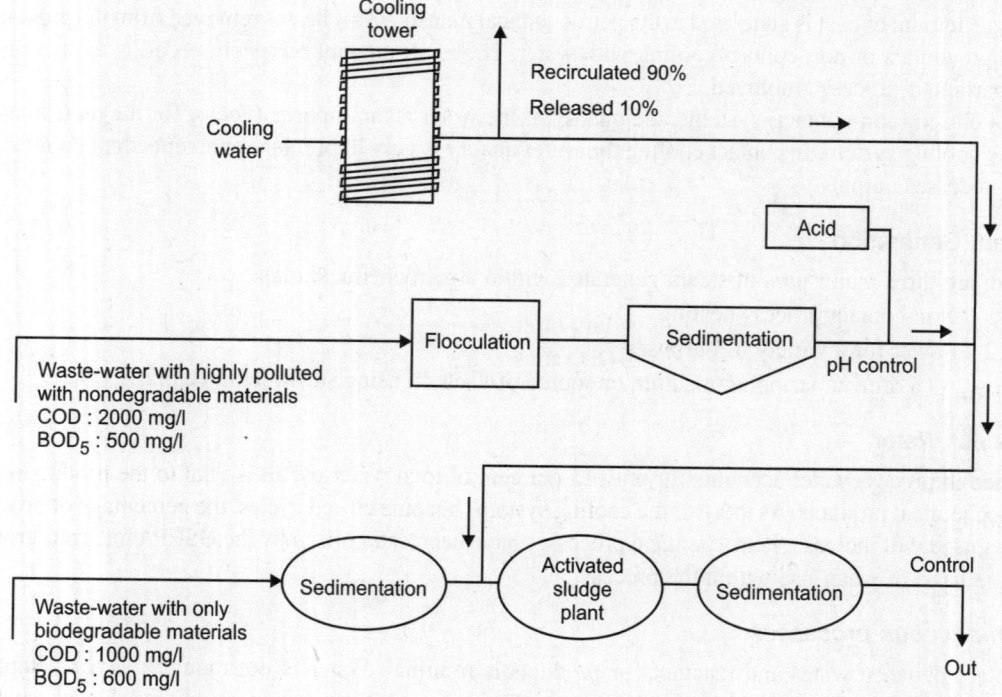

Fig. 14.3. Treatment of waste water from the organic chemicals industry.

Recycling and Reuse

Since the petrochemical industry requires a great deal of water to operate, in arid regions where water is scarce or in developed regions with conflicts over water use the amounts of intake water normally required for a petrochemical installation may have to be reduced. One way to achieve this is by alternative process technologies that require less water. However, even with the most water-efficient technologies there still may be a water supply problem. In this case, recycling and reuse of water within the petrochemical installation is an alternative.

Possible in-plant modifications that can result in reduced water intake through water recycling or reuse are as follows:
1. The substitution of noncontact heat exchangers for direct contact water cooling.
2. The use of nonaqueous quench media as a substitute for water where direct contact quench is required.
3. The reuse of process water (after treatment) as make-up water for cooling towers through which non-contact water is circulated.
4. The use of process water to produce low-pressure steam by noncontact heat exchangers in reflux condensers of distillation columns.
5. The use of nonaqueous solvents for extraction of products.
6. The recycling of process water, such as between absorber and stripper.

Sources and Characteristics of Waste-water

In the section on water use it was seen that water in the petrochemical industry is used in a variety of ways; the resulting raw waste-water streams are just as varied. The major sources of waste-water in the petrochemical industry are:
1. Raw materials themselves.
2. Products remaining in the solution after separation.
3. By-products produced during reactions.
4. Spills, washdowns and vessel clean-outs.
5. Cooling tower and boiler blow-down, steam condensate and water-treatment processes.
6. Storm water run-off.
7. Sanitary systems.

Wastes from these sources have a variety of characteristics that include high COD concentrations, high total dissolved solids, high COD/BOD ratios and contamination by heavy metals or compounds inhibiting biological treatment. Except for spills, storm-water run-off and sanitary systems the sources listed are found in the four process groups listed in the section on process water.

Waste-water Treatment Practices

The three main components of petrochemical wastes are biological substances, other chemical substances that are only slightly biodegradable or not at all, and suspended matter. The expected ranges of concentration of these components in the raw-waste streams from petrochemical production are given in Table 14.2.

One of the major issues in the treatment of petrochemical wastes is whether the waste is biodegradable. Some chemicals will actually inhibit biological treatment and cause damage to joint municipal-industrial treatment plants. It is, therefore, important to identify the compounds found in the various waste streams. Chemicals frequently found in petrochemical waste streams are listed in Table 14.3 according to biodegradability. The importance of the biodegradability of the waste stream has influenced the development of waste-treatment practices in the petrochemical industry. The first step is the reduction in quantity of effluents by recycling and reuse of water and raw materials. In addition to the six steps for recycling water listed above, the recovery of spent acids or caustic solutions for reuse and the recovery and reuse of spent catalyst solutions are possible measures to reduce waste loads.

Table 14.2. Characteristics of waste-water from the production of organic chemicals.

Product	BOD_5 (mg/l)	COD (mg/l)	Suspended matter (mg/l)
Phthalic acid anhydride and maleic acid anhydride	–	150–300	20–50
Methacrylic acid	–	7000–12000	6000–12000
Butadiene and styrene	4000–8000	800–1500	200–500
Acrylates	1000–2000	2000–3200	50–100
Ethylene and propylene	400–600	800–1200	20–40
Isocyanates	300–600	900–1600	40–75
Methyl and ethyl parathion	2000–3500	4000–6000	50–100
Acrylonitrile	200–500	600–1200	80–150
Raw materials for the pigment industry	200–400	1000–2000	80–200
Esters	5000–12000	10000–20000	20–100
Acetaldehyde	15000–25000	40000–60000	150–300
Organic acids	300–600	5000–15000	100–200
Ketones	10000–20000	20000–40000	50–100
Organic phosphate compounds	500–1000	1500–3000	200–400

Table 14.3. Biodegradable and non-biodegradable organic substances.

Biodegradable	Non-biodegradable
Aliphatic acids	Ethers
Aliphatic alcohols	Ethylene/chlorhydrin
Aliphatic primary and secondary alcohols	Isoprene
Aliphatic aldehydes	Butadiene
Aliphatic esters	Methylvinyl ketone
Alkylbenzene sulphonates	Napthalene
Amines	Various polymeric compounds
Mono- and dichlorophenols	Polypropylene benzene sulphonates
Glycols	Certain carbon hydrates, especially of aromatic structures, including alkyl-aryl compounds
Ketones	
Nitriles	Tertiary benzene sulphonate
Phenols	Tri-, tetra- and pentachlorophenols
Styrene	
Phenyl acetate	

When the possibilities for waste reduction by recycling or reuse have been exhausted, treatment of the waste is the next step. An example of an integrated waste-water treatment system for a petrochemical plant in which the waste stream is first separated into two streams, one containing only biodegradable waste and the other containing non-biodegradable waste, is given in Fig. 14.4. Preliminary treatment includes equalisation, pH adjustment, flocculation or coagulation, primary settling, flotation and oil separation.

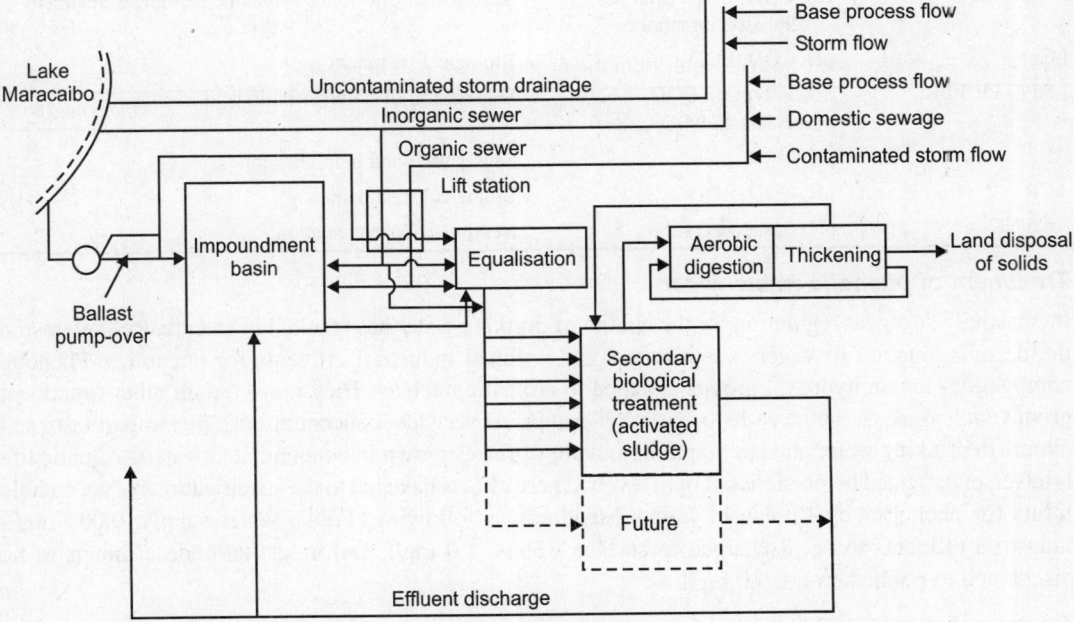

Fig. 14.4. Schematic diagram of treatment system at petrochemical complex.

Traditional biological treatment may consist of anaerobic digestion, trickle filtration, aeration (in ponds) or treatment with activated sludge. Physico-chemical treatment includes the use of activated carbon.

Advance treatment methods that are still rather new include hydrocyclone processes, ultra filtration, reverse osmosis, photo-oxidation, electrodialysis, steam stripping and special cooling techniques.

Water Pollution Control in Petrochemical Industries

Huge volume of water is required for various purposes in manufacture of petrochemicals. Water is used for cooling, process operations, fire fighting, drinking, housekeeping, steam raising, etc. Whole water after use reappears as effluents which requires treatment and disposal. Characteristics of effluents from various petrochemical plants are given in Table 14.4.

Table 14.4. Characteristics of effluents from various petrochemical plants.

Plant	Water pollutants	Mode of treatment
DMT	Total dissolved solids, pH	Segregation and neutralisation, chlorine recovery
Aromatics	pH, BOD, COD, suspended solids	Neutralisation followed by biological treatment
Olefins	pH, COD, BOD, Sulphides, phenol, hydrocarbon, oil and spent caustic, cyanides	Oil separation, segregation of fluoride effluent, biological treatment

(Contd...)

Plant	Water pollutants	Mode of treatment
LAB	pH, COD, BOD, F, oil surfactants, sulphides	Oil separation segregation of fluoride effluent, biological treatment
ACN	pH, COD, BOD, cyanides, sulphates, ammonia	Segregation of cyanide effluents, biological treatment
EG	pH, COD, BOD	Biological treatment
LDPE, PP, PBR	pH, Polymers, COD, surfactants BOD	Segregation of polymer biological treatment
AF	pH	Segregation and neutralisation
ACR	pH, COD, BOD	Biological treatment
VC/PVC	pH, Suspended solids	Neutralisation coagulation

Treatment of phenolic waste-water

Increasingly stringent regulation for the quality of drinking water has resulted in an enhanced interest in the decontamination of water, waste-waters and polluted industrial effluents for phenolics. Phenolic compounds contain hydroxyl groups attached to aromatic nucleus. They may contain other functional groups such as alkyl, methoxy, halo, carbonyl groups. At very low concentrations, they impart taste and odours in drinking water and may taint the flavour of fishes grown in contaminated waters. Aquatic life is adversely affected by phenolics at ppm levels. These effects have led to the specifications of acceptable limits for phenolics by Bureau of Indian Standards as following: Public water supply, 0.005 mg/l; industrial effluents to be discharged to surface waters, 1.0 mg/l; and industrial/trade effluents to be discharged to public sewers, 5.0 mg/l.

Sources of phenolic waste-water

Generally the level of phenolics in domestic waste-waters are low. Many industries produce phenolic waste-waters. These include petrochemical industries petroleum refineries, coal cooking and coal gasification, resin manufacturing, dye synthesis, wood preserving plants, pulp and paper mills and aircraft manufacture. Reported average phenolic concentrations for some of the industrial waste-waters are given in Table 14.5.

Treatment methodologies

Commonly used methods for the treatment of phenol bearing waste-waters include solvent extraction, physical adsorption, chemical oxidation and aerobic biological processes. Alternate biological treatment methods such as anaerobic and anoxic processes and enzymatic approach for the removal of phenolic are receiving wide attention only recently. These methods are discussed briefly with their merits and demerits.

Solvent extraction

Solvent extraction procedure is practiced for the recovery of phenols from concentrated industrial waste-waters. The criteria for choosing the appropriate solvent for extraction are: (i) low water solubility, (ii) no emulsion formation with water, and (iii) must be easily regenerable. Among many solvents which include octane, mixtures of benzene with butylacetate and isopropyl, isobutyl acetate and isopropyl ether two are extensively used for dephenolising waste-waters. Solvent extraction process can typically reduce phenol concentrations from as high as 17,000 to 920 mg/l. Dephenolised waste-water will have to be treated by other suitable methods for further removal of phenols.

Table 14.5. Industrial sources and concentrations of phenol.

Industry	Concentration, mg/l
Coking plant	
Weak ammonia liquor without dephenolisation	58010.0
Weak ammonia liquor after dephenolisation	4–332
Wash oil still wastes	30–150
Oil refineries plant	
Sour water	80–185
General waste-water	10–100
APF separator effluent	0.3–6.8
Petrochemical plant	
Benzene refinery	210
Tar distillation	300
Nitrogen	250
Plastic	600–2000
Phenolic resin	1600
Fibreboard	150
Fibreglass manufacturing	40–400
Aircraft maintenance	200–400

Adsorption method can be used for the removal of phenols from contaminated drinking water sources as well as for waste-waters that contain moderately low phenol concentrations. It is a separation process in which phenols are transferred from aqueous phase to the surface of adsorbent. Extensively used adsorbent is activated carbon. Commercially available activated carbons are derived from natural materials such as coconut shell, wood charcoal, which are carbonised at controlled conditions so as to have precise surface properties and high surface area.

These materials are costly and need to be regenerated and reactivated for further repeated use. Generally used-procedure for the regeneration of activated carbon is burning at 600°C under controlled conditions of moisture and oxygen, which is highly energy intensive. About 5–10 per cent material is lost during regeneration.

Other source materials, such as straw, used rubber tyres, fertiliser waste slurry have been tried in an attempt to produce low cost, disposable activated carbons. These materials have exhibited adsorptive properties similar to commercially available activated carbons. Activated carbon derived from fertiliser waste slurry has been successfully used for the removal of phenolics from synthetic as well as actual waste-waters from oil refineries.

Chemical oxidation

Many oxidants like ozone, H_2O_2 and permanganate have been used for the chemical oxidation of phenols. It has been reported that ozonisation can result in structural modifications of pollutants, which make them amenable for biodegradation. Requirement of around 7.5 kg of ozone per kg of phenol present is reported for the oil refineries waste-waters, if effluent quality of 0.2 mg/l is to be achieved. Reaction between phenol and hydrogen peroxide is slow under non-catalysed conditions. However in the presence of Fe(II) ions H_2O_2 oxidises phenol to hydroquinone and catechol, which are further oxidised to quinones, carboxylic acid and finally to CO_2. Fenton's Reagent [H_2O_2 = Fe (II) Catalyst, pH 2–4] is reported to be the cheapest oxidation system as compared to ozone, chlorine dioxide and potassium permanganate. If the waste-water contains phosphates, catalytic action is reduced and the oxidation of phenols cannot be achieved. Phenolic effluents from paint and, pharmaceutical manufacturing have been treated with Fenton's Reagent.

Enzymatic treatment

Enzymatic approach for the treatment of phenolic waste-waters is now attracting wide attention. Use of enzymes which include peroxidases, tyrosinase and laccases are now being explored to convert soluble phenolics to insoluble polyphenolic precipitates, which can then be removed by filtration.

Although enzymatic conversion of phenols to quinone is very rapid, nonenzymatic formation of oligomers and polyphenols is a very slow process and may require several hours. Unless phenols are converted to insoluble polyphenols, they remain in solution leading to serious practical limitations of this process. Recently use of Chitosan along with the enzymes to specifically adsorb quinones has been reported.

Chitosan is prepared from Chitin, which is an abundantly occurring natural polymer. This is present in shells of crustaceans such as crabs, prawns, etc. and is a waste product from marine food industry. There has been a considerable interest in developing chitosan based products for various applications, because of the problems associated with the disposal of Chitin rich shell. If the enzymatic approach is to be cost effective and applicable for drinking water, enzyme has to be repeatedly used and the treated water has to be free from the enzyme protein. To achieve this, the enzyme has to be immobilised on solid supports such as resins to facilitate its easy separation of the enzyme from treated water.

Advantage of the enzymatic approach are the broad specificity of the enzymes, which enable them to react with a wide range of phenols, suitablility, for dilute waste-waters and less sensitivity to the operational upsets. Present studies on the use of enzymes have been restricted to the laboratory. Although the method appears to be feasible, presently used enzymes are still costly. Thus field applications of the enzymatic method depends on the availability of cheaper microbial source.

Biological methods

Biological treatment processes are preferred over abiotic methods as they lead to the complete mineralisation of phenol to CO_2 and other inorganic components. Physical and enzymatic methods convert phenol from one form/phase to another, which still requires further treatment. Biological methods can be classified as: (i) aerobic, (ii) anoxic, and (iii) anaerobic depending on the environment present in the treatment system. Generally biological processes are used after dephenolising the concentrated phenolic waste-waters by solvent extraction. However, there are also a few reports on the application of anaerobic and anoxic biological treatment processes for treating concentrated waste-water.

Aerobic biological methods

Several micro-organisms can utilise phenols as a source of carbon and energy in the presence of oxygen. It is now well established that the presence of two phenolic hydroxyl groups is a prerequisite for microbial aromatic ring fission. Monohydric phenols, are converted to dihydric phenols in the presence of monooxygenases, which require molecular oxygen. Ring cleavage of dihydric phenols in the presence of highly specific dioxygenases leads to the formation of various aliphatic intermediates finally yielding CO_2. Complete pathway for the aerobic biodegradation of aromatic compounds has been elucidated, which have been reviewed by many researchers.

Aerobic biological methods have been extensively used for the treatment of phenolic waste-waters. Among the various processes available, activated sludge (ASP) has been the method of choice for many carbonisation industrial waste-waters. Of the several possible versions, the single stage completely mixed ASP has been generally used. It has been reported that an efficiency of 98–99 per cent reduction in phenols could be achieved when the average level of phenol in waste-waters is around 500 mg/l. It has been reported that ASP could withstand influent phenol concentration of 1000 mg/l and produced an effluent of 0.5 mg/l, when operated at 24 hours HRT. Pilotplant studies with simulated and actual (low temperature carbonisation plant) waste-waters have also yielded similar results. ASP, oxidation ditch and biological filters are being used in India to treat coking plant effluents have assessed the feasibility of modified rotating biological contactor with polyurethene foam attached to the disc as the porous support medium for microbial growth for treating petroleum refinery waste-waters. Ammonia nitrogen and phenolic removals were reported to be 99 and 85 per cent respectively. Major disadvantages of aerobic biological methods are the high energy input for the constant supply of air and large volumes of sludge production which require further treatment. These have led to the increased interest on the application of alternate biological processes for the treatment of phenolic waste-water.

Anaerobic biological methods

In the absence of oxygen, micro-organisms adopt a distinctly different pathway for the degradation of phenolics leading to the formation of methane and CO_2. Evan has reviewed the probable pathway for anaerobic biodegradation of aromatic compounds. Conversion of phenol to methane requires the coordinated participation of at least three groups of bacteria. Phenol is converted to acetate and hydrogen by phenol degrading heterotrophic bacteria, which are then converted to methane and CO_2 by two specific groups of methane forming bacteria.

Traditionally anaerobic biotechnology was used for stabilisation of sludge generated from aerobic treatment plants and concentrated animal wastes. Basic insight into the anaerobic process and the introduction of novel retained biomass reactors have led to the application of this process to domestic as well as industrial waste-waters, especially because of its advantages over aerobic treatment processes. They include low sludge production, no energy intensive aeration requirement and the production, of high Btu methane gas as one of the products. Most of the available literature on the application of anaerobic process to phenolic waste-water treatment are still from either laboratory or on pilot plant studies. Anaerobic filters, with solid supports such as rasching rings, anthracite and granular activated carbon (GAC) for microbial growth have been used for the treatment of phenolic waste-waters. It was observed that performance of biofilters with GAC was much more superior as compared to other supports. This is ascribed mainly to its adsorptive properties. Recently there have been reports in development of anaerobic granular sludge with phenolic waste-water.

Laboratory studies with anaerobic filters using carbonisation, steel plant, phenol synthesising plant, petrochemical and aircraft waste-waters have been conducted. Results show that phenolic waste-waters can be treated with a removal efficiency ranging from 70–90 per cent depending on influent concentration and the type of waste-water.

Demerits of anaerobic biotechnology are longer startup period, requirement of constant monitoring of process parameters and nonbiodegradability of some substituted phenolics.

Anoxic biological treatment

Nitrate respiration is used in the absence of oxygen by a few facultative heterotrophic anaerobic bacteria. Various organic compounds can serve as electron donors and nitrate is used by the organisms as terminal electron accepter leading to the formation of gaseous end products such as nitrogen and oxides of nitrogen. This process of denitrification is termed as 'anoxic' instead of 'anaerobic', because principal biochemical pathways for energy release are not anaerobic, but only a modification of aerobic pathway. Many phenolics are biodegraded in anoxic conditions. Aromatic ring cleavage, however follows a pathway similar to anaerobic process because of the lack of oxygen in the environment.

Anoxic or denitrification technology has always been used for the treatment of nitrogen rich wastewaters. Methanol is generally used as the source of organic carbon.

It is only recently that application of this process for the removal of organics is being explored. Major advantages of anoxic process are: (i) high concentrations of soluble electron accepter (NO_3^-) can be maintained in the reactor; (ii) denitrifying organisms are less sensitive to environmental inhibitory compounds as compared to anaerobic microbial consortia; (iii) formation of innocuous end products like nitrogen; (iv) maintaining conditions required for denitrification is easier as these organisms tolerate small amounts of oxygen; and (v) many phenolics are degraded under anoxic conditions.

Laboratory studies on the removal of phenols using anoxic filters have shown that high removal efficiencies (> 90 per cent) can be achieved using this process. Evaluation of performance of a laboratory scale upflow anoxic sludge reactor in a study has shown that a space loading rate of 13/g/l/d, more than 90 per cent removal could be achieved with an influent phenol concentration of 600 mg/l.

Several studies with coke oven waste-waters have shown the feasibility of application of anoxic process for the treatment of phenolic waste-waters. As few phenolic industrial waste-waters, especially from steel mill coke oven, are rich in nitrogen anoxic treatment of these waste-waters seem to offer an attractive option. Feasibility studies at pilot plant scale with other stimulated and actual industrial wastewaters are still needed prior to field application.

Thus appropriate removal methods for phenolics waste-waters depends upon the chemical nature of phenol and concentration. Solvent extraction is the most suitable method for waste-waters having high phenol concentrations. Dephenolised waste-water will have to be further treated by suitable methods to meet the regulatory standards. For contaminated drinking water and dilute waste-waters containing mainly phenolics, activated carbon adsorption, chemical oxidation as well as enzymatic methods are suitable. Choice of the method will be determined by the cost of treatment. It is reported that, anaerobic-aerobic sequential treatment may be more successful in the removal of certain phenolics such as chlorophenols.

Use of adapted or mutant-cultures instead of activated sludge for the improved treatment of phenolic waste-waters has been reported. These cultures, are obtained by enriching and adapting naturally occurring bacterial strains to increasing concentrations of phenols. This can be followed by induced mutation to further improve the biodegradability characteristics. Large scale production of these strains is then carried out and marketed under specific trade names. One such example is Phenobac, developed for phenol degradation. Seeding the waste-water with these cultures, considerably reduces the response time for achieving the treatment efficiencies.

SOLID WASTES POLLUTION IN PETROCHEMICAL INDUSTRIES

Solid wastes differ from other pollutants. Unlike air or water pollutants, solid wastes are yet to have seen any unpredicted growth of public concern. However, more and more concern on solid wastes too is anticipated in the coming years. Solid wastes differ from air and water pollutants in another sense too, viz. the solid wastes remain at their origin points till such a time a decision is taken by those concerned to collect and dispose them off.

Solid wastes in petrochemical industries can be in different forms: (i) toxic solids, (ii) non-toxic solids, (iii) actual solids, (iv) semi solids, (v) combustible solids, (vi) noncombustible solids, (vii) degradable solids, (viii) nondegradable solids, (ix) biodegradable solids, (x) organic solids, (xi) inert solids, (xii) dry solids, (xiii) aqueous solids, and (xiv) nonaqueous solids, etc.

The means of collection and methods of disposal of these differ from one to the other are listed below:
1. Recycling.
2. Chemical conversion.
3. Incineration.
4. Pyrolysis.
5. Landfill.
6. Ocean dumping.
7. Lagooning.
8. Salvaging.
9. Biological treatment.
10. Reclamation.
11. Open dump burning.
12. Landforming.

Basic Factors to be Considered

The following methods are applicable either alone or in combinations. Before reviewing the basic factors that are required to be considered in depth before attempting to design the relevant method of storing collecting and disposing the solid wastes, a look at the various types of such wastes will be of great interest. Some of them are: (i) lunchroom wastes, (ii) catalysts, (iii) organic chemicals, (iv) wastewater treatment sludges, (v) filter cakes, (vi) viscous solids, (vii) ashes, (viii) fly-ashes, (ix) incinerator residue, (x) plastics, (xi) ferrous metals, (xii) nonferrous metals, (xiii) water treatment sludges, (xiv) inorganic chemicals, (xv) polymers—pellets, dust, powder, (xvi) off-specification products, and (xvii) contaminated products, etc.

Considering the wide varieties of types, forms, etc. as stated above, the major factors considered while designing the facilities are discussed below.

Physical characteristics
(i) Volume, (ii) weight, (iii) density, and (iv) handling problems.

Chemical characteristics
(i) Combustibility, (ii) degradability, (iii) toxicity, (iv) potential value of materials salvaged for recycling, (v) availability of land for disposal, (vi) anticipated future effects on land due to the disposal, and (vii) effect of method on environment.

Cost of process

Wasteless processing probabilities
Almost every petrochemical industry creates some or other type of solid waste that can be categorised in any of the above listed types, etc. For their processes, many petrochemical units need water treatment facilities. The chemicals that are required for satisfactory operation of these treatment units are of

various types and nature. Flocculants, the precipitates they form, etc. too add to the complexity of the solid wastes that are generated. 'Plant trash' is the generic term coined to define miscellaneous wastes that may be found in the manufacturing facility. Trash is considered harmless if they consist only of non-combustibles. Striking examples of such a category are glass, metal scraps, waste wood, floor sweepings, etc. Trash can originate from boilers, incinerators, etc. too. The ash or residues left of fuel (solid) is trash. The incinerated material (such as sludge) may also leave trash. Power generation facility also is capable of inducing trash. The commonly identifiable scrap in any petrochemical unit is the metal chips, parts or portions of discarded, corroded or damaged vessels, tanks, pumps, etc. and are mostly ferrous in nature. This ferrous scrap contains within it the capability of getting contaminated by the exposure to toxic substances generated elsewhere in the plant. Recent day technologies play around catalysts and they have become a must for any petrochemical unit. By definition, catalysts do not take part in the reactions they catalyse and hence is a commodity with that highly beneficial and economic factor of recoverability.

Over a period of usage, they too get 'Spent'. Spent catalysts may be solid, semisolid, etc. Catalysts are generally possessing complexity in their configuration, nature and chemical characteristics. Spent catalysts of solid/semi solid configuration impart challenges in selection of treatment/disposal methods. Production processes that encompass combinations of multiple complexities end up with equally complex solid pollutants. A wide variety of other organic/inorganic chemicals have every chance of entering the waste streams a result of all these complexities.

The family of plastics, viz. polypropylene, polyester, polystyrene, polyvinyl chloride, polyurethane, etc. is present to a high degree in the solid waste group of petrochemical units. These polymers can be identified in any of the following forms: (i) pellets, (ii) chops, (iii) cleanouts, (iv) shreds, (v) dust, (vi) powder, (vii) fines, (viii) granules, (ix) emergency dumps, (x) off-specification products, and (xi) contaminated products.

Not only physical but also chemical processes are made use of to treat waste water streams emanating from any petrochemical unit. Unit operations are available to effect the process via biological routes. Methods like usage of trickling filters, aeration, stabilisation ponds and activation of sludge do work well but not without their own solid waste derivatives. Variance of production techniques or combinations of treatment processes alter the end waste characteristics by a huge variance.

Disposal/Treatment

Chemical treatment

This is a factor often neglected or overlooked as a method of useful waste management in petrochemical industries. Processes that help in chemical conversion of wastes may be enumerated as: (i) stripping, (ii) distillation, (iii) leaching, and (iv) extraction, etc. Some physical characteristics of materials also are made use of in conversion processes of wastes. They are: (i) separation by magnets, (ii) flotation, (iii) hand picking, (iv) jigging, and (v) electrical conductivity, etc.

Saleable products are possible outcomes of adoption of the right combination of selective methods in the correct sequence during waste treatment. Even if such a cleverly worked out strategy fails to derive fruitful products, they can render innocuous materials that are easier to dispose off or are not detrimental to the environment. Some possibilities in this direction, are enumerated below:

1. Organics are hydrolysable to product fuels.
2. Waste cellulose is acetylated to give cellulose acetate.

Recycling

This is a way to set up a wasteless closed circuit flow scheme. Here, a stream is returned back into the process. Recycling helps in more vigorous and effective conversions with savings in materials, energy and safety from objectionable emissions to surroundings. Usage of catalysts and solvents are acquired to the maximum extent possible in this process, when-performed at optimum conditions. Reversible reactions such as the ones represented by:

$$A + B \Leftrightarrow C + D$$

are well effected via recycling. This is more true and proves cheap especially when the cost of the reactants are high or when treatment of residual reactants before disposal, etc. is costly and/or hazardous. The cheaper reagent is used in excess and the costlier one is circulated repeatedly to get more yields of the products. Here, reactor design plays a vital role.

Incineration

This is a combustion process, conducted in controlled conditions, for burning solid wastes (in addition to the liquid and gas phase wastes), and hence is a valuable route for solid waste disposal. The common products of incineration are carbon dioxide, water, ash, etc. These are non-combustibles. The wastes are converted to a less bulky, toxic or noxious material. However, this process has its limitations also, in the sense that, compounds containing the following elements must not be incinerated as they will product oxides: (i) sulphur, (ii) nitrogen, (iii) chlorine, (iv) fluorine, (v) bromine, (vi) iodine, (vii) mercury, (viii) arsenic, (ix) selenium, (x) lead, and (xi) cadmium.

Incineration is considered useful where availability of land is scarce and methods other than land disposals are to be used. Following variables, however, have a great effect on completion of proper incineration:

1. Residence time of waste in incinerator.
2. Flammability limits of wastes.
3. Temperature inside the incinerator.
4. Ignition temperature of the wastes.
5. Turbulence in the reaction zone.
6. Combustibility of wastes being incinerated.

Reduction of operating costs of incineration can be envisaged by heat recovery through steam generation. Incineration helps not only reduce one's investment in pollution control equipments but also gives out products which are a very small fraction of the original weight and volume of the input wastes for landfill. More of organic wastes find incineration as their funeral, compared to inorganics. This also constituents a way out for harmful bacterial, viral and toxic constituents. Necessity of support fuels may be considered as the negative aspect of this process, in addition to the limitations of items that can be handled in this process. Several processes are in use for incineration of petrochemical solid wastes, such as: (i) fluidised bed reactors, (ii) dual chamber incinerators, (iii) rotary kilns, (iv) open chamber incinerators, (v) liquid burner furnaces, (vi) hearths, (vii) multiple hearth furnaces, (viii) catalytic chambers, (ix) molten salt incinerators, and (x) pyrolysis units. Factors that tend to limit the selection of incineration as the favoured method of treating solid wastes are given below:

Physical state of wastes

(i) Solid, (ii) semisolid, (iii) ease of handling, (iv) reacting capacity, (v) combustion temperature, (vi) moisture content, (vii) combustion products, (viii) ash content, (ix) combustibility, and (x) possibility of disintegration at elevated temperatures, etc. The incombustibles are the ones that require a support fuel for sustaining combustion and thus incineration. These are generally identified to be due to the following factors: (i) moisture content, (ii) inorganic salts, (iii) halogens, (iv) nitrogen, (v) sulphur, and (vi) phosphorus.

Combustibles are born for easy incineration. They do not need any support fuels and hence their incineration is cheaper, compared to the incineration process of noncombustibles. Functioning of combustibles, during incineration too depends on the following factors: (i) mixing with air, (ii) physical state, (iii) spraying (for liquids only), etc.

Dual chamber incinerators have been designed to combust solid wastes at the rate of 5700 kg/day from petrochemical units. Here one of the two chambers is for vapourisation and the other for combustion of the vapour. The latter is operated at three times higher at a temperature of the first, which operates at 400°C. Mixtures of organics with inorganics pose the toughest task to incineration. This is easy to understand by considering the fact that inorganics may contain water and that inorganic salts form oxides, carbonates, etc. unlike the organics which tend to decompose. Usage of equipments like scrubbers are necessitated.

Hazardous wastes that are given the incineration treatment as a method of disposal include organohalogen compounds, and chlorine, fluorine and bromine derivatives. Presence of halogens depresses the combustion temperatures. The organohalogen wastes are first hydrolysed to form respective acid gas that are water soluble and thus treatable through easy and well known routes like water absorption or alkaline absorption methods. Compounds that may pose explosion hazards (like organo nitrogen compounds) are treated using long residence, low temperature incineration methods. Incinerators can also handle organo phosphorus compounds and sulphur bearing compounds also, of course followed by few secondary treatment methods.

Pyrolysis

While normal incineration demands 50–100 per cent excess air over stoichiometric value, pyrolysis is a zero-air process, as this functions on indirect heating principles only. Temperatures as high as 1650°C are envisaged in an air-free or air starved chamber in pyrolysis. Basically, this resembles processes like distillation or cracking. Pyrolysis wastes are distilled to combustible forms that gets discharged from the furnace, pyrolysis poses comparatively less air pollution than incineration. This process is capable of handling large volumes. Burnable materials are degraded marginally in pyrolysis. The greatest plus point of this technique is the possibility of recovering chemicals or synthesis gas. Partial combustion of pyrolysis gas within the furnace provides the process heat. This results in using the unoxidised portions of the combustible gases as fuel elsewhere and recovering that heat. Only wastes having high calorific value are processed in this route. Problems of clinkering or fusion of ash generally encountered with the incineration process are absent here, mainly because of the application of abnormally high temperatures in pyrolysis. Waste polymers are conveniently converted to fuel oils by pyrolysis. The pyrolysis unit can be a tubular, plug flow reactor and the process adapted is thermo cracking. It is proposed in many experiments that if operating conditions are maintained properly, even other polymers can be converted into various other useful end products. Few of these are: (i) additives for gasolene, (ii) fuel oils, (iii) gaseous fuels, (iv) stain removers, (v) solvents of inferior strengths, (vi) lighter fractions, and (vii) feedstocks for few other petrochemical processes, etc.

Waste sludges of non-autogenous character are often combined with high calorific value materials to facilitate releasing of high energy levels required for pyrolysis. It is quite interesting to note here that few wastes such as spent filter media, shredded wood, torn paper wastes, unusable oils, etc. comprise the family of such 'high calorific value materials'.

Open dumping

Open dumping or open dump burning is an unacceptable method of solid waste disposal and is illegal in many countries. This demands availability of huge area of land which will become ultimately unusable. Contamination of ground water or even surface water, is also likely.

Landfill

This simple technique involves spreading and compacting solid wastes into cells and covering them. This method of 'sanitary landfill' requires low level labour and is low in cost in addition to requiring low degree of expertise and ease of operation. Wastes that are allowed to be dumped must be inert or capable of degradation of microbial attack to harmless products. Production of inflammable gases during this degradation is possible in some cases.

Land forming

Land forming is the spreading of wastes over a vast area wastes in a thin layer and is applicable for semisolids or solids mixed with liquids. On spreading, the liquid either evaporates or seeps down the earth, depending upon the type of liquid, thus leaving the solids for degradation. Predominantly paraffin wastes are given this treatment. To some extent, even aromatic sludges are land-formed. Heavy metal concentrates pose a limitation for this method. Materials having no contamination of the following metals or their salts can be disposed off using this process: (i) lead, (ii) nickel, (iii) manganese, (iv) copper, (v) chromium, and (vi) arsenic.

Lagooning

A method generally found suitable for solids, semisolids and liquid wastes generated by petrochemical industries, lagooning comprises dumping wastes into ponds or pits. Inspite of its advantages (such as easy operation, low labour requirement, low construction cost and low degree of expertise required), this process is inherent of a bad characteristic, viz. if being odouriferrous and unsightly. As degradation of wastes is limited in lagoons, this method is necessarily to be followed by a secondary treatment of concentrates the lagoon will produce. Hazards can be faced in either of the following four forms: (i) corrosive, (ii) ignitive, (iii) reactive, and (iv) toxic. At least in case of these wastes, waste management extends further to:
1. Minimise production of such wastes.
2. Investigate possibilities of using these in industries elsewhere as a desirable material.
3. Reprocessing to minimise end wastes.
4. Separate these from other common and nonhazardous wastes.
5. Treat to give a nonhazardous waste (however high that treatment may cost).
6. Dispose with care, security and caution in landfills.

Various products that produce hazardous waste streams are listed below in their descending order of quantity generated in major petrochemical units: (i) atrazines, (ii) lead alkyls, (iii) furfural, (iv) explosives, (v) perchloroethylenes, (vi) methyl methacrylates, (vii) epichlorohydrin, (viii) parathion, (ix) malathion, (x) chlorobenzene, (xi) vinyl chloride, (xii) trifluoroaniline, (xiii) ethanolamines, (xiv) toluene di-isocyanate, (xv) maleic anhydride, (xvi) chloromethane, (xvii) acrylonitrile, (xviii) nitrobenzene, (xix) fluorocarbon, and (xx) chlorotoluene, etc.

It may be hard to believe that many wastes do contain valuable substances that can be extracted from concentrated wastes economically. As pollution control facilities aim at the production of the least environmental effect processes within the constraints of cost, it is mandatory to look for indirect generation of pollutants while selecting such methods that may seem suitable and cost-effective at first sight. Means of avoiding the inadvertent creation of 'indirect pollutants' must also be derived for implementation. Some of industries and pollutants generated are given in Table 14.6.

Table 14.6. Some industries and pollutants generated.

Resins	
ABS	Acrylonitrile
	Aromatics
Acrylic	Cyanide
Latex	Acrylonitrile
	Acrolein
Alkyd	Acrolein
	Aromatics
	Poly aromatics
Epoxy	Aromatics
	Chlorinated compounds
	Phenol
Phenolics	Aromatics
	Phenol
HDPE	Aromatics
PP	Aromatics
PVC	Chlorinated paraffins
SAN	Aromatics
	Acrylonitrile
SBR	Aromatics
UPE	Phenol
	Aromatics
Fibres	
Acrylics	Acrylonitrile
Others	
Cellulose accetate	Isophones
Polycarbonates	Phenol
	Chloroaromatics
	Halomethanes
Polyester	Aromatics
	Phenol
Polystryene	Aromatics

Technologies do exist for tactful and efficient management of any waste. But they may be unaffordable. It is much more simple to assign monetary values to the long-term damage to humanity and mother earth or environment that may result from a poor management of such wastes, than to set aside the process of coining a suitable treatment and disposal method of neglecting the total issue as 'highly costly'. Regulations that are not available from governments or agencies may be another possible reason of not caring for a proper waste management system. But then, there will be no government or agency tomorrow to pose any regulation whatsoever, as we ourselves have not cared to pose self-regulation in spite of our being aware of the fact that tomorrow's world is to be made worthy of living (if not beautified) for our off-springs. After all, what is sown today is to be reaped tomorrow and nothing is costlier than an eroded environment.

Treatment of Polymers

Polymers are generally in the form of fine powder. It is a valuable product. Because of economical consideration, it has been always endeavour of entrepreneur to recover maximum product escaping along with effluent. Moreover most of polymers manufactured in the petrochemical industries has specific gravity very close to oils (hydrocarbons). Hence polymers when comes in contact with oil, emulsified scum is formed which interferes with deoiling process as well as biological treatment. Polymers are removed by under given methods.

Vibro-screen

Screen having very fine mesh size and vibrating mechanisms are used to recover fine polymer powder. Frequent chocking of screen is common phenomenon. Screen requires quite a bit of maintenance.

Side hill screen

This type of screens are used in the paper industry for recovery of fine fibre. Generally screens are inclined at 30° to 45° and normally mesh size of screen is kept at 60, effluent containing polymers is screened through for recovery of polymers. Good recovery is possible and quite economical.

Gravity separators

Polymers being light floats on the surface of water, hence slot pipe arrangement is used to recover polymers from the surface of water and water under flows for further treatment.

CONTROL OF AIR EMISSIONS IN PETROCHEMICALS

Phthalic Anhydride

The major end use for PAN is in the manufacture of plasticisers. Commercially, PAN is made by the catalytic air oxidation of either o-xylene or naphthalene (Fig. 14.5). o-Xylene is obtained from petroleum sources and typically is about 98 per cent pure, with the major impurity being p-xylene. The naphthalene is obtained from coal tar sources, and the purity typically exceeds 95 per cent. The catalyst is a vanadium pentoxide-titanium dioxide mixture containing small quantities of promoters. This active material is surface coated on a support that is inert at the operating temperature levels. The reaction is strongly exothermic.

Air emissions characterisation

Air emission sources from the process include the following:
1. Incinerator or scrubber effluent.
    (a) Oxygenated hydrocarbons.

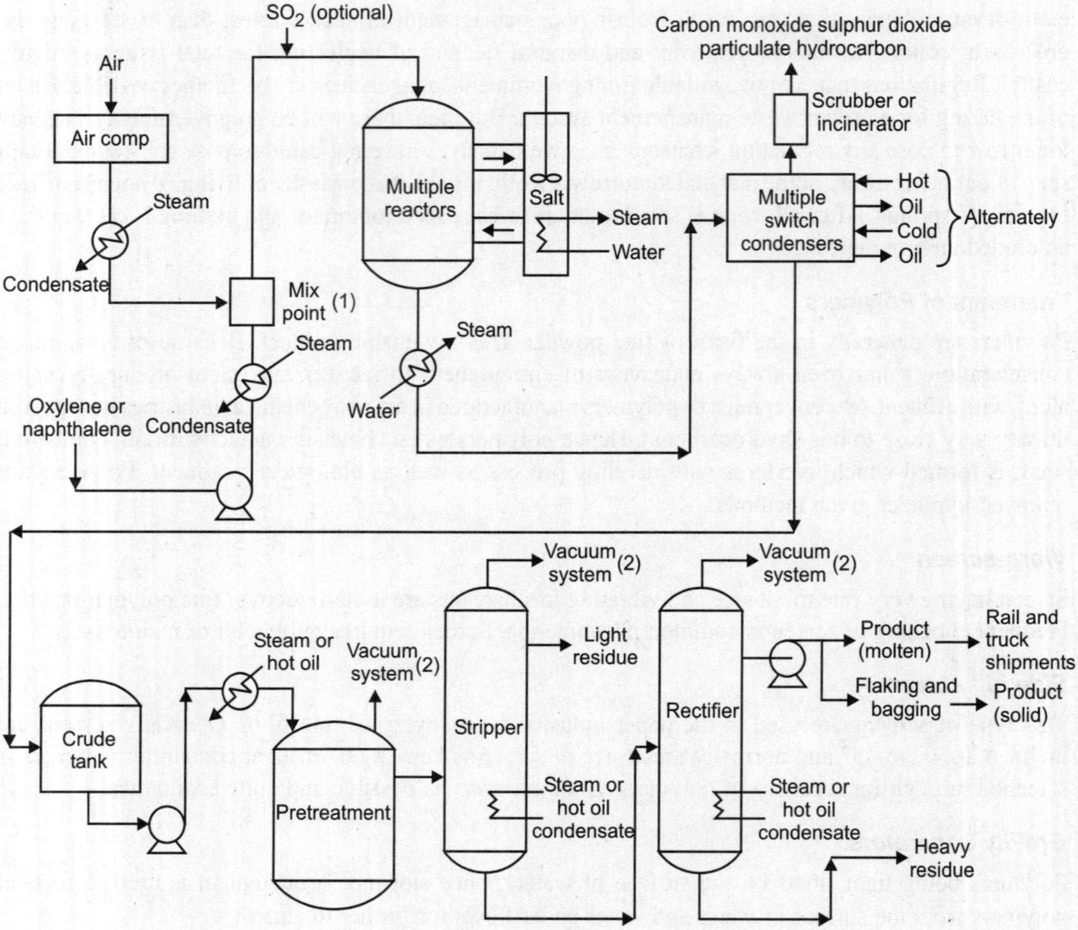

Fig. 14.5. Phthalic anhydride process.

(b) Particulates (Note: most are organic solids that eventually vapourise on exposure to air at ambient conditions): (i) primarily due to PAN, which desublimates when cooled below its freezing point of 268°F, (ii) minor quantities of ash, primarily from corrosion of carbon steel heat-transfer equipment, (iii) in the case of scrubbers, liquid particles from entrainment of the scrubber solution, and (iv) residue particulates if residue is incinerated.

(c) Sulphur dioxide (SO_2) is optionally injected into the reactor as a catalyst promoter: it passes essentially unchanged to the incinerator or scrubber. Also, SO_2/SO_3 is produced by combustion of sulphur containing fuels used in process heaters and thermal incinerators.

(d) Carbon monoxide (CO) is produced in the process reaction, as well as in thermal incinerators and process heaters.

(e) Nitrogen oxides (NO_x are not produced in the process, but are produced in the thermal incineration and in process furnaces from the combustion of the fuel used to provide the heat).

2. Vacuum system vents.
   (a) Steam jets with barometric condensers—similar to scrubber effluent, but with very little CO and maleic anhydride present.
   (b) Vents discharged to a furnace—part of furnace emissions.
   (c) Recycle to switch condensers part of switch condenser effluent.
   (d) If the pretreatment is operated above atmospheric pressure, a vacuum system is not required to vent it to, for example, the switch condenser effluent circuit.
3. Storage tanks.
   (a) Crude and product storage and transfers—particulates (PAN).
   (b) Feed tankage—o-xylene or naphthalene hydrocarbon emissions.
   (c) Heat transfer oils—weathering of light ends if degradation occurs.
4. Fugitive emissions valves, flanges, and so on. Very small quantities are involved because most of the process operates close to atmospheric pressure or under a vacuum. Values are based on heavy liquid because of the low vapour pressure of PAN. Hydrocarbon emissions also occur from the o-xylene–naphthalene feed system up to the reactor catalyst bed.
5. Emissions from flaking operations—particulates (PAN).

Emission factors

1. The 'oxidation of naphthalene,' which was largely based on the older fluid-bed processes, was deleted because PAN is currently produced only in fixed-bed units.
3. For particulates, the recovery systems have been improved with the increased production so that losses per tonne have remained constant. Incineration and scrubbing technology has been improved so that 98 per cent instead of 95 per cent destruction of the particulates is achieved.
4. Only incineration technology the best of current technology. Combinations of scrubbers and incinerators were deleted because it is currently more likely that an incinerator would be used to replace a scrubber than be added to a scrubber.
5. NO_x emissions have been added. The process produces nil NO_x, but NO_x is produced in the thermal incineration and in any process furnaces. NO_x can also be produced in catalytic incineration prior to startup. Values shown are based on fuel gas firing in a thermal incinerator.

Air Pollution Control Measures

The incinerator or scrubber is an integral part of the process because of the large volumes of air used. Although about 99 per cent of the phthalic is recovered in the switch condensers, the remaining phthalic and by-products have to be removed. Water scrubbing recovers the materials for possible sales or for combustion as a liquid solution. Scrubbing has practical limitations in the recovery of the oxygenated hydrocarbons and does not convert CO to carbon dioxide. Therefore, the preferred choice, currently, is either catalytic or thermal incineration. In thermal incinerators, the effluent is heated to about 1500°F by burning additional fuel. Residence times are about 1 second at this temperature, after which the gases are cooled, usually by generating steam. Catalytic incinerators operate in the 500–800°F temperature range and, depending on the amount of organics in the waste gas, may not require auxiliary fuel, except for startup and end-of-run operations. Both types achieve a destruction efficiency of the hydrocarbons and CO of 98 per cent. Increased destruction efficiency can be obtained in a thermal incinerator by increasing residence time and/or temperature. Similar results can be obtained in a catalytic system by increasing either inlet temperature or the amount of catalyst (i.e. residence time in the catalyst bed).

Higher temperature in a thermal incinerator increases NO_x formation and requires more steam generation to remain efficient. Catalytic incinerators are limited on catalyst bed outlet temperatures because of potential catalyst damage at elevated temperatures.

Vacuum systems

Because PAN solidifies at 268°F and reacts with water to form solid phthalic acid, vacuum pumps with water seals normally are not used to provide a vacuum. Instead, vacuum jets are used. The effluent from the jets can be treated in one of three ways: scrubbed with water and the resultant solution treated in a biox system, burned as fuel, or recycled back to the process. In the last case, the noncondensibles in the stream are eventually treated in the switch condenser effluent incinerator or scrubber. Alternatively, condensers can be installed ahead of the jets to reduce the organic levels to the point where further treatment of the jet effluent is not required.

Tank vents and transportation

Vent control consists of two types: desublimators or a vacuum system using jets (as described above and including any vents at elevated pressure that can bypass the jets and go directly to the downstream treatment steps). the simplest type of desublimator is a large sheet-metal box that provides sufficient surface area to cool the vented gases. On cooling, the PAN desublimes from the vapour and collects in the box. Suitable baffles are provided to prevent the vapour from exiting the box until cooled and for periodic removal and recycling of the solid PAN crystals. The vent gas composition corresponds to the vapour pressure of PAN at the exit temperature. Because PAN has a low vapour pressure (0.0035 mm Hg at 100°F), the concentration is low. The total amount of PAN vented will depend on tank filling rates and the rate of the gas blanketing on the tank.

Desublimators mounted on the top of the tank and equipped with heating and cooling facilities also are used. These can be operated to lower outlet temperatures depending on the cooling-medium temperature and do not require manual transfer of the recovered solid Phthalic.

MINIMAL NATIONAL STANDARDS (MINAS) FOR OIL REFINERIES

The considerations for the development of the Minimal National Standards are: (i) quality of treated effluent in refineries having treatment system comprising primary, secondary and biological unit processes; and (ii) low cost of treatment to achieve the quality. It is considered that refineries having once through cooling system should not dispose of the pollutants under the deceptive dilution offered by the cooling water. Therefore, MINAS are stipulated both in terms of concentration and quantum of pollutants worked out on the basis a waste-water generation of 700 kilo litre per 1000 tons of crude refined. The minimal standard for oil refineries are presented in Table 14.7.

Table 14.7. Minimal national standards for oil refineries.

Parameters	Maximum permissible concentration in mg/l	Maximum permissible quantum in kg/100 tons of crude processed
Oil of grease	10	7
Phenol	1	0.7
Sulphide	0.5	0.35
Biochemical oxygen demand	15	10.5
Suspended solids	20	14
pH	6	8.5

Chapter 15

Leather and Tannery

INTRODUCTION

Tanning is the chemical process that converts animal hides and skins into leather. The term hide is used for the skin of large animals (e.g. cows or horses), while skin is used for that of small animals (e.g. sheep). The hide is composed of three layers: epidermis, dermis and subcutaneous. The dermis consists of about 30 to 35 per cent protein, which is mostly collagen, with the remainder being water and fat. The dermis is used to make leather after the other layers have been removed using chemical and mechanical means. In the process of tanning, chemical reactions convert the semi-soluble protein 'collagen' present in the corium of animal skins and hides into tough, flexible and highly durable leather.

TANNING PROCESS

Leather production involves tanning which is a chemical process converting the derma, epidermis and flesh into a stable non-putrescible material known as leather. The process involves cleaning and washing off the dirt, blood, flesh, etc. from raw hides and skins after flaying before preserving with salt till they are transported to the tannery. In the tannery hides and skins are first brushed smoothly manually or in some cases mechanically to remove as much salt as possible. Then these hides and skins are subjected to various processes such as soaking, liming, deliming, fleshing, deliming and washing in case of vegetable or E.I. tanning process. When chrome tanning is done, the process is same upto deliming and then the hides and skins are subjected to bating and pickling before chrome tanning. The tanned leathers are then dyed and finished.

Pretanning Operation

Hides and skins

Skins of cows and buffaloes are called 'hides'. Skins of goats and sheep are called 'skins'. In India, 80 per cent of hides are available from fallen—those that died naturally—cows and buffaloes due to ban on cow slaughter in many parts of the country. Goat and sheep-skins are, however, by-products of the meat industry. Hides are 2 to 3 square metres in size and weight 10 to 20 kg. Skins are smaller in size, 0.4 to 0.5 square metre and lighter in weight around 1 to 2 kg. Slaughter hides and skins contain 60–70 per cent of moisture, which make them liable to bacterial attack which in turn decompose the hides and skins.

Curing

Protective treatment administered soon after the hides and skins are flayed, is called curing. Curing creates an environment for the hides and skins in which the protein destroying organism cannot function.

Its sole purpose is to ensure that the hides and skins are protected during transit from slaughter-house to the tanneries which are generally located some good distance away. It also facilitates storing. In India, curing is done by the following methods: (i) wet salting, (ii) dry salting, and (iii) drying.

In the first process, 30–40 per cent common salt is used on green weight basis to dehydrate the hides and skins. In the second process hydration is achieved by salting and natural drying. The last process achieves dehydration by natural drying without salt.

Trimming and sorting

Cured hides and skins arriving at tannery are trimmed to remove long shanks and other unwanted areas. Trimmed hides are sorted for size and weight and formed into batches ready to undergo further operations in the tannery.

Types of Tanning Process and Their Unit Operations

The various sectional operations of processing raw hides and skins into semi-finished leather and/or finished leather varies from one area to another area and tannery to tannery. From the data collected from about 1000 tanneries functioning in various parts of the country, it is observed that the tanning process can mainly be classified into 5 major types namely:

1. Processing raw hides and skins into vegetable tanned semi-finished leather (raw to E.I).
2. Processing raw hides and skins into chrome tanned semi-finished leather (raw to wet blue).
3. Processing of the raw hides and skins into finished leather by adopting chrome tanning (raw to finishing).
4. Processing the vegetable tanned semi-finished leather into finished leather by adopting chrome tanning process (E.I. to finishing).
5. Processing the chrome tanned semi-finished leather into finished leather (wet blue to finishing).

Raw hide to vegetable tanned semi-finished leather

The various stages of operations in processing of raw hides/skins into vegetable tanned semi-finished leather called East India Leather (E.I.) is shown in Fig. 15.1.

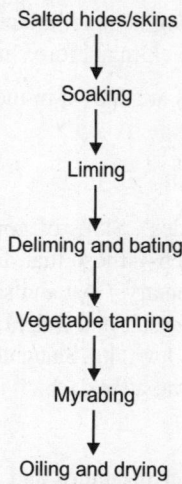

Fig. 15.1. Stages involved from raw hide to vegetable tanned finished leather.

Raw to E.I. process is an old type of tanning mostly carried out in masonry pits. This process is adopted by rural and small-scale tanneries. A few medium and large-scale tanneries, especially in Uttar Pradesh and Bihar area also adopt the vegetable tanning process for manufacturing sole leathers and industrial leathers from cow and buffalo hides.

Raw hide to chrome tanned semi-finished leather

The various stages of operations while processing raw hides/skins into chrome tanned semi-finished leather (raw to wet blue) is shown in Fig. 15.2.

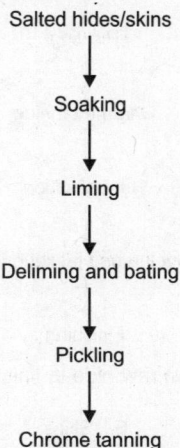

Fig. 15.2. Stages involved from raw hide to chrome tanned semi-finished leather.

Most of the small-scale and medium-scale tanneries adopt this process. These tanneries are mostly feeder units for major mechanised tanneries having facility to process the wet blue into finished leather. Beamhouse operations namely, soaking, liming, deliming and bating are carried out in pits or paddles. Tanyard operations namely, pickling and chrome tanning, are carried out in drums.

Raw hide in finished leather by chrome tanning

The various stages of operation while processing raw hides and skins into finished leather (raw to finish) by adopting chrome tanning process are shown in the Fig. 15.3.

Most of the medium-scale and large-scale existing tanneries having machines, finishing facilities and licence, adopt raw to finishing process. The tanning operations and wet finishing operations are carried out in drums.

E.I. to finished leather

The various operations of processing the vegetable tanned semi-finished leather into finished leather (E.I. to finishing) are shown in the Fig. 15.4.

This process is adopted by the medium and large-scale tanneries mostly for producing leathers for garment manufacturing. The E.I. to finishing operations are carried out in drums.

Wet blue to finished leather

The various operations of processing the chrome tanned semi-finished leather into finished leather (wet blue to finish) are shown in the Fig. 15.5.

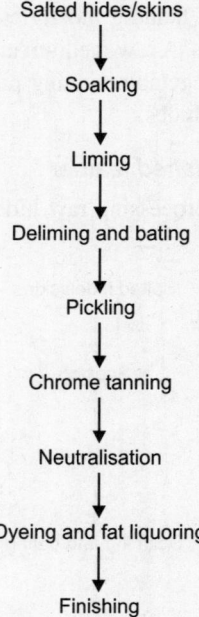

Fig. 15.3. Stages involved from raw hide in finished leather by chrome tanning.

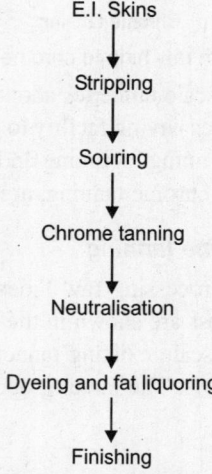

Fig. 15.4. Stages involved from E.I. to finished leather.

Most of the new tanneries who are not having licence and land space for raw to finishing adopt this process by purchasing wet blue skins and hides from the feeder tanneries. The wet blue to finishing operations are carried out in drums.

SOURCES OF WASTES

Of the wet salted hide purchased by the tannery only 30–35 per cent is potentially convertible into leather and of this amount, only some 30 per cent is eventually converted into high quality leather with further 10 per cent yielding lower grade material. The rest of the hide together with excess processing chemicals and the large volumes of water employed form the solid and liquid residues.

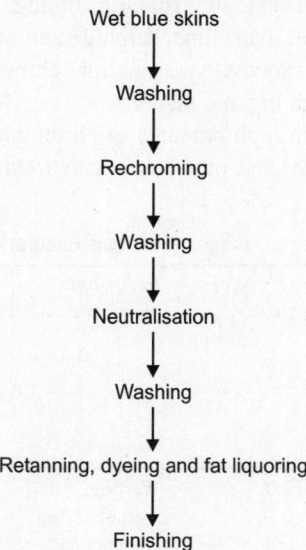

Fig. 15.5. Stages involved from wet blue to finished leather.

The composite liquid waste let out from a tannery contains large amounts of putrescible organic matter and is highly polluting in nature. Although the relatively high pH of the composite wastes tends temporarily to inhibit putrefaction, the waste-waters are inherently putrescible and eventually become highly offensive. Sludge produced during primary or biological treatment of tannery effluents can also give rise to smells.

Liquid Wastes

In the development of the tanning industry, water plays a vital role as the industry consumes large quantities of water. Approximately 30–40 litres of water is used for processing one kg of raw hide/skin into finishing leather. Most of the Indian tanneries which are located near the river banks or natural water bodies draw surface water. Groundwater from their own open wells/tube wells existing within their premises is also used by some tanneries. Most of the traditional tanneries do not have overhead water tanks for proper distribution system. Water is being pumped directly to the process and in a few tanneries, it is stored in open cement lined pits and ground level tanks.

In general, the quantity of water usage and nature of waste-water discharge varies from process to process and tannery to tannery and from time to time. Most of the discharges are intermittent. The average water usage and waste-water discharge per kg of hide/skin for different process are as follows: (i) raw to E.I. 25–30 l/kg of raw weight, (ii) raw to wet blue 25–30 l/kg of raw weight, (iii) raw to finish 30–40 l/kg of raw weight, (iv) E.I. to finish 40–50 l/kg of E.I. weight, and (v) wet blue to finish 20–25 l/kg of wet blue weight.

Most of the tanneries neither have proper drainage system for collection of the waste-water nor any effluent treatment system. The waste-water is discharged from various sectional operations intermittently and it takes its own course to the nearby low-lying area neighbouring land, pond, street, roadside, etc.

Characteristics of waste-water

Characteristics of the effluents vary from tannery to tannery and in any one tannery with respect to time. The waste-water from beamhouse process, viz. soaking, liming, deliming, etc. are highly alkaline, containing decomposing organic matter, hair, lime, sulphide and organic nitrogen with high BOD and COD. The waste-water from tanyard process, viz. pickling, chrome tanning are acidic and coloured. Vegetable tan waste-water contain high organic matter.

The chrome tanning wastes contain high amounts of chromium mostly in the trivalent form. The details of the tanning operations, water and other chemicals used, general constituents in the waste-water are furnished in Table 15.1.

Table 15.1. Details of tanning process, water usage, chemicals used and general constituents of waste-water.

Important operations in tanning process	Mode of operation waste-water discharge in M^3/ton of skin/hide processed	Approx. qty. of water used	Important chemicals	General constituents of waste-water
Soaking	Pits/paddles	9.0–12.0	Wetting, emulsifying agents and bactericidal agent	Olive green in colour, obnoxious smell, contains soluble proteins, suspended matter and high amount of chlorides
Liming	Pits/paddles	2.5–4.0	Lime and sodium sulphide	Highly alkaline, contains high amount of sulphides, ammoniacal nitrogen, suspended solids, hair, pulp and dissolved solids
Deliming	Paddles/pits/drums	2.5–4.0	Ammonium salts, enzymatic bates	Alkaline, contains high amount of organic matter and ammoniacal nitrogen
Vegetable tanning	Pits/drums	1.0–2.0	Vegetable tanning material	Highly coloured, acidic and has a characteristic offensive odour
Pickling and chrome tanning	Drums	2.0–3.0	Common salt, acid, basic chrome salt	Coloured, acidic, contain high amount of trivalent chromium, TDS and chlorides
Dyeing and fat liquoring	Drums	1.0–1.5	Dyes and fatty oils	Coloured, acidic, dyes and oil emulsions
Composite waste-water including washing (raw to finish process)	–	30.0–40.0	–	Alkaline, coloured contains soluble proteins, chromium, high TDS, chlorides, sulphides, suspended solids, etc.

Characteristics of the sectional waste-water

The characteristics of the sectional waste-water from the beam house operations, viz. soaking, liming, deliming, are given in Table 15.2. Characteristics of the sectional waste-water from tanyard operations, viz.

pickling, chrome tanning, vegetable tanning, myrob liquor are furnished in Table 15.3. The characteristics of sectional waste-water from finishing operations, viz. neutralisations, dyeing and fat liquoring are furnished in Table 15.4.

Table 15.2. Characteristics of effluents from beamhouse process.

Parametre	Soaking	Liming	Deliming and bating
pH	7.40–7.90	10.0–12.5	8.10–9.00
Alkalinity as $CaCO_3$	600–1500	8000–15000	1000–1600
BOD 5 days @ 20°C	1000–1600	5000–9000	900–1500
COD	2300–3500	15000–30000	1400–3300
Chlorides as Cl^-	15000–21000	4000–7000	700–1200
Total solids	22000–35000	30000–45000	15000–25000
Dissolved solids	19500–29500	25000–33000	13500–22000
Suspended solids	2500–5500	5000–12000	1500–3000
Sulphide as S^{2-}	–	350–700	25–40

Note: All values except pH are expressed in mg/l.

Table 15.3. Characteristics of effluents from tanyard process.

Parametre	Pickling	Chrome tanning	Vegetable tanning
pH	2.80–3.30	3.40–4.20	3.50–4.50
Acidity as $CaCO_3$	500–1200	4000–5500	2000–4000
BOD 5 days @ 20°C	250–600	400–650	12000–18000
COD	900–2500	1800–4000	25000–35000
Chlorides as Cl^-	20000–35000	11000–16000	1500–3000
Total solids	45000–65000	40000–65000	55000–80000
Dissolved solids	43000–60000	38000–60000	50000–65000
Suspended solids	2000–5000	2000–5000	5000–15000
Total chromium	–	1700–3500	

Note: All values except pH are expressed in mg/l.

Characteristics of the composite waste-water

The characteristics of the composite waste-water is governed by the following factors:
1. Intermittent discharge of waste-water from different sectional operations.
2. Wide variation in the volume and quality of waste-water from section to section.
3. Partial operations in one tannery and balance operation in another tannery.

It would be difficult to arrive at a realistic characteristic range of the composite effluents to be discharged by various tanning units.

However, from the analysis of the waste-water samples collected from various tanneries located in Kolkata and Tamil Nadu region, the general characteristics range of the composite waste-water from raw to finishing process is given in Table 15.5.

Table 15.4. Characteristics of effluents from finishing process.

Parametre	Neutralisation	Dyeing and fat liquoring
pH	4.00–6.50	3.80–4.50
Acidity as $CaCO_3$	300–1200	800–2300
BOD 5 days @ 20°C	400–1000	2200–4500
COD	2000–4000	5000–10000
Chlorides as Cl^-	600–1800	1200–3000
Total solids	7000–12000	7000–12000
Dissolved solids	6000–10000	6200–10000
Suspended solids	1000–2000	800–2000
Total chromium	25–50	Trace

Note: All values except pH are expressed in mg/l.

Table 15.5. Characteristics of composite waste-water from 'raw hide to chrome tanning finished leather'.

Parametre	Concentration range
pH	7.50–8.50
Alkalinity as $CaCO_3$	1100–2000
BOD 5 days @ 20°C	1200–2500
COD	3000–6000
Chlorides as Cl^-	4500–6500
Total solids	17000–25000
Dissolved solids	14000–20500
Suspended solids	3000–4500
Sulphides as S^{2-}	20–40
Total chromium	80–250

Note: All values except pH are expressed in mg/l.

The wide variation of BOD, COD, chromium, sulphide and other parametres exhibited in raw to finish composite waste-water is due to the variation in the process, changes in the type, quantity and quality of chemicals used for the process, fluctuations in the volume of water used for process and washings.

Impact of Liquid Wastes on Environment

Tannery effluents contain vegetable tannins and non-tannins which exert oxygen demand. They also contain high amounts of protein, especially when a hair pulping unhairing system is used. These proteins are biologically degradable and exert high BOD. About 79–80 per cent of the total dissolved volatile solids in the settled and filtered composite tannery wastes is composed of organic matter present in the form of proteins, fatty acids, ether solubles and tannins and 20–21 per cent composed of other organic compounds. Chlorides, trivalent chromium, nitrogen, phosphorous, sulphate, ammonium salts, lime, etc. are the inorganic pollutants present in significant quantities. These pollutants are of more permanent nature, unless they are in suspended state or precipitated from solution and settled. Using normal tanning techniques chromium will be present in the trivalent form which will be precipitated in the mixed effluent. The discharge of untreated tannery effluents effects streams, groundwater, land and sewer in which they are discharged.

The discharge of untreated waste-waters in water courses may affect the physical, chemical and biological characteristics of the water and deplete dissolved oxygen from the water bodies. The high oxygen demand of tannery wastes is due to proteins, fatty matter and tannins. High pH, excessive alkalinity, suspended matter, sulphides are injurious to fish and other aquatic life in streams. Sulphides present in tannery waste-waters can cause unpleasant odour problems, react with iron and other metals causing black precipitate, render the water unfit for industrial uses and affect fish and other aquatic life. The sulphide toxicity to fish increases as the pH value is lowered. Nitrogen and phosphorus from tannery effluents encourage uncontrolled growth of algae and other aquatic plants in water bodies. High amounts of chloride present in tannery waste-waters can make the receiving water less suitable for drinking, industrial and agricultural purposes.

Suspended solids both inorganic and organic present in tannery waste-waters may settle in a stream and affect fisheries by covering the bottom of the stream thereby destroying bottom fauna necessary for fish as food or reduce the spawning ground of fisheries. In general the toxic effects of chromium salts particularly hexavalent chromium salts towards aquatic life varies with species, temperature, valence of chromium and the complex synergistic and antagonistic effects due to other factors such as hardness. The effluent from vegetable tanning is coloured and contain some amounts of non-biodegradable matter. When these waste-waters are discharged into streams it is reported that the colour attributed persisted for a long time. The discharge of untreated tannery wastes into water course may also increase the turbidity of water thereby reducing light penetration and impairing photosynthetic activity of aquatic plants.

Groundwater has been found to be affected where waste-water from tanneries are ponded or lagooned spread out on land or discharged into dry river beds. The groundwater is reported to be rendered unfit for drinking and irrigation where the tanneries are concentrated together. When the tannery waste gains access to cultivable lands or when the lands are irrigated with such waste, fertility of the soil is reported to have been affected. It may change the characteristics of the soil and interfere with the intake of water by plants. When tannery waste-waters are applied to land, the soil productivity decreases.

The problems associated with discharge of tannery wastes into municipal sewers, include incrustation of sewers, sewer clogging and other forms of interference with sewage treatment. Tannery effluents are known to cause deposition of calcium carbonate with the consequent choking of the receiving sewer. Lime is converted to calcium carbonate by the carbon dioxide produced by decomposition of the organic matter present in the effluent and the hair and fleshings help to form a binder with this calcium carbonate which firmly adhere and build up gradually on the surface of sewer. Concrete sewers are likely to suffer damage when they are made to carry sewage containing a high concentration of hydrogen sulphide due to admixture with tannery wastes. If the tannery waste concentration in domestic sewage is higher, the interference with waste treatment operation could be due to:

1. Excessive alkalinity or pH.
2. Hair and fleshings which form scum.
3. Lime sludge and adhering deposits.
4. Higher concentration of sulphides and chromium.
5. Excessive loads of organic matter.

CONTROL OF WATER POLLUTION

Tannery effluents if disposed off without any treatment either on land or in inland surface waters, may create severe problems leading to damage of the environment. Tannin, trivalent chromium, proteinous matter, sulphides and high BOD/COD of the waste-water call for proper treatment.

Tannery wastes in general, can be treated to get the desired end results in the following main stages:
1. In-plant measures.
2. Primary treatment.
3. Chemical treatment.
4. Secondary treatment

In-plant Control Measures

In-plant measures include reduction of water consumption in the tannery, process modifications to reduce pollutional load of the waste being generated, segregation of soak liquor and lastly, recovery of the by-products and reuse of various process liquors.

Reduction of water usage in a tannery

To reduce the volume of the effluent, the water usage in tanneries can be considerably reduced by:
1. Better housekeeping.
2. Alteration of processes and low float systems to use less water.
3. Separation of cleaner fractions of the waste for direct reuse without treatment.
4. Recycle after complete or partial treatment.

Process modifications to reduce pollutional load

The quantities of pollutants which are inevitable in tannery waste-water are as follows:

Hide salt	150 g/kg
Hair protein	40 g/kg
Hide protein	25 g/kg
Hide fat and carbohydrate	15 g/kg
Dirt and manure	5 g/kg
Organic substances from raw	2 g/kg
Organic solids from cleaning and premises	3 g/kg

Thus the total quantity of inorganic pollutants that are inevitable is 150 g/kg of hide or skin processed and the quantity of organic solids is 90 g/kg. In a normal tannery in India the pollutants are more than the theoretical minimum. Therefore, there is a lot of scope for reducing the quantity of pollutants discharged from a tannery by adopting some of the measures, viz.
1. Waste reduction during hide preservation.
2. Waste reduction during beamhouse operations.
3. Waste reduction during vegetable tanning.
4. Waste reduction during pickling and chrome tanning.
5. Waste reduction from post chrome wet process.
6. Reuse of process liquors.

Segregation of salt laden effluents

The high concentration of cationic portion of salts or minerals is reported to exert salt toxicity, thereby seriously interferring with the biodegradation of waste.

Recovery and utilisation of by-products

The various wastes generated by the tannery have great potential for their reuse. Tannery unhairing effluent and acid are the two waste liquids individually difficult to dispose off or to utilise, but if combined together, useful products can be recovered from this combination. These precipitated products have high contents of essential amino acids such as cysteine, lysine, valine, leucine, etc.

Primary treatment

Most of the tanneries in India do not have any treatment facilities. In tanneries where treatment of effluents is carried out, it is limited to the mixing of the various effluents, followed by sedimentation. Even those operations are not carried out satisfactorily. Screening for removal of coarser impurities, hair and fleshing followed by settling for atleast 4 hours in a continuous flow settling tank form the essential primary treatment of tannery waste-waters.

A simple system of three earth lagoons, each capable of holding an entire day's flow has also been studied. They are operated in rotation on the fill and draw principle, filling during one day and discharging over the next 24 hours period, thus leaving the third day available for sludge removal. In a system of this type, removal of 80–90 per cent of the suspended solids and 40–50 per cent of the BOD has been reported. Experiments on sedimentation with controlled pH indicated that pH of 8 is the optimum for sedimentation of the suspended solids, the removal of which in the form of COD is of the order of 45 per cent. Tannin removal was also maximum at this pH although it was only 18 per cent.

Primary sedimentation is highly desirable where subsequent biological treatment of the wastes is proposed in order to reduce the size of the biological stage where space is restricted. Settling is most conveniently effected in specially designed tanks.

These tanks may be of horizontal or vertical flow type. Although more difficult to construct, vertical flow tanks have the advantage of being self-desludging when provided with side sloping at an angle of 60 degrees, sludge being removed via a valve fitted at the base of the tank.

Chemical Treatment

Tannery effluent can be treated by chemical coagulants like alum, carbon dioxide from flue gas, sulphuric acid, ferric chloride and lime. Inspite of the excellent results shown by some of these chemicals, chemical treatments are not widely adopted because of heavy cost and sludge handling problems involved in these methods.

A comparison of capital costs and annual operation costs for pre-treatment systems in class B tanneries reveals that the annual operation cost of system employing ferrous sulphate is substantially lower (0.33 times) despite higher capital investment (1.4 times) as compared to the system employing alum for comparable COD and tannin reduction (55–65 per cent and 52–68 per cent respectively).

Ferrous sulphate is not suitable for coagulation of vegetable tanning wastes, since it forms an ink with the gallic acid in the wastes giving them an intense black colour. If necessary, these wastes are treated with aluminium sulphate, which is much more effective if the wastes are previously acidified.

Experiments with potassium permanganate, alum, carbon dioxide and polymers showed that addition of alum was a cost effective means of reducing suspended solids. Alum doses of 250 mg/l can be utilised to coagulate combined beam and tan wastes and achieve 85 per cent TSS and 62 per cent BOD reduction. The large quantity of lime present in tannery waste-waters provide sufficient alkalinity for floc formation and particle entrainment. The utilisation of passive tube settlers and rapid sand filtration further enhanced solids removal.

Most discrete leather particles and hair are easily removed by simple sedimentation, but proteins, metals and lime can only be precipitated by chemical flocculation. Some of the coagulants and their removal efficiencies attained with tannery waste-water are summarised in Table 15.6.

Table 15.6. Removal efficiencies attained by some coagulants.

Coagulant	Dose mg/l per cent	TSS removal per cent	BOD removal
Ferric sulphate	100	88	77
Anionic polymer	10	90	70
Ferric chloride plus	25	92	75
Anionic polymer	1	–	–
Hexametaphosphate	50	92	–
Magnesium carbonate	50	85	–
Alum plus	100	80	–
Anionic polymer	1.5	–	–

The removal of chromium from chrome liquor could be obtained by adjusting the pH of the wastewater using lime liquor. The optimum pH for effective removal of chromium from chrome liquor is 10.5. The percentage removal of chromium at pH 10.0 is about 96 per cent. Out of various alkalies such as lime, NaOH, Na_2CO_3 and NH_4OH, lime was appeared to be the cheapest alkali for the economical removal and recovery of chromium. Using lime, about 98 per cent chromium removal was possible at a pH value of 6.6. Alum and ferric chloride for the chemical treatment of vegetable tan liquor and also of composite tannery waste gives fairly good removals of COD and tannin.

Secondary Biological Treatment

Low cost technology and conventional treatment systems (both aerobic and anaerobic or in combination) can be employed for the secondary biological treatment of equalised and settled tannery wastes. Tannery effluent can also be satisfactorily treated in admixture with sewage in a sewage treatment plant provided the proportion of tannery effluents is not high. Comparative statement for alternative type of treatment systems is shown in Table 15.7.

Table 15.7. Comparative statement for alternative type of treatment systems.

Description	Conventional aerated lagoon	Activated sludge aerated lagoon process (extended aeration lagoon system)
BOD removal, per cent	70–80	90–98
Detention time, days	3–8	0.5–1.5
Land area requirement, m^2	700–900	250–350
Equipment required	Aerators	Aerators, recycle pumps
Oxygen requirement, g/kg BOD removal	0.6–0.8	1.0–1.2
Power capacity, H.P.	2.0–3.0	7.0–1.0
Nitrification	None	Likely
Sludge handling	Manual desludging once in 5–10 years	No digestion
Maintenance	Semi-skilled operation	Skilled operation

Aerobic Systems

Attempts have been made in treating tannery effluents in an oxidation pond, but its application on full-scale operation is still awaited.

High rate biological filters provide a convenient means of removing a large proportion of the biologically oxidisable matter in a relatively small volume of plant for tannery waste-water. Biological filters filled with stones and provided with recirculation can achieve removals of 70–90 per cent BOD when loaded with rates as high as 2 kg BOD/m^3/day. Effective underdrainage is essential to collect percolated liquor and just as important the free access of air up the tower, as the process is aerobic.

Treatment of settled and 100 per cent diluted vegetable and chrome tannery effluents on trickling filter has been studied in India. The average 5-day BOD value of pre-treated vegetable tannery effluent applied to the filter was 900 mg/l and the BOD of the filter effluent was in region of 158 mg/l at a hydraulic loading of 68×106 l/ha/day or 6 MGAD and without recirculation. It was interesting to note that settlement of the filter effluent for 24 hours further brought down BOD to 56 mg/l. With the chrome tanning effluents, BOD was reduced from 821 mg/l when the filter effluent was settled for 24 hours. It was reported that practically complete removal of sulphides and chromium was effected by this treatment.

Tannery effluents can be treated successfully by activated sludge process. But until recently the principal drawbacks with this process were the long aeration period required for any appreciable reduction in BOD.

Anaerobic Systems

Some of the tanneries are using anaerobic lagoons with success. Experiments on anaerobic lagoon with the settled waste-water gave percentage reduction varied from 42.4 per cent at a BOD loading of 0.41 kg/m^3/day to 85.5 per cent at a BOD loading of 0.14 kg/m^3/day. It was found that ten days detention time would be sufficient to bring BOD reduction of about 85 per cent. If the final effluent has to be discharged on land, it would be sufficient to treat the settled waste in anaerobic lagoons for 10 days. Anaerobic lagoon treatment could reduce BOD by 88.5 per cent but colour removal could not be obtained.

Anaerobic treatment of vegetable tannery waste-water is practically feasible at shorter time using fixed film reactors. The success in treatment is strongly dependent on the high concentration of active biomass retained within the reactor and on provisions of sufficient contact time between active biomass and waste-water. The increased sludge retention enhances the rate of conversion of organic matter to methane and carbon dioxide. The reactors, having different methods of sludge retention investigated for the treatment of vegetable tannery waste-water corroborate the above facts.

Combined Process

For complete treatment to meet the standards of discharge into river, anaerobic treatment followed by aerobic process is necessary. The treatment of raw tannery waste-water having BOD concentrations of 2000–4000 mg/l is not economically feasible to treat by aerobic method due to oxygen transfer limitations. The effluent produced after anaerobic treatment of tannery waste-water has COD and BOD concentrations permissible for discharge into public sewer. The anaerobic pre-treatment with faster conversion rate at shorter time with added advantage of biogas generation will be a better approach to opt before aerobic polishing treatment. In view of the promising results obtained in the above mentioned experiments more systematic and detailed evaluation of anaerobic pre-treatment of tannery waste-water followed by aerobic treatment is necessitated to develop necessary design criteria for full scale treatment facility. To overcome the difficulties of fixed-film reactors, the feasibility of anaerobic-aerobic concept of tannery waste treatment was further studied by the use of an anaerobic contact reactor in series with an aerobic activated sludge.

Anaerobic-aerobic reactor system exhibited the ability to provide 98 per cent removal of suspended solids, 86 per cent removal of COD and 70 per cent removal of sulphides from raw combined tannery beamhouse waste-water without pre-treatment.

The conclusions drawn from anaerobic-aerobic treatment of tannery waste includes:
1. Cost required for primary treatment (chemical cost) can be eliminated by directly taking raw waste after equalisation for anaerobic treatment.
2. Production of methane gas is high enough to be of value as nonconventional energy could potentially benefit tanners to offset some of the energy costs in the tannery.
3. In aerobic reactor, aeration time is much reduced as compared to the time required for direct aerobic treatment of tannery waste-water, thereby reducing electrical energy required for aeration.

Treatment through Non-conventional Methods

Use of non-conventional methods for tannery waste treatment include:
1. Use of water hyacinth.
2. Use of fungi.
3. Sand filters.
4. Agricultural utilisation.
5. Disposal by spray irrigation.

TREATMENT OF WASTE-WATER

Primary Treatment

The primary treatment units principally comprise coarse screens, two numbers of equalisation-cum-settling tanks (used alternately) and sludge drying beds. The settling tanks, are of about 1–2 days capacity each, thereby acting also as equalisation tanks. Alternatively separate equalisation tank and settling tank/clarifier have been provided by a few tanneries.

Depending on the quality of composite effluent, addition of neutralising chemicals and coagulants like lime, alum, ferric chloride, etc. would be required for effective precipitation of chromium and removal of suspended solids in the sedimentation process.

The sludge from the settling tank/clarifier is removed and dried on sludge drying beds made up of filtering media like gravel, sand with supporting masonry structure. For operational reasons, sludge drying beds are divided into four or more compartments. The sludge drying period varies from 4 to 8 days depending upon the type of sludge, atmospheric conditions, etc. The dried sludge from the sludge drying beds can be used as manure or for landfill if it is from vegetable tannery waste. In case of chrome tannery wastes, the dried sludge should be buried or disposed off suitably as per the directions of regulatory and local bodies.

Secondary Biological Treatment

The pretreated effluent needs suitable secondary biological treatment to meet the pollution control standards. The biological treatment units generally adopted by the Indian tanneries are anaerobic lagoon, aerated lagoon, extended aeration systems like oxidation ditch, etc.

Anaerobic lagoon is a simple anaerobic treatment unit adopted by South Indian tanneries. The system is suitable for tanneries located outside town limits and having sufficient land. The depth of the lagoon may be 3–5 metres and the detention time may be 10–20 days depending upon the pollutional load and atmospheric conditions. The lagoon is provided impervious lining to prevent any subsurface infilteration of waste-water. The anaerobic lagoon is an open type digester with no provision for mixing and gas collection. No power is required for this system and its performance has proved to be efficient under South Indian conditions. pH control, sulphate reduction and atmospheric temperature are important factors in the performance of the system. pH in the range of 7.0 to 8.5, sulphates — amounts less than 500 mg/l and atmospheric temperature of 25°–40°C are favourable for anaerobic lagoon treatment.

Anaerobic contact filter and upflow anaerobic sludge blanket (UASB) are other types of anaerobic treatment system for tannery wastes under pilot scale study. These are closed type units made up of RCC or steel. These units occupy less land area since the detention time is about 1–2 days in case of contact filter and about 8 to 10 hours in case of UASB. These systems are found to be efficient for treating tannery effluent combined with domestic sewage. Though the capital cost would be high as compared to anaerobic lagoon, this system can be adopted for partial treatment of tannery wastes when adequate admixture of sewage is possible. About 60 per cent reduction in BOD is reported to be achieved by this system.

Aerated lagoon is a shallow water tight pond of about 2–3 metres depth with a detention time of about 4–6 days. Fixed or floating type surface aerators are provided to transfer oxygen from atmospheric air to the effluent for biological treatment using micro-organisms under aerobic conditions. Many tanneries in South India have provided aerated lagoons as a second stage biological treatment unit and the system is found suitable for treating low organic loads.

Extended aeration system like oxidation ditch is an improved aerobic biological treatment system occupying less land area compared to lagoon since the detention time/capacity would be only about 1–2 days. A few tanneries in Tamil Nadu have provided oxidation ditches as second stage biological treatment. These units require secondary settling and sludge recirculation arrangements. Excess sludge can be disposed off by drying and use as soil enrichener. Extended aeration system has proved to be efficient. The operational and maintenance cost is comparatively high for smaller installations, but the method is preferable for treatment capacity of 150 cubic metres and above per day.

The typical treatment system adopted by the tanneries processing raw hides and skins into finished leather is shown in Fig. 15.6. The typical treatment system adopted by the tanneries processing semi-finished leather into finished leather in shown in Fig. 15.7. The primary treatment units are mostly similar in all the treatment combinations and the secondary biological treatment combinations vary depending upon the mode of disposal and other local environmental conditions.

Many tanneries in Kanpur and in South India which are planning to provide combined effluent treatment plants have provided primary treatment units in their tanneries. Primary treatment comprising equalisation-cum-settling tanks or equalisation tank followed by separate settling tank and sludge drying beds. The pretreated effluent will be collected through a common drainage and treated in a central effluent treatment system.

Tanneries in Tamil Nadu have provided solar evaporation pans for the collection and treatment of soak liquor and particularly first soak liquor when 2 or 3 stage soaking is practised. The segregation of soak liquor is not practised in Kanpur and other North Indian tanneries due to nonavailability of land area and other constraints in installation and operating the solar evaporation pans. Segregation of soak wastes reduce the TDS concentrations in the final effluent by about 15–20 per cent.

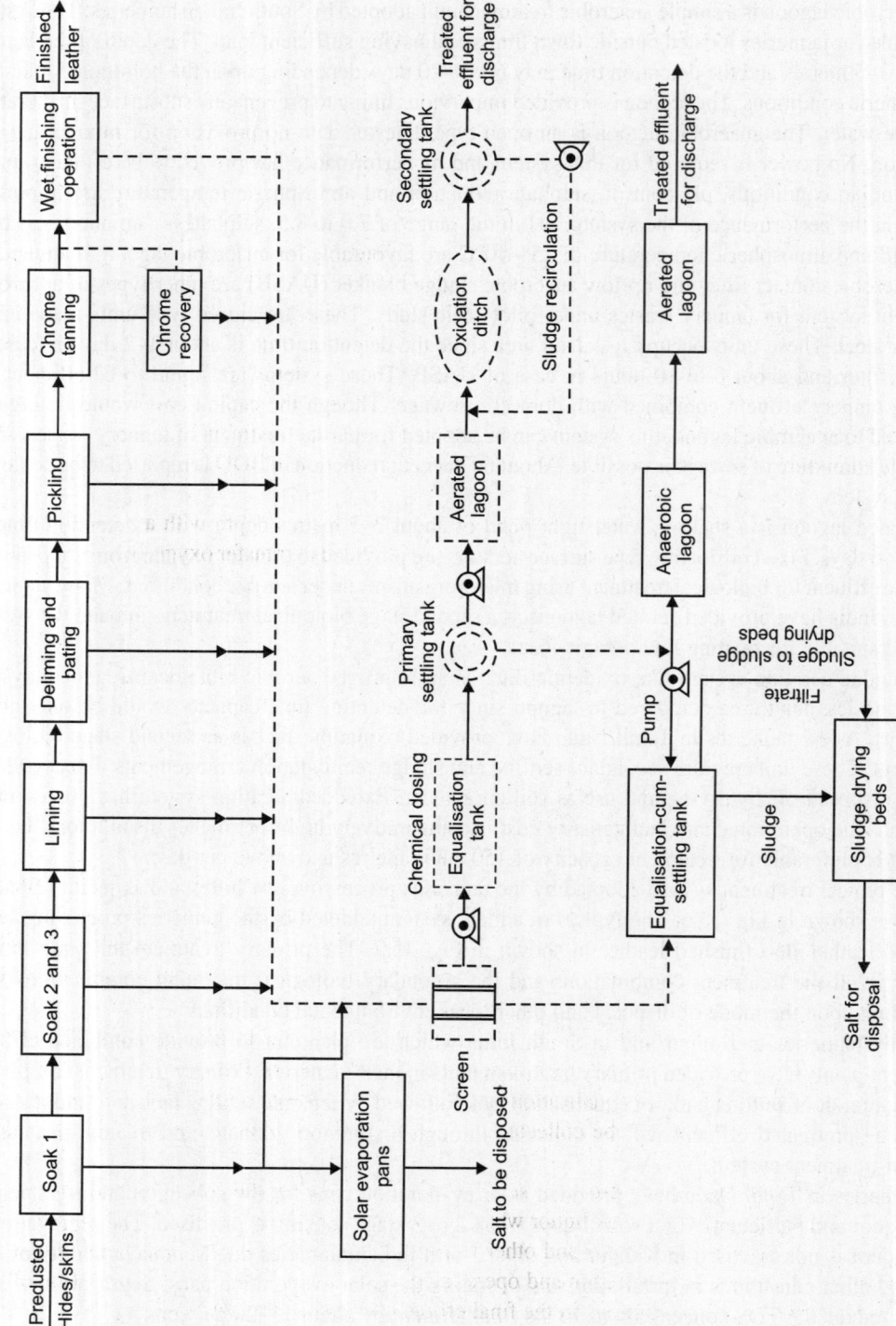

Fig. 15.6. Typical treatment system for raw to finishing tannery waste-water.

Leather and Tannery 313

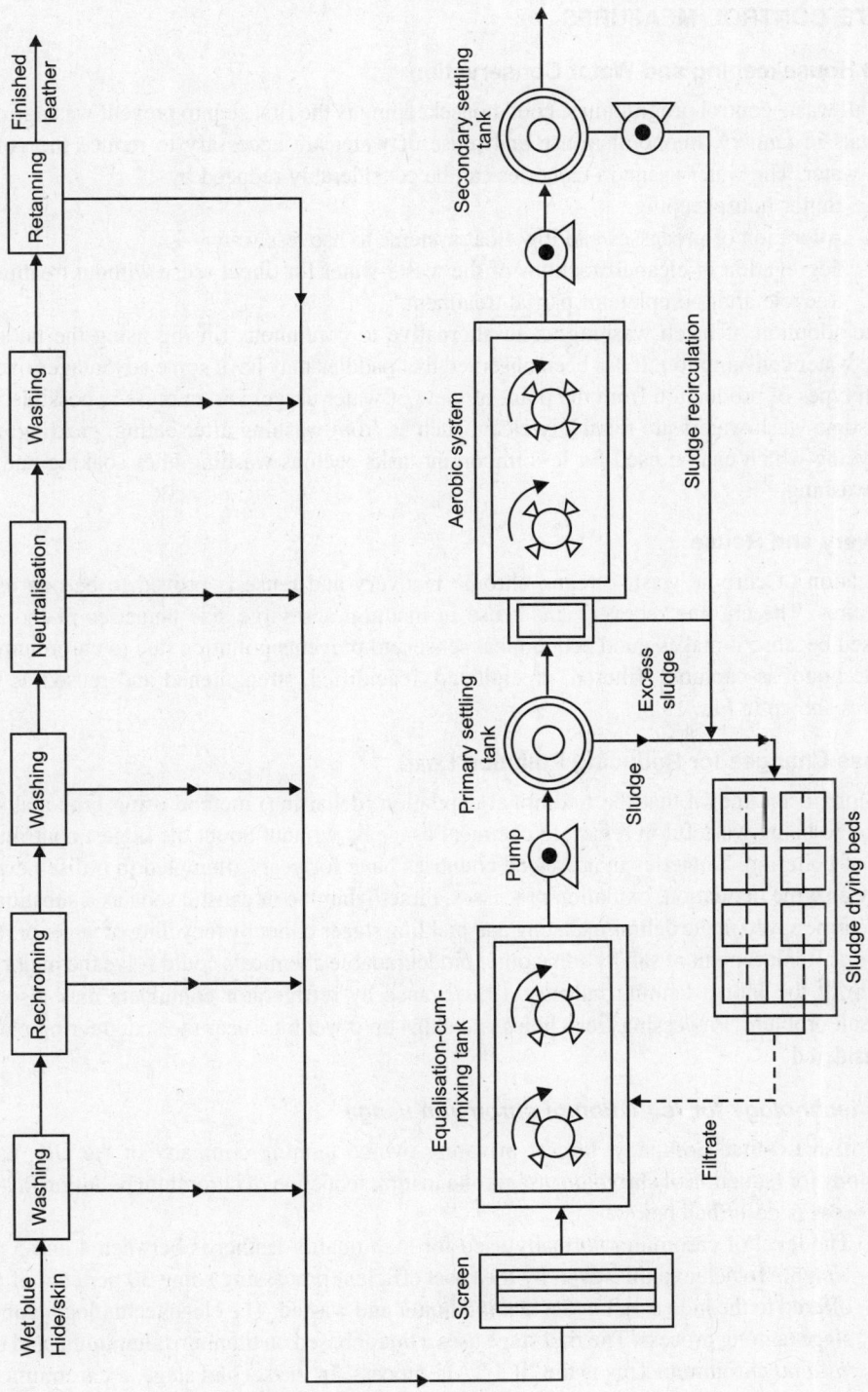

Fig. 15.7. Typical treatment system for wet blue to finishing tannery waste-water.

WASTE CONTROL MEASURES

Good Housekeeping and Water Conservation

As in all waste control programmes, good housekeeping is the first step to prevent wastage of water and materials in tannery. Economical use and reuse of water are necessary to reduce the volume of the waste-water. The water usage in tanneries can be considerably reduced by:
1. Better housekeeping.
2. Alteration of processes and low float systems to use less water.
3. Segregation of cleaner fractions of the waste-water for direct reuse without treatment.
4. Recycle after complete or partial treatment.

The adoption of batch washing as an alternative to continuous rinsing using the lattice door can reduce water consumption. It has been observed that paddles may have some advantages over drums for certain types of production from the point of view of water usage. Direct reuse is possible in a tannery since some wash waters are relatively clean, such as from washing after bating, pickling, neutralising and dyeing which can be used for less important tasks such as washing after soaking and liming and floor washing.

Recovery and Reuse

Segregation of chrome waste stream, chrome recovery and reuse is proved to be one of the viable procedures. The chrome recovery and reuse in medium and large size tanneries in India should be practised because it makes good economical sense and prevents pollution due to chromium. The spent chrome liquor is captured, filtered, precipitated, reacidified, strengthened and reused as the tanning liquor as shown in Fig. 15.8.

Process Changes for Reducing Pollution Load

It has long been known that the traditional depilation (dehairing) method using lime and sulphide, in addition to being wasteful in regard to chemical usage, is without doubt the largest contributor to high levels of pollution. Tanneries in advanced countries have for years attempted to utilise new processes, namely enzyme depilation, oxidation processes, dimethylamine or caustic soda as a substitute for lime. Water can be saved in the deliming, bating and pickling stages either by recycling of water or regeneration of liquors. Replacement of salt by some other biodegradable chemicals could solve the major pollutional problem of the Indian tanning industry. Conveyance by refrigerator containers may also be tried to avoid salt problem. Processing fresh hides and skins upto wet blue near the slaughtering place can also be considered.

Clean technology for reduction of chromium usage

The British Leather Company, largest privately owned tanning company in the UK, uses a clean technology for reduction of chromium usage, and in turn, reduction of chromium pollution in its tanneries. The process is described below:

> The level of chromium normally used for high quality leather is between 4 and 5 per cent by weight. To achieve this, even by the most efficient processing some 30 per cent of the chrome offered to the hide is left in the tanning liquor and wasted. The clean technology employs a two-stage tanning process. The first stage uses a liquor based on titanium, aluminium and magnesium with no chromium. This is the 'ICI-TAL' process. In the second stage, a chromium tan is used

with 90 per cent chromium instead of normal 17 per cent. This results in a leather with a chromium content of about 3 per cent but with characteristics comparable to traditional leather. Residual chrome in the spent liquor is reduced because less chrome is used initially and percentage uptake is greater. The overall effect is to reduce the chromium content of the spent liquor from 1200 to 350 ppm and the level in the final effluent to 10 ppm. The advantages of this process are:

1. Chromium level in the discharge is substantially reduced removing a potential constraint on production.
2. The technology requires no additional capital equipment and can be used with existing plant.
3. Economic benefits—there are modest savings in tanning reagent costs and no new capital investments.

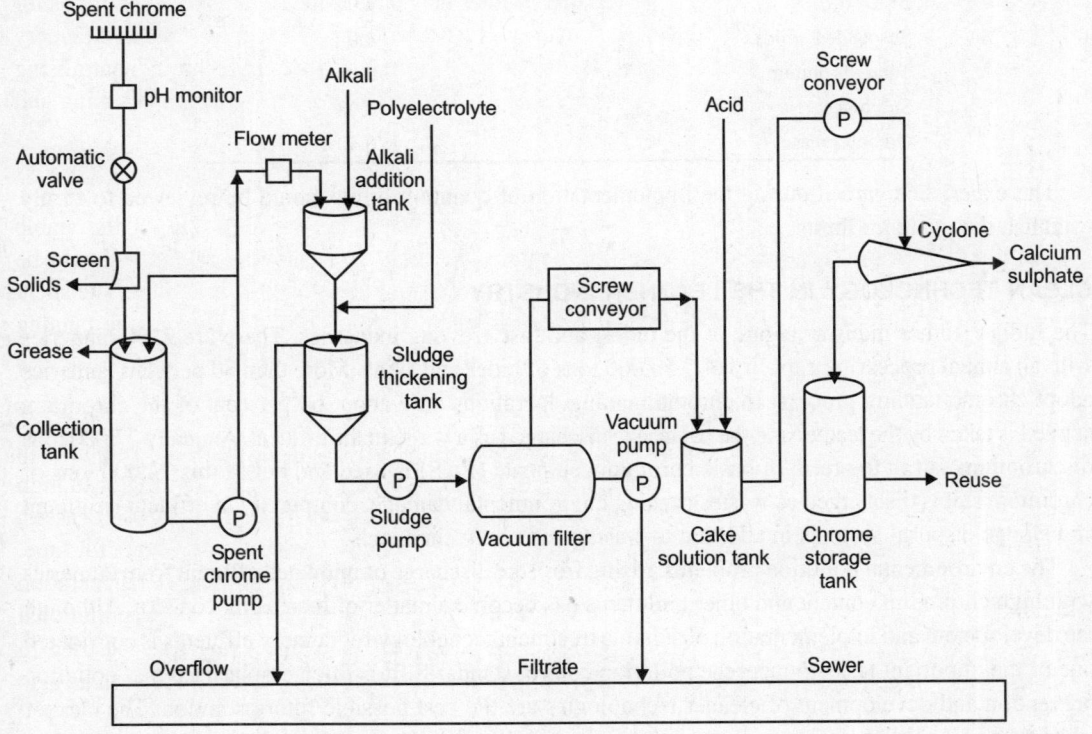

Fig. 15.8. Chrome recovery system.

GUIDELINES FOR QUANTUM LIMITS OF POLLUTANTS IN THE TREATED WASTE-WATER

It is proposed to include quantum limits of pollutants in treated waste-water in the form of guidelines. The quantum limits are not prescribed as standards due to the anticipated constraints in ascertaining the compliance of standards. There are two main factors to establish the quantum of pollutants, as under:

1. The quantity of raw hide processed.
2. The quantity of waste-water generated.

At the time of monitoring, it is not possible to firmly find out the above two factors. One has to wait for end of day or to depend on the data of preceding days.

The quantity of waste-water depends on the different processes and products made by different tanneries. The water and waste-water quantity varies from process to process and product to product as shown in Table 15.1. It is evident from the data that average figures cannot be taken for evolving the quantum limits. Because the industries which are using less water will be encouraged to use more water at the same time, industries which are consuming more water will find it difficult to reduce the water consumption.

In the light of the above, at this stage only a guideline on quantum limits are prescribed. In evolving quantum limits, waste-water generation has been considered as 40 m^3/MT of raw hide processed. Based on the above considerations the pollutant limits are as follows:

Parametre	Limit (kg/MT of hide/skin processed)
BOD	4.0
Suspended solids	4.0
Total chromium	0.08
Sulphides	0.08
Oil and grease	0.40

The experience gained during the implementation of quantum limits, could be reviewed to firmly establish the quantum limits.

CLEAN TECHNOLOGY IN THE LEATHER INDUSTRY

The Indian leather industry is one of the oldest and fast growing industries. There are 2300 tanneries with an annual processing capacity of 5,50,000 tons of hides and skins. More than 80 per cent tanneries adopt chrome tanning process. In chrome tanning operations only about 60 per cent of the chromium applied is taken by the leather and the balance is discharged as a waste in the effluent. Annually 27,000 tons of chromium salt in the form of basic chromium sulphate (BCS) is used and out of this 12,000 tons of chromium salt is discharged as wastes causing environmental damage, complexity in effluent treatment and sludge disposal systems in addition to wastage of costly chemicals.

The environmental pollution problems arising from the discharge of untreated effluent from tanneries with high chromium content and other pollutants has become a matter of increasing concern. Although the development and implementation of suitable treatment technology for tannery effluents is considered one of the important tasks to meet the pollution control standards. It is often emphasised that pollution prevention and development of cleaner technologies are the best possible future solution. The cleaner technologies including chrome recovery and reuse developed and adopted in other countries cannot be duplicated in India without modifications due to the traditional nature of the tanning process adopted, characteristics of the effluent, their discharge pattern, technical manpower capabilities in tanneries, local environmental conditions, etc. Therefore, it has become necessary to develop an appropriate technology to recover and reuse the chrome in Indian tanneries. Central leather research institute (CLRI), Chennai, India in association with M/s Haskoning and TNO/ILS—Netherlands formulated and developed a simple chrome recovery system to suit the local conditions of the traditional industry and pilot plant studies were carried out in one of the commercial tanneries at Kanpur where 160 tanneries are located in a cluster as a part of Indo-Dutch Environmental and Sanitary Engineering Project under Ganga Action Plan.

Characteristics of Waste-water

The characteristics of the waste chrome liquor and overall composite tannery waste-water are given in Table 15.8.

Table 15.8. Characteristics of the waste chrome liquor and composite tannery waste-water (Average value per tone of hides/skins processed).

Parametre	Chrome exhaust liquor	Composite tannery waste-water including chrome exhaust liquor
pH	3.2	8.6
BOD (Total) 5 at 20°C	0.5	55
COD (Total)	2.5	125
Total solids	81	700
Dissolved solids	80	600
Suspended solids	1	100
Chloride (as Cl)	20	150
Sulphate (as SO_4)	10	50
Sulphide (as S)	–	2
Chromium (as Cr)	4.0	4.5

Note: All values except pH are expressed in kilogram.

Need for Chrome Recovery and Reuse

In most of the countries including India pollution control authorities insist that the treated waste-water discharged to sewers or surface water contain less than 2 mg/l of total chromium per litre. Basically it is not difficult to remove chromium from the waste chrome tanning liquor because it is present in trivalent form which is generally insoluble at the pH range of 6–12. By mixing the chromium containing liquor with liming liquor from the pre-tanning operation followed by proper pH control and settling will be sufficient to remove the chromium. Also, during the biological treatment the residual chromium will be precipitated and/or combined with the protein containing sludge in the system.

The disposal of a large volume of chrome containing sludge, which can be estimated to be about 1,00,000 tons per year (i.e. 50 per cent dry weight basis), is a serious problem.

Choice of Chromium Recovery and Reuse System

Basically all types of alkalies such as sodium hydroxide, sodium carbonate or bicarbonate, lime, etc. are useful for chromium precipitation. Most of these alkalies are cheap. The highly reactive alkalies give a voluminous chromium sludge (more than 25 per cent by volume) which makes it necessary to separate the sludge from the liquor by a filter press. Some alkaline (sodium hydroxide) make it necessary to heat the liquor in order to obtain complete chromium precipitation. The use of lime causes a simultaneous precipitation of chromium and calcium sulphate (Plaster of Paris) which makes the reuse of the chromium problematic. Finally two systems were considered for chrome recovery. One with sodium alkalies which need the use of filter presses and the other with magnesium oxide (MgO) which, because of its low reactivity and solubility, causes the chromium to settle in a very compact way, so that separation from the liquor is merely a question of decantation of the supernatant. Dissolving of the sludge can be done instantly with sufficient sulphuric acid to obtain a reusable liquor.

Pilot Plant for Chromium Recovery

Principle of the selected system

The use of the system with MgO as alkali is considered the more appropriate for small to medium sized tanneries because of its relative simplicity and low investment costs. More than 90 per cent of the tanneries in India are small or medium-scale with processing capacity less than 10 tons per day and hence this system was chosen for the pilot plant study.

The process flow diagram of the pilot chrome recovery plant is shown in Fig. 15.9. The chromium containing waste-water including wash water is screened and collected in a treatment tank. A calculated quantity of magnesium oxide is added to the liquor and stirred. During this stirring period the pH is gradually rising to the required value of about 8. After stabilisation of the pH stirring is stopped. The chromium precipitates and settles to a very compact sludge within an hour which is only about 6 to 8 per cent by volume of the exhaust chrome liquor. The supernatant liquor is decanted. The sludge is dissolved in sulphuric acid so that again basic chromium sulphate is formed which can be reused as tanning agent.

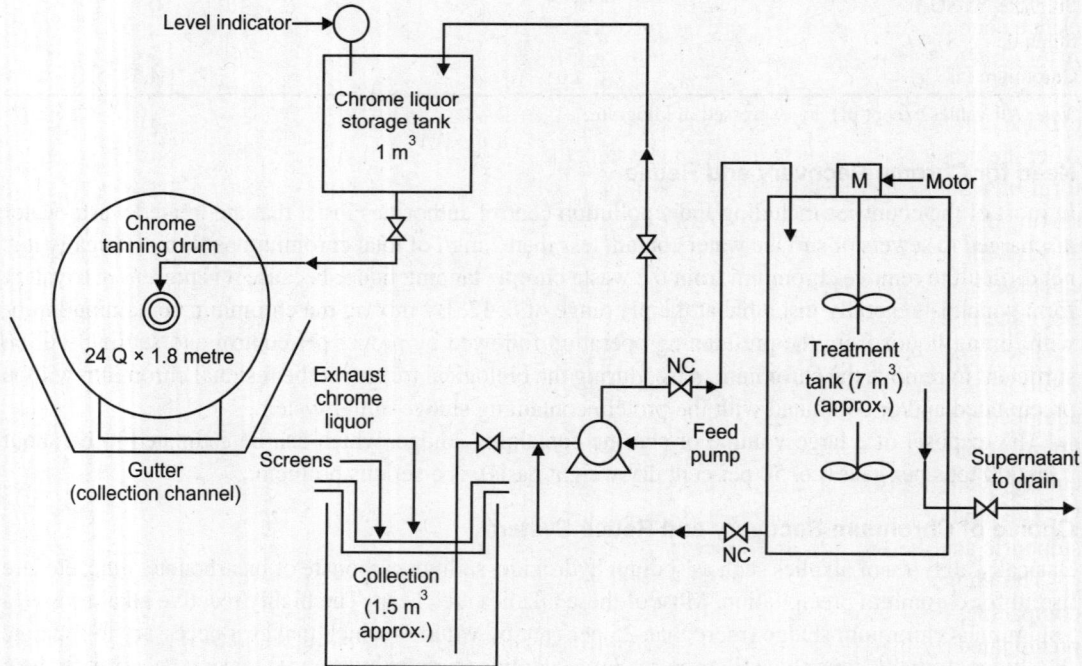

Fig. 15.9. Pilot chrome recovery plant, Kanpur.

The chemical reaction of the chrome recovery is as follows:

$$Cr_2(SO_4)_3 + 3MgO + 3H_2O \longleftrightarrow 2Cr(OH)_3 + 3MgSO_4$$
$$MgO + H_2SO_4 \longleftrightarrow MgSO_4 + H_2O$$
$$2Cr(OH)_3 + 3H_2SO_4 \longleftrightarrow Cr_2(SO_4)_3 + 6H_2O$$

To transform 0 per cent basic chromium sulphate to 100 per cent basic chromium hydroxide 790 grams MgO is needed for every kilogram of chromium oxide (Cr_2O_3).

Description of the pilot plant

The pilot chrome recovery plant has been installed in one of the commercial tanneries at Kanpur. This plant is also used as a demonstration plant for other tanneries in Kanpur area. The capacity of the plant was designed to recover the chromium from the exhaust chrome liquor from one drum initially with a provision to extend further. The same drum was used for tanning with recovered chromium. The capacity of the drum is 1000–1500 kg of pre-tanned hides (Pelt). The quantity of exhausted liquor including wash water was 3000–5000 litre per lot.

In order to collect the chrome liquor separately from other liquors and also to avoid contamination by dirt from the floor, the drum has been provided with an outlet in the form of a round hole of 15 cm diametre. This hole was covered by an easily removable lid. Inside the drum a coarse screen was provided over the hole to prevent the hides from clogging the hole and to allow the liquor to pass through into the special collection gutter. The gutter collects exclusively the liquor flowing from the hole during as well as after drumming. In this way rinsing and washing is possible with open hole. The gutter is connected by a pipe with the collection pit.

The exhaust chrome liquor contains small pieces of hide material (fleshings). In view to avoid accumulation of this material, the exhaust liquor needs to be screened before treatment. In the pilot plant a double screen has been placed on top of the collection pit. The volume of this pit is about 1500 litre. After screening, the liquor is collected in the pit, from where it is pumped into the treatment tank.

The volume of the treatment tank is 7000 litre. It is sufficient to allow for the collection of the exhaust chrome liquor as well as for washing liquors. The tank is made of steel lined with polyethylene. After the liquor is collected in the treatment tank the required quantity of MgO is added manually in the form of a slurry mixed with water. The liquor is then stirred well using a mechanical stirrer. After stirring the chromium is allowed to settle as sludge and the supernatant is decanted. Valves are provided at different heights from the bottom of the tank for decantation of supernatant. After decantation, the calculated quantity of sulphuric acid is added and after short stirring the regenerated liquor is pumped to a storage tank.

The storage tank has a volume of 1000 litre. It is made up of PVC, and is placed on such a level that addition of recovered chromium to the tanning drum is possible by gravity. The storage tank has been equipped with a level indicator readable outside the tank to measure the quantity of liquor added to the drum. Addition of recovered chromium with the required fresh chromium takes place through a pipe ending in a funnel which is connected with a pipe to the hollow axle of the drum.

The regular analysis of the waste liquor is necessary to estimate the required quantity of MgO and sulphuric acid for the recovery process.

The choice of MgO as an alkali to precipitate chromium proved to be suitable because of the compactness of chrome sludge and its easy separation from the liquor without a filter press. The precipitated chrome sludge volume is less than 10 per cent of the waste chrome liquor and the supernatant liquor which is about 90 per cent is decanted. Thus the major amount of dissolved solids and pollutional load are removed along with the decanted supernatant liquor as given in the Table 15 9.

The pilot plant studies reveal that continuous recycling without major variations in the quality of the recovered chromium is possible. It is also established that the procedure involving chromium recovery by precipitation ensure that considerable amounts of waste salts are removed and the reusable chromium is improved. After the study period the tannery is utilising the chrome recovery pilot plant as a regular chrome recovery and reuse system for the entire tannery by making suitable modifications in the chrome exhaust liquor collection drainage.

Table 15.9. Characteristics of the waste chrome liquor, precipitated chrome sludge and decanted supernatant liquor (Average value per ton of hides/skins processed).

Parametre	Waste chrome liquor from each lot	Precipitated chrome sludge using MgO	Decanted supernatant liquor
Volume in litres	1200	90	1110
pH	3.7	8.2	8.3
Chromium in kg	4.0	3.99	0.01
Dissolved solids in kg	80	6	74
Chloride in kg	20	1.5	18.5

RECYCLE, REUSE AND MANAGEMENT OF WASTE-WATER IN TANNERIES

Leather processing has taken roots in India several decades ago. The East Indies Tanning Method (viz. E.I. tanning), had been a popular mode of tanning in India. The growth of tanning industry without giving adequate development of infrastructure has resulted in the industry gaining a 'not so clean image'. Public outrage against the industry is on the rise.

The environmental problems associated with the tanning activity has been offsetting the economic benefits from the leather sector. The industry is in cross roads. It is necessary to seek technological interventions to pollution from this industry. The survival of the tanning sector is now closely circled to its ability to render leather processing not only economically viable but also environmentally sustainable.

WASTE-WATER IN LEATHER INDUSTRY

As already discussed leather making involves the stabilisation of a putrescible collagenous matrix skin, against degradation by micro-organisms and thermomechanical stress. This involves a series of operations as demonstrated with the help of a flow chart Fig. 15.10. Traditionally the leather processing has been grouped as Pretanning, Tanning and Post-tanning — aiming at preparation to tanning, the stabilisation against putrefaction and giving an aesthetic appeal.

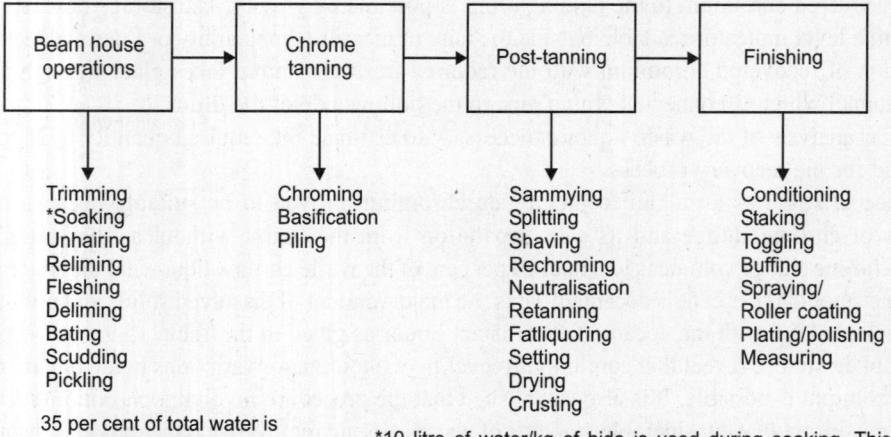

Fig. 15.10. Process sequence in leather processing.

The various chemical inputs into leather processing are demonstrated in the Fig. 15.11. Water forms the main medium of transport for the chemicals. Much of the leather processing steps depend on a large use of water. This explains in part the development of the industry along the river basins. The leather industry employs about 35–55 litres of water per kilogram of hide processed. With the present annual processing capacity of 0.9 billion kilograms of hides and skins in India, it could be estimated that nearly 30–50 billion litres of liquid effluent is generated annually. This gives rise to two major problems for the leather industry, viz. the availability for good quality water and the need for treatment of such large quantities of effluents. This leads to major investments in effluent treatment plants.

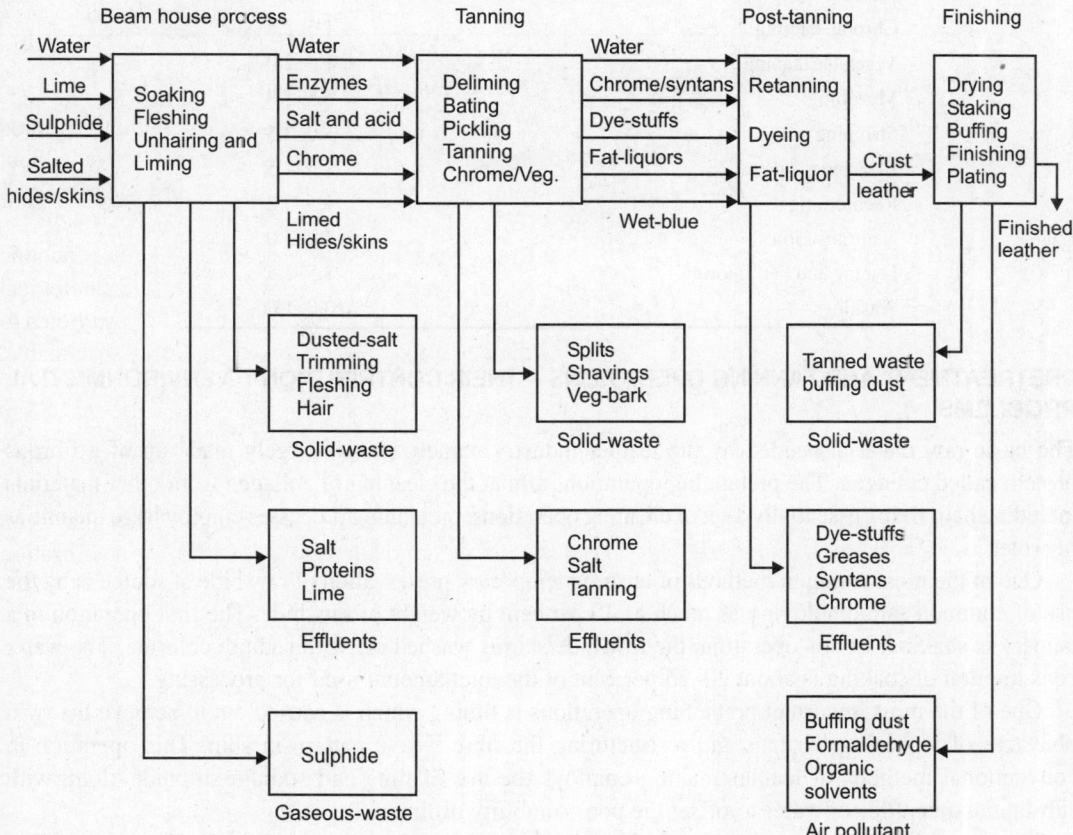

Fig. 15.11. Various chemicals used in leather processing.

The pretanning operations consume nearly 15–20 litres of water per kilogram of hide processed, while the tanning operation consumes 1.5–2.5 litres of water and post-tanning 6–8 litres per kilogram of hide processed. The operation-wise break-up of water usage is demonstrated in Table 15.10. Each of these operations give rise to characteristic pollutant loads. The various pollutant loads from each operation is presented in Table 15.11. Treatment of the large volumes of effluents and their subsequent discharge into sewers, rivers, etc. is only a short-term solution, which adds to compounded problems of investment and recurring expenditure. Other technological options to contain the usage of water, their reuse are to be considered for evolving long-term solutions for the leather industry.

Table 15.10. Quantity of waste-water discharged for 1000 kg of skin/hide processed/range values.

Operation	Quantity cu.m/day
Soaking I	4.0–5.0
Soaking II and III	5.0–6.0
Liming	4.0–6.0
Deliming	2.5–3.0
Pickling	2.0–3.0
Chrome tanning	1.5–2.5
Vegetable tanning	1.5–2.0
Myrobing	0.5–1.0
Stripping	3.0–3.5
Bleaching	2.5–3.5
Rechroming	1.5–2.2
Neutralisation	2.0–3.0
Dyeing and Fatliquoring	2.5–3.5
Washings	10.0–13.0

PRETREATMENT AND TANNING OPERATIONS—THEIR CONTRIBUTION TO ENVIRONMENTAL PROBLEMS

The basic raw material needed by the leather industry namely skin is largely made up of a fibrous protein called collagen. The pretanning operations aim at the cleaning of collagen from other materials including hair. Being essentially a set of cleaning operations, pretanning processes employ large quantities of water.

One of the most common methods of ensuing temporary preservation of raw hide at source is by the use of common salt, employing as much as 40 per cent by weight of raw hide. The first operation in a tannery is soaking. In this operation, the raw hide/skin is washed off with sodium chloride. The water consumption of soaking is about 20–25 per cent of the total amount used for processing.

One of the most important pretanning operations is liming which is carried out to achieve the twin objective of loosening the hair and restructuring the fibre weave pattern in skin. This operation in conventional methods of leather making employs the use of lime and sodium sulphide along with substantial quantities of water to offset the poor solubility of lime.

The next set of operations deliming and bating aim at removal of lime and interfibrillary matter. The liquid effluents contain nitrogenous compounds and proteinous matter.

Pickling operation aims at the preparation of the raw stock for the subsequent tanning step, employing substantial quantities of sodium chloride and sulphuric acid. This is the last of the beam-house or pretanning operation in usual leather processing activity.

Tanning is the operation in which the leather making protein is permanently stabilised against enzymatic degradation. Although there are numerous tanning agents, the commercially most common of all the tanning agents are vegetable tanning and basic chromium sulphate. The liquid wastes from chrome tannery effluent pose a major threat to tanning activity because of the fear of the pollutional hazards of chromium.

Table 15.11. Characteristics of tannery waste-waters.

Parameter	Soaking	Liming	Deliming	Pickling	Chrome tanning fatliquoring	Dyeing (including washing)	Composite
Vol. of effluent in m³/T of hides/skins	6–9	3–4	1–2	0.5–1.0	1–2	1–2	30–40
pH	7.5–8.0	10.0–12.8	7.0–9.0	2.0–3.0	2.5–3.0	3.5–4.5	7.0–9.0
BOD	1100–2500	5000–10000	1000–3000	400–700	350–800	1000–2000	1000–3000
COD	3000–6000	10000–25000	2500–7000	1000–3000	1000–2500	2500–7000	2500–8000
Total solids	35000–55000	30000–50000	4000–10000	35000–70000	30000–60000	4000–10000	15000–25000
Dissolved solids	32000–48000	24000–30000	2500–6000	34000–67000	29000–57500	3400–9000	13000–21000
Suspended solids	3000–7000	6000–20000	1500–4000	1000–3000	1000–2500	600–1000	2000–4000
Chlorides as Cl	15000–30000	4000–8000	1000–2000	20000–30000	15000–25000	500–1000	6000–9500
Total chromium as Cr	–	–	–	–	2000–5000	40–100	100–250

Note: All values except pH are in mg/l.

With mounting pressures from the environmental agencies, the need to contain the discharge of toxic substances as well as BOD, COD and TDS is much felt in recent years. The specifications for the discharge of industrial effluents are given in Table 15.12.

Table 15.12. Specifications for the discharge of industrial effluents.

Important characteristics	Tolerance limits for industrial effluents discharged		
	Into inland surface water	Into public sewers	Onland for irrigation
Colour and odour	Absent	–	Absent
pH	6.0–9.0	6.0–9.0	6.0–9.0
Suspended solids	100	600	200
BOD_5	30	350	100
COD	250	–	–
TDS	2100	2100	2100
Chlorides as Cl	1000	1000	600
Total chromium as Cr	2	2	2
Hexavalent Cr	0.1	0.1	0.1
Sulphide as S	2	2	2
Sodium, per cent	–	60	60
Boron as B	2	2	2
Oil and grease	10	20	10

Note: All values except pH and sodium are in mg/l.

Further to maintain the quality of groundwater, cleaner technological options to contain the pollutants in the tannery waste-waters as well as minimised usage of water in processing are becoming increasingly essential. Thus inplant pollution abatement measures are the need of the hour.

Approaches to Better Water Management

As the water consumption of the pretanning and tanning operations contributes to nearly 50 per cent of the total water used for leather processing, which means about 20 billion litres of water is used annually. The main unit operations which require immediate steps to evolve inplant control methods are:

1. Soaking.
2. Liming.
3. Deliming and bating.
4. Pickling.
5. Tanning.

In order to overcome the treatment cost and availability problem and also to ensure saving of chemicals, water and reduction of pollution load the approaches which are of need are:

1. Recycling and reuse of spent liquors.
2. Optimisation/rationalisation of water required for each operation.

By adopting these techniques considerable amount of water can be conserved in each of the pretanning and tanning operations.

Soaking

Recycling of soak liquor becomes necessary as there is a need to reduce the total consumption of water in leather processing. The primary soak liquor contains too high amounts of dirt and foreign matter. Recycling of this stream is a challenge and calls for removal of dirt and foreign matter and disinfection through irradiation methods prior to reuse. On the other hand the soak II and III can be recycled for the successive batches easily. Countercurrent soaking method can be advantageously employed as shown in Fig. 15.12.

This technique could provide a net saving of 60–65 per cent of the water generally used for soaking operation. In this method hides/skins move in one direction while water flows in the counter direction. With 9 l/kg of hides/skins processed employing countercurrent soaking technique, a saving of 6 l/kg of hides/skins processed is achieved. The analysis of spent soak liquors of normal soaking and countercurrent method are given in Table 15.13. From the table it is clear that no change in effluent characteristics is noticed. The physical properties of leathers obtained by this countercurrent soaking method are comparable with that of normal processed leathers.

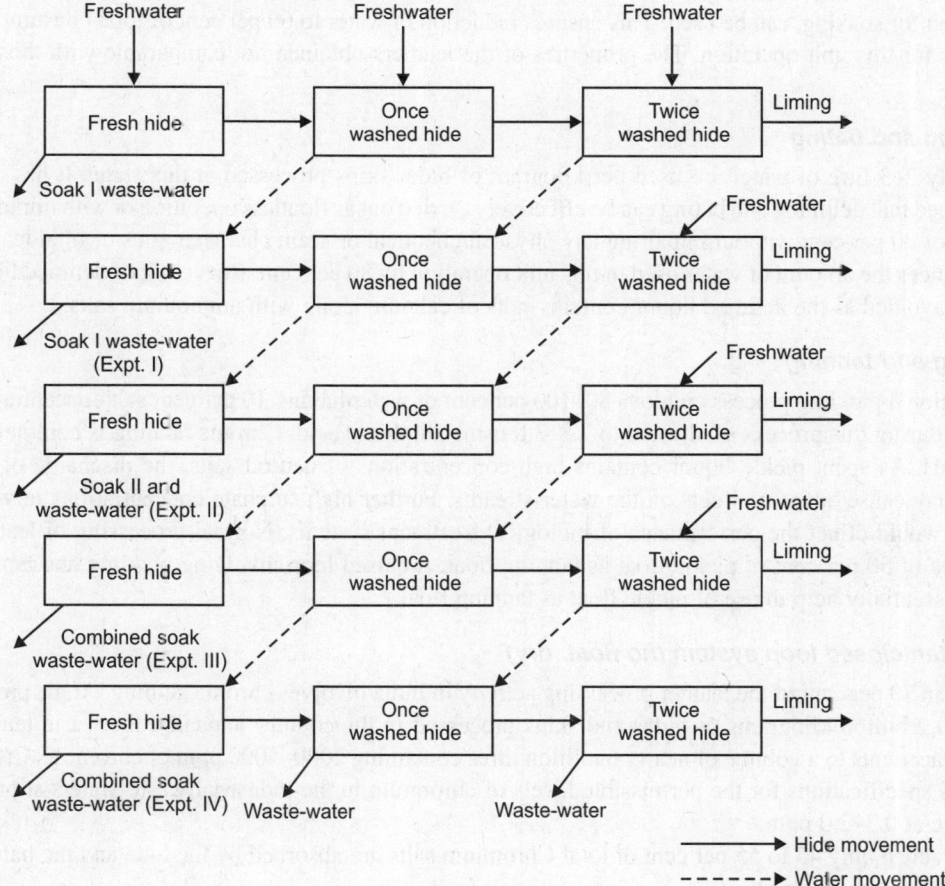

Fig. 15.12. Countercurrent soaking operation.

Table 15.13. Analysis of spent soak liquors.

Properties	Control	Experiment
pH	7.5	7.5
BOD	1383	1325
COD	5417	4898
Total solids	166545	152160
Suspended solids	32215	22255
Chlorides	73174	76694

Note: All values except pH are in mg/l.

Liming: Countercurrent approach

Liming operation utilises large quantities of water. In this operation, recycling of once used lime liquor for the next lot provides reduction in the usage of water. The similar model of countercurrent method as employed for soaking, can be used. This ensures reduction of water to 60 per cent of total consumption of water for this unit operation. The properties of the leathers obtained are comparable with those of control.

Deliming and bating

Generally 2–3 litre of water are used per kilogram of hides/skins processed at this stage. It has been established that deliming and bating can be effectively carried out as floatless operation or with minimum of float of 20 per cent without impairing any physical/chemical or grain characteristics of final leather. This reduces the amount of water used in this unit operation by 80 per cent. Recycling of delimed liquor is to be avoided as the delimed liquor contains salts of calcium along with ammonium salts.

Pickling and tanning

Conventional pickling process employs 80–100 per cent of water having 10 per cent salt concentration. The pH during this process is adjusted to 2.8–3.0 using sulphuric acid. Chrome tanning is commenced at this pH. As spent pickle liquor contains high concentrations of neutral salts, the discharge of this liquor may cause adverse effects on the water streams. Further high sulphate concentrations in waste streams would effect the performance of biological treatment systems. Normal processing of leathers make use of 50 per cent of pickle float as tanning float. A closed loop involving pickling and tanning would essentially help in use of pickle float as tanning float.

Pickle-tan closed loop system (no float, dry)

More than 70 per cent of the leather processing activity in India involves chrome tanning. At the present rate of 0.9 billion kilograms of hides and skins processed in the country annually, the waste tanning streams accounts to a volume of nearly 6 million litres containing 2000–5000 ppm of chrome as Cr(III). The BIS specifications for the permissible levels of chromium in the industrial waste-waters stipulate the range at 0.3–2.0 ppm.

In practice only 45 to 55 per cent of total Chromium salts are absorbed by the hide and the balance goes in the waste as Cr_2O_3.

Chrome recovery/reuse, adopted as an inplant pollution control measure ensures nearly 98–99 per cent chrome recovery and is easy to adopt. However, the supernatant discharged from the recovery plant contains considerable amounts of neutral salts as well as magnesium sulphate which could render the groundwater hard.

A scientific solution to overcome these problems has been to increase the absorption in tanning and minimise the wastage of chromium, if possible to near zero values. One novel method developed and proven successfully in commercial tanneries is the minimum waste high exhaust pickle-tan chrome tanning. This cleaner technological option forms the theme to meet the socio-economic problems in order to maintain the quality of the groundwater.

The novel system employs normal pickling followed by chrome tanning with a combination of Alutan[Aluminium syntan developed by CLRI and commercialised as Balsyn AL by Balmer Lawrie and Company and basic chromium sulphate (BCS)]. The pickling was done with anhydrous sodium sulphate (5.0 per cent) and the pH was maintained at 3.3–3.5 as against normal practice of 7.0 per cent sodium chloride to a pH of 2.5–3.0. Alutan was employed in tanning bath along with BCS of 5.0 per cent instead of 8 per cent BCS generally used in normal chrome tanning. The chrome absorption levels achieved are above 90 per cent during tanning and the spent solution can be reused in pickling in actual tanneries. The composition of spent chrome liquors of normal chrome tanning as well as Alutan-BCS is compared in Fig. 15.13. In other words, the recycling method leads to the reduction in BOD, COD and TDS load in the effluent.

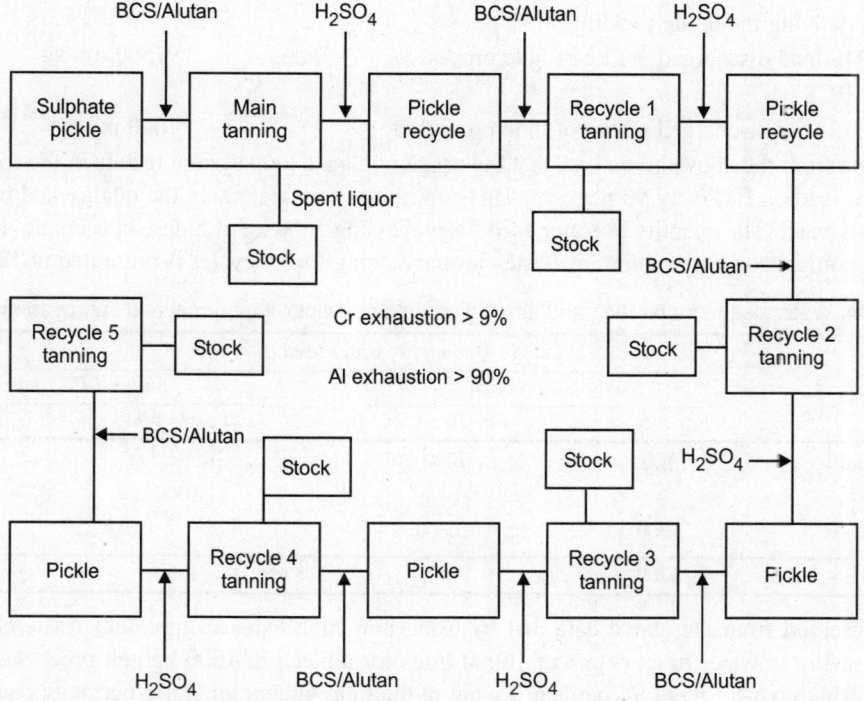

Fig. 15.13. Pickle-tan closed loop system.

The closed loop system involves the recycling of the waste tanning solution in the pickling operation. This ensures no discharge of chemicals and also saving of water and chemicals to a larger extent. No draining of pickle float as well as no additional sodium sulphate salt is added. The flow diagram is given in Fig. 15.13. The recycling could be continued for at least 10 cycles and it was well established in commercial tannery involving the normal production line that by employing closed loop system the quality of the leather comparable to that of normal tanned leather. The BOD, COD and TDS loads on the effluent treatment plants (ETP) generated from normal chrome tanning practice as well as Alutan-BCS closed loop system with 10 recycles are shown below.

BOD load on ETP's
Normal tanning including pickling
    BOD_5 load discharged per kg of hide processed   1270 mg/kg
Alutan-BCS
    BOD_5 load discharged per kg of hide processed   141 mg/kg

COD load on ETP's
Normal tanning including pickling
    COD load discharged per kg of hide processed   6200 mg/kg
Alutan-BCS
    COD load discharged per kg of hide processed   582 mg/kg

TDS load on ETP's
Normal tanning including pickling
    TDS load discharged per kg of hide processed   97500 mg/kg
Alutan-BCS
    TDS load discharged per kg of hide processed   7000 mg/kg

It is seen from the above given data that the usage of closed loop system results in the reduction of the effluent load on ETP's by 90 per cent. This contributes to maintaining the quality and minimising the usage of water. The quantity of water used for processing 1000 kg of hides/skins employing normal process in comparison with minimum waste chrome tanning for 10 cycles is presented in Table 15.14.

Table 15.14. Water usage for pickling and tanning operations employing normal and improved method.

Operation	Quantity of water used (litre)			
	Normal process		Alutan-BCS process	
	For 1 cycle	For 10 cycles	For 1 cycle	For 10 cycles
Pickling (litre)	700	7000	–	–
	–	–	1000	1000
Tanning (litre)	1000	10000	–	–
Total (litre)	1700	17000	1000	1000

It is observed from the above data that by using new high exhaust minimum waste closed loop system a saving in water to an extent of 16000 litres for a batch of 1000 kg pelt processed could be achieved, which means about 94 per cent saving in the total volume of water normally employed for these operations. The actual volume of waste-water to be treated is reduced to $\frac{1}{17}$ th of the total volume which means tremendous saving in the quantity of the water. As the availability of quality water is scarce, the cleaner technologies forms a suitable approach for industrialised nations. The use of Alutan-BCS system for tanning provides also reduction in the usage of posttanning chemicals normally employed, viz. Retanning agents: 30–40 per cent, Fatliquors: 5–7 per cent and Dyes: 10–20 per cent.

This reduced use of chemicals produce better utilisation of chemicals on one hand and savings in the overall manufacturing cost. Thus this leads to the reduction in the total load on the effluent generated. The various additional benefits of the high exhaust minimum waste closed loop system are:
1. Promises near total utilisation of chromium.
2. Promises water saving from pickling and tanning operations by 90 per cent.
3. Promises material economy.
4. Permits reduction in wet finishing chemicals, wattle: 30–40 per cent; retanning agents: 20 to 30 per cent; fatliquors: 5–10 per cent and dyes: 10–20 per cent.
5. Promises reduction of BOD, COD and TDS load by 90 per cent from the identified stages on the ETP's.
6. Affords reduction in the actual volume of effluent treated to $\frac{1}{17}$ th of the total volume.
7. Affords leathers of desired physical properties.
8. Affords fuller leathers in comparison to normal chrome tanning.

Washings

Water usage for washings in leather processing contributes to nearly 20–25 per cent of total water used. This includes material (pelts/leather) washing, floor washings, etc. Proper addition of water to wash the materials and controlled usage for floor washings could reduce the quantum of water used for washings by 30–50 per cent. The continuous washing in leather processing should be controlled and batch type washings carried out. This alone can bring down water usage for washing.

Employing the above mentioned water conservation methods, the amount of water saved in each operation during leather processing is as follows:

1. Soaking : 60–66 per cent reduction
2. Liming : 60–65 per cent reduction
3. Deliming/Bating : 80–90 per cent reduction
4. Pickling and Tanning : about 90 per cent reduction
   Overall Reduction : 72 per cent
   (Refer Table 15.15 for detailed analysis)

Table 15.15. Water reduction in various operations after employing newer approaches.

Operation	Conventional method		Recycle/ optimisation method		% Reduction
	l/kg	Bl/0.9 Bkg*	l/kg	Bl/0.9 Bkg	
Soaking	7.5–9.0	8.1	3.0	2.7	65
Liming	4.0–6.0	5.4	2.4	2.2	60
Deliming	2.5–3.0	2.7	0.6	0.54	80
Pickling and Chrome tanning	3.5–5.5	5.0	0.5	0.45	90
Total	17.5–23.5	21.2	6.5	5.85	72

* Annual processing capacity in India is 0.9 billion kg raw hides/skins (Year 2006).

It is evident from the Table 15.15 there is an overall saving of 72 per cent in the total quantity of water used for pretanning and tanning operations in leather processing. As these are the most polluting streams in leather processing, the reduction in the quantum of water used for these operations will reduce the pollution load on the effluent treatment plant. This can be attributed to the reduction in the quantity of effluent generated. With present annual processing capacity of 0.9 billion kilograms in India, the amount of water used for these operations, employing normal procedure is 21.2 billion litres. This could be reduced to 5.85 billion litres by the application of better water management. The usage of pickle tan closed loop reduces the water usage by 90 per cent. The better water management technologies brings tremendous reduction of water in leather processing.

Thus, the leather processing with recycle/optimisation approach provides a saving in the quantum of water used for pretanning and tanning operations by 72 per cent. The reduction in the usage of water with the present annual production capacity in India is 15.35 billion litres. This quantity is a phenomenal amount, which otherwise would lead to contamination of water reserves. Thus better water management approach in industrial applications play a vital role in the conservation of water for the very survival of the mankind.

GASEOUS POLLUTION

Tanneries discharge into the atmosphere odorous gases, smoke and dust. The main source of smells in a tannery are the compounds containing nitrogen and sulphur. The end products of anaerobic decomposition or putrefaction of proteins include indole, skatole, mercaptans and miscellaneous aldehydes, all of which are odorous.

The other odour producing compounds in tanneries include sulphide, fatty acids like butyric acid, valeric acid and caproic acids, solvents, lacquers, formalin and some of the chemicals used in finishing operations. Smells in tanneries intensify from unhygienic practices in skin and hide processing and delayed disposal of liquid and solid wastes. In many tanneries it is the foul odour which emanate from the putrescible solid and liquid wastes which account for much of the smell traditionally associated with the tanneries. Immediately after the hide or skin is removed from the animals, decay starts unless it is cured properly. The operations like soaking, liming, deliming, fleshing etc., are the most disagreable steps in leather manufacture. They involve use of bad smelling materials and production of putrescible organic matter like soak pit sludge, lime sludge, green fleshings, limed fleshings. During soaking, the removal of curing salt and dehydration of the skin introduces the possibility of bacterial growth and protein putrefaction. Many unhairing systems in practical use are based on balance between sodium sulphide, sulphohydrate, dimethylamine sulphate and sodium hydrosulphide. Some of these sulphides have potential to liberate hydrogen sulphide when mixed with pickling liquors. Hydrogen sulphide is an extremely bad smelling gas which can be detected in concentrations as low as 0.1 ppm. Hydrogen sulphide and organic sulphides cause odour nuisance when present in the air at concentrations of 10 to 150 times smaller than the lowest concentration of sulphur dioxide detectable by smell. There is a possibility of hydrogen sulphide generation during deliming from the sulphides sticking to hide if the limed hides are not properly washed.

Impact of Gaseous Pollutants on Environment

The problems associated with gaseous pollutants from a tannery include noxious smell, hydrogen sulphide and dust. The effect of noxious smells on people is primarily a nuisance effect. The loss of property values near poorly operated tanneries is partly a consequence of offensive odours. Dust problems normally arise in tanneries from leather buffing operations. Leather dust of finer sizes are reported to be harmful to human health and comfort. Hydrogen sulphide which is liberated during some of the tanning operations is an irritant gas and has a very bad smell. The maximum allowable concentration for 8 hours exposure in working areas is 20 ppm.

Control of Gaseous Pollutants

Source control is the most effective means of abating odour. Good sanitation practices are usually cheaper than control measures. Source control of odours can be carried out by following methods given below:
1. Drawing the odorous air from the working atmosphere by exhaust fans and diluting with relatively clean air.
2. Removal of causative impurities from the tannery.
3. Masking the odour with objectionable additives.
4. Removal of odour bearing dusts by cyclone separators.
5. Sorption of odorous gases through a granular sorbents like activated carbon.

Chemical air scrubber can be used to eliminate the odour nuisance in the tannery waste-water treatment plant. Odour masking chemicals can be added. Masking agents can also be used to control odours outdoors in such places as waste lagoons.

The smell caused by putrefaction of solid and liquid wastes generated during tannery operations can be reduced by quickly disposing them off without allowing for putrefaction. Aerobic biological methods of effluent treatment and dewatering of sludge produced by methods like vacuum filtration, drying on sand drying bed will considerably reduce the smell caused by them.

USE OF COMPUTERS IN LEATHER PROCESSING

Leather is a unique natural product. It has been associated with human civilisation since ages and continuously to beat the rough and tough of the nature and to provide comfort for the human race. In the current context, leather products serve as glamour accessories. The importance of Indian leather sector stems from two angles: employment generation and earnings through foreign exchange.

Tanning industries exist in the small-scale sector. Out of 1600 tanneries in the country, only around 10 per cent are large-scale tanneries. Most of the small-scale tanneries are practicing leather production in an unorganised manner. The operations of leather processing are controlled with the help of human skills; even administration works are carried out manually.

But introduction of personal computer with simplified operation procedure and availability of readymade software for various administration works has changed mindset of people to use computer in their company. Now tanners have started mechanising their tanneries and, to little extent, investing in automatic chemical feeding systems. But still there is a long way to go to achieve complete automation in leather processing.

Use of Computers in Technology

The use of computers in technology falls into two categories: one is traditional and another is quite novel. The first category includes the applications in which the computer carries out a set of instructions, with little or no human intervention. The second category is a new class of application, in which computer is an active partner of man. The control of chemical plants is a good example of an application in which the computer can deal with a large amount of information, monitoring many variables involved in such a way as to maintain optimum production and quality of product. The variables in a chemical process- temperature, pressure, flow, valve settings, viscosity, colour and many others are interrelated in complicated ways and usually their relation are non-linear.

If two ingredients must flow into a reaction vessel in a certain ratio and the flow rate of one ingredient is deficient for some reasons, it does no good for the computer systems to attempt to rectify the deficiency by opening the supply valve wider, if the valve is already open; instead the computer will take into account the state of affairs and close the valve on the supply line, until desired ratio of flow is achieved.

A computer is able to receive information from many measuring stations located at strategic places in the plant, make the necessary comparisons, take decision on how to monitor the control mechanism and feed command back to them in such a way as to ensure optimum operations. This capability has no limit to the complexity of the information with which the computer can deal. Industrial engineers can now devise processes so intricate that it would be difficult, if not possible, to control them with human workers.

Artificial Intelligence

Artificial intelligence (AI) is a program, which carries out job without human interference in a particular field or area. Computer scientists are trying to determine what sort of things computer can be made to do. Some researchers conceptualise their work as an exploration into the nature of human intelligence or cognition. Artificial intelligence programers aim to do the specific job either with the help of robots or by simply correlating the available value or result with set point and giving simple instruction to open or close the valve. Artificial intelligence is a vast topic.

Automation of Tannery Procedures

The advent of semi-conductors and micro-processors has facilitated the use of computers to control various tannery operations. A microprocessor-based control system has made significant impact on automatic control of tannery procedures. The microprocessor-based automatic control systems for tannery procedures are broadly classified as under:

Controlled water addition systems

1. Automatic water mixing and temperature stabilising.
2. Directing the mixed water to the appropriate drum.
3. Selection and volume metreing of water.
4. Selection and vigorous control of mixed water and temperature.
5. Programing and printing out of water addition schedule.

Controlled chemical addition systems

1. Feeding of chemicals or their formulation from storage tank farm.
2. Automatic dosing of chemicals in volumes into the chemical premixing tank.
3. Transfer of mixed chemicals to the prescribed drum.
4. Memory storage of chemicals quantities according to batch weight.
5. Printout of the dosing along with time schedule.

Process control in drum operations

1. Drum function control.
2. Recycling of tan liquor.
3. pH control and automatic load measurement.

Present Trends in Tannery Automation

Due to the advancement in technology and availability of sophisticated process control equipments at reasonable price, tannery automation is made easy. Many leading companies are providing fully computerised process control tools for tannery right from beam-house to finishing. These types of computerised process control systems eliminate human supervision and reduce batch variation. These control systems works merely by checking output signal to reference point. These systems do not possess any artificial intelligence or cannot act independently.

Creating an artificial intelligence in leather processing is a highly complicated job, because leather processing is more of an art, rather than a science. To create an artificial intelligence tool for leather processing requires more complex data for each and every point of the tannery operation, which has to be documented in the correct manner. That means human practical knowledge has to be translated into electrical signals and then stored electronically to perform any artificial intelligence operations. Process adjustments vary according to the nature of the raw materials, quality of water used, quality of chemicals, as well as behaviour of commercial leather auxiliaries produced by various chemical companies towards the leather substrate. So the task is highly complicated.

One has to select different types of measuring devices, according to the place of application and then find suitable transducers to convert mechanical measurement into electrical signals. These signals have to be processed to get appropriate result, which can be matched or correlated with stored information and then analyse the result in various dimensions. It is easy to perform tasks up to matching the result with the stored information, but to analyse the result with stored information in various dimensions is an uphill task. According to the unit operation, detailed set of information about nature of substrate, chemical reaction, input and output condition, problems and remedies, etc. should be documented and a set of databases should be created and then embedded into a chip to perform a unique task.

Electronic Information System on Leather Processing Chemicals

To create an artificial intelligence, a comprehensive set of information about leather processing is a must. It is not possible to create such information overnight; it requires a lot of paperwork, plus practical simulations. These results are to be correlated with actual production and then these findings should be debated to arrive at a final solution. To make this possible, leather processing should be divided into key areas. These include for example, quality of the basic chemicals and their usage in leather processing; quality of the commercially available chemicals and their pros and cons in leather processing; chemical modification of collagen in each and every unit operation and their impact in quality of the final leather; etc. Each module can be divided into sub-modules and various parametres required for that module are gathered. Each module is prepared individually and then combined together to get a module. The first step is to create a database for commercially available leather processing chemicals. Although a number of companies manufacture leather auxiliaries under various brand names, they are identical in chemical nature. Hence, their reactivity towards collagen and the properties of the final leather are unique. So, creating a database on leather auxiliaries is also useful for creating AI; it will also be useful to student as well as research scholars.

This database contains details, which are available in technical litreature. Due to practical difficulties production details are not appended. This database will give suitable match for a particular chemical. The main benefit of this database is that it is possible to get a correct match from other chemical companies, reduction in time for searching the chemical from litreature and less storage space.

Classification of Leather Auxiliaries

Tables 15.16 and 15.17 represent a broad classification of commercial leather processing chemicals used in leather industry. Product varies mainly due to the selection of basic raw material and preparation mode, but fall within the classification. Most chemicals are supplied by various chemical companies.

About TFEISLPC

Technologist Friendly electronic information system on leather processing chemicals (TFEISLPC) is an interactive expert system, presenting the database on leather processing chemicals in an organised manner. TFEISLPC aids the leather technologist in the tannery to classify pertinent chemicals for various processes and also for identifying alternative chemicals in case of any problem.

TFEISLPC is an attempt to create application software about tannery chemicals mostly used in the tannery. For the best human-computer interaction, TFEISLPC draws supporting knowledge from several leather technology experts functioning in various small and medium scale tanning industries. TFEISLPC generates technologist friendly intelligence for producing economical and good quality leathers from the available raw materials in the tannery.

The software aims at presenting the database on leather processing chemicals in an organised manner (Table 15.16 and 15.17).

Table 15.16. Classification of beam house auxiliaries.

Category	Chemicals
Soaking aid	Soaking enzyme based on proteolytic enzymes
Unhairing aid	Unhairing enzyme proteolytic enzyme
Liming aid	Alkylolamines (nitrogen based auxiliary), liquid molasses, glucose
Deliming aid	Nitrogen-free deliming agent
Bating aid	Bating enzyme based on proteolytic enzymes
Degreasing aid	Lipolytic enzymes
Pickling aid	Non-swelling organic acid

Table 15.17. Classification of surfactants.

Category	Examples
Wetting agents	Ethyl phenol polyglycol ether
Degreasing agent	Fatty alcohol ethoxylate, fatty alcohol polyglycol ether
Emulsifiers	Alkyl phenol polyethenoxy ether, ester sulphate

This software is created in Visual FoxPro 6.0 and it is user-friendly. It is completely menu driven and the user can get information by moving the mouse to the appropriate icon.

Salient Features

Some of the salient features of the software and the mode of operation are explained below:

On opening the software, a welcome message will appear, followed by customary statement, pressing the 'I Accept' command button; a Main menu will appear on the screen. The Main menu consists of five sub-menus, viz:

1. Basic chemicals.
2. Leather processing chemicals.
3. General information.
4. Contact address and URL.
5. Quit.

Basic chemicals menu

When a user clicks the basic chemicals menu, two sub-menus appear:
1. Elements.
2. Chemical compounds.

When the user clicks the elements menu, it displays element name, symbol, atomic weight, and atomic number of the element according to the periodic table. When the user clicks the Compounds Menu, it displays compound name, molecular formula, solubility, molecular weight. The above information will be very useful for the leather technologists, while using basic chemicals in leather processing (Tables 15.18 to 15.21).

Table 15.18. Classification of tanning agents.

Category	Examples
Mineral tanning agents	
Poly bases	Chrome, aluminium zirconium, iron, titanium tanning agent
Poly acids	Pseudo tanning agent phosphoric acid, silicic acid tanning agent
Polyaromatic tanning agents	
Vegetable tanning agent	
Hydrolysable tannin	Oakwood, Chestnut, Myrobalan, Sumac and Dividivi
Condensed tannin	Mimosa, Quebracho, Gambier
Synthetic tanning agent	
Phenol based tanning agent	Replacement tanning agent, white tanning agent, pretanning and retanning agents, nonphenolic tanning agent, auxiliary and bleach tanning agent
Aliphatic tanning agent	
Aldehyde taning agent	Formaldehyde, glutraldehyde, dialdehyde, starch
Polycondensation and polymerisation compounds	Methylol urea, methylol melamine, methylol dicyandiamide, diisocyanate, acrylate
Paraffin derivatives and fats	Paraffin sulphochloride fatty alcohol sulphonates, fish oil

Table 15.19. Classification of fat liquoring agents.

Category
Untreated oils, fats and waxes
Sulphonated oils, fats and fatty alcohols
Sulphated products (–C–O–S– bond, ester like splittable)
Sulphonated products (–C–S–bond true sulpho acid, unsplittable)
Chlorinated oils and fats: Chlorinated products, sulpho chlorinated products
Oxidation products of oils and fats
Hydrolysis products of oils and fats
Polymeric product of oils and fats

Table 15.20. Classification of dyes.

Chemical classification	Classification by application
Nitro and nitroso dyes	Anionic dyes
Azo dyes	Simple acid dyes

(Contd...)

Chemical classification	Classification by application
Monoazo dyes	Direct dyes
Polyazo dyes	Special dyes
Metal complex dyes	Dyes soluble in water and organic solvents
Stilbene dyes	Dyes soluble in organic solvents
Diphenyl and triphenyl methane dyes	Cationic dyes
Quinonimine dyes	Oxidative dyes
Oxazine dyes	Reactive dyes
Thiazine dyes	Disperse dyes
Azine dyes	Developed dyes
Sulphur dyes	Sulphur dyes, selected vat dyes
Pyridine, quinoline and acridine dyes	
Phthalocyanines	Natural and mordant dyes
Indigoide dyes indigosoles	Fat and oil soluble dyes
Anthraquinone and multiring dyes	
Natural dyes	

Table 15.21. Classification of finishing chemicals.

Category	Examples
Pigments	Organic and inorganic pigments; they are further divided into anionic or cationic pigments
Binders	Protein binders; further divided based on film hardness and charge
Resin binders	Mainly divided into acrylates, butadiene, polyurethane, etc. they are further classified according to film hardness and charge
Others	Fillers, waxes, lacquer and lacquer emulsion, slip agent, feel modifiers, anti-tacking agent, etc.

Leather processing chemicals menu

This menu is an important option of this software. After clicking this option, the user can see three options, viz.

1. Company-wise search.
2. Product-wise search.
3. Products of a particular company.

If the user choose the option 'company wise search' a list consisting of the names of chemical companies, available in the database is displayed. When the user choose a company from the list displayed, another list consisting of the nature of the product like, beam house auxiliary, surfactants, tanning agent, syntan, fatliquor, etc. will be displayed in the screen. When the user chooses any one of the classification, details about the available product, like company name, product name, classification, sub-classification, chemical composition, appearance, active matter, pH, charge of ionic nature, stability, application, etc. will be displayed.

When the user click the option 'product wise search' menu, which is the key for this software, another 'classification menu' which consists of options: Beam house auxiliary, surfactants, tanning agent, syntan, fatliquor, etc. is displayed. When the user clicks one of the above classifications, it will further display a submenu called sub-classification menu. For example, if you click fatliquor, it again displays sub-classification menus namely sulphited fatliquor, sulphated fatliquor, synthetic fatliquor, semi synthetic fatliquor, waterproofing agent and oils. If you choose one of the above sub-classifications, it will display information about the available products under different companies, under the classification and sub-classification chosen by the user available within the database. The information displayed includes company name, product name, classification, sub-classification, chemical composition, appearance, active matter, pH, charge of ionic nature, stability, application, etc. Finally when the user chooses the option 'products of a particular company', a list of chemical companies available in the database will be displayed. The user may choose any one of the companies, which, in turn, will display information about various products of a particular company available within the database. The information displayed includes company name, product name, classification, sub-classification, chemical composition, appearance, active matter, pH, charge of ionic nature, stability, application, etc.

General information menu

When the user clicks this menu, it displays a sub-menu, viz:
1. Leather calculator.
2. Quality requirements.
3. Acid base indicators.
4. pH of tannery processes.
5. Articles on leather.
6. Calendar of any year.
7. Back to main menu.

If the user chooses the option, 'leather calculator' the user can find simple and useful calculation procedures which are required for leather processing like conversion of temperature, conversion of measurement, conversion of weight, conversion of square measurement, density calculations and basicity calculations. It has one submenu called 'calculate'. If you click the sub-menu 'calculate', the following list of conversion factors will be displayed:
1. Centigrade to fahrenheit.
2. Fabrenheit to centigrade.
3. Centimetres to inches.
4. Inches to centimetres.
5. Metres to feet.
6. Feet to metres.
7. Square inches to square centimetres
8. Square centimetres to square inches.
9. Square metres to square feet.
10. Square feet to square metres.
11. Kilograms to tons.
12. Tons to kilograms.
13. Pounds to grams.
14. Grams to pounds
15. Ounce to grams
16. Grams to ounce.
17. Gallons to cubic metres.
18. Litres to cubic metres.

19. Horse power to watts.
20. Density to barkometre.
21. Density to baume.
22. Density to twaddell.
23. Barkometre to density.
24. Baume to density.
25. Twaddell to density.
26. Increase basicity in terms of Cr_2O_3.
27. Increase basicity in terms of Cr.
28. Reduce basicity in terms of Cr_2O_3.

When the user clicks one of the above conversion factors, a separate pop up window, will be opened, which will ask the user for values for conversion. When the user enters the values for conversion, the program will display the converted answer. If the user chooses the option 'quality requirements' then the properties required for upholstery hides, clothing leathers, grain upper leather, chamois leather, industrial gloves/sports gloves, shoe lining leathers are displayed, which can be utilised by the leather technologist as a ready reckoner. If the user chooses the option 'acid base indicators' then for various indicators, the two parametres, pH range in which colour changes and the colour change will be displayed.

If the user chooses the option 'pH of tannery processes', pH range for various tannery processes will appear in the screen for the use of leather technologist during tannery processes. If a user chooses the option 'articles on leather' then various articles related to leather processing would be displayed as word files. Finally, if a user chooses the option, 'calendar of any year' then a form that displays the calendar of any year is activated.

Contact address and URL

When the user clicks this menu, it displays address of the chemical companies and URL of the company, which are available within the database. One can directly access the website of the particular company.

MINIMAL NATIONAL STANDARDS (MINAS)

Generally two main aspects are taken into consideration for development of standards of waste-water discharges. One relates to the adverse effects on health and environment and the other achievability of limits of pollutants by incorporation of appropriate pollution control measures.

The latter approach aims at use of the best available and economically feasible technology. Economically feasible technology assures that the cost of pollution control measures will remain within the affordability of the industrial units. Standards developed on these principles are techno-economic standards and these standards are uniform throughout the country.

An advantage of the technology-based approach is that within a specific group of industries the extent of pollution control measures are alike. In addition, these standards serve to preserve the environmental quality in non-polluted areas without modification. The disadvantage of this approach is that these standards may become unnecessary burden on the industry where the recipient environment does not demand such control measures. This is because these standards do not relate to the actual environmental situation of the specific site.

However, it may be considered that development of standards based on the local environmental requirements is not a practicable proposition for a country like India. Therefore, it is logical to evolve industry-specific standards at the national level. To provide safeguard to the local environmental conditions, the local enforcing authorities are required to lower the limit values of pollutants (become stringent) as per case to case evaluation of the respective recipient bodies. On such exercise, these standards serve both as specific for industry and location. The parametre of relevance in tannery industries are BOD, total chromium sulphide, total suspended solids, total dissolved solids, chloride, sulphate, oil and grease, etc.

Comparative Analysis of Standards

A comparative statement of standards prescribed by BIS and Ministry of environment (EPA) is given in Table 15.22. Based on the performance study carried out, the BIS and EPA standards and consideration of the feasibility of the standards techno-economically, the MINAS has been evolved and presented in Table 15.23.

Table 15.22. Comparative statement of standards and MINAS (*BIS : 2490 - Part 3 - 1985*).

Receiving water body	Inland surface waters	Public sewers	Land for irrigation	Inland surface waters	Public sewers	Land for irrigation
Parametres						
pH	6.0–9.0	6.0–9.0	6.0–9.0	6.0–9.0	6.0–9.0	6.0–9.0
Suspended solids mg/l	100	600	200	100	600	200
BOD_5 at 20°C mg/l	30	350	100	30	350	100
COD, mg/l	250	–	–	–	–	–
Total dissolved solids, mg/l	2100	2100	2100	–	–	–
Sulphide, as S, mg/l	2.0	2.0	2.0	2.0	5.0	–
Total chromium, as Cr, mg/l	2.0	2.0	2.0	2.0	2.0	2.0
Oil and grease, mg/l	–	–	–	10	20	10
Hexavalent chromium Cr^{+6}, mg/l	–	–	–	0.1	0.2	0.1
Chlorides, mg/l	1000	1000	600	1000	1000	600
Sodium, per cent	–	–	–	–	60	60
Boron as B, mg/l	–	–	–	2.0	2.0	2.0

Table 15.23. Minimal national standards (MINAS) for tanneries.

Parametre	Limits not to exceed
pH	6.5–9.0
BOD_5, 20°C, mg/l	100
Total suspended solids, mg/l	100
Sulphides, mg/l	2
Total chromium, mg/l	2
Oil and grease, mg/l	10

Note: For effluent discharge into inland surface waters, BOD limit may be made stricter to 30 mg/litre by the concerned State Pollution Control Boards.

Chapter 16

Pulp and Paper

INTRODUCTION

India has undergone a large process of economic reforms aiming at liberalisation and attracting investments. That process, which began in the early 1990s, continues to have an important impact on the whole economy, including the pulp and paper industry.

The paper industry has an important role to play in the Indian economy. The overall paper consumption in India reached 5 million tons in 2006, making India a large market from any perspective. The potential for per capita consumption increase, originated due to economic growth, increasing purchasing power and emerging export-led industries, attracts companies to invest and modernise. The increasing demand for paper puts pressure on the supply of paper making fibres, including recycled paper, nonwood raw materials and wood.

At the same time competition in global pulp and paper markets is intensifying. This is having an increasing impact on the Indian market too. Indian pulp and paper industry consists of some 500 paper mills, mainly of small and medium size companies. There is a growing need for mill modernisation, productivity improvements and building of new capacity. If adequate measures are taken, India's competitiveness could substantially be improved and the industry shall be ready for global competition.

Today the installed capacity is over 6 million tons, out of which about 1.1 million tons is idle because of the closure of a large number of mills. The per capita consumption is still around 4.2 kg and thus the country has much potential to grow. The industry employs more than 0.3 million people directly and about 1 million people indirectly.

GRADES OF PAPER AND BOARD AND DEMAND PROJECTION

There are large number of grades of papers and boards in the market, requiring different process conditions, raw material mix and additives:

1. Writing papers—normally bleached papers.
2. Printing papers—normally bleached papers.
3. Coating base and coated papers.
4. Absorbent papers.
5. News print.
6. Office/copying/computer stationary.
7. Packaging paper and boards.
8. Speciality papers.

All the grades are available in large number of varieties. They are produced in different types and sizes of the mills. Printing and writing papers and containerboard are expected to grow most.

FIBRE SOURCES

The fibre resources used by the Indian pulp and paper industry, comes from three main sources:

Forests

Forests including bamboos and mixed hardwoods from forest fellings and Eucalyptus wood from plantations (both organised plantations and farmers' fields/agro-forestry plots). Wood and bamboo are mainly used by large integrated paper mills.

Agricultural Residues

Agricultural residues are emerging as a significant alternative raw material resource for the pulp and paper industry in India. Their share of total fibre use of the paper industry is 29 per cent (1.3 million tons). The use of agricultural residues has grown since the early 1970s partly due to the dwindling bamboo resources and partly due to the Government's industrial policy encouraging investments in agro-based paper production. Major incentives for investors included tax holidays, excise duty remissions and liberalised imports of machinery. The main agricultural residues utilised by the paper industry include bagasse, cereal straws (wheat and rice), kenaf/mesta, jute sticks, many varieties of grasses and cotton stalks.

Waste Paper

Waste paper based industry accounts for about one-third of Indian paper capacity. The recovery of waste paper has increased from 65,000 tons in 1995 to 9,00,000 tons in 2005. Most of paper is recovered, but due to alternative uses, the recovery rate for paper industry is still only about 20 per cent. This is low by international standards: Thailand (42 per cent), China (33 per cent), Germany (71 per cent). Waste paper recovery and trading are still unorganised in India. The collection is being carried out by individual wheelers and the system of sorting is unsophisticated. The Indian waste paper recovery is not keeping pace with recycled paper utilisation, resulting in increase in imports (8,00,000 tons in 2005). Multiple use of paper products (as wrapping papers, packaging applications, etc.) is common in India and often these end uses pay better price for waste paper than the paper industry.

Currently the proportions of each category in the total production of pulp and paper are 36, 29 and 35 per cent, respectively. Over the past few years, the share of forest based industries has declined.

MILL CAPACITY

There are over 500 paper mills in India. Most of the mills are small, only 34 integrated paper mills have a capacity of over 33,000 tons per annum (Table 16.1). The Indian paper machines are mostly small units. In international comparison, even the largest machines are medium-size, as large-scale machines are today in the range of 4,00,000–6,00,000 T/annum and have a trim width up to 10 metres. The following parametres are illustrative of the Indian paper industry:
1. The average capacity of paper machines is about 14000 T/annum.
2. Most of Indian paper machines have a trim width from 1.5 to 3.5 metres.

3. There are only 9 paper machines with trim width of 5 metres or more.
4. Only 14 machines have capacities 50,000 T/annum or more.

Table 16.1. Size of the paper mills in India.

Category	Capacity	Number of mills
Large paper mills	> 100 TPD	~ 34
Medium paper mills	50 to 100 TPD	~ 150
Small paper mills	< 50 TPD	+ 350
Globally–normally	> 500 TPD	

Paper mills (PM) size and speed do not thus match with economy of scale. High speed machines (>1000 m/min) are needed for cultural papers and one-layer boards. With a higher design speed, technology and paper quality will be closer to the international standard.

BASIC PROCESS STEPS IN PAPER AND BOARD MANUFACTURE

Depending upon the raw material used and product required, the various manufacturing processes are selected.

In Wood and Agro based Mills

1. Raw materials preparation—chipping, washing, cutting, depithing.
2. Pulping or cooking—direct/indirect, batch/continuous, vertical/spherical digesters.
3. Washing of pulp—weak black liquor to chemical recovery.
4. Bleaching of pulp (optional).
5. Stock preparation—blending of pulps.
6. Paper making—head box, paper machine, pressing, drying and reeling.

Chemical Recovery Operation

1. Weak black liquor evaporation (12–15 per cent to 50–60 per cent).
2. Burning of concentrated Black liquor (BL) organics to recover heat.
3. Dissolution of smelt to have green liquor.
4. Causticisation of green liquor to recover inorganics for reuse in pulping.
5. Lime mud washing and reburning for reuse of lime in causticisation.

Improvement of Paper Quality

For the improvement of paper quality, some of the options could be:
1. Good raw material preparation, pulping and washing systems should be installed to produce lower Kappa No Pulp with minimum amount of water and steam.
2. Bleaching sequences should be improved to get the higher standard ISO-brightness, with environment friendly TCF bleaching processes.
3. Domestic raw material quality and availability must be improved.
4. The level of quality measurement and control should be improved.

Characteristics of waste paper based mills is shown in Table 16.2.

Pulp available after flotation washing is blended with other pulps, as per need. It can be sent for bleaching too if required.

Table 16.2. Characteristics of water paper based mills.

Slushing in hydro pulpers	Percentage values	Remarks
Low consistency	<5%	Addition of chemicals and heat
Medium consistency	5–10%	Addition of chemicals and heat
High consistency	>10%	Addition of chemicals and heat
High density cleaner	~3–5%	Dilution and separation on density
Centri cleaner operation	~1–1.5%	Dilution and separation on density
Pressure screens	~1.5–3.5 %	Dilution and separation on size and shape
Flotation Cells	~1–2%	Dilution and separation by air flotation

EFFLUENT BENCH MARKING OF INDIAN MILLS

1. Water consumption and effluent flow are high in Indian forest and agro-based mills leading to a relatively high effluent load (Table 16.3).
2. Some of the large waste paper based mills are closer to the international standards in quality but still the effluent loads are almost double. The figures in small paper mills are still very high.
3. All loads have been calculated on per ton of air dry finished paper basis.

Table 16.3. Effluent load of Indian mills.

Integrated mill type	Effluent flow, m^3/T	TSS mg/l	COD kg/T	BOD kg/T	AOX kg/T
Indian Agro-based mills	100–180	40	47	10	–
Indian forest based mills	100–175	11	37	3.7	0.4
European forest based mills	40	1.5	12	1.5	0.1
Indian waste paper based mills	30–80	2.6	8.7	1.3	–
European waste paper based mills	10	1.0	5.0	1.0	–

The effluent load is directly related with the level of technology, raw material used and process water used. As already indicated most of the small and medium agro-based mills are not having chemical recovery systems and proper recycle of process water leading to very high consumption of raw water.

TECHNOLOGY TRENDS IN PULP PRODUCTION

In Europe the move towards total chlorine free (TCF) has almost ceased and thus not recommended for India too. TCF bleaching gives better pulp, requires less energy and capital and still the environmental standards can be met.

Enzymes are recommended to help bleaching, if the original brightness of raw material is low, chemical costs are high or final brightness low, by normal methods.

TCF bleaching and modified cooking methods require so high investment costs that they are not viable investments for small agro based mills. Some small mills have gone for oxygen delignification, with good results.

Table 16.4, shows the improvement in pulping and bleaching technology, which not only improved the paper quality but also decreased the water and other chemical inputs. As they are finally leading to lower Kappa Number, they can be treated as more environment friendly too.

Table 16.4. Improvement in bleaching technology with time.

Kappa no.	30	25	20	15	10	5	0
1970	Conventional cooking		Bleaching (CEH)		–		–
1975	Conventional cooking		O_2 Delignification		Bleaching (TCF)		–
1985	Modified batch cooking		O_2 Delignification		Bleaching (TCF)		Enzyme
2005	Dispersed super batch/RDH		O_2 Delignification		Bleaching (TCF)		Enzyme

Medium and small agro-based mills still continuing with spherical rotary digesters which are relatively inefficient and with poacher washers create very dilute black liquor, which is discharged from the mills through ETP, still causing a heavy discharge, much more than the permissible limits.

WATER USAGE AND POLICY

Government policy on water is greatly emphasising for decreased use and increased discharge quality. High water consumption in the mills is due to old/obsolete machinery leading to higher water usage. The lack of chemical recovery at smaller mills on one side increases the chemical costs but on the other side it increases the water use, poor discharge quality and large effluent discharge. Supreme court is coming heavily on such mills and they have no option but to install chemical recovery systems at the earliest. This shall also be inline with Government energy policy to go for cogeneration at pulp mills. This shall also decrease the consumption of conventional fuels such as coal. During the paper making process, the water is added or removed at a number of places, as shown in the Table 16.5.

Table 16.5. Fibre concentrations at different process conditions.

Process step	Fibre concentrations	Operation
Pulping	~10%	Dilution
Washing	~1.5–2% to 10%	Dilution/Squeezing
Bleaching	~3–5%	Dilution/Squeezing
Stock preparation	~1–4%	Dilution
Head box	~3.5–4 % to 0.9–1.3%	Dilution
Paper machine	~0.9–1.3% to 10–12%	Drainage
Press section	~10–12% to 50%	Squeezing
Dryers	~45–95 %	Evaporation

The steps involve multistage dilution, squeezing, drainage and evaporation during paper making, thus water reuse is essential for water conservation. Table 16.6 shows the average water usage of different groups of Indian Paper industry.

Table 16.6. Performance of Indian paper industry.

Water usage m^3/T of paper	Large	Medium*	Small**
80's	170–400	180–390	120–200
90's	130–250	110–280	90–120
2000's	110–200	100–220	50–100
2010 (Target)	60–100		
World average	40–50		

*Agri residue based—with no chemical recovery.
**Waste paper based—ideally it should be ~10.

The flow sheet of the processes involving straw, rag and waste paper are given in Figs 16.1 to 16.3, respectively. A generalised flow sheet adopted for making paper in small mills depicting sources of wastes is given in Fig. 16.4.

WASTE-WATER GENERATION

Sources of Waste-water

As identified earlier, in agricultural residue based units, the waste-water is released from the following sections:

1. Black liquor from cooking section.
2. Pulp wash water from pouchers.
3. Beater section specially when rags are used.
4. Bleaching section.
5. Thickener.
6. Paper machine.

Waste paper based mills generate waste-waters from sections 4, 5 and 6 only as no chemical pulping is adopted.

Black liquor is the most polluting among the different streams. Black liquor is not segregated and it ends up in pulp washing waste-water. Thus pulp washing section contributes nearly 80 per cent of the total pollution load. Further, flow and composition of pulp wash waste-water will be highly varying because the operations are carried out in batches. Only in a very few mills countercurrent washing using two stages is adopted.

Paper machine waste-water is least polluting among all the sections but contributes appreciable amounts of suspended solids. In most of the mills, paper machine waste-water either as such or after fibre recovery, is used in pulp washing, beaters, etc. The percentage pollution contributed by each section could not be worked out since flow data for individual sections are not available.

WATER POLLUTION

Big Paper Mills

Considerable work has been carried out on the characteristics of various pulp and paper mill effluents. The main polluting constituents in pulp and paper mill waste-water are suspended solids, colour, foam, inorganic such as sodium carbonate, bicarbonate, chlorides and sulphates, toxic chemicals such as mercaptans and inorganic sulphides, mercury if caustic chlorine plant forms a part of pulp and paper mill, BOD and COD.

The waste-water when discharged untreated will damage the water course and the colour in water persists for a long distance since lignin is not readily biologically degraded. 'As paper industry being scattered all over the country, it may be said that no river is spared from pollution due to the discharge of these waste-waters.'

Another important pollutant (that is yet to be assessed) in pulp and paper industry is mercury. This is so because some of the mills make their own caustic and chlorine in mercury cells and it has been reported that about 0.25 kg of mercury is lost per ton of caustic made. Further, some organic mercury compounds such as phenyl mercuric acetate (50–100 ppm of dry weight of pulp), methoxy ethyl mercury and pyridyl mercury salts are being used as slimicides on paper machines which finally find their way into the effluent. In Sweden it is reported that about 25 tons of mercury salts are used per annum.

346 Pollution Control in Chemical and Allied Industries

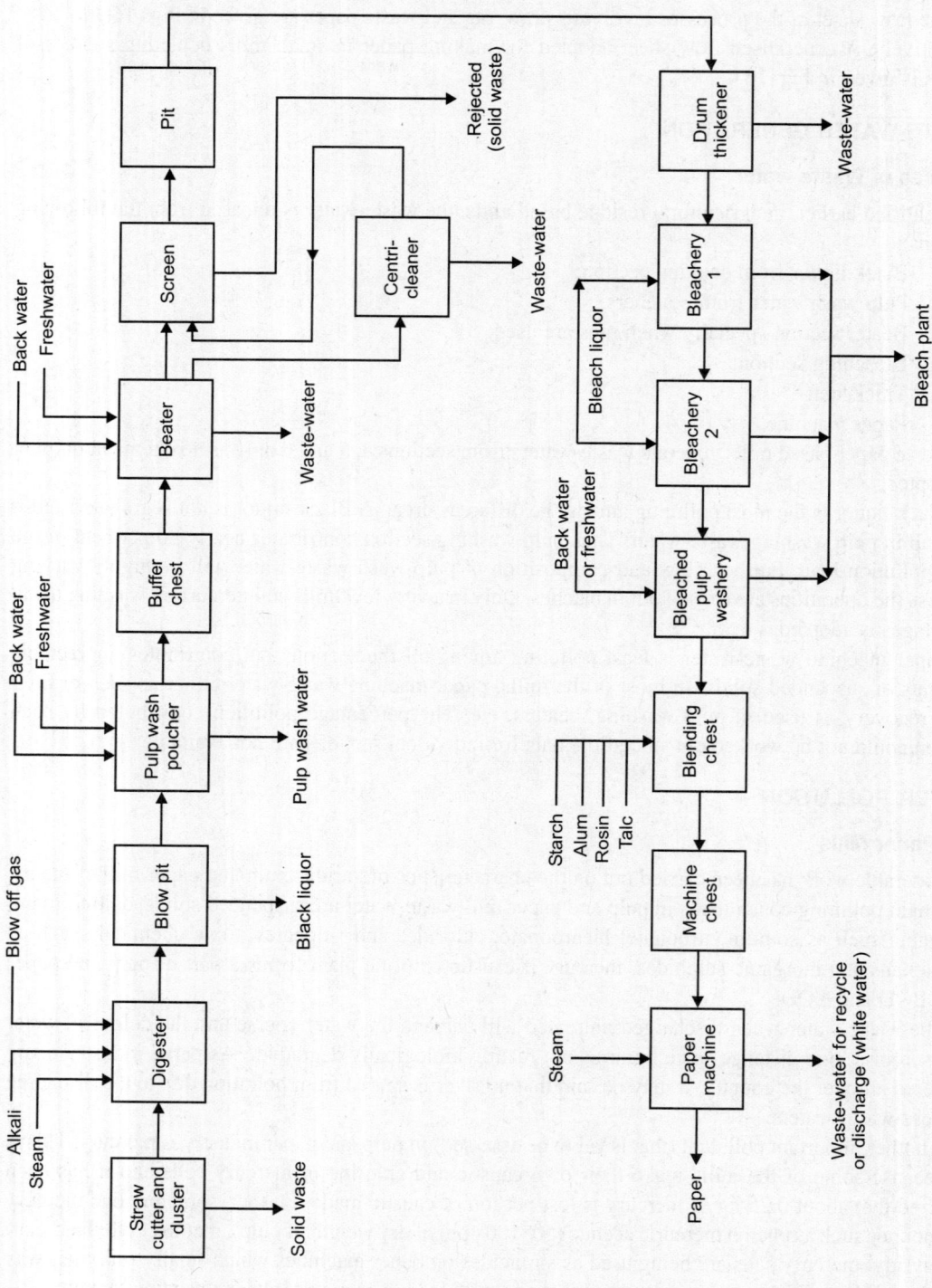

Fig. 16.1. Process flow sheet and sources of waste straw based mill.

Pulp and Paper **347**

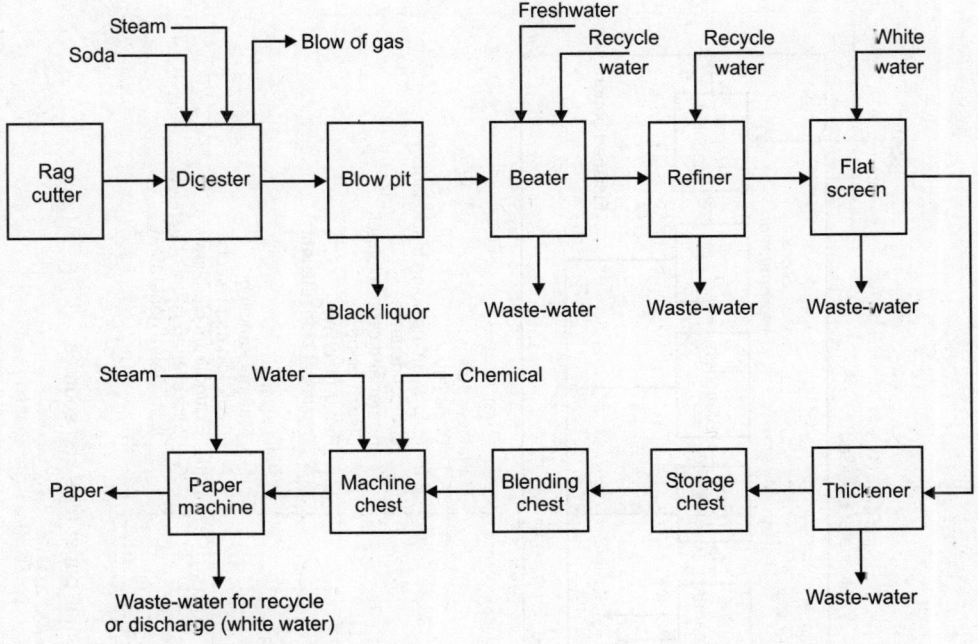

Fig. 16.2. Process flowsheet and sources of waste rag based mill.

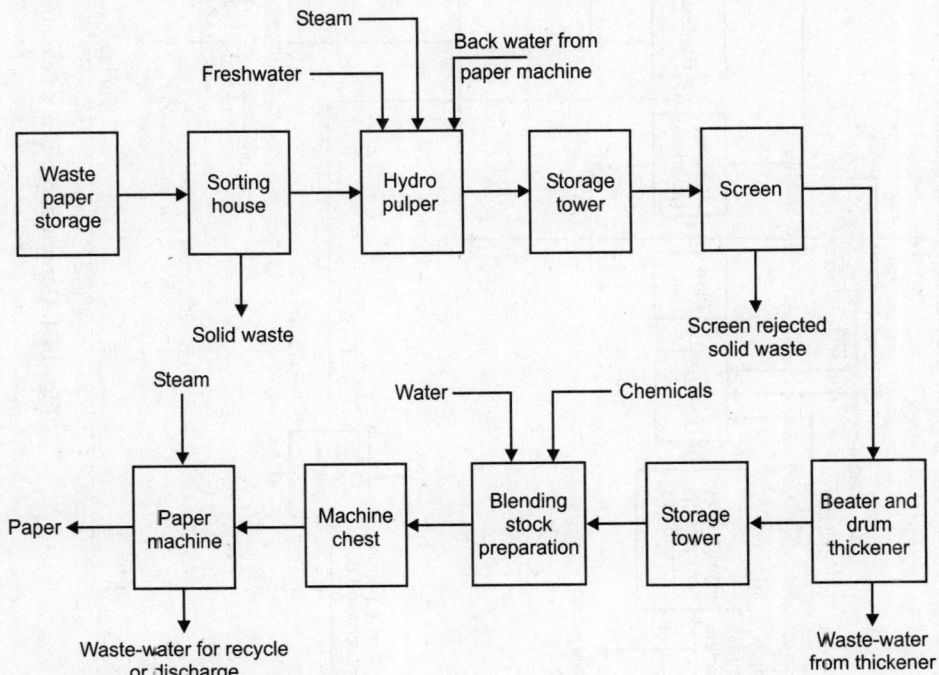

Fig. 16.3. Process flow sheet and sources of waste in waste paper based mill.

348 Pollution Control in Chemical and Allied Industries

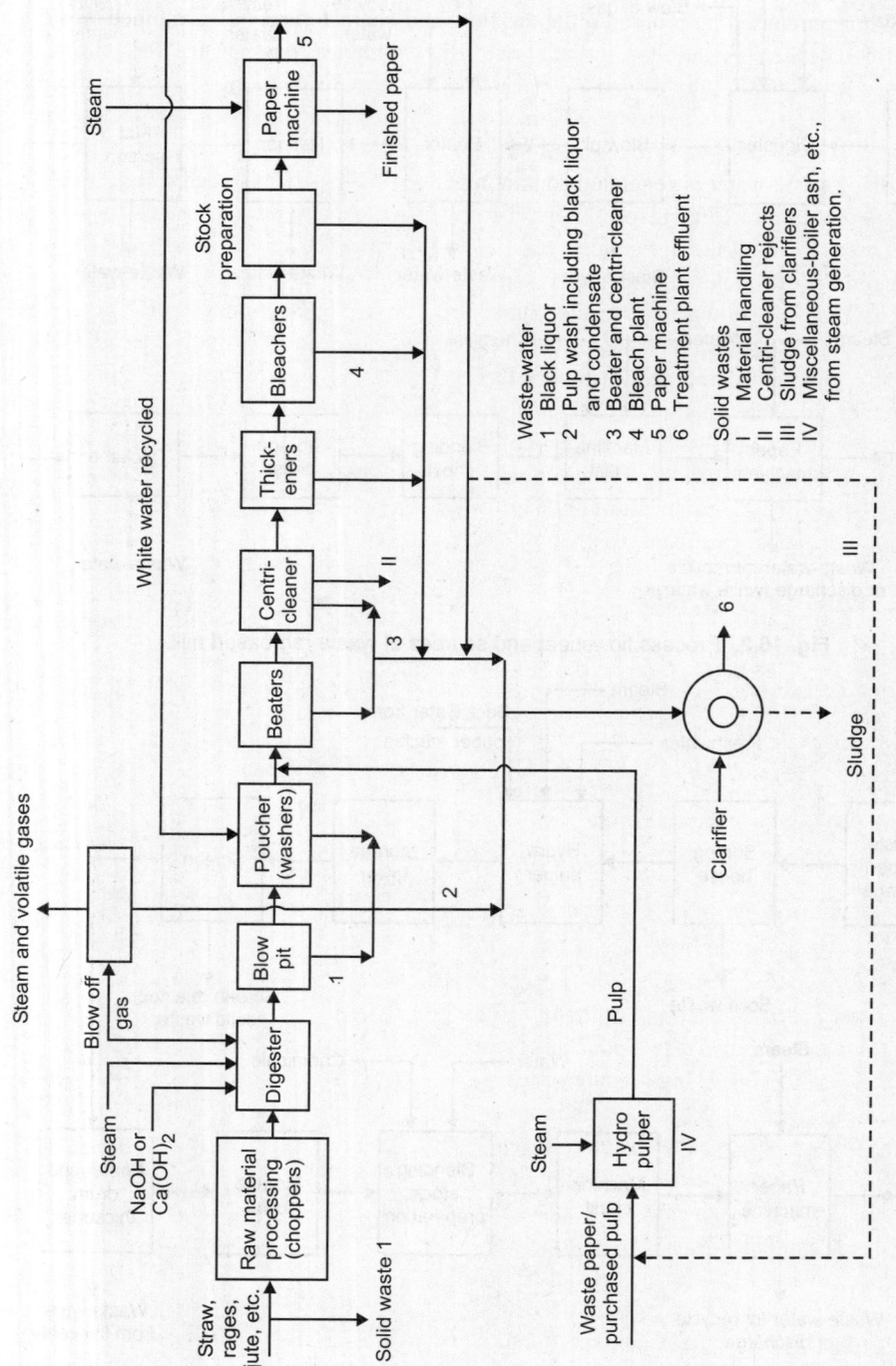

Fig. 16.4. Generalised process flow sheet of small paper mills and sources.

Waste-water
1 Black liquor
2 Pulp wash including black liquor and condensate
3 Beater and centri-cleaner
4 Bleach plant
5 Paper machine
6 Treatment plant effluent

Solid wastes
I   Material handling
II Centri-cleaner rejects
III Sludge from clarifiers
IV Miscellaneous-boiler ash, etc., from steam generation.

CHARACTERISTICS OF EFFLUENTS

Waste paper and/or purchased pulp based mills the flow and characteristics of combined waste-water indicate that although there is variation in the flow of the combined waste-water, the minimum and maximum values respectively are 0.6 and 1.5 times the average flow. There is also variation in the characteristics of the waste-water and is mainly due to the different practices adopted for fibre recovery and extent of recycle of both recovered fibre and water. On an average for each ton of paper made 107 ± 28.4 cu.m. of waste-water is generated containing 58 ± 10.7 kg of suspended solids, 20 ± 10.5 kg of BOD and 70 ± 15.2 kg of COD respectively. The proportion of agricultural residue to waste paper used varies appreciably in all the agricultural residue and waste paper based mills. Even in the same mill, depending on the raw material availability, season, type of paper made, the proportions vary even within a week. Hence, waste-water flow and characteristics are found to vary. It is not possible to generalise the flow and pollution loads contributed per ton of paper made from these mills. BOD rate constant for black liquor and combined waste-water after settling compares with that of domestic waste-water. The nonbiodegradable fraction of COD is high and is mainly due to the lignin in black liquor. In case of combined settled waste-water, 70 per cent of the nonbiodegradable COD is due to lignin and the remaining due to suspended solids, etc. Hence, it can be concluded that the COD due to lignin cannot be removed even after biological treatment. The only alternative method is to remove colour vis-a-vis lignin by chemical treatment.

TREATMENT FOR THE WASTE-WATER

For Mills Using Agricultural Residues as Raw Materials

The unit process involved in the treatment of waste-water from agricultural residue-based paper mills are:

1. Equalisation of flow from pulp wash section.
2. Primary clarification for combined waste-water.
3. Secondary biological treatment.
4. Sludge drying beds or lagoons for primary sludge depending on the availability of land.

Equalisation for pulp wash waste-water

Pulp washing section accounts for about 20–25 per cent of total waste-water and contributes around 70–80 per cent of pollution load from small paper mills using agricultural residues as raw materials. These waste-waters are discharged intermittently since, in most of the mills, washing of pulp is done in batches using pouchers. Normally two pouchers are employed and operated in series and thus generate two washes. The period of washing is more or less same in both but the quantity of water used varies appreciably and thus the first wash water is more concentrated than the second wash water. The flow variation has been observed to be fairly wide since the minimum and maximum values, respectively are 0.15 and 2.3 times the average flow. Discharge of these washes to the main waste-water stream (on an intermittent basis), as is being practised will alter the composition of the combined waste-water appreciably. This necessitates provision of equalisation to the pulp washes and discharge at a constant rate into the main sewer. Two alternatives can be considered for this purpose.

Alternative 1: It envisages flow equalisation for the first pulp wash waste-water. This waste-water is generated at a rate of 12.5 cu.m./T of paper made and is discharged at a rate of 1.56 cu.m./batch in 8 batches of 2 hours each per day. The discharge rate is fairly uniform during the 2 hours of washing period.

Capacity of equalisation tank to be provided works out to	1.0 cu.m./T paper
Adding 25 per cent extra volume including free board to meet any surge discharge	0.25 cu.m./T paper
Total volume of equalisation tank	1.25 cu.m./T paper
Rate of pumping of waste-water from the equalisation tank into sewer	0.52 cu m./hr for 24 hours

Therefore, two equalisation tanks are to be provided to facilitate cleaning and maintenance.

Alternative 2: It envisages flow equalisation of the entire pulp wash waste-water (2 washes). The total pulp wash waste-water is generated at a rate of 50 cu.m./T of paper and is discharged at a rate of 6.25 cu.m./batch in 8 batches of 2 hours each per day. As stated above, the flow rate of waste-water is fairly uniform during the 2 hours washing period.

Capacity of equalisation tank to be provided works out to	4.0 cu.m./T paper
Adding 25 per cent extra volume including free board to meet any surge discharge	1.0 cu.m./T paper
Total volume of equalisation tank	5.00 cu.m./T paper
Rate of pumping of waste-water from the equalisation tank into sewer	2.08 cu.m./hr for 24 hours

Therefore, two equalisation tanks are to be provided to facilitate cleaning and maintenance. Pumps of suitable capacity have to be provided one each for the two tanks and one as a stand-by.

Primary clarifier and sludge drying

If secondary treatment like activated sludge, oxidation ditch or rotating biological disc is used as suggested, the excess secondary biological sludge should be added to the combined waste-water before primary settling such that the settled sludge can be filtered on drying bed. The combined primary and secondary sludges do not require biological stabilisation. The dried sludge can be disposed off by burning in an incinerator or dumping in open pits in a controlled manner.

The treatment alternatives suggested are already shown in Figs. 16.5 to 16.8 for agricultural based mills. The waste-water is practically devoid of nitrogen and phosphorus. Hence, nutrients will have to be supplemented to biological treatment process. For anaerobic system BOD:N:P should be 100:2.5:0.5 and the corresponding ratio for aerobic system is 100:5:1. The treatment alternatives are detailed below:

Treatment alternative 1: The primary clarified effluent is proposed to be treated in an anaerobic lagoon after proper seeding and acclimatisation. The lagoon will have a detention time of 20 days. The anaerobic effluent will be subsequently treated in an aerated lagoon with a detention time of 4 days. The effluent leaving the aerated lagoon will be passed through a polishing pond with a detention time of 2–3 days before discharge or use on land for agriculture (Fig. 16.5a). Anaerobic treatment prior to aerated lagoon will reduce foaming problems.

Treatment alternative 2: Effluent from primary clarifier will be treated in an aerated lagoon with a detention time of 6 days. The lagoon effluent will be taken through a polishing pond with 3–5 days detention time before final discharge or use on land for irrigation after suitable correction for SAR and total dissolved solids (Fig. 16.5b).

Surface aeration of pulp mill wastes generate considerable foam which most often covers the entire surface of the aeration tanks and thus prevents or reduce oxygenation of the medium. It is observed that in square type of aeration basin or tank, the foam formed is very stable and does not disperse or break away easily. On the other hand, it is noticed that in rectangular tanks with length to breadth ratio equal to or greater than 2 to 1, although foam formation takes place, the foam stability is less and breaks away at a faster rate, i.e. more easily. Further spacing of aerators in the basin and the rotation in clockwise and anti-clockwise direction for alternate aerators will also help in reducing the foam accumulation as well as breaking away of foam formed. With these arrangements and addition of anti-foaming agents, it is likely that aerated lagoon can be used as a method of treatment. However, this system is likely to create a problem of foam which should be kept in mind.

Treatment alternative 3: Oxidation ditch operated on extended aeration principle with high MLSS (4000–5000 mg/l) can be used to obtain high degree of BOD removal. High MLSS in the system will help in destabilisation of foam. Although foam formation cannot be eliminated, it is observed in one of the mills, where this system is used, that foam breaks away as the mixed liquid flows in the channel and the liquid surface is more or less free from foam. The mixed liquor after secondary clarification will have low BOD (Fig. 16.6a).

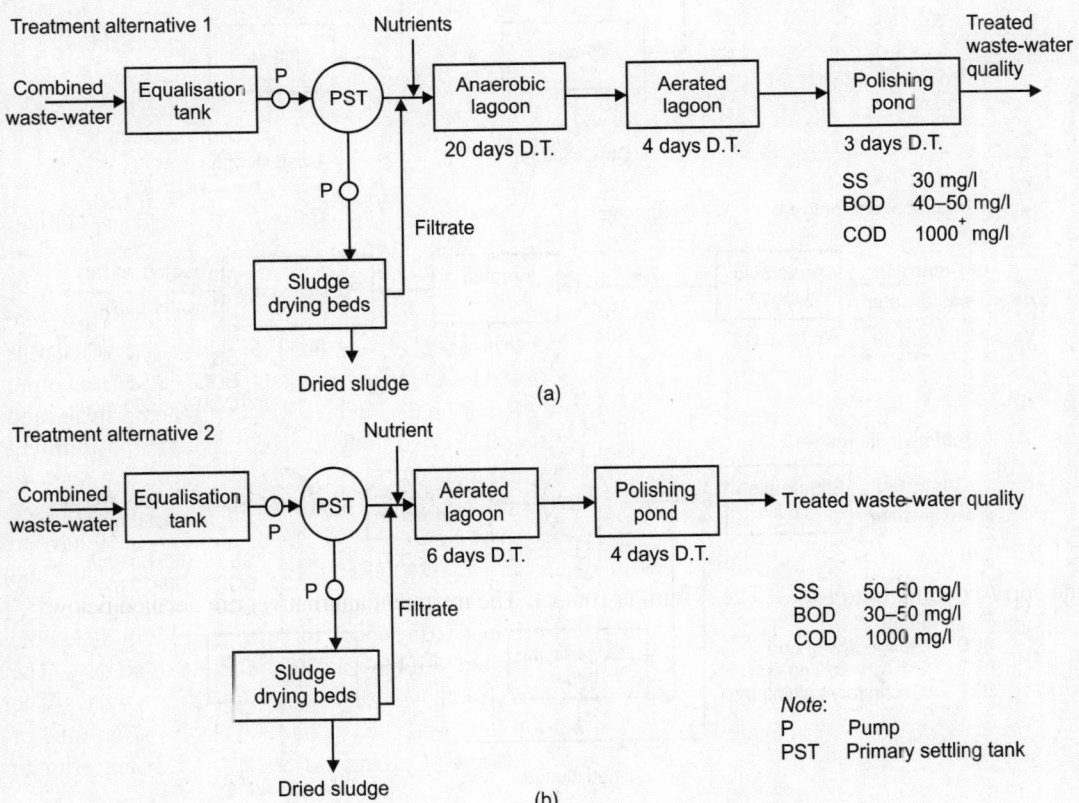

Fig. 16.5. Waste-water treatment alternatives for agricultural residue-based paper mills: (a) anaerobic lagoon; and (b) aerobic lagoon.

In place of oxidation ditch, a diffused aeration activated sludge system can also be used on the same principle. In this system, foam does form but because of diffused aeration, the oxygenation of the liquid in the tank will not be affected *vis-a-vis* biological purification. However, foam control from a physical point of view will be required. A secondary settling tank is also necessary.

Treatment alternative 4: Combined effluent without equalisation and primary settling is proposed to be treated in anaerobic lagoon after proper seeding and acclamatisation. The lagoon will have detention time of 25 days. The effluent from anaerobic lagoon will be subsequently treated in aerated lagoon with a detention time of 4 days followed by polishing pond with 4 days detention time (Fig. 16.6b). Treatment alternative 5: Same as alternative 2 except that the polishing pond is replaced by settling tank (Fig. 16.6c).

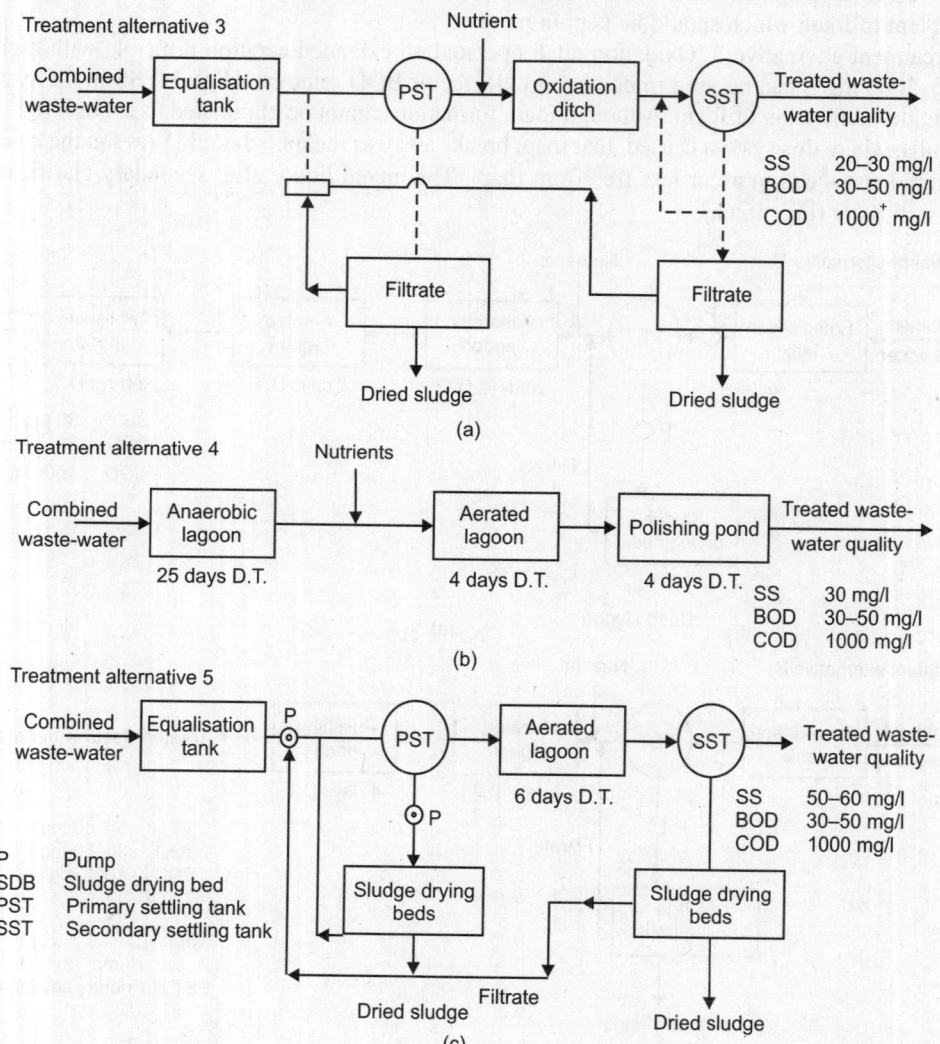

Fig. 16.6. Waste-water treatment alternatives for agricultural residue-based paper mills: (a) equalisation tank with oxidation ditch; (b) anaerobic lagoon with polishing pond; and (c) equalisation tank with aerated lagoon.

Treatment alternative 6: Rotating biological contactor (RBC) can be successfully used for the treatment of the settled combined waste-water from paper mills. This has been demonstrated in the USA and other places. Actual design data used for treating pulp and paper mill wastes by RBC are not readily available. No foam problem will exist in RBC treatment as there is no liquid and air mixing through agitation.

Thus RBC may be considered as a package plant for the treatment of the waste-water. A secondary clarifier is required to remove the biological solids which can be collected and sent to the inlet of primary clarifier and removed along with primary sludge. The design data available for domestic sewage treatment by RBC can be used for treatment plant design for pulp and paper mill wastes. Over 90 per cent BOD reduction can be expected.

Solid Wastes

Most of the solid waste from the pulp and paper industry consists of 'mill residue', which comprises bark, wood residues, refuse (pulp, paper and cardboard), ash from combustion facilities and sludge from the treatment of procesed water and de-inking. The major contributors of solid wastes in pulp and paper mill effluent are bamboo and wood dust, lime sludge, coal ash, fly ash and other rejects. Of these, lime sludge and ash are generated to the extent of 1250–1500 kg and 400–500 kg per MT respectively of the paper produced. Under high-temperature combustion, organic compounds containing carbon, hydrogen, and often chlorine are oxidised. Other substances potentially incinerated include sulphur, arsenic, and metals such as chromium mercury, and lead. Both the toxic emissions produced during the burning process and toxic ash left over after combustion are often more hazardous than the original materials. These emissions consist of products of incomplete combustion, escaping heavy metals, and new combinations of materials as a result of the burning process.

Liquid Wastes

The nature of mill effluent depends on the type of process used, but most mills release suspended solids (SS) such as fibre and bark, organic matter which increases the biological oxygen demand (BOD), various inorganic compounds (mainly aluminium, manganese and zinc), hydrocarbons from machine lubricants, resin and fatty acids and phenolic compounds. Enormous amount of waste-water (i.e. effluent) laden with organic matter is generated in a pulp and paper mill. Pulp mills are voracious water users. Their consumption of fresh water can seriously harm habitats near mills, reduce water levels necessary for fish, and alter water temperature. The conventional chlorine bleaching of Kraft pulp has been estimated to produce approximately 100–300 gram of chlorinated phenolic compounds per ton of pulp. Because of extensive chlorination, these sediments of waterbodies receive such effluents.

About 300 different chemical compounds have been identified in pulp and paper mill effluents, but these represent only 10–40 per cent of the total chlorinated organic compounds present.

For Mills Using Waste Paper and Purchased Pulp as Raw Material

Treatment of waste-water from mills using waste paper and purchased pulp may be carried out in the following processes:
1. Fibres should be recovered and recycle of recovered fibre and water to the maximum possible extent to be adopted. The dissolved air flotation system is found suitable for recovery of fibre from white water. After fibre recovery this water is recycled to the pulp section for pulp washing.
2. Primary clarification of the combined waste-water before or after fibre recovery is essential for which rectangular or circular clarifiers can be used.
3. Primary sludge can be dried on sludge drying beds or lagoons depending on the availability of land.
4. The effluent from primary clarifier needs to be further treated either in a stabilisation pond or in an aerated lagoon (½ day detention time) if land is a limiting factor (Fig. 16.7). The treated effluent will then meet the permissible effluent standards.
5. Primary clarified effluent can be used on land for crop irrigation.

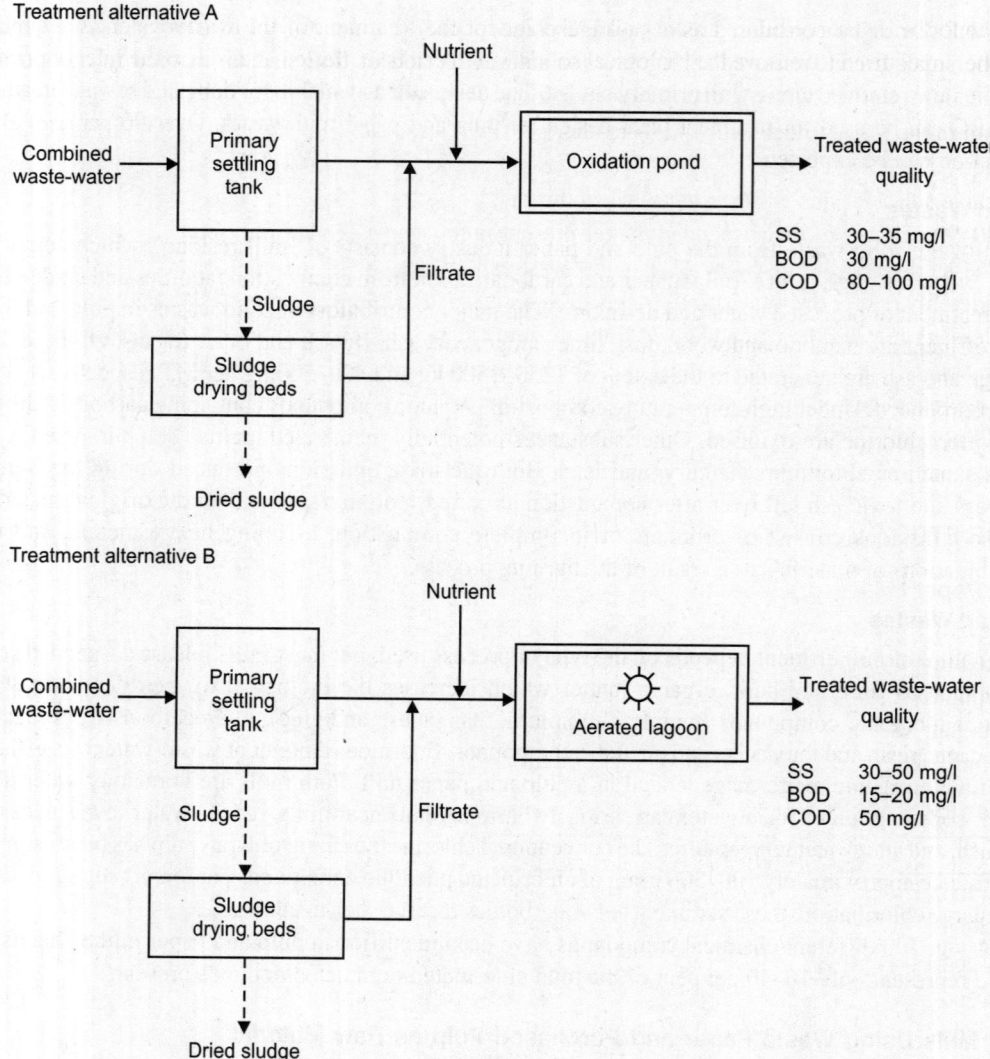

Fig. 16.7. Waste-water treatment alternatives for waste paper and purchased pulp-based paper mills.

Black Liquor Segregation and Disposal

Black liquor (BL) from the agricultural based mills constitutes the most polluting waste-water. It is not economically possible to remove or segregate black liquor as is being done in large mills where it is used for chemical recovery.

It has been reported that black liquor can be segregated by squeezing or by allowing it to drain out from a false bottom. From the information supplied by various mills, it is observed that 30–50 per cent of black liquor can be separated out from straw pulp and only 20 per cent from gunny rag. The segregated

BL cannot be discharged into stream or on land except during flood time in the river. Since its volume will be small it can be stored in lagoons for suitable period of time and discharged in a controlled manner into surface waters only during monsoon. This will help in reducing the colour, BOD and sodium in the remaining waste-water. Another possible place for segregation is the rejects from centri-cleaners. These rejects will be small in volume but contain a very large percentage of suspended and settleable solids. If this waste-water is taken through a side hill screen, it should be possible to separate out 80–90 per cent of the suspended matter present in the centri-cleaner rejected.

The sludge can be disposed of as solid waste. This method of segregation and treatment would considerably reduce the total load of suspended solids in the combined waste-water reaching the clarifier and lessens the problem of their handling in dilute form in the clarifier underflow. Paper machine waste-water contains high amount of fibre which settles rapidly. This can also be recovered by employing suitable recovery system and both the recovered fibre and water can be recycled into the process. This would not only help in reducing the cost of waste treatment but also generate revenue in the form of extra paper and reduced water requirements (Fig. 16.7).

CASE STUDY

Star Paper Mills Ltd., Saharanpur

Tables 16.7 and 16.8 show the data of Star Paper Mills, Saharanpur. This is an integrated paper mills located in northern India, which has progressed very well in all the areas in the last few years. It has not only improved its production but has reduced its specific consumption of water, energy and other chemical inputs too. Waste water generated in compressors have been recycled back for reuse at washer in the pulp mill. The condensate and white water, recirculation systems have been improved in the mill.

Table 16.7. Case study of Star Paper Mills Ltd. Saharanpur effluent load.

Year	pH	SS mg/lit	COD mg/lit	BOD mg/lit
2001–2002	7.7	68	203	25
2002–2003	7.7	68	208	27
2003–2004	7.6	70	202	25
2004–2005	7.7	80	207	25
2005–2006	7.6	75	203	23
2006–2007	5.5–8.5	100	250	30

Table 16.8. Case study of Star Paper Mills Ltd. Saharanpur (water consumption).

Year	Water consumption, m^3/t	Production, t/d
2001–2002	186	158
2002–2003	175	171
2003–2004	160	176
2004–2005	155	178
2005–2006	140	183
Target 2010	100	200

The two main sections from where the effluents are generated are—bleach wash liquor and from white water recirculation systems. The condensates from multi effect evaporator systems and Heat recovery systems, wash water from raw material washing, weak black liquor from Brown stock washers, centri-cleaner rejects, pressure screen rejects, etc. are other places from where the waste process water is generated for reuse and/or purge. Over the years the mill has given due attention in this regard which has lead to decreased consumption of raw water as is evident from the data. The basic inputs are also relatively much higher in India in comparison to developed countries, as detailed below in Table 16.9.

Table 16.9. Consumption of basic inputs.

	India	Western countries
Heat, GJ/T	15–30	4–8
Electricity kwh/T	800–1500	400–800
Water m^3/T	50–200	5–50
Chemical recovery %	88–94	95–98
Productivity T/per person	25–300	800–3500
Personnel man hr/T	10–100	0.5–2

There is always a fixed and variable part in all the inputs. This is the reason that bigger and well working mills are able to reduce specific consumptions.

Thus, the Indian paper industry has grown not only in size but in quality and specific consumption of water too. In the years to come, the average size of the mill shall have to grow with larger paper machines and more automation. It is expected that if the Indian mills continue to work properly to implement the various improvement methodologies, then only they can grow and survive globally. Agro-based mills shall have to go for chemical recovery systems or close down, in accordance with the Supreme Court orders.

CHARACTERISTICS OF WASTE-WATER MANAGEMENT IN SMALL PAPER MILLS

The quality and quantity of wastes generated depends primarily on the kind, quality and composition of various input raw materials and the process adopted for paper manufacturing.

SPPMs using agro-residues as raw materials, normally use soda process with chlorination-extraction-hypochlorite (CEH) bleaching sequence, whereas waste paper based mills use hydrapulper to produce pulps for paper manufacturing.

The mills having chemical pulping process generate heavy pollution loads during pulping, washing and bleaching operations. B/L obtained from pulp washing units contributes very high solids, BOD, COD, colour and lignin, besides creating tremendous foaming nuisance.

The effluent characteristics of different capacity small paper based on Non-conventional raw materials (NCRMs), mainly agro-chemical residues are presented in Table 16.10, which shows considerable variation in pollution load generation per ton of paper production.

In order to find out the possibility of waste-water generation minimisation/treatment at different stages, various possible sources of waste water generation are identified in a typical small paper mill, producing about 15 TPD unbleached kraft paper, mainly from NCRMs.

Various possible resources of waste-waters generation include, raw material cooking and pulp washing, stock preparation, bleaching, thickener and centri-cleaners, paper machines, freshwater tank overflow, boiler house, etc.

Table 16.10. Effluent characteristics of agro-residues based different capacity small paper mills.

Parametre	Mill A	Mill B	Mill C	Mill D	Mill E	Mill F
Production capacity (tons/day)	35	25	30	80	10–40	15
Volume, m^3/T	110	110	100	100	252	148
pH	8–8.5	9–10	6–8	8.5	5–8.5	6–8
Total suspended solids (TSS)						
mg/l	650	700	1100	1300	615	2141
kg/T	72	77	88	130	155	317
Bio-chemical oxygen demand (BOD)						
mg/l	850	800	2000	1500	698	966
kg/T	94	88	168	150	176	143
Chemical oxygen demand (COD)						
mg/l	1400	3400	4000	4800	2940	2980
kg/T	154	154	320	480	741	441

Raw material composition
Mill A = 70% bagasse + 30% hessian.
Mill B = 80% bagasse and wheat straw + 10% jute + 10% waste paper.
Mill C = 90% bagasse and wheat straw + 10% hessian + 10% waste paper.
Mill D = 70–80% bagasse and wheat straw + 20–30% jute and waste paper.
Mill E = 70–80% bagasse and wheat straw + 20–30% hessian and waste paper.
Mill F = 50% bagasse + 35% hessian + 15% waste paper.

The effluent characteristics of individual stream from a typical small paper mill using agro-residues (50 per cent), hessian (35 per cent) and waste paper (15 per cent) are presented in Table 16.11. From the in-depth study, the following observations are made:

1. For producing one ton of unbleached paper, about 140–170 m^3 fresh-water is required, of which as much as 90 per cent may be generated as effluent from various sections.
2. Large flow variations are observed due to batch operation of pulp washing units (20–60 m^3/hr) and intermittent undesirable discharges (overflows) from thickener and fresh-water tanks (16–78 m^3/hr).
3. The total pollution load in terms of total solids (TS), suspended solids (SS), BOD and COD is about 750, 320, 140 and 440 kg/T of paper respectively, of which about 90 per cent is due to pulping section (or 60–65 per cent from pulp washing unit only). The concentration of TS, SS, BOD and COD in combined effluent is of the order of 5100, 2100, 3000 and 1000 mg/l respectively.
4. Pulping as well as paper machine section effluents contain very high sodium content (85–88 per cent of total exchangeable Ca and Mg ions) and also very high sodium absorption ratios (11–12), thus making it unsuitable for irrigation purposes, unless concentration of Ca and Mg ions is enhanced.

The above effluent quality data indicate that SPMs cannot discharge their effluents without treatment to any receiving body, i.e. water or land.

Table 16.11. Effluent characteristics in a typical small paper mill from different sources.

Parametres	Poucher washer (A)	Straw beater (B)	Hessian beaters (C)	Pulp mill (D = A + B + C)	Paper machine (E)	Combined effluent (F = D + E)
Volume, m^3/T	32	4	38	74	74	148
Temperature, °C	25–40	26–30	26–30	25–36	27–31	27–33
pH range	8–10	7–8	7–8	7–9	6–7	7–9
TS, mg/l	14630	9250	3880	8813	1348	5081
kg/T	468	37	147	657	100	752
TSS mg/l	4770	5547	2568	3690	588	2142
kg/T	153	22	98	273	44	317
TDS, mg/l	9860	3703	1312	5149	760	2953
kg/T	316	15	50	381	56	437
VTS, mg/l	8755	5600	2355	5297	536	2919
kg/T	280	22	90	392	40	432
VSS, mg/l	3520	3428	1650	2568	343	1453
kg/T	113	14	63	190	25	215
VDS, mg/l	5235	2172	750	2784	197	1493
kg/T	168	9	29	206	15	221
COD, mg/l	9800	5250	2100	5908	350	2980
kg/T	314	21	80	415	26	441
BOD, mg/l	–	–	–	1800	140	966
kg/T	–	–	–	133	10	143
Na, mg/l	362	238	246	342	230	286
Na, %	90	84	86	85	88	87
SAR	15	10	10	12	11	12

Waste Minimisation/Management Options

Waste minimisation/management options for small pulp and paper mills are divided into following heads are:
1. Identification and implementation of simple internal measures.
2. Increased use of waste paper.
3. Possibility of chemical recovery.
4. Modification of pulping processes, where B/L generation is either eliminated or reduced.
5. Segregation of B/L generating stream for:
    (a) Lignin isolation.
    (b) Biogas production.
6. Waste-water treatment.

Various waste minimisation/management options are summarised in Fig. 16.8 and are discussed here briefly.

Internal measures

Various possible process modifications requiring minimum equipments, which will help in pollution, load reduction significantly, are identified and discussed. The internal measures include prevention of leakages, spills, overflows, better pulp washing operations, segregation of highly contaminated but low volume effluents and recycling of less contaminated streams with/without treatment. Though more emphasis is required in the pulping section, as it generates maximum pollution load (around 90 per cent), yet precautions are also necessary during raw material preparation and in paper machine section. The internal measures suggested in Fig. 16.8 shall reduce pollution generation as well as shall help in improving the performance of the mill.

Increased use of waste paper

Waste paper is a readily available low cost regenerative source of paper making fibre, which requires relatively simpler process, compared to other cellulosic raw materials. All grades of paper can be manufactured economically, when blended with some virgin fibres. Increased use of waste paper for agricultural residues reduces waste regeneration (air, water and solid) to a large extent along with various input material, viz. chemicals, steam, coal; power, water, etc. At 50 per cent waste paper composition (considered to be the maximum recycle rate) pollution load in terms of SS, BOD and COD can be reduced by about 30, 45 and 45 per cent respectively and at 33 per cent waste paper recycling, the corresponding reductions are about 21, 30 and 30 per cent, with around 25 per cent less effluent generation. The overall impact on environment will depend on the extent of waste paper recycle for virgin pulp.

Possibility of chemical recovery

In chemical pulping processes, lot of black liquor (B/L) containing dissolved organic and inorganic compounds is generated. The quality of dissolved solids depend upon the chemical dosage in the pulping and the degree of pulp yield obtained. Conventionally, about 1.6 tons of dissolved solids come out along with B/L in chemical pulping processes per ton of pulp, besides almost whole of the chemicals charged (soda and/or soda sulphur) which is recovered in chemical recovery systems.

Very high cost of conventional chemical recovery systems acts as limitation in its adoption for even medium scale paper mills, upto 40–50 TPD capacity. As a result, several relatively low cost, non-conventional chemical recovery systems, viz. direct alkali recovery system (DARS), wet cracking process, copeland system, wet air oxidation process, smelter cyclone evaporator system, BKMI process, detromax recovery system, etc. have been developed and are now available worldwide for chemical recovery from B/Ls in small paper mills.

Among these processes, only DARS and wet cracking process are reported to be well suited and economically viable for a small paper mills of 15–20 TPD capacity. However, further R & D efforts through pilot plant studies shall explore the techno-economic feasibility of these systems under Indian conditions.

Modifications in pulping processes

The most suited pulping processes developed for small and medium sized paper mills are capable of reducing pollution load to a large extent. These processes are categorised as where:
1. Alkali is used but higher pulp yields are obtained.
2. Alkali requirement is reduced.
3. The process does not use alkali at all.
4. Mechanical or high yield pulping processes.

Some of these techniques are also being adopted in Indian paper mills, particularly in new mills.

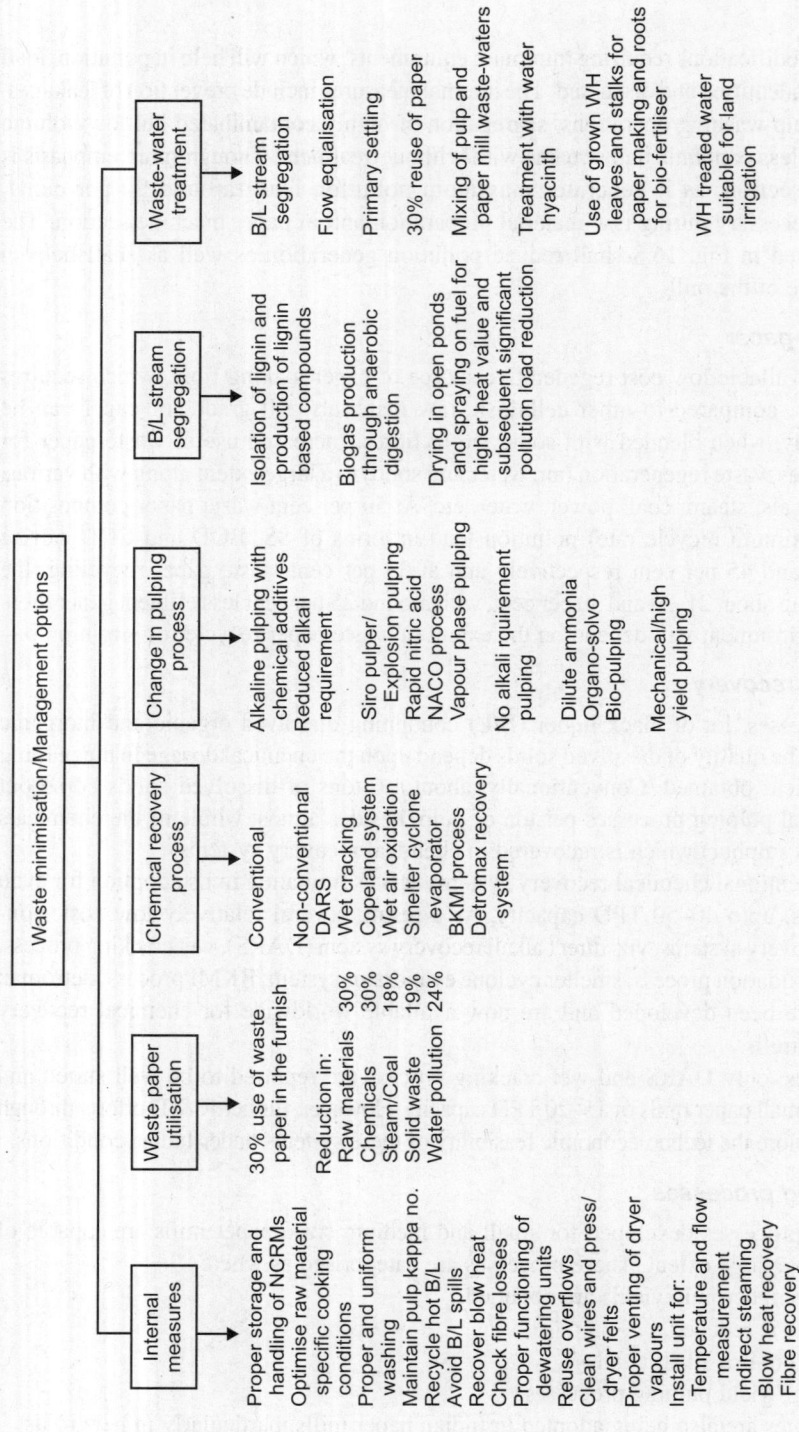

Fig. 16.8. Options of waste minimisation/management for small pulp and paper mills.

Segregation of Black Liquor Generating Stream

Lignin isolation

In the absence of chemical recovery process, B/L in SPMs is thrownout as effluent. Agricultural residues B/L, with 1–2 per cent TS concentration (sometimes 4–6 per cent or even more) essentially contain various lignin based compounds. For recovery of these compounds, first, the lignin needs to be isolated in purest form from B/L. But, the difficulty lies in the separation of pure lignin from large volumes of B/L, economically. However, the performance of lignin isolation unit can be improved considerably, if B/L is rich in total solids concentration. For the purpose, a screw press can be installed before pulp washing to extract most concentrated 15–18 per cent (TS) hot liquor from the pulp. This will further reduce the size of the equipment required for acid precipitation and also the energy required for concentrating B/L. Though, upto 86 per cent lignin can be isolated through acid precipitation, yet there is a need for inexpensive methods of pure lignin separations. The expenses incurred in the lignin isolation process will be offset by producing other useful lignin based chemicals, which have wide commercial application.

Biogas production

Alternatively, segregated highly concentrated (20–25 per cent TS or more) B/L stream containing high organic contents (about 70 per cent), BOD (20–35 g/l) and COD (50–90 g/l) can be used for gas production (methane content 65–70 per cent) through any of the anaerobic digestion processes, viz. fixed bed, fluidised bed, contact process or UASB system. About 70–75 per cent reduction in COD and 80–85 per cent reduction in BOD can be achieved in the process with biogas yield of 0.45–0.50 m^3 per kg of COD destroyed.

If B/L could not be used for lignin and its derivatives manufacturing, it can be used at least for its fuel value. B/L can be collected in separate small ponds for few days and after concentration to some higher total solids, can be sprayed over the fuel (rice husk or coal), being used in the boiler. This, besides tremendous pollution load reduction, will result in higher calorific value of the fuel. Few Indian small paper mills are practicing this technique to cope with pollution and energy problems to some extent.

Waste-water treatment

Any pulp and paper mill with best internal measures will find it difficult to discharge effluents meeting standards without recourse to end of the pipeline treatment. Though, through internal measures, it is possible to reduce the pollution load by about 40 per cent, yet to meet MINAS levels, the waste-water needs external treatment.

Waste-water treatment strategy

Various effluent treatment alternatives suggested by Central Pollution Control Board require high capital and operational costs, which SPMs of the order of 10–15 TPD cannot afford. Further, it would not be economically viable to treat whole of the effluent from SPMs which will require large land, high cost of machinery and the construction work. Therefore, looking into the practical problems of whole of the effluent treatment, the following simple strategy has been suggested, which will certainly relax the small industrial sectors.

The overall waste-water treatment strategy for agro-residues based small pulp and paper mills is presented in Fig. 16.9 along with effluent characteristics at different stages and potentiality of water hyacinth utilisation. The salient features of which are:
1. Segregation of the most polluted poucher washed (pulp washing) stream.
2. Grit, debris removal and flow equalisation unit separately for B/L and remaining pulp mill streams.
3. Settling of waste-waters from straw and hessian pulp beaters with little addition of alum.
4. Reduction in SS–70 per cent, TS–75 per cent, BOD–40 per cent and COD–40 per cent.
5. Mixing of surplus paper machine effluent.
6. 30–40 per cent recycle/reuse of paper machine backwater, overflows for pulp washing, stock dilution or pulp beating.
7. Treatment of clarified effluent with water hyacinth (WH) yielding 80 per cent reduction in SS and 70 per cent in COD to grow water hyacinth.
8. Harvesting of cultivated water hyacinth (average 1.35 T/ha/d) and separation of roots from the plants for bio-fertiliser manufacture.
9. Use of dried stalks and leaves (under sun for 2–3 days and upto 40–45 per cent moisture content) as a source of raw material for paper manufacturing/bio-gas production.
10. Use of WH treated effluent for land irrigation (SS–141 mg/l and COD 251 mg/l).

The potential of WH in effluent treatment, paper making and biogas generation has been highlighted with a view to draw attention of the industrial organisations towards this otherwise nuisance creating valuable waste material.

Potentiality of Water Hyacinth

In waste-water treatment

It is reported that under favourable conditions, one hectare area of land produces over 150 tons of dry plant material annually, assimilating over 17 tons of nitrogen and over 3.75 tons of phosphorus from domestic waste-water. The same area of WH would assimilate about 50 kg of phenol, 220–400 kg of silver, strontium, cobalt, cadmium and nickel and about 100 grams of lead and mercury from industrial waste-waters.

Therefore, effluent treatment in natural lagoons in the presence of WH will offer a low cost and simple method to small paper mills, where it is possible to reduce SS and COD load by 80 per cent and 70 per cent, respectively, within 15 days detention time. The influent SS and COD concentrations in the range of 600–700 mg/l and 800–900 mg/l respectively, shall result effluent SS and COD concentrations in the range of 120–140 mg/l and 240–270 mg/l respectively, which are near to MINAS levels.

In addition, substantial colour reduction can also be achieved. About 90 per cent COD reduction can be achieved for paper machine effluents, only within 6 days detention time. WH can grow in very wide range of pH (4–10) and it has pH reduction capability also. The optimum temperature range for growth of WH is 26° to 40°C. Under favourable conditions, WH growth may be in the range of 0.9–1.8 T/ha/day. But, the growth of WH reduces linearly with the water temperature in the range of 30° to 20°C.

In paper making

The whole WH plant contains about 24–32 per cent roots, 42–56 per cent stalks and rest 12–34 per cent leaves. The whole plant of WH cannot be used for paper making, as the roots do not contain any fibrous value. Therefore, the roots need to be separated from the plants for manufacture of bio-fertiliser.

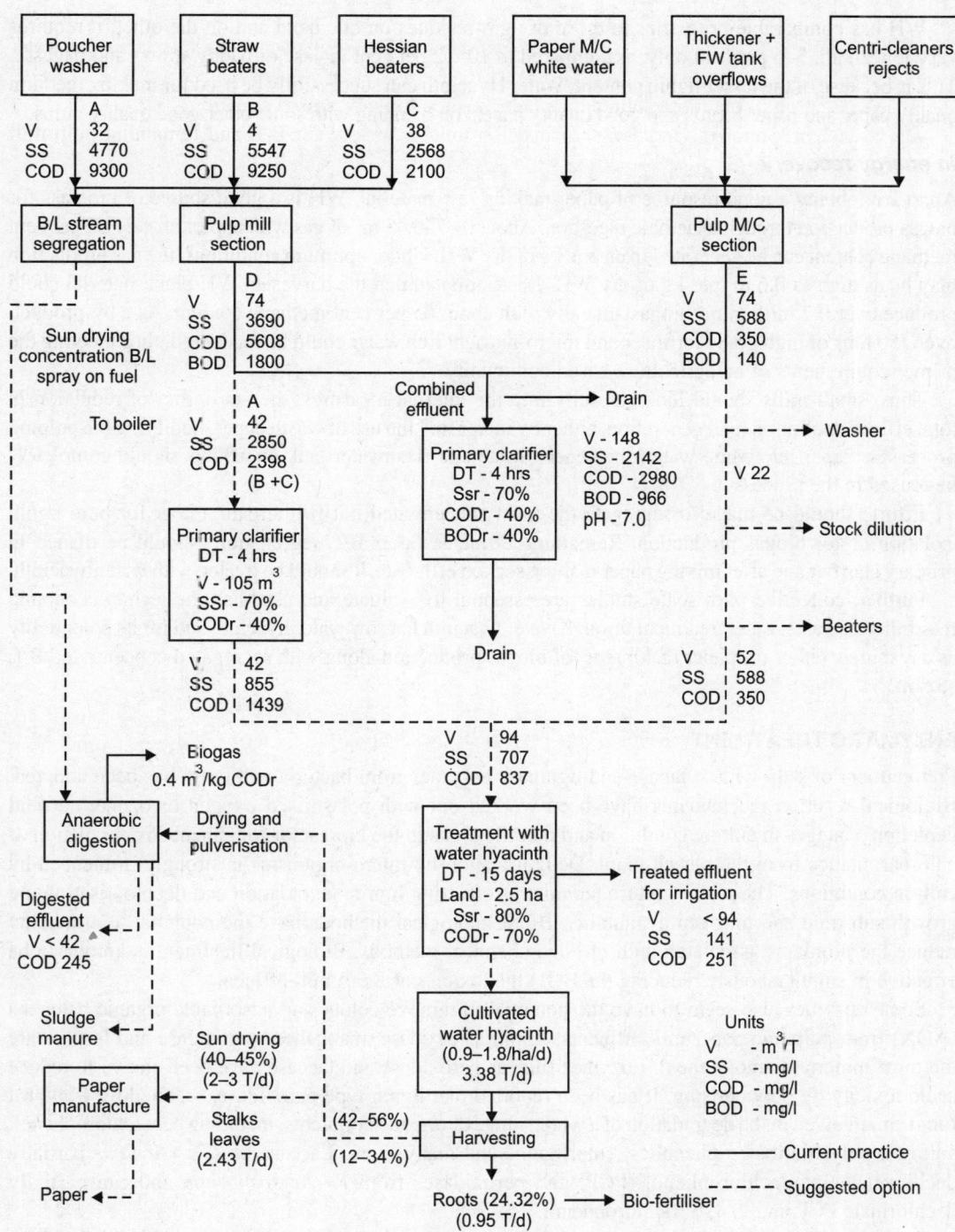

Fig. 16.9. Waste-water treatment strategy for agro-residues based small pulp and paper mills.

WH has comparable properties to those of agro-residues on one hand and on the other, it requires very low alkali, 5–6 per cent only, as compared to 10–12 per cent in case of paddy straws and bagasse. This is because of the lesser lignin content. Water Hyacinth can successfully be used for making medium quality paper and paper boards and good quality papers on blending with some other good quality pulps.

In energy recovery

Apart from being a good resource of paper making raw material, WH is a good source of biomass for biogas production through anaerobic digestion. About 0.37–0.45 m^3 of gas with approximately 65 per cent methane content can be generated from one kg of dry WH. Under optimum conditions, the gas production may be as high as 0.6 m^3 per kg of dry WH. It is estimated that the harvested WH plant material could produce over 0.7 million m^3 biogas annually with about 65 per cent methane content. As a by-product, over 150 tons of high grade fertiliser and micro-nutrient rich water could be recovered, thus meeting the prime requirements of an agriculture based community.

Thus, small mills should look carefully into the internal measures, the possibility of reduction in total effluent/pollution load generation, either by increasing the use of waste paper modifying the pulping processes. Paper m/c white water, thickener filtrate and freshwater tank overflows should completely be reused in the process.

Efforts should be made to segregate the most concentrated hot B/L stream, either for pure lignin isolation or for biogas production. Remaining effluents (after B/L segregation) should be treated in primary clarifier and after mixing paper making section effluents, it should be treated with water hyacinth.

Further, collective pilot scale studies are essential to evaluate and establish the techno-economic feasibility of waste-water treatment through water hyacinth to grow water hyacinth and for its potentiality as a resource, either for paper making or for biogas production along with segregated concentrated B/L stream.

ENZYMATIC TREATMENT

Pretreatment of pulp with xylanase and ligninase enzymes from bacteria and fungi has been adopted. Biological reaction mechanisms have been worked out with pressurised oxygen or ozone, nutrient depletion, changes in culture condition and modifications in the bioreactor for enhanced degradation of chloroaromatics from the bleach plant. Delignification by micro-organisms is strongly influenced by culture conditions. The major culture parametres affecting lignin degradation and decolourisation are growth substrate and nitrogen availability. Biotechnological methods have the potential to eliminate/ reduce the problems associated with physico-chemical methods. Biological treatment is known to be effective in simultaneously reducing the BOD the toxicity of Kraft mill effluent.

Some enzymes also seem to have the potential to remove colour and adsorbable organic halogens (AOX) from pulp and paper mill effluents. Lignin peroxidase, manganese peroxidase and laccase are the most important among them. Enzymes such as peroxidase and laccase have been shown to reduce acute toxicity by polymerising. It has been reported that lignin type peroxidase, secreted by white rot fungi are involved in the degradation of a whole range of organic pollutants, including pentachlorophenol, anilines, polychlorinated phenolics, chlorlignins and many more. Laccase of *T. versicolour* partially dechlorinates pentachlorophenol (PCP) and peroxidases from *P. chrysosporium* and can partially dechlorinate PCP and 2, 4, 6 trichlorophenol.

An enzyme-based treatment process might be a viable approach for the removal of phenolics from foul condensates. Peroxidase enzymes utilise hydrogen peroxide (H$_2$O$_2$) to oxidise a wide range of phenols and aromatic amines to their corresponding radicals. These subsequently combine to form polymers, which can be removed from the aqueous phase through coagulation and precipitation, followed by filtration or sedimentation. Horseradish peroxidase (HRP) is a promising candidate for industrial application due to its broad substrate specificity and stability. The quantity of solids produced from the enzymatic process is very small in comparison with the sludge produced during biological treatment. Enzyme-derived polymers seem to exhibit similar mineralisation kinetics as naturally occurring lignins and humic acids. Enzyme, Lignozyme GmbH, Germany, has been purified on a commercial scale and applied for treatment. However, methods for their safe disposal have not been investigated adequately so for.

GASEOUS EMISSIONS AND AIR POLLUTION

Air Pollution from Big Paper Mills

Air pollution caused by the pulp and paper industry results from the use of mechanical processes, and the combustion equipment required for heating the fibre and pulp. The use of combustion equipment is common to both mechanical and chemical processes. Therefore, the industry is responsible for emissions of suspended particulates and pollutants, specifically related to fossil fuel combustion, carbon monoxide (CO), carbon dioxide (CO_2), sulphur dioxide (SO_2) and nitrogen oxides (NO_x). Pulp mill emissions contain very high levels of NO_x, which is responsible for the formation of acid rain and photochemical smog (an ozone-depleting gas). Pulp mills also emit large quantities of CO_2, a greenhouse gas. Mills using chemical pulping technology produce various sulphur oxides (SO_x), including SO_2, and a number of foul-smelling compounds, such as hydrogen sulphide (H_2S) and thiols. Sulphur dioxide (SO_2) is a major atmospheric pollutant that contributes to the formation of acid rain, precipitation and acid smog. This gas, which is almost completely absorbed between the nose and pharynx, is irritating to the respiratory tract. Sulphur dioxide is also fairly damaging to plants, with concentrations around 0.03 ppm causing acute lesions on foliage. Air pollution from pulp mills has not been well studied. Mills usually do not monitor the range of air emissions, such as particulate matter, carbon dioxide, sulphur dioxide, hydrogen sulphide, volatile organic compounds, chlorine, chloroform, and chlorine dioxide. The general gaseous wastes are digester relief gases, digester blow tank and water hood vent gases, chemical recovery furnace emissions, smelt dissolving tank gases and power/stream boiler emissions. The emissions of pulp and paper mills, using sulphate method of pulping, contain malodorous gases like mercaptans and hydrogen sulphide. Therefore, odour problem exists in all mills. An increase in temperature of the chlorination results in an increase in pollution loadings in chlorination and extraction effluent. An increase in the end pH and temperature of chlorination and chlorine charge, increases the formation of chloroform, potential mutagens and mutagens in the chloroform liquid. Big paper mills are those which produce abvoe 10,000 tons of paper per annum.

Air Pollution from Small Paper Mills

Production capacity of small paper mills

Small paper mills are those which produce 10,000 tons per annum of paper without chemical recovery.

Air pollution, due to release of gaseous emissions into atmosphere, occurs mainly from two sources in small paper mills, viz. (i) digesters, and (ii) steam boilers. The third source could be the captive power generation facilities provided in the mills.

In agricultural residue based mills, after the raw material digestion with caustic soda and/or lime is completed, the pressure in the digester is released after the digester attains a temperature of about 90°C. During the process about 1.4 tons of steam per ton of pulp escape containing volatile organics released during digestion process. The escaping gases have characteristic odour and cause aesthetic pollution problems. These mills unlike the conventional sulphate mills (kraft mills) do not generate hydrogen sulphide and mercaptans (organic sulphides) since sodium sulphide is not used in pulping chemicals. The pollution is mostly confined to the surroundings of the mill, intermittent due to batch process adopted and can be felt at the time of digester gas release.

Coal is the commonly used fuel for generating steam required in all the small paper mills. It is reported that about 3.4 tons of coal is required per ton of paper made in agricultural residue based mills and the corresponding coal requirement in waste paper based mills is 1.5 tons per ton of paper. In all these boilers, coal lumps of 2.5 to 5.0 cm diametre are fired. Very few boilers use powdered coal. Besides coal, paddy husk, diesel oil, etc. are also used in a very few mills. Indian coals are reported to contain 0.5 to 0.8 per cent sulphur. Since the quantity of steam generation is small, the steam boilers are not provided with any air cleaning equipment in these mills. In a study conducted by NEERI on air pollution problems from steam boilers, it is observed that a boiler plant producing 57 tons steam per hour, and provided with an electro-static precipitator (ESP) would release the following quantities of pollutants. The ESP is reported to be working with 80 percent efficiency.

Suspended particulate matter	620–1030 mg/N cubic metre
Sulphur dioxide, SO_2	360–390 mg/N cubic metre
Oxides of nitrogen, NO_x	90–120 mg/N cubic metre

Captive power generation is reported in 5 mills among the 93 mills from which information were obtained. Four of the mills exclusively used diesel oil while the fifth one uses coal and diesel as fuels. The main purpose for providing captive power is to meet sudden power shut-offs and these units are put into operation for specific period whenever such exigency arises. The capacity of power generating units ranged from 110 KW to 2170 KW.

Air pollution is also caused by the delignification of wood by alkalis producing alkyl sulphides, like dimethyl sulphide, and mercaptans, like methyl mercaptan. These products are volatile and, therefore, escape to the atmosphere. Such compounds are not only toxic, but also possess offensive odour. However, as for as pulping is concerned, the air pollution problem is a relative minor one, being restricted to the areas in the immediate vicinity of the mills.

Treatment of air pollutants

A high concentration of contaminants are emitted from the stacks of significant sources like boilers, furnaces, incinerators, and electricity generating units. Particulates can be major air pollutants in many industries, and improved removal can reduce the impact of industrial pollution. Electrostatic precipitators, thermal precipitators and bioscrabbers are used for the removal of air pollutants from pulp and paper mills. Wet scrubbers are applied to provide a more effective removal of fine particles, in the size range of 0.1 to 2.0 m radius from an exhaust gas stream. The research is focused on ways to improve drop size and distribution in a venturi scrubber to optimise particle removal. Postcyclones, which use the energy in an exhaust stream leaving a cyclone to further remove fine particles, may provide a cost effective means of improving particle removal. The research involves both laboratory tests and computer modelling air and particulate movement through the cyclones.

MINIMAL NATIONAL STANDARDS (MINAS)

The MINAS for pulp and paper mill effluent have been evolved after looking into practical difficulties, limitations, techno-economic feasibility and economic impact on the industry.

The basic considerations that went into the development of the MINAS were, therefore, as follows:
1. Characteristics of effluent from small pulp and paper mills without chemical recovery systems.
2. Achievability and techno-economic feasibility of various waste-water treatment alternatives.
3. Maintained ratio of annualised cost to the turn over of the industry.

The tolerance limits for effluents from small pulp and paper industries after considering the above mentioned aspects have been evolved as presented in Table 16.12.

Table 16.12. MINAS for small pulp and paper industry.

Parametre	Concentration
pH	6.0–9.0
Suspended solids	100 mg/l
Bio-chemical oxygen demand	50 mg/l

Permissible limits for chemical oxygen demand and lignin for, which no suitable economic technology is presently available, are not prescribed at the moment. The COD limits shall be introduced in MINAS as and when the suitable economic treatment systems for removal of colour/lignin would be available.

The implementation of above prescribed limits will be made in phased manner. In the first instance, the BOD and suspended solids of treated effluent should be reduced by 90 per cent of the total BOD and SS load in the effluent by June, 2010.

Chapter 17

Soap and Detergent

INTRODUCTION

In modern times, the soap and detergent industry, although a major one, produces relatively small volumes of liquid wastes directly. However, it causes great public concern when its products are discharged after use in homes, service establishments and factories.

The water pollution resulting from the production or use of detergents represents a typical case of the problems that followed the very rapid evolution of the industrialisation that contribute to the improvement of the quality of life after World War II. Prior to that time, this problem did not exist. The continuing increase in consumption of detergents (in particular, their domestic use) and the tremendous increase in production of syndets are the origin of a type of pollution whose most significant impact is the formation of toxic or nuisance foams in rivers, lakes and treatment plants.

CLASSIFICATION OF SURFACTANTS

Soaps and detergents are formulated products designed to meet various cost and performance standards. The formulated products contain many components, such as surfactants to tie up unwanted materials (commercial detergents usually contain only 10–30 per cent surfactants), builders or polyphosphate salts to improve surfactant processes and remove calcium and magnesium ions and bleaches to increase reflectance of visible light. They also contain various additives designed to remove stains (enzymes), prevent soil redeposition, regulate foam, reduce washing machine corrosion, brighten colours, give an agreeable odour, prevent caking and help processing of the formulated detergent.

The classification of surfactants in common usage depends on their electrolytic dissociation, which allows the determination of the nature of the hydrophilic polar group, e.g. anionic, cationic, nonionic and amphoteric.

Anionic surfactants: These produce a negatively charged surfactant ion in aqueous solution, usually derived from a sulphate, carboxylate or sulphonate grouping. The usual types of these compounds are carboxylic acids and derivatives (largely based on natural oils), sulphonic acid derivatives (alkylbenzene sulphonates LAS or ABS and other sulphonates) and sulphuric acid esters and salts (largely sulphated alcohols and ethers). Alkyl sulphates are readily biodegradable, often disappearing within 24 hours in river water or sewage plants. Because of their instability in acidic conditions, they were to a considerable extent replaced by ABS and LAS, which have been the most widely used of the syndets because of their excellent cleaning properties, chemical stability and low cost. Their biodegradation has been the subject of numerous investigations.

Cationic surfactants: These produce a positively charged surfactant ion in solution and are mainly quaternary nitrogen compounds such as amines and derivatives and quaternary ammonium salts. Due to their poor cleaning properties, they are little used as detergents; rather their use is due to their bacteriocidal qualities. Relatively little is known about the mechanisms of biodegradation of these compounds.

Nonionic surfactants: These are mainly carboxylic acid amides and esters and their derivatives, and ethers (alkoxylated alcohols) and they have been gradually replacing ABS in detergent formulations (especially as an increasingly popular active ingredient of automatic washing machine formulations). In nonionic surfactants, both the hydrophilic and hydrophobic groups are organic, so the cumulative effect of the multiple weak organic hydrophils is the cause of their surface-active qualities. These products are effective in hard water and are very low foamers.

Amphoteric surfactants: These presently represent a minor fraction of the total surfactants production with only speciality uses. They are compounds with both anionic and cationic properties in aqueous solutions, depending on the pH of the system in which they work. The main types of these compounds are essentially analogues of linear alkane sulphonates, which provide numerous points for the initiation of biodegradation and pyridinium compounds that also have a positively charged N-atom (but in the ring) and they are very resistant to biodegradation.

SOURCES OF DETERGENTS IN WATERS AND WASTE-WATERS

The detergent concentrations that actually find their way into waste-waters and surface water bodies have quite diverse origins: (i) soaps and detergents, as well as their component compounds, are introduced into waste-waters and water bodies at the point of their manufacture, at storage facilities and distribution warehouses and at points of accidental spills on their routes of transportation; (ii) the additional industrial origin of detergent pollution notably results from the use of surfactants in various industries, e.g. textiles, cosmetics, leather tanning and products, paper, metals, dyes and paints, production of domestic soaps and detergents and from the use of detergents in commercial/industrial laundries and dry cleaners; (iii) the contribution from agricultural activities is due to the surface runoff transporting of surfactants that are included in the formulation of insecticides and fungicides; and (iv) the origin with the most rapid growth since the 1950s comprises the waste-waters from urban areas and it is due to the increased domestic usage of detergents and, equally important, their use in cleaning public spaces.

IMPACTS ON WASTE-WATER TREATMENT PROCESSES

The effect of surfactants on waste-water oils and greases depends on the nature of the latter, as well as on the structure of the lipophilic group of the detergent that assists solubilisation. As is the case, emulsification could be more or less complete. This results in a more or less significant impact on the efficiency of physical treatment designed for their removal. On the other hand, the emulsifying surfactants play a role in protecting the oil and grease molecules from attacking bacteria in a biological unit process.

In water treatment plants, the coagulation/flocculation process was found early to be affected by the presence of surfactants in the raw water supply. In general, the anionic detergents stabilise colloidal particle suspensions or turbidity solids that, most times, are negatively charged. Langelier reported problems with water clarification due to syndets, although according to Nichols and Koepp and Todd concentrations of surfactants on the order of 4 to 5 ppm interfered with flocculation. The floc, instead of settling to the bottom, floats to the surface of sedimentation tanks. Other studies, such as those conducted by Smith and Cohen, indicated that this interference could be not so much due to the surfactants themselves, but to the additives included in their formulation, i.e. phosphate complexes. Such interference was observed both for alum and ferric sulphate coagulant, but the use of certain organic polymer flocculants was shown to overcome this problem.

Surfactants are only partially biodegraded in a sewage treatment plant, so that a considerable proportion may be discharged into surface water bodies with the final effluent. The shorter the overall detention time of the treatment plant, the higher the surfactant concentration in the discharged effluent.

INDUSTRIAL OPERATION AND WASTE-WATER

The soap and detergent industry is a basic chemical manufacturing industry in which essentially both the mixing and chemical reactions of raw materials are involved in production. Also, short- and long-term chemicals storage and warehousing, as well as loading/unloading and transportation of chemicals are involved in the operation.

Manufacture and Formulation

This industry produces liquid and solid cleaning agents for domestic and industrial use, including laundry, dishwashing, bar soaps, speciality cleaners and industrial cleaning products. It can be broadly divided into two categories: (i) soap manufacture that is based on the processing of natural fat, and (ii) detergent manufacture that is based on the processing of petrochemicals.

Numerous processing steps exist between basic raw materials for surfactants and other components that are used to improve performance and desirability and the finished marketable products of the soap and detergent industry.

Inorganic and organic compounds such as ethylene, propylene, benzene, natural fatty oils, ammonia, phosphate rock, trona, chlorine, peroxides and silicates are among the various basic raw materials being used by the industry. The final formulation of the industry's numerous marketable products involves both simple mixing of and chemical reactions among compounds such as the above.

SOAP MANUFACTURE AND PROCESSING

Manufacturing of soap consists of two major operations: the production of neat soap (65–70 per cent hot soap solution) and the preparation and packaging of finished products into flakes and powders, bar soaps and liquid soaps. Many neat soap manufacturers also recover glycerine as a by-product for subsequent concentration and distillation. Neat soap is generally produced in either of two processes: the batch kettle process or the fatty acid neutralisation process, which is preceded by the fat splitting process.

Neat Soap Manufacture and Waste Streams

Batch kettle process

It consists of the following operations: (i) receiving and storage of raw materials, (ii) fat refining and bleaching, and (iii) soap boiling. The major waste-water sources, as shown on the process flow diagram, are the washouts of both the storage and refining tanks, as well as from leaks and spills of fats and oils around these tanks. These streams are usually skimmed for fat recovery prior to discharge to the sewer.

The fat refining and bleaching operation is carried out to remove impurities that would cause colour and odour in the finished soap. The waste-water from this source has a high soap concentration, treatment chemicals, fatty impurities, emulsified fats and sulphuric acid solutions of fatty acids. Where steam is used for heating, the condensate may contain low-molecular-weight fatty acids, which are highly odourous, partially soluble materials.

The soap boiling process produces two concentrated waste streams: sewer lyes that result from the reclaiming of scrap soap and the brine from Nigre processing. Both of these wastes are low-volume, high pH with BOD values up to 45000 mg/l.

Fatty acid neutralisation

Soap manufacture by the neutralisation process is a two-step process:

$$\text{Fat + water} \longrightarrow \text{fatty acid + glycerine (fat splitting)}$$
$$\text{Fatty acid + caustic} \longrightarrow \text{soap (fatty acid neutralisation)}$$

Fat splitting

The manufacture of fatty acid from fat is called fat splitting. Washouts from the storage, transfer and pre-treatment stages are the same as those for Batch kettle process. Process condensate and barometric condensate from fat splitting will be contaminated with fatty acids and glycerine streams, which are settled and skimmed to recover the insoluble fatty acids that are processed for sale. The water will typically circulate through a cooling tower and be reused.

Occasional purges of part of this stream to the sewer release high concentrations of BOD and some grease and oil.

In the fatty acid distillation process, waste-water is generated as a result of an acidification process, which breaks the emulsion. This waste-water is neutralised and sent to the sewer. It will contain salt from the neutralisation, zinc and alkaline earth metal salts from the fat splitting catalyst and emulsified fatty acids and fatty acid polymers.

Fatty Acid Neutralisation

Soap making by this method is a faster process than the kettle boil process and generates less waste-water effluent. Because it is faster, simpler and cleaner than the kettle boil process, it is the preferred process among larger as well as small manufacturers.

Often, sodium carbonate is used in place of caustic. When liquid soaps (at room temperature) are desired, the more soluble potassium soaps are made by substituting potassium hydroxide for the sodium hydroxide (lye). This process is relatively simple and high-purity raw materials are converted to soap with essentially no by-products. Leaks, spills, storm runoff and washouts are absent. There is only one waste-water of consequence: the sewer lyes from reclaiming of scrap. The sewer lyes contain the excess caustic soda and salt added to grain out the soap. Also, they contain some dirt and paper not removed in the strainer.

Glycerine Recovery Process

The process consists of three steps: (i) pre-treatment to remove impurities, (ii) concentration of glycerine by evaporation, and (iii) distillation to a finished product of 98 per cent purity.

There are three waste-waters of consequence from this process: Two barometric condensates, one from evaporation and one from distillation, plus the glycerine foots or still bottoms. Contaminants from the condensates are essentially glycerine with a little entrained salt. In the distillation process, the glycerine foots or still bottoms leave a glassy dark brown amorphous solid rich in salt that is disposed of in the waste-water stream. It contains glycerine, glycerine polymers and salt. The organics will contribute to

BOD, COD and dissolved solids. The sodium chloride will also contribute to dissolved solids. Little or no suspended solids, oil and grease or pH effect should be seen.

Glycerine can also be purified by the use of ion-exchange resins to remove sodium chloride salt, followed by evaporation of the water. This process puts additional salts into the waste-water but results in less organic contamination.

Production of Finished Soaps and Process Wastes

The production of finished soaps utilises the neat soap produced to prepare and package finished soap. These finished products are soap flakes and powders, bar soaps and liquid soap.

Flakes and powders

Neat soap may or may not be blended with other products before flaking or powdering. Neat soap is sometimes filtered to remove gel particles and run into a crutcher for mixing with builders. After thorough mixing, the finished formulation is run through various mechanical operations to produce flakes and powders. Since all of the evaporated moisture goes to the atmosphere, there is no waste-water effluent.

Some operations will include a scrap soap reboil to recover reclaimed soap. The soap reboil is salted out for soap recovery and the salt water is recycled. After frequent recycling, the salt water becomes so contaminated that it must be discharged to the sewer. Occasional washdown of the crutcher may be needed. The tower is usually cleaned down dry. There is also some gland water that flows over the pump shaft, picking up any minor leaks. This will contribute a very small, but finite, effluent loading.

Bar soaps

The procedure for bar soap manufacture will vary significantly from plant to plant, depending on the particular clientele served. The amount of water used in bar soap manufacture varies greatly. In many cases, the entire bar soap processing operation is done without generating a single waste-water stream. The equipment is all cleaned dry, without any wash ups. In other cases, due to housekeeping requirements associated with the particular bar soap processes, there are one or more waste-water streams from air scrubbers.

The major waste streams in bar soap manufacture are the filter backwash, scrubber waters, or condensate from a vacuum drier and water from equipment washdown. The main contaminant of all these streams is soap that will contribute primarily BOD and COD to the waste-water.

Liquid soap

In the making of liquid soap, neat soap (often the potassium soap of fatty acids) is blended in a mixing tank with other ingredients such as alcohols or glycols to produce a finished product or the pine oil and kerosene for a product with greater solvency and versatility. The final blended product may be and often is, filtered to achieve a sparkling clarity before being drummed. In making liquid soap, water is used to wash out the filter press and other equipment. According to manufacturers, there are very little effluent leaks. Spills can be recycled or handled dry. Washout between batches is usually unnecessary or can be recycled to extinction.

DETERGENT MANUFACTURE AND WASTE STREAMS

Detergents, as mentioned previously, can be formulated with a variety of organic and inorganic chemicals, depending on the cleaning characteristics desired. A finished, packaged detergent customarily consists of two main components: the active ingredient or surfactant, and the builder. The processes discussed in the following section will include the manufacture and processing of the surfactant as well as the preparation of the finished, marketable detergent. The production of the surfactant is generally a two-step process: (i) sulphation or sulphonation, and (ii) neutralisation.

Surfactant Manufacture and Waste Streams

Oleum sulphonation/sulphation

One of the most important active ingredients of detergents is the sulphate or sulphonate compounds made via the oleum route. In most cases, the sulphonation/sulphation is carried out continuously in a reactor where the oleum (a solution of sulphur trioxide in sulphuric acid) is brought into contact with the hydrocarbon or alcohol and a rapid reaction ensues. The stream is then mixed with water, where the surfactant separates and is then sent to a settler. The spent acid is drawn off and usually forwarded for reprocessing, and the sulphonated/sulphated materials are sent to be neutralised.

This process is normally operated continuously and performs indefinitely without need of periodic cleanout. A stream of water is generally played over pump shafts to pick up leaks as well as to cool the pumps. Waste-water flow from this source is quite modest but continual.

Air-SO_3 sulphation/sulphonation

This process for surfactant manufacture has many advantages and is used extensively. With SO_3 sulphation, no water is generated in the reaction. In addition, there are usually several airborne sulphonic acid streams that must be scrubbed, with the waste-water going to the sewer during sulphation.

SO_3 solvent and vacuum sulphonation

Undiluted SO_3 and organic reactant are fed into the vacuum reactor through a mixing nozzle. This system produces a high-quality product, but offsetting this is the high operating cost of maintaining the vacuum. Other than occasional washout, the process is essentially free of waste-water generation.

Sulphamic acid sulphation

Sulphamic acid is a mild sulphating agent and is used only in very specialised quality areas because of the high reagent price. Washouts are the only waste-water effluents from this process as well.

Neutralisation of sulphuric acid esters and sulphonic acids

This step is essential in the manufacture of detergent active ingredients as it converts the sulphonic acids or sulphuric acid esters into neutral surfactants. It is a potential source of some oil and grease, but occasional leaks and spills around the pump and valves are the only expected source of waste-water contamination.

Detergent Formulation and Process Wastes

Spray-dried detergents

In this segment of the processing, the neutralised sulphonates and/or sulphates are first blended with builders and additives in the crutcher. The slurry is then pumped to the top of a spray tower of about 4.5 to 6.1 metre (15–20 ft) in diameter by 45 to 61 metre (150–200 ft) high, where nozzles spray out detergent slurry. A large volume of hot air enters the bottom of the tower and rises to meet the falling detergent. The design preparation of this step will determine the detergent particle's shape, size and density, which in turn determine its solubility rate in the washing process.

The air coming from the tower will be carrying dust particles that must be scrubbed, thus generating a waste-water stream. The spray towers are periodically shut down and cleaned. The tower walls are scraped and thoroughly washed down. The final step is mandatory since the manufacturers must be careful to avoid contamination to the subsequent formulation. Waste-water streams are rather numerous and include many washouts of equipment from the crutchers to the spray tower itself. One waste-water flow that has high loadings is that of the air scrubber which cleans and cools the hot gases exiting from this tower. All the plants recycle some of the waste-water generated, while some of the plants recycle all the flow generated. Due to increasingly stringent air quality requirements, it can be expected that fewer plants will be able to maintain a complete recycle system of all water flows in the spray tower area. After the powder comes from the spray tower, it is further blended and then packaged.

Liquid detergents

Detergent actives are pumped into mixing tanks where they are blended with numerous ingredients, ranging from perfumes to dyes. From here, the fully formulated liquid detergent is run down to the filling line for filling, capping, labelling, etc. Whenever the filling line is to change to a different product, the filling system must be thoroughly cleaned out to avoid cross contamination.

Dry detergent blending

Fully dried surfactant materials are blended with additives in dry mixers. Normal operation will see many succeeding batches of detergent mixed in the same equipment without anything but dry cleaning. However, when a change in formulation occurs, the equipment must be completely washed down and a modest amount of waste-water is generated.

Drum-dried detergent

This process is one method of converting liquid slurry to a powder and should be essentially free of the generation of waste-water discharge other than occasional washdown.

Detergent bars and cakes

Detergent bars are either 100 per cent synthetic detergent or a blend of detergent and soap. They are blended in essentially the same manner as conventional soap. Fairly frequent cleanups generate a waste-water stream.

WASTE-WATER CHARACTERISTICS

Waste-waters from the manufacturing, processing and formulation of organic chemicals such as soaps and detergents cannot be exactly characterised. The waste-water streams are usually expected to contain trace or larger concentrations of all raw materials used in the plant, all intermediate compounds produced during manufacture, all final products, coproducts and by-products and the auxiliary or processing chemicals employed. It is desirable, from the viewpoint of economics, that these substances not be lost, but some losses and spills appear unavoidable and some intentional dumping does take place during housecleaning and vessel emptying and preparation operations.

According to a study by Zimmerman presenting estimates of industrial waste-water generation as well as related pollution parameter concentrations, the waste-water volume discharged from soap and detergent manufacturing facilities per unit of production ranges from 0.3 to 2.8 gal/lb. (2.5–23.4 l/kg) of product. The reported ranges of concentration (mg/l) for BOD, suspended solids, COD and grease were 500 to 1200, 400 to 2100, 400 to 1800, and about 300, respectively.

These data were based on a study of the literature and the field experience of governmental and private organisations. The values represent plant operating experience for several plants consisting of 24 hours composite samples taken at frequent intervals. The ranges for flow and other parametres generally represent variations in the level of plant technology or variations in flow and quality parametres from different subprocesses. In particular, the more advanced and modern the level of production technology, the smaller the volume of waste-water discharged per unit of product. The large variability (up to one order of magnitude) in the ranges is generally due to the heterogeneity of products and processes in the soap and detergent industry.

WASTE-WATER CONTROL AND TREATMENT

In-Plant Control and Recycle

Significant in-plant control of both waste quantity and quality is possible particularly in the soap manufacturing subcategories where maximum flows may be 100 times the minimum. Considerably less in-plant water conservation and recycle are possible in the detergent industry, where flows per unit of product are smaller.

The largest in-plant modification that can be made is the changing or replacement of the barometric condensers. The waste-water quantity discharged from these processes can be significantly reduced by recycling the barometric cooling water through fat skimmers, from which valuable fats and oils can be recovered and then through the cooling towers.

The only waste with this type of cooling would be the continuous small blowdown from the skimmer. Replacement with surface condensers has been used in several plants to reduce both the waste flow and quantity of organics wasted.

Significant reduction of water usage is possible in the manufacture of liquid detergents by the installation of water recycle piping and tankage and by the use of air rather than water to blowdown filling lines. In the production of bar soaps the volume of discharge and the level of contamination can be reduced materially by installation of an atmospheric flash evaporator ahead of the vacuum drier. Finally, pollutant carry-over from distillation columns such as those used in glycerine concentration or fatty acid separation can be reduced by the use of two additional special trays.

In a recent document presenting techniques adopted by the French for pollution prevention, a new process of detergent manufacturing effluent recycle is described. As shown in Fig. 17.1, the washout effluents from reaction and/or mixing vessels and washwater leaks from the paste preparation and pulverisation pump operations are collected and recycled for use in the paste preparation process. The claim was that pollution generation at such a plant is significantly reduced and although the savings on water and raw materials are small, the capital and operating costs are less than those for building a waste-water treatment facility.

Besselievre reported in a review of water reuse and recycling by the industry that soap and detergent manufacturing facilities showed an average ratio of reused and recycled water to total waste-water effluent of about 2:1. That is, over two-thirds of the generated waste-water stream in an average plant was being reused and recycled. Of this volume, about 66 per cent was used as cooling water and the remaining 34 per cent for the process or other purposes.

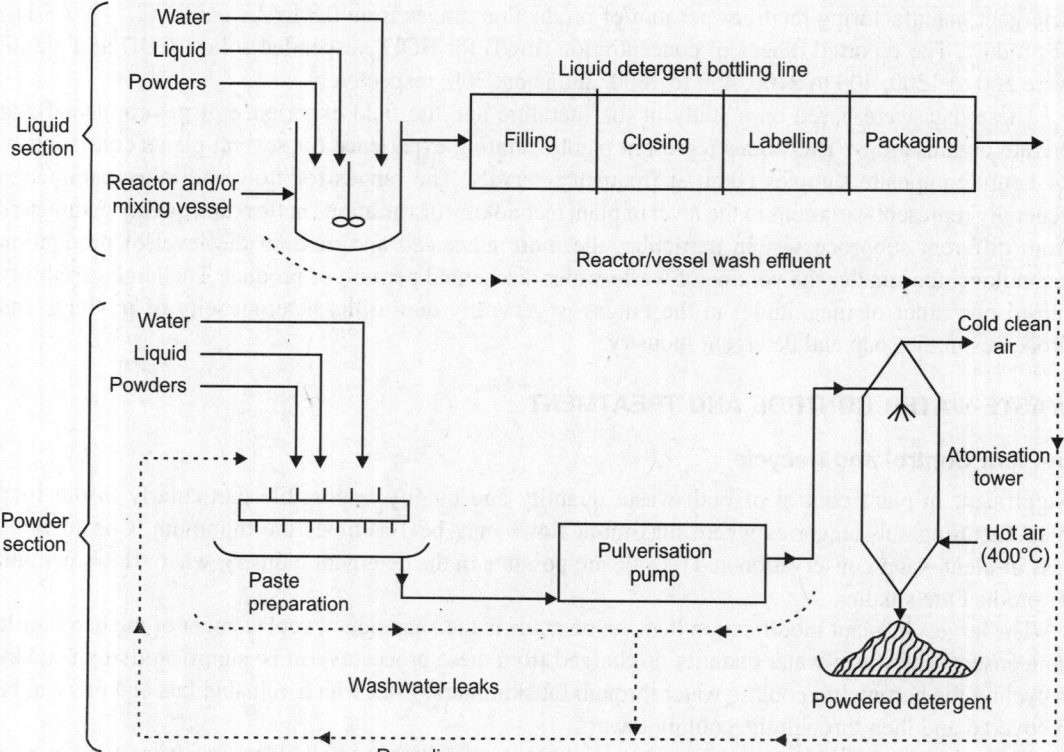

Fig. 17.1. Process modification of waste-water recycling in detergent manufacture.

Waste-water Treatment Methods

The soap and detergent manufacturing industry makes routine use of various physico-chemical and biological pre-treatment methods to control the quality of its discharges. A survey of these treatment processes is presented in Table 17.1 which also shows the usual removal efficiencies of each unit process on the various pollutants of concern. According to Nemerow the origin of major wastes is in washing and purifying soaps and detergents and the resulting major pollutants are high BOD and saponified soaps (oily and greasy, alkali and high-temperature wastes), which are removed primarily through air-flotation and skimming and precipitation with the use of $CaCl_2$ as a coagulant.

Table 17.1. Treatment methods in the soap and detergent industry.

Pollutant and method	Efficiency (percentage of pollutant removed)
Oil and grease	
API-type separation	Up to 90% of free oils and greases
	Variable on emulsified oil.
Carbon adsorption	Up to 95% of both free and emulsified oils.
Flotation	Without the addition of solid phase, alum or iron, 70-80% of both free and emulsified oil. With the addition of chemicals, 90%.

(Contd...)

Pollutant and method	Efficiency (percentage of pollutant removed)
Mixed-media filtration	Up to 95% of free oils. Efficiency in removing emulsified oils unknown
Coagulation/sedimentation with iron, alum or solid phase (bentonite, etc.)	Up to 95% of free oil. Up to 90% of emulsified oil.
Suspended solids	
Mixed-media filtration	70–80%
Coagulation/sedimentation	50–80%
BOD and COD	
Bioconversions (with final clarifier)	60–95% or more
Carbon adsorption	Up to 90%
Residual suspended solids	
Sand or mixed-media filtration	50–95%
Dissolved solids	
Ion exchange or reverse osmosis	Up to 90%

Figure 17.2 presents a composite flow diagram describing a complete treatment train of the unit processes that may be used in a large soap and detergent manufacturing plant to treat its wastes. As a minimum requirement, flow equalisation to smooth out peak discharges should be utilised even at a production facility that has a small-volume batch operation. Larger plants with integrated product lines may require additional treatment of their waste-waters for both suspended solids and organic materials' reduction.

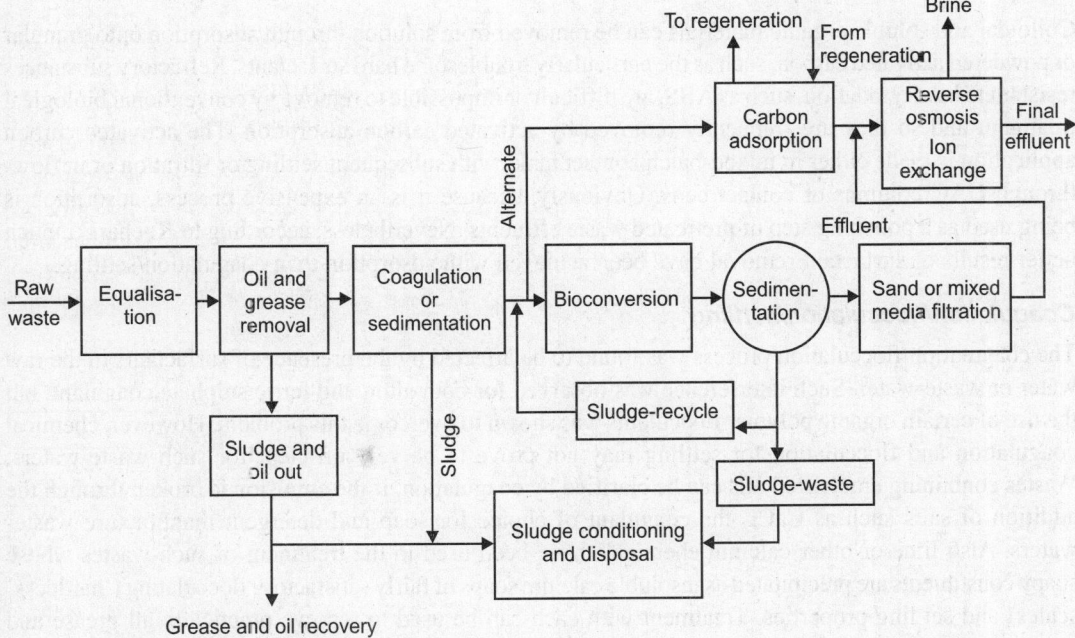

Fig. 17.2. Composite flowsheet of waste treatment in soap and detergent industry.

Coagulation and sedimentation are used by the industry for removing the greater portion of the large solid particles in its waste. On the other hand, sand or mixed-bed filters used after biological treatment can be utilised to eliminate fine particles. One of the biological treatment processes or, alternatively, granular or powdered activated carbon is the usual method employed for the removal of particulate or soluble organics from the waste streams. Finally, as a tertiary step for removing particular ionised pollutants or total dissolved solids (TDS), a few manufacturing facilities have employed either ion exchange or the reverse osmosis process.

Flotation or foam fractionation

One of the principal applications of vacuum and pressure (air) flotation is in commercial installations with colloidal wastes from soap and detergent factories. Waste-waters from soap production are collected in traps on skimming tanks, with subsequent recovery floating of fatty acids. Foam separation or fractionation can be used to an extra advantage. Not only do surfactants congregate at the air/liquid interfaces, but other colloidal materials and ionised compounds that form a complex with the surfactants tend to also be concentrated by this method. An incidental, but often important, advantage of air flotation processes is the aerobic condition developed, which tends to stabilise the sludge and skimmings so that they are less likely to turn septic. However, disposal means for the foamate can be a serious problem in the use of this procedure. It has been reported that foam separation has been able to remove 70 to 80 per cent of synthetic detergents, at a wide range of costs. Gibbs reported the successful use of fine bubble flotation and 40 minutes detention in treating soap manufacture wastes, where the skimmed sludge was periodically returned to the soap factory for reprocessing. According to Wang the dissolved air flotation process is also feasible for the removal of detergents and soaps from water.

Activated carbon adsorption

Colloidal and soluble organic materials can be removed from solution through adsorption onto granular or powdered activated carbon, such as the particularly troublesome hard surfactants. Refractory substances resistant to biodegradation, such as ABS, are difficult or impossible to remove by conventional biological treatment and so they are frequently removed by activated carbon adsorption. The activated carbon application is made either in mixed-batch contact tanks with subsequent settling or filtration or in flow-through GAC columns or contact beds. Obviously, because it is an expensive process, adsorption is being used as a polishing step of pretreated waste effluents. Nevertheless, according to Kucharski much better results of surfactant removal have been achieved with adsorption than coagulation/settling.

Coagulation/flocculation/settling

The coagulation/flocculation process was found to be affected by the presence of surfactants in the raw water or waste-water. Such interference was observed for both alum and ferric sulphate coagulant, but the use of certain organic polymer flocculants was shown to overcome this problem. However, chemical coagulation and flocculation for settling may not prove to be very efficient for such waste-waters. Wastes containing emulsified oils can be clarified by coagulation, if the emulsion is broken through the addition of salts such as $CaCl_2$ the coagulant of choice for soap and detergent manufacture waste-waters. Also lime or other calcium chemicals have been used in the treatment of such wastes whose soapy constituents are precipitated as insoluble calcium soaps of fairly satisfactory flocculating ('hardness' scales) and settling properties. Treatment with each can be used to remove practically all grease and suspended solids and a major part of the suspended BOD. Using carbon dioxide (carbonation) as an auxiliary precipitant reduces the amount of calcium chloride required and improves treatment efficiency. The sludge from $CaCl_2$ treatment can be removed either by sedimentation or by air or vacuum flotation.

Ion exchange and exclusion

The ion-exchange process has been used effectively in the field of waste disposal. The use of continuous ion-exchange and resin regeneration systems has further improved the economic feasibility of the applications over the fixed-bed systems. One of the reported special applications of the ion-exchange resins has been the removal of ABS by the use porous anion exchanger that is a strong base and depends on a chloride cycle. This resin system is regenerated by removing a great part of the ABS absorbed on the resin beads with the help of a mixture of HCl and acetone. Other organic pollutants can also be removed by ion-exchange resins and the main problem is whether the organic material can be eluted from the resin using normal regeneration or it is economically advisable to simply discard the used resin. Wang and Wood successfully used the ion-exchange process for the removal of cationic surfactant from water.

The separation of ionic from non-ionic substances can be effected by the use of ion exclusion. Ion-exchange can be used to purify glycerine for the final product of chemically pure glycerine and reduce losses to waste, but the concentration of dissolved ionisable solids or salts (ash) largely impacts on the overall operating costs. Economically, when the crude or sweet water contains under 1.5 per cent ash, straight ion exchange using a cation-anion mixed bed can be used, whereas for higher percentages of dissolved solids, it is economically feasible to follow the ion-exchange with an ion-exclusion system. For instance, waste streams containing 0.2 to 0.5 per cent ash and 3 to 5 per cent glycerine may be economically treated by straight ion-exchange, while waste streams containing 5 to 10 per cent ash and 3 to 5 per cent glycerine have to be treated by the combined ion-exchange and ion-exclusion processes.

Biological Treatment

Regarding biological destruction, as mentioned previously, surfactants are known to cause a great deal of trouble due to foaming and toxicity in municipal treatment plants. The behaviour of these substances depends on their type, i.e. anionic and non-ionic detergents increase the amount of activated sludge, whereas cationic detergents reduce it and also the various compounds decompose to a different degree. The activated sludge process is feasible for the treatment of soap and detergent industry wastes but, in general, not as satisfactory as trickling filters. The turbulence in the aeration tank induces frothing to occur and also the presence of soaps and detergents reduces the absorption efficiency from air bubbles to liquid aeration by increasing the resistance of the liquid film.

On the other hand, detergent production waste-waters have been treated with appreciable success on fixed-film process units such as trickling filters. Also, processes such as lagoons, oxidation or stabilisation ponds and aerated lagoons have all been used successfully in treating soap and detergent manufacturing waste-waters. Finally, Vath demonstrated that both linear anionic and nonionic ethoxylated surfactants underwent degradation, as shown by a loss of surfactant properties, under anaerobic treatment.

AIR EMISSIONS

Slurry Preparation

The formulation of slurry for detergent granules requires the intimate mixing of various liquids, powdered, and granulated materials. The soap crutcher is almost universally used for this mixing operation.

380 Pollution Control in Chemical and Allied Industries

Premixing of various minor ingredients is performed in a variety of equipment prior to charging to the crutcher or final mixer. The slurry, mixed in batch operations, is then held in surge vessels for continuous pumping to the spray dryer.

Air emissions characterisation

The receiving, storage, and hatching of the various dry ingredients create dust emissions. Pneumatic conveying of fine materials causes dust emissions when conveying air is separated from the bulk solids. Many detergent products require raw materials with a high percentage of fines. Typical specifications for some raw materials include the following percentage of fine materials passing a 200 mesh screen: sodium sulphate, 12 per cent; TSPP, 74 per cent; STP, 53 per cent.

The storage and handling of the liquid ingredients, including the sulphonic acids, sulphonic salts, and sulphates, do not cause emission problems other than mild odours. In the batching and mixing of the fine dry ingredients to form slurry, dust emissions are generated at scale hoppers, mixers, and the crutcher. Liquid-ingredient addition to the slurry creates no visible emissions, but may cause odours.

Air pollution control measures

Conrol of dusts generated from pneumatic or mechanical conveying or from discharge of fine materials into bins or vessels can be controlled by fixing bag filters. No unique problems occur in hooding or exhaust systems for controlling dust emissions from conveying and slurry preparation.

Baghouses are employed not only to reduce and eliminate the dust emissions, but also for the salvage of raw materials. None of the dusts causes any serious corrosion problems. Filter fabrics should be selected that have good resistance to alkalis. Filter ratios for baghouses with intermittent shaking cleaning mechanisms should be under 3 cfm/ft^2. Continuous process for fatty acids and soaps is shown in Fig. 17.3.

Spray Drying

Process description

All spray-drying equipment designed for detergent granule production incorporates the following components: spray-drying tower, air heating and supply system, slurry atomising equipment, slurry pumping equipment, product cooling equipment, and conveying equipment. The towers are cylindrical with cone bottoms and range in size from 12 to 24 ft in diametre and 40 to 125 ft in height. Single towers may vary in diametre, being larger at the top and smaller at the bottom. Air is supplied to the towers from direct-heated furnaces fired with either natural gas or fuel oil. The products of combustion are tempered with outside air to lower temperatures and then blown to the dryer under forced draft. The towers are usually maintained under slightly negative pressure, between 0.05 and 1.5 inch of water column, with exhaust blowers adjusted to provide this balance.

Most towers designed for detergent production are of the countercurrent type, with the slurry introduced at the top and the heated air introduced at the bottom. A few towers of the concurrent type are used for detergent spray drying, with both hot air and slurry introduced at the top. Some towers are equipped for either mode of operation, as illustrated in Fig. 17.4.

In most towers today, the slurry is atomised by spraying through a number of nozzles rather than by centrifugal action. The slurry is sprayed at pressures of 600–1000 psi in single-fluid nozzles and at pressures of 50–100 psi in two-fluid nozzles. Steam or air is used as the atomising fluid in the two-fluid nozzles.

Soap and Detergent 381

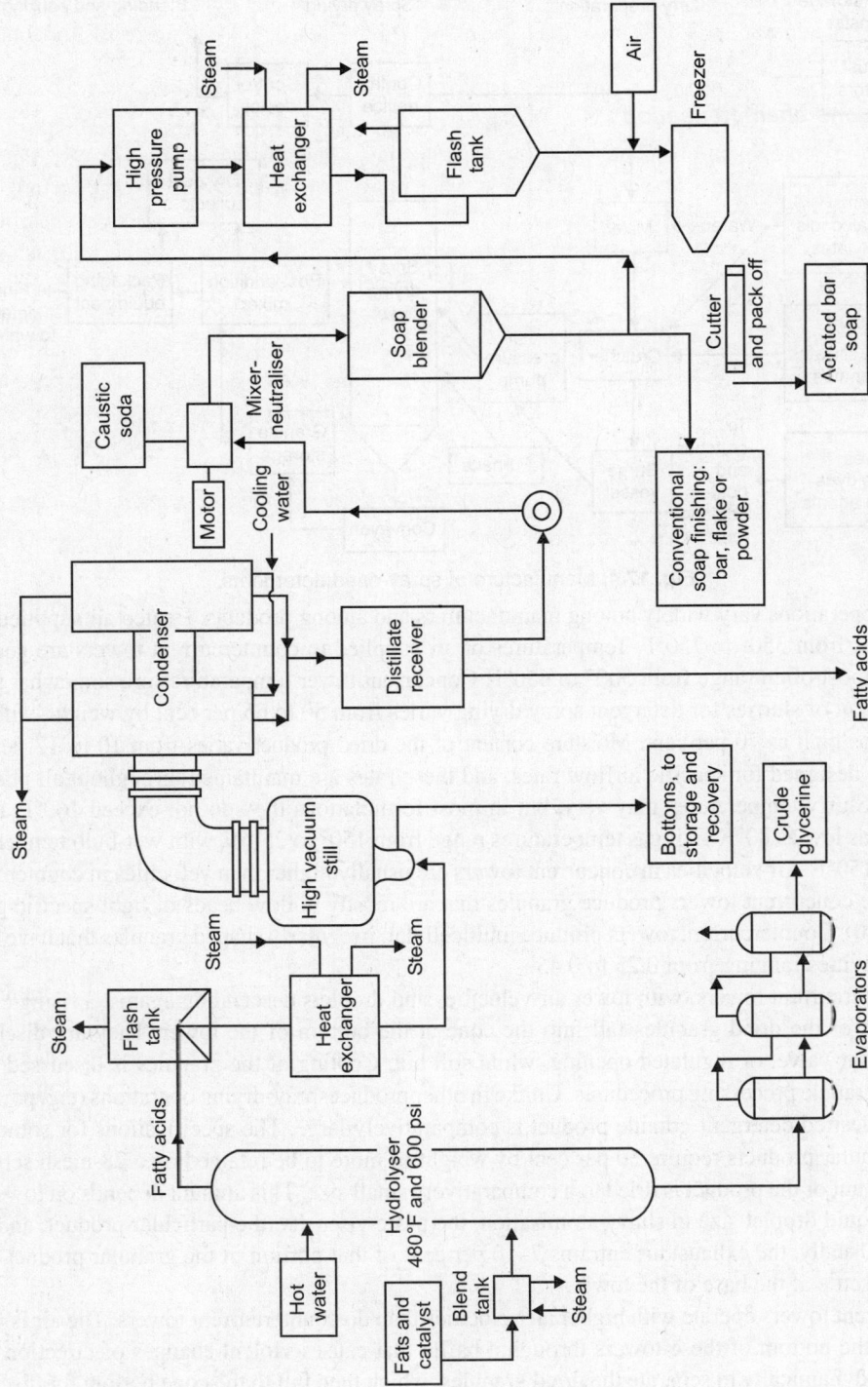

Fig. 17.3. Continuous process for fatty acids and soaps.

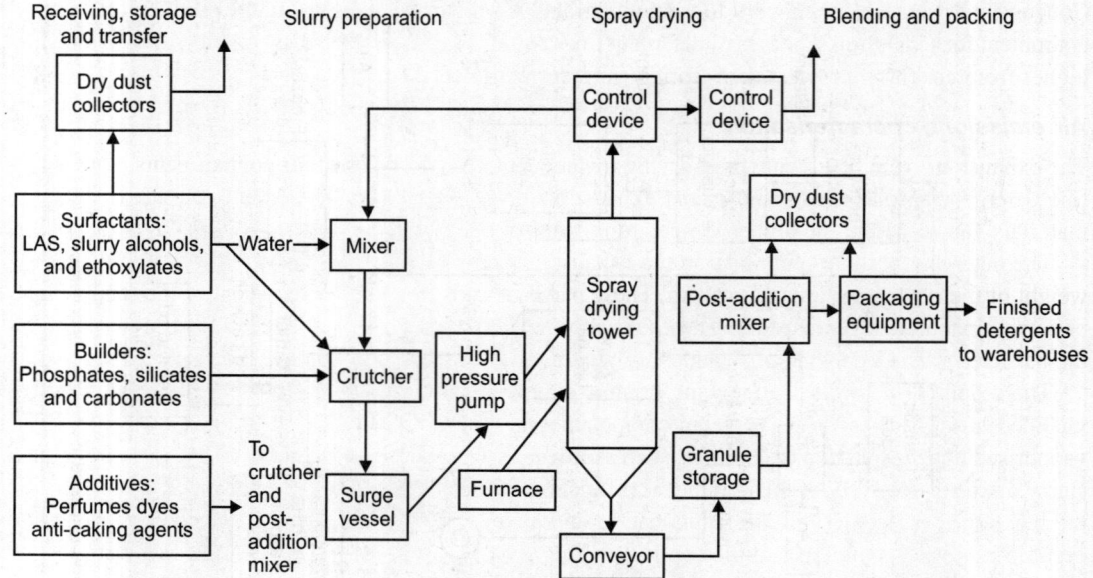

Fig. 17.4. Manufacture of spray-dried detergents.

Tower operations vary widely among manufacturers and among products. Heated air supplied to the tower varies from 350° to 750°F. Temperatures of air supplied to countercurrent towers are generally lower and most often range from 500° to 650°F. Concurrent tower temperatures are somewhat higher. Solids content of slurries for detergent spray drying varies from 50 to 65 per cent by weight, with some operations as high as 70 per cent. Moisture content of the dried product varies from 10 to 17 per cent. Towers are designed for specific airflow rates, and these rates are maintained throughout all phases of operation. Slurry temperatures may vary, but in most formulations they do not exceed 160°F and are frequently as low as 80°F. Exit gas temperatures range from 150° to 250°F, with wet-bulb temperatures of 120° to 150°F. Air velocities in concurrent towers are usually higher than velocities in countercurrent towers. The concurrent towers produce granules that are mostly hollow beads of light specific gravity (0.05 to 0.20). Countercurrent towers produce multicellular, irregularly shaped granules that have higher specific gravities ranging from 0.25 to 0.45.

In countercurrent towers, with lower air velocities and droplets descending against a rising column of air, most of the dried granules fall into the cone at the bottom of the tower. They are discharged through a star valve, or regulated opening, while still hot. Cooling of the granules is discussed below with other granule processing procedures. Unlike in other product spray-drying operations (e.g. powdered milk), the desired detergent granule product is comparatively large. The specifications for some well-known granular products require 50 per cent by weight or more to be retained on a 28-mesh screen. A certain amount of the product is dried to a comparatively small size. This amount depends on tower feed rates, the liquid droplet size in slurry atomisation, the paste viscosity, the particular product, and other variables. Usually, the exhaust air entrains 7–10 per cent of that portion of the granular product that is too fine to settle at the base of the tower.

Concurrent towers operate with higher air velocities than do countercurrent towers. The air is vented just above the bottom of these towers through a baffle that causes violent changes of direction to the exhaust air dynamically to separate the dried granules, which then fall to the cone bottom for discharge.

Concurrent towers producing very-low-gravity granules vent air still conveying the product to auxiliary equipment for separation. The loss of detergent fines entrained in the exhaust air stream will be somewhat higher from concurrent towers than from countercurrent towers.

Air emissions characterisation

The exhaust air from detergent spray-drying towers contains two types of air contaminants. One is the fine detergent particles entrained in the exhaust air discussed above; the second consists of organic materials vapourised in the higher-temperature zones of the tower.

The detergent particles entrained in the exhaust air are relatively large in size. Over 50 per cent by weight of these particles are over 40 μm. These particles constitute over 95 per cent of the total weight of contaminants in the exhaust air. They consist principally of detergent compounds, although some of the particles are uncombined phosphates, sulphates, and other mineral compounds.

The second type of air contaminant, organic compounds, originates primarily from the surfactants included in the slurry. Various organic components in the slurry vapourise in the tower. The amount vapourised depends on many variables, such as tower temperatures and volatility of the organics in the slurry mixture. Volatility of the organics is a function of the chain length of surfactant compounds and the reaction completeness in the sulphonation and sulphation of anionic surfactants. The vapourised organic materials condense upon cooling in the tower exhaust air stream into micrometer- and submicrometer-sized droplets or particles.

The variety of possible detergent compounds is almost infinite, and manufacturers are continually introducing new formulations or reformulating older ones. It is not always possible to predict how certain organic compounds in a slurry mixture will affect stack emission. Work in the early 1970s indicated that if amides are present in the slurry in amounts greater than 0.5 per cent by weight, emission problems will occur. Source tests of exhaust air from an air-pollution control scrubber with amides present in the slurry being dried in the spray tower revealed 0.08 grain of organic particulate per standard cubic foot of exhaust gas. The presence of this relatively low concentration of submicrometer-sized aerosols causes water vapour plumes to persist for long distances. Following the break or end of the water vapour plume, a highly visible contaminant plume persists for even greater distances. The amide emission rate increases as tower operating temperatures rise. Many tower operating variables affect air contaminant emissions. The most significant variables are the formulation's final granule temperature and moisture, the dryer inlet air temperature, and temperature profiles in the dryer.

Slurry formulations containing alcohol ethoxylate surfactants or alcohol sulphates with high levels of unsulphated alcohols cause similar aerosol emissions. Source tests of the aerosol leaving the scrubber indicate a particle size range of 0.2 to 1 μm, where ethoxylated alcohol surfactants were added to the crutcher. It is believed that very small amounts of unreacted alcohols coat water droplets like amides, inhibiting the evaporation of the droplets and resulting in gray water vapour plumes that persist for long distances.

Air pollution control measures

The collection of air contaminants not only provides for the economic return of detergent fines to the process, but it also provides for control of submicrometer particles and aerosols to ensure compliance with air pollution prohibitory rules.

Manufacturers producing detergent granules have developed several separate approaches for capturing the detergent fines in the spray-dryer effluent for return to process. Dry cyclones and cyclonic impingement scrubbers are the primary approaches used. Cyclonic impingement scrubbers return a slurry to the crutcher, while dry cyclones return detergent fines to the crutcher. The dry cyclone separators can remove 90 per cent or more by weight of the detergent product fines in the tower exhaust air. The detergent dust remaining in the effluent vented from the cyclones consists of particles 95 per cent by weight less than 10 μm. Particulate concentrations vary from 0.1 to 1.0 g/scf. The cyclones are designed for relatively high efficiencies and operate at pressure drops of from 8 to 10 inches of water column.

Secondary collection equipment is used to collect fine dust and organic aerosols that pass the primary collectors. Mist eliminators are used after cyclonic impingement scrubbers. Cyclones are generally followed by scrubbers, scrubber/precipitators, or fabric filters. Several types of scrubbers can be used following the cyclone collectors. Venturi scrubbers operating at 8 to 10 inches of water pressure drop (using water at 8 to 10 psig distributed through nozzles in the throat, velocities of 8500 fpm, and water supplied to the throat at a ratio of 4.5 to 5.0 gal per 1000 ft^3 of effluent) have been used following the cyclones. Venturi scrubbers have been replaced largely by packed bed scrubbers operating at 0.5 to 2.0 inches of water pressure drop and water rates of 1 to 3 gal per 1000 acfm. Packed-bed scrubbers are usually followed by wetpipe-type electrostatic precipitators constructed immediately above the packed bed in the same vessel. Fabric filters are sometimes used after cyclones, but their application is limited to spray dryers having low drying loads and drying products with low surfactant concentrations. On efficient spray dryers with exhaust conditions near the dew point drying products with more than 10–15 per cent surfactant, condensing water vapour and condensing organic aerosols will bind the filter fabrics.

Typical control efficiencies and emission factors for various control options are given in Table 17.2. As previously mentioned, small amounts of organic, especially paraffin alcohols and amides, can result in a highly visible plume that can persist after the condensed water vapour plume has dissipated. The precipitator/scrubber appears to provide the best control for these compounds, although no conclusive data exist. Differences in formulation among manufacturers and differences in drying systems make comparison impossible. Fabric filters operating at higher temperatures will not collect organic aerosols, some of which will be caught on the filter.

Table 17.2. (Metric and English units) particulate emission factors for detergent spray drying[a] emission factor rating: E[b].

Control device	Efficiency (%)	Particulate	
		kg/mg of product	lb/ton of product
Uncontrolled	NA	–	–
(SCC 3–01–099–01)	–	45	90
Cyclone	85	7	14
Cyclone with			
Spray chamber	92	3.5	7
Packed scrubber	95	2.5	5
Venturi scrubber	97	1.5	3
Wet scrubber	99	0.544	1.09
Wet scrubber/ESP	99.9	0.023	0.046
Packed bed/ESP	99	0.47	0.94
Fabric filter	99	0.54	1.1

[a]Some type of primary collector, such as a cyclone, is considered integral to a spray-drying system. ESP, electrostatic precipitator; SCC, Source Classification Code.
[b]Emission factors are estimations and are not supported by current test data.
[c]Emission factor has been calculated from a single source test. An efficiency of 99 per cent has been estimated.

Opacity and the organics emitted are influenced by granule moisture and temperature at the end of drying, temperature profiles on the dryer, and formulation of the slurry. An alternative method for controlling visible emissions from the dryer caused by volatile organics in the slurry is to reformulate the slurry to eliminate these offending organic compounds. Sometimes the purity of raw materials such as alcohol ethoxylates can be improved by stripping short-chain molecules from the raw material. When amide compounds were identified as causing the emission problems, some manufacturers developed other formulations or methods for adding the amides to the spray-dried granules after the dryer to achieve a comparable product.

When reformulation is not possible, the tower production rate may be reduced, permitting operation at lower air inlet temperatures and lower exhaust gas temperatures. When tower temperatures are reduced, lower temperatures may result in lower granule temperature and higher granule moisture, thus reducing organic emissions.

Granule Handling

Process description

Many manufacturers discharge hot granules from the spray tower into mixers where dry or liquid ingredients are added. The granules are usually mechanically conveyed away from the tower or mixer discharge and then are air-conveyed to storage and packaging. Air conveying cools the granules and elevates them for gravity flow through further processing equipment to storage and packaging. Air conveying of low-density granules usually is designed for 50–75 scfm air per pound of granules conveyed. At the end of the conveyor, gravity or centrifugal separators remove granule product from the conveying air. Some manufacturers mechanically lift the granules from the spray tower to aeration bins, where they are cooled or aged by injecting air at the bottom of the bin. This air percolates upward through the scrubber.

The cooled granules are screened to deagglomerate the large granules and to remove undersize or oversize particles. Further mixing or blending may be performed to add heat sensitive compounds such as enzymes to the detergent products. Many manufactures do not store the finished granules, but convey them directly to packaging equipment. Some detergent products are held in storage, in either large fixed bins or small-wheeled buggy bins, and then are charged to packaging equipment. Packaging is done with either scale or volumetric filling machines.

Air emissions characterisation

Conveying, mixing, packaging, and other equipment used for granules can cause dust emissions. The granule particles, which are hollow beads, are crushed during mixing and conveying and generate fine dusts. With continuous exposure dusts emitted from screens, mixers, bins, mechanical conveying equipment, and air-conveying equipment are quite irritating to eyes and nostrils. Some additive materials, such as enzymes, bleaches, bleach activators, and brighteners, also can cause health problems. Equipment involving enzymes requires very efficient ventilation in addition to proper dust collection. Dust emissions in most cases represent a significant product loss and their collection and return to process (usually as an ingredient of the slurry to be spray dried) is necessary for cost-efficient plant operation, as well as for air pollution control.

Air pollution control measures

Dust generated by granule processing, conveying, and storage equipment does not create unique air pollution control problems. Usually, baghouses provide the best control. Collection efficiencies for baghouses are high; efficiencies can exceed 99 per cent. No extreme conditions of temperature or humidity have to be met, but filter fabrics selected must show good resistance to alkaline materials. Baghouses utilising intermittent shaking mechanisms should not have filtering velocities exceeding 3 fpm. Baghouses with continuous cleaning mechanisms may have filtering velocities as high as 6 fpm.

Current developments

The recent introduction of superconcentrated detergents and a new push for more environmentally friendly products and packaging may cause a change equal to or greater than the shift to liquids sparked by phosphate bans. Manufacturers are developing more biodegradable surfactants from natural oils. The introduction of polymers for builder systems offers improved performance for zero- and low-phosphate detergents. Superconcentrated powders offer the environmental benefit of reducing packaging material and shipping energy by eliminating fillers and compressing the detergent to achieve two-to-one and three-to-one volume reductions. These factors may push the market back toward powder detergents. While this may increase the demand for spray-drying capacity, newer formulations might also allow other manufacturing processes with less pollution potential.

Chapter 18

Pesticide

INTRODUCTION

Pesticide is a composite term that includes all chemicals that are used to kill or control pests. In agriculture, this includes herbicides (weeds), insecticides (insects), fungicides (fungi), nematicides (nematodes), and rodenticides (vertebrate or rodent poisons). A fundamental contributor to the Green Revolution has been the development and application of pesticides for the control of a wide variety of insectivorous and herbaceous pests that would otherwise diminish the quantity and quality of food produce.

However, there is overwhelming evidence that the agricultural use of pesticides has a major impact on water quality and this has led to serious environmental consequences.

Indiscriminate and nonjudicious use of pesticides has led to acute contamination of food and food products, groundwater and air. Misuse of pesticides is a serious problem in most of the developing countries, including India.

Availability and sale of banned and outdated pesticides is quite common. Precautionary measures to be taken by the personnel involved in spraying operations are hardly followed, thereby leading to serious health problems in farm workers. The waiting period recommended for different pesticides is hardly even followed before the product is harvested.

Pesticide manufacturing is energy-intensive. Most pesticides are derived from ethylene and propylene, which are obtained by catalytic cracking of crude petroleum oils, or from methane from natural gas. Some pesticides are more energy-intensive than others; however, pesticides also vary in their energy use, per unit area of application. The trend in pesticide manufacturing is towards the production of pesticides that are more energy-intensive per unit, but are applied at a very low rate per acre.

The environment is being polluted by industrial wastes, animal and plant wastes, automobile exhausts and agricultural chemicals, which include pesticides and fertilisers. The major source of environmental contamination by pesticides is the deposits resulting from the application of these chemicals to control agricultural pests and pest causing public health problems. What concerns the society the most is the biological effects of these pesticide derivatives, which at various stages of environmental alteration can come in contact with many biological systems. Two aspect of the problem appear to be particularly important:
1. Effect on man and domestic animals which may take up pesticides mostly through contaminated food and feeds.
2. Effect on wild life, where certain species can severely be affected by absorption or accumulation of pesticides through the food web, disturbing the balance of the ecosystem in nature.

Pesticides are generally applied to the soil, plants, water bodies and human settlements. There would be least environmental contamination if these fall exactly on the target and get degraded to harmless compounds.

This situation does not happen and a large portion of the pesticides applied drift into environment, thus affecting the structure and function of the ecosystem and communities.

Fears have been expressed about possible pollution of natural waters and reservoirs by high concentrations of nitrogen, potassium and phosphate nutrients from the run-off of modern chemical fertilisers from agricultural land.

CLASSIFICATION OF PESTICIDES

By definition, a pesticide is a pest-killing agent. The term usually refers to one or more materials developed and used to destroy a broad range of specific pests. In legal terminology, pesticides may be defined as 'any substance used for controlling, preventing, destroying, repelling, or mitigating any pest'. Hence, even chemicals that do not actually kill pests may, for practical and legal reasons, be considered pesticides. Included under the term are compounds used as repellents, attractants, antifeedants etc. A pesticide product consists of an active material in a certain concentrated formulation, such as the above emulsifiable concentrate, seed dispersing in water, designed to enable its safe and effective application. A great many pesticides are available today for the control of unwanted organisms. The insecticides, fungicides, herbicides and microbial agents given below are classified by chemical nature, which stresses the relationship in chemical structure; this shows the principle involved for these well-known compounds.

Pesticides/Fertilisers Causing Pollution

Pesticides, by their very nature, are toxic compounds and, as such, besides controlling pests and diseases, they also have potentialities of affecting life and the environment adversely. Pesticides are classified as insecticides, herbicides, rodenticides, fungicides and molluscides.

Organochlorine insecticides

Organochlorine insecticides includes:
1. Chlorinated ethane derivatives, of which DDT and Methoxychlor are the best known examples.
2. Cyclodienes, which include chlordane, aldrin, dieldrin, endrin, heptachlor, endosulphan and toxaphene.
3. Hexachlorocyclohexanes, such as lindane.

These are absorbed from the intestinal tract and, if it occurs in the air in the form of a very fine aerosol or dust, it may enter the alveoli of the lung from which it is absorbed readily.

Organophosphorus insecticides

These are absorbed by the skin as well as by the respiratory and gastrointestinal tracts. These insecticides, act by the inhibition of the enzyme acetylcholinesterase (AchE). The inhibition of AchE result in accumulation of endogenous acetylcholine (Ach) in nerve tissue and the effectors organs. The AchE content of various tissues is not equally affected in the same poisoned animal, and the level in all tissues, including even the brain, can be lowered markedly from the pre-poisoning level without seriously affecting normal function, especially if the reduction is gradual. A sudden marked depression may lead to critical and frequent fatal poisoning.

Carbamate insecticides

Carbamates are absorbed by all portals including the skin. These insecticides are reversible inhibitor of AchE. The reversal is so rapid that, unless special precautions are taken, measurements of blood AchE of people or animal exposed to carbamates are likely to be inaccurate and always in the direction of appearing to be normal. Intoxication may lead to blurred vision, increased blood pressure and profuse sweating.

Diphenyl herbicides

These may be absorbed, if taken by mouth or, presumably, if inhaled, but have very little cumulative effect. Skin absorption is slight. Acute doses in man lead to fibrillary twitching of muscles hyporeflexia and urinary inconvenience. Ventricular fibrillation may cause death.

Heterocyclic herbicides

It is rapidly absorbed and eliminated in the animal body primarily in the urine. On continuous, long term exposure it leads to goiter, probably as a result of continuous inhibition of peroxidase activity.

Bipyridinium herbicides

It is rapidly absorbed and excreted, mainly unchanged in the urine. It can delay the maturation stages of spermatogenesis and at higher dose level it can produce teratogenic effects in animals.

Fungicides

The mercury containing fungicides are responsible for many deaths or permanent neurological disability. Another compound, hexachlorobenzene, is also a highly toxic compound and can lead to severe skin manifestation, including hypersensitivity.

Fumigants

Fumigation is an important method for soil sterilisation as a means of killing insects, nematodes and metabolised when absorbed, but has relatively high chronic toxic, because it can produce liver tumors in several species of animals.

Rodenticides

The majority of poisons effective against mammalian pests are also very toxic to man and domestic animals. Accidental poisoning of sodium cyanide may lead to epistaxis, bleeding gums, and occasionally paralysis due to cerebral hemorrhage.

Fertilisers

Fertiliser nutrients can be lost from farmlands in three different ways:
1. By drainage water percolating through soil, leaching soluble plant nutrients.
2. By the inefficient return to the land of the excreta of stock.
3. By the erosion of surface soils or the movement of fine soil particles into subsoil drainage systems.

Nitrogen fertilisers are readily converted into nitrates, which are soluble and these pose more serious problems. Potassium has limited mobility in soils, however, whilst phosphorus is virtually immobile and therefore neither is leached out very easily nor appears to have any serious effects on the natural environment or on human health.

Fate of Pesticides in the Environment

Pesticides are generally applied to the soil, plant, water bodies and human settlements by man-mounted equipment, by tractors or by aircrafts, either as liquids, dusts or granules. If these pesticides hit only the target species, there would be least pollution. Often, as little as 25–50 per cent of the pesticide formulations land in the crop area when applied by aircraft and the remaining drift in the atmosphere to contaminate very remote areas of the abiotic environment. The migration of pesticides and their fate, and the routes for the loss of pesticides in the environment are given in Fig. 18.1 and 18.2.

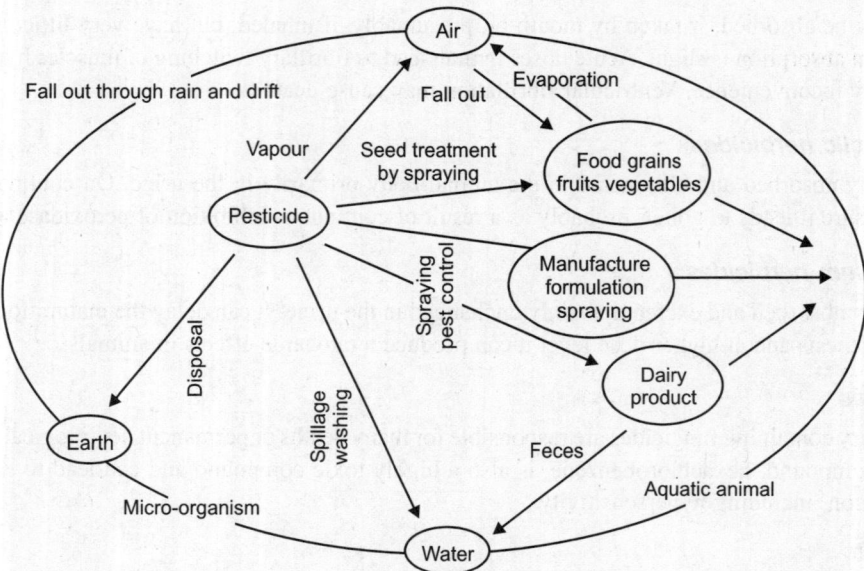

Fig. 18.1. Pesticide cycle in the environment.

Although most of the problems arising from pesticide drift are confined to within a few kilometers of their application sites, large scale effects are invariably produced either directly or indirectly.

Furthermore, the persistence of pesticides in the environment is dependent upon many factors such as the nature of the pesticide, temperature, light, humidity, air movement, activity of micro-organism and factors which influence the breakdown and mechanical dispersion of these chemicals.

Effects of Pesticides in the Environment

Effects on soil

The hazards of pesticides in soil largely depend on their persistence. The longer they persist, the greater are the chances for contamination of environment. For persistent organo-chlorine insecticides, it has also become a dogma that residues of these insecticides, gradually increases as they pass from lower to higher trophic levels through a process of bio-magnification.

Under tropical conditions, these persistent insecticides are lost from the soil into atmosphere in large amounts because of hot and humid weather. Among pesticides, fumigants and fungicides have definitiely shown a detrimental effect on the soil microbial population. However, these effects are often temporary and soon the microbes recover and resume their activities. Invariably, their application leads to increase in NH_4^+–N, but nitrification is generally inhibited. Herbicides, on the contrary, enhance the nitrification rates, especially at low of application. Some organophosphate and carbonate insecticides inhibit soil enzyme activity, when they are absorbed on soil particulates.

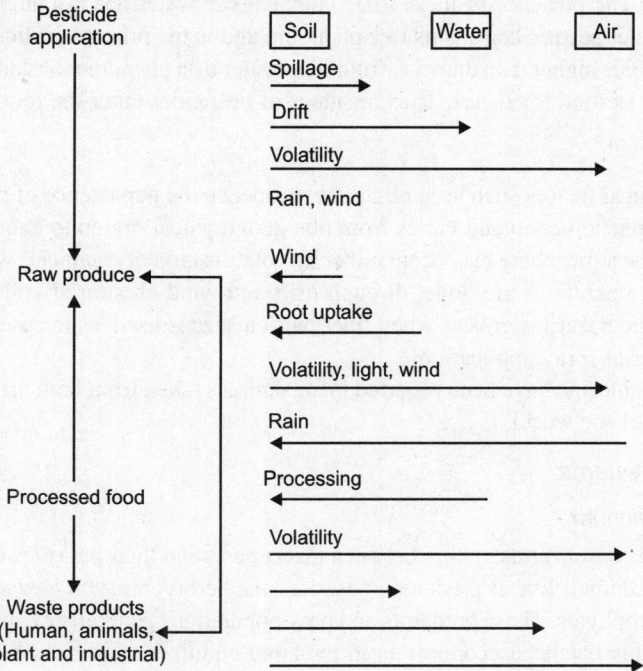

Fig. 18.2. Routes for the loss of pesticides to the environment.

Pesticides migrating from soil are important because it leads to contamination of drinking water or food crops grown on such soils. Generally, more residues are found in crops grown on soils treated with organo-chlorinated insecticides; for example, potatoes and carrots grown on soils treated with aldrin at 0.5 kg/acre are reported to contain 0.03 ppm and 0.05 ppm, respectively. The most significant harmful effect of pesticides residues in soil is that they are taken up by plants, which besides being toxic to the consumers, also causes phytotoxicity to the plants.

Effects on water

The factors which influence the persistence of pesticides in water are solubility, bottom mud, organic matter, temperature and pH.

The pesticides that are of importance in connection with water quality include chlorinated hydrocarbons and their derivatives, persistent herbicides, soil insecticides, pesticides that are easily leached out from soil, and pesticides that are systematically added to water supplies for disease vector control for other purpose. Of these compounds, only the chlorinated hydrocarbon persist in the environment and are known to have drifted over thousands of kilometers and from water melted from Antarctic snow. Traces of chlorinated hydrocarbon pesticides in water may accumulate progressively in different steps of a food chain; for example, DDT can accumulate in fish to levels more than 10,000 times the concentration present in the surrounding water.

Guideline values are recommended for several chlorinated hydrocarbon pesticides because of the possibility of their presence in water. These guideline values are derived from the acceptable daily intake (ADI) values with the assumption that is not more than 1 per cent of the ADI would be derived from drinking water. The presence of these toxic chemicals in water has also significance as they are picked up by unicellular aquatic organisms like plankton, and in the process pesticides get accumulated in body tissues manifolds higher than that of surrounding water by a phenomenon called bioconcentration. This plankton serves as food for fishes. Thus, residues of pesticides enter the food chain.

Effects on air

Several physio-chemical factors such as sunlight etc. influence the persistence of pesticides in air. This is a highly complex phenomenon and varies from one geographical region to another.

The pollution of the atmosphere may occur either by volatilisation of chemicals where filling, loading, mixing and spraying operations are done, through drift and wind erosion of soil particles laden with absorbed residues. The hazard is greater when finer particle size is used, as in case of ultra low volume application and in aerial spray application.

Traces of organo-chlorine have been reported in air samples taken from both urban and rural areas of developed countries of the world.

Effects on biotic systems

Disturbance in equilibrium

Pesticides disturb the equilibrium existing between insect pests and their parasites or predators. Perhaps 90 per cent of all the animals live as predators (parasites and herbivores) and feed on living matter; only a few live as true saprophytes. These predators and prey populations generally evolve a balanced supply-demand economy. This balanced economy is in dynamic equilibrium, which, however, if altered by pesticides, will disturb the equilibrium between predators and prey populations.

Increased disease susceptibility

Pesticides are also known to increase the disease susceptibility in hosts. For example, 2, 4-D increased the size of tobacco mosaic virus (TMV) lesions on hyper-sensitive tobacco.

Bioaccumulation

It is an interesting phenomenon that has significant impact on the dynamic equilibrium that exists between predators and prey. Both animals and plants have the ability to concentrate many types of pesticides in their body tissues. The chlorinated hydrocarbon insecticides have the greatest tendency for their biological concentration. The other pesticides having tendency for bioaccumulation are diazinon, mirex, diquate, paraquate and 2,4-D.

Fish (Golden shiners) are reported to take DDT at 0.265 ppt in water and concentrate it 1,00,000 times in their bodies. In general, however, the capacity for bioaccumulation is usually not as great as this, but when the procedure is repeated at each successive link in the food chain, extremely high concentrations of toxicant occurs in the species at the top of the food chain.

Development of tolerances

Due to intense selective pressure from pesticides, a wide variety of insects, mites, animals and plants have evolved significant levels of tolerance. Of about 2000 peat insects and mite species, a total of 225 species have been reported to show resistance to pesticides. Of these groups, 121 are crop pests, 97 man and animal pests, 6 stored produce pests and 10 forest pests. In at least one instance, the level of increased resistance is reported to be 25000 folds. Amphibians (cricket frogs) from heavily DDT treated cotton field have evolved resistance to this insecticide.

Disturbance in reproductive physiology

Several studies have reported a decrease in the thickness of egg shells in Europe and North America. This partly explained the increased incidence of egg breakage and egg disappearance observed in several species. Many more studies could be cited all showing the above mentioned association between DDT (in particular its main metabolite, DDE) and egg shell thinning and reduced reproductive success. The results from control experiments indicated that DDE is involved in egg shell thinning and reduced reproductive success in certain species of birds.

Effects on behaviour

Another significant change that takes place is in pattern of behaviour of non-targeted species. The patterns of behaviour in animals determine their survival in the habitat in which they live. Each animal responds to environmental cues to help it to select its food and shelter, avoid natural enemies, select a mate and so forth. Hence, any change in an animal's behavioural pattern may significantly influence the survival of the population.

For example, the herbicide 2,4-D made beetles to become sluggish in their behaviour. When this happens they are less successful as predators than unaffected beetles and some of their prey escape capture.

Effects on wildlife, fish and other aquatic organisms

Wildlife biologists believe that DDT and other pesticide residues are potentially toxic to fish and other aquatic organisms. Known effects of pesticides on fish mainly refer to cases of mortality which may occur after accidental spillage or after the use of fish toxic pesticides in places from which the compounds could enter waterways, fish ponds, lakes or important habitats for fish such as rice paddies.

Two pesticides, in particular, have become well known in this respect, viz. Endrin and Endosulfan. However, there are more pesticides, which may kill fish or birds when they are applied or disposed of in the neighborhood of aquatic habitats. Examples of these are DDT, dieldrin, toxaphase, methyl mercury and niclosimine. Indiscriminate spraying of broad spectrum pesticides such as phosphamidon against spruce budworms and application of DDT, Dieldrin and endosulfan may affect a wide variety of animal groups as was shown in Africa where mammals, birds, reptiles, amphibians, fish and invertebrate died in often large numbers during the tsetse control operation.

Effects on plants

Together with target pests, pesticides also alter the chemical make up of plants and, in turn, affect the animals' dependent upon them for food. The changes that occur are specific for both the plants and the pesticides involved. For instance, heptachlor in soil has caused significant changes in the elements N, P, K, Ca, Mg, Mn, Fe, Cu, Al and Zn. Similarly, the protein content of wheat and the KNO_3 content of sugar beet was increased after 2,4-D. This nitrate level was highly toxic to cattle.

Another chemical changes is the increased sugar content seen in plants like ragwort, a weed toxic to cattle, exposed to sub-lethal doses of 2,4-D. The high sugar content makes it attractive to cattle and sheep, with disastrous results because the toxicity of the plant remains high.

Reduced seed production has been reported after the use of pesticides. For example, when the herbicide 2,4-D was applied, a 10–15 per cent reduction in seed production has been reported.

Effects on foods

Pesticides residues have been detected in food commodities. Low level of residues in foods may result from growing of crops, especially root crops, in soil containing residues of persistent pesticides.

A number of pesticides have an affinity for lipid materials and get accumulated in animal systems, thus contaminating animal products with residues.

Levels of DDT and BHC have been reported at more than the tolerance limits prescribed by WHO and FAO. Apart from these, residues of endrin, heptachlor and endosulphan have been detected in major food crops.

Pesticide residues as high as 35 ppm DDT and 60 ppm BHC in spinach; 36 ppm DDT in Brinjal; 50 ppm DDT in cauliflower, cabbage and radish and nearly 26 ppm in oil have been detected in the Delhi market. Many farmers even spray pesticides and vegetables just before harvesting them, in order to give them a fresh appearance; such causes are more alarming.

The pesticide residue levels detected in meat and poultry, etc. are equally high and often higher. According to a study, the average daily intact (ADI) of DDT and BHC for nonvegetarian needs is respectively 24 and 10 per cent higher. Even this average DDT intake exceeds the maximum ADI of 5 mg/kg of body per day. Thus, there is probably nothing that an Indian eats, which is not contaminated by pesticides.

Possible Health Hazards of Pesticides Residues

Since our food articles contain excessive amount of DDT and BHC, it is not surprising that the Indian population has an unusually high amount of these insecticides in the body. Other sources such as air, water and direct contact with contaminated surface etc. may also contribute to some extent to the DDT or BHC burden in human body.

BHC has been shown to cause higher incidence of liver tumours than DDT and the life span of individual exposed to these insecticides is reduced. The most likely effect of low level residues of insecticides is the impact on enzymes.

DDT is known to alter the efficacy of some drugs in the body, by interfering with enzymatic activity of the drug detoxification mechanism. High infant mortality was observed in Guatemala where high residues of DDT were found in mother's milk.

Effect of Fertilisers on the Environment

Although fertiliser loss is economically important to the farmers, the more important concern is its effect on the environment.

There are two major problems that have been associated with high nutrient levels.
1. Algal blooms have been attributed to soluble nitrates and phosphates in water. The causes of such growth are incompletely understood, but the levels of nutrients which it is claimed limits algal growth are below 0.3 ppm N (nitrogen) and 0.01 ppm P (phosphorus), respectively. In most low land water in the UK, nutrient concentrations have been above these limits for decades, but blooms have been rare and short lived. It is thought that factors such as water, temperature, CO_2 concentrations and the presence of organic matters are important and, in general, it seems unlikely that marginal increase in nutrient levels would initiate algal growth.
2. Increased nitrate levels can cause a health hazard in drinking water for very young babies (up to age of about three months). The nitrate in the water can be reduced in the baby's stomach to nitrate, which combines with the haemoglobin to form methaemoglobin, preventing the transport

of oxygen around the body. This is one of the causes of 'blue babies' and about 10 cases of methaemoglobinaemia have been confirmed since the condition was recognised 20 years ago. These have all been associated with ground waters, usually with water from shallow wells, which may have been contaminated with sewage.

Large scale use of nitrate fertilisers could also lead to depletion of ozone. It has been estimated that a 10-fold increase in the use of nitrate fertilisers by the year 2010 could generate enough additional nitrate (N_2O) to reduce the ozone cover significantly. Reduction ranging from 1–20 per cent has been suggested for this source.

POLLUTION PREVENTION AND CONTROL

Every effort should be made to replace highly toxic and persistent ingredients with degradable and less toxic ones. Recommended pollution prevention measures are as follows:
1. Control the quantities of active ingredients to minimise wastage.
2. Reuse of by-products from the process as raw materials or as raw material substitutes of other processes.
3. Use automated filling to minimise spillage.
4. Use 'closed' feed systems for batch reactors.
5. Use nitrogen blanketing where appropriate pumps, storage tanks, and other equipment minimise the release of toxic organics.
6. Give preference to non-halogenated and non-aromatic solvents, wherever feasible.
7. Use high-pressure hoses for equipment cleaning to reduce waste-water.
8. Use equipment wash down waters and other process waters (such as leakages from pump seals) as makeup solutions for subsequent batches.
9. Use dedicated dust collectors to recycle recovered materials.
10. Vent equipment through a recovery system.
11. Maintain losses from vacuum pumps (such as water ring and dry) at low levels.
12. Return toxic materials packaging to the supplier for reuse or incinerate/destroy in an environmentally acceptable manner.
13. Minimise storage time of off-specification products through regular reprocessing.
14. Find productive uses for off-specification products to avoid disposal problems.
15. Minimise raw material and product inventory to avoid degradation and wastage that could lead to the formation of inactive but toxic isomers or by-products.
16. Label and store toxic and hazardous materials in secure, bonded areas. A pesticide manufacturing plant should prepare a hazard assessment an operability study, and also prepare and implement an emergency preparedness and response plan that takes into account neighbouring land uses and the potential consequences of an emergency. Measures to avoid the release of harmful substances should be incorporated in the design, operation, maintenance, and management of the plant.

WASTE-WATER GENERATION IN PESTICIDES INDUSTRY

The entire manufacturing process for a particular product is a combination of various unit operations. The unit operations, where water is used and waste-water is generated, are to be identified so as to accurately identify the characteristics of the waste-water and the quantum of pollution load generated. A schematic diagram of an unit operation is shown in Fig. 18.3. The waste-water flow lines from process to treatment to final disposal are to be identified so that the entire status of the waste-water generation from the factory is depicted at one place. The waste-water flow lines in a typical industry are shown in Fig. 18.4. The network for monitoring waste-water may be fixed so as to analyse the characteristics of each of the waste-water streams and to evaluate the performance of the pollution control systems.

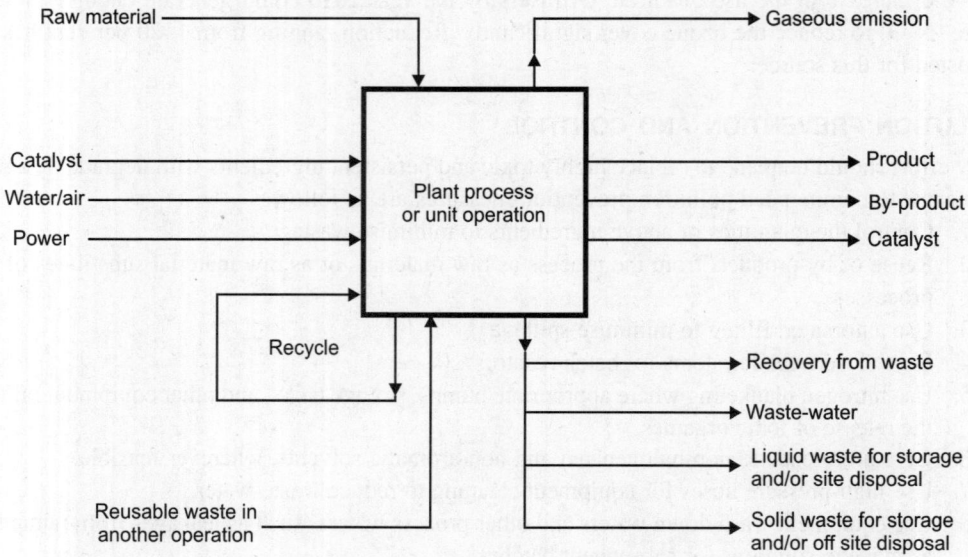

Fig. 18.3. A typical unit operation indicating waste generation routes.

The observations on waste-water generation and its characteristics are given below:
1. The possible sources of pollutants are raw material used, in pesticide synthesis in excess of their stoichiometric requirements, impurities in raw materials, solvent used as a carrier medium, solvent used as extraction medium, impurities in solvents, catalysts, manufacturing products, etc.
2. There is a great variation in quality and quantity of waste-water generated per unit of product. The biochemical oxygen demand (BOD) and chemical oxygen demand (COD) are widely varying in waste-water from product to product having no fixed correlation and has to be determined only on case to case basis. In general, BOD is in the range of 1000–5000 mg/l. The variation in the waste characteristics is hourly, daily and seasonal. The observations on manufacturing processes mentioned in above, also contribute to these variations. In general, the waste-water is toxic and not easily biodegradable in nature.
3. Some process streams are inorganic and do not need biological treatment.
4. Some streams are organic and/or inorganic and have very high dissolved solids (TDS). These cannot be treated biologically.
5. Details of various other pollutants in the waste-water are given below:
    (a) Volatile organics: Benzene, toluene, chlorobenzene, etc. which are used as raw materials.
    (b) Halomethanes: Methylchloride, chloroform, carbon tetrachloride, etc. which are used as raw materials and extraction solvents.

Pesticide 397

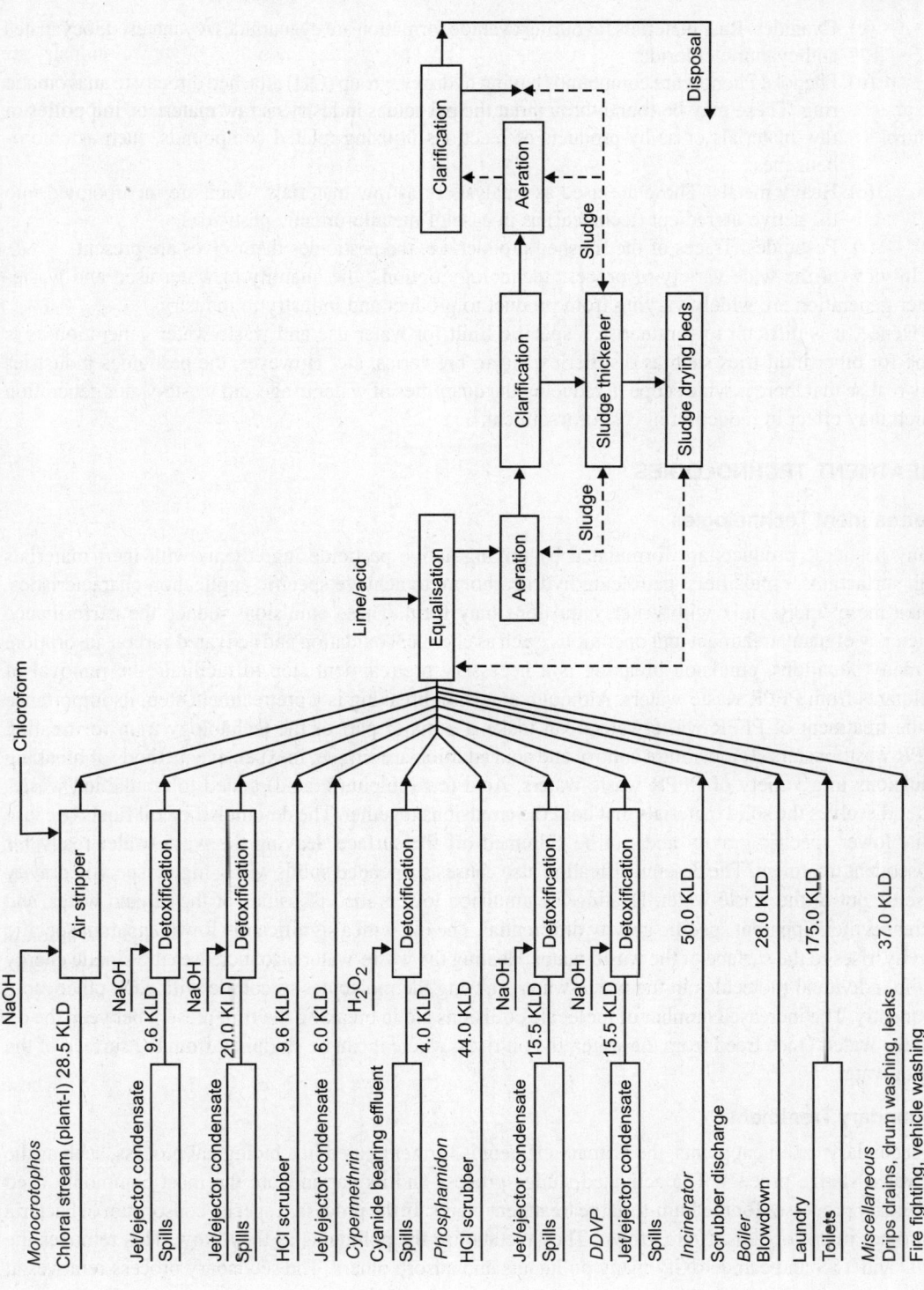

Fig. 18.4. Waste-water flow-lines in a typical industry.

(c) Cyanides: Raw materials favouring cyanide formation are cyanamides, cyanates, thiocyanates and cyanuric chloride.
(d) Phenols: Phenols are compounds having hydroxyl group (OH) attached directly to an aromatic ring. These may be found throughout the pesticides industry as raw materials, impurities in raw materials or as by-products of reactions utilising related compounds, such as chlorobenzenes.
(e) Heavy metals: These are used as catalysts or as raw materials which are incorporated into the active ingredient (technical) as in case of metallo-organic pesticides.
(f) Pesticides: Traces of the finished product, i.e. the pesticides themselves are present.

In view of the wide variety of process technology options, the quantity of water used and waste-water generation are widely varying from product to product and industry to industry.

Hence, it is difficult to summarise a specific limit for water use and waste-water generation as is done for other industries such as distilleries, sugar, breweries, etc. However, the pesticides industries may realise that there is wide scope in reducing the quantities of water usage and waste-water generation which may effect in reduction in cost of treatment.

TREATMENT TECHNOLOGIES

Pretreatment Technologies

Many pesticide products are formulated by mixing active pesticide ingredients with inert materials (e.g. surfactants, emulsifiers, petroleum hydrocarbons), to achieve specific application characteristics. When these 'inerts' mix with water, emulsions may form. These emulsions reduce the performance efficiency of many treatment unit operations, such as chemical oxidation and activated carbon adsorption. In many situations, emulsion breaking is a necessary pretreatment step to facilitate the removal of pollutants from PFPR waste-waters. Although emulsion breaking is a pretreatment step, its importance in the treatment of PFPR waste-waters can make it a major part of the technology train for treating PFPR waste-waters. Temperature control and acid addition are simple, inexpensive methods of breaking emulsions in a variety of PFPR waste-waters. Acid (e.g. sulphuric acid) added to emulsified waste-water dissolves the solid materials that hold the emulsions together. The de-emulsified oil floats because of its lower specific gravity and can be skimmed off the surface, leaving the waste-water ready for subsequent treatment. The de-emulsification also causes suspended solids with a higher specific gravity to settle out of the waste-water. Heating the emulsion lowers the viscosities of the oil and water, and increases their apparent specific gravity differential. The oil, with a significantly lower apparent specific gravity, rises to the surface of the waste-water. Heating the waste-water also increases the kinetic energy of the individual molecules in the waste-water, causing the molecules to collide with each other more frequently. The increased number of molecule collisions aid in breaking the film present between the oil and the water. Once freed from the water, the oil rises, where it can be skimmed from the surface of the waste-water.

Secondary Treatment

At secondary treatment plants, the primary effluent is further treated in a biological process, such as the activated sludge process. The activated sludge process and its variants are the most commonly used secondary processes for medium-to-large treatment plants. In the process, a special consortium of bacteria and other micro-organisms are grown. They metabolise the pollutants as they grow. This removes the BOD and TSS and can detoxify many pollutants and adsorb others. The secondary process removes at least 85 per cent of the BOD and TSS, and well-designed and operated plants may remove 95 per cent of the BOD and TSS. The effluent is clear with only a few pinhead sized suspended solids. It is not potable and still contains pathogens. It may contain trace metals and hard-to-treat organics compounds, such as pesticides. Nevertheless, it is usually suitable to be discharged into many receiving waters.

Tertiary Treatment Technologies

Activated carbon adsorption

Activated carbon effectively removes organic constituents from waste-water through the process of adsorption. The term 'activated carbon' refers to carbon materials, such as coal or wood, that are processed through dehydration, carbonisation, and oxidation to yield a material that is highly adsorbent, due to a large surface area and high number of internal pores per unit mass. As waste-water flows through a bed of carbon materials, molecules that are dissolved in the water may become trapped in these pores. In general, organic constituents (including many pesticide active ingredients) with certain chemical structures (such as aromatic functional groups), high molecular weights, and low water solubility are amenable to activated carbon adsorption. These constituents adhere to the stationary carbon material, so the waste-water leaving the carbon bed has a lower concentration of pesticide than the waste-water entering the carbon bed. Eventually, as the pore spaces in the carbon become filled, the carbon becomes exhausted and ceases to adsorb contaminants. Spent carbon may be regenerated or disposed of; cost and/or other regulatory factors (e.g. RCRA) generally determine the choice.

Carbon adsorption depends on process conditions, such as temperature and pH and process design factors such as carbon/waste-water contact time and the number of the carbon columns. If performed under the right conditions, activated carbon adsorption can be an effective treatment technology for PFPR industry waste-waters. Carbon adsorption capacity depends on the characteristics of the adsorbed compounds, the types of compounds competing for adsorption, and the characteristics of the carbon itself. If several constituents that are amenable to activated carbon adsorption are present in the waste-water, they may compete with each other for carbon adsorption capacity. This competition may result in low adsorption or even desorption of some constituents.

Activated carbon comes in two sizes: powdered carbon which has a diameter of less than 200 mesh, and granular carbon which has a diameter greater than 0.1 millimetre. While granular carbon is more commonly used in waste-water treatment; powdered carbon is used less frequently because the small particle size creates regeneration and design problems. Activate carbon is obtained from vendors in bulk or in a variety of container sizes. At smaller facilities, the container in which the carbon is sold is intended to be used as the carbon bed, with influent waste-water passing into one end of the container and treated effluent water passing out of the opposite end. At larger facilities, carbon is purchased and added to a column that is installed at the facility.

Carbon is regenerated by removing the adsorbed organic compounds through steam, thermal, or physical/chemical methods. Thermal and steam regeneration are the most common methods to regenerate carbon used for waste-water treatment. These methods volatilise the organic compounds that have adsorbed onto the carbon.

Afterburners are required to ensure the destruction of the organic vapours; a scrubber may also be necessary to remove particulates from the air stream. Physical/chemical regeneration uses a solvent, which can be a water solution, to remove the organic compounds. Carbon is usually shipped back to the vendor for regeneration, although some facilities with larger carbon beds may find it economical to regenerate carbon on-site.

Chemical oxidation

Chemical oxidation modifies the structure of pollutants in waste-water to similar, but less harmful, compounds, through the addition of an oxidising agent. During chemical oxidation, one or more electrons transfer from the oxidant to the targeted pollutant, causing its destruction. One common method of chemical oxidation, referred to as alkaline chlorination, uses chlorine (usually in the form of sodium hypochlorite) under alkaline conditions to destroy pollutants, such as cyanide and some pesticide active ingredients. However, facilities treating waste-water, using alkaline chlorination should be aware that the chemical oxidation reaction may generate toxic chlorinated organic compounds, including chloroform, bromodichloromethane, and dibromochloromethane, as by-products. Adjustments to the design and operating parameters may alleviate this problem, or an additional treatment step (e.g. steam stripping, air stripping, or activated carbon adsorption) may be required to remove these by-products.

Chemical oxidation can also be performed with other oxidants (e.g. hydrogen peroxide, ozone, and potassium permanganate) or with the use of ultraviolet light. Although these other methods of chemical oxidation can effectively treat PFPR waste-waters, they typically entail higher capital and/or operating and maintenance costs, greater operator expertise, and/or more extensive waste-water pretreatment than alkaline chlorination.

Chemical precipitation

Chemical precipitation is a treatment technology in which chemicals (e.g. sulphides, hydroxides, and carbonates) react with organic and inorganic pollutants present in waste-water to form insoluble precipitates. This separation treatment technology is generally carried out in the following four phases:
1. Addition of the chemical to the waste-water;
2. Rapid (flash) mixing to distribute the chemical homogeneously throughout the waste-water.
3. Slow mixing to encourage flocculation (formation of the insoluble solid precipitate).
4. Filtration, settling, or decanting to remove the flocculated solid particles.

These four steps can be performed at ambient conditions and are well suited to automat control. Hydrogen sulphide or soluble sulphide salts (e.g. sodium sulphate) are chemicals commonly used in the PFPR industry during chemical precipitation. These sulphides are particularly effective in removing complexed and heavy metals (e.g. mercury, lead, and silver) from industrial waste-waters. Hydroxide and carbonate precipitation can also be used to remove metals from PFPR waste-waters, but these technologies tend to be effective on a narrower range of contaminants.

Hydrolysis

Hydrolysis is a chemical reaction in which organic constituents react with water and break into smaller (and less toxic) compounds. Basically, hydrolysis is a destructive technology in which the original molecule forms two or more new molecules. In some cases, the reaction continues and other products are formed. Because some pesticide active ingredients react through this mechanism, hydrolysis can be an effective treatment technology for PFPR waste-water. The primary design parameter considered for hydrolysis is the half-life, which is the time required to react 50 per cent of the original compound. The half-life of a reaction generally depends on the reaction, pH and temperature and the reactant molecule (e.g. the pesticide active ingredient). Hydrolysis reactions can be catalysed at low pH, high pH or both, depending on the reactant molecule. In general, increasing the temperature increases the rate of hydrolysis. Identifying the best conditions for the hydrolysis reaction results in a shorter half-life, thereby reducing both the size of the reaction vessel required and the treatment time required.

SEPARATION TECHNOLOGIES FOR REMOVAL OF ORGANIC AND PESTICIDAL CHEMICALS FROM WASTE-WATER

The following unit operations have been grouped under the Separation Technology for removal of organic and pesticidal chemicals from waste-water:

1. Absorption.
2. Adsorption including bubble adsorption
3. Centrifugation
4. Clathration.
5. Coagulation.
6. Coalescence.
7. Condensation.
8. Cyclonic action.
9. Desorption.
10. Dialysis.
11. Diffusion process.
12. Electrophoresis.
13. Evaporation.
14. Extraction.
15. Filtration.
16. Flash expansion.
17. Floatation.
18. Foam fractionation.
19. Gravity settling.
20. Impringement.
21. Membrane permeation.
22. Precipitation.
23. Reverse osmosis.
24. Scrubbing.
25. Stripping.
26. Ultra-filtration.

Toxicity of Chemical Pesticides

Based on the toxicological evaluations, the chemical pesticides are classified into extremely hazardous, highly hazardous, moderately hazardous and slightly hazardous groups. Most of the presently used chemical pesticides in India fall in the fist two categories. Table 18.1.

Table 18.1. Classification of some selected pesticides.

Hazard class	Pesticide
Extremely hazardous	Aldicarb, Methoyl, Monocrotophos, Phorate, Parathion, Phosphamidon
Highly hazardous	Aldrin, Carbofuran, Diehlorovos, Dimethoate, Endosulphan, Ethion, Fenthion, Fenvalerate, Methyl Parathion, Phosalone, Triazophos
Moderately hazardous	Chlorpyriphos, Cypemethrin, Lindane, Fenitrothion
Slightly hazardous	Carbendazim, B.T., Trichogramma, Azadirachin, N.P.V.

MANAGEMENT

As a consequence of the environmental hazards associated with use of pesticides, most advanced countries of the world have enacted laws enforcing safe use. In these countries, maximum residues limits have statutory authority. If any commodity is found to contain residues exceeding the legally permitted levels, it is taken off the market and legal proceedings are drawn against trader/cultivators.

In India, the following regulations have been enacted to regulate pesticide levels.
1. 1948: The Factories Act (Pollution and Pesticide).
2. 1954: Assam Agricultural Pests and Disease Act.
3. 1954: Prevention of Food Adultration Act.
4. 1968: The Insecticide Act.
5. 1980: Air Pollution Prevention and Control Act.

Considering the hazards associated with use of pesticides, most developed and some developing countries have enforced a system of monitoring pesticide residues in commodities aimed to measure the level of toxic pollutants.

The futility of a unidirectional approach to pest control has become pronounced after nearly four decades of total reliance on chemical pesticides. Alternate pest control methods that are more in harmony with nature are being developed.

Traditional methods like crop rotation, soil and water management and biological control that employ natural enemies of pests are gaining respectability. Such tactics can represent a valid approach to pest management, provided natural enemies are given optimum opportunity to work and selective pesticides are employed on a need basis.

Pest control requires the input of many disciplines and the cooperation of pest control experts with those who understand more thoroughly the sensitivities of the natural environment. Bringing together the needed expertise is essential for utilisation of our resources to the best advantage of man and the friendly inhabitants that surround him in the environment.

Risks of inhalation of pesticides by field workers can be reduced if safe re-entry interval before the workers enter treated crop areas are provided. Exposure would be reduced by almost 50 per cent if entries to treated crops are delayed by about 12 hours. People involved in mixing and spraying pesticides should wear respirators during operations.

To achieve effectiveness of pesticides and minimise environmental pollution problems, newer formulation application techniques are being introduced. Slow release granule, encapsulation, fogging and ULV preparation not only reduce per acre cost, but also ensure safety to enduser. Granular or encapsulated formulations are necessary for highly toxic chemicals like Phorate.

Algal stripping entails growing a dense crop of algae in reservoirs so that at harvesting a considerable proportion of the nitrate nitrogen in the water could be removed in the algal bodies.

Bacterial denitrification using anaerobic denitrifying bacteria, which reduce nitrates to gaseous nitrogen, are also added to water, either free floating in a deep pond or reservoir or preferably attached to a filter at the bottom.

Either organic or slow acting nitrogen fertilisers would be effective in controlling pollution. There are, at present, some organic chemicals in the market which inhibit nitrification of ammonium ion.

ENVIRONMENTALLY SAFE LAYOUT CODE FOR MANUFACTURING UNITS

1. An environmentally safe layout plan takes care of material loss, cost of collection, disposal, recycle and treatment which are parts of the process itself, and consequently of the layout arrangement.

2. This layout code postulates that environment protection is a factor for designing any equipment, reaction vessel, material transfer arrangement, storage tank, and service support to operate the production system.
3. All places of storage of solid and liquid materials are to be diked without drains. Any spillage is to be wiped out and cannot be washed out.
4. Each vessel should have its own catchpit to collect spills.
5. Each pump must be mounted on its own catchpit; a suction line of the pump should be connected to empty the pit, periodically or regularly or continuously.
6. As losses of materials take place during charging of the reaction vessels, discharging of produce and dripping of outlet valves, and as materials may be either solid or solid slurry or liquid, care needs to be exercised to prevent the losses, if necessary by changing the charging/discharging and transfer devices.
7. In order to collect spills from a particular vessel before the spilled materials get a chance of contamination with spills from another nearby vessel, the two vessels must be installed at sufficient distance so that inter-contamination cannot take place. The extra distance, 'noncontaminating distance' is to be provided for recycle of materials.
8. Flange joints should be avoided wherever avoidable.
9. Corrosion-prone areas and construction materials liable to atmospheric and process induced corrosion should be given special attention for finding better replacement material and stricter preventive maintenance frequency.
10. Exhaust ducts and fan outlets are sources of pollution, if the thrown out air is contaminated with pollutants. These may be treated before vented. Any vapour line should be connected with either a recovery system or an absorption system.
11. The engineering code for the operation of pressurised systems and the established practice for preventive maintenance are consistent with the protection of the environment. These systems are fitted with pressure release valves, and in many cases with rupturable discs. The present practice is to allow the released materials to the atmosphere. To be environmentally, safe, these lines ought to be connected to recovery/adsorption/absorption arrangements. The rupturing of safety discs is accompanied with sudden release of high pressure; the design of the recovery arrangement of the released materials should be befitting the sudden emerging conditions of high temperatures/pressure/volumes.
12. New units will build floors with expanded metals, slotted angles, steel grills, steel grates, prefabricated industrial floor gratings, and the like which will make floor washing redundant.
13. If the plant layout demands that vessels should be installed in upper floors, arrangements should be simultaneously made to spill avoidance/collection. Vulnerable points of leakage should be taken special care of. This is necessary not only for pollution control but also for the safety of plant personnel working in lower floors.
14. Storage tanks of raw materials for supply to the production vessels, should be installed on a separate structure located just outside the main plant building, with arrangement for holding spills and overflow. Level alarms should be installed where possible; where the same is not feasible because of the nature of the liquid, two overflow pipes at two different levels of the tank should be fitted.
15. Plant management should evolve its own code for washing equipment, where a particular equipment is used for the manufacture of different products. Dry scraping of equipment surface

followed by moping with wet cloth should be carried out before hosing operation. This will reduce the quantity of contaminants and waste-water volume.
16. All channels be fitted with waste-water measuring devices, half barrier for the separation floating immiscible liquid and in-built separation/sedimentation basins for withholding settleable particulate matters. This provision may be treated as compulsory for waste-water channels in the immediate vicinity of waste-water generating units.
17. All water usages that do not come ill contact with chemicals, should have no opportunity to mix with process water. Uncontaminated water should have separate outlets from the plant and if recycle is not possible, should be drained out through separate channels, without any change of getting contaminated.
18. This proposed layout code recognises the solid waste generated in the process of manufacture must find a place within the factory premises. It will be stored on land/lagoon which will be lined with compatible geotextile materials.
19. The detoxification operation is to be carried out outside the main production plant, and provision has to be kept for the same.
20. Storm water drains should be segregated from process water drains. The former may be used for the removal of cooling water and non-process water.

Standards for pesticide industry (manufacturing and formulation)

Parameter	Concentration not to exceed mg/l (except pH and bio-assay)
Compulsory parameters:	
pH	6.5 – 8.5
BOD	100 mg/l
(5 days 20°C)	
Oil and grease	10 mg/l
Total suspended solids bioassay test	100 mg/l
Bioassay test	Minimum 50 per cent survival after 96 hours with fish or crustacean at 10 per cent dilution, i.e. with 90 per cent effluent
Optional parameters:	
Heavy metal	
Copper	1.0 mg/l
Managnese	1.0 mg/l
Zinc	0.1 mg/l
Mercury	0.01 mg/l
Tin	0.1 mg/l
Any other like Nickel	Shall not exceed 5 times the drinking water standards (BIS)
Organics	
Phenol and phenolic compounds as C_6H_5OH	1.0 mg/l
Inorganics	
Arsenic as As	0.2 mg/l
Cyanide as CN	0.2 mg/l
Nitrate as NO_3	5.0 mg/l
Phosphate as P	5.0 mg/l

(Contd...)

Parameter	Concentration not to exceed mg/l (except pH and bio-assay)
Specific pesticides	
Benzene haxachloride	10 µg/l
DDT	10 µg/l
Dimethoate	450 µg/l
Copper oxychloride	9600 µg/l
Ziram	1000 µg/l
2–4D	400 µg/l
Paraquat	23000 µg/l
Propanil	7300 µg/l
Nitrofen	780 µg/l
Other pesticides individually	100 µg/l
Insecticides	
Aluminium phosphide	Lindane Pyrethrem extract
Dichlorovos	Malathion Quralphos
EDTC mixer	Methyl bromide Mkonocrotophos
Ethylene dibromide	Nicotine sulphate Carbaryl
Ethion	Oxydemetion methyl Endosulfan
Fenitorthion	Methyl parathion phosphamidon Fenvalerate phoraste
Lime-sulphur	
Temephos	
Fungicides	
Aureofungin	Organomercurials (MEMC and PMA)
Barium polysulphide	Sulphur (colloidal wettable and dust)
Cuprous oxide	Streptocycline
Ferbam	Thiram
Mancozeb	Carbendazim
Manab	Carbendazim
Nickel chloride	Tridemorph
Redenticides	
Comfury	
Warfarin	
Zinc phosphide	
Nematicide	
Metham N sodium	
Weedicides	
Fluchloralin	
Isoproturon	
Butachlor	
Anilophos	
Plant growth regulants	
Chloromequat chloride	
Napthalene acetic acid	

Notes

1. Limits should be complied with at the end of the treatment plant before any dilution.
2. From the optional parameters only those relevant (based on the raw materials used and products manufacatured) parameters may be prescribed.
3. Bioassay test should be carried out with as per BIS 6582–1971.
4. State Boards may prescribe the optional parameter after case by case study.
5. No limit for COD is prescribed. If the COD in a treated effluent is persistently greater than 250 mg/l such industrial units are required to identify chemicals causing the same. In case these are found to be toxic as defined in Hazardous Rules, 1989 in Schedule I, the State Boards in such cases shall direct the industries to instal tertiary treatment stipulating time limit. Otherwise COD may not be stipulated. This may be done case by case studies.
6. Solar evaporation followed by incineration is a recognised practice, provided the guidelines of solar evaporation are followed.

MINIMAL NATIONAL STANDARDS

The Minimal National Standards (MINAS) for the pesticide industry are evolved considering achievability through the recommended treatment alternatives, environmental sensitivity, cost aspect in respect of ozonation and the various pollutants that emanate from the industry.

Each of the industrial units should make a feasibility study for treatment options taking into consideration the specificity of its manufacturing processes and waste-water generation.

The units must immediately equip themselves with the capability to determine the concentrations of various parameters listed in the MINAS.

The MINAS are to be met after treatment but before any dilution with cooling water or fresh water, or neutralisation water.

The MINAS shall be met in respect of each and every parameter; partial treatment to satisfy a few of the parameters is not acceptable.

Minimal national standards for pesticide manufacturing and formulation industry

Characteristics	Limiting concentration
General	
Temperature	Shall not exceed 5°C above the receiving water temperature
pH	6.5 to 8.5
Fat, oil and grease (FOG value)	10 mg/l
Oxygen absorption test (4 hours in acid potassium permanganate at 27°C)	60 mg/l
Non-pesticidal suspended solids (inert formulating materials, biological sludge)	30 mg/l
Total pesticide including its metabolites and isomers	*10 µg/l
Organic solvents and compounds (in original or aqueous phase as carbon-chloroform extract)	*30 µg/l

(Contd...)

Characteristics	Limiting concentration
Raw material and process intermediates	Shall not be present in their original form
Bio-assay Test	Standard method for the static bio-assay test is to be applied. Test should be separately carried out with suitable species of fish and crustacean at 10 per cent dilution. 50 per cent survival at the end of 96 hours should be attained at that dilution and for each test specimen. Bio-accumulation and bio-magnification phenomena are not relevant for the plant test.
Optional	
Heavy metals	
Copper	1 mg/l
Manganese	1 mg/l
Zinc	1 mg/l
Mercury	0.01 mg/l
Tin	0.1 mg/l
Any other like nickel etc.	Shall not exceed the drinking water 5 times standards
Organics	
Organic chlorine	Amount of any one shall not exceed the stoichiometric equivalent of allowable discharge of pesticide organic solvents and compounds
Organic nitrogen	
Organic phosphorus	
Organic sulphur	–
Acetate	–
Phenol and phenolic compounds as C_6H_5OH	1 mg/l
Inorganic chemicals	
Arsenic as As	0.2 mg/l
Cyanide as CN^-	0.2 mg/l
Nitrate as NO_3^-	50 mg/l
Dissolved phosphates as P	5 mg/l
Sulphate as SO_4^{--}	1000 mg/l

*μg = microgram.

Notes

1. MINAS are suggested for disposal of treated waste-waters to flowing waters. Land application has to be selective.
2. No MINAS are proposed for sludges. These are to be incinerated either in cement kilns or in a suitably designed incinerator, irrespective of their toxicity.
3. No Total Organic Carbon (TOC) or Dissolved Organic Carbon (DOC) or COD limit has been specified, which can be done only after establishing correlationship with relevant parameters. In the determination of COD of pesticide industry effluent care must be taken to minimise loss of volatile organics.
4. MINAS do not cover emission of vapour and gas to the air environment.

5. For any test or a set of tests, waste-water sampling has to be carried out proportionate to the volume rate of low. A 2-hourly compositing over 24 hours period is stipulated.

To sum up pesticides are, in general, low volume and highly biologically active environmental contaminants. Unlike fertilisers, pesticides don't physically alter the environment, as such, in quantity. Their potential hazards, therefore, must be judged in terms of their biological effects and studies of metabolic degradation in the environment should be primarily concerned with the detection of potential terminal residues.

It is also essential that only plant protection schedules developed and recommended for control of various pests involving the use of pesticide cleared from the point of persistence be adopted for use. These may not leave any toxic residues for long and degrade to safe levels by harvest.

Certain river and well water has occasionally high nitrate levels and a careful and constant watch should be maintained.

Systematic research to test and evaluate potential hazards of agricultural chemicals to man or to his environment will help in management of environmental pollution.

Chapter 19
Fertiliser

INTRODUCTION

Fertilisers can be classified as nitrogenous fertilisers, phosphatic fertilisers and complex fertilisers. Plants may be producing nitrogenous fertilisers, like urea, ammonium sulphate, ammonium nitrate and ammonium chloride, or phosphatic fertilisers like super- phosphates; there are plants where complex fertilisers containing both nitrogen and phosphates, like ammonium phosphate and ammonium sulphate phosphate are produced. Also, some fertiliser units are only involved in mining and undertake no other activity.

MANUFACTURING PROCESS

Ammonia is the principal intermediate in the manufacture of all nitrogenous fertilisers. So, except when the by-product ammonia is available from a coke-oven, raw materials for nitrogenous fertiliser production are carbonaceous material, which are required for making ammonia. Thus, all nitrogenous fertiliser plants essentially have an ammonia production unit and a reactor where the synthetic ammonia is reacted with other chemicals to produce the final product. The plant may have auxiliary units to produce reacting chemicals also.

Steps in Manufacture

Basic process steps in the manufacture of urea from carbonaceous raw materials like naphtha are:
1. Reaction of the carbonaceous material with steam and air to form a mixture of hydrogen and carbon monoxide, known as synthesis gas.
2. Reaction of carbon monoxide with steam over a catalyst to form more hydrogen and carbon dioxide.
3. Separation and purification of carbon dioxide.
4. Removal of residual carbon monoxide from gas mixture.
5. Synthesis of ammonia by reacting hydrogen and nitrogen over a catalyst (nitrogen is supplied as air in an earlier step).
6. Synthesis of urea by treating ammonia with carbon dioxide in a reactor at higher temperature and pressure.

Plants using by-product ammonia from other manufacturing plants (like coke oven) have to produce carbon dioxide separately for the production of urea.

Ammonium sulphate may be produced by reacting anhydrous ammonia with sulphuric acid, usually obtained as a by-product from other manufacturing plants. Ammonium sulphate may also be manufactured from gypsum or from calcium sulphate sludge, obtained from the phosphatic fertiliser plant, using ammonia and carbon dioxide obtained from ammonia plant. In this process, calcium sulphate is reacted with ammonium carbonate solution to produce ammonium sulphate.

Ammonium nitrate is produced when ammonia is reacted with nitric acid. Normally, the required quantity of nitric acid is produced in the same plant, by oxidising ammonia.

Superphosphate is produced by merely mixing the phosphate ore (commonly known as phosphatic rock) with sulphuric acid to convert the phosphate to 'monocalcium phosphate'. The by-product calcium sulphate of this process may be used in the manufacture of ammonium sulphate.

Ammonium phosphate is made by treating phosphoric acid with ammonia; the phosphoric acid production process involves the following steps:
1. Dissolving phosphate rock in enough sulphuric acid.
2. Holding the mixture until the calcium sulphate crystals grow to adequate size.
3. Separation of the phosphoric acid and calcium sulphate by filtration.
4. Concentration of acid to the desired level.

WASTE-WATER FROM FERTILISER PLANTS

A variety of waste streams are discharged from fertiliser plants in the form of:
1. Processing chemicals like sulphuric acid.
2. Process intermediates like ammonia, phosphoric acid, etc.
3. Final products like urea, ammonium sulphate, ammonium phosphate, etc.

In addition, oil bearing wastes from compressor houses of ammonia and urea plants, and some portion of cooling water and wash water from the scrubbing towers for the purification of gases, also come as waste.

Wash water from the scrubbing towers may contain toxic substances like arsenic, monoethanolamine, potassium carbonate, etc. in a nitrogenous fertiliser plant, while that in a phosphatic fertiliser plant may contain a mixture of carbonic acid, hydrofluoric acid and fluosilicic acid. Both alkaline and acidic wastes are also expected from the boiler feed water treatment plant, the wastes being generated during the regeneration of anion and cation exchanger units.

Previously, waste cooling water contained toxic elements like chromates, zinc, etc. which were used for corrosion control. The development of non-chromate technology using quaternary ammonium compounds has eliminated these toxic substances. Additional pollutants like phenol and cyanide will be introduced in the list of pollutants in a fertiliser plant where ammonia is derived from the waste ammoniacal liquor of coke ovens.

Effects of Wastes on Receiving Streams

All the components of waste from the fertiliser plants induce adverse effects in streams. Acids and alkalies can destroy normal aquatic life. Arsenic, fluorides, and ammonium salts are found to be toxic to fish. Amines are not only toxic to fish but also exert a high oxygen and chlorine demand.

Presence of different types of salts renders the stream unfit for use as a source of drinking water in the downstream side. Nitrogen and other nutrient content of the waste encourages growth of aquatic plants in the stream.

Treatment of Fertiliser Waste-water

Major pollutants in fertiliser waste-water for which treatment is necessary include oil, arsenic, ammonia, urea, phosphate and fluoride. Oil is removed in a gravity separator. Arsenic containing waste is segregated and after its concentration, the solid waste is disposed of in a safe place. Phosphate and fluoride bearing wastes are also segregated and chemically coagulated by lime; clarified effluent, which still contains some amount of phosphate and fluoride, is diluted by mixing with other wastes. Several alternatives are there for the treatment of ammonia bearing wastes, including:

1. Steam stripping.
2. Air stripping in towers.
3. Lagooning after pH adjustment.
4. Biological nitrification and denitrification.

For all practical purposes, 'steam stripping' for ammonia removal from fertiliser wastes has been found to be uneconomical. Removal of ammonia gas from the solution in an air stripping tower, packed with red wood stakes, is found to be a very efficient method.

Very encouraging results are obtained from some laboratory and pilot plant studies conducted by national environmental engineering research institute (NEERI) in the removal of ammonia by simply lagooning the waste. It was found that considerable reduction in the ammonia content can be accomplished just by retaining the ammoniacal waste in an earthen tank, about 1 metre deep, for a day or two, after pretreatment of the waste by lime to increase the pH to 11.0. However, no reduction in urea content was observed within this period in wastes containing urea; thus waste containing both urea and ammonia required to be retained in the lagoon for a longer period, to allow urea to decompose to ammonia first.

Biological nitrification involves oxidation of ammonia to nitrate, via nitrite under aerobic conditions; this is followed by denitrification of the nitrified effluent under anaerobic condition, in which gaseous N_2 and N_2O are the end products and are released into the atmosphere. The denitrification requires addition of some quantity of carbonaceous matter in the reactor.

In all the ammonia removal methods described above, urea remains untouched. If urea removal is required, wastes containing urea must be retained for a sufficiently long time in an earthen lagoon to allow it to decompose first to ammonia. The effluent treatment of a complex fertiliser plant are given in Fig. 19.1.

Pollution control aspects related to single superphosphate (SSP) fertiliser, straight nitrogenous fertiliser and complex fertilisers are discussed here.

Pollution Control in Single Superphosphate (SSP) Plant

Single superphosphate (SSP) is the most popular and the cheapest phosphatic fertiliser among Indian farmers. There is still a wide gap between the total installed capacity and the actual production. Fluorine emitted as a result of rock phosphate–acid reaction is a major pollutant in an SSP plant.

The importance of single superphosphate (SSP) as an important source of phosphatic fertiliser in our country has already been accepted. This importance is further increased as it provides some important micro-nutrients like sulphur and calcium in addition to P_2O_5. It is all the more crucial that SSP production, which is going to have increasing share of growing phosphatic demand in our country should be allowed in the plants which have taken adequate precautions to protect the environment around them by treating emissions from the plant to the extent that technology can economically support.

Single superphosphate is mainly manufactured by a continuous process. Ground rock phosphate and diluted sulphuric acid are mixed in a Broadfield mixer and passed through a closed slow moving conveyer

(Den). Most Indian SSP manufacturers are using the above mentioned process. During the acidulation process, some acidic fumes containing fluorine are emitted from the Den and the mixer. They are to be scrubbed to remove toxic gases. Hydrogen silicofluoride formed in the scrubber can be used to make sodium-silico-fluoride, a useful by-product.

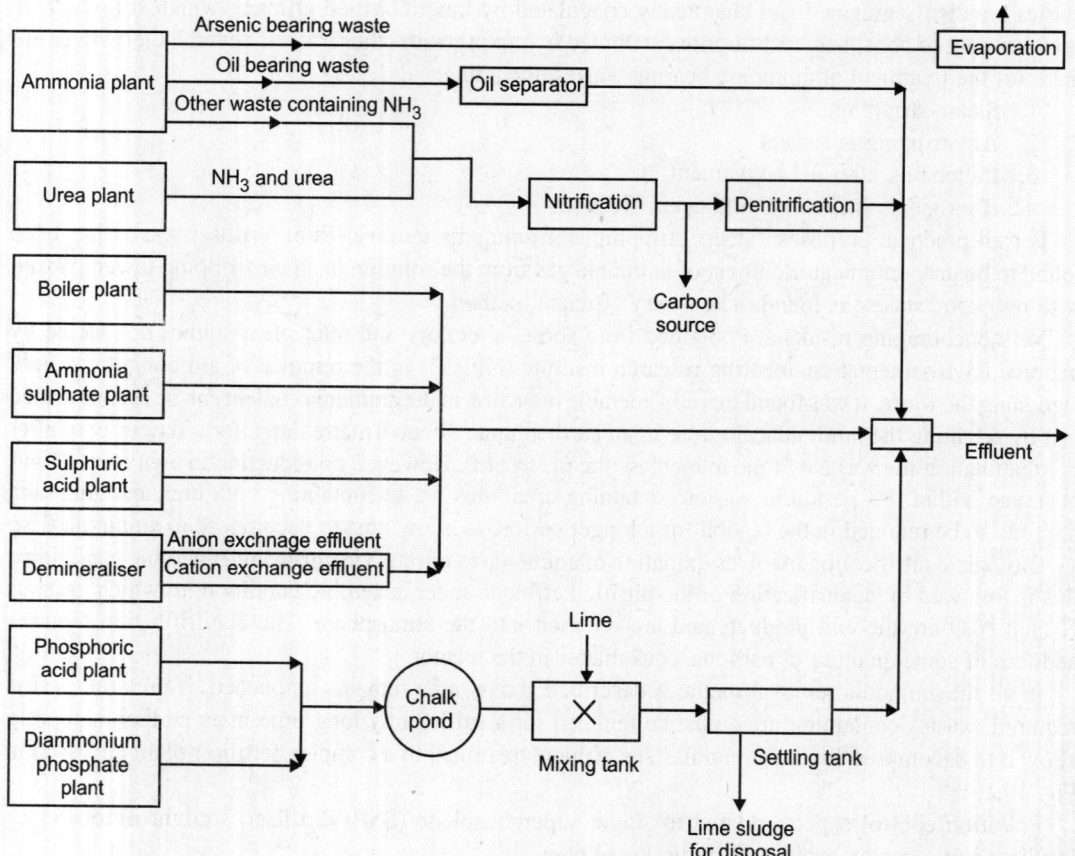

Fig. 19.1. Flow diagram for effluent treatment of a complex fertiliser plant.

Minimal national standards for straight phosphatic fertiliser (MINAS)

The minimal national standards for the straight phosphatic fertiliser units are based on the available treatment system and achievable quality of effluent by the technology in use today. The minimal national standards are presented in Table 19.1.

Straight Nitrogenous Fertilisers

Urea and salts of ammonia are referred to as straight fertilisers and if combined with other nutrients such as phosphates and potash, they are called complex/mixed fertilisers.

Nitrogenous fertiliser plants use a large quantity of water mainly for process cooling, steam generation and process use, resulting in waste-water generation at various points in the manufacturing process. In an ammonia plant, if partial oxidation process is used, then the carbon of hydrocarbon feedstock is not completely combusted. As such, unburnt carbon as soot is generated. In most plants, inplant provisions are incorporated for recycle of 80 per cent of the carbon generated. The rest 20 per cent of carbon produced is required to be disposed of.

Table 19.1. Minimal national standards for straight phosphatic fertiliser industry.

Parameter	Concentration not to exceed, mg/l (except pH)
pH	7.0–9.0
Fluorides as F	10
Phosphate as P	5
Chromium as Cr	
Total	2
Hexavalent	0.1
Oil and grease	10
Suspended solids	100

Further, cyanide is formed due to a side reaction during gasification. Both carbon and cyanide come out in the washwater. Inplant system is usually provided for removal of major quantity of cyanide in the stripper, where the stripping is done by steam and acidic gases. However, the stripper bottom effluent containing some residual cyanide is continuously discharged. In case of fuel oil feedstock, the effluent discharged contains suspended carbon, sulphide, formate, ammonia and metals like vanadium, nickel, iron, etc. along with cyanide. Where the fuel oil feedstock contains high sulphur, a sulphur recovery plant is incorporated.

Where coal is used as feedstock, considerable quantity of fly ash is discharged in the wash water of the gasification section. This washwater contains a small quantity of cyanide also.

In the steam reformation process, steam in excess of stoichiometric requirement is supplied in the primary reformer. When the gas is cooled, the excess steam condenses and forms process condensate, which contains ammonia, methanol and some organics. In a modern plant, built-in system is provided for removal of major quantity of ammonia by stripping. This makes the condensate fit for use as boiler feed water, with or without polishing. However, when the contaminants in the condensate are high, the condensate is unfit for use as boiler feed water and is discharged as effluent.

In the carbon dioxide recovery section, though the absorption towers are operated in closed circuit, some absorbent chemicals find their way out in the cooler condensate, sludge (formed in the process) and in leakage and spillage from the system. The type of pollutant depends on the absorbent solution used.

In the urea plant, the urea solution with associated ammonia and carbon dioxide is concentrated in vacuum concentrators. During evaporation, the vapours condense to form contaminated condensate. This condensate contains urea, ammonia and carbon dioxide. This is the main source of pollution in a urea plant. In modern urea plants, a built-in facility—thermal hydrolyser stripper is provided whereby a major quantity of contaminants are removed and recycled before the discharge of the condensate.

Another effluent stream is generated from the carbon dioxide compression section, which mainly contains oil. The ammonia and carbamate solution plunger pump leakage also form an effluent stream, which contains small quantity of ammonia, urea and oil.

Disposal of the final treated effluent and monitoring

Disposal of the final treated effluent to the receiving water system is an important aspect in the pollution control system. The detailed procedure to be adopted for disposal system is given as under:

1. It is desired that all the effluents after treatment shall be routed to a properly lined guard pond for equalisation and final control.
2. The guard pond should have two compartments, each of at least four hours capacity. All the effluent streams shall be connected to these compartments by a parallel connection system. One compartment of the guard pond shall be used for the routine disposal of effluent, while the other compartment will remain empty and be utilised when effluent do not conform the limits.
3. In the guard pond, an automatic monitoring system for flow and the relevant pollutants shall be provided with high-level alarm system. The parameters necessary for automatic monitoring are pH, ammoniacal nitrogen, nitrate nitrogen and hexavalent chromium. Monitoring of nitrate nitrogen is applicable to the industries where nitric acid is produced and used for production of fertilisers, and the monitoring of hexavalent chromium is applicable to the industries where chromate-based inhibitors are used for cooling water conditioning.
4. When pollutants in the final effluent exceed the stipulated figures as indicated by the alarm system and the effluent stream responsible cannot be identified, all the effluent streams shall flow the empty compartment of the guard pond. The effluent thus stored should be treated before discharge.
5. The area around the guard pond shall be developed with proper road connection and lighting system so that it can be approached easily at any time.
6. Till the continuous monitoring system as indicated above are not installed, the industry shall collect grab samples at a four-hour interval and analyse these for pH and ammoniacal nitrogen.
7. The other parameters as relevant to the industry concerned such as total kjeldahl nitrogen, hexavalent chromium, total chromium cyanide, nitrate nitrogen, vanadium, arsenic, suspended solids and oil and grease should be analysed in the grab sample collected once a day at fixed hours.
8. For effective appraisal of the performance of treatment units, the industry shall monitor the concerned parameters at least once a shift before and after treatment.

Minimal national standards for straight nitrogenous fertiliser industry

The minimal national standards for the straight nitrogenous fertiliser units are based on the available technology of inplant pollution control, end on treatment systems and achievable quality of effluent based on the technology in use today. These are presented in Table 19.2.

Table 19.2. Minimal national standards for straight nitrogenous fertiliser industry.

Parameter	Concentration not to exceed, mg/l (except pH)
pH	6.5 to 8.0
Ammoniacal nitrogen	50
Total Kjeldahl Nitrogen (TKN)	100
Free ammoniacal nitrogen	4
Nitrate nitrogen	10

(Contd...)

Parameter	Concentration not to exceed, mg/l (except pH)
Cyanide (as CN)	0.2
Vanadium (as V)	0.2
Arsenic (as As)	0.2
Hexavalent chromium	0.1
Total chromium	2.0
Suspended solids	100
Oil and grease	10

PHOSPHATE FERTILISER

Industrial Operation and Waste-water

The phosphate manufacturing and phosphate fertiliser industry is a basic chemical manufacturing industry, in which essentially both the mixing and chemical reactions of raw materials are involved in production. Also, short- and long-term chemical storage and warehousing, as well as loading/unloading and transportation of chemicals, are involved in the operation. In the case of fertiliser production, only the manufacturing of phosphate fertilisers and mixed and blend fertilisers containing phosphate along with nitrogen and/or potassium is presented here.

Regarding waste-water generation, volumes resulting from the production of phosphorus are several orders of magnitude greater than the waste-waters generated in any of the other product categories. Elemental phosphorus is an important waste-water contaminant common to all segments of the phosphate manufacturing industry, if the phossy water (water containing colloidal phosphorus) is not recycled to the phosphorus production facility for reuse.

Categorisation in phosphate production

The phosphate manufacturing industry is broadly subdivided into two main categories: phosphorus-derived chemicals and other nonfertiliser phosphate chemicals. For the purposes of raw waste characterisation and delineation of pretreatment information, the industry is further subdivided into six subcategories. The categorisation system of the various main production streams and their descriptions are shown in Table 19.3. It will be used in the following discussion to identify process flows and characterise the resulting raw waste.

Table 19.3. Categorisation system in phosphorus-derived and nonfertiliser phosphate chemicals production.

Main category	Subcategory
Phosphorus-derived chemicals	Phosphorus production
	Phosphorus consuming
	Phosphate
Other nonfertiliser phosphate chemicals	Defluorinated phosphate rock
	Defluorinated phosphoric acid
	Sodium phosphates

The manufacture of phosphorus-derived chemicals is almost entirely based on the production of elemental phosphorus from mined phosphate rock. Ferrophosphorus, widely used in the metallurgical industries, is a direct by-product of the phosphorus process. Over 85 per cent of elemental phosphorus production is used to manufacture high-grade phosphoric acid by the furnace or dry process as opposed to the wet process that converts phosphate rock directly into low-grade phosphoric acid. The remainder of the elemental phosphorus is either marketed directly or converted into phosphorus chemicals. The furnace-grade phosphoric acid is marketed directly, mostly to the food and fertiliser industries. Finally, phosphoric acid is employed to manufacture sodium tripolyphosphate, which is used in detergents and for water treatment and calcium phosphate, which is used in foods and animal feeds.

On the other hand, defluorinated phosphate rock is utilised as an animal feed ingredient. Defluorinated phosphoric acid is mainly used in the production of animal foodstuffs and liquid fertilisers. Finally, sodium phosphates, produced from wet process acid as the raw material, are used as intermediates in the production of cleaning compounds.

Phosphorus and phosphate compounds

Phosphorus production

Phosphorus is manufactured by the reduction of commercial-quality phosphate rock by coke in an electric furnace, with silica used as a flux. Slag, ferrophosphorus (from iron contained in the phosphate rock) and carbon monoxide are reaction by-products. The standard process, consists of three basic parts: phosphate rock preparation, smelting in an electric furnace and recovery of the resulting phosphorus. Phosphate rock ores are first blended so that the furnace feed is the uniform composition and then pretreated by heat drying, sizing or agglomerating the particles and heat treatment.

Waste-waters are generated in the process of scrubbing contaminants from gaseous effluent streams. This water requirement is of significant volume and process conditions normally permit the use of recirculated contaminated water for this service, thereby effectively reducing the discharged waste-water volume.

Leaks and spills are routinely collected as part of process efficiency and housekeeping and, in any case, their quantity is minor and normally periodic.

Defluorinated phosphoric acid

One method used in order to defluorinate wet process phosphoric acid is vacuum evaporation. The concentration of 54 per cent P_2O_5 acid to a 68 to 72 per cent P_2O_5 strength is performed in vessels that use high-pressure (30.6–37.4 at or 450–550 psi) steam or an externally heated Dowtherm solution as the heat energy source for evaporation of water from the acid. Fluorine removal from the acid occurs concurrently with the water vapour loss.

A second method of phosphoric acid defluorination entails the direct contact of hot combustion gases (from fuel oil or gas burners) with the acid by bubbling them through the acid. Evaporated and defluorinated product acid is sent to an acid cooler, while the gaseous effluents from the evaporation chamber flow to a series of gas scrubbing and absorption units. Finally, aeration can also be used for defluorinating phosphoric acid. In this process, diatomaceous silica or spray-dried silica gel is mixed with commercial 54 per cent P_2O_5 phosphoric acid. Hydrogen fluoride in the impure phosphoric acid is converted to fluosilicic acid, which in turn breaks down to SiF_4 and is stripped from the heated mixture by simple aeration.

The major waste-water source in the defluorination processes is the wet scrubbing of contaminants from the gaseous effluent streams. However, process conditions normally permit the use of recirculated contaminated water for this service, thereby effectively reducing the discharged waste-water volume.

Waste-water Characteristics and Sources

Waste-waters from the manufacturing, processing and formulation of inorganic chemicals such as phosphorus compounds, phosphates and phosphate fertilisers cannot be exactly characterised. The waste-water streams are usually expected to contain trace or large concentrations of all raw materials used in the plant; all intermediate compounds produced during manufacture; all final products, coproducts and by-products; and the auxiliary or processing chemicals employed. It is desirable from the viewpoint of economics that these substances not be lost, but some losses and spills appear unavoidable and some intentional dumping does take place during housecleaning, vessel emptying and preparation operations.

Few fertiliser plants discharge waste-waters to municipal treatment systems. Most use ponds for the collection and storage of waste-waters, pH control, chemical treatment and settling of suspended solids. Whenever available retention pond capacities in the phosphate fertiliser industry are exceeded, the waste-water overflows are treated and discharged to nearby surface water bodies. The range of waste-water characteristics and concentrations for typical retention ponds used by the phosphate fertiliser industry are given in Table 19.4.

Table 19.4. Raw waste-water characteristics of phosphate fertiliser industry retention ponds.

Quality parameter	Phosphate
Suspended solids (mg/l)	800–1200
pH	1–2
Ammonia (mg/l)	450–500
Sulphate (mg/l)	4000
Chloride (mg/l)	58
Total phosphate (mg/l)	3–5 M
Fluoride (mg/l)	6–8.5 M
Aluminium (mg/l)	110
Iron (mg/l)	85
Radium 226 (picocuries/l)	60–100

M = Thousand.

The specific types of waste-water sources in the phosphate fertiliser industry are: (i) water treatment plant wastes from raw water filtration, clarification, softening and deionisation, which principally consist of only the impurities removed from the raw water (such as carbonates, hydroxides, bicarbonates and silica) plus minor quantities of treatment chemicals; (ii) closed-loop cooling tower blowdown, the quality of which varies with the makeup of water impurities and inhibitor chemicals used (the only cooling water contamination from process liquids is through mechanical leaks in heat exchanger equipment. Table 19.5 highlights the normal range of contaminants that may be found in cooling water blowdown systems); (iii) boiler blowdown, which is similar to cooling tower blowdown but the quality differs as given in Table 19.6; (iv) contaminated water or gypsum pond water, which is the impounded and reused water that accumulates sizable concentrations of many cations and anions, but mainly fluorine and phosphorus [concentrations of 8500 mg/l F and in excess of 5000 mg/l P are not unusual; concentrations of radium 226 in recycled gypsum pond water are 60 to 100 picocuries/l and its acidity reaches extremely high levels (pH 1–2)]; (v) waste-water from spills and leaks that, when possible, is reintroduced directly to the process or into the contaminated water system; and (vi) nonpoint source discharges that originate from the dry fertiliser dust covering the general plant area and then dissolving in rain water and snowmelt that become contaminated.

Table 19.5. Range of concentrations of contaminants in cooling water.

Cooling water contaminant	Concentration (mg/l)
Chromate	0–250
Sulphate	500–3000
Chloride	35–160
Phosphate	10–50
Zinc	0–30
TDS	500–10000
TSS	0–50
Biocides	0–100

Table 19.6. Range of concentrations of contaminants in boiler blowdown waste.

Boiler blowdown contaminant	Concentration (mg/l)
Phosphate	5–50
Sulphite	0–100
TDS	500–3500
Zinc	0–10
Alkalinity	50–700
Hardness	50–500
Silica (SiO_2)	25–80

In the specific case of waste-water generated from the condenser water bleedoff in the production of elemental phosphorus from phosphate rock in an electric furnace, Horton reported that the flow varies from 10 to 100 gpm (2.3–23 m³/hr), depending on the particular installation. The most important contaminants in this waste are elemental phosphorus, which is colloidally dispersed and may ignite if allowed to dry out and fluorine that is also present in the furnace gases. The general characteristics of this type of waste-water (if no soda ash or ammonia were added to the condenser water) are given in Table 19.7.

As previously mentioned, fertiliser manufacturing may create problems within all environmental media, i.e. air pollution, water pollution and solid wastes disposal difficulties. In particular, the liquid waste effluents generated from phosphate and mixed and blend fertiliser production streams originate from a variety of sources and may be summarised as follows: (i) ammonia-bearing wastes from ammonia production, (ii) ammonium salts such as ammonium phosphate, (iii) phosphates and fluoride wastes from phosphate and superphosphate production, (iv) acidic spillages from sulphuric acid and phosphoric acid production, (v) spent solutions from the regeneration of ion-exchange units, (vi) phosphate, chromate, copper sulphate and zinc wastes from cooling tower blowdown, (vii) salts of metals such as iron, copper, manganese, molybdenum and cobalt, (viii) sludge discharged from clarifiers and backwash water from sand filters, and (ix) scrubber wastes from gas purification processes.

Table 19.7. Range of concentrations of contaminants in condenser waste from electric furnace production of phosphorus.

Quality parameter	Concentration or value
pH	1.5–2.0
Temperature	120°–150°F
Elemental phosphorus	400–2500 ppm
Total suspended solids	1000–5000 ppm
Fluorine	500–2000 ppm
Silica	300–700 ppm
P_2O_5	600–900 pp
Reducing substances as (I_2)	40–50 ppm
Ionic charge of particles	Predominantly positive (+)

Considerable variation, therefore, is observed in quantities and waste-water characteristics at different plants. The most important factors that contribute to excessive in-plant materials losses and therefore, probable subsequent pollution are the age of the facilities (low efficiency, poor process control), the state of maintenance and repair (especially of control equipment), variations in feedstock and difficulties in adjusting processes to cope and an operational management philosophy such as consideration for pollution control and prevention of materials loss. Because of process cooling requirements, fertiliser manufacturing facilities may have an overall large water demand, with the waste-water effluent discharge largely dependent on the extent of in-plant recirculation. Facilities designed on a once-through process cooling flowstream generally discharge from 1000 to over 10,000 m^3/hr waste-water effluents that are primarily cooling water.

According to research results reported by Fuller, the removal of semicolloidal matter in settling areas or ponds seems to be one of the primary problems concerning water pollution control. The results of dissolved oxygen (DO) and BOD surveys indicated that receiving streams were actually improved in this respect by the effluents from phosphate operations. On the other hand, no detrimental effects on fish were found, but there is the possibility of destruction of fish food aquatic micro-organisms and plankton under certain conditions.

The waste-water characteristics vary from one production facility to the next and even the particular flow magnitude and location of discharge will significantly influence its aquatic environmental impact. The degree to which a receiving surface water body dilutes a waste-water effluent at the point of discharge is important, as are the minor contaminants that may occasionally have significant impacts. Fertiliser manufacturing wastes, in general, affect water quality primarily through the contribution of nitrogen and phosphorus, whose impacts have been extensively documented in the literature. Significant levels of phosphates assist in inducing eutrophication and in many receiving waters they may be more important (growth-limiting agent) than nitrogenous compounds. Under such circumstances, programmes to control eutrophication have generally attempted to reduce phosphate concentrations in order to prevent excessive algal and macrophyte growth.

In addition to the above major contaminants, pollution from the discharge of fertiliser manufacturing wastes may be caused by such secondary pollutants as oil and grease, hexavalent chromium, arsenic and fluoride. As reported by Beg in certain cases, the presence of one or more of these pollutants may have adverse impacts on the quality of a receiving water, due primarily to toxic properties or can be inhibitory to the nitrification process. Finally, oil and grease concentrations may have a significant detrimental effect on the oxygen transfer characteristics of the receiving surface water body.

Phosphate Fertiliser Manufacture

The limitations applied to process waste-water, which establish the quantity of pollutants or pollutant properties that may be discharged by a point source into a surface water body after the application of various types of control technologies. The total suspended solids limitation is waived for process waste-water from a calcium sulphate (phosphogypsum) storage pile runoff facility, operated separately or in combination with a water recirculation system, which is chemically treated and then clarified or settled to meet the other pollutant limitations. The concentrations of pollutants discharged in contaminated nonprocess waste-water, i.e. any water including precipitation runoff that comes into incidental contact with any raw material, intermediate or finished product, by-product or waste product by means of precipitation, accidental spills or leaks and other non-process discharges, should not exceed as per details given below:

Effluent characteristic	Maximum for any 1 day	Average of daily values for 30 consecutive days shall not exceed
Contaminated nonprocess waste-water		
Total phosphorus (as P)	105	35
Fluoride	75	25

Waste-water Control and Treatment

The pollution control and treatment methods and unit processes used are discussed in more detail in the following section.

In-plant control, recycle and process modification

The primary consideration for in-plant control of pollutants that enter waste streams through random accidental occurrences, such as leaks, spills and process upsets, is establishing loss prevention and recovery systems. In the case of fertiliser manufacture, a significant portion of contaminants may be separated at the source from process wastes by dedicated recovery systems, improved plant operations, retention of spilled liquids and the installation of localised interceptors of leaks such as oil drip trays for pumps and compressors. Also, certain treatment systems installed (i.e. ion-exchange, oil recovery and hydrolyser-stripper systems) may, in effect, be recovery systems for direct or indirect reuse of effluent constituents. Finally, the use of effluent gas scrubbers to improve in-plant operations by preventing gaseous product losses may also prevent the airborne deposition of various pollutants within the general plant area, from where they end up as surface drainage runoff contaminants.

Cooling water

Cooling water constitutes a major portion of the total in-plant wastes in fertiliser manufacturing and it includes water coming into direct contact with the gases processed (largest percentage) and water that has no such contact. The latter stream can be readily used in a closed-cycle system, but sometimes the direct contact cooling water is also recycled (after treatment to remove dissolved gases and other contaminants and clarification). By recycling, the amount of these waste-waters can be reduced by 80 to 90 per cent, with a corresponding reduction in gas content and suspended solids in the wastes discharged to sewers or surface water.

Phosphate manufacturing

Significant in-plant control of both waste quantity and quality is possible for most subcategories of the phosphate manufacturing industry. Important control measures include stringent in-process abatement, good housekeeping practices, containment provisions and segregation practices. In the phosphorus chemicals industry, plant effluent can be segregated into noncontact cooling water, process water and auxiliary streams comprising ion-exchange regenerants, cooling tower blowdowns, boiler blowdowns, leaks and washings. Many plants have accomplished the desired segregation of these streams, often by a painstaking rerouting of the sewer lines. The use of once-through scrubber waste should be discouraged; however, there are plants that recycle the scrubber water from a sump, thus satisfying the scrubber water flow rate demands on the basis of mass transfer considerations while retaining control of water usage.

The containment of phossy water from phosphorus transfer and storage operations is an important control measure in the phosphorus-consuming. Although displaced phossy water is normally shipped back to the phosphorus-producing facility, the usual practice in phosphorus storage tanks is to maintain a water blanket over the phosphorus for safety reasons. This practice is undesirable because the addition of makeup water often results in the discharge of phossy water, unless an auxiliary tank collects phossy water overflows from the storage tanks, thus ensuring zero discharge. A closed-loop system is then possible if the phossy water from the auxiliary tank is reused as makeup for the main phosphorus tank.

Another special problem in phosphorus-consuming is the inadvertent spills of elemental phosphorus into the plant sewer pipes. Provision should be made for collecting, segregating and by-passing such spills and a recommended control measure is the installation of a trap of sufficient volume downstream of reaction vessels.

Process modifications

The following are possible process modifications and plant arrangements that could help reduce waste-water volumes, contaminant quantities and treatment costs: (i) in ammonium phosphate production and mixed and blend fertiliser manufacturing, one possibility is the integration of an ammonia process condensate steam stripping column into the condensate boiler feedwater systems of an ammonia plant, with or without further stripper bottoms treatment depending on the boiler quality makeup needed; (ii) contaminated waste-water collection systems designed so that common contaminant streams can be segregated and treated in minor quantities for improved efficiencies and reduced treatment costs; (iii) in ammonium phosphate and mixed and blend fertiliser production, another possibility is to design for a lower-pressure steam level (i.e. 42–62 atm) in the ammonia plant to make process condensate recovery easier and less costly; and (iv) when possible, the installation of air-cooled vapour condensers and heat exchangers would minimise cooling water circulation and subsequent blowdown.

Recently new techniques have been adopted by French company for pollution prevention, for a new process modification for steam segregation and recycle in phosphoric acid production in which, raw water from the sludge/fluorine separation system is recycled to the heat-exchange system of the sulphuric acid dilution unit and the waste-water used in plaster manufacture. Furthermore, decanted supernatant from the phosphogypsum deposit pond is recycled for treatment in the water filtration unit. This process modification permits an important reduction in pollution by fluorine and that it makes the treatment of effluents easier and in some cases allows specific recycling. Finally, the new process produced a small reduction in water consumption, either by recycle or discharging a small volume of polluted process water downstream and required no particular equipment and very few alterations in the mainstream lines of the old process.

Waste-water Treatment Methods

Phosphate manufacturing

Nemerow summarised the major characteristics of wastes from phosphate and phosphorus compounds production (i.e. clays, slimes and tall oils, low pH, high suspended solids, phosphorus, silica and fluoride) and suggested the major treatment and disposal methods such as lagooning, mechanical clarification, coagulation and settling of refined waste-waters. The various waste-water treatment practices for each of the six subcategories (Table 19.3) of the phosphate manufacturing industry are summarised.

Phosphate fertiliser production

Contaminated water from the phosphate fertiliser is collected in gypsum ponds and treated for pH adjustment and control of phosphorus and fluorides. Treatment is achieved by 'double liming' or a two-stage neutralisation procedure, in which phosphates and fluorides precipitate. The first treatment stage provides sufficient neutralisation to raise the pH from 1 to 2 to a pH level of at least 8. The resultant effectiveness of the treatment depends on the point of mixing of lime addition and on the constancy of pH control. Fluosilisic acid reacts with lime and precipitates calcium fluoride in this step of the treatment.

The waste-water is again treated with a second lime addition to raise the pH level from 8 to at least 9 (where phosphate removal rates of 95 per cent may be achieved), although two-stage dosing to pH 11 may be employed. Concentrations of phosphorus and fluoride with a magnitude of 6,500 and 9,000 mg/l, respectively, can be reduced to 5 to 500 mg/l P and 30 to 60 mg/l F. Soluble orthophosphate and lime react to form an insoluble precipitate, calcium hydroxy apatite. Sludges formed by lime addition to phosphate wastes from phosphate manufacturing or fertiliser production are generally compact and possess good settling and dewatering characteristics and removal rates of 80 to 90 per cent for both phosphate and fluoride may be readily achieved.

The seepage collection of contaminated water from phosphogypsum ponds and reimpoundment is accomplished by the construction of a seepage collection ditch around the perimeter of the diked storage area and the erection of a secondary dike surrounding the first. The base of these dikes is usually natural soil from the immediate area and these combined earth/gypsum dikes tend to have continuous seepage through them. The seepage collection ditch between the two dikes needs to be of sufficient depth and size to not only collect contaminated water seepage, but also to permit collection of seeping surface runoff from the immediate outer perimeter of the seepage ditch. This is accomplished by the erection of the small secondary dike, which also serves as a backup or reserve dike in the event of a failure of the primary major dike.

The sulphuric acid plant has boiler blowdown and cooling tower blowdown waste streams, which are uncontaminated. However, accidental spills of acid can and do occur and when they do, the spills contaminate the blowdown streams. Therefore, neutralisation facilities should be supplied for the blowdown waste streams, which involves the installation of a reliable pH or conductivity continuous-monitoring unit on the plant effluent stream. The second part of the system is a retaining area through which noncontaminated effluent normally flows. The detection and alarm system, when activated, causes a plant shutdown that allows location of the failure and initiation of necessary repairs. Such a system, therefore, provides the continuous protection of natural drainage waters, as well as the means to correct a process disruption.

Mixed fertiliser treatment technology consists of a closed loop contaminated water system, which includes a retention pond to settle suspended solids. The water is then recycled back to the system. There are no liquid waste streams associated with the blend fertiliser process, except when liquid air scrubbers are used to prevent air pollution. Dry removals of air pollutants prevent a waste-water stream from being formed.

Phosphate and fluoride removal

Phosphates may be removed from waste-waters by the use of chemical precipitation as insoluble calcium phosphate, aluminium phosphate and iron phosphate. The liming process has been discussed previously, lime being typically added as a slurry and the system used is designed as either a single or two-stage one. Polyelectrolytes have been employed in some plants to improve overall settling and clariflocculators or sludge-blanket clarifiers are used in a number of facilities. Alternatively, the dissolved air flotation process is also feasible for phosphate and fluoride removal.

A number of aluminium compounds, such as alum and sodium aluminate, have also been used as phosphate precipitants at an optimum pH range of 5.5 to 6.5, as have iron compounds such as ferrous sulphate, ferric sulphate, ferric chloride and spent pickle liquor. The optimum pH range for the ferric salts is 4.5 to 5 and for the ferrous salts it is 7 to 8, although both aluminium and iron salts have a tendency to form hydroxyl and phosphate complexes. As reported by Ghokas, sludge solids produced by aluminium and iron salts precipitation of phosphates are generally less settleable and more voluminous than those produced by lime treatment.

According to Sprecht, in the two-step process to remove fluorides and phosphoric acid, water entering the first step may contain about 1700 mg/l F and 5000 mg/l P_2O_5 and it is treated with lime slurry or ground limestone to a pH of 3.2 to 3.8. Insoluble calcium fluorides settle out and the fluoride concentration is lowered to about 50 mg/l F, whereas the P_2O_5 content is reduced only slightly. The clarified supernatant is transferred to another collection area where lime slurry is added to bring the solution to pH 7 and the resultant precipitate of P is removed by settling. The final clear water, which contains only 3 to 5 mg/l F and practically no P_2O_5, is either returned to the plant for reuse or discharged to surface waters. The two-step process is required to reduce fluorides in the water below 25 mg/l F, because a single-step treatment to pH 7 lowers the fluoride content only to 25 to 40 mg/l F. In the process where the triple phosphate is to be granulated or nodulised, the material is transferred directly from the reaction mixer to a rotary dryer and the fluorides in the dryer gases are scrubbed with water.

In making defluorinated phosphate by heating phosphate rock, one method of fluoride recovery consists of absorption in a tower of lump limestone at temperatures above the dewpoint of the stack gas, where the reaction product separates from the limestone lumps in the form of fines. A second method of recovery consists of passing the gases through a series of water sprays in three separate spray chambers, of which the first one is used primarily as a cooling chamber for the hot exit gases of the furnace. In the second chamber, the acidic water is recycled to bring its concentration to about 5 per cent equivalence of hydrofluoric acid in the effluent, by withdrawing acid and adding freshwater to the system. In the final chamber, scrubbing is supplemented by adding finely ground limestone blown into the chamber with the entering gases. Hydrochloric acid is sometimes formed as a by-product from the fluoride recovery in the spray chambers and this is neutralised with NaOH and lime slurry before being transferred to settling areas.

Ammonium Phosphate Fertiliser and Phosphoric Acid Plant

The fertiliser industry is plagued with a tremendous problem concerning waste disposal and dust because of the very nature of production that involves large volumes of dusty material. Two types of problems are associated with waste from the manufacture of ammonium phosphate: (i) wastes from combining ammonia and phosphoric acid and the subsequent drying and cooling of the products; and (ii) wastes from the handling of the finished product arising primarily from the bagging of the product prior to shipping. Because the ammoniation process has to be 'forced' by introducing excess amounts of ammonia than the phosphoric acid is capable of absorbing, there is high ammonia content in the exhaust air stream from the ammoniator. Since it is neither economically sound nor environmentally acceptable to exhaust this to the atmosphere an acid scrubber is employed to recover the ammonia without condensing with it the steam that nearly saturates the exhausted air stream.

The filtered gypsum cake from phosphoric acid production, slurried with water to about 30 per cent, is pumped to the settling lagoons, from where the clarified water is recycled to process. To provide a startup area, approximately two acres (or 8100 m^2) of the disposal area were black-topped to seal the soil surface against seepage and the gypsum collected in this area was worked outward to provide a seal for enlarging the settling area. A dike-separated section of the disposal area was designated as a collection basin for all drainage waters at the plant site. From this basin, after the settling of suspended solid impurities, these waters are discharged to surface waters under supervision from a continuous monitoring and alarm system that guards against accidental contamination from any other source.

Air streams from the digestion system, vacuum cooler, concentrator and other areas where fluorine is evolved are connected to a highly-efficient absorption system, providing extremely high volumes of water relative to the stream. The effluent from this absorption system forms part of the recycled water and is eventually discharged as part of the product used for fertiliser manufacture. Some plants require a constant recirculating water load in excess of 3000 gpm (11.4 m^3/min), but multiple use and recycle reduce makeup requirements to less than 400 gpm (1.5 m^3/min) or a mere 13 per cent of total water use.

The second major waste-water discharge (about 1.2 m^3/min or 318 gpm) is from the bottom of the scrubbers used after a dry dust collector cyclone to reduce the dust concentration in the effluent air stream from the phosphate dryer. The underflow of the scrubber contains a high concentration of dust and mud, laden with tiny phosphate organic and inorganic silica and clay particles and is disposed of in a nearby stream.

The main pollutant in the flow to the stream is P_2O_5 at a concentration of nearly 1200 mg/l and it remains mostly as suspended particles.

Furnace Wastes from Phosphorus Manufacture

The electric furnace process for the conversion of phosphate rock into phosphorus, the phosphate, the major source of waste-water is the condenser water bleed off from the reduction furnace, the flow of which varies from 10 to 100 gpm (2.3–22.7 m^3/hr) and its quality characteristics are already given in Table 19.4.

Phossy water, a waste product in the production of elemental phosphorus by the electric furnace process, contains from 1000 to 5000 mg/l suspended solids that include 400 to 2500 mg/l of elemental phosphorus, distributed as liquid colloidal particles. These particles are usually positively charged, although this varies depending on the operation of the electrostatic precipitators. Furthermore, the chemical equilibrium between the fluoride and fluosilicate ions introduces an important source of variation in suspended solids that is a pH function. Commonly used coagulants such as alum or ferric chloride were unsatisfactory for waste-water clarification because of the positively charged particles, as were inorganic polyelectrolytes despite their improved performance. However, high-molecular-weight protein molecules at a suitable pH level (which varies for each protein) produced excellent coagulation and were highly successful in clarifying phossy water.

Horton investigated or attempted such potential treatment and disposal methods as lagooning, oxidising, settling (with or without prior chemical coagulation), filtering and centrifuging and concluded that the best solution appears to be coagulation and settling. The pilot units installed to evaluate this optimum system are shown in Fig. 19.2, together with a summary of the experimental results. The proper pH for optimum coagulation with proteins alone was obscured at higher pH levels by the formation of silica, which tended to encrust the pipe lines.

It was found that the addition of clay, as a weighting agent with the coagulant, eliminated the scale problem without decreasing the settling rates. Finally, it was concluded that in the pilot plant it was possible to obtain a 40 fold concentration of suspended solids (or 25 per cent solids) by a simple coagulation and sedimentation process.

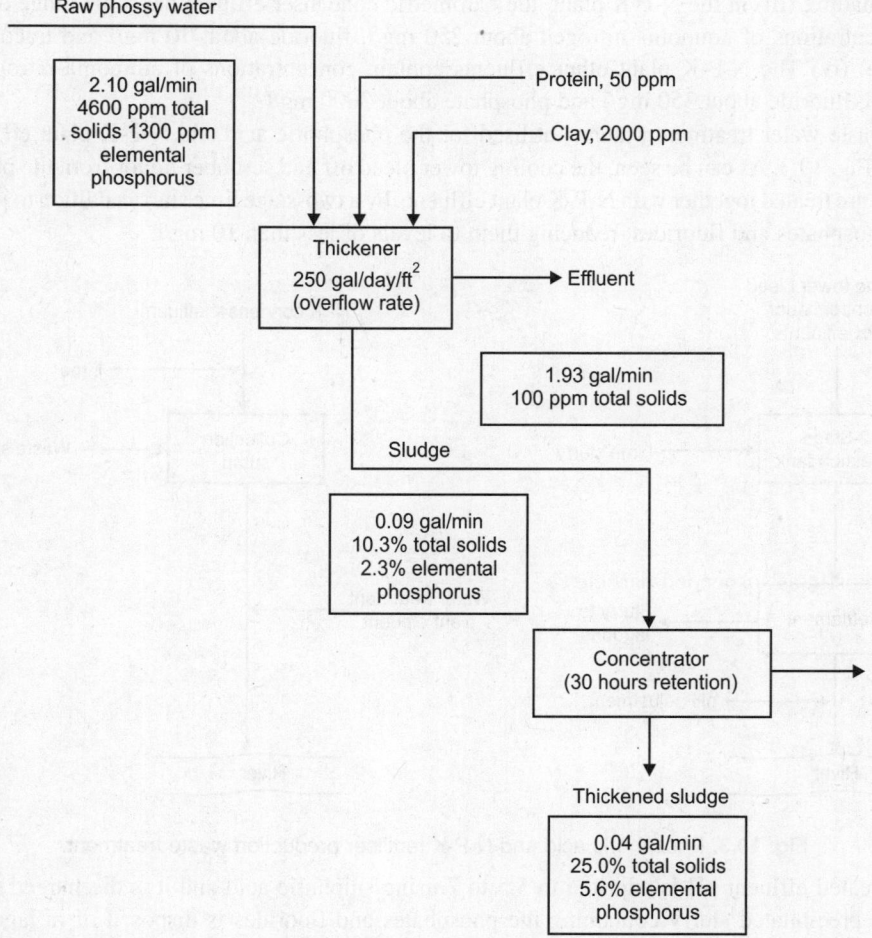

Fig. 19.2. Summary of materials balance in a pilot plant for recovery of phosphorus from phossy water.

Phosphoric Acid and N-P-K Fertiliser Plant

The heterogeneous nature of fertiliser production plants precludes the possibility of presenting a 'typical' case study of such a facility. Nevertheless, Smith presented information regarding waste-water flows and characteristics as well as the treatment systems for a phosphoric acid and N-P-K fertiliser plant, which was part of a large fertiliser manufacturing facility. The full facility additionally included an ammonia plant, a urea plant, a sulphuric acid plant and a nitric acid plant. The typical effluent flows were 183 m^3/hr (806 gpm) from the phosphoric plant and 4.4 m^3/hr (20 gpm) from the water treatment plant associated with it, whereas in the N-P-K plant they were 420 m^3/hr (1850 gpm) from the barometric condenser and 108 m^3/hr (476 gpm) from other effluent sources.

These waste-water effluents had quality characteristics that could be described as follows: (i) in the phosphoric acid plant, the contributing sources of effluent are the cooling tower bleed off and the scrubber liquor solution that contains concentrations ranging for phosphate from 160 to 200 mg/l and for fluoride from 225 to 7000 mg/l; (ii) in the water treatment plant, the waste-water effluent is slightly acidic in nature; (iii) in the N-P-K plant, the barometric condenser effluent has a pH range of 5.5 to 8 and concentrations of ammonia-nitrogen about 250 mg/l, fluoride about 10 mg/l and trace levels of phosphate; (iv) The N-P-K plant other effluents contain concentrations of ammonia-nitrogen about 2000 mg/l, fluoride about 350 mg/l and phosphate about 3000 mg/l.

The waste-water treatment systems utilised for the phosphoric acid and N-P-K plant effluents are shown in Fig. 19.3. As can be seen, the cooling tower bleed off and scrubber liquor from the phosphoric acid plant are treated together with N-P-K plant effluents by a two-stage lime slurry addition to precipitate out the phosphates and fluorides, reducing them to levels of less than 10 mg/l.

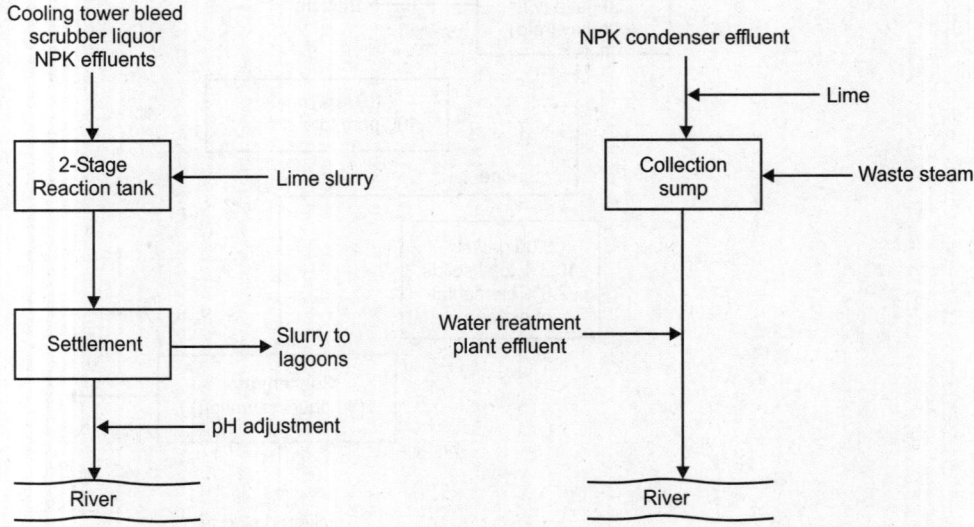

Fig. 19.3. Phosphoric acid and N-P-K fertiliser production waste treatment.

The treated effluent pH is adjusted to 5.5 to 7 using sulphuric acid and it is discharged to a river, while the precipitated slurry containing the phosphates and fluorides is disposed of in lagoons. The effluent of the barometric condenser has its pH adjusted to 11 by adding lime to remove residual ammonia-nitrogen, subsequently waste steam is introduced to remove free ammonia and the final effluent is mixed with the water treatment plant effluent prior to discharge in a river (Fig. 19.3).

Environmentally Balanced Industrial Complexes

Unlike common industrial parks where factories are selected simply on the basis of their willingness to share the real estate, environmentally balanced industrial complexes (EBIC) are a selective collection of compatible industrial plants located together in a complex so as to minimise environmental impacts and industrial production costs. These objectives are accomplished by utilising the waste materials of one plant as the raw materials for another with a minimum of transportation, storage and raw materials preparation costs. It is obvious that when an industry neither needs to treat its wastes, nor is required to import, store and pretreat its raw materials, its overall production costs must be reduced significantly. Additionally, any material reuse costs in an EBIC will be difficult to identify and more easily absorbed into reasonable production costs.

Such EBICs are especially appropriate for large, water-consuming and waste-producing industries whose wastes are usually detrimental to the environment, if discharged, but they are also amenable to reuse by close association with satellite industrial plants using wastes from and producing raw materials for others within the complex. Examples of such major industries that can serve as the focus industry of an EBIC are fertiliser plants, steel mills, pulp and paper mills and tanneries Nemerow presented the example of a steel mill complex with a phosphate fertiliser and a building materials plant as the likely candidates for auxiliary or satellite industries (Fig. 19.4).

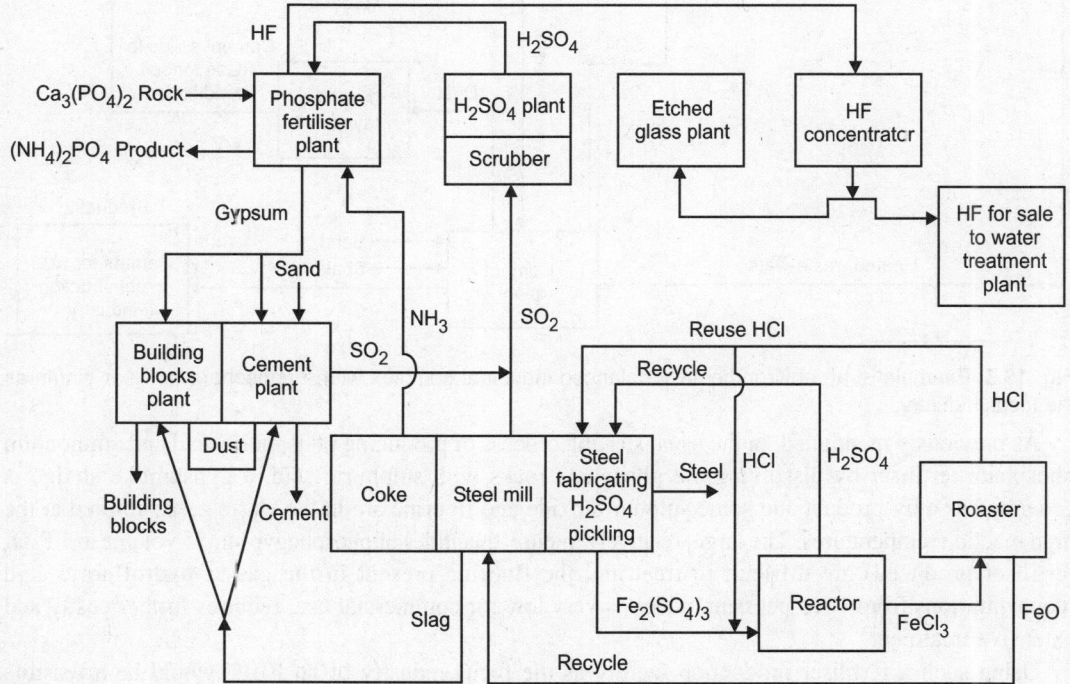

Fig. 19.4. Example of environmentally balanced industrial complex centered about a steel mill plant.

A second example presented was an EBIC centered about a phosphate fertiliser plant, with a cement production plant, a sulphuric acid plant and a municipal solid wastes composting plant (its product to be mixed with phosphate fertiliser and sold as a combined product to the agricultural industry) as the satellite industries (Fig. 19.5).

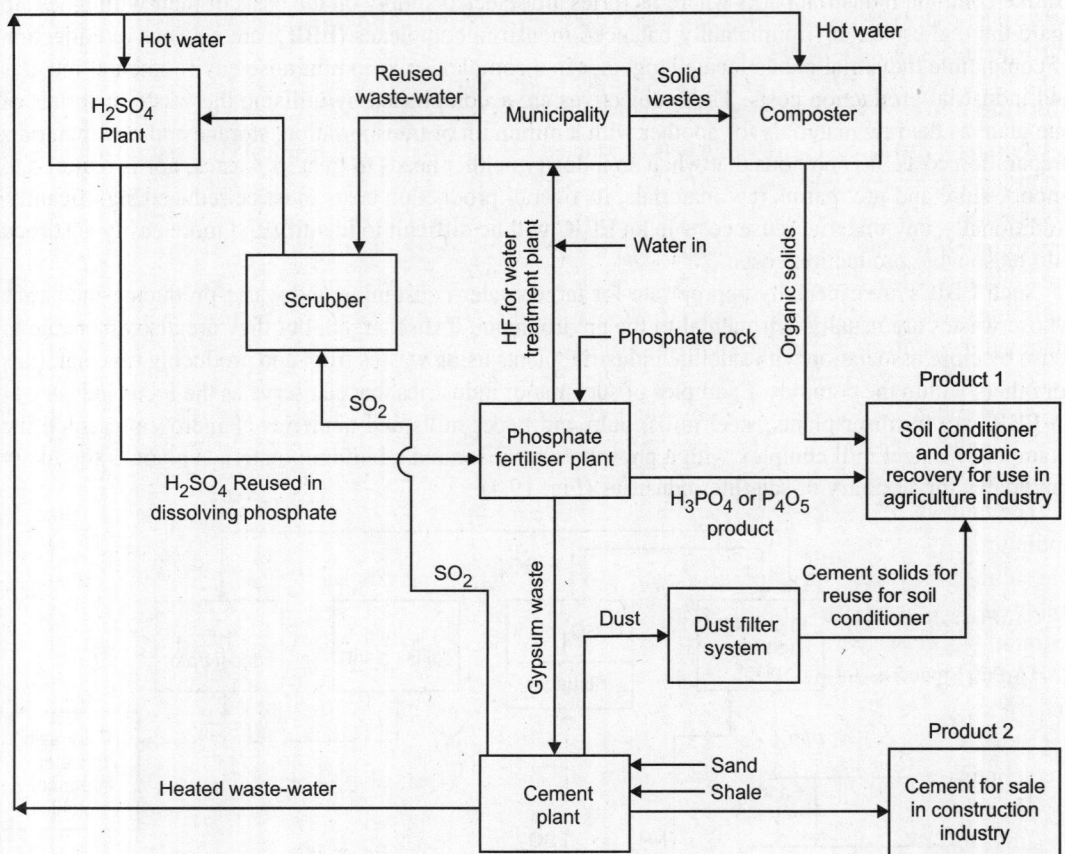

Fig. 19.5. Example of an environmentally balanced industrial complex with a phosphate fertiliser plants as the focus industry.

As previously mentioned, in the usual starting process of producing phosphoric acid and ammonium phosphate fertiliser by dissolving the phosphate rocks with sulphuric acid, a gypsumlike sludge is generated as a by-product and some sulphur dioxide and fluorine are in the waste gases emitted at the high reaction temperatures. The large, relatively impure, quantities of phosphogypsum (5 volume to 1 P_2O_5 fertiliser produced) are difficult to treat and the fluorine present in the gas as hydrofluoric acid (concentrations from 1–10 per cent, which is very low for commercial use) requires further costly and extensive treatment.

Using such a fertiliser production facility as the focus industry of an EBIC would be a feasible solution to the environmental problems if combined, i.e. with such satellite industries as: (i) a sulphuric acid plant to feed its products to the phosphoric acid plant and to use some of the hot water effluent from the cement plant and the effluent from the SO_2 scrubber of the phosphate fertiliser plant; (ii) a municipal solid wastes composting plant utilising hot water from the cement plant, serving as disposal facility for the garbage of a city and producing composted organic solids to be used as fertiliser; and (iii) a cement and plasterboard production plant utilising the phosphogypsum waste sludge in the manufacture of products for the construction industry, producing hot water effluent to be used as mentioned above and waste dust collected by a dust filter and used as a filler for the soil fertiliser produced by mixing composted garbage and phosphate fertiliser.

Fluoride and Phosphorus Removal from a Fertiliser Complex Waste-water

A laboratory-scale treatability study was conducted for the Mississippi chemical corporation to develop a physico-chemical waste-water treatment process for a fertiliser complex waste-water to control nitrogen, phosphorus and fluoride and to recover ammonia. The removal technique investigated consisted of precipitation of fluorides, phosphorus and silica by lime addition, a second stage required for the precipitation of ammonia by the use of phosphoric acid and magnesium and a third stage for further polishing of the waste-water necessary to remove residual phosphate. The waste-water quality parameters included the following concentrations (mg/l):

> Fluoride 2000, ammonia 600, phosphorus as P_2O_5 (P) 145 and an acidic pH level of 3.5. Ammonia removals of 96 per cent were achieved and the insoluble struvite complex produced by the ammonia removal stage is a potentially commercial-grade fertiliser product, whereas the fluoride and phosphorus in the effluent fell below 25 and 2 mg/l, respectively.

The multistage treatment approach was to first remove the fluorides by lime precipitation, with the optimum removal (over 99 per cent) occurring with a two-step pH adjustment to about 10.4 (removal was more a function of lime dosage, rather than a pH solubility controlled phenomenon). Each step was followed by clarification and the required lime equivalent dosage was 180 per cent of the calcium required stoichiometrically.

The effluent from the first stage had an average fluoride content of 135 mg/l (93 per cent removal) and a phosphorus content of less than 5 mg/l, whereas the hydraulic design parameters were a 15 minutes per stage reaction time and a 990 gpd/ft^2 (40.3 m^3/m^2/day) clarifier overflow rate. The resulting precipitated solids underflow concentration was 7.7 per cent by weight.

The second-stage (ammonia removal) effluent contained unacceptable levels of F and P and had to be subjected to third-stage lime treatment. It raises the pH from 8.5 to 11.4 and produces an effluent with concentrations of F and P equal to 25 and 2 mg/l, respectively. The hydraulic design parameters were a 15 minutes reaction time and a 265 gpd/ft^2 (10.8 m^3/m^2/d) clarifier overflow rate. The resulting precipitated solids underflow concentration was 0.6 per cent by weight. In all three stages, an anionic polymer was used to aid coagulation, solids settling and effluent clarification.

The solids resulting from the first- and third-stage treatment consisted of calcium fluoride, calcium phosphate and fluorapatite-type compounds. Typically, in the fertiliser industry, such sludges are disposed of by lagooning and subsequent landfilling. Other studies have investigated the recovery of the fluoride compounds, such as hydrofluoric and fluosilisic acids, for use in the glass industry and in the fluoridation of drinking water supplies.

AIR EMISSIONS

The major air emissions in fertiliser industry are given in Table 19.8.

Table 19.8. Air emissions in fertiliser industry.

Pollutants	Source
SO_2	Sulphuric acid plants, boilers using coal, fuel oil, LSHS, etc.
NO_x	Nitric acid plants, NP/NPK plants
	Particulate matter rock grinding in the phosphoric and NP/NPK plants. Prilling/granulation in urea, NP/NPK/SSP/TSP plants
Fluorine	Phosphoric acid, single super phosphate, triple super phosphate, NP/NPK plants

Control of Air Emissions

Over the years, the technology adopted has been upgraded on the basis of improved systems available in advanced countries. With the energy crisis, greater attention is being paid to this aspect along with stricter emission controls adopted due to increasing environmental consciousness. All the plants, therefore, have incorporated the most modern pollution control systems, minimising emissions to practicable limits with the best of technologies available to meet the standards. However, there are many old plants which have adopted older technologies where the emission levels are higher and so also is energy consumption.

Control of sulphur dioxide and acid mist emissions from sulphuric acid plants

Sulphuric acid is produced by the oxidation of sulphur to sulphur dioxide (SO_2) and conversion of this to sulphur trioxide (SO_3). This is absorbed to produce sulphuric acid. All plants utilise a vanadium catalyst to convert SO_2 to SO_3 according to the contact process.

In the past 20 years all new sulphuric acid plants that have come up are of DCDA type which gives a conversion efficiency of 99.5–99.7 per cent. Table 19.9 shows the amount of SO_2 emitted by sulphuric acid plants operating at several gas strengths and conversion levels.

Table 19.9. Sulphur dioxide emission from sulphuric acid plant.

	Exit SO_2, ppm		
% Conversion	6% SO_2 in feed	8% SO_2	10% SO_2
96	2626	3616	4672
97	1972	2716	3510
98	1316	1813	2344
99	658	907	1174
99.5	329	454	587
99.7	197	272	352

In a typical sulphuric acid plant, the conversion of SO_2 to SO_3 is achieved in a four bed vanadium pentoxide catalyst bed. After fourth bed, when SO_2 content has been reduced from 10 to 0.06 per cent corresponding to 600 ppm, the sulphur trioxide is absorbed in 98.5–99 per cent sulphuric acid in the final absorption tower (FAT). Before being vented to the atmosphere, the gases pass through a brink mist eliminator located after the intitial absorption tower (IAT) and FAT where acid mist is removed. The catalyst bed temperature, catalyst activity, pressure drop and concentration of the circulating acid are vital parameters affecting the performance.

In spite of adopting double conversion double absorption (DCDA) system, during initial start up or start up after a shut down of more than 4 hours when catalyst beds are cold or relatively cold the required temperature for conversion of SO_2 to SO_3 is not present resulting in higher emissions of SO_2 during start up. Sulphuric acid plants are therefore equipped with either a start up heater to increase catalyst bed temperature to the required level or a start up scrubber (which is a packed tower using caustic soda to absorb the SO_2). In the case of start up scrubbers Na_2SO_3/$NaHSO_3$ is formed depending on the pH of the scrubbing solution.

Control of sulphuric acid mist

Acid mist comprises of liquid droplets which range in size from 10 microns down to 0.07 microns. It is a well known problem in sulphuric acid production. This not only produces a visible and persistent plume in the stack gases but also causes equipment corrosion. Of course, this is a pollution hazard. Mist is formed when water vapour in the air combines with SO_3 in the convertor and in the absorption circuit. By and large, this is controlled by drying the air thoroughly before feeding it to the convertor and by proper maintenance of temperatures and acid concentration in the absorption towers.

Efficient drying of air is achieved by circulating 93–99 per cent acid so that water vapour in the outlet gas is around 5–100 mg/nm^3. At higher concentration of acid higher operating temperatures can be tolerated while the reverse is true for lower acid concentrations. The type of mist eliminator used in the drying tower are either the impaction type fibre bed mist eliminator or the mesh pad type. Wire mesh pad require periodic replacement in drying towers. Although plastic mesh pad offer better resistance to corrosion there have been cases where back flow of hot gases during a shut down past a leaking shut off damper have melted the plastic mesh pad.

The absorption of SO_2 in acid gives rise to acid mist. The acid concentration should be maintained between 98–99 per cent to keep the H_2SO_4 + H_2O + SO_3 vapour pressure at a minimum. At lower concentrations, the partial pressure of water vapour is high enough to give rise to production of acid mist, by combination of this water vapour with the sulphur trioxide in the gas steam. Typical mist loadings in the IAT are 50–1500 mg/nm^3 with an average particle size diameter of 1–2 μm. Brownian diffusion filter elements are hollow cylinders which look like candles containing hydrophobic glass mattress packed between two stainless steel cages. Most mist particles 1 μm in size and larger are removed upto 96 per cent of those smaller than 1 μm are collected. The comparative efficiencies of filters are given in Table 19.10.

Table 19.10. Comparative efficiencies of various type of filter.

Mist size (μ)	Single mesh pad	Double mesh pad	Single fibre glass pad	Candle filter
5	100%	100%	100%	100%
5 to 2	90–100%	100%	100%	100%
3 to 2	50%	95%	90%	99%
2 to 1	–	90%	–	98–99%
1	–	–	–	96%

Advances are being made in mist collection equipments. High efficiency fibre beds are being developed with different efficiencies for various submicron particle sizes and various companies offer mist eliminators for sulphuric acid plant service (Table 19.11)

Table 19.11. Efficiency of fibre beds.

Parameter	High efficiency	High velocity
Efficiency on particle > 3 μm	95–99%	90–98%
Efficiency on particles < 3 μm	100%	100%
Pressure drop (inches water)	5–15	6–8

Control of particulate matter

Urea prilling tower dust

Urea is produced by reacting NH_3 and CO_2 at elevated temperatures and pressures. Traditionally urea has been transformed into solid form by prilling techniques. The urea solution leaving the synthesis section has a concentration of 72–76 per cent and is treated in a vacuum concentration section to obtain a urea melt to feed to the prilling systems.

It is during this prilling that particulate matter is emitted. The conversion of urea to solid form is achieved in a prilling tower by cooling the solution with air to expel its water content. The quantity of air required has been estimated to be 10–1200 nm^3/T of urea produced. In a typical 1000 tpd plant the total quantity of air is 5,00,000 nm^3/hr and it contains around 200–500 mg/nm^3 of urea dust which has to be controlled, otherwise it would cause air pollution and result in loss of the product. Most of the plants use wet scrubbers for particulate removal. Dust laden air rising through the prilling tower enters the annular duct. Air is drawn from this annular duct by a ring of liquid jets consisting of nozzles arranged in annular space which provides the energy required to overcome the pressure drop in the system. The liquid droplets act as a spherical collectors for the urea dust whose size ranges from 2–200 microns. The water is sprayed at a pressure of 4.5–6 kg/cm^2. It is finally discharged into the atmosphere after passing through demister packings. Fresh make up water is sprayed into the demisters for washing purposes while the jet nozzles are fed with the urea solution which collects in the annular basin. After the concentration of urea in the recirculating solution reaches around 10–15 per cent, it is drained and sent to a tank for further processing (Table 19.12).

Table 19.12. Absorption efficiency of wet scrubbers.

Urea concentration in recirculating solution %	Absorption efficiency %
0.0	99
5.0	95
20.0	92

The outlet dust concentration is 30–50 mg/nm^3.

Control of particulate matter in complex fertiliser plants

Dust is emitted during rock grinding, product cooling and granulation operations. Normally, fabric filters are used for rock grinding and wet scrubbers like venturi scrubbers for granulation (i.e. drying of slurry in an equipment called the 'Spherodiser'). The hot gases are sucked out by an exhauster and enter the Venturi scrubber whose essential feature is the presence of a constricted cross-section or 'throat', through which the gas is forced to flow at high velocity. Water is introduced ahead of the 'throat' and the atomised liquid drops act as collectors. The concentration of dust at the outlet is 60 mg/nm^3.

Dust is emitted during the grinding of rock phosphate. Bag houses are commonly used for dust control. The dust concentration to the inlet of the bag house is normally in the range of 20–30 gms/nm^3 which is brought down to less than 150 mg/nm^3. The fabric usually used is polypropylene or polyester with a special coating. Pulse jet cleaning is adopted when the pressure drop is around 6″ (150 mm) of water.

Control of oxides of nitrogen

Nitric acid plants

Nitric acid is commercially produced by reacting ammonia with air to produce nitrogen oxides which are then absorbed in water to yield the acid. Despite the many variations in operating details among the plants producing nitric acid from ammonia, three basic steps are common to all:
1. Reaction of ammonia with air over a catalyst (platinum, radium) at a high temperature and moderate pressure to produce nitric oxide.
2. Oxidation of the nitric oxide by oxygen remaining in the gas stream to produce nitrogen dioxide.
3. Absorption of the nitrogen dioxide in water to produce nitric acid releasing additional nitric oxide which must be reoxidised.

After absorption, the tail gas containing, namely, nitrogen, water vapour and oxides of nitrogen (commonly known as NO) is normally preheated and expanded through a turbine to recover energy, which is used to power the nitric acid plant's compressor.

The NO_x content in the tail gas varies from 1000–3000 ppm of which approximately 60 per cent is NO_2 and the rest is NO. The NO_2 component imparts a yellowish colour to the tail gas.

In general, there are three types of processes which can be used to reduce or control emissions:
1. Catalytic reduction (selective and nonselective): In selective catalytic reduction an improved process is adopted in some of the plants where selecting reduction of NO_x (i.e. NO and NO_2) in the gas system is done with slight excess of stoichiometric quantities of ammonia over a mixed catalyst. The catalyst is selective and the ammonia consumption is almost stoichiometric and virtually no ammonia reacts with oxygen present in the gas which results in much lower exit gas temperatures allowing use of simpler and cheaper equipment. Two catalysts are used in the reaction, the first is based on precious metals mixed with metallic oxides, noble or iron group or V_2O_5 on alumina which promotes the NO_x reduction and the second is platinum based which destroys or converts any excess ammonia fed to the system into harmless nitrogen. The exit temperature is raised by 20°–30°C due to exothermic reaction. The only care to be taken is to see that ammonia gas to the reactor is not fed when the tail gas temperature is less than 200°C, in order to avoid the formation of ammonium nitrite and nitrate. The NO_x could be brought down to less than 300 ppm to NO_x. This technique has become widely used in recent years.

In nonselective catalytic reduction the oxides of nitrogen are formed during the combustion of ammonia are absorbed in water to form nitric acid. The unabsorbed portion of oxides of nitrogen and inerts are emitted to atmosphere which generally contain 0.2–0.4 per cent of NO_x and 2–3 per cent free oxygen. The tail gas leaving the absorber is mixed with a fuel gas like natural gas (or refinery gas or purge gas from the ammonia plant) and the mixture is passed over a catalyst (platinum vanadium, iron oxide or titanium based) bed. This converts the NO_2 to NO rendering the exit gas colourless. This step is called the decolourisation reaction. This alone may not be sufficient and complete destruction of NO by converting it to N_2 may be required. To achieve this the oxygen present in the tail gas (2–3 per cent) must first react with the fuel which must be present in slight excess of the stoichiometric amount. The reaction with oxygen results in a large amount of heat of combustion which results in temperatures of nearly of 650°–700°C.

2. Extended absorption: The extended absorption processes aims to continuing the process of absorption of the NO_x in water beyond the level at which it normally ends, i.e. 2000–3000 ppm NO_x. The extended absorption is made possible by provision of large absorption columns.

   The additional absorber produces nitric acid. Owing to the large size of the absorbers which have to be necessarily of stainless steel construction, the investment costs are the highest as compared to the other NO_x control processes. However, there is no additional operating cost involved.

3. Chemical absorption: The tail gas could be cleaned by scrubbing with a liquid containing caustic soda or urea. In case of the caustic, it results in the formation of $NaNO_2$ and $NaNO_3$. A means must be found to recover and reuse the $NaNO/NaNO_3$ from the solution otherwise it could be a source of pollution. Urea also reacts with the nitrogen oxides resulting in the liberation of N_2/CO_2 and formation of ammonium nitrate in solution. The scrubbing yields a steamy but colourless plume. Here again, the by-product ammonium nitrate solution must be utilised otherwise, it would result in the total use of urea as well as a source of water pollution. Liquid scrubbing systems are best adopted in combination with extended absorption. After cleaning the tail gas substantially in additional absorber, liquid scrubbing could be adopted to reduce the pollution levels. This would result in considerable reduction in the absorber volume while, at the same time, keeping chemical consumption reasonably low.

Control of NO_x in the nitrophosphate plants

Rockphosphate (fluoropatite $3Ca_3 (PO_4) CaF_2$), which is naturally available contains phosphorus in the form of tricalcium phosphate which is not soluble and hence not available to the plants. To convert this phosphorus to the soluble forms, i.e. dicalcium phosphate and monocalcium phosphate the CaO/P_2O_5 ratio has to be reduced.

This is done either by removing part of the calcium nitrate formed (after addition of nitric acid to rock phosphate) by chilling and crystallisation or by the additional external P_2O_5 or by both. By this fertiliser of required grade containing nitrogen and phosphorous (NP) is produced.

The emission of NO_x takes place during the treatment of rockphosphate with nitric acid. In theory no NO_x should be formed as the nitric acid is being used purely as an acid to dissolve phosphate rock. In practice, however, the rock phosphate contains some oxidisable impurities such as sulphides and organic material and it is these which reacts with nitric acid to produce NO_x.

The NO_x emission can be minimised but not eliminated by maintaining proper temperature in the acidulation reactor. To combat the emission of NO_x, urea is added at the digestion stage to yield nitrogen, ammonium nitrate and water.

The operation and maintenance of this system is very simple provided urea addition is continuously done. This is achieved through an urea solution tank equipped with a pump to add urea at the rock digestion stage. Good results have been obtained with apparently no adverse effects on product quality.

Absorption

The absorption of NO_x on a fixed bed of solids has not been extensively commercialised as other processes. Normal (dry) absorption on activated carbon or molecular sieves is very efficient but the absorbents must be regenerated either by thermal desorption or by pressure reduction.

Control of fluorine emissions

Fluoropatite ores are used for the production of fertilisers such as single superphosphates (SSP), triple superphosphate or converted to wet process phosphoric acid and there on to other fertilisers. Depending upon the source of origin the fluorine content in the rock varies between 1 and 5 per cent which is released as silicon tetrafluoride (SiF_4) and hydrogen fluoride (HF) in the presence of active silica contained in the rock when rock is acidulated and ammoniated for fertiliser manufacture. To circumvent this problem it has become common to fix volatile fluorine compounds as H_2SiF_6 (hydrofluorisilicic acid) by scrubbing with water which then forms a base for production of a variety of fluorine related compounds.

Fluorine is released during the attack on rock by sulphuric or phosphoric acids. In the manufacture of SSP 12–25 per cent of total fluorine in the rock is evolved as gaseous vapours in the form of hydrogen fluoride and silicon tetrafluoride. The hydrogen fluoride liberated during acidulation reacts with silica to produce SiF_4 which reacts with the water vapour generated during acidulation and with water in the scrubber towers to form hydrofluoride acid and finely divided silica. In the TSP manufacture 5–15 per cent of the total fluorine contained in the acid and rock is evolved during the acid attack and 20–40 per cent of the total fluorine during drying. The reaction is nearly the same with a part of fluorine getting converted to H_2SiF_6 and the rest remaining as HF and SiF_4 which is removed by further scrubbing. Phosphoric acid manufactured by action of sulphuric acids on rock at 70°–80°C evolves fluorine during the acidulation and concentration steps. Total volatalisation of flourine (F) during acidulation and concentration of phosphoric acid is approximately 40 per cent. The fluorine emissions are controlled by scrubbing these gases in either Venturi scrubbers, packed towers, crossflow scrubbers or a spray chamber. Water is used as scrubbing liquor which converts the HF and SiF_4 to H_2SiF_6. Efficient scrubbing system can operate upto 25 per cent H_2SiF_6 but above this point the SiF_4 vapour pressure rises steeply. Many towers operate at 15–20 per cent H_2SiF_6 or less.

One of the commonly encountered problems is deposition of silica when concentration of phosphoric acid is less than 50 per cent. This leads to choking of scrubber nozzles. Hence nozzles have to be periodically cleaned and specially designed for this service.

Chapter 20
Paints and Dyes

INTRODUCTION

In order to identify the sources of generation of pollutants, understanding of the manufacturing process of various paints is an imperative prerequisite. An attempt is, therefore, made to study the manufacturing process and to identify the sources of pollution. Grouping of paint industries can be done on the basis of products manufactured, which in turn can be based on the raw materials used and manufacturing process involved. Paint is a general term used for oil paints, enamels, emulsions, lacquers, etc. Broadly, it can be said that the type of binder or vehicle used defines the type of paint. When oil is the base it is oil paint. With synthetic resins as the base, the product is synthetic enamel. Lacquers are nitro-cellulose base paints. Acrylic or PVC base paints are acrylic emulsions and so on. Depending on the quality and consistency of the products slight changes are made in the manufacturing process such as, the temperature maintained during manufacture, the additives used, reaction period, etc. All oil-based paints can be grouped together. The raw materials used for these are oils as vehicles or binders, besides pigments and solvents. The general process steps followed are mixing, grinding and thinning.

MANUFACTURING PROCESS

Paint Manufacture

In general, the manufacture of paint requires the following steps:
> Dry pigments are weighed, mixed and fed through hoppers or chutes into mills where they are dispersed in an appropriate resin vehicle. The milled pigment is transferred to a mixer where thinners, drying agents, etc. may be added to adjust consistency, viscosity, colour and drying time. When mixing is complete, the paint is filtered through filter cloth and packed into containers (Fig. 20.1).

Varnish Manufacture

Varnish is an unpigmented resinous surface coating. In traditional varnish is manufactured in an open kettle which is set over a fire in which copal or other natural gum is heated and dissolved in hot oil. Other ingredients such as driers are added and after cooling the varnish is thinned out to a workable consistency with solvent. It is then clarified by gravity settling, straining or centrifugation. Modern manufacture takes place in jacketted enclosed kettles. The decomposition products of resin and oil released as fumes are removed through an exhaust system where the fumes are scrubbed by jets of water spray. The fume oil is separated from the water by decantation and the water is recirculated.

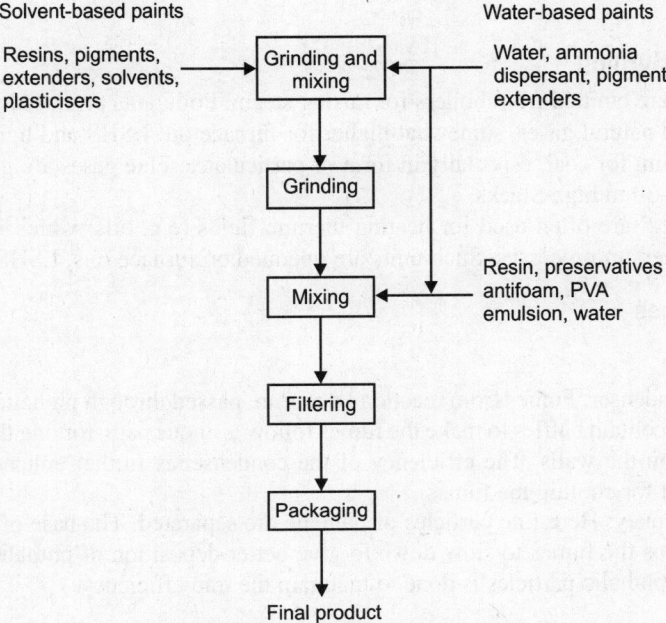

Fig. 20.1. Flow diagram for manufacturing paint.

Lacquer Manufacture

Lacquers are coatings that dry only by evaporation. The pigments are first dispersed in ball mills with plasticisers and natural or synthetic resins are added. Solvents are added to achieve required consistency. This may be done either by cold churning or by gentle warming and mixing.

Distemper Manufacture

Powdered raw materials together with additives go to pug mill in which soft water is used. Other raw materials are emulsified with stand oil prior to entering the pug mixer. Further grinding may be done before packing.

Manufacture of Resins and Emulsions

Other items of interest in paint manufacture are resins and emulsions for the manufacture of which various chemicals/substances are reacted at controlled temperatures in special reactors.

Bituminous Paint

Bituminous paint is manufactured in synthetic resin plants. The process followed is almost identical to that of resin manufacturing process. The bitumen (generally asphalt grade of tar is used) is melted in a closed, jacketted kettle. Heating is done indirectly by circulating thermic fluid around the kettle, at a temperature of 150°–180°C. Bitumen is melted in the kettle in 2 hours' time. During the heating process fumes are given out which are passed through a water scrubber. This water is let out into the drains from where it is finally collected and treated in the effluent treatment plant.

Melted bitumen is then thinned down to the required consistency by mineral turpentine, filtered and cooled.

AIR POLLUTION

Emission—Fuel Burning

1. Flue gases are emitted from boilers for raising steam. Pollutants produced are minimal in case of LPG and natural gases, somewhat higher for furnace oil, LSHS and light diesel oil (LDO), and maximum for coal, especially in form of particulates. Flue gases are generally discharged through 20–30 m high Stacks.
2. 'Thermopaes' are often used for heating thermic fluids (e.g. oils) which in turn are used for heating in certain processes. Such units are operated on furnace oils, LSHS or LDO.

Emission—Process

Resin production

1. Phthalic condenser: Fumes from reaction kettles are passed through phthalic condensers. These condensers contain baffles to make the fumes follow a sinous path, forcing the phthalic particles to deposit on the walls. The efficiency of the condenser is further enhanced by providing a water jacket for cooling the fumes.
2. Special chimney: Here fine particles of phthalic are separated. The base of the chimney being conical, helps the fumes to slow down to give better deposition of phthalic particles. Weekly removal of phthalic particles is done to maintain the trap efficiency.

Varnish production

Decomposition products of resin and oils are released as fumes. These fumes are scrubbed by jets of water spray.

Bituminous paint production

During bitumen melting, fumes are given out and scrubbed by water.

Air extraction

Air extractors and exhaust fans are provided with suitably designed hoods at key positions to help improve ventilation in working areas. The air is dispersed through stacks.

The flue gas emission from fuel sources should meet the standards prescribed in the emission regulation of the central pollution control board and Bureau of Indian standards.

WASTE-WATER GENERATION

Waste-water Quality

On close examination of manufacturing process of various paints, it is observed that water is not a constituent in any of the production processes of resins, varnishes, lacquers, etc. However, water is required for the manufacture of stiff or water-based paints and cooling of the ball mills or sand mills in manufacture of oil paints. Water is used for washing of floors which are littered by spillage of powders, solvents, etc. when a particular batch is finished. In cases where there is vast difference in the colour shade of next batch, the mills and containers etc. have to be cleaned. This cleaning is done by xylene which is recovered and reused. Once in a while the vessels are cleaned with caustic soda. This cleaning constitutes the major part of the waste-water generated which is highly alkaline in nature. Other source is the cooling tower blow down. The mills used for grinding powders and the solvents used in mixing have to be maintained at room temperature. They are cooled by circulating cooling water. Cooling tower blow down generally enters the effluent stream. In summary, it can be stated that waste-waters from the paint manufacturing industries generally tend to be alkaline, contain some oil and grease and biochemical oxygen demand (BOD), chemical oxygen demand (COD) and suspended solids (SS). The waste-waters can be assumed to contain small amounts of the products.

The BOD and COD values give only a gross measure of organics in the wastes. Some of the organic and inorganic compounds used in the manufacturing operations are classified as toxic and hazardous. Chromium, copper, lead, zinc, ethyl benzene, di-(2-ethylhexyl) phthalate, tetrachloroethylene and toluene were found in high concentrations.

In all the industries surveyed, raw materials used have very less chances of entering the effluent stream. This is because, the majority of them being toxic are carefully used. These toxic powders, containing chromium, copper, lead, zinc, titanium, etc. cannot be traced as such in the effluent because during the manufacturing process they are mixed and blended with a variety of other substances and solvents. The only toxic substance traced was phenols. Phthalic anhydride is a common raw material widely used in paint industry. This, after reacting with other substances, sets phenols free as a reaction product. These are at a high temperature and when condensed are traced in the condensate water. They are in high concentrations in the condensate water but get diluted further when mixed with other effluent streams.

Waste-water Quantity

Widely varying results of waste-water quantity were obtained from the paint units studied. The consumption of water ranges from 26 cu.m^3/day to 150 cu.m^3/day depending upon the type of products manufactured at the time of study and the extent of water consumed in the cooling towers and floor washings. Thus no direct co-relation between products manufactured and water consumed could be established. The industrial waste-water generation was assessed to be between 10 and 120 cu.m/day with one exception of 3200 cu.m/day. The domestic waste-water discharge was both industrial and domestic sources ranged between 25–224 cu.m/day.

Waste-water Characterisation

The process and productwise characteristics of the major waste-water streams are as follows:
1. Caustic cleaning operations.
2. Resin house.
3. Stiff paint.
4. Laboratory/R & D.
5. Liquid paint.

The data collected during investigation are summarised in Table 20.1.

Table 20.1. Process and productwise effluent characteristics.

Parameters mg/l except pH	Caustic cleaning operations	Resin house	Stiff paints	Laboratory/ R & D	Liquid paint
pH	8.5–13.5	3.2–6.3	6.5–13.5	5.1–6.7	13.5
SS	200–600	240–400	400–700	20–200	40
COD	1100–3800	240–78,000	1215–6000	150–200	560
BOD	475–2400	225–60,000	380–980	18–45	60
Oil and grease	32–150	14–25	252	–	6
Phenolic content	12.5	6–86	6.4–100	0.84–5.5	5.5

Waste-water characteristics from main sections

Waste-waters from caustic cleaning

It was observed during survey that caustic cleaning resulted in maximum production of waste-water in the paint industry. Vessels used in manufacturing processes have to be cleaned before change of product. This is done using water and alkali. Thus, the waste-water pH is in range of 8.5 to 13.5. This waste-water consists of suspended solids in the range of 200 to 600 mg/l. BOD varies from 475 to 2400 mg/l and the COD is between 1100–3800 mg/l. The oil and grease is within a range of 32–150 mg/l. The phenolic concentration of this waste-water is 12.5 mg/l.

Waste-water from resin house

The waste-water from resin house is that which accumulates from the condenser into the separator. The water layer is disposed off in the waste-water stream and the upper layer of solvents is reused. The water from separator is generally acidic with pH range from 3.2 to 6.3. The suspended solids range from 240 to 400 mg/l. The BOD and COD of waste-water vary from 225 to 60,000 mg/l and 240 to as high as 78,000 mg/l respectively. The phenolics are between 6 and 86 mg/l. The oil and grease is within a range of 14–25 mg/l.

Waste-water from stiff paints

Stiff paints are water base paints and the main product is distemper. Cleaning of hoppers, grinders and containers consume large amount of water and contributes 400 to 700 mg/l of suspended solids, a large fraction of which is settleable. Waste-water contributed by stiff paint section is very less in quantity because of less production in this section and is not generated continuously. As observed during the survey, characteristics of the waste-water stream are COD 1215 to 6000 mg/l, BOD 380 to 980 mg/l and phenolic content varies from 6.4 to 100 mg/l. The oil and grease content of the waste-water is 252 mg/l.

Combined waste-water

The characteristics of combined waste-water is presented in Table 20.2.

Table 20.2. Combined waste-water characteristics.

Parameters mg/l, except pH	Range
pH	6.5–10.5
SS	220–1200
COD	300–5700
BOD	1700–3100
Oil and Grease	22–138
Phenolics	18–55

Waste-water Treatment

There is considerable variation in the nature of waste-water normally generated in paint manufacturing industry. Before evolving waste-water treatment scheme, the following aspects should be looked into:
1. Segregation of waste-waters based on characteristics and strength.
2. Reduction of volume and strength of waste-waters by adopting in-plant control measures.
3. Study of treatability of various waste-water streams to decide the best combination of treatment system.

Segregation

Segregation of wastes is, however, recommended to reduce capital costs, improve the treatment efficiency and reduce chemical consumption. Effluents generated from paint industry can be segregated into the following streams:
1. Caustic cleaning effluent.
2. Stiff paint effluent.
3. Effluent from remaining units.
4. Domestic waste.

Caustic cleaning effluent is highly alkaline and needs neutralisation. Effluents from other units comprise thermopac burner cleanings, resin house waste, etc. which contain oil and has to be separated before mixing with the other effluents. Stiff paint effluent contains easily settleable solids which are settled in the primary clarifier and then the waste-water is given further treatment. For a typical factory producing 100 cu.m^3/day effluent, the segregated quantities may be of the following order:

Caustic cleaning effluent	83 cu.m^3/day
Effluent from stiff paint section	6 cu.m^3/day
Effluent from other units	11 cu.m^3/day
Total	100 cu.m^3/day

To the above quantity of waste-water, domestic waste-water at the rate of 100 litres per worker per day may be added.

Reduction of Waste-water

Schemes for reducing the generation of waste-water at source should be practised. This is to reduce the effluent load rather than finding methods to treat it. Unnecessary use of water not only adds to the quantity of effluent and the cost of treating it, but also increases the wastage of heat, power and/or product in the effluent. Steps that can be taken to reduce the generation of waste-water are discussed below:

Good housekeeping

Good housekeeping reduces generation of both waste-waters and solid wastes. The cooling water is usually uncontaminated and thus, should be collected and reused. It could be used for floor washing or discharged separately into the receiving water bodies rather than mixing with polluted water and discharging into the treatment plant. Accidental spills and leakages should be reduced to a minimum through proper maintenance of equipment and training of personnel. In case of caustic cleaning, instead of washing away the caustic solution, it can be collected, stored and used for further cleaning. In case of stiff paints, water from first cleaning should be collected and used later as process water for a similar type of batch. Waste-water volume can also be reduced through reuse of rinse water for preparation of alkali solution. The above procedures can reduce the quantity of generation of caustic cleaning water significantly.

Recovery of wastes

A large number of solvents are used in paint manufacturing and a majority of them are recovered and therefore not lost in the waste-water streams. In case of oil paints, solvents are added in grinders which are closed units, therefore, loss of solvents through evaporation is considerably reduced. High temperature is maintained in resin and varnish manufacture, resulting in evaporation of solvents added. These solvent vapours, along with the water vapours generated through chemical reactions are condensed and collected in a separator. The solvent layer is removed and reused in the next batch. Figure 20.2 shows the paint production process along with sources of liquid and solid waste generated.

442 Pollution Control in Chemical and Allied Industries

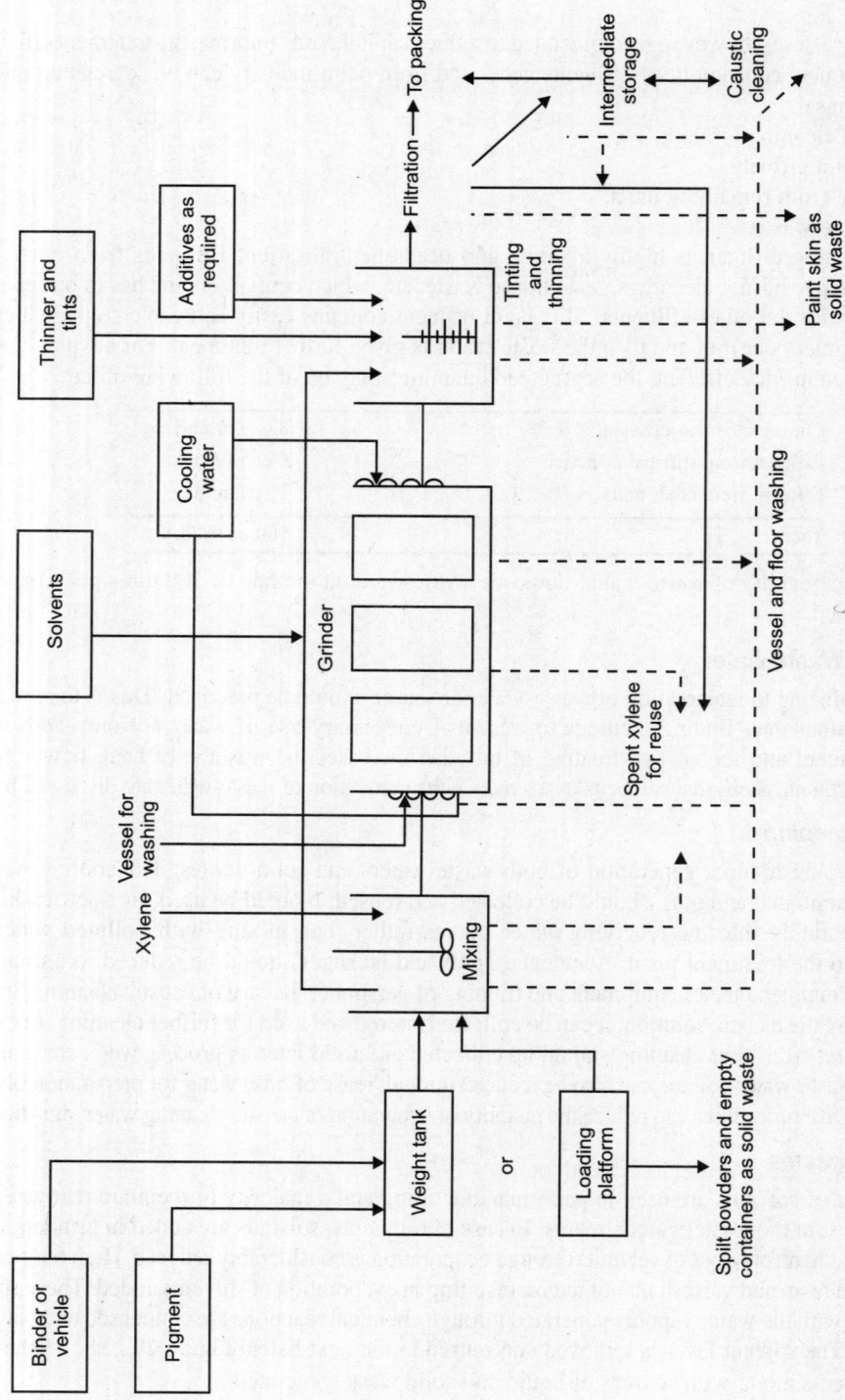

Fig. 20.2. Flow diagram for manufacturing paints with general sources of solid and liquid waste generation and possible solvent recovery steps.

Status of Waste-water Treatment in India

Although physico-chemical processes are commonly used for the waste-water treatment in paint industry, it is noticed during survey that biological treatment also is employed. It has been investigated that if waste-water contains a high amount of settleable solids and the waste-water quantity is low, primary treatment followed by treatment in oxidation pond may result in reasonably high BOD removal.

But for a higher quantity of waste-water primary treatment should be followed by secondary treatment. On the other hand, removal of COD always demands tertiary treatment.

Treatability Aspects of Waste-water

Combined effluent from paint industries can be satisfactorily treated using the usually physico-chemical and/or biological treatment methods. The treatment consists of coagulant addition and adjustment of pH to an optimum level for maximum precipitation, the precipitated material is removed by gravity separation, either on batch basis or in a continuous flow tank. Ultra filtration and activated carbon adsorption have also been tried but have not been found to be cost effective. Using physico-chemical processes, removals of 90 per cent or greater can be achieved for some significant toxic pollutants. Even after this treatment if some toxic pollutants remain, biological treatment can reduce their concentration. Some of the organics which remain in the supernatant after physico-chemical treatment are biodegradable and can be removed through biological treatment.

Waste-water Treatment

Considering the practice of waste-water treatment followed in the country and the performance efficiency of each operation, the best alternative is evolved. The treatment system consists of physico-chemical treatment units followed by biological treatment units (Fig. 20.3).

Primary treatment

1. Oil and grease removal: Effluents from all units except stiff paint section and caustic cleaning waste are passed through an oil and grease removal device.
2. Equalisation-cum-neutralisation: Effluent from caustic cleaning operation is highly alkaline in nature and requires neutralisation prior to further treatment. An equalisation-cum-neutralisation tank is provided with an agitator. Effluent from stiff paint is mixed with the neutralised waste-water, dosed with a coagulant and sent to flash mixer. The effluent is then subjected to clariflocculation.
3. Clariflocculation: The effluent is clarified in clariflocculator and subjected to biological treatment. Sludge generated in this unit is carried to the sludge drying beds for dewatering.

Secondary treatment

1. Extended aeration: Domestic waste-water from the factory premises is mixed with the supernatant from clariflocculator and is biologically treated by extended aeration process.
2. Secondary clarification: Mixed liquor from the aeration tank overflows to the secondary clarifier. The settled sludge is recycled continuously through return sludge pumps to the aeration tank and excess sludge is discharged to sludge drying beds. Effluent from the secondary clarifier is fit for discharge to the environment.
3. Sludge drying: Sludge from oil and grease trap, clariflocculator and secondary clarifier is dewatered on sludge drying beds. Filtrate from these beds is returned to equalisation-cum-neutralisation tank.

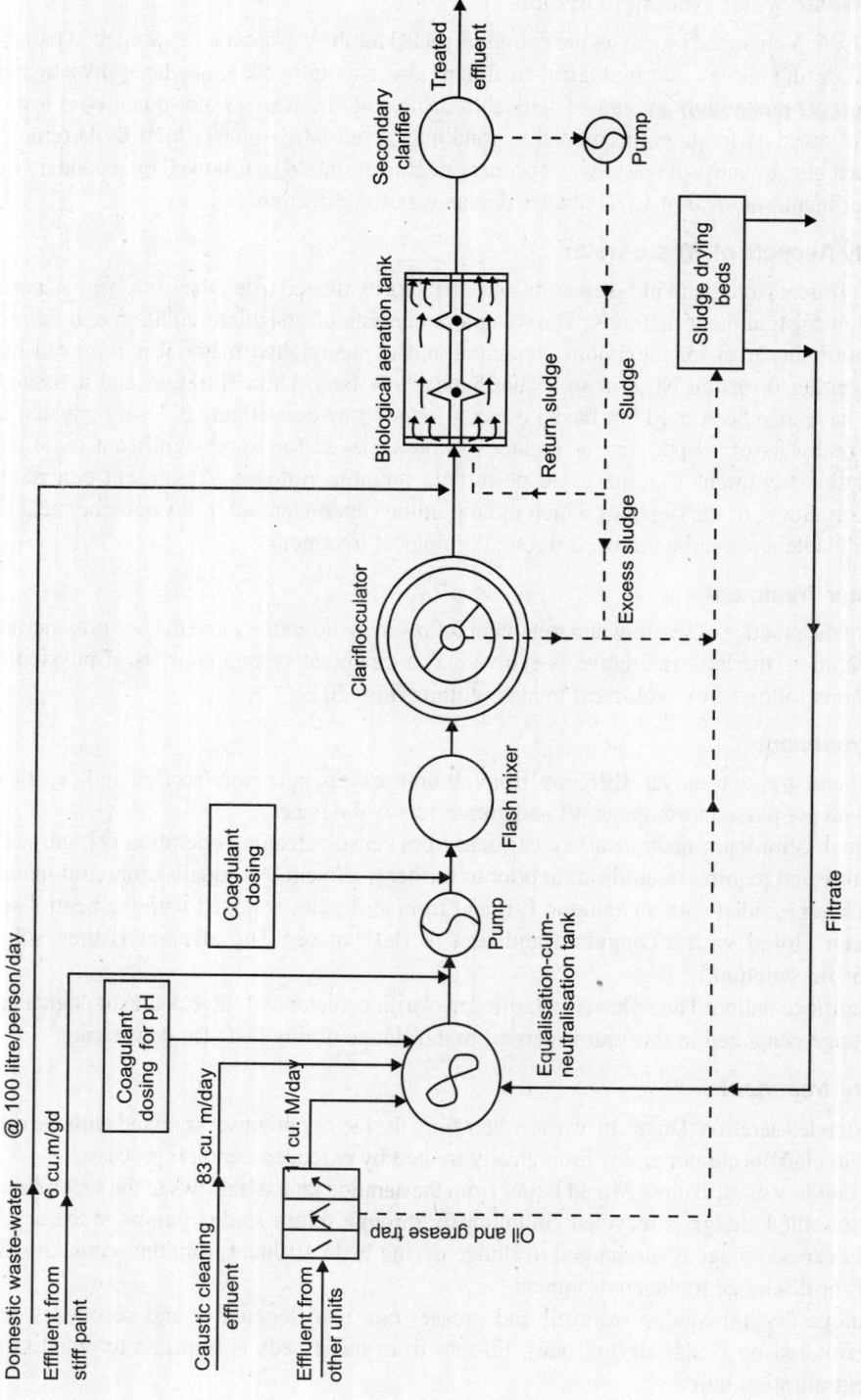

Fig. 20.3. Typical waste scheme for a factory producing 100 m³/day industrial waste-water and domestic waste-water @ 100 l/person/day.

The above treatment process is expected to achieve 90–95 per cent efficiency of removal of pollutants and thus acceptable to the recipient environment.

Recommended treatment system

The following minimum steps are recommended for waste-water treatment:
1. Adjustment of pH.
2. Removal of oil and grease.
3. Removal of suspended solid.
4. Removal of toxic substance.

The proposed waste-water treatment system should be supported by segregation, reduction of generation and recycling of waste-water.

Standards to Satisfy Environmental Requirements

The concerned authorities have to modify the MINAS consistent with the local conditions. The receiving body has much to do with such standards. The following factors should be taken into account in prescribing such standards:
1. Degree of dilution available in the receiving system.
2. Biotic species that will have to be protected.
3. Mean tolerance limit in respect of pollutants to the identified biotic species.
4. Application factor in respect of the mean tolerance limit.

In developing the limits of pollutants, it is considered that the treated waste-waters will be diluted ten times in the receiving water streams. It is ensured that the treated waste-water thus diluted is not likely to cause adverse effects on the quality of receiving water stream. The drinking water standards (IS-10500–1983) and treated water standards (IS-2490, part I-1981) are also compared while arriving at the limits of pollutants.

The volume of waste-water generated is generally low. Thus the quantity of pollutants in the treated waste-water are also reasonably low. In case the treated waste-water is discharged on land with proper management practice, the groundwater is likely to be affected. However, as a precaution groundwater should be monitored.

The heavy metal parameters are kept optional as the heavy metals were not found to be present in the waste-water during the study. The State Pollution Control Boards may include heavy metal parameters as essential parameters to be compiled with, if the individual situation demands so. It was considered that synergistic effect of the combined heavy metals may cause adverse effects on recipient water body. Therefore, limit of total heavy metal is prescribed. The MINAS as evolved is presented in Table 20.3.

Table 20.3. Waste-water discharge standard (MINAS) for paint industry.

Parameters	Limit, not to exceed mg/l, except pH and bioassay test
Essential	
pH	6.0–9.0
Suspended solid	100
BOD_5 at 20°C	50
Phenolics as C_6H_5OH	1.0

(Contd...)

Parameters	Limit, not to exceed mg/l, except pH and bioassay test
Oil and grease	10.0
Bioassay test	90% survival in 96 hours
Optional	
Lead as Pb	0.1
Chromium as Cr	
Hexavalent	0.1
Total	2.0
Copper as Cu	2.0
Nickel as Ni	2.0
Zinc as Zn	5.0

Note: (i) BOD is relaxable upto 250 mg/l, if the waste-water is discharged into town sewer where secondary treatment of sewer effluent is carried out; (ii) state pollution control boards may stipulate quantum limits after case to case study; and (iii) total heavy metals should not exceed 7.0 mg/l.

DYE AND DYE INTERMEDIATES

A substance may be called a dye, if it satisfies certain conditions as:
1. It must have a suitable colour.
2. It must be capable of being fixed to the fabric directly or with the help of certain reagents, called mordants.
3. When fixed, it must not be fugitive, i.e. the colour must be fast to light and resistant to soap and water, and to a certain extent to dilute acids and alkalies.

Thus dyes are substances capable of colouring fabrics in such a manner that the colour cannot be removed by rubbing or washing. It should also be noted that every coloured substance is not a dye. For example, azobenzene is of orange red colour, but it is not a dye, because it is not capable of colouring the fibre.

Basic Operations in Dyeing

The basic operation of the dyeing process involves the following important steps:

Preparation of the fibre

The raw fibres are generally associated with foreign materials such as oils, waxes, lubricants, etc. used during spinning.

Preparation of the dye bath

The dye bath can be prepared by adding requisite amounts of chemicals and other necessary ingredients to the solution of water soluble acidic, basic and direct dyes.

In addition to dye, some other ingredients such as wetting agents, carriers, salts, retarders, etc. have also been added to the dye bath to improve certain properties.

Application of the dye

The fibre is dyed by immersing in the dye bath for a specified period and at optimum temperature.

Dyeing can be carried out either by hand operation or on a machine. In hand operation, the fibrous material is moved in an open vat containing the dye colour. The process on a machine is continuous and two types of dyeing machines are generally used. In one type of continuous dyeing process, the dyeing bath is maintained stationary and the fabric is moved in it.

Finishing

Dyed fibres are finally finished by making use of a number of finishing processes. These processes are used to produce lustre, resistance to shrinkage, crease, and other desirable qualities of feel and appearance.

Manufacturing

Manufacturing process generally adopted involves conversion of simple organic products like benzene, toluene, xylene, naphthalene, anthracene, etc. into a vast number of complex chemical intermediates and dye through several steps of operation like: (i) chemical conversion, (ii) sulphonation, (iii) neutralisation, (iv) fusion, (v) chlorination, (vi) nitration, (vii) reduction, (viii) animation, (ix) acetylation, (x) hydrolysis, and (xi) carboxylation.

Each process requires different operations like charging, reflux, distillation, filtration, washing, drying, and grinding, etc.

The major types of dyes, their main use and the process chemicals used as raw material are given in Table 20.4.

Table 20.4. Types of dyes, application and process chemicals required.

Type of dye	Main use	Process chemicals
Acid	Wool, nylon	H_2SO_4, CH_3COOH, Na_2SO_4, surfactants
Azoic	Cotton	Metal, HCHO, NaOH, $NaNO_2$, acids, and salts
Basic	Acrylic	CH_3COOH, softening agents sodium
Direct	Cotton, synthetics	salts, fixing agents, metal salts (copper or chromates)
Disperse	Polyester	Carrier, NaOH, $NaHSO_3$ chromium
Mordant	Wool	and other metal salts
Reactive	Cotton, wool	NaCl, NaOH, $NH_2CH-CHNH_2$
Sulphur	Cotton, synthetics	sodium sulphide and other salts, acetic acid
Vat	Cotton, synthetics	NaOH, $NaHSO_3$ and other salts, surfactants

Waste Generation

The water usage in the industry is mainly for the following purposes:
1. Synthesis of the dyes and dye intermediates.
2. Steam generation and cooling system.
3. Washing and rinsing of reaction kettles, filter press, floors, etc.
4. Domestic and other miscellaneous activities.

The water consumption pattern varies widely from one industry to another. In the same industry the rate of water consumption often changes due to frequent changes of the feed material synthesis reaction and desired products. The change of product pattern needs cleaning and washing which consumes a substantial quantity of water. Thus water requirement of a dye and dye intermediate industry depends on the following factors:

1. Type of dye produced.
2. Number of products.
3. Gross production.
4. Pattern of working of factory, i.e. continuous or in one shift only.
5. Frequency of change of product pattern, etc.

In-depth study reveals that in general, process water consumption is highest, next to it is the cool and boiler make-up water requirement. The water needs for domestic purposes is the lowest. It is found that for production of one kg of vat dye, water consumption is the largest (1528 to 10345 l/kg) whereas naphthol dye consumes the least quantity of water (6 to 17 l/kg.). The generation of waste-water follows the trend of consumption of water. In Table 20.5 comparative quantities of water consumption and waste-water generation are provided.

Table 20.5. Water consumption and waste-water generation during production of various types of dye products.

Type of product	No. of units	Range of water consumption	Litre/kg of product waste-water generation
Direct dyes	15	2.5–667	1.0–644
Reactive dyes	11	2.0–186	2.0–157
Basic dyes	4	60–4200	50–200
Azo dyes	8	90–400	8.0–213
Vat dyes	2	1528–10345	1389–7980
Dye intermediates	6	36–230	9.0–74
Naphthol dye	2	6.0–17	5.0–8.0
Pigments	3	93–923	7.0–7.85
Indigosol colours	1	529	429
Disperse dyes	1	70	12–42.5
All varieties of dyes/intermediates	10	13–2300	11–1146

The waste-water quantity varieties according to the number of batches of products manufactured in a day, week or month, the duration of synthesis of the dye in the reactor vessel and the duration of the washing and rinsing operations.

The frequency of discharge may be intermittent in the small dye units, operating one or two batches per day, but in the case of the medium and larger units, which operate a number of reactors simultaneously, waste-water discharge is continuous. The waste-water generation sources are as follows:

1. Process waste-water including left out mother liquor.
2. Washing and rinsing wastes.
3. Sanitary and other miscellaneous waste-water.

Characterisation of Waste Effluents

Some of the effluents while manufacture dye are: Dyes produced across the world generate 10 per cent effluent, in which 2 per cent is generated from manufacturing, while 8 per cent is from colouring.

Colour matter in waste water comes from makers and users of colouring matter i.e. dyestuffs, pigments, textiles, dyeing units and tanneries. The pulp and paper sector and distilleries which use raw materials with colour as a by-product, also discharge colour matter into waste water. Effluent left after dyeing contains unused dyes in the form of organic and inorganic compounds, toxic metals, suspended and dissolved solids.

Air emission

The principal air pollutants from dye manufacturing are volatile organic compounds (VOCs), nitrogen oxides (NO_x), hydrogen chloride and hydrogen bromide (HCl and HBr), sulphur oxides (SO_x), ammonia (NH_3), chlorine and bromine gas (Cl_2 and Br_2).

Solid hazardous wastes

Hazardous solid wastes are in the form of ETP sludge, iron sludge, organic sludge, heavy metal sludge, inorganic sludge etc. The main heavy metals present in dye effluent are copper, nickel, chromium, cobalt, lead, mercury and molybdenum.

Liquid effluents

Liquid effluents from dye manufacturing are mainly waste-water (biodegradable and non-biodegradable) or high COD waste-water. About 15 per cent of the dye that is manufactured and used by the industry is discharged into the water. Of this, approximately 2–3 per cent is discharged by the dye manufacturing industry and about 14–15 per cent by the textile dyeing industry.

Waste-water characteristics

The process waste-water is mainly the mother liquor left over after the product is isolated and separated by filter press. This waste-water is of smaller volume and highly concentrated in terms of pollutants. The vessel washings also contain similar type of pollutants but with lower concentration. It has been identified that the waste-waters of the industries have the following characteristics:

1. High levels of BOD and COD.
2. High acidity.
3. High TDS.
4. Deep colour of different shades.
5. High levels of chlorides and sulphates.
6. Presence of phenolic compounds.
7. Presence of heavy metals, e.g. copper, cadmium, chromium, lead, manganese, mercury, nickel and zinc.
8. Presence of oil and grease.

The dye industry waste-waters, if derived from naphthalene and anthracene bases are resistant to biodegradation. The colour removal is also not adequate by the conventional chemical and biological treatment.

The characteristics of the combined waste-water of a dye industry is presented in Table 20.6. In Tables 20.7 and 20.8 the characteristics of combined waste-water of two dye industries engaged in the production of other products are given.

It may be seen that the industry which manufactures dye products only generates combined waste-water which is dark brown in colour having BOD of 503 mg/l (Table 20.3).

Table 20.6. Waste-water treatment in dyestuff Industries.

Characteristics, mg/l except pH, temperature and colour	Waste-water Untreated	Treated	Efficiency % removal
pH	3.4	7.2	–
Temperature°C	45	30	–
Colour	Dark brown	Dark brown	–
TDS	5455	5658	–
SS	96	65	32
BOD	503	49	90
COD	1518	824	44
Chloride	1952	1927	–
Sulphate	1675	1028	–
Cyanide	0.075	0.04	47
Oil and grease	19.3	7.2	62.5

Note: Treatment system: Equalisation, neutralisation, flocculation, sedimentation, extended aeration and clarification. The sludges are dried in sludge drying bed.

Table 20.7. Waste-water treatment in dye industry.

Characteristics, mg/l except pH	Waste-water Untreated	Treated	Efficiency % removal
pH	1.9	6.0–8.5	–
TDS	3670	6650	–
SS	370	200	46
BOD	350	275	22
COD	930	500	46
Chloride	1125	800	–
Sulphate	1985	–	–
Phosphate	23	–	–
Total Nitrogen	17	–	–
Oil and grease	20	10	50
Phenol	10.8	4.0	66

Note: Treatment system: Equalisation and neutralisation by aeration adopting batch process.

Table 20.8. Waste-water treatment in dye-intermediate(s) industry.

Characteristics, mg/l except pH, temperature and colour	Waste-water Untreated	Treated	Efficiency % removal
pH	3.0	7.5	–
Temperature °C	29	29	–
Colour	Brown yellow	Pale yellow	–
TDS	1000	500	–
SS	200	50	75

(Contd...)

| Characteristics, mg/l except | Waste-water | | Efficiency |
pH, temperature and colour	Untreated	Treated	% removal
BOD	1000	100	90
COD	1500	400	73
Cyanide	3	Nil	100
Oil and grease	10	1.0	90

Note: Treatment system: Oil and grease removal, homogenisation, neutralisation, primary clarification, biological treatment and secondary clarification.

Reduction of Waste-water

In-plant control

It is essential to have proper in-plant control measures before going for waste-water treatment. Some of the relevant measures in-plant are summarised below.

Reduction of waste

The volume of waste-water can be reduced by proper control of freshwater consumption. The cooling water blow-down may be reduced by raising the concentration factor. Timely maintenance of the units may be done to prevent leakage, spillage, etc. Minimal usage of water for washing and rinsing may be practised. The last wash water may be recycled for first washing. Application of counter-current washing may also reduce waste-water generation. Dry cleaning of the floor is preferred. When floor washing is absolutely necessary, the treated waste-waters may be used. The pollution load can be reduced by recovery of chemicals and solvents as far as practicable. Spills, leakages, overflows, etc. may not be allowed to join the waste-water stream.

Segregation

Storm waters need segregation to reduce the volume of waste-water. Similarly the uncontaminated and less contaminated waste-water streams like cooling water blow-down, boiler blow-down, condensate, etc. are to be segregated and should not be permitted to join the process waste-water stream. The highly contaminated and coloured mother liquors should be segregated and collected separately. Sometimes strong wastes of the reaction vessels are required to be discharged. These unforeseen strong wastes are to be collected in separate holding tanks and drawn into the waste-water treatment plant at a regulated rate.

Process modification

The production process equipments should be modified so as to generate less wastes. The raw materials used in the synthesis may be substituted by choice of more readily biodegradable chemicals.

Effluent control

The waste-water streams originating as mother liquor and strong waste-waters as mentioned earlier should be segregated and collected in two separate collection tanks. The rest of the process waste-waters may be diverted to the equalisation pond.

Oil and grease and the floating matters present in the waste-water streams are to be arrested by putting up appropriate control measures at the boundary limit of each plant. The waste-water collected in equalisation pond will be monitored to ascertain the concentration of pollutants. Based on the data obtained, the waste-waters of holdings, ponds are to be diverted to the equalisation pond at regulated rates so that the concentration of pollutants are acceptable to the biological treatment system.

The equalisation of waste-waters is followed by pH adjustment and clarification in primary clarifier. Clarifier may be of sludge blanket type. Ferrous sulphate and lime may be used for precipitation of heavy metals. The overflow of the clarifier after pH adjustment will flow to the extended aeration type biological treatment unit. The effluent of biological treatment unit will be clarified in convenient type clarifier using coagulant. Figure 20.4 gives the scheme of the proposed integrated treatment of waste-water.

The reduction of colour may be achieved by segregating and controlled discharge of the mother liquor, which also contains most concentrated form of the chemicals left after dye synthesis. It is envisaged that substantial colour reduction of the waste streams may be achieved by the three-stage treatment system as proposed above. However, if the desired limit of colour (400 Hazen unit) is not achieved by the above treatment then one or both of the following steps are to be adopted:

1. Evaporation of the segregated mother liquor and washings of vessels by solar evaporation or by indirect heating using steam.
2. Powered activated carbon adsorption of the treated waste-water.

Environmental Impact of Dye and its Intermediates

Use of synthetic dyes involves the release of a large amount of hazardous chemicals in the environment during their production, and also during their subsequent use. Within the dyestuff industry, it was realised that the ecotoxicological problems ahead could be tackled more effectively by the deployment of the large but nonetheless limited scientific and technical resources on a collaborative basis. This led to the formation of an international association, ecological and toxicological association of the dyestuffs manufacturing industry (ETAD) in 1974.

The notified dye and intermediates are a stable powder at room temperature and pressure. The granular powder has a propensity to form dust. The notified chemical contains a number of impurities with unknown toxicity. However, as the chemical has been tested in its impure form, hazards associated with the impurities should not warrant any additional concerns. The chemical exhibits low oral and dermal toxicity. Inhalation toxicity is also expected to be low, based on the low toxicity by the oral and dermal routes. The chemical is a moderate eye and skin irritant. Chronic exposure to high levels of chemical may result in kidney and liver effects. The chemical is expected to be genotoxic. As the notified chemical is a reactive dye, and respiratory sensitisation has been associated with reactive dyes, this chemical should be considered a possible respiratory sensitiser. The major toxicological hazards associated with the chemical will be skin and eye irritation, and possible respiratory effects. During weighing, transferring and sampling operations, the potential for skin contact is high. Eye and inhalation exposure, however, will only result if dust is generated, or in the event of spills. The engineering controls which the applicant describes (exhaust ventilation in colour kitchens and negative pressure ventilation in weighing and mixing areas) should be adequate to reduce eye and inhalation exposure to safe levels. The use of impervious gloves and protective clothing will reduce dermal exposure to safe levels.

The presence of dyes in watercourses can also cause many water-borne diseases like nausea, haemorrhage, ulceration of skin and mucous membrane, dermatitis, perforation of nasal septum and severe irritation of the respiratory tract. The people also may come in contact with the yarn or fibre products dyed with the notified chemical. However, as the dyestuff is chemically fixed to the fibre, this exposure is expected to be negligible. The notified chemical is, therefore, considered not to constitute a significant health risk when used in the proposed manner. Dark colour dyes block sunlight from entering the water, thus inhibiting photosynthesis and killing aquatic flora and fauna. Impact of dyes on human health and experiments carried out on fish and related animals have led to regulatory actions in several countries.

Paints and Dyes 453

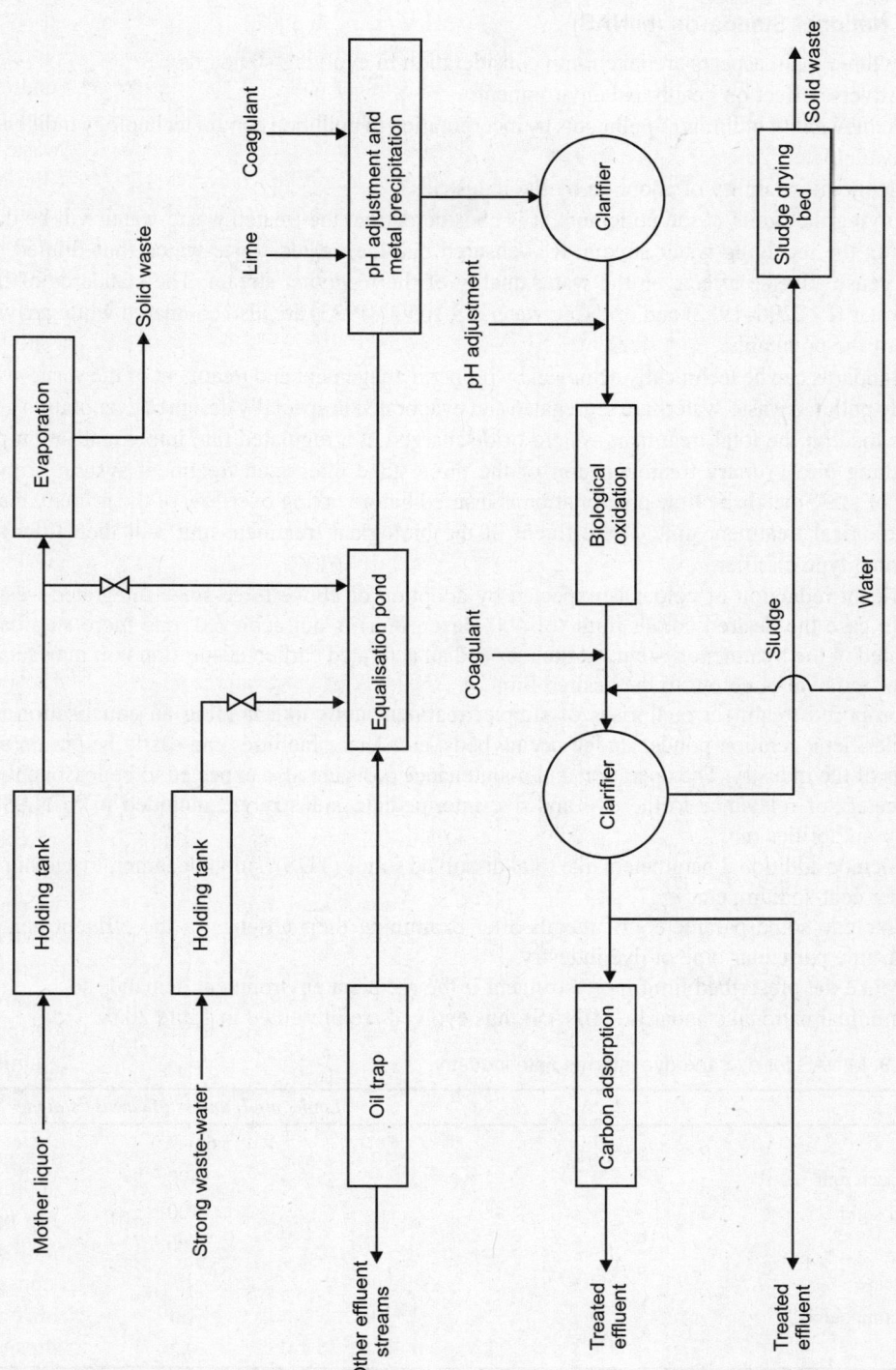

Fig. 20.4. Treatment scheme for dye manufacturing industry effluent.

Minimal National Standards (MINAS)

Generally three main aspects are taken into consideration in evolving standards:
1. Adverse effect on health and environment.
2. Achievability of limits of pollutants by incorporation of pollution control technology indigenously available.
3. Economic viability of adoption by the industries.

In evolving the limits of the pollutants, it is considered that the treated waste-water will be diluted ten times in the receiving water stream. It is ensured that the treated waste-water, thus diluted is not likely to cause adverse effects on the water quality of the receiving stream. The standards of inland surface water (IS 2296–1982) and drinking water (IS 10500–1983) are also compared while arriving at the limit of the pollutants.

The standards can be technically achieved by proper management and treatment of the waste-waters. The highly polluted waste-waters are segregated and evaporated in specially designed evaporation system to reduce load on the total treatment system or discharged at a regulated rate into equalisation ponds before taking into primary treatment unit of the three-stage integrated treatment system proposed. Removal of toxic metals by lime precipitation is insured before taking overflow of the primary clarifier to the biological treatment unit. The effluent of the biological treatment unit will then flow to the conventional type clarifier.

Significant reduction of colour is expected by adoption of above three-stage integrated treatment system. In case the desired colour limit (of 400 Hazen unit) is not achieved, one more step may be incorporated in the treatment system. At the rear end an activated carbon adsorption unit may serve the purpose of removal of colour to the desired limit.

The proposed treatment comprises of simple treatment units like holding an equalisation tanks, pumps, clarifiers, aeration ponds, sludge drying beds, etc. These facilities can easily be put up within the means of the industry. The operation and maintenance costs are also expected to be reasonably low.

Parameters of relevance to the dye and dye intermediate industry are included in MINAS. The regulatory authorities may:
1. Include additional parameters like total dissolved solids (TDS), sulphide, ammoniacal nitrogen, per cent sodium, etc.
2. Exclude some parameters of metals after examining their absence in the effluent generated from a particular type of dye industry.
3. Make the prescribed limit more stringent if the recipient environment demands so.

The minimal national standards (MINAS) thus evolved are presented in Table 20.9.

Table 20.9. MINAS for dye and dye intermediate industry.

Parameter	Limits mg/l, unless otherwise stated
pH	6.0–9.0
Colour, Hazen unit	400
Suspended solid	100
BOD_5	100
Oil and grease	10
Phenolic compounds	1.0
Cadmium	0.2

(Contd...)

Parameter	Limits mg/l, unless otherwise stated
Copper	2.0
Lead	0.1
Mercury	0.1
Manganese	2.0
Nickel	2.0
Zinc	5.0
Chromium	
Hexavalent	0.1
Total	2.0
Bioassay test	90 per cent survival of the test animals after 96 hours in treated waste-water (1:10 dilution)

Note: The limit of colour may be made stringent by the local authorities considering the dilution available in the recipient water body and water usage pattern of the recipient water body.

Chapter 21

Drugs and Pharmaceuticals

INTRODUCTION

The pharmaceutical industry, although a strong and important entity in itself, is frequently considered to be part of the chemical industry. The pollution problems of the pharmaceutical industry, particularly those of the major companies, parallel those of the chemical industry. Because the problems of the industry are not unique, this chapter will be concerned with emphasising the problems, where they should be looked for, and some of the more interesting solutions to solve these. The pharmaceutical companies facing major problems in connection with both air and water pollution are the ones that operate fermentation plants, principally in the production of antibiotics (penicillin, streptomycin, etc.).

PRODUCTION PROCESS

The production process involved in this industry can be divided into two categories:
1. Manufacture of basic drug.
2. Formulation.

A majority of the pharmaceutical industries in India are formulation units, whereas the multinational units and some of the large-scale units in public sector have both manufacturing and formulating facilities in the same complex.

Pharmaceutical manufacturing industries generally employ batch operations for manufacture of most basic drugs and their derivatives. Formulation units mainly employ physical operations for preparation of tablets, capsules, syrups, injections, liquid preparations, ointments, etc. However, the industry is so large and the products are so diversified that it is beyond the scope report of this chapter to describe manufacturing processes for individual drugs. The manufacturing processes are broadly classified and described in the following sections.

PHYSICAL METHODS

Formulation: Formulation products are prepared by physical methods such as mixing, grinding, sieving, filtration, washing, drying, milling, encapsulation, packing, etc. Different types of capsules, tablets, injectables, liquid tonics, syrups, ointments, etc. are prepared by these methods.

Extraction: Extraction is also a physical method and it is involved in the separation of a useful constituent from crude or partially refined basic drugs. Suitable solvents like water, alcohol, ether, acetone or steam are used in the separation process. Sometimes the extraction is done by mechanical force like maceration, centrifugation, fractional or pressure percolation, etc.

Fermentation: Fermentation is a bio-chemical reaction within a reactor in the presence of selected active microbes or enzymes. Reactions are carried out under mild chemical and physical conditions. Various drugs like antibiotics, enzymes, hormones, Vitamin B12, vaccines, etc. are manufactured by the process of fermentation.

Antibiotics, Vitamins and Enzymes

The following steps are involved in a typical fermentation process:
1. Selection of proper microbial strain.
2. Isolation of the strain in a pure culture.
3. Inoculation and germination in a fairly complex nutrient medium.
4. Fermentation under controlled aerobic and/or anaerobic conditions.
5. Separation of the mycellium from the fermentation broth.
6. Recovery and purification of the final product from fermentation broth.
7. Style and finishing of the final product.

Raw materials like sugar, starch, corn steep liquor, soyabean meals, fish or whale solubles, etc. are used as raw materials. Vitamins and minerals are used as growth factors. The end products are often less than 10 per cent of the total preparation. After separation of the mycellium, the final product is recovered by various techniques like adsorption, filtration, precipitation, ion-exchange, concentration, decolourisation, vacuum drying, etc.

Organic synthesis

A large number of pharmaceuticals are manufactured by organic synthesis. This involves several steps like oxidation, reduction, nitration, sulphonation, halogenation, amination, aminolysis, Friedel Craft's acytilation, alkylation, esterification, crystallisation, hydrogenation, precipitation, etc. Products produced by organic synthesis are like chloramphenicol, sulpha-drugs, quinolines, dexamethasone, antidiobatic, antihelmintic, antifilarial, antileprotic, anti-malarial, anti-T.B., anti-pyretic, analgesics, vitamins, etc. A wide range of chemicals are used in different processes for production and purification of various drugs. Generating a product may involve many steps and many intermediates. After the final product is formed, it is separated from the intermediates, solvents and reactants, etc.

WASTE GENERATION

Waste generation from the basic drug manufacturing houses are higher than the formulating units. The nature and the quantity of waste generated vary depending upon the processes involved.

In general, the waste and emissions generated during manufacturing of pharmaceuticals depend on the raw materials and equipment used, as well as the manufacturing, compounding and formulation process employed. In designing bulk manufacturing processes, consideration is given to the availability of the starting materials and their toxicity, as well as the wastes (e.g. mother liquors, filter residues, and other by-products) and the emissions generated. When bulk manufacturing reactions are complete, the solvents are physically separated from the resulting product. Due to purity concerns, solvents are not often reused in a pharmaceutical process. They may be sold for non-pharmaceutical uses, utilised for fuel blending operations, recycled, or destroyed through incineration.

Air Emissions

Both gaseous organic and inorganic compounds, as well as particulates, may be emitted during pharmaceutical manufacturing, separation, purification and formulation. Some of the volatile organic compounds (VOCs) and inorganic gases that are emitted may be hazardous. The various sources of emission in the order of their priority are distillation units, storage and transfer of materials, filtration, extraction, centrifugation, and crystallisation.

During reaction: VOC emissions from reactor vents, material loading and unloading, acid gases (halogen acid, sulphur dioxide, nitrous oxides); fugitive emissions from pumps, sample collections, valves, tanks, etc.

During separation: VOC emissions from filtering systems which aren't contained and fugitive emissions from valves, tanks and centrifuges

During purification: Solvent vapours from purification tanks, fugitive emissions.

During formulation: Tablet dusts, other particulates

Effluent Generation

Pharmaceutical manufacturers use water for process operations, as well as for other non-process purposes. However, the use and discharge practices and the characteristics of the waste-water will vary, depending on the operations conducted at the facility. Additionally, in some cases, water may be formed as part of a chemical reaction.

Solid Waste Generation

Both nonhazardous and hazardous wastes are generated during all three stages of pharmaceutical manufacturing. These wastes can include obsolete raw materials or products, spent solvents, reaction residues, used filter media, bottom still, used chemical reagents, dust from filtration or air pollution control equipment, raw material packaging wastes, laboratory wastes, spills, as well as wastes generated during packaging of the formulated product.

Human and Animal Contribution

Last but not the least is the contribution from humans and animals. Drugs are consumed by humans and animals, alike. These drugs are not completely utilised in the body. The residues and the metabolites are excreted in urine and faeces.

Physical Processes

Formulation and extraction operations come under this section. In a formulation unit the major pollution load comes from washing of vessels, floor and equipments. The pollutants are carbohydrates and inert formulating materials. The effluents coming from different sections of a medium size formulation unit and their average qualities and quantities are furnished in Table 21.1. Certain harmful organic impurities may come out from extraction of natural ingredients.

Table 21.1. Waste-water generation from a pharmaceutical formulation unit.

Source of waste-water generation	Average flow of waste-water (m^3/hr)	Range of average effluent characteristics in mg/l except pH	
Sterile products	1.0 to 10	BOD	50 to 100
		TSS	20 to 260
Syrup preparation	10 to 25	COD	150 to 2500
		Cl^-	20 to 150

(Contd ...)

Source of waste-water generation	Average flow of waste-water (m^3/hr)	Range of average effluent characteristics in mg/l except pH	
Malt preparation	1 to 5	BOD	300 to 2000
		SO_4^-	20 to 200
Pastilles preparation	10 to 25	BOD	2000 to 2500
		Heavy metals	1 to 20
		TSS	100 to 300
		pH	4.0 to 8.0

Considerable amounts of pollutants may generate from extraction of herbs and natural raw materials. Waste-waters are generated from washings before extraction which may contribute grit of organic impurities. The flow of waste-water ranges between 1 to 10 m^3/kg of the desired product.

Chemical Methods
Cyanide destruction
Several cyanide destruction treatment technologies are currently used in the pharmaceutical manufacturing industry, including alkaline chlorination, hydrogen peroxide oxidation, and basic hydrolysis.
1. The alkaline chlorination treatment process involves reacting free cyanide with hypochlorite (formed by reacting chlorine gas with an aqueous sodium hydroxide solution) to form nitrogen and carbon dioxide. The reaction is a two-step process and is normally performed separately in two reactor vessels. Because treatment is normally performed in batches, it is necessary to use an additional equalisation tank to store accumulated waste-water during treatment. The reactors need to be equipped with agitators, and both reaction steps require close monitoring of pH and oxidation reduction potential (ORP). These reactions are normally performed at ambient temperatures.
2. Hydrogen peroxide treatment involves adding hydrogen peroxide to cyanide-bearing waste-water to convert free cyanide to ammonia and carbonate ions. This treatment is normally performed batch-wise, in a reaction vessel. The treatment process consists of heating the waste-water to approximately 125°C and adjusting the pH in the reaction vessel to approximately 11. Hydrogen peroxide is added to the vessel and is allowed to react for approximately one hour. Equipment required for this process includes reaction vessels, storage vessels for hydrogen peroxide and a pH adjustment compound (typically sodium hydroxide), an equalisation tank, and feed systems for hydrogen peroxide and sodium hydroxide.
3. Hydrolysis treatment involves reacting free cyanide with water under basic conditions to produce ammonia. This process requires approximately one hour and is typically performed at a temperature between 170° and 250°C, at a pH of 9 and 12. Hydrolysis is normally performed in a reactor vessel equipped with a heat exchanger and a system to store and deliver sodium hydroxide (or other basic compounds).

Oxidising agents like ozone, UV rays and hydrogen peroxide
Ozone gas is highly reactive and is used in the treatment of waste-water. Ozone is effective in oxidising antibiotics, betablockers, antiphlogistics, lipid regulator metabolites, the antiepileptic drug, carbamazepine and the natural estrogen, estrone from STP. Also, the combined treatment of hydrogen peroxide and UV radiations reduces the amount of diclofenac present in waste-water.

Semiconductor photocatalysis

In semiconductor photocatalysis, the samples are exposed to radiation in the presence of a semiconductor, generally TiO_2. TiO_2 is remarkably active, cheap, non-toxic and chemically stable over a wide pH range and it is not subject to photo corrosion. In general, the goal of the application of photocatalysis in water treatment is the transformation, deactivation and finally minimisation of environmentally persistent compounds. Carbamazepine, clofibric acid and iomeprol showed higher degradation tendency, when subjected to semiconductor photocatalysis. In this, TiO_2 suspension was freshly prepared by suspending TiO_2 in ultra pure water and then was sonicated for at least 30 minutes. Before starting the irradiation, the pharmaceutical or contrast media solution was equilibrated with humidified air in order to get defined concentrations of oxygen and bicarbonate in all samples, then mixed with the sonicated TiO_2 suspension and adjusted to a defined pH value. Then the samples were irradiated.

Fermentation

Waste-waters from a fermentation unit emanate mainly from the recovery and purification of the final product and also from washings of floor, vessels, equipments, etc. About 5 m^3/kg of final product is the waste-water generated. The main sources of wastes generation from a penicillin fermentation plant are listed in Table 21.2.

Table 21.2. Waste-water generation from an antibiotic plant.

Source of waste-water generation	Nature of waste-water and solid waste	Ranges for average characteristics of combined effluent in mg/l except pH	
Fermentation block	Floor and equipment washings,	pH	4.0 to 8.0
	leakages of valves/ machines, contaminated batches,	TSS	100 to 1000
	cooling waters, laboratory and utility wastes	BOD	500 to 6,000
Filtration/centrifugation	Mycellium cakes, filter washings, floor washings	BOD	Upto 10,000
Recovery and purification block	Faecal wastes, acid and alkali wastes (from regeneration of ion-exchangers), floor washings laboratory wastes	CN^- Heavy metals Oil and grease	0.1 to 1 1 to 5 20 to 50
Style and finishing block	Floor and equipments, washings and other utility wastes	Phenol (after mycellium cake separation)	1 to 5

The major waste load is contributed by the mycellium cake. The wet cake has a BOD value in the order of 40,000 to 70,000 ppm and solid content is in the order of 30,000 to 50,000 ppm. After filtration the BOD load in the filtrate may be as high as 10,000 mg/l. The spent liquor from the process has an average BOD of 2500 mg/l and a maximum of 6,000 mg/l. The average characteristics furnished in Table 21.2 are after the separation of mycellium from the fermentation broth.

Vaccination, microbial suspension antitoxin preparation

The waste generated from these sections includes animal tissues, hides, leather, blood, fat, egg, fluid and shells, spent grains, biological culture media, solvents, salts, antiseptic agents, bactericides, equipment and floor washings. The waste-water characteristically contains high BOD, COD and bad odours. High total solids and chlorides are also expected as a result of washing operations. Antiseptic and anti-bacterial agents may contribute to toxicity.

Organic synthesis

Classified descriptions of various kinds of effluents generated from organic synthesis plants are presented below.

Acidic and alkaline effluents

Highly acidic or alkaline waste-waters are discharged from certain processes. Mineral acids like HCl, H_2SO_4 and organic acids like acetic, formic, sulphanilic, etc. are present in waste-water. The discharge of acids in waste-water may range upto several thousand kilograms per day and may give a shock load to the effluent treatment plant. Dissolved salts are produced in the process of mixing acidic and alkaline effluents. The most predominant salts are sulphates and chlorides in the concentration range of 2,500 to 30,000 mg/l and flow upto 20 m^3/hr.

General process liquors

Varying composition, concentration and toxicity may add problems in the treatment of pharmaceutical and fine chemical process effluents. However, the flows of such streams are small and they are partly biodegradable.

Strong process liquors

Such waste-waters are generated from batch process and purification units which are mostly toxic. These contribute high COD and organic load in the treatment plant.

Emulsified effluents

Various physical and chemical operations may give rise to effluents containing oils, emulsions and other extractable materials which often form emulsions in waste-water and reduce the efficiency of the biological treatment system.

WASTE-WATER

A general flow diagram for the treatment of waste-waters from various types of pharmaceutical products are presented in Fig. 21.1. Depending upon the nature of the waste-water, selection/omission of the specific treatment units can be made.

Before selection of a particular treatment system for effluents of pharmaceutical industries the following aspects are required to be considered:
1. Good housekeeping practices.
2. Segregation of certain waste-water streams.
3. Process and equipment modifications.
4. Recovery of by-products and recycle possibilities.

Certain waste-waters containing pathogenic micro-organisms from fermentation particularly vaccine section need autoclaving before discharge into common treatment plant.

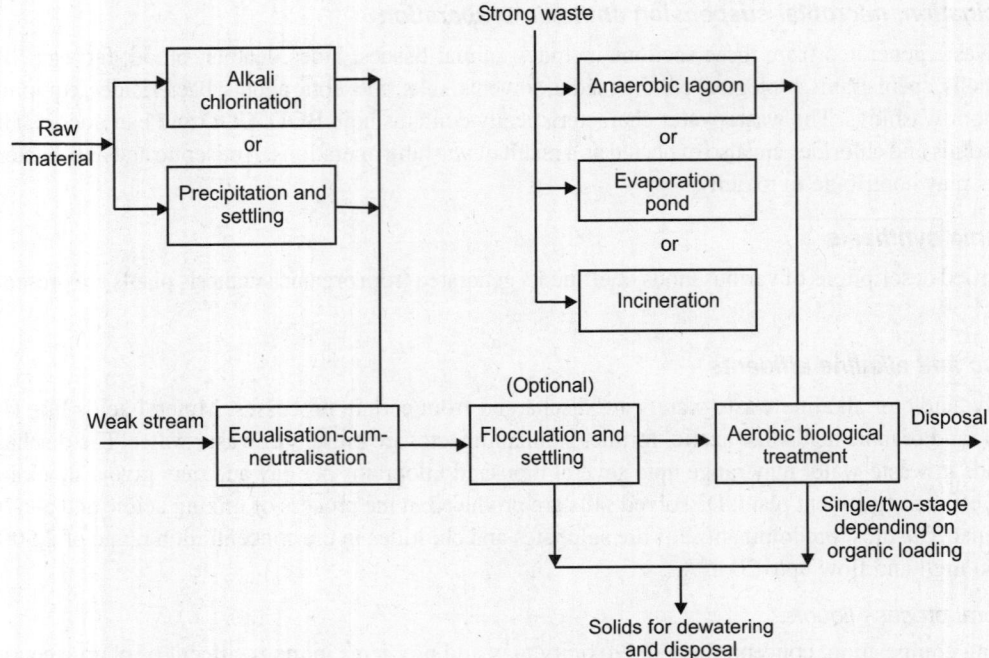

Fig. 21.1. A general flow diagram for the treatment of waste-waters from pharmaceutical industries.

Various combinations of unit processes are involved in effective treatment of pharmaceutical industry effluent.

The schematic flow diagrams suitable for waste-water treatment of bulk drug manufacturing unit and for formulation unit are presented in Fig. 21.2 and Fig. 21.3 respectively. Description of various treatment processes are presented in subsequent sub-sections.

Physical Treatment

The common physical treatments are plain settling, dissolved air flotation, adsorption, solar evaporation. These methods can be adopted individually or in combination depending upon the waste-water quality and quantity.

Plain settling

Water insoluble compounds can be removed from the waste-water by plain settling. The settling characteristics can be improved with the addition of suitable coagulants. The system consists of a settling basin with adequate volume to provide sufficient settling time for the impurities.

Dissolved air flotation

Suspended, colloidal and emulsified impurities can be removed from waste-water by air flotation technique.

Air dissolved in the waste-water under pressure when released in the flotation tank forms bubbles which trap the lighter impurities and lift them to the surface. The materials form a floating layer which is removed by a skimmer mechanism.

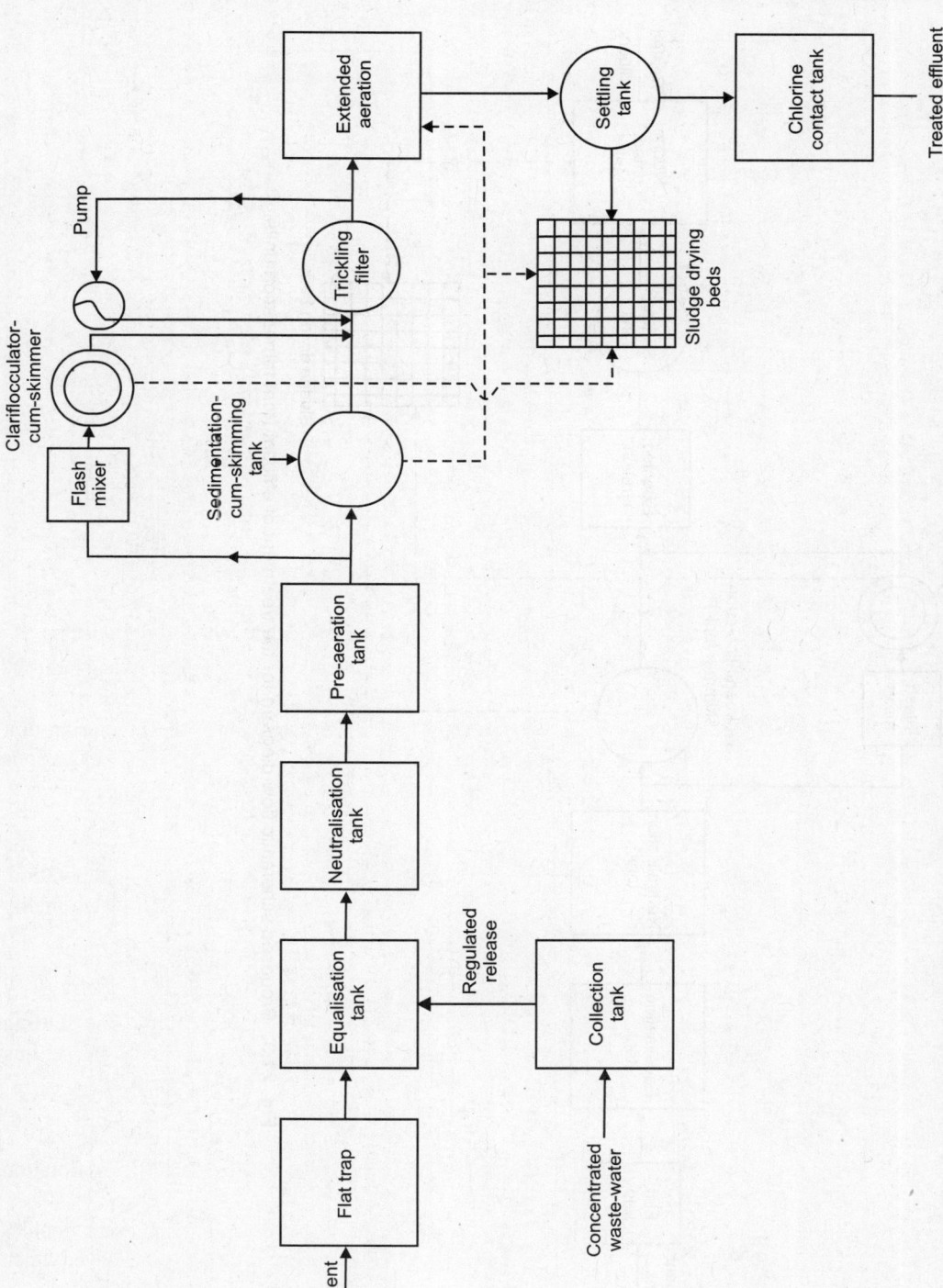

Fig. 21.2. Proposed schematic flow diagram for the treatment of effluent from bulk drug units.

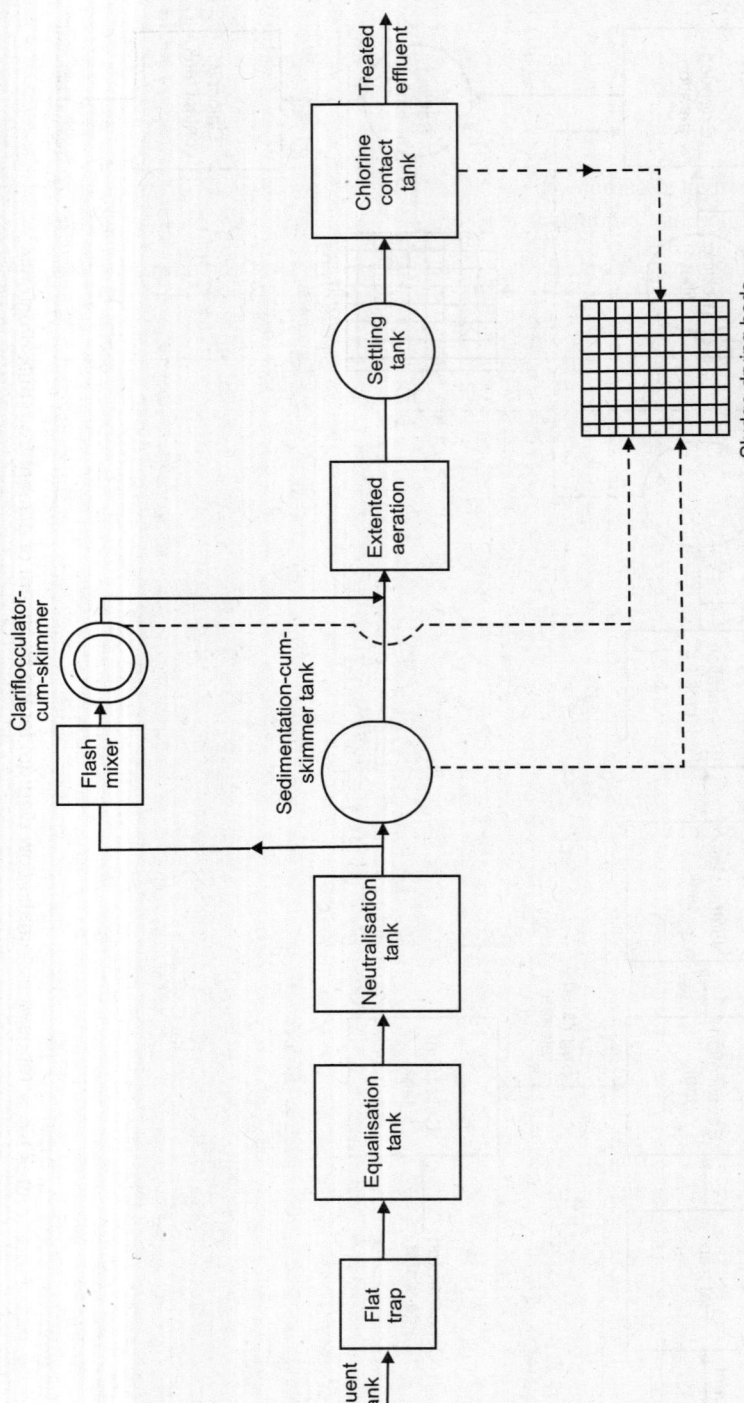

Fig. 21.3. Proposed schematic flow diagram for the treatment of effluent from formulation units.

Adsorption

Activated carbon and activated clay are used for the adsorption of refractory chemicals which are either non-degradable or inhibit the degradation of other organic compounds. The process is carried out in columns or tanks. Carbon treatment can be employed as a polishing unit to further improve the quality of the biologically treated waste-water.

Summary of a study using activated carbon on biologically treated waste-water from an industry manufacturing basic drugs and microbial products is presented in Table 21.3.

Table 21.3. Summary of adsorption studies using activated carbon on biologically treated waste-water.

Carbon dosage mg/l	Sample	Chemical minimum	Oxygen average	Demand (mg/l) maximum
25	Influent	216	325	448
	Effluent	184	295	396
	% Removal	6.3	10.3	21.2
50	Influent	216	325	448
	Effluent	192	276	392
	% Removal	4.3	15.6	27.5
100	Influent	216	325	448
	Effluent	168	263	344
	% Removal	9.1	21.9	28.8

Evaporation pond

When waste-water quantity is less and contains high organic and inorganic impurities, it can be disposed through solar evaporation. This system is feasible where adequate area is available. Area required for effective evaporation depends upon the local annual evaporation rate.

Primary Treatment/Chemical Treatment

Segregation of different waste streams is an important step for economic design of a treatment plant. The following streams may be segregated for this purpose:
1. Strong process liquors.
2. Streams containing cyanide, heavy metals, toxic chemicals.
3. Condensate and cooling waters.
4. Acidic and alkaline streams.

Various treatment alternatives for strong process liquors are incineration, solar evaporation and treatment by anaerobic filter. The toxic effluents either can be incinerated or treated by suitable technologies like carbon adsorption, ion-exchange, chemical precipitation, reverse osmosis, etc. Condensate and cooling waters can be recycled and reused. The acidic and alkaline waste streams can be either treated separately with acid/alkali for pH correction or may be combined suitably with other waste streams. The effluent containing chemical sludge, settleable solids and high oil concentration (over 50 mg/l) can be treated by coagulation, flocculation and settling after neutralisation. Coagulants like alum, $FeSO_4$, $FeCl_3$, etc. with/without polyelectrolytes can be used. The coagulation process also breaks oil emulsions and nullifying the zeta potential. Pre-aeration for 2 to 3 hours by means of diffused air may help to bring down the BOD load of about 30 to 40 per cent. Diffused aeration is known to bring oily and fatty matters in suspension form in the waste-water.

Secondary Treatment

Various factors are responsible to select a suitable treatment system, i.e. quality and quantity of influent to be treated, desired degree of treatment, site conditions, change of products and overall economics. Secondary biological treatments employed are mainly aerobic and in some cases anaerobic followed by aerobic treatment. Trickling filter, extended aeration and conventional activated sludge systems are generally practised. Anaerobic filter and anaerobic lagooning are also being used for treatment of pharmaceutical industry waste-waters.

Anaerobic Filter and Lagooning

Anaerobic filter can operate with an influent COD concentration upto 2000 mg/l for a period upto six months. The hydraulic retention time is 35 to 40 hours at an operating temperature of around 35°C. It can operate with an efficiency of COD removal upto 80 per cent and BOD removal of 90 to 95 per cent. However, it is not suitable for shock loads of high suspended solids and sulphate concentration. It cannot operate at all climatic conditions in certain parts of the country. Anaerobic lagooning is not suitable for effluents having higher COD load and lesser biodegradability. For these reasons, aerobic treatment systems are preferred to anaerobic systems which are described below:

Trickling filter

This is widely employed for treatment of effluents of pharmaceutical industry in India and abroad. The major advantage of trickling filter is that it is capable of handling higher shock loads of BOD and toxic waste materials than any other biological treatment system. It can function with a load variation of ±20 per cent without affecting the performance. Recirculation is usually practised upto a ratio of 8:1 with an input BOD load of about 2500 mg/l. In pharmaceutical industry, the type and load of pollutants change widely depending on the local market demand. This aspect necessitates putting up additional appropriate treatment in the downstream of trickling filter.

Conventional activated sludge

Conventional activated sludge is commonly practised in pharmaceutical manufacturing industry. This requires lesser space compared to oxidation ditch or aerobic pond and easily operates with a high efficiency of BOD removal of more than 90 per cent. The dynamic behaviour of the process is complicated by the presence of mixed population of microbes, varying physical and chemical characteristics of organic loads, fluctuation of influent flow and limiting substrate concentration. The presence of nickel in effluent has toxic effects on nitrifying bacteria in the activated sludge process. This system needs high rate of aeration and sludge recirculation.

Extended aeration

Extended aeration treatment system has all the advantages of conventional activated sludge as described earlier. In addition, it takes care of complete sludge stabilisation and remaining BOD in the sludge as well as in the waste-water. But the disadvantages are it can handle low organic load and unable to tackle shock loads of BOD fluctuations and toxic compounds. It requires an aeration time of 35 to 40 hours. But the system is suitable for treating effluent from a pharmaceutical formulation unit and bulk drug manufacturing unit. Since this unit will be in the downstream of the trickling filter, major portion of the shock loads of pollutants will be absorbed in the trickling filter.

Tertiary treatment

Desired effluent quality is the basis for the selection of any treatment scheme. In most cases, pharmaceutical industry effluents are not suitable for land disposal for farming due to the presence of high concentration of dissolved salts. Tertiary treatments are required to kill virus and bacteria and to remove other impurities like colour, bad smell, etc. Chlorination and sand filtration may generally be practised for tertiary treatment.

Biological treatment

Various factors are responsible for the selection of biological treatment system, i.e. quality and quantity of waste-water to be treated, desired degree of treatment, site conditions and overall economics. Biological treatment systems employed for pharmaceutical wastes are mainly aerobic in nature. In some cases where the organic loading is high, anaerobic treatment units are used prior to aerobic system.

The widely used anaerobic treatment unit for pharmaceutical wastes is anaerobic lagoon and for aerobic treatment, trickling filter, aerated lagoon or aeration tank followed by clarifier are used.

PERFORMANCE OF EXISTING TREATMENT SYSTEMS

For the development of minimal national standards (MINAS), a number of existing treatment plants of bulk drug manufacturing as well as formulation units were monitored. The performance of selected treatment plants are presented in Table 21.4.

Table 21.4. Performance of existing treatment systems.

Type of industry	Flow m^3/day	Unit steps of treatment system	Parameters	Performance of treatment plants		
				Conc. in mg/l in per cent		
				(Influent)	(Effluent)	Removal
Bulk drug manufacturing and formulation	540	Neutralisation, settling tank, alum flocculation, Second stage settling, plain aeration, extended aeration, final settling tank and sludge drying	TSS	200	20.0	90
			BOD	975.0	22.0	97.9
			COD	1648.0	240.0	85.4
			Mercury	0.085	Nil	–
Bulk drug manufacturing and formulation	450	Oil and grease removal, equalisation-cum-neutralisation, clariflocculator, extended aeration	TSS	551.0	80.0	85.5
			BOD	350.0	36.0	89.7
			COD	4120.0	428.0	89.6
			NH_3-N	11.6	1.9	83.4
			Phenol	1.28	0.32	75.0
Bulk drug manufacturing and formulation	750	Equalisation-cum-neutralisation, first aeration tank,	TSS	280.0	18.0	93.6
			BOD	475.0	76.0	84.0
			COD	2720.0	80.0	97.0

(Contd...)

Type of industry	Flow m³/day	Unit steps of treatment system	Performance of treatment plants			
			Parameters	Conc. in mg/l in per cent		Removal
				(Influent)	(Effluent)	
		clariflocculator, second aeration tank, clarifier and sludge drying	Mercury	0.083	0.036	56.6
Antibiotics	1600	Equalisation-cum-	pH	7.59	7.75	–
		neutralisation,	TSS	2200.0	100.0	95.5
		aeration tank,	BOD	195.0	79.0	59.5
		clariflocculator and	COD	1094.0	428.0	61.7
		sludge drying	Oil and grease	60.0	40.0	33.3
Antibiotics	600	Homogenisation,	pH	7.1	8.1	–
		first aeration tank,	TSS	990.0	85.0	91.4
		second aeration tank	BOD	600.0	25.0	95.8
		and sludge drying	COD	1753.7	155.2	91.1
			NH_3-N	32.0	5.0	84.3
Formulation	75	Oil and grease	pH	6.52	7.37	–
		removal and	TSS	167.0	68.0	59.3
		sedimentation tank	BOD	265.0	116.0	56.2
			COD	676.0	266.0	60.6
			Oil and grease	34.0	7.0	79.4

Cost of Waste-water Treatment

Capital cost and economics for operation and maintenance of the treatment plant are the deciding factors to opt for a treatment system. Various alternatives of unit processes may be practised for primary treatment, depending on raw waste characteristics and desired quality of final effluent. The cost of secondary or biological treatment may be calculated on the basis of the following alternatives:

1. Single stage biological treatment consisting of extended aeration or conventional activated sludge or trickling filtration.
2. Two stage biological treatment for BOD value above 1500 mg/l, consisting of trickling filtration followed by extended aeration or conventional activated sludge process.

AIR EMISSIONS CHARACTERISATION

This section describes the characteristics of emissions associated with bulk pharmaceutical xhemicals (BPC) manufacturing processes. Subsections presented include: emissions sources, types of emissions, emission estimates, and estimation techniques. Guidance is provided on estimating emission rates.

Emission Sources

A summary of typical material inputs and emissions for BPC manufacturing processes is presented in Table 21.5. Sources of emissions at BPC manufacturing plants vary widely from facility to facility, depending on the products manufactured. Sources of emissions associated with the major BPC manufacturing unit process operations are described below.

Reactors

Emissions from reactors may occur by several mechanisms, including: (i) displacement of air during the charging of reactants, (ii) inert gas flows during the reaction cycle, (iii) opening of the reactor during operation to collect samples or charge reactants, (iv) venting of emissions during refluxing, (v) solvent vapourisation during cleanup operations, and (vi) the relieving of pressure between cycles (where a), reactor is operated under pressure). Emissions are maximised when reactions involve highly volatile chemicals and elevated temperatures. In contrast, emissions are minimised with less volatile chemicals, low temperatures, and/or sometimes subatmospheric pressures and when water-based reactions are involved.

Distillation units

Emissions occur from the condensers used to recover the evaporated liquids and from the vents as a result of vapour displacement during startup.

Extractors

Emissions from extractors may occur from the: (i) displacement of air while charging, emptying, and cleaning the extractor, (ii) agitation of liquids during the extraction process, and (iii) filling and emptying of associated surge tanks.

Table 21.5. Summary of typical material inputs and outputs in BPC processes.

Process	Inputs (examples of some commonly used chemicals provides)	Air emissions
Chemical synthesis reaction	Solvent, catalysts, reactants, e.g. benzene, chloroform, methylene chloride, toluene, methanol, ethylene glycol, methyl isobutyl ketone (MiBK), xylenes, hydrochloric acid, etc.	VOC emissions from reactor vents, manways, material loading and unloading, acid gases (halogen acids, sulphur dioxide, nitrous oxides); fugitive emissions, from pumps, sample collections, valves, tanks
Separation	Separation and extraction solvents, e.g. methanol, toluene, hexanes, etc.	VOC emissions from filtrering systems that are not contained; fugitive emissions from valves, tanks and centrifuges
Purification	Purification solvent, e.g. methanol, toluene, hexanes, etc.	Solvent vapours from purification tanks; fugitive emissions
Drying	Finished active drug(s) or intermediates	VOC emissions from manual loading and unloading of dryers
Natural product extraction	Plants, roots, animal tissues, extraction solvents, e.g. ammonia, chloroform, phenol, toluene, etc.	Solvent vapours and VOCs from extraction chemicals
Fermentation	Inoculum, sugars, starches, nutrients, phosphates, fermentation solvents, e.g. ethanol, amyl alcohol, methanol, MiBK, acetone, etc.	Odouriferous gases, extraction solvent vapours, particulates

Fermentors

Emissions occur during the fermentation process as a result of displacement of large volumes of solvent from one vessel to another, and recovery of product from the concentrated solvent by crystallisation, filtration, and drying of the solid product.

Centrifuges

A significant source of emissions is the vapourisation of organic compounds from the 'wet' solids as they are removed from the centrifuge and transported to the next operation. In addition, emissions can occur as the centrifuge is opened, as the filtrate is discharged into a holding tank as inert gases are vented. Some older facilities may have open-top centrifuges, where direct vapourisation to the ambient air can occur.

Filters

Emissions typically occur when a filter is opened to remove the collected solids or when the filter is purged with steam or inert gas prior to cleaning.

Crystallisers

Emissions from crystallisation operations are usually not significant, unless supersaturation is achieved through solvent evaporation. These emissions may be vented directly to the atmosphere or through a capture device (e.g. a condenser).

Dryers

In the pharmaceutical industry, drying is the unit operation with generally the highest VOC emissions. With a convective dryer, the exhaust gas is a high-flow, relatively dilute stream, which traditionally has often been vented directly to the atmosphere without emission control. With a conductive dryer, the vent gas is much more concentrated; so the flow rate is low. In the case of vacuum drying, VOC emissions occur downstream from the vacuum pump, where the air in-leakage is exhausted from the system.

Storage and transfer

The vapour space in a tank will in time become saturated with the stored organics. During tank filling, vapours are displaced causing an emission or a 'working loss'. Some vapours also are displaced as the temperature of the stored VOC rises, such as from solar radiation, or as atmospheric pressure drops; these are 'breathing losses'. The amount of loss depends on several factors: type of VOC stored, size of tank, type of tank, diurnal temperature changes, and tank throughput.

In-plant transfer of VOC is done mainly by pipeline, but also may be done manually (e.g. loading or unloading 55-gal drums). Raw materials are delivered to the plant by tank truck, rail car, or in 55-gal drums. These storage devices are susceptible to breathing and working losses, the magnitude of which depends on the type of compound stored, the size and design of the tank, ambient temperature and diurnal temperature changes, and tank throughput. Emissions may also be associated with manual material transfer operations within the facility and with the transfer of liquids from tanker cars, trucks, and rail tank cars.

Types of Emissions

Pharmaceutical manufacturing facilities are similar to many chemical-manufacturing operations in that both gaseous and particulate emissions can be significant.

VOCs

Usually of greatest concern at pharmaceutical facilities are emissions of VOCs. Many of these compounds are photochemically reactive and contribute to tropospheric (lower-level) ozone formation (a concern for US facilities located in ozone nonattainment areas), while others may deplete the stratospheric (upper-level) ozone layer. An added concern is that some compounds are air toxics or HAPs, categorised by the EPA as possible or probable human carcinogens, thereby making long-term human exposure undesirable.

Finally, some compounds used in the industry may be acute toxicants, where short-term exposure, including emergency releases, can lead to adverse health effects.

Particulates

Emissions of particulates also may be of concern. One reason is that particulate emissions, especially those associated with formulation and packaging operations, may include biologically active ingredients. Additionally, some of these emissions may be in the respirable size range. Within this range, a significant fraction of the particulates may be inhaled directly into the lungs, thereby enhancing the likelihood of being absorbed into the body and damaging lung tissues. Fortunately, because of concern about worker health and product contamination, particulate emissions are often closely controlled, especially where toxic compounds are in use. Consequently, particulate emissions at most pharmaceutical manufacturing facilities are insignificant.

Odours

Additionally, odours can be a problem, especially where fermentation takes place. While odours can create a community nuisance, they generally do not represent a serious air pollution problem.

Other

Another class of gaseous emissions is acid gases; these are principally sulphur dioxide, nitrous oxides, and hydrochloric acid. The oxides of sulphur and nitrogen are typically associated with the large-scale combustion of fossil fuels are utilised for power generation or steam production by large chemical facilities.

Because of the generally smaller-scale operations of the pharmaceutical industry, acid gases are usually not an issue. The exception may be halogen acids (e.g. hydrochloric), which are formed as by-products of chemical synthesis or as products of the combustion of halogenated compounds.

AIR POLLUTION CONTROL MEASURES

Information on air pollution control measures that apply to pharmaceutical processes has been divided into the following three subsections: emission control considerations, process design considerations, and operational practices.

Emission control systems are commonly used to reduce emissions of VOCs (including odours) and particulates. Because some VOCs and particulates may contain toxic components (HAPs), these control systems are also effective in reducing emissions of air toxics. Indeed, air-toxics control practices have been followed in the industry for many years as a result of the biologically active and high-value materials handled by some operations. 'Air emissions characterisation' VOC emissions are of primary concern at pharmaceutical facilities and are thus our focus here.

The primary focus of the remainder of this discussion is the control of VOC emissions.

A variety of control systems are available for use in the pharmaceutical industry, depending on the particular application and on process operating parameters.

A listing of the types of controls that are in use or available for use at pharmaceutical facilities is presented below.

Control category	Available control systems
Add-on controls	Condensers
	Absorbers
	Adsorption units
	Thermal destruction
	Vapour containment
Process design	VOC minimisation
	Solvent substitution
Operational practices	Cleanup operations

Information is presented in the following on the use of these pollution control concepts: add-on controls, process design, and operational practices.

Emission Control Considerations

Becauses of the batch nature of many pharmaceutical process unit operations, it is common for critical emission parameters to vary throughout a process operating cycle. These parameters include temperature, pressure, flow rate, VOC composition and concentration, and particulate loading.

To control these variable emissions streams effectively, emission controls must be designed to handle both peak and nonpeak flow conditions. Additionally, the controls must be designed to be started up and stopped numerous times over a year and to be effective over short operational cycles. Consequently, it is very important that control equipment be sized correctly and designed to handle a range of operating parameters and conditions. Add-on controls available for use at pharmaceutical manufacturing facilities include condensation, absorption, carbon adsorption, thermal destruction, and vapour containment.

Condensation

The principle of operation of a condenser in emission control service is that a gas stream containing a (VOC can be cooled to below the saturation temperature ('dew point'), forming a second phase consisting of the VOC liquid. Separating the liquid from the gas (e.g. by gravity in the condenser combined with a demister pad to trap entrainment) removes a fraction of the VOC from the emission stream. The amount of VOC that remains in the gas stream as a vapour is a function of the temperature and the vapour–liquid equilibrium, of the VOC species. In general, the lower the temperature, the lower will be the VOC content in the exit gas stream. In theory, any desired level of VOC removal (control efficiency) approaching 100 per cent is possible with a condenser operating at a low enough temperature. There are some very real practical limitations, however.

Low condensation temperatures are required to control high-volatility VOCs or low concentrations of VOCs with any significant degree of volatility. In the latter case, most of the cooling duty of the condenser is to cool the non-condensible gas to the saturation temperature. Because the heat-transfer coefficient is low under these circumstances, the required heat-transfer area is large, increasing the cost of the condenser.

Producing low temperatures requires energy to drive the refrigeration system. Lower temperatures have progressively higher energy consumption (and thus cost) per unit quantity of heat removed. One scheme to raise the required condensation temperature is to compress the vent stream to a higher pressure so that the dew-point temperature will be raised. This approach is particularly useful where high control efficiencies of low-boiling-point materials (e.g. methylene chloride) are required. It is sometimes possible to raise the dew-point temperature sufficiently so that the problem with water vapour to be described can be avoided.

The presence of water vapour in the emission stream along with the VOC is common in actual practice. If room air sweeps or air purges are used, as is often the case in pharmaceutical manufacture, normal humidity will be present. If the process contains water as well as organic solvents, then even isolated vent streams will contain water vapour. Even conditioned 'plant air' has traces of water, since the design operation of typical plant air dryers is to lower the water content to a dew point of only about $-40°C$ ($-40°F$). If a gas stream containing water passes through a condenser that is operating below the freezing point, frost (rime) will form on the tubes, thereby effectively decreasing the capacity of the condenser by blocking efficient heat transfer.

Low-temperature condensation of volatiles from a gas, stream containing water can be accomplished with a series of two or more condensers with decreasing temperatures. The first condenser(s) removes the water vapour, while successive, lower-temperature units remove the organic compounds. A concern with this scheme is that the condensate containing the water may also have a substantial organic content (by condensation) and thus may pose a water pollution control problem.

Because a vent condenser can be a modest cost-control device, it is often recommended. Based on the foregoing discussion, there are clearly some situations that are well suited to condensation. For example, when there is little or no water vapour in the stream and a reasonable concentration of a medium-or high-boiling-point VOC (10 per cent or more), the use of a condenser is particularly appropriate. When the VOC is a low-boiling-point material, when there is water vapour present, or when the VOC concentration is low (less than 1 per cent) the applicability of condensers is marginal, and multiple units operated at low temperatures may be required. When the VOC concentration is quite low (at the fractional percent level), condensers are probably not appropriate, as the achievable level of control will be unacceptably low.

Absorption

Absorption is the selective transfer of one or more components of a gas mixture (solute) into a solvent liquid. For any given solvent, solute, and set of operating conditions (pressure and temperature), there is an equilibrium ratio of solute concentration in the gas mixture-to-solute concentration in the solvent. If the solute concentration is higher than the equilibrium value, there is a 'driving force' for mass transfer of that solute into the solvent, leading to its removal from the gas stream. If the equilibrium concentration in the gas is low compared with the concentration in the liquid solvent, the solute is said to have high solubility. Conversely, if a high concentration of the solute in the gas phase is in equilibrium with the liquid, the solute is said to have low solubility.

The apparent solubility of a solute species may be enhanced if there is a chemical reaction between it and components of the solvent solution. This reaction has the practical effect of minimising the concentration of the actual solute species in the liquid by changing the solute species into another, more soluble species.

In emission control service for VOC removal, the solvents are chosen for their high organic solubility. These solvents often include water, high-boiling-point mineral oils, and sometimes-aqueous solutions of sodium carbonate and sodium hydroxide.

Emission control devices that are based on the principle of absorption are commonly called scrubbers and include spray towers, venturi scrubbers, packed columns, and trayed (plate) columns. The basic design concept of all these devices is to provide the most economical means of achieving a high degree of gas-liquid contact in order to promote the desired mass transfer. The degree of control achieved (i.e. the fraction of the solute that is transferred from the gas to the liquid) is a function of residence time, mass transfer area, and physical and thermodynamic properties of the VOC species involved.

Spray towers are inherently simple devices consisting of an open column up which the gas stream flows and a group of atomisation nozzles to distribute the solvent as a spray of fine drops. Spray towers are versatile in that particulate matter also can be removed, but they have low mass transfer coefficients and typically are used only with high-solubility gases.

Venturi scrubbers, which are slightly more complex mechanically than spray towers, have a high degree of gas-liquid mixing and high particulate removal efficiency, but have relatively short residence times that limit the mass transfer efficiency. They are typically used only with high-solubility gases. For controlling organic compound emissions, packed and trayed columns are generally preferred. Packed columns are usually more economical than trayed columns for small diameter (less than 0.6 metre) units and can be readily designed for low-gas-phase pressure drop and for handling corrosive materials and liquids that tend to foam or plug.

Trayed columns are preferred for larger-diameter units, where internal cooling is required or where low liquid flow rates would be insufficient to wet the packing.

Scrubbers have been applied particularly to control water-soluble inorganic gases (e.g. sulphur dioxide, hydrogen chloride, hydrogen sulphide, and ammonia) in airstreams using water as the solvent. In the case of acid gases, using an alkaline scrubbing medium increases the effectiveness even more.

In the pharmaceutical industry, scrubbers are also used to remove VOCs. Solvent selection, equipment design parameters, and operating conditions all determine the control efficiency. To absorb organic compounds that have relatively high water solubility (e.g. most alcohols, organic acids, aldehydes, ketones, amines, and glycols), water is the preferred solvent. For organic compounds with low water solubility, another organic liquid (usually one with low vapour pressure) may be chosen. However, in the case of low solubility, water may still be selected, and the equipment will be designed to handle the high flow rate required.

Increasing the depth of the packing or the number of trays increases the control efficiency by increasing the amount of mass transfer area. Decreasing the operating temperature may result in a more favourable gas-liquid equilibrium and achieve higher control efficiency with a given piece of equipment.

Absorption can be used effectively with a wide range of concentrations, ranging from several per cent to as low as 200 or 300 ppm. Control efficiencies typically range from 60 to 95 per cent or better. However, because of the hydraulic behaviour of packing and trays, a relatively steady gas flow is required to maintain good efficiency. Turndown ratios of approximately 50 per cent are standard, thus limiting the ability of absorption to handle the intermittent flow of vents from batch processes.

Adsorption

The principle of operation for adsorption and the application of adsorption to VOC removal from gas streams are discussed below.

Principle of operation

Gas adsorption is the selective transfer of one or more components (solutes) of a gas mixture onto the surfaces of the pores of a microcrystalline solid sorbent. The selectivity is most pronounced in a monomolecular layer next to the solid surface, although selectivity may persist to a height of three or four molecules. Both natural and synthetic materials are suitable for use as sorbents if they have a microcrystalline structure. The large number of pores in these materials results in extremely high surface areas (e.g. more than a square kilometer per kilogram, or 0.2 square mile per pound, of solid in some instances). Adsorbents in large-scale commercial use include carbons, silicates, aluminas, and aluminosilicates (molecular sieves).

Generally, adsorption is reversible: The effective adsorption capacity (performance) of a sorbent for a particular solute increases with the concentration of the solute in the gas stream and decreases with increasing temperature. This means that the sorbent may be repeatedly 'regenerated'; that is to say, the solute can be desorbed by passing a lower-concentration or higher-temperature gas stream over the bed. Reversible adsorption is desired for VOC removal because it allows for reuse of the adsorbent. However, in some situations, the adsorption of a solute from a gas is irreversible. The capacity of a solid sorbent is nearly constant and independent of the gas stream composition, but because the solute is so tightly bonded to the sorbent, it cannot be reused.

Activated carbon is the preferred absorbent material for removing VOCs from gas streams. It has a high affinity for nonpolar compounds, with an even greater capacity for high-molecular-weight materials. The high-molecular-weight compounds can pose some problems, as they tend to be irreversibly adsorbed and effectively decrease the capacity of the carbon in a reversible system. A further problem is that high-molecular-weight compounds can displace lower molecular-weight materials already adsorbed.

The performance of a particular type of activated carbon for a particular solute species must be determined from equilibrium behaviour. Where only one species of solute is present in the gas stream, the equilibrium behaviour is conveniently represented by a simple curve that plots solute concentration in the solid phase as a function of solute concentration in the gas phase. These curves are usually only valid at a single temperature and are known as isotherms. Isotherms for common organic vapours being adsorbed on activated carbon are available in the open literature and from carbon suppliers. They may represent strictly experimental data or, more commonly, experimental data fitted to established algebraic formulas (e.g. Langmuir or freundlich isotherms). In some rare cases, they may represent purely theoretical calculations of equilibrium based on the molecular statistics of the solute–surface interactions.

Application to VOC removal from gas streams

In designing a carbon adsorption pollution control device in which the inlet VOC concentration is known and a certain outlet concentration is desired, it is possible to determine from the isotherm and material balance calculations how much carbon will be required. The process design is finalised by applying experience factors in terms of 'ageing' of the carbon beds (i.e. loss of capacity with time as a result of pore plugging or some other problem) and safety factors. The desired outlet concentration of the solute in the gas phase is used to determine the concentration of the solute on the activated carbon. When the entire mass of carbon reaches the 'saturation' level, the solute no longer will be adsorbed from the gas stream to the desired level, and 'breakthrough' is said to have occurred. The carbon then must be either regenerated (in the case of reversible adsorption) or disposed of (in the case of irreversible adsorption).

To allow vent streams or other gas streams containing VOCs to come into contact with the activated carbon, the carbon granules are usually arranged in a fixed-bed arrangement in either a vertical or horizontal cylindrical vessel. Small units are manufactured with the carbon already in place (e.g. a canister) with the intent that the entire unit will be replaced when the capacity limit of the carbon is reached. The supplier may regenerate the unit, or it may be disposed of. Larger units are constructed so that the carbon granules are loaded after installation of the vessel. In these units, the carbon may be periodically regenerated in place, or it can be removed and shipped off to a central location for regeneration on a contract basis.

Fixed-bed units are operated 'batchwise'; that is, the carbon is gradually saturated with the VOC and must be periodically regenerated, usually in place. This regeneration in place is done by having two or more beds installed and manifolded together so that one bed is in use while the other bed is undergoing regeneration. For very large adsorption operations, 'continuous' processes have been designed in which the carbon bed is continuously moved (by a fluidised bed, conveyor belt, or other mechanical means) from an adsorption zone through a regeneration zone and back to the adsorption zone. Continuous adsorption generally does not apply to the smaller scale operations at pharmaceutical plants.

Activated carbon beds in VOC removal service usually are regenerated using steam. The steam serves as a gas stream with zero concentration of the VOC solute (so that the solute tends to be desorbed) and acts to raise the temperature of the bed so that the equilibrium for desorption will be favoured. A much smaller volume of steam is required for regeneration compared with the volume of gas that is treated during the adsorption portion of the cycle. Following desorption, the steam is condensed (further reducing its volume) and sent to a waste-water treatment system to remove the organic content. In some cases, depending on the species and loading on the carbon bed, the VOC concentration in the water is sufficient such that two liquid layers form with the steam condensate. The second layer, with the higher organic compound concentration, can be readily decanted and either recycled or disposed of in a more efficient manner than waste-water.

In the pharmaceutical industry, because of the final product purity requirements, there are limited opportunities for in-plant recycling of recovered solvents. However, where practiced, recovery of a separate organic stream is desirable, since there may be opportunities to sell recovered solvents to another industry or to incinerate them for in plant heat recovery.

Activated carbon adsorption is currently in use in pharmaceutical plants to remove VOCs from vent streams, back up condensers, and control odours. Removal efficiencies of 95 per cent are common with values as high as 98–99 per cent seen in selected cases. In odour-control applications, the malodorous organic compounds are often fairly large molecules, and thus carbon adsorption may be substantially irreversible. In these cases, the used disposable carbon devices (such as canisters) may be favoured. A related issue is the problem of displacement: If the carbon bed is used to adsorb the larger molecules, it is unlikely that it also will be very effective at simultaneously removing the lighter-molecular-weight compounds. Thus, the removal efficiency of the lighter-weight compounds will be poor.

It should be noted that potential fire hazards are associated with using carbon adsorption on air vents or vent streams containing oxygen and flammable hydrocarbons. Whereas the gas streams may be sufficiently dilute in flammables as to be well below the lower flammability limit, the carbon bed concentrates the flammable materials, so that a potentially flammable mixture may be formed inside the bed. Furthermore, the carbon itself, with a high surface area, is combustible. For these reasons, special precautions for early fire detection or bed cooling may be required in some instances or an alternative to carbon adsorption may need to be used.

Thermal destruction

There are four main types of thermal destruction systems in general use: flares, boilers and process heaters, thermal incinerators, and catalytic (oxidisers) incinerators.

Flares

Flares burn flammable or combustible vapours with an open flame and can be either tower or ground mounted. Nozzle design, steam injection, and air injection are some of the design aspects used to enhance destruction efficiency. Flares are well suited for use with gas streams having a significant fuel value (more than 7.45 MJ/nm^3 or 200 Btu/scf) and are particularly well suited for handling a large, intermittent flow such as might occur from an emergency vent header. They have only limited capability for handling chlorine-containing compounds owing to corrosion of the burner elements and the lack of postcombustion emission control devices (for acid gas removal). The use of flares in the pharmaceutical industry for emission control is very limited; more typical applications are in petroleum refining or large chemical complexes.

Boilers and process heaters

Boilers and process heaters represent an opportunity to burn a waste stream while simultaneously recovering the stream's heating value. Typical applications are combusting a high-fuel-value liquid waste stream (uncommon in pharmaceutical operations) or using a pollutant-laden airstream as a source of combustion air. A fairly high-fuel-value vent gas stream is required for the boiler or process heater to function without added fuel. As with flares, boilers and process heaters operating with waste streams are more typically found in petroleum refining or large chemical complexes.

Thermal incinerators

Thermal incinerators are controlled combustion devices where fuel and air are added to a combustion chamber to maintain a high minimum operating temperature. Because of these high temperatures, thermal incinerators are able to handle dilute vent streams with high destruction efficiency. Gases with heating values below about 1.86 MJ/nm^3 (50 Btu/scf) can be handled with the addition of fuel. Combustion chamber temperatures range from 700° to 1300°C (1300°–2400°F). Packaged, single-unit thermal incinerators are available in a wide range of sizes and are able to handle flow rates ranging from 0.1 nm^3/sec (200 scfm) to about 24 nm^3/sec (50000 scfm).

The control efficiency achieved by thermal incineration is typically 98 per cent destruction or an exit gas concentration of 20 ppm by volume, whichever is less stringent. Thus, an inlet stream with a VOC concentration of 2000 ppm by volume or higher (which, with the typical 1:1 air dilution, becomes 1000 ppm by volume at the inlet to the combustion chamber) will have 98 per cent of the incinerator inlet VOC content removed. Inlet streams with lower VOC concentrations will result in outlet gas concentrations of 20 ppm by volume of unburned organics, but with lower destruction efficiencies. Thermal incinerators can handle chlorine and sulphur containing compounds but may require a scrubber on the outlet to control acid gases.

Due to the wide range of sizes available and the ability to handle a wide range of compounds in vent gas streams, thermal incineration is well suited for emission control in the pharmaceutical industry. It is, however, a capital- and energy-intensive approach.

Catalytic incinerators

Catalytic incineration is a variation of thermal incineration. A catalyst is used to promote oxidation of the inlet gas stream at lower temperatures than are required in standard thermal combustion. Catalytic units usually operate over a range of 320°–650°C (600°–1200°F), and these lower operating temperatures reduce energy requirements. Catalytic incineration units may be smaller than standard thermal incineration units, but the cost of the catalyst may tend to offset any potential savings in the capital investment. Two additional design constraints are that: (i) high organic concentrations may cause catalyst failure, and (ii) the catalyst may be poisoned by sulphur-containing compounds, heavy metals, and halogens.

The control efficiency of catalytic incineration is equivalent to that of thermal incineration, and a wide range of packaged unit sizes is available.

As with thermal incineration, catalytic incineration is well suited for emission control in the pharmaceutical industry.

Vapour containment

Transferring volatile organic liquids from delivery tank trucks and tank cars to storage tanks and from storage tanks to process vessels displaces gas from the headspaces that contain some fraction of VOC vapour. While these intermittent vent streams may be treated by one of the add-on control devices described above vapour containment may in some cases be the preferred option. In this technology, additional piping is provided so that as the liquid is transferred from the supply tank to the receiving tank, the displaced gas and vapour from the receiving tank are returned in a separate line to the supply tank. With a properly designed system, very little gas/vapour escapes into the atmosphere, and VOC emissions are minimised. This concept can be extended to filling reactors from drums by using a drum pump and vapour return line rather than picking up the drum and pouring.

Other unit operations that can be improved in terms of vapour containment are those designed to operate at atmospheric pressure. Because of the relatively unrestricted mixing of the process vapours with the atmosphere, reactor and condenser vents open to the atmosphere are difficult-to-control sources of emissions. An option is to operate them as closed systems, at a slight pressure or vacuum, so that the flow of non-condensible gas (air or inert gas) is restricted. This approach is likely to decrease emissions, and the restricted flow rate stream that occurs is easier to control.

Bulk quantities of volatile organic liquids typically are stored in atmospheric pressure tanks with vents to relieve excessive pressure and to break any vacuum that may form. However, diurnal temperature fluctuations will result in 'breathing losses' where the headspace vapour, approximately saturated with volatile organic vapours, is expelled when the temperature rises. When the temperature drops, fresh air (or an inert blanketing gas) is drawn into the tank, from which it may later be expelled with some degree of saturation. Installing 'conservation vents' raises, by a small amount, the pressure or vacuum threshold at which this event occurs, thereby reducing the emission stream volume. Isolated, especially high-pressure, rises or unusually low vacuums will still open the conservation vents, but normal daily fluctuation will not, so that the total volume vented to the atmosphere is substantially reduced.

Process design considerations

To use add-on controls, including vapour containment, a certain amount of process design must be undertaken. However, in the case of new products or revamps of existing processes, it may be possible to use a different pollution control concept—designing a process that has an inherently lower potential for air emissions. One strategy is to minimise the use of VOCs, to substitute lower-volatility (and/or less toxic) compounds, and to keep these compounds contained.

For example, in the case of drying emissions in pharmaceutical manufacture, a modified process design could use and transfer intermediates as solutions or slurries, rather than as dry products. Where solvents are required, a lower-vapourpressure solvent would tend to minimise evaporative losses and increase recovery efficiency when using condensers. With some processes, a reaction or formulation step traditionally conducted in a solvent possibly could be carried out in an aqueous medium. One ultimate goal may be the practice of a 'solventless' synthesis, where the reaction is conducted 'neat'. In this case, a customised reactor design may be required to handle potential heat transfer and rheological problems associated with the resulting more concentrated reaction mixture.

Although there are often physical or chemical reasons why a process design must incorporate a particular volatile compound, there are many other instances where, with ingenuity, VOC emissions can be minimised by eliminating the offending material altogether.

Another strategy in a modified process design is to minimise the transfer of materials from process vessels by having a particular unit perform more than one unit operation. An example is a filter-dryer combination. These units, increasingly being used in the pharmaceutical industry, are batch pressure filters that, instead of discharging a wet filter cake, remain closed and convectively dry the cake with a recirculating gas stream.

For maximum effectiveness with some add-on controls, process redesign may be required. An example is the use of closed-loop drying systems.

Acknowledging that a convective dryer is potentially a large source of emissions, a condenser, or a condenser combined with gas compression, is often a good choice. Instead of venting the drying gas, a further refinement of this approach is to recirculate the gas from the discharge of the condenser through the heater and back to the dryer. During the drying cycle there are, other than fugitives, essentially no emissions from the closed loop.

Operational practices

This last control concept deals with the potential for changing workplace practices in such a way as to minimise emissions. Much of the everyday operation of a pharmaceutical synthesis facility is governed by standard operating procedures (SOPs). With proper engineering and management review, the SOPs can sometimes be revised and modified to reduce emissions. One area where emission reductions are particularly possible is in the use of solvents in cleanup practices. Typical procedures may call for the filling and subsequent emptying and air drying of process vessels. The air-drying step results in significant emissions regardless of the vapour pressure of the solvent involved. As an improved SOP, it may be feasible to eliminate the drying step, leaving residual solvent present before the start of the next cycle. Alternatively, a low-VOC aqueous cleaner might be substituted.

A similar emission source involves flushing and blowing transfer lines (piping). Again, it may be possible to change the SOPs so that the lines are left filled with solvent rather than blown dry to the atmosphere.

MINIMAL NATIONAL STANDARDS (MINAS)

Objective

The Minimal National Standards (MINAS) are developed based on achievability and environmental requirements. The objective is to make a general approach for minimisation of pollution and good water quality management by achievable technology. General standards are prescribed irrespective of products and processes of manufacture.

Location specific standards may be prescribed by State Boards taking into account the characteristics of recipient system. The following compulsory and optional standards are stipulated:

Compulsory parameters	*Limiting concentration in mg/l except pH*
pH	5.5–9.0
Oil and grease	10
Total suspended solids	100
Bio-chemical oxygen demand (5 days at 20°C)	30
Bio-assay test	90 per cent survival after 96 hours. Test should be carried out with suitable species of fish at 100 per cent effluent concentration

Optional parameters   *Limiting concentration (in mg/l)*

Optional parameters	Limiting concentration (in mg/l)
Mercury	0.01
Arsenic	0.2
Chromium (hexavalent)	0.1
Lead	0.1
Cyanide	0.1
Phenolics (as C_6H_5OH)	1.0
Sulphides (as S)	2.0
Phosphate (as P)	5.0

Notes
1. The optional parameters are applicable to pharmaceutical manufacturing units depending upon the process and products. Formulators are exempted from the optional parameters.
2. State Boards may prescribe limit for chemical oxygen demand correlated with BOD limit.
3. State Board may prescribe limit for total dissolved solids depending upon uses of receipient water body.
4. Limits should be complied with at the terminal of the treatment unit before letting out of the boundary limit.
5. For the compliance of limits, analysis should be done in the composit sample collected every hour for a period of 8 hours.

Environmental requirements vary from one site to another. It is not possible to develop uniform environmentally acceptable standards for the whole country especially in case of waste-water discharge. The receiving body is most important to decide such standards. The concerned regulatory agency can modify the MINAS depending on the local conditions. While prescribing such standards, the following factors are to be noted:
1. Degree of dilution available in receiving system.
2. Protection of important biotic species.
3. Mean tolerance limit for pollutants to the identified biotic
4. Application factors in respect of mean tolerance limit.

Chapter 22

Sugar, Distillery and Fermentation Products

INTRODUCTION

Today, the sugar industry is the third largest industry in the country and ranks next only to the iron and steel, and cotton industries. The effluent from sugar industries contains a large amount of dissolved organic matter. This organic matter is readily decomposed by biological action. Consequently, discharge of these liquors to surface water causes serious damage to aquatic life. One good example that provides a cleaner technology scenario can be found in the sugar industry. India is one of the large sugar producing countries in the world and there are about 425 sugar mills in operation at present. This is a seasonal industry and operates for 6–8 months in a year.

SUGAR

Bagasse Utilisation

It is well known that sugarcane (Fig. 22.1) is the basic raw material for the extraction of sugar in India. After crushing of sugarcane, the juice is squeezed and the left-over cellulosic bagasse, which otherwise is a solid waste for the industry, is rich in fibre contents. Since, such a fibrous substance is a good raw material for the pulp and paper manufacturing industries, it can easily be sold to these industries.

This not only solves the solid waste disposal problems, (as disposal would lead to some extra cost), but also brings in some extra money. Use of sugarcane bagasse as a raw material in paper industries will certainly cut down the felling of a large number of trees and, ultimately, conserve our forest reserves. Since bagasse also possesses a reasonably good amount of fuel value, majority of industries use it as a substitute of coal in the boiler house for steam raising. This is a major step in the conservation of fuel resources. However, this process is not free from air pollution.

Press mud—a source of energy

The extracted sugarcane juice is then subjected to a series of purification processes to make it free from impurities, such as dirt, colour and organic matter. The purification process is termed as 'sulphitation process' whereby at raised temperatures, juice is reacted with lime solution and sulphur dioxide gas in a reactor and the contents are either vacuum filtered or press filtered. The filter cake, thus separated, is called 'press mud' and is rich in organic matter. As such, this component is again a form of solid waste which requires safe disposal. Due to its richness in organic matter, some farmers use it as manure and thus, can be sold to them.

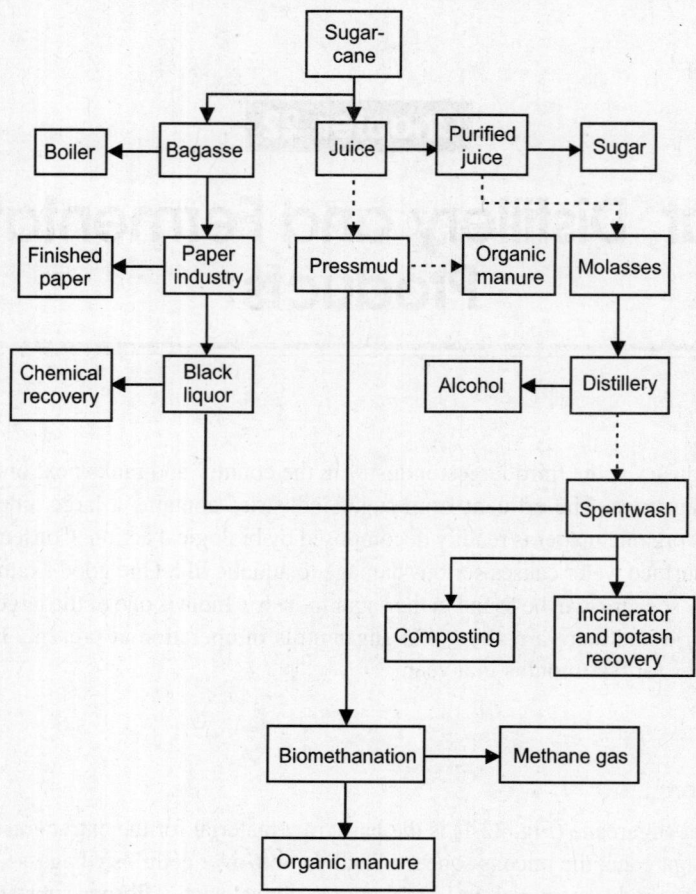

Fig. 22.1. Sugar industry—an example of waste utilisation

Recently, it has been recognised that press mud is a useful substrate for biogas production. Based on the studies conducted by the Department of Agriculture of the Annamalai University, Tamil Nadu, the Ministry of Nonconventional Energy Sources, selected Ugar Sugar Works Ltd. Karnataka, for a demonstration plant to generate methane gas out of press mud, and a biogas distribution system to 120 families residing in the factory premises. The results indicated that as high as 80 Nm^3 biogas/T of fresh press mud could be produced. The effluent slurry is also useful as a nutrient rich biofertiliser. This clearly reflects wealth recovery out of waste by-products.

Hot condensate

After the purification of the juice, it is transferred to evaporator pans for concentration prior to crystallisation of sugar. In the evaporator pans, the juice is vacuum evaporated so that its water content is reduced. These water vapours are, in fact, one of the purest forms of water when condensed, without any carry over of sugar, and can be used as boiler-feed water.

In this regard, the 'polybaffle entrainment catcher' in pan evaporation for sugar recovery provides an alternative over the conventional pan evaporation technique. Such water will not only reduce the scaling in the boiler tubes, but also conserve large quantities of freshwater to be used. In addition, this hot condensate needs far lesser amount of energy (when compared with the freshwater) to be transformed into steam for obvious reasons, resulting in money saving to the industry. More recently, it is emerging that if hot condensate is used to extract the last traces of sugar from the crushed sugarcane, not only does this leads to increased recovery, but it also saves large quantities of freshwater that is used otherwise. It also requires less heat energy for the sulphitation process. According to an estimate by the Central Pollution Control Board, in general, approximately 1500 to 2000 litres of water is used per ton of cane crushed. The recycling and reuse of hot condensate water can reduce this consumption to as low as 100 to 200 litres per ton of cane crushed. Proper housekeeping, periodic checking and maintenance of pipe joints, valves and glands can further reduce this. Ideally speaking, the industry can be brought down to zero effluent discharging; if not, the small quantities of waste-water can easily be treated in the end-of-pipe treatment, requiring least amount of energy and resources.

Molasses—The Key Raw Material for Distillery

The last step in the manufacturing of sugar is crystallisation, wherein sugar crystals are removed by centrifuging process and a dark brown viscous substance called 'molasses' is produced as a waste by-product. This contains very high concentration of organic matter and some sugars that escape crystallisation. As such, it is a highly polluting substance which can consume vast amount of energy and resources for treatment. But thanks to technology, molasses constitute the key raw material for the extraction of alcohol by distilleries. The left-out sugars in the molasses are fermented with the help of yeast into alcohol and subsequently collected through distillation process. As a matter of fact, once the molasses is sold to distilleries, not only are the pollutional problems of sugar industry eliminated, but it also earns revenues from the sale of its wastes.

Sugar Mill Wastes

In countries like Cuba, Jamaica and India, sugar is produced from sugarcanes, while in many other places, beetroots are used as the raw material. In India, most of the sugar mills are situated in the countryside and operate for about 4 to 8 months a year, just after the harvesting of sugarcanes. A large volume of waste of organic nature is produced during the period of production and normally they are discharged onto land or into the nearby water courses, usually small streams, practically without pre-treatment.

The condition becomes worse as the stream flow reaches a very low level and eventually enough dilution water is not available during the period of operation of the mills (early November to end of May or June). Putrefaction of the polluted stream water caused by the heavy discharge of organic waste, resulting in the odour nuisance near the sugar mills is a very common phenomenon.

In fact, all the concerned bodies, both sugar industry and pollution control agencies, are aware of these problems and are trying to find an economical means to check the nuisance created by sugar mill effluents.

Manufacturing Process

Sugarcane is normally harvested manually in India, which eliminates the carriage of soil and trashes to the factory. Sugarcane is cut into pieces and crushed in a series of rollers to extract the juice in the 'mill house'. The milk of lime is then added to the juice and heated, when all the colloidal and suspended impurities are coagulated; much of the colour is also removed during this lime treatment. The coagulated juice is then clarified to remove the sludge. The clarifier sludge is further filtered through filter presses and then disposed of as solid waste. The filtrate is recycled to the process, and the entire quantity of clarified juice is treated by passing sulphur dioxide gas through it. The process is known as 'sulphitation process'; colour of the juice is completely bleached out due to this process. The clarified juice is then preheated and concentrated in evaporators and vacuum pans. The partially crystallised syrup from the vacuum pan, known as 'massecuite' is then transferred to the crystallisers, where complete crystallisation of sugar occurs. The massecuite is then centrifuged, to separate the sugar crystals from the mother liquor. The spent liquor is discarded as 'black strap molasses'. The sugar is then dried and bagged for transport. The fibrous residue of the mill house, known as 'bagasses' may be burnt in the boilers, or may be used as raw-material for the production of paper products. The black strap molasses may be used in the distilleries. A flow diagram of the process of manufacturing in a typical sugar mill is given in the Fig. 22.2.

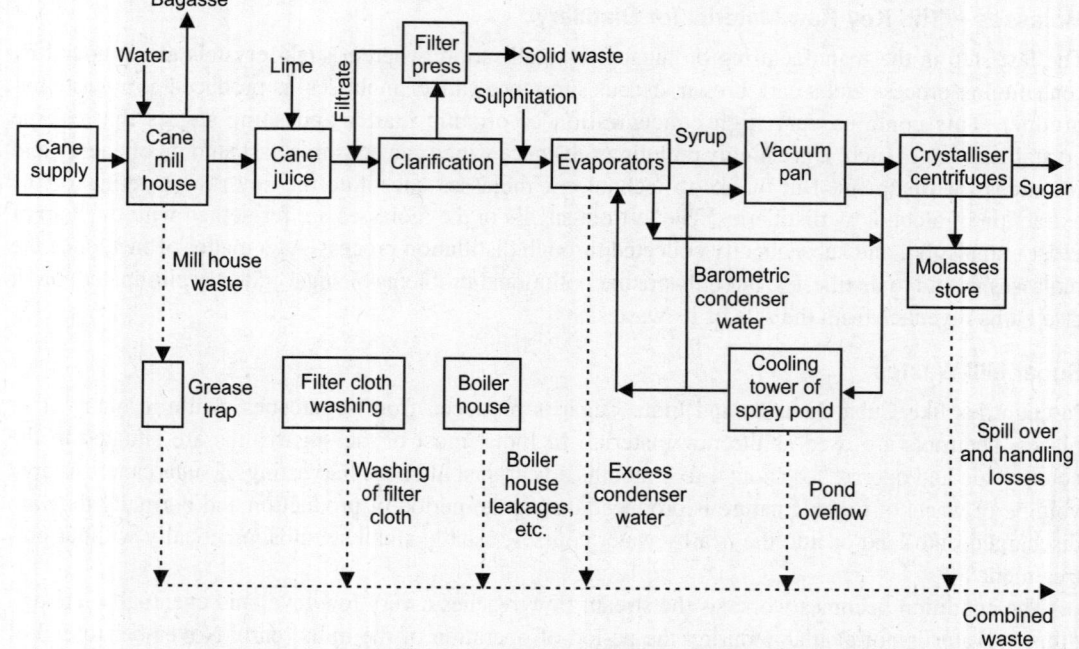

Fig. 22.2. Flow diagram for the sugar manufacturing process.

Sources of Waste-water and the Characteristics of the Wastes

Wastes from the mill house include water used as splashes to extract maximum amount of juice, and that used to cool the roller bearings. As such, mill house waste contains high BOD due to the presence of sugar, and oil from the machineries. The filter cloth, used for filtering the juice, needs occasional cleaning. The wash water thus produced, though small in volume, contains high BOD and suspended solids.

A large volume of water is required in the barometric condensers of the multiple effect evaporators and vacuum pans. The water is usually partially or fully recirculated, after cooling through a spray pond. This cooling water gets polluted as it picks up some organic substances from the vapour of the boiling syrup in the evaporators and vacuum pans. The water from spray pond when it overflows, becomes a part of the waste-water, and is usually of low BOD in a properly operating sugar mill. But because of poor maintenance and bad operating conditions, a substantial amount of sugar may entrain in the condenser water; this polluted water, instead of being recirculated, is discarded. These discharges contribute substantially to the waste volume and moderately to BOD in many sugar mills.

Additional waste originates due to the leakages and spillages of juice, syrup and molasses in different sections, and also due to the handling of molasses. The periodical washings of the floor also contribute a lot to the pollution load. Though these wastes are small in volume and are discharged intermittently, they have a very high BOD. The periodic blow off of the boilers produce another intermittent waste discharge. This waste is high in suspended solids, low in BOD and is usually alkaline.

Characteristics of wastes from the different sections of common sugar mills and that of the combined waste are given in Table 22.1. Composition of the waste and its volume varies widely from mill to mill, depending upon the availability of water.

Table 22.1. Characteristics of sugar mill wastes.

Characteristics	Mill-house waste	Filter cloth washings	Condenser water	Boiler house and floor washings	Combined waste (excluding condenser water)
Rate of flow litres per ton of cane crushed	730	360	1640	230	–
pH	6.7	9.5	–	7.2	4.6–7.1
COD, mg/l	–	–	–	–	600–4380
BOD, mg/l (5-day 20°C)	210	1765	–	5150	300–2000
Total solids, mg/l	1760	6970	–	5130	870–3500
Total volatile solids, mg/l	–	–	–	–	400–2200
Total suspended solids, mg/l	910	4000	–	120	220–800
Total nitrogen mg/l	–	–	–	–	10–40
COD/BOD ratio	–	–	–	–	1.3–2.0

Effects of the Waste on Receiving Water

The fresh effluent from a sugar-mill decomposes rapidly after few hours of stagnation. It has been found to cause considerable difficulties when their effluent gets an access to water courses, particularly the small and non-perennial streams in the rural areas. The rapid depletion of oxygen due to biological oxidation followed by anaerobic stabilisation of the waste causes a secondary pollution of offensive odour, black colour, and leads to fish mortality. No question of the discharge of this waste into the sewers arises, as most sugar mills are situated in the unsewered rural areas.

Treatment of the Wastes

Like any other industry, the pollution load in sugar mills can be reduced with a better water and material economy practised in the plant. Judicious use of water in various plant practices and its recycling, wherever practicable, will reduce the volume of waste to a great extent. Volume of mill house waste can be reduced by recycling the water used for splashing. Dry cleaning of floors or floor washings using controlled quantity of water will also reduce the volume of waste to certain extent. The organic load of the waste can only be reduced by a proper control of the operations.

Overloading of the evaporators and the vacuum pans, and the extensive boiling of the syrup lead to a loss of sugar through condenser water; this, in turn, increases both volume and strength of the waste effluent. Disposal of the effluent on land as irrigation water is practised in many sugar mills, but it is associated with odour problems.

The reasonable COD/BOD ratio of mill effluents indicate that the waste is amenable to biological treatment. However, generally it is found that the aerobic treatment, with conventional activated sludge process and trickling filters, is not too efficient, even at a low organic loading rate. A maximum BOD reduction of 51 per cent was observed in a pilot plant study at Kanpur, where both trickling filter and activated sludge process were tried.

In view of the high cost of installation and supervision of the treatment units, and the seasonal nature of the operation of this industry, it is generally observed that the conventional aerobic treatment will not be economical in India. Anaerobic treatment of the effluent, using both digesters and lagoons, have been found to be more effective and economical. A BOD reduction of about 70 per cent was observed in a pilot plant study with an anaerobic digester, where BOD loading was 0.65 kg/m^3/day with a detention time of 2.4 days at a controlled temperature of 37°C. In the same study, the BOD reduction was found to be 60 per cent in an anaerobic lagoon, where a BOD loading of 0.23 kg/m^3/day and detention time of 7 days was provided. In another laboratory study, a BOD reduction of 88.9 per cent was observed in an anaerobic digester, with a detention time of 2 days and controlled temperature of 36 ± 1°C, at a higher BOD loading of 1.79 kg/m^3/day. The effluents of the anaerobic treatment units are found to contain sufficient nutrients (nitrogen and phosphorus). As such, further reduction of BOD can be accomplished in aerobic waste stabilisation ponds.

Where sufficient land is available, two-stage biological treatment, with anaerobic lagoons followed by aerobic waste stabilisation ponds, is recommended for Indian conditions. The mill effluent, however, is to be pre-treated primarily in bar screens and grease trap. With the process parameters as indicated in Fig. 24.2, it is expected that the BOD reduction in the anaerobic process will be in the order of 60 per cent, while overall BOD reduction may be in the order of 90 per cent.

DISTILLERY INDUSTRY

Process

The distillery industry uses sugar cane molasses, cereal and other agro products to produce alcoholic beverages. Typical primary products of a grain alcohol distillery include the beverages: whiskey, gin and vodka. Cooked slurries of ground whole grains, called 'mash', are fermented, without prior filtration, using a yeast that operates rapidly at about 95°F. Ratios of the various grains are varied within specified limits and malted barley furnishes saccharifying enzymes. Complete conversion of the grain starch to alcohol is sought in a fermentation time of 2–3 days. The alcohol content of a finished fermentation is about 7 per cent.

Small quantities of amyl alcohol (fusel oil), lactic acid and glycerol are formed during fermentation. The slurry is sent to the stills without separation of the solid material. Extraction of botanicals and redistillation complete the gin process. Whiskey attains its colour and significant parts of its flavour from oak wood while ageing in charred barrels. By-product feed recovery includes the use of screens, centrifuges, vacuum evaporators and rotary and drum dryers.

Distilleries use different kinds of raw material such as sugar cane juice, sugar cane molasses, sugar beet molasses, wine or corn, for the production of alcohol. The cycle of raw materials starts from the farm, through a product formation, and fermentation, up to distilling, for production of alcoholic beverages, and finally spent wash. The sugar manufacturing process broadly involves the extraction, clarification and concentration of sugarcane juice. Finally, the concentrated juice is crystallised and dried. The manufacturing processes produce molasses, bagasse and press-mud as waste. The manufacturing process in a distillery involves dilution of molasses with water, followed by fermentation. The product is then distilled to obtain rectified spirit or neutral alcohol. The distilling process begins with raw material like grains, which are initially ground in order to make a mash, at which point the starch is converted into sugar. This process then undergoes: (i) grinding, (ii) saccharification of the starch, (iii) fermentation process, and (iv) distillation.

Treatment of Distilling Industry Effluents

Most of the distilleries are facing the environmental issue of treatment and disposal of its spent wash. A wide range of technologies have been tried for the treatment of spent wash, however, none of these methods have been found to be effective and economically viable, in order to achieve the standard norms set by the Central Pollution Control Board, Government of India.

Primary treatment

Primary treatment methods selectively remove materials, which could interfere with the physical operation and subsequent treatment processes. The processes are based on exploitation of the physical properties of the contaminants, and are generally used at the initial stages of effluent treatment. It helps in improving the treatment efficiencies by reduction of surface area.

Screening, flow equalisation, comminution, mixing, flotation, flocculation and sedimentation are the methods used during the initial stages of effluent treatment. Chemicals are also used in conjugation with physical treatment, in which sedimentation is performed, by the addition of coagulants and other additives such as alum, lime, ferric chloride, activated charcoal etc. Different chemical treatment methods were investigated, based on chemical properties of the pollutants, and as a result, chemical treatment with Fe(III) and anionic polymer, followed by H_2O_2 +Fenton reagent was selected as the best option because of its performance and viability.

Strong oxidants like ozone and hydrogen peroxide were used. Though they result in decolourisation, they are not accepted at present, because of economic reasons. It has been observed that inorganic flocculants such as iron sulphate, iron chloride and aluminium sulphate do aid the separation of colouring matter. Oxidation by ozone combined with hydrogen peroxide is found to be very effective in the degradation of some aromatics present in waste-water and removal of colour. Due to generation of sludge, accumulation of metals, cost, and related contaminants, chemical methods are not successful.

Secondary treatment

Secondary treatment is also known as biological treatment, in which soluble and colloidal form of organic matter is removed. Both aerobic and anaerobic micro-organism is used in secondary treatment. Anaerobic digestion is the most suitable option for the treatment of high strength organic effluents. The presence of biodegradable components in the effluents, coupled with the advantages of the anaerobic process over other treatment methods, makes it an attractive option. In addition, a modification in the existing reactor designs for improving the efficiency of digestion has also been suggested.

Anaerobic and aerobic treatment systems can remove most of the biologically removable organics, CODs and colour. Biologically treated distillery waste-water is nonbiodegradable. Anaerobic treatment of effluents from different alcohol-producing industries over a long-term period is performed. Since most of the COD and colour in biologically treated distillery waste-water is nonbiodegradable, identification and optimisation of biotechnological treatment methods are a necessity today.

Tertiary or alternative treatment

A variety of treatment methods and strategies, like thermal pre-treatment, wet air oxidation, concentration-incineration, anaerobic treatment, etc. have been suggested or tested for the treatment of distillery waste-water. As individual schemes, these are either incomplete, impractical or unviable. Thus, there is an urgent need to assess the possibility of combining the available partial treatment schemes, for the complete treatment.

Wastes

Typical wastes from a beverage grain alcohol distillery, are shown in Table 22.2, based on BOD data. The total population equivalent per bushel of incoming grain is about 3.0. Good correlation exists between in-plant survey totals and periodic effluent surveys and between these surveys and production efficiency accounts. The portions of waste originating from 4 groups of unit processes are shown in the table. Each waste entry includes start-up, normal operating and cleanup activities. Substrate preparation (cooking) and fermenting operations contribute 12.3 per cent of the plant's BOD.

Table 22.2. Waste distribution in a grain distillery.

Total effluent: kg of BOD per bushel	0.51
Population equivalents per bushel	3.0

Operation	Per cent of total BOD
Cooking and fermenting	12.3
Distilling	1.8
Power plant	2.5
Feed recovery	83.4

Distillation of the fermented slurry adds 1.80 per cent and powerhouse losses account for 2.5 per cent. Losses from maturing, blending and packaging the final products of a grain distillery, in contrast to brewery practice, are so small as to warrant omission from this summary. Wastes amounting to 83.4 per cent of the total come from operations performed on residues from the stills after the desired primary products are obtained. This sizable portion of the waste continues to present a challenge for further waste abatement measures in modern grain distilleries.

A major purpose of a summary such as Table 16.1 is to indicate the best possibilities for further improvement. For example, the subtotals that make up the figure of 12.3 per cent for substrate preparation and fermenting wastes, though not shown in Table 16.1, are useful. The 12.3 per cent breaks down as less than 2 per cent resulting from routine operations of cooking and fermenting, over 7 per cent from entrainment from flash cooling the mash and about 3 per cent from equipment cleanups.

Improvements made recently have included lowering the operating and cleanup losses by a system of improved spray washers, so that prepared fermentables may go with greater efficiency to fermentors rather than to cleanup wastes. Other possible improvements would be better reclamation of weekend scourings and design and operating changes to reduce flash cooler losses. The cost of equipment changes and the possibility of loss of cooling efficiency by changing vacuum cooler cycles enter the picture. The value of additional by-products recovered from better retention of scouring would not be great enough to control the decision and the final question would relate more to the cost of waste abatement than to additional by-product profit.

Although powerhouse wastes do not form a significant portion of grain distilling wastes, the nature of many power plant effluents is such as to cause possible errors in waste analysis. Powdered coal and fly-ash washings, resulting from air pollution abatement equipment, have a much greater COD than BOD. Useful correlations on composite waste streams are possible only if power plant washings are excluded.

By-product recovery

Further improvement in stillage recovery may be possible within the narrowing limits of low profit or nonprofit expenditures. Subtotals within the 83.4 per cent of total BOD shown in Table 16.1 include about 50 per cent from stillage evaporator condensate, 32 per cent from dry entrainment from feed dryers and coolers and 1 per cent from cleanup. Pilot tests have indicated that changes in control of the steam and vacuum may reduce evaporator losses. However, all abatement measures pointed out by in-plant surveys have not been developed so simply.

By-product reclamation in order to lessen industrial waste is not novel, but it is noteworthy that reclamation procedures, in an industry previously identified with heavy organic wastes, now permit operation of large installations with minimum emphasis on waste treatment. The major potential waste of the grain distilling industry is the stillage or residual portion of substrate not used by the fermentation process.

At first, only the suspended solids trapped by screening were reclaimed. This product, known as distillers' dried grains, decreased the population equivalent of the waste from 53 to 43. Heavy BOD still remains in the soluble matter which passes through the screens. Concentration of the screened effluent in vacuum evaporators has proved successful. Centrifuging or filtering prior to evaporation, to remove residual suspended matter, permits evaporation to 50 per cent solids concentration.

This placed two more products on the market. Distillers' dried solubles is prepared by drum or spray drying of the evaporated concentrate. If the concentrate is mixed with the distillers' grains for drying, the resulting mixed grains and solubles product is called distillers' dark grains.

A portion of the screened effluent has also been used as the base or medium for a secondary microbial process which synthesises riboflavin. On drying, this yields a feed rich in riboflavin.

Breweries

In-plant methods of waste abatement have been applied successfully in the brewing industry, using about the same techniques as in the grain distilling industry.

Processes required for the primary products of the brewery have yielded by-products different from distilleries, but the waste reduction accomplished has, been a factor in the trend to large economical brewing plants.

In making beer, a water extract is made from the malt, adjunct materials and hops. This wort is filtered and inoculated with a yeast that works slowly in a cool environment. After fermenting and blending over a period of a few weeks, the product is clarified, the carbonation is adjusted and the beer is bottled or canned. Selection of raw materials and control of fermentation is such that certain residual carbohydrates and protein flavour components remain in the product.

Variations in the amount of substrate preparation done on the premises, cleanup routines and the amount of spillage of beer are the most significant variables. Some surveys have reported up to 75 per cent of brewery BOD losses originating from spillage and washing operations. Nutrient values are different from distillers' feeds because the brewing process does not take the grains through the fermentation.

Wineries

The wine industry presents some of the most difficult or complicated waste control problems in the fermentation industries. The problems stem from the traditional production patterns of the industry, many of which are based on European and South African wine industry practice. To meet the current market, many types of wine are needed, each requiring a different process.

Three different types of waste are recognised at wineries. These are not necessarily produced at the same plant; indeed, trends to relocate and consolidate parts of the wine production are steps in the abatement of waste. Some disposal possibilities involve supplying the wastes of one group to another, but most of these possibilities would still be available if the operations were at separate locations.

The first waste results from the pressing of fermentable juice from the grapes, leaving a canning industry type of waste called 'pomace'. Of course, wastage of fermentable juice at these installations is also a potential source of BOD. Pomace corresponds to the intentional wastes from substrate preparation noted in discussing other fermentations. A second type of waste comes from the fermenting, racking decanting, blending and bottling of the wines. This group of wastes may be likened to brewery wastes in that yeast residues and the spillage of fermentables and final product are included.

A third waste comes from the brandy plant and may be compared with the dealcoholised stillage of grain or molasses alcohol distilling plants. Separate study of these three types of waste is necessary in contemplating either reclamation or treatment. As standards of practice develop, along with more firmly set pollution limits three different criteria will be necessary for these three groups of unit operations.

As dumped from the presses, pomace contains 30–40 per cent water and on this wet basis amounts to about 12 per cent of the original grape weight. High fibre content lowers its potential value for animal feed and fertiliser. Residual fermentables vary with the thoroughness of the pressing.

Hard pressings yield a heavier wine or a raw material for brandy and enter a different market than first-press juice. Other types of wine call for fermentation 'on the skins', which brings a pomace residue problem to the fermenting section of the industry. Pomace, because of its residual sugar content, may be handled by fermenting and subsequent extraction in a pomace brandy still. A promising sequence for waste control calls for as complete as possible a pressing, then extraction of the residues with several recyclings of hot water. The extract would be available for various types of fermentation. Fibre content can be reduced by removal of seeds and extraction of the seed oil, if economic boundaries permit. Dried pomace with seeds contains about 35 per cent fibre; the seeds contain over 50 per cent fibre.

The lees (residues from fermentation) resulting from winemaking are available for distillation to brandy and can be readily transported from winery to brandy plant if these are at different locations. Spillage losses in the fermentation division are 'housekeeping and operating problems'. Tartrate (cream of tartar) is a marketable by-product of winery operation and has been obtained from each of the three classes of wine waste, including stillage when the market demanded it.

The best quality of tartrate, reclaimed with the least effort, crystallises in wine maturing vessels as a white deposit called 'argol'. Tartrate from this source enters the domestic market profitably, although tartrate from other wine wastes now meets difficult competition from imports from other wine producing countries. Removal of tartrate reduces the BOD of winery waste-waters directly.

In evaluating brandy stillage, the source must be considered. Brandy can be made from wine, which forms a low solids substrate; from lees, with a medium amount of residue sustained; or from pomace, which gives a ratio of BOD to alcohol produced similar to grain stillage. Mixtures of these substrates may also be used for brandy production.

Cleaner Technology for Distillery Industry

In distilleries, after alcohol is separated, the remaining contents are called 'spentwash'. This is a non-toxic, biodegradable effluent, with BOD values as high as 40,000 to 50,000 mg/l and, obviously, needs huge resources for treatment and disposal. The cleaner technology has also provided a number of ways to recover wealth from the spentwash. The first and the foremost way is 'biomethanation' of spentwash.

The flow sheet for the complete treatment of the sugar mill waste is shown in Fig. 22.3.

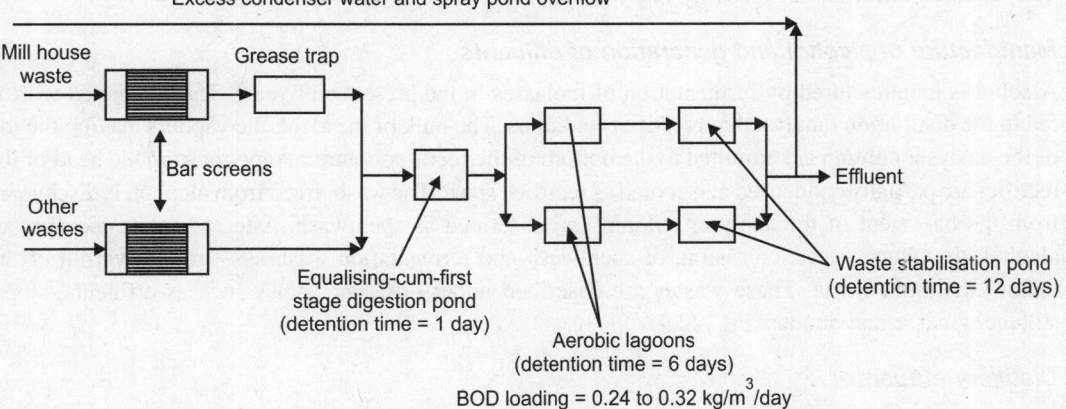

Fig. 22.3. Flow diagram for complete treatment of sugar mill waste.

Biomethanation

Biomethanation is a process involving decomposition of organic matter (and to some extent, inorganic matter) in the absence of molecular oxygen, resulting in methane-rich gas that has good fuel value. The generation of biogas can meet more than 60 per cent of the fuel requirement of the distillery.

During such bioconversion, approximately 80 to 90 per cent of BOD is also removed from the spentwash. Over the years, a number of distilleries in the country have adopted bio-methanation for the treatment of spentwash. Although this is a cost-intensive technology, yet the investment made in this system is usually recovered within 4 to 5 years. For a 30 kld alcohol producing industry, generally the spentwash generated is of the magnitude of 450 kld and, on an average, the total biogas production is about 12,000 Nm^3/day. Based on the net calorific value, the coal saving is as high as 12 tpd, which amounts to a net saving of Rs. 72 lakh a year as calculated in the year 2006.

Composting

In yet another process, the spentwash is mixed with filler materials, such as press mud, rice husk, wood chips, bagasse and pith. The mixture is seeded with cowdung or specially developed micro-organisms to hasten the process. The process is aerobic and it takes 12 to 14 weeks to be completed, resulting in black compost, which can be used as manure.

Incineration and potash recovery

The third possible alternative is based on the principle that the spentwash, when concentrated to 60 per cent w/w, develops sufficient calorific value and can burn by itself without any external input of energy. This process results in potash rich ash which can again be used as a fertiliser.

Utilisation of distillery effluents

Shortage of raw materials for industry and inputs for agriculture is the major problem facing the country today. Energy crisis has aggravated the situation. Rapid industrialisation programmes have resulted in the generation of liquid, solid and gaseous wastes in such huge quantities that a serious threat is likely to be posed to the 'quality' of life in the years to come. All these problems can be solved only through a change in our philosophy and attitude to consider wastes not a nuisance or a disposal problem but as new resource materials for recovery of by-products.

Manufacture of alcohol and generation of effluents

Alcohol is manufactured by fermentation of molasses in the presence of yeast. The fermented wort is fed to the distillation (analyser and rectifier) columns. The bulk of the alcoholic vapours leaving the top of the analyser column are admitted to the bottom of the rectifier column. Vapours from the head of the rectifier are partially condensed and cooled as rectified spirit. The wash, freed from alcohol, is discharged from the basement of the analyser column and is termed as spentwash. After complete recovery of alcohol, the plant wastes, consisting of spentwash and fermentation washings, are thrown out of the factory as liquid waste. These wastes are described under different names such as effluents, slops, stillage, vinasse and dunder (Fig. 22.4).

Distillery effluents

In India, there are around 130 distilleries with a total installed capacity of 735 million litres of alcohol per annum. For every litre of rectified spirit (ethanol) manufactured, about 12–15 litres of effluents are discharged which pose a serious pollution problem, particularly due to their generation in substantial quantities. They are characterised by reddish brown to dark colour, with pH around 4.0, temperature about 90°C, having a smell of burnt or caramelised sugar.

The effluents have a high biological oxygen demand (BOD), of the order of 40,000 mg/l, mainly due to their organic load, and when allowed to decompose in open ponds, produce foul smell and favour breeding of mosquitoes, thus rendering the surroundings completely unhygienic and polluted.

They are 60–70 times more hazardous than domestic sewage when discharged into rivers, streams, lakes and groundwater. They deplete the dissolved oxygen completely and destroy aquatic life. The disposal of distillery wastes is easy when the plant is situated near the sea, but those which are in the interior have to face many difficulties.

Treatment and disposal methods

The treatment and disposal of effluents has become a major technical and economic problem for all the distilleries. This is mainly due to a variety of factors which tend to defy the development of a successful utilisation pattern based upon recovery of by-products—BOD, temperature, acidity, volume, retention time, capital cost, operational and maintenance cost, land and water requirements for dilution and application, energy input for drying and incineration, sulphide content and disposal of sludge.

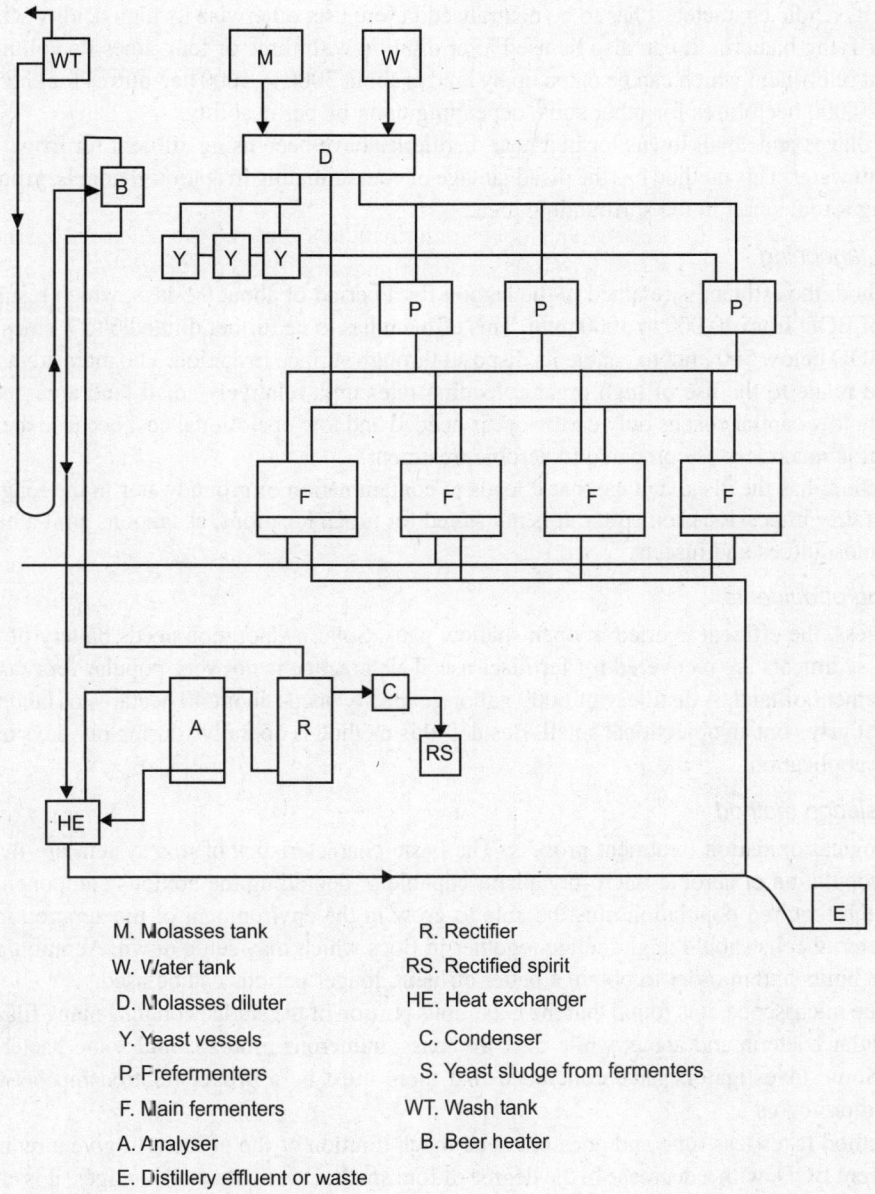

Fig. 22.4. Distillery process for ethyl alcohol from molasses.

M. Molasses tank
W. Water tank
D. Molasses diluter
Y. Yeast vessels
P. Prefermenters
F. Main fermenters
A. Analyser
E. Distillery effluent or waste

R. Rectifier
RS. Rectified spirit
HE. Heat exchanger
C. Condenser
S. Yeast sludge from fermenters
WT. Wash tank
B. Beer heater

Following are some of the methods used in various distilleries:

Disposal of irrigation

This is the most economical method, provided sufficient water and land are available nearby. The fertilising value of 100 hectolitres of effluents has been estimated at one ton of farmyard manure. On account of its acidic character, it has to be neutralised before use; otherwise its high acidity will inhibit the soil nitrifying bacteria. It can also be used after dilution with three or four times its volume water. The amount of effluent which can be taken up by land is about 3000 to 4000 hectolitres for clay soil and 10,000 to 20,000 hectolitres for other soils, depending upon its permeability.

Some villages and small towns located near distilleries have been using effluent for irrigation after dilution with water. This method has the disadvantage of contaminating irrigation channels, groundwater and creating a foul smell in the surrounding area.

Anaerobic lagooning

In this method, the effluent is retained in the lagoon for a period of about 90 days, which results in the reduction of BOD from 40,000 to 3000 ppm. This effluent has to be further diluted 6 to 7 times to bring down the BOD below 500 ppm to enable its disposal through surface irrigation. The main advantages of this method relate to the use of high organic loading rates on a relatively small land area because of larger depth, low capital cost as only earthwork is needed and low operational cost because the nutrient requirement is much less as compared to aerobic treatment.

This method has the disadvantage that it leads to contamination of groundwater in the long run and also needs fairly impervious soil strata. It is not suited for urban locations, as lagoons emit a bad odour and breed mosquitoes and insects.

Solar drying of effluents

In this process, the effluent is dried in open shallow pans. Solar evaporation needs battery of trenches wherefrom sediments are recovered for fertiliser use. This practice is not very popular for reasons of a vast requirement of land. A distillery of 6000 gallons capacity needs about 40 hectares of land for solar drying and it gives out an objectional smell. Besides, this method is operative during dry days only, thus limiting its application.

Activated sludge method

It is a biological oxidation treatment process. The basic characteristics of this system are three-fold: (i) mixed population of aerobic micro-organisms capable of degrading the noxious components of the waste, (ii) the required population must be able to grow in the environment of the aeration tank, and (iii) the bacterial cells should agglomerate together in flocs which may settle down. Aeration period is usually 4–8 hours and in order to obtain a better effluent, longer periods can be used.

Under the microscope, it is found that the gelatinous portion of the sludge contains many filamentous and unicellular bacteria and algae, while over the mass, numerous protozoa and some bacteria crawl and feed. Some investigators have concluded that there must be a proper relationship between the bacteria and protozoa.

This method takes less time and does not need much dilution of the effluent to give a reduction of 90–95 per cent BOD, with a decrease in the degree of foul smell. Due to these advantages, it is employed in industrial and sewage waste treatment. Despite these advantages, there are some disadvantages. Activated sludge process needs a high capital cost and the time required to build up activated sludge and the quantum of electricity consumed may prove to be obstacles to flexible operation. The choice of a suitable organism is a matter of great importance in aerobic fermentation.

Anaerobic digestion and methane gas generation

A large amount of data on the anaerobic digestion of sewage for methane gas production has accumulated and considerable details have been worked out. Pilot plant studies have shown that methane gas can be increased by recirculating carbon dioxide. High capital cost is one of the many restricting factors discouraging plants to make use of this potentiality.

Fermentation also leaves behind a sludge of BOD value around 500 ppm. The disposal of this sludge is not without its own problems.

Ammonification and nitrification of effluents

In this process, the effluent is first treated with ammonifying bacteria anaerobically. Pilot scale trials were carried out by the National Sugar Institute (NSI), Kanpur, in collaboration with Somaiya Organo Chemicals, Ltd.

The final BOD reached was 2,500–3,000 ppm, which needs further dilution before being discharged through irrigation. The main advantages of this process are a low capital cost and fast treatment of the waste without giving chance for seepage or foul odour; however, this method also suffers from the same problems as anaerobic lagooning.

Cultivation of torula yeast

A process has been patented in France where distillery slops have been used as a raw material for cultivation of torula yeast that can be used as a cattle feed. The yeast is developed under vigorous aeration and as a result, BOD is reduced by about 50 per cent. Due to high ash content and other impurities in distillery waste, this method has not been found suitable in India. In Brazil, the average yield of dry cellular matter based on total soluble solids was found to be 18.8 per cent.

In this process, there is a need for a large amount of investment and power requirement is high; the residual water needs treatment to reduce pollution.

Recovery of potash

Distillery effluents contain about 11.5 per cent potash which can be recovered by multiple effect evaporators. The main problem in making use of this valuable plant nutrient lies in high capital investment as well as high energy needs involved in evaporation, drying and transportation.

Recovery of vitamin B_{12}

The synthesis of vitamin B_{12} was studied during treatment of distillery waste at NEERI, Nagpur. It has been found that it is worthwhile to recover the vitamin from the sludges, using plain water at boiling temperature. Addition of 1 mg/l cobalt increased vitamin synthesis by 50 per cent.

A good deal of research work has been initiated at a number of institutes/centres; namely, National Sugar Institute, Kanpur; National Environmental Engineering Research Institute, Nagpur, on anaerobic lagooning, anaerobic digestion in closed digesters with methane recovery, ammonification, drying, incineration, potash recovery, and direct use for irrigation purposes; but the high cost involved in the treatment and recovery of by-products has been one of the main constraints.

In order to tackle this problem, there is a need to adopt an integrated system (Fig. 22.5) which may help recover energy and water for inplant utilisation, water and potash for farmlands and feed for animal nutrition, consistent with minimum pollutional hazards.

As such, keeping in view the prospects and problems in the utilisation of distillery wastes, it is highly important to constitute a planning group to prepare a feasibility report and a coordinated programme.

Efficient utilisation of effluents may not only help provide sanitary conditions all around the factory areas but also open up a vast potential to generate new resource materials.

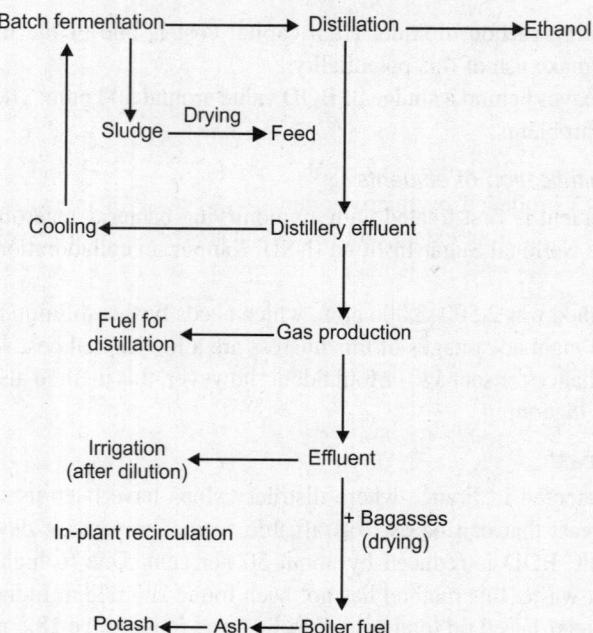

Fig. 22.5. Recycling distillery effluents for fuel, ethanol, animal feed and potash.

FERMENTATION PRODUCTS

Fermentation is a natural process around which several industries have been built. Traditional uses of the process have expanded over the past quarter-century and many new uses for microbiological processes have developed, mostly in the pharmaceutical industry. In the same period, the competitive techniques of chemical synthesis has replaced fermentation as the preferred production method for certain compounds.

The interlocking of these industries is significant. One view of fermentation shows the process as a link in the stepwise degradation of biologically decomposable agricultural materials.

The interdependent economy of raw materials, processing needs and products allows surplus by-product material of some of these industries to be fermented for profit, rather than to be disposed of directly as waste material. For example, corn steepwater is currently used as a major component of industrial microbial media. Also, corn processors now prepare and market ready-mix liquid brewers adjuncts; this sequence leaves the industrial wastes from adjunct preparation as a problem of the corn processor rather than the brewer.

Alternate animal feed markets for molasses, with its residual sugar intact, are being expanded. As these markets shift, applications of the fermentation process also shift and the responsibility for disposal of the industrial wastes associated with these industries moves from one industry to another.

Many of the processes used in cheese making, meat curing and other food manufacturing operations may correctly be called fermentation processes. Use of the fermentation process to produce ethanol from sulphite pulp waste has been investigated as to economics and technology by the paper industry, primarily as a means of pollution control.

Industrial Fermentations

Apparently, fermentation processes were operating as a part of the world's ecology long before man began to modify nature's processes to serve his needs. The fermentation of wines and other beverages was a commercial pursuit for many centuries before the Christian Era.

About the time of World War II, the biosynthesis of penicillin gave the medical profession the first of a series of antibiotics produced by micro-organisms. The use of industrially practical microbial processes has expanded to such an extent that review of the meaning of the word 'fermentation' has become necessary. Indeed, the meaning is undergoing change or expansion beyond the concepts of the older schools of microbiologists. The Latin word itself, *fermentare*, to boil, would appear to cover any bubbling reaction.

Pasteur's description of the 'life without air' of certain microbes limits the definition of fermentation to anaerobic microbiology, with yeast conversion of carbohydrate to alcohol as the process in mind.

Expansion of the concept includes the production by *Clostridiun* of acetone and butanol from molasses carbohydrate, usually considered a typical fermentation process. More lenient usage includes the commercial production of lactic acid and propionic acid, the conversion of wine to vinegar and the processes for producing citric acid and antibiotics.

Wastes versus by-products

Descriptions of waste composition are of little value if based on the averages of composite wastes from several microbial process industries. If the descriptions are limited to the fermentation industries under the strictest definition of the word or even if they are limited to the distilling or brewing industry alone, they must still be qualified and applied carefully.

This logic is based on the fact that microbial processes are merely unit processes: they are supported by other processes which must vary, not only with the product desired, but also with the raw materials available under the interlocked economy of these industries. Therefore, the type and sequence of supporting processes used in industries employing the fermentation process form an important part of the study of fermentation wastes.

The economic downgrading or degradation of agricultural products has been stated in the sequence: food, feed, fertiliser, fuel. Any of the last three of these, economically significant though it may be, might have been discharged as a waste material from any category of higher rank. The manufacturer's decision on when to omit steps in this value sequence, such as to go directly to conventional waste treatment with the production of methane and sludge, is based on the economy relating recovery and marketing at the step where the material is produced.

The material remaining after the desired products of a fermentation have been obtained is usually of the value level of animal feedstuff, with the cost of recovery as feed near the market value. Final wastes from these plants then include the wastes from by-product processing, but the potential waste of the fermentation residues is reduced to the extent permitted by efficiency of the recovery operation and market for the feed.

The practice of in-plant waste control has been so successful in some large units of the fermentation industry that final effluent can be discharged directly to the river or combined with city sewage. Thus, the unit processes used to accomplish most of the waste abatement in these industries are not of the conventional waste treatment type but, insofar as practical, are at the feedstuff or perhaps the fertiliser level. Because of the fermentation industry's reliance on non-conventional in-plant abatement of pollution, the materials that might be wastes become by-products instead.

Improvement of the manufacturing processes and perfection of means for patrolling the improvements have been the procedures preferred by these industries in attaining waste abatement and pollution control.

Process

Applied research on the fermentation process has been pursued since ancient times. Pasteur's work advanced this endeavour tremendously and was a beginning of basic research. Knowledge is presently increasing concerning the intermediate reactions that occur between the glucose and ethanol stages and between other substrates and the desired compounds of microbial processes. Application of the research on biochemical pathways may lead to additional improvement of fermentation yields. However, the knowledge will probably prove more valuable in raising the efficiency of the newer uses of micro-organisms rather than in improving alcohol fermentations, for which satisfactory efficiencies have been achieved. Research using chromatography and other sensitive analytical techniques is finding answers to questions on the specific compounds that make up whiskey, wines, beer and antibiotic structures. Improvement of the flavour qualities of alcoholic beverages will become possible when this information can be combined with the new knowledge of biochemical reaction routes. More intelligent control may then be exercised over the biosynthesis of important flavour compounds, which are produced as side reactions during the fermentation.

Several tools for control of fermentations are available and used. The nature of the substrate material, whether a grain, a fruit or a cane product, is especially significant. The fermentation may or may not be conducted in the presence of the protein, pectin and pentosan portions of the raw material; these substances are converted during fermentation to compounds affecting colour, flavour and other properties of the product. The pH, temperature and dilution of the medium in which the yeast must work and the time allowed for fermentation are further controls on the process. The efficiency and the timing of supplementary enzymes supplied to the yeast also control reaction sequences. Research on the fermentation process is aimed at more precise control of the synthesis route and the end products, as well as the quantitative efficiency of the fermentation.

For some industrial microbial reactions, the efficiency with which the desired product is produced may not be directly related to the degree to which the biochemical oxygen demanding substances of the substrate are consumed. Although alcohol is produced by the substantially complete metabolic activities of yeast operating anaerobically, other desired products of micro-organisms may be accumulated by intermediate blocking or side-tracking of a metabolic or cell-structuring reaction.

Cell material itself is not a desired product of the microbial processes, except in the production of yeast or the experimental production of foods by micro-organisms. The disposal of mycelia, production of which is incidental to an antibiotic process, is one of the troublesome waste problems in this industry.

Raw materials

To put fermentation to efficient and precise use requires attention to preparation of the substrate and to product and by-product recovery as well as to establishment of proper conditions for the biological process.

The type and quantity of the products and wastes produced by fermentation, from a given amount of raw material, are functions of the efficiency attained in the supporting operations as well as of the inherent efficiency of the yeast or other biological system.

Fermentations for beverage products require raw materials such as grain, malted barley, fruit and molasses. For the entire range of commercial microbiological processes, it is estimated that over 200 strains of organisms are under controlled culture. The satisfying of so many diets requires supplements such as corn steepwater, meat and fish by-products and surplus animal fat. Protein and carbohydrate raw material supplied in these several forms obviously includes unwanted fibres and other materials that are rejected by the fermenting organisms and that remain for reclamation as another by-product or for disposal as waste-water.

These are predetermined or premeditated waste materials and are not a function of process efficiency in the manner that loss of desired fermentable material would be.

The wastes incidental to substrate preparation and product recovery, as well as the amount and type of the predetermined waste, can be controlled to a significant extent by careful selection of the raw materials. The cheapest material containing the necessary substrate nutrients is not always the material of lowest over all cost when fermentation product value, by-product value, recovery cost and pollution control cost are all considered. The amount of substrate preparation carried out at the fermentation plant varies from one plant to another and has a major effect on the waste potential.

If the preparation is done elsewhere, the discarded portions of the material are wastes belonging to someone else. As may be concluded from the interlocking nature of the economy surrounding the fermentation industry, a considerable amount of shopping for raw materials is possible and proper. In the past, this privilege has been neglected to a surprising extent by the older fermentation industries.

Cleaning, tempering, sprouting and drying of barley are usually accomplished off the premises of the main brewery or distillery. Malt sprouts, a low-value feed ingredient, remain with the maltster. In addition to ground malt, the brewing recipe may include corn grits, brewer's rice in ready-milled form or one of the new liquid adjuncts prepared to the brewer's specifications by the corn refining industry. The preparation of brewing wort is completed by extraction with hot water at atmospheric or slightly elevated pressure.

Distillery mash is cooked at higher pressures. Winery fermentations are of several types, some conducted in a lightly pressed juice, some in juice containing grape skin components and others with the skins as part of the substrate.

Auxiliary Operations

Methods of obtaining the desired primary products from the fermented material are varied. A brewery ferments a cleanly filtered wort; after the fermentation is halted, the product needs little more than adjustment of clarity and carbonation before packaging for sale. Grain distilleries in the United States ferment a whole-grain slurry, then distil to yield whiskey distillates containing about 60 per cent alcohol or spirit distillates of up to 95 per cent alcohol. Wineries are concerned with the direct products of fermentations, in the form of wine and also with distilled brandies.

The desired products of some pharmaceutical fermentations are commonly produced by the microorganisms in concentrations of a few milligrams per litre and processes of adsorption, solvent extraction, ion exchange and filtration are required to recover the marketable product from the broth.

The final packaging of the products of the fermentation industries is the subject of research and quality control effort. Advancement of design and maintenance of package quality is important in the competitive beverage market. With pharmaceuticals, fine packaging is necessary to protect the expensive products.

Wastes from packaging at breweries, mainly from washings and spillage, may form a significant portion of the waste load. The packaging wastes of distilleries are almost zero because at this stage these products are at full tax-paid value. Distillery containers are not reused, so there is no washing loss.

Marketing of by-products

The marketability and cost of reclamation of by-products left in the fermentation residues are among the most strategic points in determining the degree of waste control to be sought. Under favourable market conditions, the profit on feed by-products from a completely integrated distillery may almost pay the cost of the grain.

To accomplish this requires not only production efficiency but also competent research to substantiate the values of the by-product and to develop its market. Some microbial processes yield residues that are difficult or impossible to market. In these processes, much can be gained by selecting low residue raw materials, such as refined sugars or other pure compounds, as sources of soluble nutrient.

Commonly used, recovery operations include screening, filtering, centrifuging, evaporation and drying of the substrate residue. By these steps, breweries reclaim yeast and enzyme products as well as animal feed; distilleries have found the animal feed market most profitable.

Beverage Alcohol

The oldest and largest segment of the fermentation industry is the manufacture of products containing ethyl alcohol, for beverage use. The major subdivisions are distilleries, breweries and wineries, each of which is discussed later in this chapter. The processes and the wastes that result from the manufacture of industrial fermentation products may be studied in the light of experience and standards of practice developed in the grain distilling industry. Identification of the wastes from several of the unit processes of the grain distilling industry and description of the survey methods illustrate typical fermentation industry wastes and pollution control practices.

By limiting the presentation of quantitative waste survey data to the well-established procedures of the distilling industry, the nature of the wastes and the techniques of abatement can be illustrated more specifically than if necessarily wide ranges of data are used to show the quantities of waste expected from all commercial applications of the fermentation process. This procedure also avoids estimating the closely held production data of the pharmaceutical industry and avoids fermentations for which waste abatement standards are being determined. A study of winery processes would similarly provide a good illustration of fermentation industry wastes; certainly the wine industry is one of the oldest applications of fermentation, but modern winery procedures and wastes are not as neatly standardised. This is not a fault but a characteristic of the industry, because the market demands many types of wine and brandy, necessitating varied fermentation and recovery techniques. So wide a field would not make a concise, illustration.

Wastes from a fermentation plant have a high potential biochemical oxygen demand per ton of raw material or per unit of product, caused by residual oxidisable organic material normally remaining in the spent fermentation medium and in unwanted material screened or filtered out when preparing the medium. Also, both the prepared medium and the product characteristically have high BOD values and, if small amounts are wasted through accident or inefficient processing, a heavy load is added to the effluent. For these wastes, BOD remains the most practical and useful parameter.

However, residue on evaporation, its fixed and volatile fractions and chemical oxygen demand furnish useful support to the BOD analysis, especially in the evaluation of test runs or attempts to improve operating efficiency of a single process. Useful correlation of COD and BOD analyses is possible if comparison of the two analyses is limited to wastes from an individual unit process. Final summaries of the wastes may rightly consider only BOD, frequently expressed as the corresponding population equivalent. The BOD is, after all, the nature of the pollution that may be sustained from these wastes.

OTHER INDUSTRIES

The manufacture of alcohol, rum or other microbial products from molasses is, in reality, a waste control operation of the sugar refining industry. Molasses is a seasonally supplied and priced material. It is important to distinguish black strap molasses from high test molasses; the former is a by-product of sugar refining, the latter is a whole molasses, containing the sugar, made available in times of sugar surplus.

Industrial Alcohol

In the preparation of molasses for industrial alcohol production, most of the non-carbohydrate material is separated and left at the sugar mill, near the cane or beet fields. A variable degree of further removal of protein and pectin material is accomplished by heat precipitation and lime treatment. After adjustment of pH and balancing of auxiliary nutrients, a suitable yeast accomplishes practically complete conversion of fermentable carbohydrate to alcohol. Molasses is used as a substrate for microbial production of several other commercially important compounds.

Stillage from other fermentations may be handled in a similar manner. The stillage carries about 8 per cent solids and a BOD of approximately 25,000 mg/l. When evaporated and dried in much the same manner described for grain distilling operations the resulting product contains approximately 30 per cent ash, 10 per cent sugars which are mostly non-fermentable but available for BOD, 8 per cent protein and lesser amounts of glycerine and organic acids. The high ash content is not encountered with grain stillage and must be taken into account in feed formulation.

Secondary fermentations to produce a feed ingredient with high riboflavin content have survived feed market fluctuations somewhat better than similar products from the grain distilling industry. A syrup of 50 per cent solids, from the evaporation of molasses stillage, has found a good market as a feed ingredient. It is not subject to spoilage at this concentration, it is acceptable to feed blenders and its production reduces molasses distillery operating waste by eliminating wastes incidental to the drying process.

Antibiotics

Microbial processes have been widely adopted by the pharmaceutical, industry for the production of a spectrum of antibiotic agents. The first commercially significant item was penicillin, developed during World War II.

The ratios of waste to production, which are of considerable value in the fermentation industries described earlier, are more difficult to apply to the manufacture of antibiotics. A major reason, of course, is the high potency and the small yield, on a weight basis, of the active product. In other words, the ratio of waste to raw material is nearly 100 per cent, unless primary wastes can be converted to useful by-products.

The ratios can be used, however, in reporting reclamation costs or waste reduction accomplishments. Modifications in the fermentation process, which may be of little significance in relation to product yield, may be important with respect to material recovery and pollution control.

Antibiotic concentrates have entered the feed market for their feed value as well as their antibiotic content. Determination of recovery costs for borderline profit items is important when exploiting new markets.

Heavy mycelia from streptomycin filtration is difficult to treat for any value greater than that of fertiliser; sometimes burial in garbage dumps is the simplest means of disposal. The relative advantages or disadvantages of various dewatering units could be shown by unit cost accounting and perhaps good accounting of the costs of mycelia production might promote new research on the primary process, to improve antibiotic production and decrease cell production in the fermentation.

Chemical solvents used in certain antibiotic extractions must be of high purity to gain the efficiencies required. For recovery and reuse, stripping of the solvent in waste reclaiming operations must be highly efficient if the effort is to prove more economical than the purchase of new solvent accompanied by conventional waste treatment for the old. The potency of solvent waste is readily calculated from BOD values of the pure compounds. Isolation and selection of waste streams for simultaneous application of different types of conventional treatment, solvent recovery, feed reclamation or composting may provide a practical choice of processes if accurate waste accounting points the way.

Treatment of Non-recoverable Waste

Knowledge of the types of waste expected from industrial fermentation processes permits setting general guides for conventional waste treatment, when this is necessary. The wastes remaining after application of in-plant abatement procedures are of about the same chemical type as were the heavy loads from these plants in the days before good in plant control. They exhibit a strong demand for oxygen.

The main difference lies in the dilution factor. Among the troubles encountered are too much dilution, deficiencies of certain components needed to nourish a healthy sewage flora and frequently more organic acid than is desirable.

Fermentation wastes are amenable to treatment by most of the modern conventional biological treatment processes.

In-plant measures may still be valuable when treatment is added as an auxiliary process. The plant surveys provide definitions of the wastes for which treatment is contemplated. Knowledge of the amounts of waste emanating from various parts of the plant is needed in order to select portions of the waste for treatment and to select different portions for different treatments. The size and cost of the treatment facility depend to a considerable degree on these final applications of the in-plant approach.

The fermentation industry is fortunate in another aspect of waste treatment planning, in that the technologies required by the fermentation itself and by the waste treatment are similar. Both rely on biological processes and employ techniques of bioengineering. Advances in aeration techniques, for example, have been applied both to activated sludge and to microbial processing.

If exploited, the common knowledge can eliminate any remaining barriers between fermentation plant and treatment plant and the entire plant approach can really be applied.

Summarising waste abatement planning, it should be possible to define the concentration and the total quantity of that portion of the waste for which treatment is contemplated, then to decide which portion to treat and to consider all applicable types of treatment in order to select the most economical, not forgetting combined treatment with municipal sewage to balance the nutrients. In-plant abatement measures of borderline economic advantage may then be reevaluated, after treatment costs have been estimated.

Chapter 23

Miscellaneous Process Industries

INTRODUCTION

This chapter discusses the miscellaneous process industries chemical products and raw materials and the nature of the waste-water problems encountered and techniques for solving them.

The effluents discussed in most of the other chapters are characterised by similarity. Individual plants within these industries produce the same general product or series of products from the same raw materials and by approximately the same methods. Techniques and methods of waste-water treatment applicable to one plant are generally usable in another plant of the same industry without substantial change.

The manufacture of inorganic chemicals can hardly be called a unit industry. The products are of an inorganic nature, but the individual plants are characterised by dissimilarity. A great variety of products are manufactured and any given product may be made by a number of processes, from different raw materials and with effluents of entirely different characters.

INORGANIC CHEMICAL INDUSTRY

To indicate the diversity of products manufactured by the inorganic chemical industry, Table 23.1 lists the major products. This list includes only manufactured chemicals; materials which are mined and purified, such as NaCl and KCl, are not included. It is interesting to note that among the top 10 inorganic chemicals from the standpoint of tonnage, only sodium carbonate and phosphoric acid have liquid waste problems of significance.

Multiproduct plants are common in the industry sometimes by the nature of the process and sometimes because of utilisation of by-product streams which would otherwise be wasted. A plant producing chlorine by electrolysis of salt brine also produces sodium hydroxide in a fixed ratio. A soda ash plant may advantageously operate in conjunction with a calcium products plant to utilise calcium values in the side streams.

Inorganic-organic industry combinations are frequent; such halogenated hydrocarbons as tetrachloroethane and metallo-organic compounds like sodium methylate and chromium carbonyl are produced by the same plants that make inorganics.

The availability of additional raw material over and above that required for a particular product leads to consideration of other products that could be made from the same source. In one plant, a large supply of salt and limestone led to the complex: soda ash, sodium bichromate and chromic acid from incremental soda ash; cement from additional limestone; chlorine and caustic from incremental salt;

pure calcium products from soda ash side streams; a variety of chlorinated products from incremental chlorine; several sodium silicate compounds from soda ash; and coke with coke by-products from a coke plant to supply the soda ash operation.

Table 23.1. Inorganic chemicals.

Product
Sulphuric acid
Ammonia
Sodium carbonate (soda ash, including trona)
Sodium hydroxide (caustic soda)
Chlorine
Nitric acid
Ammonium nitrate
Phosphoric acid
Oxygen
Ammonium sulphate
Calcium carbide
Hydrochloric acid
Carbon dioxide
Aluminum sulphate
Nitrogen solutions
Sodium sulphate
Sodium tripolyphosphate
Sodium silicate
Titanium dioxide
Phosphorus
Calcium phosphate
Hydrofluoric acid
Sodium bichromate
Potassium hydroxide
Sodium metal
Sodium chlorate

Raw Materials

The inorganic chemical industry relies to a great extent on naturally occurring elements and compounds for its raw materials. Salt deposits of high purity are found in numerous locations in India and other parts of the world. With salt as a starting material, sodium carbonate is obtained by the ammonia-soda process; sodium hydroxide and chlorine by electrolysis. Salt may be mined by either dry or wet mining; in the latter, water is pumped down into salt deposits of high purity and the concentrated brine forced to the surface. Some salt is made by evaporation of sea water in ponds.

Natural deposits are the main source of sulphur in the United States and other developing countries. In the Frasch process, used to mine sulphur, superheated water pumped into the deposit melts the sulphur, which is then forced to the surface and allowed to solidify. The major portion of sulphuric acid is made with sulphur as a beginning material. Minor portions are produced by recovery or regeneration of spent acids, burning of iron pyrites, roasting of sulphide ores in metallurgical operations and oxidation of hydrogen sulphide recovered from gas processing. From sulphuric acid, numerous sulphate compounds, both organic and inorganic, are obtained by a variety of processes.

The atmosphere furnishes oxygen and nitrogen by the fractionation of liquid air. Originally used primarily in cutting and welding operations, oxygen is increasingly used in the chemical industry as a tonnage chemical. Oxygen plants located adjacent to chemical plants are common. Nitrogen is used in the catalytic synthesis of ammonia and hence indirectly in the manufacture of numerous ammonium compounds, including fertilisers. Another major use is in the manufacture of nitric acid by a catalytic vapour phase reaction.

Phosphate rock processing yields phosphoric acid and various phosphate compounds. Primary applications are in fertilisers, soaps and detergents.

Many inorganic chemicals are found in natural deposits, requiring only refining and purification. Mixture of sodium carbonate and bicarbonate is purified and calcined to produce sodium carbonate. Potassium compounds are becoming important commercially because of fertiliser demands. Sylvinite (42.7 per cent KCl, 56.6 per cent NaCl) is mined, dissolved and the KCl separated by fractional crystallisation or flotation. From KCl, a number of potassium compounds are made.

Waste Characteristics

An outstanding characteristic of wastes from inorganic chemical manufacture is their variability. As has been pointed out above, a particular product may be manufactured by completely different processes using completely different raw materials. This situation usually gives rise to dissimilar effluents from the manufacturing plants. The dissimilarity involves differences in either waste composition or concentration. Waste components in the industry cover a wide range and normally contain raw materials, intermediates and products, as well as processing chemicals. Concentrations cover a wide range, from almost zero to saturated solutions and suspensions or slurries. A large production capacity does not indicate a large waste problem, because many of the tonnage chemicals produced do not have an appreciable liquid waste. On the other hand, tonnage production which does involve a waste problem may give rise to such quantities of waste that the volume itself becomes troublesome, as in the production or soda ash.

Biological inactivity

In general, wastes from inorganic chemical manufacture are themselves inorganic. As such, they are not amenable to biological degradation as a general approach to effluent treatment. Anions such as chlorides, sulphates and phosphates, together with typical cations such as sodium, potassium, calcium and magnesium are most often dealt with by chemical rather than biological processes.

Anions

Chloride is an example of a troublesome inorganic waste. Sodium chloride, a typical compound, is not biologically degradable, cannot be destroyed by aeration or a combination of aeration and natural processes, does not break down in the stream and can be precipitated only with a few expensive reagents such as silver and platinum salts. Chlorides are common in the inorganic plant effluent, but are not economical to recover. Sodium chloride can be purchased more cheaply than it can be recovered from even concentrated mixtures, hence recovery is not practiced. The only practical technique for handling chloride effluents is dilution, by impoundment and regulated discharge to the stream. Discharge rate is determined by measurement of stream flow and an agreed-upon maximum chloride concentration in the stream. In addition to chloride, the sulphate, nitrate, phosphate, carbonate and hydroxide anions are ordinary components of inorganic effluents. Mixtures are common and concentrations cover a wide range.

Cations

The most common cations found in inorganic effluents are sodium, calcium, potassium, ammonium, iron and magnesium. In addition, the presence of chromium, zinc, copper, nickel, cadmium and other heavy metals is often noted. Mixtures and wide concentration ranges may be expected.

Effects on receiving stream

The usual sanitary engineering parameters of dissolved oxygen, biochemical oxygen demand and oxygen depletion in the stream have little or no significance when considering inorganic effluents. Aside from a few oxygen carriers such as nitrates, oxygen levels are not affected directly by inorganics. Occasionally, natural metabolic processes of the stream biota may be halted by inorganics of sufficient concentration, through actual sterilisation of the stream. Inorganic effluents must be studied from the standpoint of cationic and anionic constituents, with consideration of the chemical characteristics of each ion.

Some components of the inorganic effluents may be toxic, depending on the material and concentration. Reference to literature data on toxicity and conferences with regulatory agencies are in order when toxicity is suspected. In particular cases, bioassays are necessary to establish toxicity levels. Fortunately, inorganic toxicity has been reasonably well evaluated and the critical materials are fairly well-known, although existing data may be contradictory as to concentration levels toxic to aquatic life. A complete discussion of toxicity is beyond the scope of this book, but the waste engineer must recognise a possible problem with some inorganic materials.

Taste and odour comprise another area which must be approached cautiously, especially where a high concentration level or proximity to water treatment plants is involved. Taste is the important aspect; in-organics generally do not present an odour problem. Taste mechanisms are not yet well understood; reactions of individuals vary widely to a given concentration of specific material. Regulatory limits for taste-producing substances are usually based on an average reaction, with a safety factor sometimes included. Among inorganics, chlorides have been singled out for attention in establishing taste limits. Several agencies have set 250 mg/litre as the maximum stream concentration permissible.

Other effects on receiving streams which must be considered in dealing with a specific inorganic effluent are hardness, concentration, corrosion, chemical reaction, solids and colour. Each effluent must be considered in the light of regulatory requirements, economics and general feasibility of treatment.

Waste Disposal Techniques

Ideally, the problem of waste disposal should be considered an integral part of the manufacturing process. Engineers carryout considerable effort on consideration of alternatives in process design in terms of equipment or process steps. As finally evolved, the design should represent the optimum balance between equipment selection and economic process alternatives. Too often, no attention may be given to the problem of waste disposal until this point; or it may even wait until the plant is under construction. Occasionally, the expertly designed plant is found to have resulted in a waste which is either untreatable or treatable only at considerable expense. The burden added by waste treatment then may affect the over-all economics of the process and, in the extreme, may preclude any possible profit from the operation. Proper consideration should be given to waste disposal during the design phase, in order that the lowest over-all manufacturing cost can be achieved. This can be accomplished primarily by adequate communication between the design engineer and the waste control engineer, from the beginning of the project.

Stream requirements

The obvious first step in a proper waste control programme is to establish the conditions that must be met in the receiving stream. This can be done by conferences with regulatory agency personnel or by reference to published regulations and requirements where these are available. Beyond this, the waste control engineer may aim for concentration levels which his experience indicates are reasonable and practically attainable. In addition to the legal requirements, attention must be given to the moral aspects of waste discharge. These considerations are part of being a 'good neighbour' in the community; when properly evaluated and dealt with, good waste control makes future public relations much easier.

Attention should also be given to probable trends in requirements and allowances made for changes in permissible levels of contaminants in the stream.

Having established the conditions that must be met, the engineer is now ready to consider means of achieving these levels. In the new project design stage, experience and the ability to visualise and solve difficulties are the waste engineer's best assets. In the already existing plant, ingenuity must be exercised to solve problems within the confines of the operating installation.

In-plant procedures and control

Money spent on waste treatment is usually not revenue-producing capital; and the burden imposed, in both investment and operating costs, must be borne by the general process economics. Any reduction in the amount of waste to be treated or in the complexity of the treatment process yields direct benefits in the form of cost reduction and therefore enhances the over-all profit possibilities of the product.

Reduction in the amount of waste to be treated can be accomplished in several ways. Possibilities range from control of incidental spillage to a major change, such as the addition of a by-product recovery plant. On a new project, proper consideration of waste treatment presumes that the question of by-product recovery has already been answered; in an existing plant such a major change must be evaluated on its economic merits. In evaluation, the feasibility considerations must extend to sales and distribution problems. The new product should be compatible with the company's present line and all costs of by-product utilisation compared with all costs of waste treatment. Because of the nature of the inorganic chemical industry, many opportunities exist for alternative processes or raw materials.

Water reuse

In complex chemical processes, opportunities exist for utilisation of slightly contaminated streams in less critical portions of the process. Such reuse decreases the total over-all liquid throughput and hence the volume to be finally treated. Inorganic values in these contaminated process streams may sometimes be utilised. In one plant producing sodium silicate, the entire waste stream was used as process makeup water; this completely eliminated the pollution problem, in addition to recovering all silicate values.

One possibility of water reuse is the application of closed-cycle cooling systems. Cost of the necessary facilities and the effect of concentration buildup through reuse must be balanced against the cost of once-through water. No generalised recommendation can be given; specific circumstances determine the best course for each plant.

Good housekeeping

A substantial reduction in waste load can be accomplished through the collective contributions of a number of minor improvements in operation. Leaks in equipment, spillage, sewering of residues left in batch equipment, floor wash down, tank overflows and general sloppy operation of a process can measurably increase the waste load. Elimination of these factors can be accomplished solely by a cooperative attitude on the part of equipment operators and never-ending vigilance on the part of supervision. Proper equipment maintenance, of course, is a necessity, as is proper education of operators. Minor additions, such as drip collectors or float controls for tank levels are helpful.

Waste segregation

Inorganic pollutants are more easily treated in concentrated solutions. Occasionally, two wastes may be segregated because one is more easily treated than the combination of the two. Where one waste component may be much worse in pollutional characteristics than others, it may be segregated and treated at the source in small volume, rather than attempting treatment of the entire plant waste volume.

Waste blending

Often, the opposite approach is indicated-blending wastes to take advantage of self-neutralisation or to achieve dilution of a troublesome component. Possible synergistic or antagonistic effects of various components must be taken into account in blending. The decision to segregate or to blend is determined by specific waste characteristics. In general, waste streams are initially segregated and kept segregated as long as high concentration and ease of treatment is an advantage. The various waste streams should be blended to achieve dilution when the advantages of segregation have been exhausted.

Chemical Techniques

After application of proper design and pretreatment techniques have reduced the waste load as much as possible, there remains an irreducible residue which must be treated. As pointed out previously, there is no general approach to treatment in the inorganic chemical industry. The waste engineer has available to him a number of chemical and physical techniques for treatment, reviewed briefly below. Proper utilisation of these techniques should result in an acceptable plant effluent.

Neutralisation

Regulatory agency requirements for pH normally range between 5.5 and 8.5, with more restrictive limits for particular streams. Many compounds can be used for pH adjustment. Most can be eliminated on the basis of cost or practicality, leaving relatively few in common usage. Selection of an alkali or an acid for pH adjustment is based on cost and availability, reaction rate, sludge production and ease of application. Where neutralisation is accompanied by precipitation, difficulties encountered in separation and disposal of the solids must be considered. Alkalies commonly in use are caustic soda, soda ash, lime and limestone. Use of each of these must be evaluated in the circumstances of the specific waste.

Similarly, many acids could be used, but the most common are sulphuric and hydrochloric. Carbon dioxide can be used where the pH change required is not too great. As flue gas, this is probably the cheapest acid value available; it is used by bubbling the gas through the solution or by contact in a spray tower.

Where several waste streams varying in pH are concerned or where the plant effluent swings frequently from acid to alkaline, equalisation basins can be used advantageously, thus eliminating some of the chemical cost involved in neutralisation.

Precipitation

Inorganic wastes frequently lend themselves to precipitation as a method of treatment. This may be accomplished by appropriate mixing of streams or by selection of a suitable precipitant. The one requirement is that the end product be insoluble or of such limited solubility that the polluting component is sufficiently removed to meet stream requirements. Consideration could be given to recovery of the precipitate as a by-product, but usually the material is not sufficiently valuable to justify separation and recovery. The waste engineer must also recognise the problem involved, in solids separation and disposal. Large land areas are required for solids disposal if the precipitate is bulky. Attention should be paid to the residues from the reagents added, as these could be more pollutional in character than the original component.

With the numerous relatively insoluble substances found in inorganic chemistry, both precipitation as a method of treatment and specific precipitants for a given component are dictated by the circumstances of each plant.

Oxidation-reduction

Oxidation and reduction processes are less common in the inorganic chemical manufacturing industry than in other inorganic industries such as metal finishing. However, these processes should not be overlooked where they are applicable.

Common oxidising agents are atmospheric or manufactured oxygen, ozone, chlorine, hypochlorites and occasionally permanganates, chromates and nitrates.

Reducing agents are even less commonly used in the inorganic field. One application is the reduction of hexavalent chromium from the manufacture of chromic acid and bichromates.

Miscellaneous

Inactivation of particular waste components by complexation with a chelating agent is possible but is usually considered too expensive as a general means of treatment. Calcium, magnesium and other multibasic cations can be complexed with certain phosphates and organic sequestering agents. Such agents are expensive and find application only in special situations.

Ion exchange is a possibility where the waste material is fairly valuable and may be recovered. This technique is occasionally practiced on chromates and copper, nickel and other metal cations.

Physical Techniques

Physical treatment methods are generally less expensive than chemical treatment methods because the equipment is simple and no chemical additives are involved.

Flocculation

Finely divided solid materials are generally removed by flocculation and settling. Precipitates of ferric hydroxide, aluminum hydroxide and silica have the property of forming large flocs of high surface area. These flocs tend to remove fine suspended particles both by adsorption and by mechanical sweeping action as they move through the solution.

The long settling times associated with finely divided solids can be shortened appreciably by flocculation and flocs may assist in removing colloids that otherwise would not settle at all. Basic requirements are time for floc formation and mild agitation for thorough mixing. Flocculator equipment is common and may be obtained from a number of suppliers.

Settling

Solids separation by settling, with or without flocculation, is common in the industry. Settling tanks are available from suppliers in a variety of designs, shapes and sizes or may be built at the site. Frequently, settling basins rather than tanks are used where volumes are large and settling rates low. In soda ash plants, the basins may cover several hundred acres. Sludge from the settling operation can be filtered for further concentration of solids or can be conducted through a shallow lagoon for additional dewatering and ultimate disposal.

Filtration

Filtration is frequently resorted to because of the crystalline nature of the solids encountered. Many types of filters are available; equipment choice is dictated by volume and nature of the solids. Plate-and-frame and leaf filters are commonly used in batch applications. For continuous operation, several types of rotary filters are available, featuring provision for washing, for cake removal by strings or by scraper and frequently the use of vacuum. For materials difficult to filter, filter aids may be used. Other types of filters include sand filters and plain screens.

Miscellaneous

If a plant effluent contains only few components, evaporation and fractional crystallisation are sometimes utilised to recover valuable materials. However, this is not frequent; in fact, economics is usually against this procedure. Even with multiple effects, evaporation is a relatively expensive operation, particularly where dilute concentrations are being handled. Where evaporation is practiced, it is sometimes difficult to distinguish whether it is a by-product recovery process or a waste treatment process.

Solvent extraction is a possible treatment operation, but it is much more frequent in organic manufacture than in inorganic. The cost of solvents other than water usually precludes extensive use of this technique for inorganic wastes.

Solids disposal

The wastes engineer must give attention to the problem of disposing of solid residues after filtration, settling or clarification. Where large volumes of bulky solids are involved, solids disposal can be a greater problem in some respects than the original pollution characteristics of the component. A sufficient quantity of land must be available and must be so located that the disposal area does not become an eyesore. The amount of land necessary, of course, varies with the amount and characteristics of the solids. Allowance must be made for foreseeable future operations. Provision must be made for conveying or trucking to the site, unless the material is conveyed as a slurry which can be further dewatered at the site. Occasionally, solids separated during processing are stored; this occurs where the solids contain valuable materials uneconomic to extract by present technology but where a future breakthrough may make recovery attractive.

Impoundment and regulated discharge

Some materials such as chlorides are, by their very nature, not amenable to any practical sort of treatment. The only reasonable technique thus far devised for handling such wastes is impoundment, with regulated discharge to control the concentration in the receiving stream.

Dilution

A very obvious and important means of disposal is by dilution. Where the receiving stream or body of water is large in relation to the waste flow, dilution of the waste can sometimes be achieved so that waste concentrations are negligible or at least reduced many times. Where untreatable wastes such as chlorides are involved, plant location next to large volumes of dilution water may eliminate an otherwise difficult pollution problem. Occasionally it may be desirable to combine waste flows with such relatively uncontaminated flows as cooling water in order to take advantage of dilution factors available.

Caustic Soda Industry

The caustic soda industry, among the other chemical industries, has of late attracted particular attention of pollution control authorities world over. The concern is due to the release of mercury into the environment from the industrial units using the mercury cell process for the manufacture of caustic soda. Caustic soda is manufactured either by electrolytic or chemical processes. In the electrolytic process either mercury or perforated steel plates are used as cathodes to which sodium ions migrate. The former is called mercury cell process and the latter diaphragm process because the plates are wrapped with either asbestos or synthetic membrane diaphragm. The diaphragm allows sodium ions to pass through but hinder diffusion of products. In the chemical process, sodium carbonate and calcium hydroxide are reached to produce caustic soda and calcium carbonate.

Manufacturing process

A flow diagram depicting the unit processes in the mercury cell electrolytic method of manufacture of caustic soda is presented in Fig. 23.1. Caustic soda is produced by the electrical decomposition of the solution of sodium chloride in water known as brine, the preparation of which is the first step in the manufacturing process. A saturated solution contains 310 grams of sodium chloride per litre.

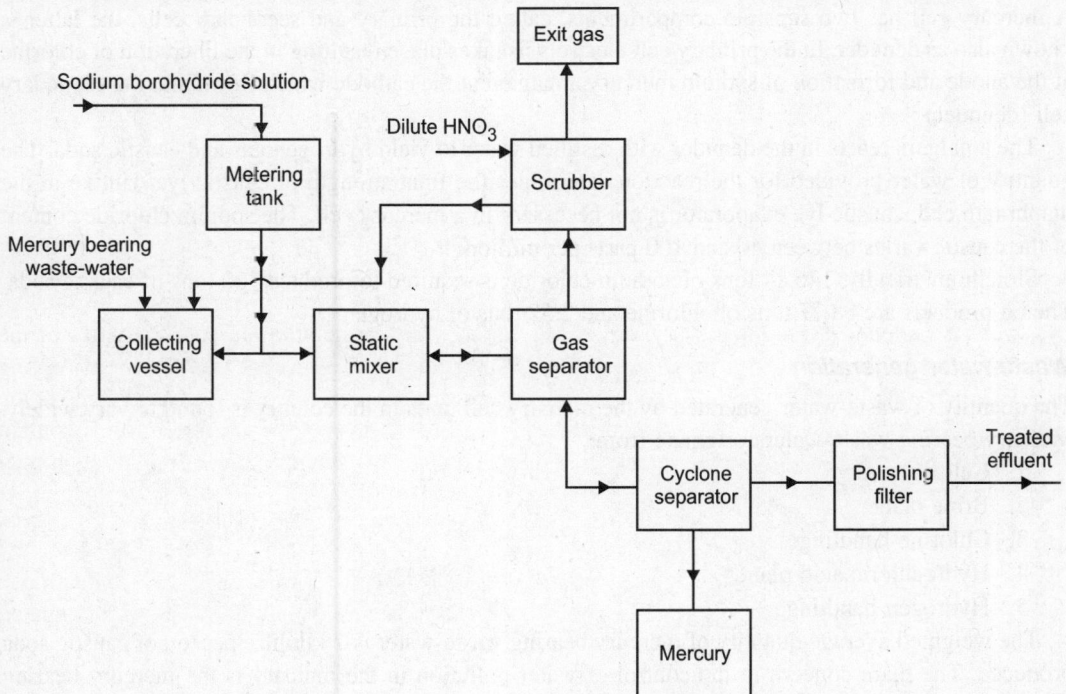

Fig. 23.1. Schematic diagram of reduction process using sodium borohydride.

However, in the electrolytic cell only 35 to 40 gram/litre of sodium chloride is decomposed. The spent brine or depleted brine is reused after replenishment with common salt. The salt supplied to the industry contains impurities. The Bureau of Indian specification (BIS) 791:1967 specifies salt quality for use in chemical industries. However, the BIS grade–1 salt is not manufactured in the country with the result purification of the supplied salt has to be carried out at the caustic soda plants. The purification is aimed at the removal of calcium and magnesium compounds, a higher degree of purification being required in the mercury cell plants than in diaphragm plants.

The purification is effected by the addition of sodium carbonate, barium carbonate, hydralated lime, caustic soda and settling agents such as glue or synthetic coagulant chemicals resulting in the precipitation of impurities. It is estimated that sludge produced in the purification process is 0.03 ton per ton of caustic soda produced which is double the international figure. The difference is due to the quality of salt made available to the industry. This aspect has a bearing on the consumption of mercury in the mercury cell plants because the mercury present in the returned spent brine gets into the sludge, during the purification process. The more the sludge produced, the more is the mercury loss with the sludge. An analysis of purification sludge in a typical mercury cell plant yielded the result of 2.4 mg of mercury per gram of sludge on dry weight basis. Based on the result the estimated mercury loss with the sludge is about 72 gram/T of caustic soda produced ([2.4 mg of Hg/gm of dry sludge] × [30 kg of dry sludge/T of caustic soda produced]).

This is a serious situation because disposal of the sludge is not being done with the care it demands. Surface runoff will carry the sludge, if not carefully disposed of, to the nearest water course.

The purified brine is electrolysed between the stationary carbon anode and mobile mercury cathode. A mercury cell has two separate compartments, called the primary and secondary cells, the latter is known also as denuder. In the primary cell electrolysis takes place resulting in the liberation of chlorine at the anode and formation of sodium mercury amalgam at the cathode which flows into the secondary cell (denuder).

The amalgam reacts in the denuder with distilled water to yield hydrogen gas and caustic soda. The quantity of water provided for the reaction determines the final strength of caustic lye. Unlike in the diaphragm cell, caustic lye evaporator is not necessary in a mercury cell. The sodium chloride content of the caustic varies between 40 and 100 parts per million.

Stoichiometrically, 146.14 tons of sodium chloride is required to produce 100 tons of caustic soda. The co-products are 88.77 tons of chlorine and 2.52 tons of hydrogen.

Waste-water generation

The quantity of waste-water generated by the mercury cell units in the country is found to vary widely. Mercury bearing waste-waters originate from:

1. Cell house.
2. Brine plant.
3. Chlorine handling.
4. Hydrochloric acid plant.
5. Hydrogen handling.

The weighted average quantity of mercury bearing waste-water is 7 kilolitre per ton of caustic soda produced. The main concern in the control of water pollution in the industry is the mercury bearing waste-water. It has to be segregated and treated for mercury removal before it is allowed to mix with waste-water from other sections of the industrial unit.

Average mercury consumption in the industry is calculated at 394 grams per ton of caustic soda produced based on mercury supplied to the industry. It is considered very high, even conceding that there are plausible causes for such a high loss.

The estimated loss of mercury along with the brine purification mud is significant accounting for 64.5 per cent of the total loss assuming an average mercury concentration of 8.5 mg/gm of mud on dry basis. The loss through the water route is only 0.3 per cent. However the possible transformation of metallic mercury into toxic methyl mercury in aquatic environment with subsequent biomagnification renders the route a cause for concern.

Significant loss of mercury with the brine mud is attributed to the poor quality of salt made available to the industry. An immediate measure being considered is to identify safe dumping place where the mud be disposed of. The ultimate solution lies in producing alkali grade salt at the salt works by installing washeries and purification units. Regarding loss of mercury through other routes the concerned agencies will have to take appropriate measures at the earliest.

Suspended and dissolved substances in the waste-water are stable inorganic chemicals and do not exert chemical or biological oxygen demand. It is the presence of mercury which is of primary concern. The quality of waste-water is subject to temporal variation. In many works caustic soda plants are part of a large industry such as rayon, pulp or textile.

In such situations the waste-water from the caustic soda unit invariably gets mixed with waste-waters from other manufacturing sections, thus providing dilution.

As a result concentration of mercury gets reduced but the mass of mercury, obviously, remains unchanged. The effect of mercury in the food chain is related to the mass of mercury in the ambient water. Therefore, dilution is no solution to mercury pollution. The characteristics of combined waste-water from a mercury cell plant are presented in Table 23.2.

Table 23.2. Typical analysis of waste-water and water consumption pattern in a mercury cell plant.

(UNITS: mg/l except for pH and temperature)

Parameter	Minimum	Maximum
pH	10.4	13.5
Suspended solids	82	216
Temperature, °C	20	30
$Ca(OH)_2$	37	1887
$CaCO_3$	100	1540
Calcium	1040	2920
Magnesium	200	300
Chloride	2942	11670
Sulphate	160	5023
Free chlorine	14.2	195.2
Available chlorine	111	2041
Mercury	2.0	4.8
BOD	–	14.92
COD	43.2	60.5
Dissolved oxygen	6.1	8.5

(Contd...)

	Water consumption pattern	
Consumption point	Quantity kilo litre/hr.	Pollutants
Salt washing	0.100	NaCl
Brine filter washing	10.000	NaCl
Cell washing	0.600	Hg, NaOH
Purification sludge	0.830	Ca, Mg, Fe, Ba, Hg
Bleach liquor sludge	12.000	Ca compounds
Chlorine drying unit	0.023	dilute H_2SO_4
Cooling water	24.547	Oil, acid, alkali
	48.100	

Unutilised chlorine is either emitted into the atmosphere or absorbed in lime slurry. In the latter case, bleach liquor containing free chlorine is formed. Its disposal into a stream should not be permitted.

Storage of the liquor in lagoons may cause groundwater pollution and therefore, careful watch has to be maintained.

Spillage and leakage of the liquor will introduce suspended and dissolved solids and chlorine in combined waste-water emanating from the industry.

The other pollutants that may enter the combined waste-water from the mercury cell units are sodium chloride as a result of handling loss in the factory and from brine filter washing, condensed acidic chlorine water containing dissolved mercury and dilute sulphuric acid from chlorine drying unit. Cooling water may carry oil and grease. Overflow from brine purification tank and spillage, if any, in the area may also find their way into the effluent. Magnesium chloride from salt washing operation may enter the combined wastestream.

Inplant control measures to minimise mercury loss in waste-water

Mercury is not part of the products and hence any replenishment requirement indicates discharge of mercury into the environment through the many routes. The following inplant control measures are suggested to minimise mercury loss through the waste-water route:

1. The brine in the mercury cell may contain 15 to 50 mg/l of mercury. Therefore, preventive maintenance of fittings vulnerable to leakage is necessary to avoid leakage of brine resulting in mercury reaching the waste-water stream. Prevention of overflow from brine storage and purification tanks is necessary for the same reason.
2. Steady supply of electricity would ensure nonoccurrence of mercury-butter phenomenon resulting in high mercury ion in the spent brine. If direct current to the cell fails the mercury circulation pumps should be run by alternate power supply. Provision should also be made to feed the cells with polarising direct current.
3. Recirculation of cell and end box wash water should be done.
4. Prevention of leakage of mercury bearing condensate from the cooled hydrogen line.
5. Sloping of the cell room floor towards mercury collection pits and rendering the flow smooth with epoxy paint.
6. Mercury in the waste-water emanating from the cell room should be kept in solution by the addition of sodium hypochlorite or by any other means.
7. Where activated carbon is used, the waste-water may be reused for brine preparation.

8. Use of metal anodes in cells in place of graphite anodes minimises the number of times the cells have to be opened and thus reduces mercury loss through vapourisation. It also eliminates the loss of mercury with graphite particles.

Waste-water treatment schemes

Segregation of mercury bearing waste-water is a prerequisite to waste-water management. Before that it is necessary to explore all possibilities of preventing mercury getting into the waste-water. It is claimed that total control systems encompassing water, air, land and product pollution by mercury are available. System ensuring total recycle of waste-water is also said to be available. The uniqueness of mercury in caustic soda manufacture is that theoretically no replenishment of mercury is necessary. Therefore, thrust in mercury pollution control has to be on the prevention of mercury loss to the environment. Treatment of waste-water to reduce concentration of mercury is secondary. However, methods to achieve the former are patented.

Treatment methods available to reduce concentration of mercury in waste-water are:
1. Reduction process.
2. Sulphide treatment.
3. Ferrous chloride treatment.
4. Magnetic ferrites.
5. Ion exchange.
6. Ion exchange followed by chelating resin.

The reduction process consists of reduction of all mercury compounds to the metallic state which may be followed by filtration and is suitable for small volumes of concentrated effluent. It can be accomplished in two different ways:
1. By treating the effluent with a less noble metal such as copper, iron, zinc and aluminium. Mercury is recovered as an amalgam or as droplets coalescing on the surface of the metal. It can be recovered in a pure state by electrolytic method.
2. By treating the effluent with reducing chemicals like hydrazine, hydroxylamine, hypophosphorous acid, formaldehyde and sodium borohydride. Mercury is recovered by coalescence and/or filtration. A 5-micron filter reduces the mercury concentration to below 0.1 mg/l. The process is illustrated in Fig. 23.1.

Sulphide treatment consists of treatment of mercury bearing waste-water with either sodium hydrosulphide (NaHS) or sodium sulphide (Na_2S) and a flocculent. The following reaction takes place in the treatment:

$$Hg \text{ (metal)} + Hg \text{ (divalent)} + Hg \text{ (monovalent)} + S = Hg + HgS$$

The interpretation of this equation is that metallic mercury remains unaffected but mercurous and mercuric compounds react to form sulphides. They are insoluble and settle down like mercury metal and can be recovered by filtration or settling or both. The drawback of this process is that an addition of excess solution of sulphide forms soluble salt complex, particularly at high pH. It is necessary therefore to keep pH restricted between 7 and 7.5. It is achieved by adding sodium carbonate at 2 gm/l. However, such strict pH control is not necessary if sodium hydrosulphide were used instead of sodium sulphide. A careful operation ensures 0.05 to 0.06 mg/l of mercury concentration in the effluent. A schematic diagram of the process is illustrated in Fig. 23.2. The waste-water thus treated, may further be treated with hydrazine (N_2H_4) to convert the remaining dissolved mercury into finely dispersed metallic form.

Further treatment through carbon precoated filter will produce an effluent with mercury content in the range of 0.02 to 0.03 mg/l.

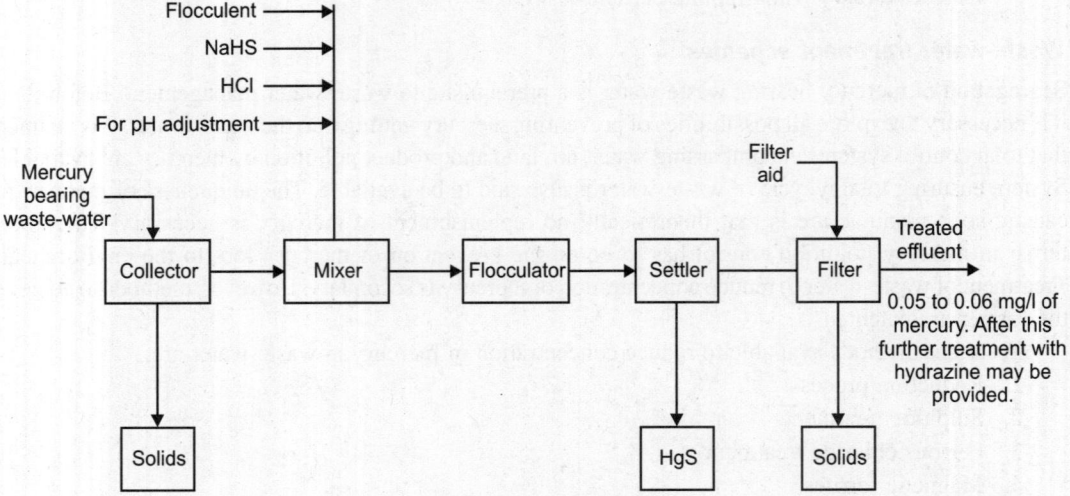

Fig. 23.2. Schematic diagram of hydrosulphide treatment.

In the ferrous chloride treatment mercury salts in waste-water are reduced to insoluble compounds as shown in the ionic equation:

$$2\ Fe^{++} + 2Hg^{++} + 8(OH)^- = 2Fe(OH)_3 + Hg_2O + H_2O$$

Insoluble ferric hydroxide and mercuric oxide are formed which jointly precipitate. The reaction takes place after pH is adjusted between 9 and 9.5 and a 50 mg/l excess alkali is added. Ferrous chloride should be added after this alkaline condition is attained by waste-water. The precipitate can be settled which takes several days or it can be filtered. The effluent contains 0.005 to 0.006 mg/l of mercury.

In the magnetic ferrites process for every mole of mercury present in water, 2 moles of ferrous sulphate is added and the water is neutralised with alkali when a dark green complex of hydroxide is formed, as shown by equation:

$$x M + Fe_{(3-x)} + 6(OH) = M_x Fe_{(3-x)} (OH)_6$$

Oxidation of the complex with air follows during which a black ferrite is formed which is represented as below:

$$M_x Fe_{(3-x)} (OH)_6 + \tfrac{1}{2} O_2 = M_x Fe_{(3-x)} (O_2 + 3H_2O$$

(*Note*: M means any heavy metal like mercury and x means the number of atoms).

A magnetic separator removes the insoluble ferromagnetic ferrite from the solution either for disposal or possible use in the manufacture of microwave transmission products.

Performance data show that the mercury content was reduced from 6 mg/l before treatment to 0.005 mg/l in the treated effluent. Ferrite does not redissolve as do simple metal hydroxides and is easy to separate because of their big particle size.

In treatment methods using ion-exchange and chelating resin mercury contamination of water, brine mud and slurries can be reduced from 20 mg/l to 0.005 mg/l. The method is in operation in Japan and Sweden for over six years. There are two steps in the demercurisation process as shown in Fig. 23.3.

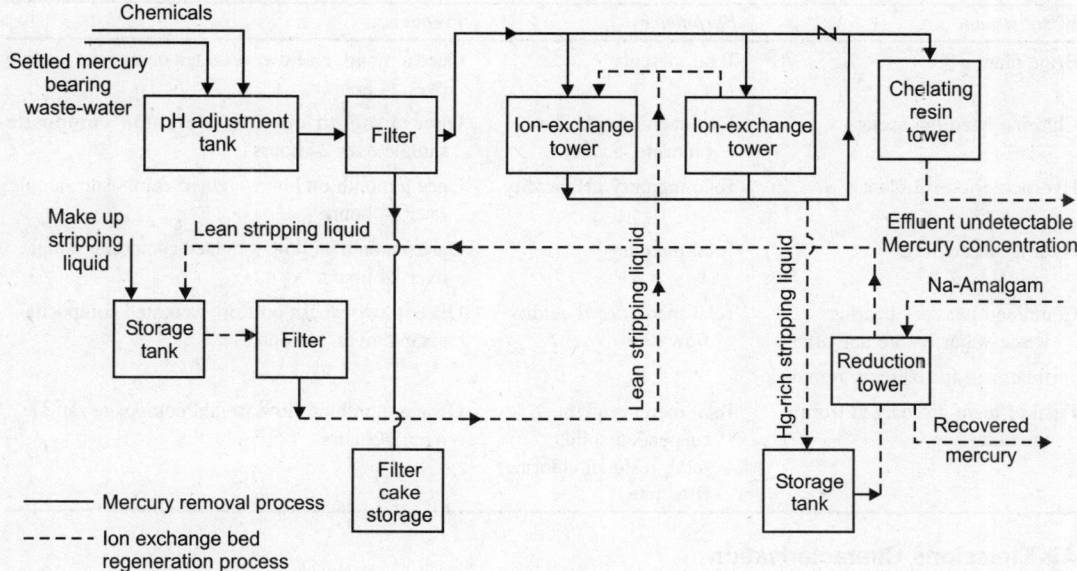

Fig. 23.3. Mercury removal by ion-exchange resin and chelating resin beds.

Step 1: Waste-water containing 20 mg/l mercury is allowed to settle in large tanks for metallic mercury to sink. Its pH is thereafter, adjusted to acidic range to remove free chlorine contamination. It is filtered to remove insolubles and then passed through an ion-exchange resin column which is selective to mercury. The concentration of mercury gets reduced to 0.1 to 0.15 mg/l. The resin is unaffected by the presence of iron, vanadium, chromium and heavy metal hydroxide but gets poisoned by sodium hydroxide and free chlorine. It is regenerated by an inorganic salt solution of proprietary composition. During regeneration, mercury is removed from the resin bed and recovered in the metallic form. The chemical identify of the resin and the regenerating agent is not disclosed.

Step 2: If required, the waste-water may be further treated in another tower containing a patented chelating resin, coded as MR which brings down mercury to an undetectable level. Break through of the resin is considered to have taken place when mercury content in the effluent reaches 0.005 mg/l and then it is replaced by fresh resin. A few modifications of the process are available.

Monitoring

Monitoring individual waste streams and the combined effluent before and after treatment in respect of relevent parameters is necessary from the point of view of pinpointing source of loss of mercury and of compliance with the standards. A monitoring schedule identifying the waste streams to be monitored, pollution parameters for analysis and frequency of sampling is presented in Table 23.3. Measurement of quantity of waste-water flow is necessary for which appropriate measuring devices will have to be installed.

Table 23.3. Monitoring schedule for caustic soda (mercury cell) industry.

Waste stream	Parameter	Frequency
Cell house	Total mercury	Once a month on flow weighted composited sample over 24 hours

(Contd...)

Waste stream	Parameter	Frequency
Brine plant	Total mercury	Once a month on flow weighted composite sample over 24 hours
Chlorine handling section	Total mercury pH, free chlorine, acidity	Once a month on flow weighted composite sample over 24 hours
Hydrochloric acid plant	Total mercury pH, acidity	Once a month on low weighted composite sample over 24 hours
Hydrogen handling	Total mercury	Once a month on flow weighted composite sample over 24 hours
Combined mercury bearing waste-water before and after treatment for mercury removal	Total mercury pH acidity flow rate	Once in two weeks on flow weighted composite sample over 24 hours
Final effluent discharged from the unit	Total mercury pH, suspended solids, total, residual chlorine, flow rate	Once a month on flow weight composite sample over 24 hours

Air Emissions Characterisation

The air emissions to be controlled from diaphragm-, mercury-, and membrane-cell chlorine plants include chlorine gas, mercury vapours, asbestos, and hydrogen.

Chlorine

Chlorine gas is listed as an extremely hazardous substance under, and any person handling in excess of 100 pounds of the gas must take certain emergency precautions.

The potential sources of the various emissions are:
1. From the cell flanges, connections, and relief devices.
2. From the low-pressure headers during normal operations, startups and shutdowns, and emergency operations.
3. From chlorine dissolved in the sulphuric acid and then released to the atmosphere.
4. From chlorine compressor seals.
5. From high-pressure piping and liquefaction equipment.
6. Tail gas from recovery units and scrubbers.
7. From tanks and equipment being loaded.

Mercury

The potential sources of the various emissions are:
1. From the cells during operation and maintenance.
2. In the vent gas from the cells.
3. In the hydrogen from the cells.
4. From piping leaks.

Asbestos

The potential sources of the various emissions are:
1. Receiving and handling.

2. Diaphragm manufacture.
3. Disposal methods.

Hydrogen

This very light gas does not appear on the regulatory lists, and spill and release reporting is not required. Hydrogen is an extremely flammable gas, and extreme care and good equipment design are required in its handling to avoid fires and explosions. Among the potential sources of the various emissions are:
1. From the cells during operation.
2. From vent leak-through.

Air Pollution Control Measures

The control and elimination of fugitive and point-source emissions depend on plant design, operating philosophy, and the maintenance programme.

Chlorine

Emissions from diaphragm cells and low-pressure headers are controlled or eliminated by the use of corrosion-resistant materials, the use of elastomer gaskets retained by bolts or heavy compression springs, and operation at atmospheric pressure or under a very slight vacuum. All headers and major equipment are vented through safety relief systems to caustic scrubbers during startups, shutdowns, or emergency conditions to eliminate releases to the atmosphere. These scrubbers must have sufficient capacity to neutralise full production capacity for the time necessary to start up with inerts or to shut down and evacuate the system.

A control system capable of maintaining a consistent pressure throughout the system under varying rates and stream compositions is extremely important to successful operation. Operation of the low-pressure system above atmospheric pressure can result in the release of fugitive chlorine, and operation at deep vacuums can result in the intrusion of atmospheric air, which causes high noncondensible flows in the high-pressure system. Cells, piping headers, and major equipment are evacuated to a vacuum system and purged prior to opening them for repairs.

The design and operation of the sulphuric acid drying system are extremely important to long-term successful operation. The rate of corrosion of steel used in the high-pressure system is accelerated to unacceptable levels as the moisture level increases. The sulphuric acid used to dry the chlorine will contain dissolved chlorine, which must be recovered and recycled by vacuum stripping or neutralised with sulphides to prevent release to the atmosphere after discharge and neutralisation of the spent acid.

Most manufacturers now use multistage centrifugal compressors, although reciprocal and sulphuric acid wetted-lobe compressors have been used. Double seals with a dry air purge between the seals are used to prevent seal leakage. It is important to design all high-pressure piping and equipment with the minimum number of flanges, and no leakage is acceptable. Because of the rapid corrosion of steel when it is exposed to wet chlorine, any small leak will rapidly become a large leak. High-pressure equipment is also vented through safety relief valves to a caustic scrubber with sufficient capacity to handle full production rates for the period required to shut down and evacuate the equipment.

The high-pressure chlorine is condensed in liquefaction equipment at temperatures ranging from $+75°$ to $-50°F$, pending on the operating pressure of the system and the chlorine removal desired. The tail gas from the system, which consists of the noncondensibles and residual chlorine, can be sent to chlorination processes that can use low-concentration streams or to recovery systems. Recovery systems are capable of complete recovery and recycling of the chlorine in the tail gas. The design of recovery systems must recognise the presence of low levels of hydrogen that can react violently with the chlorine in the system. The hydrogen level must be monitored closely and steps taken when required to ensure that it is kept below the explosive range. These systems consist of multiple stages of scrubbing, absorption/reaction, and refrigeration of the stream, with the chlorine being recovered by desorption/reaction from the scrubbing medium. The capability to divert the tail gas to a caustic scrubber should be included in the process to handle upsets or outages.

The liquid chlorine is transferred to scale tanks or spheres for storage prior to shipment. All liquid lines are equipped with expansion chambers between the valves to allow for the expansion of trapped liquid without pipeline rupture. The storage tank vapours are equalised to the drying tower system for recovery. Vapours from transportation equipment (i.e. tank cars, tank trucks, and barges) are recovered by a vacuum system. Pumps are equipped with double seals with a dry air pad to eliminate chlorine leakage at the seals. Valves and cocks are of a special leak-free design with bellows or multiseal construction.

Mercury

The control of mercury emissions is dependent on equipment and facility design, a good preventive maintenance programme, and observation of appropriate operations procedures.

The primary cell is operated at a slight vacuum imposed by the chlorine compressor to eliminate emissions. The end boxes, which are the mercury inlet and outlet chambers, should have vapour-tight covers. In addition, water is added to the end boxes to wash, cool, and control vapour emissions from the mercury. A slight vacuum is pulled on the end boxes by a fume system further to reduce the possibility of mercury release. These vent streams are cooled, scrubbed, and/or passed through treated carbon to ensure a mercury-free discharge. Mercury pump tanks, pumps, and seal pots are treated in the same way: vacuum operation, cover water, and tight covers.

Cell maintenance procedures and programme quality are extremely important to the prevention of the escape of mercury vapours. Leaks in pipes, flanges, and other equipment must be repaired immediately, and a preventive maintenance programme should be in place to prevent their occurrence. Any time that the cell cover or top must be removed for maintenance, the cell should be cooled and vented and, when possible, the bottom kept covered with water or brine to reduce emissions. The time that the cell is open and the frequency of such openings should be held to a minimum.

A potentially major source of mercury loss from the system is with the hydrogen generated in the decomposer or denuder cell. The hydrogen from the cells should be cooled at each individual cell and the pipes arranged to return condensed mercury directly to the cell, thus reducing handling of the mercury in later recovery steps. The hydrogen product is further cooled in refrigerated coolers and/or scrubbed with chemical scrubbers in series to recover the maximum quantity of mercury. Liquid mercury recovered is recycled directly to the cells, and the mercury recovered in scrubbers can be returned with the brine.

The floors in the cell room area should be smooth with no cracks and all seams sealed, so that no mercury or mercury solutions can be held for later release. Trenches should be sloped to prevent mercury puddles and a layer of water kept on sumps. Where possible, waste-waters are recycled, and when this is not possible they are treated to remove mercury before they are discharged.

A good inspection programme involving daily inspections and floor washdown can further limit emissions. Special vacuum cleaners equipped with carbon filters are available for the cleanup of mercury droplets discovered by the inspection programme.

Operator training and compliance with operating standards are an important part of the maintenance of minimum emissions. They involve housekeeping, spill control, personal hygiene, and control of operating parameters that affect emission levels. Well-designed systems can reduce emissions to less than 100 g/d from end-box systems and to less than 10 g/d from the hydrogen system. Actual testing is required to meet the 1000 grams per day emission limit for point sources.

Asbestos

The control of asbestos (friable) emissions starts with the receipt of the asbestos, which arrives in airtight plastic bags, shrink wrapped with plastic, on pallets in boxcars or trailer trucks. The pallet loads are inspected before removal is started. If any tears or spills are noted, they are repaired with tape and the spilled material cleaned up using a vacuum cleaner. Workers wear special clothing and masks during all phases of unloading. The asbestos bags should be stored in a dedicated area with appropriate warning signs. If water is used for the cleanup of spills, care must be taken to limit pressure to ensure that fibres are not broadcast into the air.

Periodically, a pallet of asbestos bags is transferred to the cell repair area. Access to this area is limited, and appropriate warning signs are posted. When it is necessary to add asbestos to the liquid bath used to produce the asbestos diaphragms, a bag is placed inside a glove box through a hinged Plexiglas door. The box is equipped with rubber gloves hermetically sealed through openings into the box. These gloves are used for all operations inside the glove box. A vacuum is pulled on the box system through the vacuum mix tank, which ensures that no fibres escape the box system. The asbestos bags can be opened inside the glove box and the asbestos transferred to the mix tank through a transfer pipe. From this point on, the asbestos is handled in the wet (nonfriable) state, eliminating exposure. The empty asbestos bags are transferred to a clean disposal bag attached to an opening in the glove box. Before adding the next bag, the box can be cleaned with a vacuum wand located inside the box. All spills of asbestos slurry are hosed down as required to prevent any later generation of fibres.

Cell diaphragms eventually must be replaced because of the gradual plugging caused by trace impurities in the brine. At that time, the wet asbestos is washed from the cathode into a sump. It is then recovered in a filter operation and, while still in the wet form, is placed in plastic-lined boxes for secure landfill. Cleanliness is important, and any spilled material should be washed to the sump for recovery.

Hydrogen

Hydrogen is a very light flammable gas that does not appear on regulatory lists. Hydrogen from mercury cells can contain mercury, for which the relevant control measures are as discussed under 'mercury'. The major emphasis for hydrogen is on control systems to provide safety in the handling of this extremely flammable gas. Cell and suction header pressures are maintained near atmospheric pressure to reduce losses from flanges. Because of the difficulty in preventing hydrogen from leaking through isolation and vent valves, the use of water seals is recommended where pressure permits. Double-block and bleed arrangements are recommended for higher-pressure service. High-pressure valves can be checked for leak-through by venting through seal tanks that can be filled with water to determine valve tightness.

Disposal of Wastes from Specific Industries

Sulphuric acid

From the standpoint of tonnage, sulphuric acid is the leading inorganic chemical. The major portion is made by the contact process, in which sulphur is burned to sulphur dioxide, purified if necessary and a mixture of dry sulphur dioxide gas and air passed through a preheater and a reactor where it is catalytically oxidised to sulphur trioxide. After cooling, the gas is partially absorbed by oleum and finally scrubbed with a 97 per cent acid. The products are oleum and 98 per cent sulphuric acid. The process is such that no significant liquid wastes are formed and hence no water pollution problem is encountered. Air pollution problems may exist.

Ammonia

The large production of ammonia results from its demand by the fertiliser industry as a source of fixed nitrogen. Ammonia may be applied to the land directly as liquid, in solution or as an ammonium compound.

Ammonia is produced by the reaction of nitrogen and hydrogen gases in the presence of a nickel catalyst at high temperature and pressure. The principal sources of nitrogen and hydrogen are air and natural gas, although other raw materials are currently used. Where other special sources of synthesis gas are employed, operating conditions and catalysts are different. No stream pollution problems of any significance are encountered in the manufacture of ammonia.

Sodium carbonate

The basic method used in the manufacture of sodium carbonate or soda ash is the ammonia-soda process perfected by Solvay. The chemistry involved is fairly complex and is discussed in most elementary chemistry texts as a classic example of inorganic reaction. Raw materials for the process are salt, limestone and coke; products and by-products are sodium carbonate, calcium chloride and carbon dioxide. Carbon dioxide and ammonia are recycled, with the ammonia entering into a number of intermediate reactions. The process consists of the following chemical reactions:

$$NH_4OH + CO_2 \longrightarrow NH_4HCO_3 \qquad \ldots (23.1)$$
$$NH_4HCO_3 + NaCl \longrightarrow NaHCO_3 + NH_4Cl \qquad \ldots (23.2)$$
$$2NaHCO_3 \longrightarrow Na_2CO_3 + CO_2 + H_2O \qquad \ldots (23.3)$$
$$2NH_4Cl + Ca(OH)_2 \longrightarrow 2NH_3 + CaCl_2 + 2H_2O \qquad \ldots (23.4)$$

Initially, a strong sodium chloride solution is purified to remove calcium, magnesium and other heavy metal ions. The brine is ammoniated and contacted with carbon dioxide in a carbonating tower. In the tower, reactions 23.1 and 23.2 occur, producing sodium bicarbonate and ammonium chloride. Because the bicarbonate has limited solubility in this solution, it crystallises and is separated on vacuum filters. The bicarbonate is calcined in a sealed rotary kiln, where reaction 23.3 occurs, forming light soda ash which is cooled and stored. Carbon dioxide from the calciner is recycled to the carbonating tower; additional CO_2 is produced by burning lime and by combustion of carbon. The filtrate from bicarbonate separation is treated with lime in a free-ammonia still, causing reaction 23.4 and the ammonia is returned to the process. The calcium chloride formed and the unreacted sodium chloride and other solids are waste products.

A typical analysis of the waste stream called 'distiller's waste' from a soda ash plant is given below:

Component	Concentration, g/litre
$CaCl_2$	85–95
NaCl	45–50
$CaCO_3$ •	6–15

(Contd...)

Component	Concentration, g/litre
$CaSO_4$ •	3–5
$Mg(OH)_2$	3–10
CaO	2–4
Fe_2O_3 and Al_2O_3 •	1–3
SiO_2 •	1–4
NH_3	0.006–0.012

The components marked (•) by an asterisk are relatively insoluble and are removed from the waste stream before discharge.

Standard practice in the industry is to remove the solids by settling in large lagoons or waste lakes, several hundred acres in extent. Impounding on this large scale is expensive and can be troublesome, especially if land availability at the plant site is limited. During its life, a soda ash plant may require several waste lakes, with new areas opened as old ones are filled. When the solids are settled, the dissolved chloride problem remains. The ammonia-soda process is only about 75 per cent efficient in terms of sodium and uses none of the chloride present in the original brine. Production of 1 ton of soda ash results in about 1½ tons of soluble chlorides as waste.

Typically, this quantity of chloride will be contained in about 2600 gal of liquid. Since chlorides generally are soluble, do not break down and are not subject to bacterial degradation, treatment possibilities are few. Part of the calcium chloride can be recovered from the waste by separation from the sodium chloride, although recovery is generally not economical at the low market price. In addition, market demand could absorb only a fraction of the calcium chloride available from soda ash production. The major use of calcium chloride is for ice and dust control on streets and roads; some minor uses have been found, as in concrete, but no processes or industries requiring substantial amounts of chloride have as yet developed. If all calcium chloride were recovered from the waste, one soda ash plant of medium size could supply the entire present demand.

A number of solutions to the soda ash waste problem have been investigated over a period of years. These include recovery of calcium chloride with separation and reuse of the accompanying sodium chloride, separation by means of ionic membranes and alternative methods of manufacture. Despite years of intensive effort by the industry, none of these possibilities has found widespread use, because of economical or technical reasons. Thus far, the ammonia-soda process has resisted substitute methods of manufacture.

Minor improvements in operating conditions and in equipment have been found and adopted, but have not resulted in significant improvement in the waste problem. In summary, there appears to be no alternative process that would produce a smaller volume and concentration of wastes and there appears to be no promising method of treatment of the waste produced.

Many of the plants originally built to produce soda ash were located on inland waters close to raw materials and markets. Later plants were established on or near oceans, where the effects of chloride discharges are negligible.

The ideal solution to this problem, with present technology, is removal of solids in a waste lake and discharge of the effluent to salt or brackish water. However, discharge of soda ash waste to freshwater is not a major pollution problem where ample dilution water is available. For plants located on the headwaters of an inland river system, detention ponds and regulated discharge to correspond with watershed runoff patterns minimise the pollution effect. Even though this practice does not reduce the total amount of chloride waste reaching the stream, detrimental effects of chloride are reduced because of the lower and more uniform chloride levels achieved.

Nitric acid

Nitric acid may be prepared by the catalytic oxidation of ammonia or by treating a nitrate salt with sulphuric acid. Most of the nitric acid is produced from ammonia. Anhydrous ammonia is mixed with air, heated and reacted in the vapour phase in the presence of a platinum-rhodium catalyst under moderately high temperatures and pressures. The nitric oxide formed is cooled and absorbed in water to yield nitric acid. The effluent from the process is primarily cooling water and presents no significant pollution problem.

Phosphoric acid

Phosphoric acid is made by treating phosphate rock with sulphuric acid or by burning elemental phosphorus produced in an electric or blast furnace. In the 'wet process' method of manufacturing phosphoric acid, phosphate rock and sulphuric acid are mixed and their reaction produces phosphoric acid and calcium sulphate.

The calcium sulphate precipitate is filtered and washed to remove as much phosphoric acid as possible and the filtrate is recycled to the process. The weak acid produced by the process is evaporated to concentrate it to about 50 per cent phosphoric acid. During evaporation, gaseous fluorides are evolved and are collected by water scrubbing. The scrub water is treated with lime to precipitate the fluorides and to raise the pH. This is ordinarily done in combination with similar wastes in an integrated phosphate plant.

Ammonium compounds

A number of ammonium compounds are produced by treating ammonia with the corresponding acid, such as nitric acid or sulphuric acid to produce ammonium nitrate and ammonium sulphate. Most ammonium sulphate, however, is a by-product recovered from coke oven operations. No significant pollution problems are presented in the manufacture of these compounds *per se*, although problems do arise from other aspects of the operation, particularly in coke ovens.

Oxygen

Commercial oxygen is produced by the liquefaction and subsequent fractionation of air. Nitrogen is produced as a coproduct and may be recovered or wasted. Additional fractionation and processing yields the rare gases argon, neon, helium, krypton and xenon. With the low temperatures involved in the production of oxygen, liquid effluents are not a problem.

Thus the inorganic chemical industry is made up of a large number of individual plants manufacturing individual products from specific raw materials and bearing little relation to each other from the standpoint of waste treatment.

In general, the products are made from raw materials that are mined for the purpose or from raw materials which are only one or two chemical steps from a naturally occurring deposit. The processes are old and well understood and reflect the most economic utilisation of materials found in nature. Until cheaper sources of raw materials are found, such as waste from other processes, it is unlikely that any of the present processes will be supplanted.

Air Emissions Characterisation

There are a number of emission sources within the natural sodium carbonate industry. Those judged most significant include calciners, bleachers, dryers, and predryers. These are process emission sources that have the potential to emit large quantities of particulate matter.

Calciners

Calciners are the largest source of particulate emissions from plants using the monohydrate process. These particulates consist of sodium carbonate and inerts. The exit gas from coal-fired calciners also contains fly ash. Particulate emissions from calciners are affected by the gas velocity and the particle size distribution of the ore feed. As the gas velocity increases, the rate of increase in the particulate emissions steadily increases. Thus, coal-fired calciners may have higher particulate emissions than gas-fired calciners because of higher gas flow rates. Additionally, small particles are more easily entrained in a moving stream of gas than are larger particles.

Sulphur oxides are produced from fuel combustion. The quantities produced depend upon the sulphur content of the fuel. Organics are also emitted from calciners, some of which are present in the feed in the form of oil shale. At the calcination temperatures used, organics may vaporise or be partially combusted. In addition, some organics may result from partial or incomplete combustion of the fuel.

Nitrogen oxides are emitted from direct-fired process heating units such as calciners and soda ash dryers, but very little emission test data are available for these processes. Reported emission factors are 0.0008 kg/mg of feed (0.016 lb/T) for one calciner with a cyclone and scrubber, and 0.056 kg/mg (0.111 lb/T) from another calciner with a cyclone and ESP.

Dryers, predryers, and bleachers

Three types of dryers are used for product drying in the monohydrate and direct carbonation processes: rotary steam tube, rotary gas fired, and fluid-bed steam tube. Sodium carbonate fines are emitted from each of these dryers.

Particulate emissions from dryers are affected by the gas velocity of the feed. As the gas velocity increases, the rate of increase in the total emission rate of particulates steadily increases. Therefore, because of higher gas flow rates, and higher gas velocities, fluid-bed steam tube dryers and rotary gas-fired dryers have higher emission rates than rotary steam tube dryers.

In the direct carbonation process, rotary steam-heated predryers are used to lower the water content of wet sodium bicarbonate crystals before they are calcined. Particulates of sodium bicarbonate are the primary type of emissions from predryers.

Emissions from bleachers consist mainly of particulates of sodium carbonate. Small amounts of compounds formed from the reactions of sodium nitrate may also be present in the particulates.

Air Pollution Control Measures

Particulate emission control techniques applicable to sources in sodium carbonate plants include:
1. Centrifugal separation.
2. Wet scrubbing.
3. Electrostatic precipitation.
4. Fabric filtration.

Centrifugal separators, or cyclones, rely on centrifugal forces to effect particulate separation from the gas stream. Cyclones are frequently used upstream of scrubbers or electrostatic precipitators (ESPs).

Scrubbers rely mainly on inertial impaction of particles with water droplets to effect particulate separation from the gas stream. Particles are contacted with wetted surfaces or atomised liquid droplets. As the gas stream diverges to pass such obstructions, the inertia of particles in the gas stream carries the particles into the water droplets or wetted surfaces. The particulate laden liquid is then separated from the gas stream, and either recycled to the production process or discharged as waste.

Electrostatic precipitators generate an electrical field by applying a high voltage to a discharge electrode system consisting of rows of vertical wires. The strength of the field depends in part on the gas composition. The subsequent migration of the charged particles to the collected plates depends on the particle size, resistivity, gas velocity and distribution, rapping, and field strength. The collecting electrodes are rigid plates that are baffled. Electromagnetic or pneumatic hammers are used to rap the electrodes, dislodging the collected particles, which then fall into hoppers. Baffling on the collecting electrodes provides shielded air pockets that reduce reentrainment of particles after rapping.

Fabric filtration is a process where dust particles in a gas stream are filtered out and collected.

Ore and product handling

Particulate emissions from ore and product handling operations, such as conveyor transfer points, crushing, and product sizing, are typically controlled by either venturi scrubbers or fabric filters (baghouses). These control devices are an integral part of the manufacturing process, capturing raw materials and product for economic reasons.

Calciners and bleachers

Calciners and bleachers are typically controlled by cyclones in series with ESPs. Venturi scrubbers, also used to control emissions from calciners, typically achieve lower removal efficiencies than ESPs. Higher removal efficiencies may be achieved with sufficient pressure drop. It appears that a pressure drop of 154 cm (60 inches) of water may be required to achieve a removal efficiency comparable to that achieved with a four-stage ESP, based on the removal efficiency achieved from EPA test data from a gas-fired calciner with a scrubber pressure drop of about 85 cm (33.5 inches) of water.

Fabric filters (baghouses) have not been reported to be in use to control emissions from calciners or bleachers. The sticky, hygroscopic nature of sodium carbonate could lead to bag blinding and/or caking.

Dryers and predryers

Venturi scrubbers are the devices used to control emissions from rotary steam tube dryers. Cyclones in series with venturi scrubbers are used to control emissions from fluid-bed steam tube dryers and rotary steam-heated predryers. Both venturi scrubbers and ESPs have been used to control emissions from gas-fired dryers.

The exhaust gas from both rotary and fluid-bed steam tube dryers and predryers is well suited to control by wet scrubbing. The removed sodium carbonate particles are quite soluble and hygroscopic. These characteristics enhance the removal of sodium carbonate particles in wet scrubbers. However, these characteristics coupled with the high water content of the dryer exit gas can result in operating problems for ESPs or baghouses.

Moisture in the exit gas can condense in ESPs, or baghouses. Wet, sticky dust adheres to the electrodes and hoppers of ESPs or blinds and cakes the bags in baghouses. ESPs have been used to control emissions from rotary gas-fired dryers, but the exit gas from these dryers is at a higher temperature and lower relative humidity than gas from steam tube dryers.

Recent and Future Industry Trends

Sodium carbonate facilities are potential major sources of particulate emissions. These emissions can be effectively controlled through the proper use and maintenance of conventional add-on particulate control techniques. Tests conducted at sodium carbonate plants along with industry data contributed to the selection of ESPs as the best system of emission reduction for calciners and bleachers and venturi scrubbers as the best system for dryers and predryers.

Small amounts of sulphur oxides and organics are also emitted from direct-fired calciners; however, source tests have indicated that these emissions are very low relative to uncontrolled particulate emissions.

Environmental issues have contributed to the phaseout of the production of sodium carbonate by the solvay process in the United States; for example, substantial quantities of aqueous waste, containing high concentrations of calcium chloride, are produced in the solvay process. Another reason for the decline in the use of the solvay process has been high fuel costs, since the Solvay process is more fuel intensive than any of the natural processes.

Most of the new sodium carbonate process lines constructed in Wyoming in the past decade use the monohydrate process. This process is likely to continue to be used in the future. Because of the proximity of the Wyoming trona mines to western coal mines, future plants are expected to make greater use of coal than older existing plants.

MINIMAL NATIONAL STANDARD (MINAS)

Generally three main aspects are taken into consideration in evolving the standards:
1. Adverse effect on health and environment.
2. Achievability of limits of pollutants by incorporation of pollution control measures indigenously available.
3. Economic viability of adoption by the industries.

Parameters of relevance to the industry are included in MINAS. The regulatory authorities may: (i) include additional parameters like total dissolved solid (TDS), sulphide, ammonical nitrogen, per cent sodium, etc.; (ii) exclude some parameters of metals after examining their absence in the industry; and (iii) make the prescribed limits more strigent if the recipient environment demands so.

In developing the limits of the pollutants it is considered that treated waste-waters will be diluted ten times in the receiving waters stream. It is ensured that the waste-water, thus diluted is not likely to cause any adverse effect on the water quality of receiving stream. The drinking water standards (BIS 10500–1988) and waste-water standards (BIS 2490 part I–1981) and also compared while arriving at the limits of the pollutants.

The volume of waste-water generated is low. Thus the quantity of the pollutants in the treated waste-water is also reasonably low. In case, the treated waste-waters are discharged on land, ground water is not likely to be affected. However, in case of land application ground water should preferably be monitored for the presence of heavy metals.

In general, the proposed treatment system is basically chemical precipitation of heavy metals to attain minimum solubility of metals. The separation of precipitates is to be carried out by settling is settlers/clarifiers. It is considered that such simple treatment system is not costly with respect to the investment. Cost for operation and maintenance is fairly low. Thus the treatment system is economically affordable by the industries.

The minimal national standards (MINAS) as evolved are presented in Table 23.4.

Table 23.4. MINAS for inorganic metallic compounds of chromium, manganese, nickel copper, zinc, cadmium, lead and mercury manufacturing industries.

Parameters	Not to exceed mg/l except pH
pH	6.0–9.0
Chromium	
Hexavalent	0.1
Total	2.0
Manganese	2.0
Nickel	2.0
Copper	2.0
Zinc	5.0
Cadmium	0.2
Lead	0.1
Mercury	0.0
Cyanide as CN	0.2
Oil and grease	10.0
Suspended solids	30.0

Note:
1. Additional parameters as necessary may be incorporated by state boards.
2. Some of the parameters relating to metals may be avoided if those metals are not present in the effluent of the industry.
3. Total concentration of the above metals should preferably be limited to 7.0 mg/l.

Chapter 24

Textile

INTRODUCTION

The textile industry actually represents a range of industries with operations and processes as diverse as its products. It is almost impossible to describe a 'typical' textile effluent because of such diversity. After fabrics are manufactured, they are subjected to several wet processes collectively known as 'finishing' and it is in these finishing operations that the major waste effluents are produced. These finishing processes are complex and ever-changing. This is a fact of life that is reflected in the variety of chemicals that find their way into textile finishing waste-waters.

WATER POLLUTION FROM BOILERS

Boilers meant for raising steam by combustion of fuels like coal, fuel oil, etc. contribute substantially in causing pollution in two ways:
  1. Water pollution caused due to discharge of boiler blowdown water.
  2. Air pollution caused due to emission of chimney gases into the atmosphere.

In order to ensure that concentration of dissolved solids in the boiler water does not exceed the permissible limit, some amount of boiler water has to be discharged from the boilers either continuously or intermittently. The boiler blowdown water so discharged contains very high amount of dissolved solids ranging from as low as 3500 mg/litre for water tube boilers to as high as 10,000 mg/litre for Lancashire boilers.

In fact in some cases, the blowdown water from the Lancashire boilers are known to contain as much as 60,000 mg/litre of dissolved solids. Whereas raw water with high carbonate alkalinity is fed into the boilers without softening, hardness precipitates inside the boilers giving rise to high level of suspended solids in the blowdown.

The alkalinity of the blowdown is also very high and is generally about 15 per cent of the dissolved solids. Likewise, pH of the blowdown is also about 10.5 to 11.0. Thus the blow-down adds to the water pollution in terms of dissolved solids, suspended solids, alkalinity and pH. Table 24.1 gives the characteristics of boiler blowdown water.

WATER POLLUTION FROM WATER-TREATMENT PLANTS

It is required that the water used for processing of textiles and in generation of steam in the boilers should not be hard. The type of softening treatment generally imparted to the hard water is ion-exchange softening process. There is no water pollution from the softening process as such. However, when the ion-exchange resin is regenerated with concentrated solution of common salt, its effluent will contain very high amount of sodium ions apart from high dissolved solids. In case of partial demineralisation process, the effluent contains weak acid. Some mills have water treatment plant of precipitation type. Here lime is added to precipitate carbonate hardness and soda ash is added to remove noncarbonate hardness. Some coagulants like alum or sodium aluminate, etc. are added to accelerate the precipitation process. In this type of treatment, some amount of water is drained out from the bottom of the reaction vessel to get rid of the precipitates. This wastewater contains high amount of suspended solids and adds to the water pollution when it joins the composite stream.

Table 24.1. Characteristics of boiler blowdown water.

pH	10.5 to 11.5
TDS	18000 mg/l
Suspended solids	1500 mg/l
Temperature	95° to 100°C
Chlorides as Cl	4500 mg/l
Sulphate as SO_4	1700 mg/l
Total alkalinity as $CaCO_3$	8000 mg/l

OPERATIONS INVOLVED IN FINISHING

It is important to know the various operations involved in finishing before the problems of treatment of waste-waters from these operations can be appreciated. It is beyond the scope of this chapter to discuss all these operations in detail. The finishing operations for cotton, wool and synthetic fibre processing are discussed very briefly in the following sections.

Cotton

Slashing

Slashing is the first process in which liquid treatment is involved. In this process, the wrap yarns are coated with 'sizing' in order to give them textile strength to withstand the process exerted on them during the weaving operation. The common sizes are starch, polyvinyl alcohol, carboxy methyl cellulose, polyacrylic acid and polyester size. Starch is commonly used especially in combination with other sizes such as polyvinyl alcohol. Starch is very high in BOD. Most other sizes exert very little BOD even though they may contribute to COD. Sizes represent the single largest group of chemicals used in the industry.

Desizing

Desizing removes the substance applied to the yarns in the slashing operation, by hydrolysing the size into a soluble form. Acid desizing and enzyme desizing are the two common methods. The method of desizing is dependent on the actual size used.

Scouring

Noncellulosic components of the cotton are removed by soap solution or hot alkaline detergents. Solvent processes, on the other hand, will use little water. Even though the projected solvent recovery is between 90 and 95 per cent, a significant amount of solvent may reach the atmosphere and the waste stream.

Bleaching

Bleaching is performed to remove the yellowish colour of the fibre and render it white. Sodium hypochlorite, hydrogen peroxide and sodium chlorite are the three bleaches most commonly used for cotton.

Mercerisation

Mercerisation uses sodium hydroxide, water and an acid wash. This process increases the dye affinity, tensile strength and surface luster.

Dyeing

Dyeing in most cases is performed in dye houses or colour shops when the goods are sent there after mercerising. A few large plants may do the dyeing in the same facility where the other finishing operations are performed. Depending on the type of dye used, the waste will vary. The general process is to use rolling machines that contain a paste of dye, thickener, hygroscopic substances, dyeing assistants, water and other chemicals.

Wool

In the processing of woolen fibres, five sources of pollution load exists: scouring, dyeing, washing after fulling, neutralising and bleaching. Scouring removes impurities from the woollen fibres. Wool scouring produces one of the strongest industrial wastes in terms of BOD by contributing 55 to 75 per cent of the total BOD load in the wool finishing. The BOD load contributed by dyeing may represent a small per cent of the mill's total BOD load, but again, this percentage may vary depending on the plant. Washing after fulling may be the second largest source of BOD in wool finishing waste-waters. Neutralisation after carbonising the fabric does not significantly contribute to the BOD load. The amount of wool fabric bleached is very small because bleaching is required for white and lightly shaded fabrics only. Hence, bleaching (like neutralisation) does not significantly contribute to effluent BOD.

Synthetic Fibres

The demand for synthetics has experienced phenomenal growth in the past several decades. Synthetic textile fibres may be cellulosic or noncellulosic. Rayon and cellulosic acetate are examples of cellulosic fibres. The major noncellulosic fibres are acrylics, nylon, polyester, etc. Different aqueous processes are involved in the production of these fibres that lead to varying pollution loads. Desizing, scouring and dyeing are the chief sources of water pollution from finishing processes for synthetic fibres.

Dyes

In this section, a discussion of dyes has been included since colour is a perplexing problem in treating textile mill waste-waters. Dyeing of fabrics is a common process for all fabrics, as discussed in the earlier sections.

It is important to note that all coloured substances are not necessarily dyes. A dye is a coloured substance that can be applied in solution or dispersion to a substrate (e.g. a textile fibre, paper, or foodstuff), thus giving it a coloured appearance. The substrate has a natural affinity for appropriate dyes and readily absorbs them from solutions or aqueous dispersion. Concentrations of dyes, temperature and pH are important factors in controlling the dyeing process. The presence of auxiliary substances may be necessary to control the rate of dyeing.

Dyes are classified according to either their chemical structure or method of application. The colour index (CI) includes a comprehensive classification of known commercial dyes according to usage and chemical constitution. There are over 7000 dye structures listed in the CI and the major application classes are well known. The dyes are not readily classified under conventional chemical headings because of structural variety.

Dyes consists of a chromogen and an auxochrome. The chromogen contains the chromophore (colour giver) and is represented by chemical radicals such as azo group, nitroso group, nitro group, ethylene group, carbonyl group, etc.

The auxochromes normally contain —NH_2, —OH, —COOH or —SO_3H.

On the basis of their chromophoric system, the dyes may be classified as nitroso, nitro, azo, azoic, anthraquinone, indigoid, etc. According to their application, the dyes may be classified as acid, direct disperse, vat, reactive, etc. Abrahart has described these classifications in detail.

Shaul reported that azo dyes constitute a significant portion of submissions to the premanufacture notification (PMN) process under the toxic substances control act (TSCA). Azo dyes are characterised by the presence of one or more azo groups (—N=N—). Some of the azo dyes, dye precursors or their degradation products such as aromatic amines are suspected to be, carcinogenic.

CHARACTERISTICS OF TEXTILE WASTE-WATERS

From the preceding sections, it is apparent that the characteristics of the waste-waters from a textile plant will depend on the specific operations in the plant. It is misleading to speak of a typical textile effluent. The type of fibre involved and machinery employed are the main factors determining the type and quantities of chemicals present in the textile waste-waters.

Finishing processes discussed earlier may be either batch or continuous. For batch processes, the discharge is intermittent, with the interval between discharged depending on the operations. All the waste-waters from a batch process are likely to come from the same operation, the first being the most heavily contaminated and the last rinse the most dilute. For continuous processes, a steady flow of effluents with moderate concentrations is expected.

Developing a database from sampling the textile effluent discharges at frequent intervals should lead to establishing reliable average values. In addition to frequent sampling, the other approach to determining the characteristics of textile waste-waters is to study the process, its waste components and volumes. For example, the amount of organic matter that is removed from a fabric in the course of normal textile processing can be visualised when one considers that about 10 per cent of the gross weight of a cotton fabric consists of natural impurities that may be removed in processing. For a firm that manufactures 20 tons of fabric per week, about 2 tons per week of impurities are discharged to the sewer. With known discharge volumes, the concentrations of the impurities in the effluent may be estimated.

Kremer and others reviewed the pollutants generated by the textile industry. They divided the pollutants into four general groups, i.e. sizes, detergents, dyes and priority pollutants. They reported that most of the priority pollutants contained in textile industry effluents are aromatics, halogenated hydrocarbons and heavy metals.

Cotton and linen contribute organic matter from the noncellulosic materials that are present in the natural fibres, whereas wool contains sand and grease that are removed during scouring. Synthetic fibres may contain spinning oils and antistatic dressings. Textile wastes are generally coloured, highly alkaline and high in BOD, suspended solids and temperature. The raw waste-water (pH = 9) of a bleachery had 660 mg/l of BOD, 2080 mg/l of COD, 34 mg/l of oil and 2700 mg/l of TDS. Randall and King reported that waste-waters from textile dyeing and finishing operations may be characterised as high in organic matter, both biodegradable and nonbiodegradable, total solids and colour and very variable in pH. In addition, they tend to be high in surfactants and contain potentially significant concentrations of oils, phenols and heavy metals such as chromium, zinc and copper. Randall and King reported the characteristics of raw waste-waters from three plants. The BODs varied between 260 and 560 mg/l. The colour (APHA units) varied between 1000 and 1335. The hue was brown, red to black or yellow to black. Kertell and Hill reported the characteristics of the waste-water from a dye and finishing company. The average BOD was 371 mg/l and the average colour was 113. About 70 mg/l of oil and grease were present. The total solids concentration was about 480 mg/l. Troxler and Hopkins reported that the introduction of continuous dyeing machines had significantly increased the strength of carpet finishing waste-water. This is due to the use of natural bean gum thickeners as a viscosity modifier for the dye solutions. The average BODs and COD from 'beck dyeing' waste-water were 232 and 943 mg/l, respectively. The average BODs and COD from continuous dyeing waste-water were 930 and 2912 mg/l, respectively.

Davis reported that the BODs of a dyehouse waste-water varied between 20 and 1250 mg/l with an average of 634 mg/l. The colour (APHA units) varied between 7,700 and 13,100 mg/l. The average flow of this plant was 1.58 mgd.

A wide variety of methods have been used for reporting results related to colour-removal processes. These include the use of American public health association (APHA) colour units, transmittance, hue, intensity, etc. The current EPA standards are, however, based on a colour analysis procedure developed by the American dye manufacturer's institute.

TREATMENT OF TEXTILE WASTE-WATERS

To solve the problems of treatment of waste-waters from a textile plant, several alternatives should be included. The alternatives are the following:

1. In-plant control for waste reduction.
2. Treatment to 'reuse standard' on an external (end-of-line) basis or by closed-loop recycle systems.
3. Direct discharge to municipal waste treatment systems (i.e. POTWs).
4. 'On-site treatment' of textile waste-waters at POTWs before combining with municipal waste-waters.
5. 'Pretreatment' of textile waste-waters at the plant before discharging to sewer.

POLLUTION CONTROL

In-Plant Control

Conservation of water

A major portion of the waste load is inherent in the methods of textile processing and is independent of the efficiency of the processing plants. For example, the chemicals used for sizing to wrap yarns must be taken off before subsequent bleaching and dyeing. In the industry the normal practice is washing the material with a good flow of running water. This washing operation is the most water-intensive in any textile mills and considerable quantity of water is consumed in this process. Since the concentration level of pollutants in the combined waste-water are not generally high, at the cost of increase in concentration at the inlet level of the treatment plant, water can be saved by judicious use of water. The biological treatment units are designed on the basis of organic loading and hence increase in concentration will not affect the size of the biological unit. On the contrary size of the units designed on hydraulic loading can be reduced apart from saving precious water.

Recently suitable enzymes have been developed which can act upon the starch applied during sizing rapidly at high temperatures ensuring faster degradation of the starch without affecting the material to be treated. With this the desizing operation can be made continuous using only a fraction of water originally required.

Chemicking is a treatment with a weak solution of sodium hypochlorite. Often excessive alkalinity of the bath detracts from bleaching. For satisfactory treatment, the bath alkalinity should not be more than 10 per cent of the available chlorine. The spent bleach can be further replenished at the required time thus saving about 2 to 3 litres of water per kg of fabric for each chemicking operation. Similarly the scouring bath also can be used 8 to 10 times after replenishing.

In mercerising 20 per cent caustic soda is used for treating the fabric and the treatment period ranges from 30 seconds for fabrics to 2 minutes for yarns. Caustic soda on cellulose material is a very tenacious chemical to wash and industries use larger quantity of water than required. This causes not only loss of water but also requires unduly long time to recover caustic from wash waters. The measures by which water can be conserved in mercerising are:

1. Use higher temperature for the wash water.
2. Use counter-counter system of washing whereby caustic soda content in the fabric is progressively reduced.
3. Reduce the concentration of the caustic soda to the optimal limit of 20 per cent strength.
4. Increase the time of washing and the force of the wash water.

Most of the caustic used in textile mills comes from the kiering and mercerising sections. Dialysis and evaporation methods have been employed to purify and reuse the caustic soda from these waste streams.

Jones also reviewed the various ways to reduce textile wastes by in-plant control. In addition to in-plant measures to reduce waste-water volume, the importance of reduction of process chemicals, recovery and reuse of process chemicals, process modifications and substitution of chemicals were discussed.

Treatment to Reuse

Groves and Buckley evaluated membrane separation technology for the reuse of textile effluents. They studied two pilot-plant applications: (i) high-temperature ultrafiltration of desizing effluent for polymer size recovery and water reuse; and (ii) hyperfiltration (reverse osmosis) of mixed cotton/synthetic fibre dyehouse effluents for water reuse. The membrane separation processes may offer potential for the recovery of various chemicals like sizing agents. Groves and Buckley concluded that the use of closed-loop recycle systems is technically feasible for textile waste-waters. They also discussed the 'fouling' problems and requirements of cleanings for the restoration of the design flux.

Davis reported that ultrafiltration has several applications for the recovery of textile sizes. Also, latex recovery can be accomplished using ultrafiltration membranes. Tinghul showed how the science of reverse osmosis offers a basis for the choice of membrane materials for use in reverse osmosis applications involving the separation of dyes in aqueous solutions.

The reuse of a textile effluent may be economic only if the plant faces an acute shortage of water. Complete reuse will probably be unrealistic under any circumstances for many more years.

Direct Discharge to POTWs

For many textile plants, direct discharge to POTWs may be the best alternative. A mill's waste-water may be clean enough or low enough in volume to be treated by the POTW at little or no extra cost. Even if preliminary or primary treatment is required, the cost to the mill may be much less than if a complete treatment facility were required. Jones listed three advantages of combined treatment, i.e. the direct discharge of textile waters to POTWs. Potential economy of operation is the first advantage. Textile waste-waters may not contain enough nutrients (nitrogens and phosphorus) required for biological treatment. Hence, combining such waste-waters with nutrient-rich domestic waste-waters appears to be another advantage. Finally, in combined treatment, the dilution of highly concentrated textile wastes can be achieved, which prevents shock loads of toxic materials from killing the bacteria in the treatment plant. Newlin studied the economic feasibility of treating textile wastes in municipal treatment plants. The general conclusion, based on 26 municipalities serving some 100 textile mills, was that problems and conditions were so diverse that each case must be given individual attention. Three of the six textile mills covered by the study would have saved money by direct discharges, as compared with the costs of treating their own wastes. However, three plants were saving money by treating their own waste-waters. In general, the mills with small amounts of waste-waters were better off paying their service charges to the POTWs.

Jones concluded that the findings regarding the cost of waste-water treatment in relation to total textile manufacturing costs in the southeastern United States raise doubts about the significance of waste-water treatment requirements as an important factor in a competitive strategy as long as the effluent standards do not change drastically. On the other hand, some old plants may face treatment costs that are high enough to create financial problems for them.

On-site Treatment of Textile Waste-waters

Figure 24.1 shows schematic diagrams for alternatives (a) and (b). It is important to note how 'on-site treatment' and 'pretreatment' are defined for this discussion.

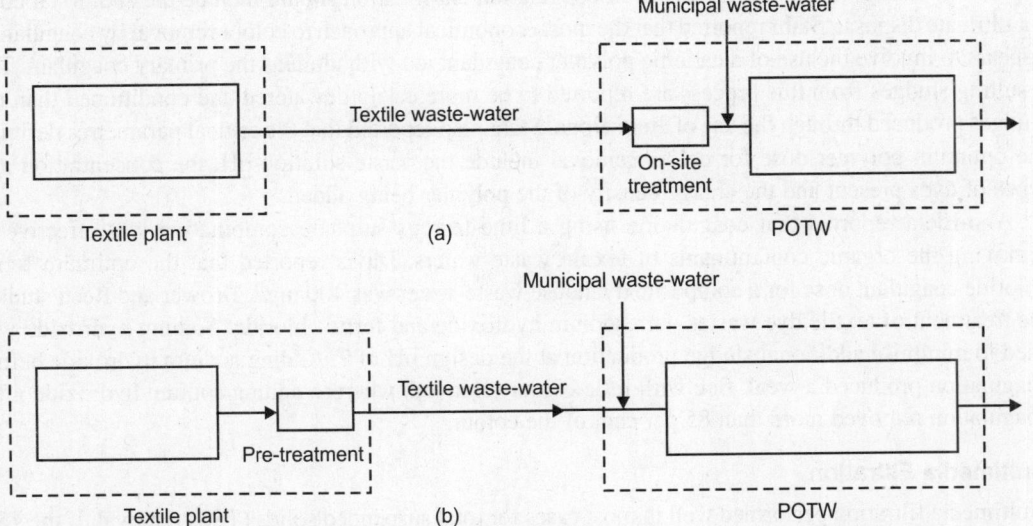

Fig. 24.1. Systemic flow diagrams for: (a) on-site treatment; and (b) pretreatment of textile waste-waters.

On-site treatment refers to any additional treatment at the POTW before combining the textile effluents with the municipal waste-waters for the subsequent treatment. Such additional treatment of the textile waste-waters at the POTW may be physico-chemical or biological. On-site treatment appears to be feasible when several small, closely located textile plants discharge the waste-waters to the same POTW.

Pretreatment of Textile Waste-water

Pretreatment refers to the treatment of waste-waters by the textile plants before discharging to the sewer (Fig. 24.1). Again, such treatments may be physico-chemical or biological. A combination of physico-chemical and biological (both anaerobic and aerobic) processes may also be feasible. Such pre-treatment appears to be feasible for large textile plants.

PHYSICAL/CHEMICAL TREATMENT

Textile waste-waters may be treated using physical/chemical processes either at the POTW (on-site) or at the plant (pre-treatment). Experience has demonstrated that chemical processes remove biodegradable organic matter. Some of the physical/chemical processes are coagulation clarification, multimedia filtration, granular carbon adoption, dissolved air flotation and ozonation.

Coagulation/Clarification

Coagulation/clarification is an effective process for textile waste-water treatment. This method may be, especially effective for colour removal. Typical coagulants are alum, ferrous sulphate, ferric sulphate and ferric chloride with lime or sulphuric acid for pH control. Other widely used coagulants are cationic, anionic and nonionic organic polymers.

For effective coagulation, the experimental determination of the optimum dosage is required. Some waste-waters may require very high coagulant dosage. Chemical dosages in the range of 500 to 1000 mg/l are not uncommon for textile waste-waters. The addition of large amounts of chemicals result in the production of significant quantities of waste solids. The ultimate disposal of these wastes may be very expensive. Hence, the cost calculations of coagulation/clarification should include the additional costs for ultimate disposal. Stahr reported that the most economical approach to colour removal by coagulation appears to involve the use of a cationic polymer coagulant aid with alum as the primary coagulant. The resulting sludges from this process are reported to be more easily dewatered and conditioned than the sludges produced through the use of alum alone. Stahr also reported that the critical parametres defining the optimum polymer dose for colour removal include the waste solution pH, the concentration and types of dyes present and the charge density of the polymer being added.

Abo-Elela reported that coagulation using a lime-ferrous sulphate combination was effective in removing the organic contaminants of textile waste-waters. Davis reported that the optimum ferric chloride coagulant dose for a composite dyehouse waste-water was 400 mg/l. Brower and Reed studied the treatment of textile dye wastes with sodium hydroxide and ferric chloride. Sodium hydroxide was used to minimise additional sludge production at the design pH of 7. Adding sodium hydroxide before coagulation produced a weak floc with little colour removed, whereas adding sodium hydroxide after coagulation removed more than 85 per cent of the colour.

Multimedia Filtration

Multimedia filtration performed well in most cases for total suspended solids (TSS) removal. If the TSS value of a waste-water was approximately 100 mg/l or less, multimedia filtration can be expected to be effective as an initial process for TSS reduction. Multimedia filtration was also an effective process for reducing TSS after coagulation/clarification.

Granular Carbon

Granular carbon adsorption worked well for some textile waste-waters, whereas for others it was found that a portion of the organic removal occurred from physical filtering rather than an adsorption mechanism. McKay evaluated a model to explain the adsorption of selected dyes on activated carbon. The feasibility of activated carbon treatment of dye house waste-waters frequently depends on costs associated with the regeneration of spent carbon. Thermal regeneration has been the primary means of regenerating granular activated carbon. Posey and Kim studied the feasibility of solvent regeneration of exhausted activated carbon using methanol as the organic solvent. They found that for the three dye compounds tested, solvent regeneration was not cost-competitive with thermal regeneration because of the large amounts of methanol required.

Dissolved Air Flotation

Dissolved air flotation is an effective pre-treatment process for textile mill effluent. The US environmental protection agency (EPA) has conducted a survey and compiled removal data on dissolved air flotation. Table 24.2 indicates the removal data of dissolved air flotation for treatment of a textile mill effluent. The treated waste-water was from a woven fabric operation with a flow rate of 1730 m^3/day. Cationic polymer was dosed as the flotation aid.

It is important to know that BOD removal was over 50 per cent. Suspended solids and phenol removals were 85 and 72 per cent, respectively. The removals of lead, *bis*(2-ethylexyl) phtalate, di-*n*-butyl phthalate and naphthalene were all over 92 per cent.

Table 24.2. Treatment of textile mills effluent by dissolved air flotation.

Pollutant/parameter	Concentration		Per cent removal
	Influent	Effluent	
Classical pollutants (mg/l)			
BOD (5)	400	< 200	> 50
COD	1000	720	28
TSS	200	32	84
Total phenol	0.092	0.026	72
Toxic pollutants (mg/l)			
Copper	320	81	75
Lead	14	ND	> 99
Bis(2-ethylexyl) phthalate	570	45	92
Di-*n*-butyl phthalate	13	ND	> 99
Pentachlorophenol	37	30	19
Phenol	94	26	72
Benzene	18	12	33
Ethylbenzene	460	160	65
Toluene	320	130	59
Naphthalene	250	ND	> 99

Ozonation

Ozonation is a very effective oxidation process for colour removal. Beszedits reviewed several studies on ozonation for colour removal. A combination of ozonation with other processes like ultraviolet light or fluidised bed systems was found to be more efficient than the individual processes.

Other physical/chemical processes especially applicable to colour removal include chlorination, photochemical degradation, treatment by irradiation, ion-exchange, liquid-liquid extraction, etc. These processes may show special advantages for specific, small-volume, dye waste-waters.

BIOLOGICAL TREATMENT

Dyes have to be more and more resistant to ozone, nitric oxides, light, hydrolysis and other degradative environments to be successful in the commercial market. It is not surprising that most studies on the biological degradation of dyestuffs yield negative results when dyes are designed to resist this type of treatment. Of those dyes that are known to undergo biodegradation, the azo dyes are perhaps the most commonly studied, although they tend not to be readily biodegradable in sewage treatment works. Shaul and others reported that some azo dyes (e.g. acid yellow 151 and acid red 337) are moderately adsorbed but not biodegraded by the activated sludge process. Azo dyes like acid blue 113 appeared to be strongly adsorbed and possibly biodegraded; acid orange 7 and acid red 88 were moderately adsorbed and significantly biodegraded. They also reported that of the various dye removal mechanisms in an activated sludge process, adsorption and/or biodegradation appeared to be the only two removal mechanisms.

In most cases, the biodegradation of dyes does not occur spontaneously in aerobic conditions. Lengthy adaptation periods are generally necessary. It is assumed that despite the presence of oxygen, the first step of degradation is a reduction of the azo bridge to the corresponding amines. In long sewers and primary settling tanks anaerobic conditions occur and hence, azo reduction is likely. The obligate anaerobic organisms are assumed to be generally capable of reducing azo compounds, but this assumption is yet to be systematically verified for the common azo dyes. It is, however, fairly well established that most of the azo dyes are not toxic to the biological systems. Grady reported that anaerobic processes are effective in decolourising many dyes, suggesting that they could be used in the treatment of specific dye streams. The mechanism of colour removal was not determined and could have been due to surface adsorption, reduction reactions, microbial activity, or combinations of the three. Kremer reported that anaerobic digestion is an effective means for the removal of selected monoazo dyes (acid red 88 and acid orange 7), up to 10 mg/l in the presence of a supplemental carbon source. The metabolites of acid red 88 were naphthionic acid, 2-naphthol, 1,2-naphthoquinone and isoquinoline.

An aerobic process following the anaerobic process should be able to biodegrade the intermediates formed by the anaerobic degradation of the dyes. Such a combination of anaerobic/aerobic processes appears to be a very promising concept that has yet to be proven.

The fibres in the textile waste-waters caused operational problems in several of the municipal plants studied by Troxler and Hopkins. These fibres tend to float and form dense mats. Problems caused include fouling of mechanical aerators, plugging of centrifugal pumps and the formation of dense blankets of floating fibres in an anaerobic digester. A good pretreatment programme to remove these fibres at the factory and routine maintenance at the POTWs can reduce these problems.

From the preceding discussion on the physical/chemical and biological treatments of textile waste-waters, it appears that a combination of these two processes should be more efficient than either of the individual processes. Interestingly, several researchers have studied the advantages and disadvantages of such combinations. Randall and King studied three full-scale treatment plants treating relatively similar dyeing and finishing waste-waters. The first plant consisted of only an extended aeration activated sludge system plus sand dyeing beds. The other two plants combined biological and chemical (lime or aluminium chloride was the primary coagulant) processes in an attempt to attain complete treatment of the textile waste-waters. One design placed the chemical system ahead of the biological unit, whereas the other design positioned the chemical/physical processes after the biological system. Randall and King concluded that both lime and aluminium chloride are satisfactory primary coagulants and anionic polymers are best as secondary coagulants for the removal of colour from textile dyeing and finishing waste-waters. They also concluded that combined physical/chemical and biological processes are required for high removals of both BOD, colour and suspended solids. Randall and King additionally reported that biological systems preceding chemical processes should be designed to minimise biological solids carry-over in an effort to minimise operating problems.

LAND TREATMENT OF TEXTILE WASTE-WATERS

Overcash and Rendall studied the feasibility of land treatment of textile waste-waters. They concluded that from an investment perspective, wool scouring with dyeing and finishing and woven fabric dyeing and finishing require the greatest expenditure for land treatment. At the other extreme, knit fabric dyeing and finishing involve the smallest investment.

To sum up due to the variability of textile waste-water characteristics, treatability studies should be conducted on a case-by-case basis in order to identify and confirm the required design parametres. For textile waste-water treatment, physical/chemical, biological or a combination of both processes may be suitable. A combined aerobic/anaerobic treatment may be shown to be a feasible process for the treatment of dye waste-waters. Land treatment may be suitable for some dye wastes. After identifying the technically feasible processes, the final selection should be based on the economics of the processes.

AIR EMISSIONS CHARACTERISATION

Possible emissions from textile processing include:
1. Oil mists and organic emissions produced when textile materials containing knitting and lubricating oils, plasticisers, and other materials that can volatilise or be thermally degraded into volatile substances, are subjected to heat. Processes that can be the sources of oil mists include tentering, calendaring, heat setting, drying, and curing.
2. Acid mists produced during the carbonising of wool.
3. Solvent vapours released during and after solvent processing operations such as dry cleaning.
4. Dust and lint produced by the processing of natural fibres and synthetic staple prior to and during spinning, as well as by napping and carpet shearing.

AIR POLLUTION CONTROL MEASURES

Emissions from finishing operations, including drying, curing, and heat setting, have been and continue to be one of the more significant air pollution problems in the textile industry. The blue haze and odour that are characteristic of these emissions have posed both technical and economic problems for large and small facilities alike. The amount of air pollution depends on the finishing processes used. Various kinds and amounts of oils and finishing resins can vapourise from the cloth in tenter frames (commonly used for drying, curing, and heat setting) operating at temperatures typically in the 300°–400°F range. Some exhausts also can carry a significant amount of lint.

There are process modifications that can reduce some of this air pollution. For example, most of the oils can be removed by effectively prescouring the cloth. Alternatively, the selection of 'less polluting' finishing chemicals may help, but cloth finishing quality and economics often limit the ability of such changes to correct the problem.

Over the longer term, new technology, such as solvent-free, radiation-curable coatings, may prove to be a factor in overcoming the air pollution problem. Such technology not only would eliminate the problem, but would result in significant energy savings, higher line speeds, and greatly reduced floor space. Unfortunately, lack of availability of radiation-curable coatings and high cost have been the major limitations to commercialising this technology.

Historically, the solution to the characteristic blue haze and odour problem has been the use of air pollution control equipment. Wet (two-stage) electrostatic precipitators, fabric-type (glass fibre-wool) mist eliminators, and fume incinerators have traditionally been the equipment of choice. Wet precipitators and mist eliminators have been plagued by continuous and costly operation and maintenance problems. In general, their performance has been strongly dependent on the effectiveness of gas stream cooling prior to treatment in the control device.

Experience has shown that temperatures generally should not exceed 120°F. Fume incinerators operating at 1200°–1400°F with approximately a 0.3 second gas residence time have been effective in eliminating blue haze and odour problems; however, capital and operating costs may have a significant impact on the facility.

Air pollution problems resulting from greige processing of cotton consist predominantly of dust generation in the high-speed mechanical operations, such as spinning, drawing, carding, and twisting. Dry dust collectors of the vacuum type are well suited to maintain a dust-free working environment.

Finely divided lint is present in the air in the vicinity of various operations associated with spinning the yarn and weaving the cloth. Most modern mills have air tunnels built under the floor to effect a downward flow of air around the machines that produce the largest volumes of waste lint (carding, roving, spinning, twisting, weaving, etc.). In the opening and picking rooms in a cotton plant, stock is usually handled by overhead ductwork. Lint and dust are collected by the vacuum system as close as possible to their point of origin. Automatic travelling vacuum cleaning systems are frequently used in conjunction with the underfloor system.

The suspended lint and dust are conveyed by air movement through ducts to a centrally located filter system for removal. Automatic drytype travelling disposable air filters are extensively used for this purpose. Since any filtering medium tends to hold the lint fibres tenaciously, a low-cost, disposable paper medium is often used. The fumes from wool carbonising pose another problem that must be dealt with. This process generates very fine carbon particles that appear as smoke, as well as some fumes and odours. These fumes probably include residual sulphur oxides (if sulphuric acid has been used for carbonising), in addition to organic decomposition products, and are generally very corrosive. Corrosion-resistant stainless steels and/or plastics have been used in the collection systems, which usually exhaust the fumes to the atmosphere.

Since a significant quantity of particulate matter is in the submicrometre range, a visible emission is usually apparent unless there is a very efficient air pollution control system. The severity of the opacity problem will depend to a large extent on location, topography, and local meteorological conditions. Incinerators, wet scrubbers, and fabric filters have been used to resolve the problem. Wet scrubbing systems have the added advantage of reducing residual gases such as sulphur oxides, as well as particulate emissions.

SECTION III

Pollution Control in Food Processing Industries

25.	**Food Processing Industry: A Review**	543
26.	**Dairy**	547
27.	**Bakery**	555
28.	**Seafood Processing and Meat Products**	562

Chapter 25

Food Processing Industry: A Review

INTRODUCTION

Agro-industries or food-processing industries, are concerned primarily with the production of edible goods for human and animal consumption from raw agricultural products. These industries include: (i) canning, (ii) dairy, (iii) brewing and distilling, (iv) meat packing and rendering (including poultry and feedlots), (v) sugar refining, (vi) soft drinks and beverages, and (vii) miscellaneous, including coffee, seafood, rice, grains, and bakeries.

The industries are characterised by the following basic production steps:
1. Washing the raw materials.
2. Removing inedible portions.
3. Preparing the foodstuff.
4. Packaging.

The processes that make up these steps in the various industries of the agro-industries sector are diverse. The water quality and quantity and the waste stream characteristics, vary from one industry to the next. This section provides an over view of water use and treatment practices of the agro-industries sector as a whole and does not focus on anyone of the subsector industies.

WATER USE

Variety is the one word to describe water use in the agro-industries. Not only are there thousands of products, but agro-industries are well established in many countries with different availabilities of water resources. The industry is not as heavily concentrated in the developed countries as other industries since food processing is nearly a universal need.

SOURCES AND CHARACTERISTICS OF WASTE-WATER

Water is in contact with raw or finished products in most processes in the agro-industries. This contact results in wastes containing organic matter, in dissolved or colloidal states and in varying degrees of concentration. The sources of these waste-waters are water that contacts spoiled raw material or finished products, rinsing or washing water, transporting water, process water, cooling water, spills and water used for cleaning equipment.

The waste streams exhibit extreme ranges among the industries. The effluent BOD can range from 100 to 1,00,000 ppm. Suspended solids can range from nearly zero to 1,20,000 ppm and pH values range from 3.5 to 11.0. The volumes of the waste also vary greatly. Table 25.1 shows data that provide a brief summary of waste-water problems facing the food industries.

Table 25.1. Characteristic water pollution control problems of the agro-industries.

Industry	Characteristics attending water pollution control	Industry	Characteristics attending water pollution control
Canning, frozen and dehydrated fruits and vegetables, soup, potato chips, speciality items, baby food, etc.	Large seasonal volumes Variation in effluent strength and volume Highly biodegradable effluents Some soluble organics difficult to remove chemically Water colouration by strong pigments in some raw products Liquid wastes highly putrescible and cannot be stored for long periods of time	Coffee	Evaporation and other effluents
		Chocolate	Suspended fats in effluents
		Fish and seafood	Liquid wastes highly putrescible and cannot be stored for long periods of time Wastes have water-colouring properties
		Red meat	Highly biodegradable effluents Liquid wastes highly putrescible and cannot be stored for long periods of time Relatively large volumes of waste-water
Edible oils	High concentrations of fats, oils and greases; BOD_5; suspended solids, dispersed organics; and dissolved solids Fats, oils and greases difficult to remove to acceptable level for direct discharge to waterways Highly biodegradable effluents Relatively large volumes of waste-water	Poultry	Highly organic effluents, high in suspended solids and floating materials such as grease Highly biodegradable effluents Relatively large volumes of waste-water Fats and grease in high concentration
Dairy	Highly biodegradable effluents Variation in flow rates and characteristics Whey from cheese production	Milling	Highly organic effluents, high in suspended solids Highly biodegradable effluents Large volumes of waste-water
Pickle	Brine, high dissolved solids in effluents		
Tea	Evaporation effluent	Sugar	Liquid wastes highly putrescible and cannot be stored for long periods of time

WASTE-WATER TREATMENT PRACTICES

Since the primary component of agro-industries' effluents is organic wastes, biological forms of treatment are effective. The alternatives for the treatment of agro-industries' waste are in-plant treatment with the effluent discharged directly to the water environment or the effluent discharge to a municipal waste treatment plant with or without in-plant pretreatment. Discharge to municipal treatment facilities usually requires pretreatment owing to higher concentrations of organics. In-plant treatment for direct discharge or as pretreatment will most likely use aerobic or anaerobic biological treatment, with the most effective methods being activated sludge, biological filtration, anaerobic digestion, oxidation ponds, lagoons and spray irrigation. The selection of a method will depend upon the degree of treatment required, nature of the organic wastes, concentration of organic matter, volume of wastes and local costs.

Table 25.2 suggests treatment methods for various types of waste that may be encountered in the agro-industries.

Table 25.2. Waste-water treatment methods of removing various categories of pollutants in the effluents of the agro-industries.

Pollutant	Treatment method
Free and emulsified oils and greases	Gravity separation
	Coagulation and sedimentation
	Carbon adsorption
	Granular media filtration
	Flotation
	Impressed current
Suspended solids	Sedimentation
	Coagulation and sedimentation
	Granular media filtration
Dispersed organics	Biological conversion
	Carbon adsorption
Dissolved solids (inorganic)	Evaporation
	Ion exchange
	Reverse osmosis
	Electrodialysis
Acidity and alkalinity	Neutralisation
Sludge from processes	Anaerobic digestion
	Aerobic digestion
	Centrifuging
	Incineration
	Stabilisation pond
	Thickening
	Vacuum filtration
	Wet oxidation
	Landfill
	Fertiliser or soil conditioner
	Animal feed

A new system has been developed that can treat the caustic waste separately installed. The system is composed of land treatment for the noncaustic waste and an aerobic/anaerobic pond system with a sand filter and pH adjustment system (Fig. 25.1).

The performance of combined system is given in Table 25.3.

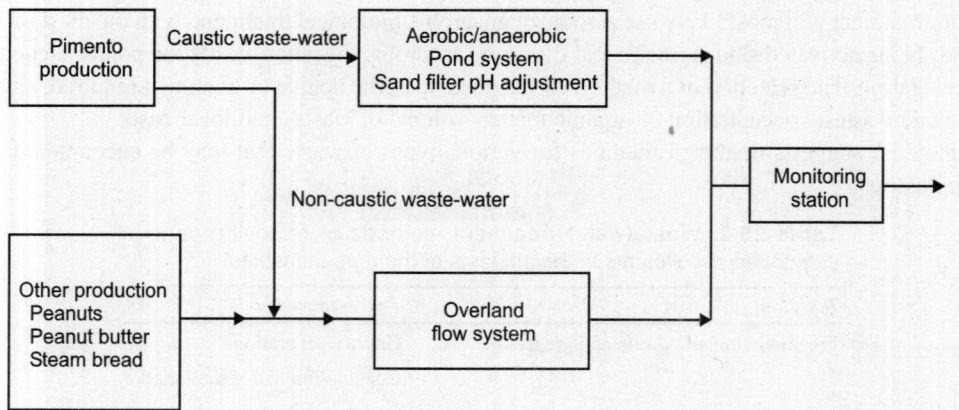

Fig. 25.1. Different treatment for different food-processing waste-water.

Table 25.3. Combined treatment performance of a food-processing plant.

Nature of waste-water	BOD_5 in (mg/l)	BOD_5 out (mg/l)	Fraction of BOD_5 removed (%)	Total suspended solids in (mg/l)	Total suspended solids out (mg/l)	Fraction of suspended solids removed (%)
Caustic	13640	85	99	3680	300	94
Noncaustic	3600	200	94	600	40	93

AIR POLLUTION

In food processing industries there are no process emissions except for combustion product from ovens, and utilities like boilers and DG sets.

Chapter 26

Dairy

INTRODUCTION

The rapid expansion of dairy industry is one of the significant developments in the food processing field during this period. The subject dairy and food engineering occupies a major importance in the food science curriculum and this emphasis is likely to continue.

In dairy industry, water has a multipurpose use. Water used for the purpose of processing, cleaning and other general uses should be of potable standard and absolutely free from microbial contaminants. Use of contaminated water in the plant, results in milk and milk products to be unsafe for human consumption.

The dairy industry has to meet the rapidly intensifying quality standards with regard to microchemical/microbiological parameters. These quality standards can be achieved by having an adequate system which covers quality of raw milk, water supplies, cleaning materials, and procedures and solid/liquid waste management system, without any compromise on aesthetic, public health, and stipulated technical standards.

An adequate 'safety screen' approach requires that:
1. Potential contamination of milk and milk products with any of the group factors like bacteria, heavy metal, and viruses, etc. should be minimised through deliberate 'point source' control with stringent inspection.
2. A secondary protective screen system be installed at critical unit process where water loops may be in contact with products.

They should adequately remove all potentially questionable factors, usually by subtractive filtration, absorption methods, appropriate for each factor in type, size, and frequency. These can usually be designed with ample safety factor to meet present and future needs. It is necessary to carefully design and build proper supply and operating systems of water for various uses in the plant based on sound engineering principles. The milk industry is one of the most widely spread of all industries.

SOURCES OF WASTE

Wastes from milk products manufacture contain milk solids in a more or less dilute condition, but in varying concentration. These solids enter the wastes from almost all of the operations.

In general the wastes and their sources may be classified as follows:
1. Spoiled raw or manufactured products.
2. By-products (buttermilk, skim milk, and whey).

3. Spillage or overflow due to inefficient equipment and careless operations.
4. Rinsings and washings from cans, equipment and floors.
5. Condenser water and condensate from vacuum pans.
6. Water from coolers, ice machine, boilers and roof drains.

BOD of Whole Milk and By-products

Milk solids are composed of the three general classifications of organic material, namely, fats, proteins and carbohydrates. Roughly, the BOD of one kg of milk fat is 0.90 kg. The milk solids content of whole milk and hence that of skim milk, buttermilk and whey varies to some extent.

Waste Prevention

Waste disposal in the milk industry may be divided into two programmes, first, waste prevention or saving, and second, waste treatment. The utilisation of by-products and a waste-saving programme will materially reduce the loss of milk solids and simplify the requirements for treatment. Such a programme should always precede the design of treatment facilities.

The first step in the programme is to segregate all possible clean water from the water containing milk solids. Segregation necessitates changes in the drain system of the plant in order to provide a separate line for cooling water, ice machine water, boiler blow-down, roof drains and vacuum pan water. The condenser water from the vacuum pan will contain entrained solids and will be discussed later, but because of its large volume it must be segregated from the plant wastes.

After as much of the clean water has been segregated as can be economically accomplished, a weir box containing a device for measuring the rate of flow and an automatic sampler is installed on the waste line. The laboratory is then provided with facilities and instructed in the procedure for the BOD test. A regular programme of sampling and analysis is initiated before any waste prevention activities are started and is continued until losses are reduced to a minimum. After this point has been reached occasional measurements and analyses are necessary to prevent a return to careless operations. The prevention programme consists of a study of the various sources of waste, and the initiation of good housekeeping methods, careful operations, by-product utilisation, employee education and adequate and efficient facilities.

Disposal of Spoiled Products

Spoiling usually occurs during periods of hot weather when cooling facilities may prove to be inadequate. More attention has been given in recent years to the handling of milk by the producer and spoilage is not as great as it was at one time. Spoiling may occur in the plant due to prolonged power failure, breakdown of equipment or lack of adequate storage.

It is a general rule that spoiled products are not to be dumped into the drain system, except perhaps in very large cities where the quantity of sewage is so large that the spoiled material will have no apparent significance. Provision should be made to prevent spoiling by installing adequate equipment and emergency power. When products do spoil, they should be returned to the producer for feeding purposes.

Utilisation of By-products

It should be a general rule in all plants where milk is processed that by-products should not be allowed to enter the drain systems. The quantity of these by-products is not always amenable to processing for use as food products or animal feed. In these cases adequate provision must be made to return the entire amount of by-product which cannot be sold as such to the farms for feeding purposes.

Where the volume is sufficiently large to warrant processing, adequate provision must be made to take care of the by-product at peak season. In some cases, it may be feasible for several plants to combine for processing by-products, if the length of haul is not excessive. There are numerous plants designed especially for by-product processing which take the material from a fairly wide area. In general, these processes consist of removing the water and recovering the solids in a semi-solid or dry condition.

Methods of Treatment

All floor wastes should pass through a simple combination sand trap, fat trap, screen tank. This is desirable for all plants even if the waste is discharged to a large stream or a municipal sewer. The simplest means of waste disposal for the milk plant is the discharge of the waste to a municipal sewer system. However, a sewer service charge can be expected which should be adequate to permit the municipality to construct the additional waste treatment facilities and pay for their operation.

Some of the effluent can be used directly for the irrigation or to spread them over an area of grassland or cultivated land. The composition of the effluent does, however, impose limits of a maximum of 200 litres per m^2 per year. Furthermore, the irrigation must be done at least 200 metres away from an inhabited area. Return of by-products waste to farmers is the most economical method of disposal. Farmers use such by-products as whey for feeding purposes. Effluent may be treated by mechanical, chemical and biological means. A combination of these methods is often used.

Mechanical treatments are screening, filtration, sedimentation, or flotation. The latter process is aided by gas bubbles to which the particles of impurities becomes attached. Only insoluble constituents can be removed by these methods. Reverse osmosis would also remove dissolved constituents, but this method is not suitable for large scale use at the present time because large volume of effluent produced makes the process too costly to run.

Chemical treatment consists of the precipitation of dissolved substances by means of suitable precipitating agents such as iron sulphate, iron chloride, aluminium sulphate, and lime, etc. A sedimentable coagulum is formed which also contains suspended solids. The coagulum may be subsequently separated from the water by mechanical means. Chemical purification is not sufficient for dairy effluents because it does not remove the dissolved lactose.

The most suitable method of treating dairy effluents is biological purification. Dissolved or colloidally suspended organic compounds are decomposed by oxidation with the aid of aerobic bacteria. Oxygen must be supplied by means of artificial ventilation. Organic substances can also be decomposed by reduction of anaerobic bacteria in a septic tank.

Percolating filters

A percolating filter contains broken pieces of pumice or plastic materials especially made for the purpose which serve as support for the growth of bacteria, yeasts, fungi, protozoa, and nematodes. These media are arranged to form a thick porous layer through which the effluent trickles, fed from a storage tank (Fig. 26.1). The slimy skin of micro-organism, growing on the filter medium absorbs the organic constituents of the effluent and decomposes them aerobically. A stream of air is often passed through the filter medium. Decomposition produces gases, liquids and solid substances. By aeration, the cells are destroyed by their own metabolism (endogenous oxidation) thus greatly reducing the sludge volume.

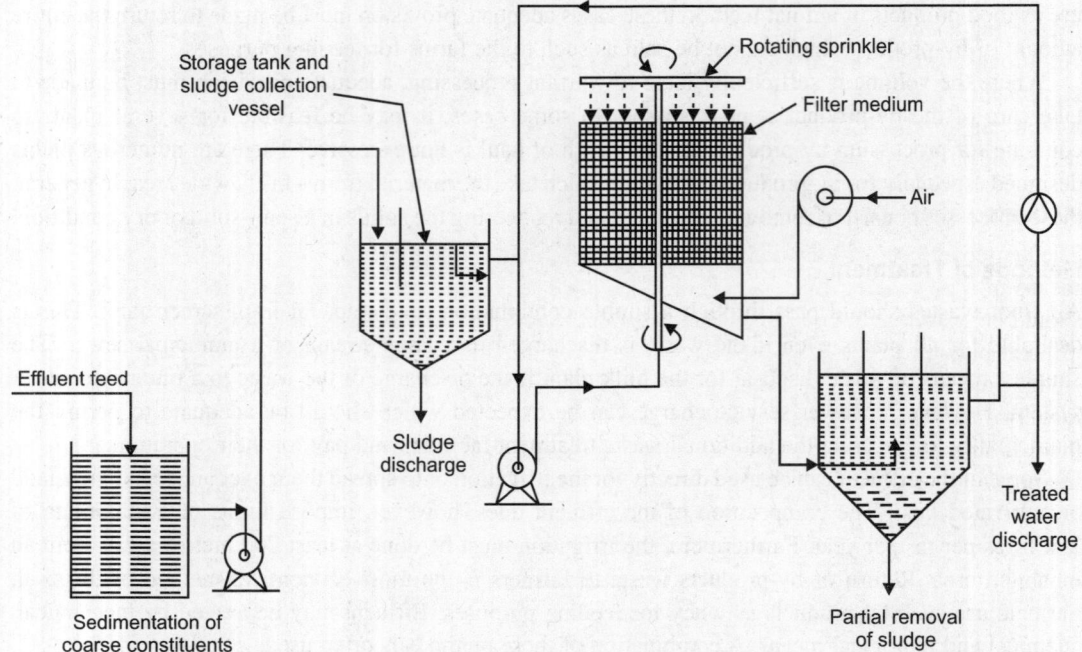

Fig. 26.1. Schematic diagram of a percolating filter.

A single passage of the effluent through the percolating filter column would not give a satisfactory result. The effluent must be passed several times over the film of micro-organisms. This is indicated by the return in Fig. 26.1. Excessive deposition of sludge is avoided because the circulation of the effluent produces a rising effect. Nevertheless, deposits of sludge have to be removed from time to time. The settling tanks are provided with tapered bottom leading to sludge-collection sump, a central inlet and constant-overflow weir. Possible nuisance due to foul odours from the filter media is avoided by a closed construction. The air from the filter is passed into clarifying tanks arranged behind the filter where it is de-odourised. The forced ventilation is of advantage for the oxygen supply of the filter medium.

A satisfactory degradation of dairy effluent can be achieved by means of percolating filters provided that: (i) the number of passes through the plant is sufficiently large, (ii) the plant is loaded lightly, (iii) a strongly fluctuating load is treated in aerated storage tanks, and (iv) too much sludge deposition is avoided.

It has been found that 92 per cent purification (BOD_5) of dairy effluent can be achieved in a percolating filter installation operation intermittently (circulation time about 20 hours at any one time). However, iron sulphate should be added to the effluent.

If a percolating filter is operated with a high load it is of advantage to rinse it to reduce the accumulation of sludge which blocks the filter. The effect of rinsing may also be achieved by an appropriately high number of recirculation passes. A useful modification is a combination of large aerated mixing and equalising tank with two percolating filters arranged in series with settling tanks between them. The order of filters is changed (e.g. every week) when the first one begins to be blocked by sludge. By this means the filter medium is freed of sludge in a relatively short time.

The BOD_5 value of the effluent leaving the filter is, however, still relatively high. It can be reduced by passing it through a low loaded activated sludge plant (oxidation pond), situated behind the filter. Additional installations arranged in series behind the filter are; final clarifying tank, effluent filter, equipment for concentrating the sludge to about 18 per cent solids consisting of a perforated drum and a twin band press. Although the running costs of such a plant are higher its operation is reliable and degree of purification produced is high.

Activated sludge method

The degree of purification obtainable with the activated sludge method is higher than that obtained by means of a percolating filter. To achieve complete biological purification, however, the reaction time should be greater than 12 hours and the loading should be low. The degree to which the volume of an effluent plant can be loaded depends on the amount of sludge the water is carrying. In the activated sludge process a flocculent sludge is formed which is kept aerated, takes up and decomposes the organic matters present in the incoming wastes. The sludge is formed in liquid suspension and it is maintained there in an active state by vigorous aeration. The individual processing steps in the activated sludge treatment are shown in Fig. 26.2.

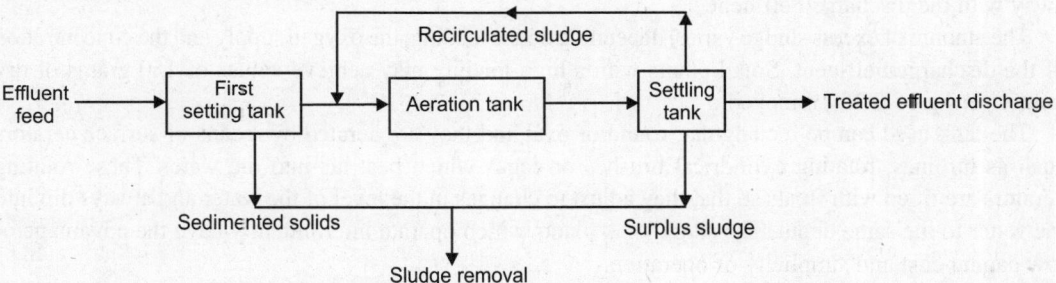

Fig. 26.2. Schematic representation of the activated sludge process.

The first step is to allow sedimentable impurities to sediment in a settling tank. This first clarification tank should, moreover, be large enough to allow the large fluctuation in pollution load and in pH values which occur in dairy effluents to be equalised. Sometimes a part of the recirculated sludge is passed into this tank which will then have to be aerated. The actual biological purification takes place in the aeration tank. These tanks are either oblong, round or oval and air is incorporated either by means of pressure jets or by surface aerators consisting of rotating cylindrical brushes or cages. The water is also agitated to prevent the flocculated material from settling. The amount of air incorporated should be sufficient to raise the oxygen content in the aeration tank to at least 1 to 4 mg/l. About 5–15 per cent of oxygen contained in the air blown into the tank is utilised, depending on the size of the bubbles. The optimum temperature for aerobic digestion is about 30°C and at 10°C the rate of fermentation is practically stationary.

In the final settling tank the flocculated material settles within two to four hours from a layer of sludge at the bottom. This activated sludge is largely pumped back into the aeration tank. The sludge should not remain for too long in the tank (not more than 6 hours) as it can start to decompose by putrefaction and the gases evolved can make it float to surface. Since a part of the organic compounds absorbed by the micro-organisms is transformed into microbial matter, excess sludge is produced which has to be removed continuously. The ratio of excess sludge to recirculated sludge should be about 1:10.

Surplus sludge has a solid content of about 1 per cent. It can be concentrated by means of a decanting centrifuge or by filtration. The activated sludge process is rather expensive to install and needs skilled attention for its successful operation.

Low load activated sludge method — effluent lagoon

This method is specially suitable for dairy effluents. Such a plant consists of a single tank which functions as a storage tank, biological oxidation tank and final settling tank. As the BOD_5 of the feed generally varies between 400 and 3000 mg BOD_5/hr and the supply may be intermittent a whole day's load should be collected in the tank. The tank is first of all, strongly aerated but the aeration can be reduced during the night when the oxygen consumption is lower. The reason is that about $\frac{2}{3}$ of the organic compounds, particularly lactose, are decomposed relatively rapidly within the first one of four hours. Organic acids are produced which lower the pH value somewhat. The protein decomposition which follows is considerably slower and produces a rise in pH.

Depending on the depth of the tank the sludges settles in 3 to 5 hours. About one hour after aeration has been stopped, the first of the purified effluent is discharged. This is done by lowering an overflow weir slowly (over a period of 2–3 hours) so that the sludge particles may not be disturbed and carried away with the discharged effluent.

The amount of excess sludge varies, depending on the loading, the oxygen supply and the concentration of the discharged effluent. Small plants with a high loading may achieve values of 150 grams of dry matter per kg of BOD_5 removed.

The tank used can be rectangular, round or oval and they are aerated by means of surface aerators such as turbines, rotating cylindrical brushes or cages which beat air into the water. These rotating aerators are fitted with floats so that they adjust to changes in the level of the water and always dip into the water to the same depth. The single tank plants which operate intermittently have the advantage of low capital cost and simplicity of operation.

Low loaded single tank plants are not only able to deal with large fluctuations in concentration, but they are also insensitive to fluctuations in acidity and alkalinity over the relatively wide pH range of 3 to 12.5. The time of exposure of pH values outside the normal range (6–8) should not be too long, of course, as otherwise the activity of the flora may be adversely affected. For this reason care should be taken that spent HNO_3— and NaOH— containing cleaning solutions are mixed and added in small doses to the effluent. Complete neutralisation of the effluent is not necessary. Dairies which uses mostly alkalies for cleaning usually produce effluents whose pH is too high. A part from acids (e.g. HNO_3, H_2SO_4, HCl) flue gases are also frequently used for neutralisation.

Anaerobic purification — septic tank method (digestion method)

In anaerobic purification, bacteria decompose organic subtances in a septic tank in the absence of air. The first reaction is the decomposition of lactose by reduction during which process oxygen is split off from water and hydrogen is liberated. In the process of acid fermentation, lactose is converted to lactic acid and fats are decomposed to low molecular weight fatty acids and other organic compounds. This takes about 24 hours at 20°C. The pH falls a little. Subsequently, methane fermentation converts proteins within two days into ammonia with the liberation of CO_2, H_2S, CH_4, and various other gases. The pH rises slowly again (sometimes upto pH 7.5).

Modern large dairy plants producing large volumes of effluent often find the results unsatisfactory, particularly in cold weather, and there is often the problem of unpleasant odours. This is the reason that septic tanks are used only rarely now-a-days. The loading of a septic tank should be less than 100 grams BOD_5/m^3 day if methane fermentation is to be maintained. In winter loads should be less than 50 grams BOD_5/m^3 day.

Turning the sludge over increases the vigour and efficiency of the digestion. The sludge is thereby mixed well with the raw effluent feed, and settling of the sludge and the formation of layers is prevented. This is shown in Fig. 26.3.

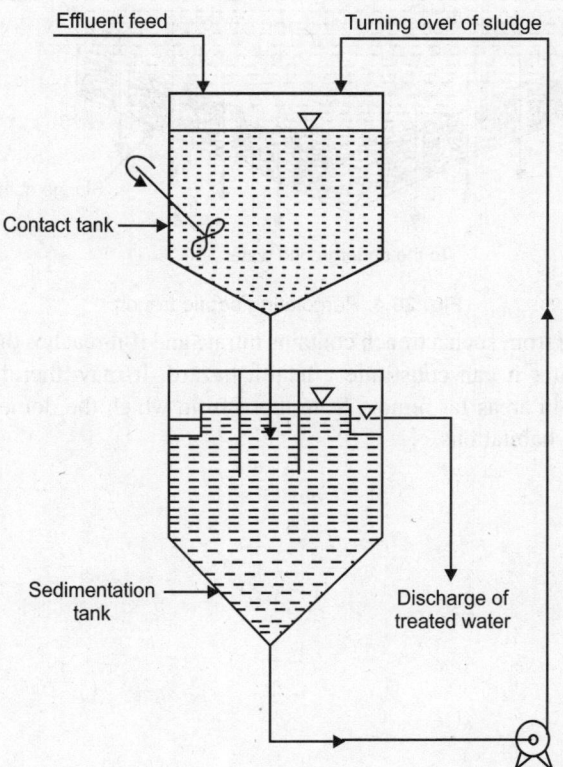

Fig. 26.3. Fermenting sludge contact method.

The use of higher temperature (e.g. 30°C) has the advantage that a greater reduction in BOD_5 can be achieved at a constant load or the load can be increased at an unchanged reduction in BOD_5 values. Higher temperatures do, however, lead to increased gas production which makes the discharged effluent turbid. The effluent has moreover a slight putrid smell. The contact method is, therefore, unsuitable for the complete biological purification of dairy effluents. It can, at best, be used as the first stage in a two-stage process.

A variation of the septic tank method is a trench from which the fermenting sludge percolates into the ground. This is shown in Fig. 26.4. The mean retention time in the trench is 20 to 30 days at a rate of filtration of about 4 cm^3/day. The thicker the layer of sludge the better is the clarification, but lower the rate of percolation. A sludge layer 10 cm thick permits a percolation rate of about 2 cm^3/day.

A high concentration of lactose (e.g. in whey) is unfavourable because the decomposition produces organic acids which lowers the pH too much (sometimes to 4.5) that further decomposition is inhibited.

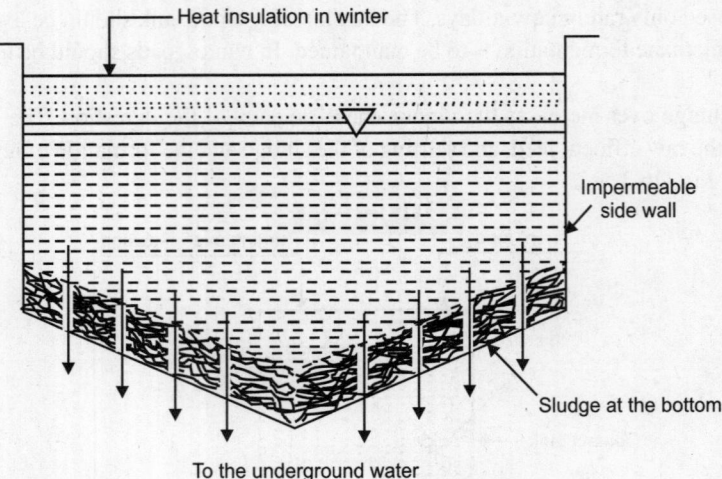

Fig. 26.4. Percolating septic trench.

The water percolating from such a trench contains nitrate and if it reaches the domestic water supply via the underground water it can constitute a health hazard. It may, therefore, be passed into the underground water only in areas far remote from those from which the domestic supply is taken and which are far away from habitations.

Chapter 27

Bakery

INTRODUCTION

The bakery industry is one of the world's major food industries and varies widely in terms of production scale and process. Traditionally, bakery products may be categorised as bread and bread roll products, pastry products (e.g. pies and pasties) and speciality products (e.g. cake, biscuits, donuts and speciality breads).

The major equipment includes miller, mixer/kneading machine, bun and bread former, fermentor, bake ovens, cold stage and boilers. The main processes are milling, mixing, fermentation, baking and storage. Fermentation and baking are normally operated at 40°C and 160°–260°C, respectively. Depending on logistics and the market, the products can be stored at 4–20°C.

WASTE-WATER

Waste-water in bakeries is primarily generated from cleaning operations including equipment cleaning and floor washing. It can be characterised as high loading, fluctuating flow and contains rich oil and grease. Flour, sugar, oil, grease and yeast are the major components in the waste.

The ratio of water consumed to products is about 10 in common food industry, much higher than that of 5 in the chemical industry and 2 in the paper and textiles industry. Normally, half of the water is used in the process, while the remainder is used for washing purposes (e.g. of equipment, floor and containers).

Typical values for waste-water production are summarised in Tables 27.1–27.3. Different products can lead to different amounts of waste-water produced. As shown in Table 27.1, pastry production can result in much more waste-water than the others. The values of each item can vary significantly as demonstrated in Table 27.2. The waste-water from cake plants has higher strength than that from bread plants. The pH is in acidic to neutral ranges, while the 5-day biochemical oxygen demand (BOD_5) is from a few hundred to a few thousand mg/l, which is much higher than that from the domestic waste-water. The suspended solids (SS) from cake plants is very high. Grease from the bakery industry is generally high, which results from the production operations. The waste strength and flow rate are very much dependent on the operations, the size of the plants and the number of workers. Generally speaking, in the plants with products of bread, bun and roll, which are termed as dry baking, production equipment (e.g. mixing vats and baking pans) are cleaned dry and floors are swept before washing down. The waste-water from cleanup has low strength and mainly contains flour and grease (Table 27.2). On the other hand, cake production generates higher strength waste, which contains grease, sugar, flour, filling ingredients and detergents.

Table 27.1. Summary of waste production from the bakery industry.

Manufacturer	Products	Waste-water production (litre/ton production)	COD (kg/tonne production)	Contribution to total COD loading (%)
Bread and bread roll	Bread and bread roll	230	1.5	63
Pastry	Pies and sausage rolls	6000	18	29
Speciality	Cake, biscuits, donuts and Persian breads	74	–	–

Due to the nature of the operation, the waste-water strength changes at different operational times. As demonstrated in Table 27.2, higher BOD_5, SS, total solids (TS) and grease are observed from 1 to 3 a.m., which results from lower waste-water flow rate after midnight.

Table 27.2. Waste-water characteristics in the bakery industry.

Type of bakery	pH	BOD_5 (mg/l)	SS (mg/l)	TS (mg/l)	Grease (mg/l)
Bread plant	6.9–7.8	155–620	130–150	708	60–68
Cake plant	4.7–8.4	2240–8500	963–5700	4238–5700	400–1200
Variety plant	5.6	1600	1700	–	630
Unspecified	4.7–5.1	1160–8200	650–13430	–	1070–4490

Bakery waste-water lacks nutrients; the low nutrient value gives BOD_5:N:P of 284:1:2. This indicates that to obtain better biological treatment results, extra nutrients must be added to the system. The existence of oil and grease also retards the mass transfer of oxygen. The toxicity of excess detergent used in cleaning operations can decrease the biological treatment efficiency. Therefore, the pretreatment of waste-water is always needed.

BAKERY WASTE TREATMENT

Generally, bakery industry waste is nontoxic. It can be divided into liquid waste, solid waste and gaseous waste. In the liquid phase, there are high contents of organic pollutants including chemical oxygen demand (COD), BOD_5, as well as fats, oils and greases (FOG) and SS. Waste-water is normally treated by physical, chemical and biological processes.

PRE-TREATMENT SYSTEMS

Pre-treatment or primary treatment is a series of physical and chemical operations, which precondition the waste-water as well as remove some of the wastes. The treatment is normally arranged in the following order: screening, flow equalisation and neutralisation, optional FOG separation, optional acidification, coagulation-sedimentation and dissolved air flotation. The pretreatment of bakery waste-water is presented in Fig. 27.1.

In the bakery industry, pretreatment is always required because the waste contains high SS and floatable FOG.

Pre-treatment can reduce the pollutant loading in the subsequent biological and/or chemical treatment processes; it can also protect process equipment. In addition, pretreatment is economically preferable in the total process view as compared to biological and chemical treatment.

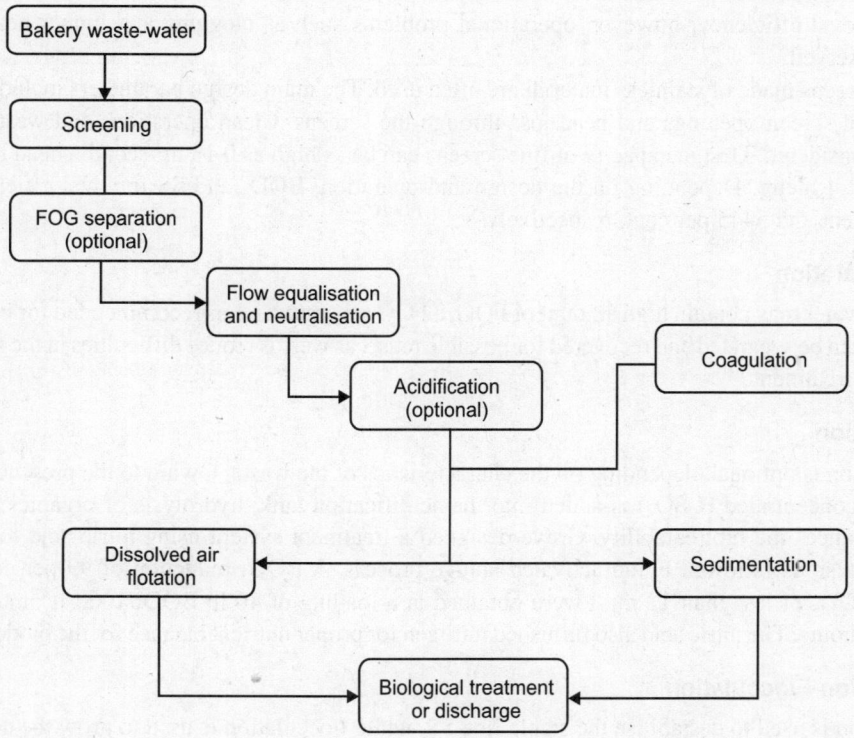

Fig. 27.1. Bakery waste-water pre-treatment system process flow diagram.

Flow Equalisation and Neutralisation

In bakery plants, the waste-water flow rate and loading vary significantly with the time as illustrated in Table 27.3. It is usually economical to use a flow equalisation tank to meet the peak discharge demand. However, too long a retention time may result in an anaerobic environment. A decrease in pH and bad odours are common problems during the operations.

Table 27.3. Average waste characteristics at specified time interval in a cake plant.

Time interval	pH	BOD_5 (mg/l)	SS (mg/l)	TS (mg/l)	Grease (mg/l)
3 am–8 am	7.9	1480	834	3610	428
9 am–12 am	8.6	2710	1080	5310	457
1 pm–6 pm	8.1	2520	795	4970	486
7 pm–12 pm	8.6	2020	953	3920	739
1 am–3 am	8.9	2520	1170	4520	991

Screening

Screening is used to remove coarse particles in the influent. There are different screen openings ranging from a few μm (termed as microscreen) to more than 100 mm (termed as coarse screen). Coarse screen openings range from 6–150 mm; fine screen openings are less than 6 mm. Smaller opening can have a better removal efficiency; however, operational problems such as clogging and higher head lost are always observed.

Fine screens made of stainless material are often used. The main design parameters include velocity, selection of screen openings and head loss through the screens. Clean operations and waste disposal must be considered. Design capacity of fine screens can be as high as 0.13 m^3/sec; the head loss ranges from 0.8–1.4 metre. Depending on the design and operation, BOD$_5$ and SS removal efficiencies are 5–50 per cent and 5–45 per cent, respectively.

FOG Separation

As waste-water may contain high amount of FOG, a FOG separator is thus recommended for installation. The FOG can be separated and recovered for possible reuse, as well as reduce difficulties in the subsequent biological treatment.

Acidification

Acidification is optional, depending on the characteristics of the waste. Owing to the presence of FOG, acid (e.g. concentrated H$_2$SO$_4$) is added into the acidification tank; hydrolysis of organics can occur, which enhances the biotreatability. Grove designed a treatment system using nitric acid to break the grease emulsions followed by an activated sludge process. A BOD$_5$ reduction of 99 per cent and an effluent BOD$_5$ of less than 12 mg/l were obtained at a loading of 40 lb BOD$_5$/1000 ft^3 and detention time of 87 hours. The nitric acid also furnished nitrogen for proper nutrient balance for the biodegradation.

Coagulation-Flocculation

Coagulation is used to destabilise the stable fine SS, while flocculation is used to grow the destabilised SS, so that the SS become heavier and larger enough to settle down. The Coagulation-flocculation process can be used to remove fine SS from bakery waste-water. It normally acts as a preconditioning process for sedimentation and/or dissolved air flotation.

The waste-water is preconditioned by coagulants such as alum. The pH and coagulant dosage are important in the treatment results. Liu and Lien reported that 90–100 mg/l of alum and ferric chloride were used to treat waste-water from a bakery that produced bread, cake and other desserts. The waste-water had pH of 4.5, SS of 240 mg/l and COD of 1307 mg/l. Values of 55 per cent and 95–100 per cent for removal of COD and SS, respectively, were achieved. The optimum pH for removal of SS was 6.0, while that for removal of COD was 6.0–8.0. It was also found that FeCl$_3$ was relatively more effective than alum. Yim used coagulation-flocculation to treat a waste-water with much higher waste strength. Table 27.4 gives the treatment results. Owing to the higher organic content, SS and FOG, coagulants with high dosage of 1300 mg/l were applied. The optimal pH was 8.0. As shown, removal for the above three items was fairly high, suggesting that the process can also be used for high-strength bakery waste. However, the balance between the cost of chemical dosage and treatment efficiency should be justified.

Table 27.4. Comparison of different bakery pre-treatment methods.

Coagulant	BOD$_5$		SS		FOG	
	Influent (mg/l)	Removal (%)	Influent (mg/l)	Removal (%)	Influent (mg/l)	Removal (%)
Ferric sulphate	2780	71	2310	94	1450	93
Alum	2780	69	2310	97	1450	96

Sedimentation

Sedimentation, also called clarification, has a working mechanism based on the density difference between SS and the water, allowing SS with larger particle sizes to more easily settle down. Rectangular tanks, circular tanks, combination flocculator-clarifiers and stacked multilevel clarifiers can be used.

Dissolved Air Flotation (DAF)

Dissolved air flotation (DAF) is usually implemented by pumping compressed air bubbles to remove fine SS and FOG in the bakery waste-water. The waste-water is first stored in an air pressured, closed tank. Through the pressure-reduction valves, it enters the flotation tank. Due to the sudden reduction in pressure, air bubbles form and rise to the surface in the tank. The SS and FOG adhere to the fine air bubbles and are carried upwards. Dosages of coagulant and control of pH are important in the removal of BOD_5, COD, FOG and SS. Other influential factors include the solids content and air/solids ratio. Optimal operation conditions should be determined through the pilot-scale experiments. Liu and Lien used a DAF to treat a waste-water from a large-scale bakery. The waste-water was preconditioned by alum and ferric chloride. With the DAF treatment, 48.6 per cent of COD and 69.8 per cent of SS were removed in 10 minutes at a pressure of 4 kg/cm^2 and pH 6.0. Mulligan used DAF as a pretreatment approach for bakery waste. At operating pressures of 40–60 psi, grease reductions of 90–97 per cent were achieved. The BOD_5 and SS removal efficiencies were 33–62 per cent and 59–90 per cent, respectively.

Biological Treatment

The objective of biological treatment is to remove the dissolved and particulate biodegradable components in the waste-water. It is a core part of the secondary biological treatment system. Micro-organisms are used to decompose the organic wastes.

With regard to different growth types, biological systems can be classified as suspended growth or attached growth systems. Biological treatment can also be classified by oxygen utilisation: aerobic, anaerobic and facultative. In an aerobic system, the organic matter is decomposed to carbon dioxide, water and a series of simple compounds. If the system is anaerobic, the final products are carbon dioxide and methane.

Compared to anaerobic treatment, the aerobic biological process has better quality effluent, easier operation, shorter solid retention time, but higher cost for aeration and more excess sludge. When treating high-load influent (COD > 4000 mg/l), the aerobic biological treatment becomes less economic than the anaerobic system. To maintain good system performance, the anaerobic biological system requires more complex operations. In most cases, the anaerobic system is used as a pretreatment process.

Suspended growth systems (e.g. activated sludge process) and attached growth systems (e.g. trickling filter) are two of the main biological waste-water treatment processes. The activated sludge process is most commonly used in treatment of waste-water. The trickling filter is easy to control and has less excess sludge. It has higher resistance loading and low energy cost. However, high operational cost is its major disadvantage. In addition, it is more sensitive to temperature and has odour problems. Comprehensive considerations must be taken into account when selecting a suitable system.

AEROBIC TREATMENT

Activated Sludge Process

In the activated sludge process, suspended growth micro-organisms are employed. A typical activated sludge process consists of a pretreatment process (mainly screening and clarification), aeration tank (bioreactor), final sedimentation and excess sludge treatment (anaerobic treatment and dewatering process). The final sedimentation separates micro-organisms from the water solution. In order to enhance the performance result, most of the sludge from the sedimentation is recycled back to the aeration tank(s), while the remaining is sent to anaerobic sludge treatment. A recommended complete activated sludge process is given in Fig. 27.2.

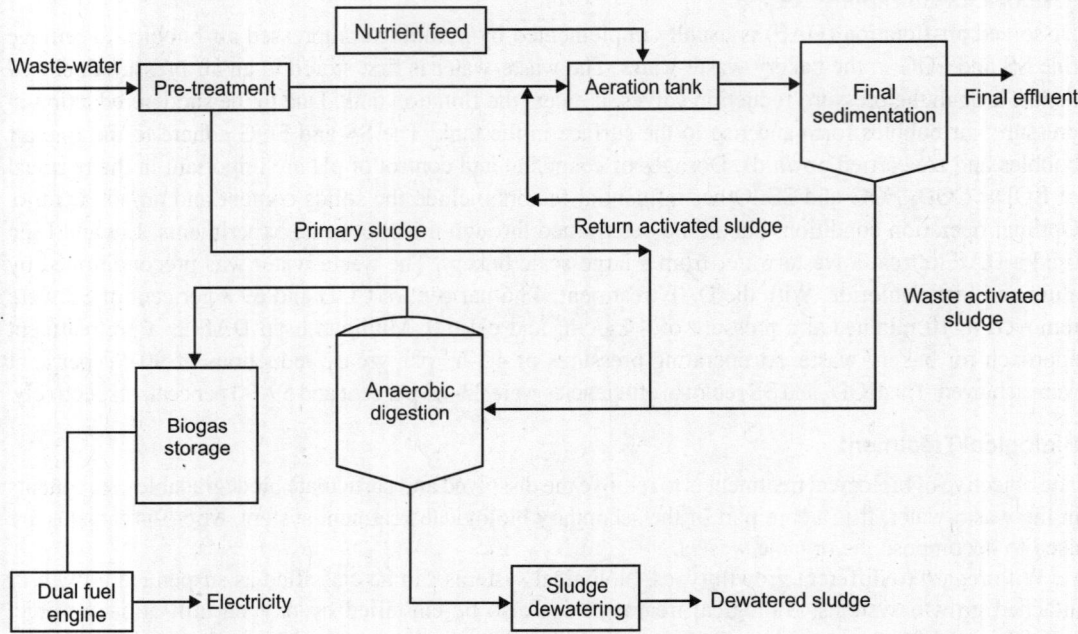

Fig. 27.2. Process flow diagram of activated sludge treatment of bakery waste-water.

The activated sludge process can be a plug-flow reactor (PFR), completely stirred tank reactor (CSTR), or sequencing batch reactor (SBR). For a typical PFR, length-width ratio should be above 10 to ensure the plug flow. The CSTR has higher buffer capacity due to its nature of complete mixing, which is a critical benefit when treating toxic influent from industries. Compared to the CSTR, the PFR needs a smaller volume to gain the same quality of effluent. Most large activated sludge sewage treatment plants use a few CSTRs operated in series. Such configurations can have the advantages of both CSTR and PFR.

The SBR is suitable for treating noncontinuous and small-flow waste-water. It can save space, because all five primary steps of fill, react, settle, draw and idle are completed in one tank. Its operation is more complex than the CSTR and PFR; in most cases, auto operation is adopted.

The performance of activated sludge processes is affected by influent characteristics, bioreactor configuration and operational parameters. The influent characteristics are waste-water flow rate, organic concentration (BOD_5 and COD), nutrient compositions (nitrogen and phosphorus), FOG, alkalinity, heavy metals, toxins, pH and temperature. Configurations of the bioreactor include PFR, CSTR, SBR, membrane bioreactor (MBR) and so on. Operational parameters in the treatment are biomass concentration [mixed liquor volatile suspended solids concentration (MLVSS) and volatile suspended solids (VSS)], organic load, food to micro-organisms (F/M), dissolved oxygen (DO), sludge retention time (SRT), hydraulic retention time (HRT), sludge return ratio and surface hydraulic flow load. Among them, SRT and DO are the most important control parameters and can significantly affect the treatment results. A suitable SRT can be achieved by judicious sludge wasting from the final clarifier. The DO in the aeration tank should be maintained at a level slightly above 2 mg/l.

Owing to the high organic content, it is not recommended that bakery waste-water be directly treated by aerobic treatment processes.

Chapter 28

Seafood Processing and Meat Products

INTRODUCTION

The seafood industry consists primarily of many small processing plants, with a number of larger plants located near industry and population centers. Numerous types of seafood are processed, such as mollusks (oysters, clams, scallops), crustaceans (crabs and lobsters), saltwater fishes and freshwater fishes. As in most processing industries, seafood-processing operations produce waste-water containing substantial contaminants in soluble, colloidal and particulate forms. The degree of the contamination depends on the particular operation; it may be small (e.g. washing operations), mild (e.g. fish filleting) or heavy (e.g. blood water drained from fish storage tanks).

The chapter focuses on biological treatments for seafood meat and poultry industry processing waste-water, namely aerobic and anaerobic treatments. The most common operations of biological treatments are also described in this chapter.

The chapter also discusses the physico-chemical treatments for seafood meat and poultry industry processing waste-water. These operations include coagulation, flocculation and disinfection. Potential problems in land application are highlighted. Economic considerations are always the most important factors that influence the final decision for selecting processes for waste-water treatment. The economic issues related to waste-water treatment process are also discussed.

TREATMENT OF SEAFOOD PROCESSING WASTES

Waste-water from seafood-processing operations can be very high in biochemical oxygen demand (BOD), fat, oil and grease (FOG) and nitrogen content. Literature data for seafood processing operations showed a BOD production of 1–72.5 kg of BOD per ton of product. White fish filleting processes typically produce 12.5–37.5 kg of BOD for every ton of product. BOD is derived mainly from the butchering process and general cleaning and nitrogen originates predominantly from blood in the waste-water stream.

Seafood-processing Waste-water Characterisation

Seafood-processing waste-water characteristics that raise concern include pollutant parameters, sources of process waste and types of wastes. In general, the waste-water of seafood-processing waste-water can be characterised by its physico-chemical parameters, organics, nitrogen and phosphorus contents. Important pollutant parameters of the waste-water are five-day biochemical oxygen demand (BOD_5), chemical oxygen demand (COD), total suspended solids (TSS), fats, oil and grease (FOG) and water usage. As in most industrial waste-waters, the contaminants present in seafood-processing waste-waters are an undefined mixture of substances, mostly organic in nature.

Primary Treatment

In the treatment of seafood-processing waste-water, one should be cognizant of the important constituents in the waste stream. This waste-water contains considerable amounts of insoluble suspended matter, which can be removed from the waste stream by chemical and physical means. For optimum waste removal, primary treatment is recommended prior to a biological treatment process or land application. A major consideration in the design of a treatment system is that the solids should be removed as quickly as possible. It has been found that the longer the detention time between waste generation and solids removal, the greater the soluble BOD_5 and COD with corresponding reduction in by-product recovery. For seafood-processing waste-water, the primary treatment processes are screening, sedimentation, flow equalisation and dissolved air flotation. These unit operations will generally remove up to 85 per cent of the total suspended solids and 65 per cent of the BOD_5 and COD per cent in the waste-water.

Screening

The removal of relatively large solids (0.7 mm or larger) can be achieved by screening. This is one of the most popular treatment systems used by food-processing plants, because it can reduce the amount of solids being discharged quickly. Usually, the simplest configuration is that of flow-through static screens, which have openings of about 1 mm. Sometimes a scrapping mechanism may be required to minimise the clogging problem in this process.

Fish solids dissolve in water with time; therefore, immediate screening of the waste streams is highly recommended. Likewise, high-intensity agitation of waste streams should be minimised before screening or even settling, because they may cause breakdown of solids rendering them more difficult to separate. In small-scale fish-processing plants, screening is often used with simple settling tanks.

Sedimentation

Sedimentation separates solids from water using gravity settling of the heavier solid particles. In the simplest form of sedimentation, particles that are heavier than water settle to the bottom of a tank or basin. Sedimentation basins are used extensively in the waste-water treatment industry and are commonly found in many flow-through aquatic animal production facilities. This operation is conducted not only as part of the primary treatment, but also in the secondary treatment for separation of solids generated in biological treatments, such as activated sludge or trickling filters.

Flow equalisation

A flow equalisation step follows the screening and sedimentation processes and precedes the dissolved air flotation (DAF) unit. Flow equalisation is important in reducing hydraulic loading in the waste stream. Equalisation facilities consist of a holding tank and pumping equipment designed to reduce the fluctuations of the waste streams. The equalising tank will store excessive hydraulic flow surges and stabilise the flow rate to a uniform discharge rate over a 24-hour day. The tank is characterised by a varying flow into the tank and a constant flow out.

Separation of oil and grease

Seafood-processing waste-waters contain variable amounts of oil and grease, which depend on the process used, the species processed and the operational procedure. Gravitational separation may be used to remove oil and grease, provided that the oil particles are large enough to float towards the surface and are not emulsified; otherwise, the emulsion must be first broken by pH adjustment. Heat may also be used for breaking the emulsion but it may not be economical unless there is excess steam available.

Flotation

Flotation is one of the most effective removal systems for suspensions that contain oil and grease. The most common procedure is that of dissolved air flotation (DAF), which is a waste treatment process in which oil, grease and other suspended matter are removed from a waste stream.

Biological Treatment

To complete the treatment of the seafood-processing waste-waters, the waste stream must be further processed by biological treatment. Biological treatment involves the use of micro-organisms to remove dissolved nutrients from a discharge. Organic and nitrogenous compounds in the discharge can serve as nutrients for rapid microbial growth under aerobic, anaerobic or facultative conditions. The three conditions differ in the way they use oxygen. Aerobic micro-organisms require oxygen for their metabolism, whereas anaerobic micro-organisms grow in absence of oxygen; the facultative micro-organism can proliferate either in absence or presence of oxygen although using different metabolic processes.

The biological treatment processes used for waste-water treatment are broadly classified as aerobic and anaerobic treatments. Aerobic and facultative micro-organisms predominate in aerobic treatments, while only anaerobic micro-organisms are used for the anaerobic treatments.

If micro-organisms are suspended in the waste-water during biological operation, this is known as a 'suspended growth process', whereas the micro-organisms that are attached to a surface over which they grow are said to undergo an 'attached growth process'.

Biological treatment systems are most effective when operating continuously 24 hours/day and 365 days/year. Systems that are not operated continuously have reduced efficiency because of changes in nutrient loads to the microbial biomass. Biological treatment systems also generate a consolidated waste stream consisting of excess microbial biomass, which must be properly disposed. Operation and maintenance costs vary with the process used.

The principles and main characteristics of the most common processes used in seafood-processing waste-water treatment are explained in this section.

Aerobic process

In seafood processing waste-waters, the need for adding nutrients (the most common being nitrogen and phosphorus) seldom occurs, but an adequate provision of oxygen is essential for successful operation. The most common aerobic processes are activated sludge systems, lagoons, trickling filters and rotating disc contactors.

Activated sludge systems

In an activated sludge treatment system, an acclimatised, mixed, biological growth of micro-organisms (sludge) interacts with organic materials in the waste-water in the presence of excess dissolved oxygen and nutrients (nitrogen and phosphorus). The micro-organisms convert the soluble organic compounds to carbon dioxide and cellular materials. Oxygen is obtained from applied air, which also maintains adequate mixing. The effluent is settled to separate biological solids and a portion of the sludge is recycled; the excess is wasted for further treatment such as dewatering.

Aerated lagoons

Aerated lagoons are used where sufficient land is not available for seasonal retention or land application and economics do not justify an activated sludge system. Efficient biological treatment can be achieved by the use of the aerated lagoon system.

Stabilisation/polishing ponds

A stabilisation/polishing ponds system is commonly used to improve the effluent treated in the aerated lagoon. This system depends on the action of aerobic bacteria on the soluble organics contained in the waste stream. The organic carbon is converted to carbon dioxide and bacterial cells. Algal growth is stimulated by incident sunlight that penetrates to a depth of 1–1.5 metre. Photosynthesis produces excess oxygen, which is available for aerobic bacteria; additional oxygen is provided by mass transfer at the air-water interface.

Trickling filters

The trickling filter is one of the most common attached cell (biofilm) processes. Unlike the activated sludge and aerated lagoons processes, which have biomass in suspension, most of the biomass in trickling filters are attached to certain support media over which they grow.

Rotating biological contactors (RBC)

Increasingly stringent requirements for the removal of organic and inorganic substances from waste-water have necessitated the development of innovative, cost-effective waste-water treatment alternatives in recent years. The aerobic rotating biological contactor is one of the biological processes for the treatment of organic waste-water. It is another type of attached growth process that combines advantages of biological fixed-film (short hydraulic retention time, high biomass concentration, low energy cost, easy operation and insensitivity to toxic substance shock loads) and partial stir. Therefore, the aerobic RBC reactor is widely employed to treat both domestic and industrial waste-water.

Anaerobic treatment

Anaerobic biological treatment has been applied to high BOD or COD waste solutions in a variety of ways. Treatment proceeds with degradation of the organic matter, in suspension or in a solution of continuous flow of gaseous products, mainly methane and carbon dioxide, which constitute most of the reaction products and biomass. Its efficient performance makes it a valuable mechanism for achieving compliance with regulations for contamination of recreational and seafood-producing wastes. Anaerobic treatment is the result of several reactions: the organic load present in the waste-water is first converted to soluble organic material, which in turn is consumed by acid-producing bacteria to produce volatile fatty acids, plus carbon dioxide and hydrogen. The methane-producing bacteria consume these products to produce methane and carbon dioxide. Typical micro-organisms used in this methanogenic process are *Metanobacterium*, *Methanobacillus*, *Metanococcus*, and *Methanosarcina*.

Digestion systems

Anaerobic digestion facilities have been used for the management of animal slurries for many years, they can treat most easily biodegradable waste products, including everything of organic or vegetable origin. Recent developments in anaerobic digestion technology have allowed the expansion of feedstocks to include municipal solid wastes, biosolids and organic industrial waste (e.g. seafood-processing wastes).

Physico-chemical Treatments

Coagulation/Flocculation

Coagulation or flocculation tanks are used to improve the treatability of waste-water and to remove grease and scum from waste-water. In coagulation operations, a chemical substance is added to an organic colloidal suspension to destabilise it by reducing forces that keep them apart, that is, to reduce the surface charges responsible for particle repulsions. This reduction in charges is essential for flocculation, which has the purpose of clustering fine matter to facilitate its removal. Particles of larger size are then settled and clarified effluent is obtained.

In seafood processing waste-waters, the colloids present are of an organic nature and are stabilised by layers of ions that result in particles with the same surface charge, thereby increasing their mutual repulsion and stabilisation of the colloidal suspension. This kind of waste-water may contain appreciable amounts of proteins and micro-organisms, which become charged due to the ionisation of carboxyl and amino groups or their constituent amino acids. The oil and grease particles, normally neutral in charge, become charged due to preferential absorption of anions, which are mainly hydroxyl ions.

Several steps are involved in the coagulation process. First, coagulant is added to the effluent and mixing proceeds rapidly and with high intensity. The purpose is to obtain intimate mixing of the coagulant with the waste-water, thereby increasing the effectiveness of destabilisation of particles and initiating coagulation. A second stage follows in which flocculation occurs for a period of up to 30 minutes. In the latter case, the suspension is stirred slowly to increase the possibility of contact between coagulating particles and to facilitate the development of large flocs. These flocs are then transferred to a clarification basin in which they settle and are removed from the bottom while the clarified effluent overflows.

Several substances may be used as coagulants. The pH of several waste-waters of the proteinaceous nature can be adjusted by adding acid or alkali. The addition of acid is more common, resulting in coagulation of the proteins by denaturing them, changing their structural conformation due to the change in their surface charge distribution. Thermal denaturation of proteins can also be used, but due to its high energy demand, it is only advisable if excess steam is available. In fact, the 'cooking' of the blood-water in fishmeal plants is basically a thermal coagulation process.

Another commonly used coagulant is polyelectrolyte, which may be further categorised as cationic and anionic coagulants. Cationic polyelectrolytes act as a coagulant by lowering the charge of the waste-water particles, because waste-water particles are negatively charged. Anionic or neutral polyelectrolyte are used as bridges between the already formed particles that interact during the flocculation process, resulting in an increase of floc size.

Since the recovered sludges from coagulation/flocculation processes may sometimes be added to animal feeds, it is advisable to ensure that the coagulant or flocculant used is not toxic.

Electrocoagulation

Electrocoagulation (EC) has also been investigated as a possible means to reduce soluble BOD. It has been demonstrated to reduce organic levels in various food- and fish-processing waste streams. During testing, an electric charge was passed through a spent solution in order to destabilise and coagulate contaminants for easy separation.

Disinfection

Disinfection of seafood-processing waste-water is a process by which disease-causing organisms are destroyed or rendered inactive. Most disinfection systems work in one of the following four ways:

(i) damage to the cell wall, (ii) alteration of cell permeability, (iii) alteration of the colloidal nature of protoplasm, and (iv) inhibition of enzyme activity.

Disinfection is often accomplished using bactericidal agents. The most common agents are chlorine, ozone (O_3) and ultraviolet (UV) radiation, which are discussed in the following sections.

Chlorination

Chlorination is a process commonly used in both industrial and domestic waste-waters for various reasons. In fisheries effluents, however, its primary purpose is to destroy bacteria or algae or to inhibit their growth. Usually the effluents are chlorinated just before their final discharge to the receiving water bodies. For this process either chlorine gas or hypochlorite solutions may be used, the latter being easier to handle. In waste solutions, chlorine forms hypochlorous acid, which in turn forms hypochlorite.

Ozonation

Ozone (O_3) is a strong oxidising agent that has been used for disinfection due to its bactericidal properties and its potential for removal of viruses. It is produced by discharging air or oxygen across a narrow gap with application of a high voltage. An ozonation system is presented in Fig. 28.1.

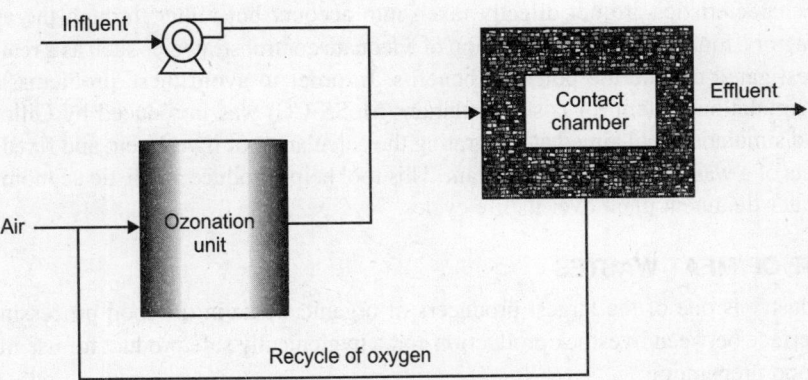

Fig. 28.1. Simplified diagram of an ozonation system.

Ozonation has been used to treat a variety of waste-water streams and appears to be most effective when treating more dilute types of wastes. It is a desirable application as a polishing step for some seafood-processing waste-waters, such as from liquid-processing operations, which is fairly concentrated.

Ozone reverts to oxygen when it has been added and reacted, thus increasing somewhat the dissolved oxygen level of the effluent to be discharged, which is beneficial to the receiving water stream. Contact tanks are usually closed to recirculate the oxygen-enriched air to the ozonation unit. Advantages of ozonation over chlorination are that it does not produce dissolved solids and is affected neither by ammonia compounds present nor by the pH value of the effluent. On the other hand, ozonation has been used to oxidise ammonia and nitrites presented in fish culture facilities.

Ozonation also has limitations. Because ozone's volatility does not allow it to be transported, this system requires ozone to be generated onsite, which requires expensive equipment. Although much less used than chlorination in fisheries waste-waters, ozonation systems have been installed in particular in discharges to sensitive water bodies.

Ultraviolet (UV) radiation

Disinfection can also be accomplished by using ultraviolet (UV) radiation as a disinfection agent. UV radiation disinfects by penetrating the cell wall of pathogens with UV light and completely destroying the cell and/or rendering it unable to reproduce.

ECONOMIC CONSIDERATIONS OF SEAFOOD-PROCESSING AND MEAT PROCESSING WASTE-WATER TREATMENT

Economic considerations are always the most important parameters that influence the final decision as to which process should be chosen for waste-water treatment. In order to estimate cost, data from the waste-water characterisation should be available together with the design parameters for alternative processes and the associated costs. Costs related to these alternative processes and information on the quality of effluent should also be obtained prior to cost estimation in compliance with local regulations.

During the design phase of a waste-water treatment plant, different process alternatives and operating strategies could be evaluated by several methods. This cost evaluation can be achieved by calculating a cost index using commercially available software packages. Nevertheless, actual cost indices are often restrictive, since only investment or specific operating costs are considered. Moreover, time-varying waste-water characteristics are not directly taken into account but rather through the application of large safety factors. Finally, the implementation of adequate control strategies such as a real-time control is rarely investigated despite the potential benefits. In order to avoid these problems, a concept of modelbased simulation system for cost calculation (MoSS-CC) was introduced by Gillot, which is a modelling and simulation tool aimed at integrating the calculation of investment and fixed and variable operating costs of a waste-water treatment plant. This tool helps produce a holistic economic evaluation of a waste-water treatment plant over its life cycles.

TREATMENT OF MEAT WASTES

The meat industry is one of the largest producers of organic waste in the food processing sector and forms the interface between livestock production and a hygienically safe product for use in both human and animal food preparation.

The first stages in meat processing occur in the slaughterhouse (abattoir) where a number of common operations take place, irrespective of the species. These include holding of animals for slaughter, stunning, killing, bleeding, hide or hair removal, evisceration, offal removal, carcass washing, trimming and carcass dressing. Further secondary operations may also occur on the same premises and include cutting, deboning, grinding and processing into consumer products.

Processing Facilities and Wastes Generated

As a direct result of its operation, a slaughterhouse generates waste comprised of the animal parts that have no perceived value to the slaughterhouse operator. It also generates waste-water as a result of washing carcasses, processing offal and from cleaning equipment and the fabric of the building.

Waste-water flow

Water is used in the slaughterhouse for carcass washing after hide removal from cattle, calves and sheep and after hair removal from hogs. It is also used to clean the inside of the carcass after evisceration and for cleaning and sanitising equipment and facilities both during and after the killing operation. Associated facilities such as stockyards, animal pens, the steam plant, refrigeration equipment, compressed air, boiler rooms and vacuum equipment will also produce some waste-water, as will sanitary and service facilities for staff employed on site: these may include toilets, shower rooms, cafeteria kitchens and laboratory facilities. The proportions of water used for each purpose can be variable.

The quantity of waste-water will depend very much on the slaughterhouse design, operational practise and the cleaning methods employed. Waste-water generation rates are usually expressed as a volume per unit of product or per animal slaughtered and there is a reasonable degree of consistency between some of the values reported from reliable sources for different animal types.

Waste-water characteristics

Effluents from slaughterhouses and packing houses are usually heavily loaded with solids, floatable matter (fat), blood, manure and a variety of organic compounds originating from proteins. As already stated the composition of effluents depends very much on the type of production and facilities. The main sources of water contamination are from lairage, slaughtering, hide or hair removal, paunch handling, carcass washing, rendering, trimming and cleanup operations. These contain a variety of readily biodegradable organic compounds, primarily fats and proteins, present in both particulate and dissolved forms. The waste-water has a high strength, in terms of biochemical oxygen demand (BOD), chemical oxygen demand (COD), suspended solids (SS), nitrogen and phosphorus, compared to domestic waste-waters. The actual concentration will depend on in-plant control of water use, by-products recovery, waste separation source and plant management. In general, blood and intestinal contents arising from the killing floor and the gut room, together with manure from stockyard and holding pens are separated, as best as possible, from the aqueous stream and treated as solid wastes. This can never be 100 per cent successful, however and these components are the major contributors to the organic load in the waste-water, together with solubilised fat and meat trimmings.

The aqueous pollution load of a slaughterhouse can be expressed in a number of ways. Within the literature reports can be found giving the concentration in waste-water of parameters such as BOD, COD and SS. These, however, are only useful if the corresponding waste-water flow rates are also given. Even then it is often difficult to relate these to a meaningful figure for general design, as the unit of productivity is often omitted or unclear. These reports do, however, give some indication as to the strength of waste-waters typically encountered and some of their particular characteristics, which can be useful in making a preliminary assessment of the type of treatment process most applicable. At best it can be concluded that slaughterhouse waste-waters have a pH around neutral, an intermediate strength in terms of COD and BOD, are heavily loaded with solids and are nutrient-rich.

The waste-water contains a high density of total coliform, fecal coliform and fecal streptococcus groups of bacteria due to the presence of manure material and gut contents. Numbers are usually in the range of several million colony forming units (CFU) per 100 ml. It is also likely that the waste-water will contain bacterial pathogens of enteric origin such as *Salmonella* sp., and *Campylobacter jejuni*, gastrointestinal parasites including *Ascaris* sp., *Giardia lamblia* and *Cryptosporidium parvum* and enteric viruses. It is, therefore, essential that slaughterhouse design ensures the complete segregation of process wash water and strict hygiene procedures to prevent cross-contamination. The mineral chemistry of the waste-water is influenced by the chemical composition of the slaughterhouse's treated water supply, waste additions such as blood and manure, which can contribute to the heavy metal load in the form of copper, iron, manganese, arsenic and zinc and process plant and pipework, which can contribute to the load of copper, chromium, molybdenum, nickel, titanium and vanadium.

Waste-water Minimisation

As indicated previously, the overall waste load arising from a slaughterhouse is determined principally by the type and number of animals slaughtered. The partitioning of this load between the solid and aqueous phases will depend very much upon the operational practices adopted, however, and there are measures that can be taken to minimise waste-water generation and the aqueous pollution load.

Minimisation can start in the holding pens by reducing the time that the animals remain in these areas through scheduling of delivery times. The incorporation of slatted concrete floors laid to falls of 1 in 60 with drainage to a slurry tank below the floor in the design of the holding pens can also reduce the amount of washdown water required. Alternatively, it is good practice to remove manure and lairage from the holding pens or stockyard in solid form before washing down. In the slaughterhouse itself, cleaning and carcass washing typically account for over 80 per cent of total water use and effluent volumes in the first processing stages. One of the major contributors to organic load is blood, which has a COD of about 4,00,000 mg/l and washing down of dispersed blood can be a major cause of high effluent strength. Minimisation can be achieved by having efficient blood collection troughs allowing collection from the carcass over several minutes. Likewise the trough should be designed to allow separate drainage to a collection tank of the blood and the first flush of wash water. Only residual blood should enter a second drain for collection of the main portion of the wash water. An efficient blood recovery system could reduce the aqueous pollution load by as much as 40 per cent compared to a plant of similar size that allows the blood to flow to waste.

The second area where high organic loads into the waste-water system can arise is in the gut room. Most cattle and sheep abattoirs clean the paunch (rumen), manyplies (omasum) and reed (abomasum) for tripe production. A common method of preparation is to flush out the gut manure from the punctured organs over a mechanical screen and allow water to transport the gut manure to the effluent treatment system.

Typically the gut manure has a COD of over 1,00,000 mg/l, of which 80 per cent dissolves in the wash water. Significant reductions in waste-water strength can be made by adopting a 'dry' system for removing and transporting these gut manures. The paunch manure in its undiluted state has enough water present to allow pneumatic transport to a 'dry' storage area where a compactor can be used to reduce the volume further if required. The tripe material requires washing before further processing, but with a much reduced volume of water and resulting pollution load.

The small and large intestines are usually squeezed and washed for use in casings. To reduce water, washing can be carried out in two stages: a primary wash in a water bath with continuous water filtration and recirculation, followed by a final rinse in clean potable water. Other measures that can be taken in the gut room to minimise water use and organic loadings to the aqueous stream include ensuring that mechanical equipment, such as the hasher machine, are in good order and maintained regularly.

Other methods can also be employed to minimise water usage. These will not in themselves reduce the organic load entering the waste-water treatment system, but will reduce the volume requiring treatment and possibly influence the choice of treatment system to be employed. For example, high-strength, low-volume waste-waters may be more suited to anaerobic rather than aerobic biological treatment methods. Water use minimisation methods include:

1. The use of directional spray nozzles in carcass washing, which can reduce water consumption by as much as 20 per cent.
2. Use of steam condensation systems in place of scald tanks for hair and nail removal.
3. Fitting washdown hoses with trigger grips.

4. Appropriate choice of cleaning agents.
5. Reuse of clear water (e.g. chiller water) for the primary washdown of holding pens.

Waste-water Treatment Processes

The degree of waste-water treatment required will depend on the proposed type of discharge. Wastewaters received into the sewer system are likely to need less treatment than those having direct discharge into a watercourse. In the European countries, direct discharges have to comply with the urban waste water treatment directive and other water quality directives. In the United States the EPA has proposed effluent limitations guidelines and standards (ELGs) for the meat and poultry products industries with direct discharge. These proposed ELGs will apply to existing and new meat and poultry products (MPP) facilities and are based on the well-tested concepts of 'best practicable control technology currently available' (BPT), the 'best conventional pollutant control technology' (BCT), the 'best available technology economically achievable' (BAT) and the 'best available demonstrated control technology for new source performance standards' (NSPS). In summary, the technologies proposed to meet these requirements use, in the main, a system based on a treatment series comprising flow equalisation, dissolved air flotation and secondary biological treatment for all slaughterhouses; and require nitrification for small installations and additional denitrification for complex slaughterhouses.

There is some potential, however, for segregation of waste-waters allowing specific individual pretreatments to be undertaken or, in some cases, bypass of less contaminated streams. Depending on local conditions and regulations, water from boiler houses and refrigerating systems may be segregated and discharged directly or used for outside cleaning operations.

Primary and secondary treatment

Primary treatment

Grease removal is a common first stage in slaughterhouse waste-water treatment, with grease traps in some situations being an integral part of the drainage system from the processing areas. Where the option is taken to have a single point of removal, this can be accomplished in one of two ways: by using a baffled tank or by DAF. A typical grease trap has a minimum detention period of about 30 minutes, but the period need not to be greater than 1 hour. Within the tank, coagulation of fats is brought about by cooling, followed by separation of solid material in baffled chambers through natural flotation of the less dense material, which is then removed by skimming.

Secondary treatment

Secondary treatment aims to reduce the BOD of the waste-water by removing the organic matter that remains after primary treatment. This is primarily in a soluble form. Secondary treatment can utilise physical and chemical unit processes, but for the treatment of meat wastes biological treatment is usually favoured.

Physico-chemical secondary treatment

Chemical treatment of meat-plant wastes is not a common practice due to the high chemical costs involved and difficulties in disposing of the large volumes of sludge produced.

Biological secondary treatment

Using biological treatment, more than 90 per cent efficiency can be achieved in pollutant removal from slaughterhouse wastes. Commonly used systems include lagoons (aerobic and anaerobic), conventional activated sludge, extended aeration, oxidation ditches, sequencing batch reactors and anaerobic digestion.

A series of anaerobic biological processes followed by aerobic biological processes is often useful for sequential reduction of the BOD load in the most economic manner, although either process can be used separately. As noted above, slaughterhouse waste-waters vary in strength considerably depending on a number of factors. For a given type of animal, however, this variation is primarily due to the quantity of water used within the abattoir, as the pollution load (as expressed as BOD) is relatively constant on the basis of live weight slaughtered. Hence, the more economical an abattoir is in its use of water, the stronger the effluent will be and *vice versa*. The strength of the organic degradable matter in the waste-water is an important consideration in the choice of treatment system. To remove BOD using an aerobic biological process involves supplying oxygen (usually as a component in air) in proportion to the quantity of BOD that has to be removed, an increasingly expensive process as the BOD increases. On the other hand an anaerobic process does not require oxygen in order to remove BOD as the biodegradable fraction is fermented and then transformed to gaseous endproducts in the form of carbon dioxide (CO_2) and methane (CH_4).

Anaerobic treatment

Anaerobic digestion is a popular method for treating meat industry wastes. Anaerobic processes operate in the absence of oxygen and the final products are mixed gases of methane and carbon dioxide and a stabilised sludge. Anaerobic digestion of organic materials to methane and carbon dioxide is a complicated biological and chemical process that involves three stages: hydrolysis, acetogenesis, and finally methanogenesis. During the first stage, complex compounds are hydrolysed to smaller chain intermediates. In the second stage acetogenic bacteria convert these intermediates to organic acids and then ultimately to methane and carbon dioxide via the methanogenesis phase (Fig. 28.2).

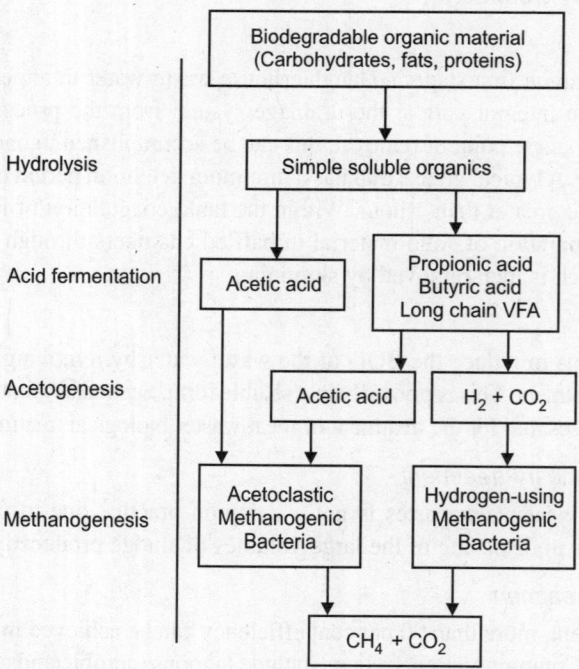

Fig. 28.2. The microbial phases of anaerobic digestion.

In most of the countries, anaerobic systems using simple lagoons are by far the most common method of treating abattoir waste-water. These are not particularly suitable for use in the heavily populated regions of western Europe due to the land area required and also because of the difficulties of controlling odours in the urban areas where abattoirs are usually located. The extensive use of anaerobic lagoons demonstrates the amenability of abattoir waste-waters to anaerobic stabilisation, however, with significant reductions in the BOD at a minimal cost.

The anaerobic lagoon consists of an excavation in the ground, giving a water depth of between 10 and 17 ft (3–5 metres), with a retention time of 5–15 days. Common practice is to provide two ponds in series or parallel and sometimes linking these to a third aerobic pond. The pond has no mechanical equipment installed and is unmixed except for some natural mixing brought about by internal gas generation and surface agitation; the latter is minimised where possible to prevent odour formation and re-aeration. Influent waste-water enters near the bottom of the pond and exits near the surface to minimise the chance of short-circuiting. Anaerobic ponds can provide an economic alternative for purification. The BOD reductions vary widely, although excellent performance has been reported in some cases, with reductions of up to 97 per cent in BOD, up to 95 per cent in SS and up to 96 per cent in COD from the influent values.

Anaerobic lagoons are not without potential problems, relating to both their gaseous and aqueous emissions. As a result of breakdown of the waste-water, methane and carbon dioxide are both produced. These escape to the atmosphere, thus contributing to greenhouse gas emissions, with methane being 25 times more potent than carbon dioxide in this respect. Gaseous emissions also include the odouriferous gases, hydrogen sulphide and ammonia. The lagoons generally operate with a layer of grease and scum on the top, which restricts the transfer of oxygen through the liquid surface, retains some of the heat and helps prevent the emission of odour. Reliance on this should be avoided wherever possible, however, since it is far from a secure means of preventing problems as the oil and grease cap can readily be broken up, for example, under storm water flow conditions.

The use of fabricated anaerobic reactors for abattoir waste-water treatment is also well established. To work efficiently these are designed to operate either at mesophilic (around 95°F or 35°C) or thermophilic (around 130°F or 55°C) temperatures.

Anaerobic filters have also been applied to the treatment of slaughterhouse waste-waters. These maintain a long SRT by providing the micro-organisms with a medium that they can colonise as a biofilm. Unlike conventional aerobic filters, the anaerobic filter is operated with the support medium submerged in an upflow mode of operation. Because anaerobic filters contain a support medium, there is potential for the interstitial spaces within the medium to become blocked and effective pre-treatment is essential to remove suspended solids as well as solidifiable oils, fats and grease.

The third type of high-rate anaerobic system that can be applied to slaughterhouse waste-waters is the upflow anaerobic sludge blanket reactor (UASB). This is basically an expanded-bed reactor in which the bed comprises anaerobic micro-organisms, including methanogens, which have formed dense granules. The mechanisms by which these granules form are still poorly understood, but they are intrinsic to the proper operation of the process. The influent waste-water flows upward through a sludge blanket of these granules, which remain within the reactor as their settling velocity is greater than the up flow velocity of the waste-water. The reactor therefore exhibits a long sludge retention time, high biomass density per unit reactor and can operate at a short HRT.

Aerobic treatment

Aerobic biological treatment for the treatment of biodegradable wastes has been established for over a hundred years and is accepted as producing a good-quality effluent, reliably reducing influent BOD by 95 per cent or more. Aerobic processes can roughly be divided into two basic types: those that maintain the biomass in suspension (activated sludge and its variants) and those that retain the biomass on a support medium (biological filters and its variants). There is no doubt that either basic type is suitable for the treatment of slaughterhouse waste-water and their use is well documented in works such as Broils and Broughton, where aerobic processes are compared with anaerobic ones. In selecting an aerobic process a number of factors need to be taken into account. These include the land area available, the head of water available, known difficulties associated with certain waste-water types (such as bulking and stable foam formation), energy efficiency and excess biomass production. It is important to realise that the energy costs of conventional aerobic biological treatment can be substantial due to the requirement to supply air to the process.

It is, therefore, usual to only treat to the standard required, as treatment to a higher standard will incur additional cost. For example, in order to convert ammonia to nitrate requires 4.5 moles of oxygen for every mole of ammonia converted. In effect this means that a 1 mg/l concentration of ammonia has an equivalent BOD of 4.5 mg/l. It is, therefore, only usual to aim for the conversion of ammonia to nitrate when this is required.

The most common aerobic biological processes used for the treatment of meat industry wastes are biological filtration, activated sludge plants, waste stabilisation ponds and aerated lagoons.

Biological filters

Biological filters can also be used for treating meat industry wastes. In this process the aerobic micro-organisms grow as a slime or film that is supported on the surface of the filter medium. The waste-water is applied to the surface and trickles down while air percolates upwards through the medium and supplies the oxygen required for purification (Fig. 28.3). The treated water along with any microbial film that breaks away from the support medium collects in an under-drain and passes to a secondary sedimentation tank where the biological solids are separated. Trickling filters require primary treatment for removal of settleable solids and oil and grease to reduce the organic load and prevent the system blocking.

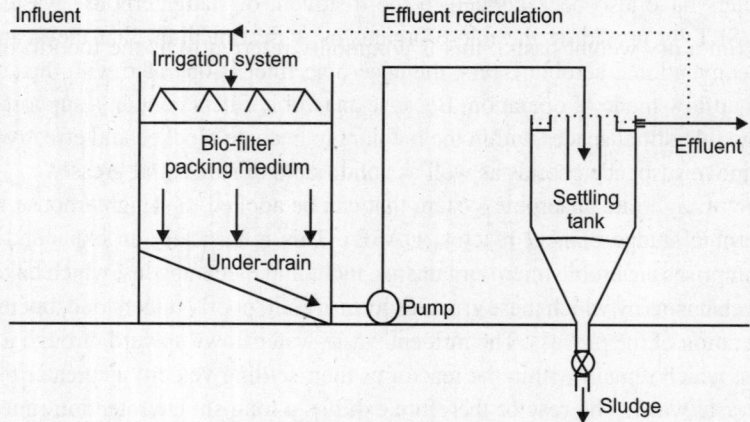

Fig. 28.3. Typical biological filtration treatment system.

Rock or blast furnace slag have traditionally been used as filter media for low-rate and intermediate-rate trickling filters, while high-rate filters tend to use specially fabricated plastic media, either as a loose fill or as a corrugated prefabricated module. The advantage of trickling filters is their low energy requirement, but the disadvantage is the low loading compared to activated sludge, making the plant larger with a consequent higher capital cost.

Because of the relatively high strength of slaughterhouse waste-water, biological filters are more suited to operation with effluent recirculation, which effectively increases surface hydraulic loading without increasing the organic loading. This gives greater control over microbial film thickness.

Biological filters have not been widely adopted for the treatment of slaughterhouse waste-waters despite the lower operating costs compared with activated sludge systems. Obtaining an effluent with a low BOD and ammonia in a single-reactor system can provide conditions suitable for the proliferation of secondary grazing macro-invertebrate species such as fly larvae and this may be unacceptable in the vicinity of a slaughterhouse. There is also the need for very good fat removal from the influent waste-water flow, as this will otherwise tend to coat the surface of the biofilm support medium.

Activated sludge

The activated sludge process has been successfully used for the treatment of waste-waters from the meat industry for many decades. It generally has a lower capital cost than standard-rate percolating filters and occupies substantially less space than lagoon or pond systems. In the activated sludge process the waste-waters are mixed with a suspension of aerobic micro-organisms (activated sludge) and aerated. After aeration, the mixed liquor passes to a settlement tank where the activated sludge settles and is returned to the plant inlet to treat the incoming waste. The supernatant liquid in the settlement tank is discharged as plant effluent. Air can be supplied to the plant by a variety of means, including blowing air into the mixed liquor through diffusers; mechanical surface aeration; and floor-mounted sparge pipes. All the methods are satisfactory provided that they are properly designed to meet the required concentration of dissolved oxygen in the mixed liquor (greater than 0.5 mg/l) and to maintain the sludge in suspension; for nitrification to occur it may be necessary to maintain dissolved oxygen concentrations above 2.0 mg/l.

The activated sludge process can be designed to meet a number of different requirements, including the available land area, the technical expertise of the operator, the availability of sludge disposal routes and capital available for construction. The first step in the design of an activated sludge system is to select the loading rate, which is usually defined as the mass ratio of substrate inflow to the mass of activated sludge (on a dry weight basis); this is commonly referred to as the food to micro-organism (F:M) ratio and is usually reported as lb BOD/lb MLSS day (kg BOD/kg MLSS day). For conventional operation the range is 0.2–0.6; the use of higher values tends to produce a dispersed or nonflocculent sludge and lower values require additional oxygen input due to high endogenous respiration rates. Systems with F:M ratios above 0.6 are sometimes referred to as high rate, while those below 0.2 are known as extended aeration systems. The latter, despite their higher capital and operating costs are commonly chosen for small installations because of their stability, low sludge production and reliable nitrification. Because of the stoichiometric relationship between F:M ratio and mean cell residence time (MCRT), high-rate plants will have an MCRT of less than 4 days and extended aeration plants of greater than 13 days. Because of the low growth rates of the nitrifying bacteria, which are also influenced markedly by temperature, the oxidation of ammonia to nitrates (nitrification) will only occur at F:M ratios less than 0.1. It is also sometimes useful to consider the nitrogen loading rate, which for effective nitrification should be in the range 0.03–0.08 lb N/lb MLSS-day (kg N/kg MLSS day).

Conventional plants can be used where nitrification is not critical, for example, as a pre-treatment before sewer discharge. One of the main drawbacks of the conventional activated sludge process, however, is its poor buffering capability when dealing with shock loads. This problem can be overcome by the installation of an equalisation tank upstream of the process, or by using an extended aeration activated sludge system. In the extended aeration process, the aeration basin provides a 24–30 hours (or even longer) retention time with complete mixing of tank contents by mechanical or diffused aeration. The large volume combined with a high air input results in a stable process that can accept intermittent loadings. A further disadvantage of using a conventional activated sludge process is the generation of a considerable amount of surplus sludge, which usually requires further treatment before disposal. Some early work suggested the possible recovery of the biomass as a source of protein, but concerns over the possible transmission of exotic animal diseases would make this unacceptable in Europe. The use of extended aeration activated sludge or aerated lagoons minimises biosolids production because of the endogenous nature of the reactions. The size of the plant and the additional aeration required for sludge stabilisation does, however, lead to increased capital and operating costs. Considering the high concentrations of nitrogen present in slaughterhouse waste-water, ammonia removal is often regarded as essential from a regulatory standpoint for direct discharge and increasingly there is a requirement for nutrient removal. It is, therefore, not surprising that most modern day designs are of an extended aeration type so as to promote reliable nitrification as well as to minimise sludge production. Efficient designs will also attempt to recover the chemically bound oxygen in nitrate through the process of denitrification, thus reducing treatment costs and lowering nitrate concentrations in the effluent.

Design criteria and loadings for activated sludge treatment have been widely reported and reliable data can be found in a number of reports.

In recent years, a great deal of interest has been shown in the use of sequencing batch reactors (SBRs) for food-processing waste-waters, as these provide a minimum guaranteed retention time and produce a high-quality effluent. A batch process also often fits well with the intermittent discharge of an industrial process working on one or two shifts. Advantages are an ideal plug flow that maximises reaction rates, ideal quiescent sedimentation and flow equalisation inherent in the design. Decanting can be achieved using floating outlets and adjustable weirs, floating aerators are commonly employed and an anoxic fill overcomes problems of effluent turbidity as well as providing ideal conditions for denitrification reactions.

SECTION IV

Special Topics

29. **Biotechnology for Pollution Prevention in Chemical Industry**   579
30. **Instrumental Techniques for Environmental Pollution Monitoring**   586
31. **Environmental Management in the Chemical Process Industries**   605

Chapter 29
Biotechnology for Pollution Prevention in Chemical Industry

INTRODUCTION

Biotechnology comprises of integrated application of theoretical and practical knowledge of biochemistry, microbiology, physiology, genetics and chemical engineering to exploit the properties of microbes and cell cultures for various beneficial technological uses. Environmental biotechnology is a very broad field which includes environmental monitoring and safety, waste treatment and recovery, restoration of envrionmental quality, substitution of nonrenewable resource-based with renewable resources research and development of various processes for the benefit of mankind with due regard to socio-economic, legal and environmental safety considerations.

Environmental biotechnology has a great potential in transformation and enhancement of resource-base, reduction of energy consumption, pollution abatement, enhancement of biomass production and conservation of nonrenewable resources, generation of useful organic chemicals from bioconversion of biomass, etc.

Some of these applications have indirect beneficial effects on wider problems such as global warming, acid rain and climatic changes.

For removing specific types of pollutants selectively, new types of proteins, peptides and biomimetic adsorbers are being developed using biotechnology. Important applications of biotechnology in pollution control and waste management include:
1. Improvement of existing processes.
2. Treatment of toxic wastes using genetically improved organisms.
3. Recovery of useful products from waste materials.
4. Development of new catalysts, new bioreactors, novel biosensors and automation of waste-water treatment processes.

Environmental biotechnology can offer cheap, compact and effective processes instead of bulky, expensive and space wasting ones. Its philosophy is linked with conservation and by-product recovery, and it is not stimulated by market pressures. While initial costs are high, the treatment may well be less costly overall.

Its full potential is not realised, and laboratory and field successes have not translated into applications. Importantly low-value products, if any, are obtained (e.g. alcohol, methane vs hormones and interferons). Biotechnology can become effective if key technical, legal, economic, business and market issues are successfully tackled. Table 29.1 highlights eco-friendly biotechnology.

Table 29.1. Eco-friendly biotechnology.

Other technologies (chemical)	Biotechnology
Raw materials used cause drain on natural resources, e.g. crude oil, petroleum	Renewable resources used in most of the cases, e.g. starch, cellulose
Processes are hazardous	Processes occur at normal temperature, pressure, etc. and are less hazardous
Products are not eco-friendly	Products are eco-friendly
Wastes are more harmful	Less harmful wastes, sometimes even recyclable wastes
Production methods are more polluting	Production methods are less polluting
Pollution control methods only concentrate, convert pollutant to another form	Pollution control methods are for total or elimination and not mere conversion
There is drain on energy/resources (petroleum fuels used)	Process consume less energy
Example of options	Examples of options
(a) Fertilisers	(a) Biofertilisers
(b) Pesticides	(b) Biopesticides
(c) Plastics	(c) Bioplastics
(d) Mining	(d) Biomining-bioleaching
(e) Petroleum fuels	(e) Biofuels

INDUSTRIAL POLLUTION

Industries differ in the type of pollution that can possibly be caused by them. Liquid effluents, solid wastes and gaseous emissions are common, resulting in impact on water-bodies, land and air, but thermal, radioactive and noise pollution are also associated with a few industries. Industries face a number of problems in their pollution abatement programmes. These include use of outdated technology, plant and equipment, financial burden and nonavailability of suitable technology.

Industrial effluents vary in load, concentration of pollutants, toxic materials and are often nutritionally unbalanced. They are complex in composition and may contain biodegradable or non-biodegradable constituents, or both. The effluent of a unit may also show seasonal variation related with production and fluctuations within a day. Non-biodegradable components have to be tackled by physico-chemical methods, while biodegradable components can be tackled otherwise. Pollutants can be classified as:
1. Readily degradable organics (e.g. from food industry).
2. Complex organics (e.g. from organochemical industry, pesticide industry).
3. Reactable inorganics (e.g. from heavy metal industries and plating).
4. Inert inorganics (e.g. from coal mining and quarrying).

Objectives of pollution treatment include reduction of BOD/COD, heavy metal removal, hazardous wastes disposal, removal of nitrogen/phosphorus, removal of oil and grease, deodourisation, removal of air pollutants, solid waste disposal, etc. However, very often much of the effort is focussed on reduction of total organic carbon (TOC) or reduction in BOD/COD.

BIOTECHNOLOGY FOR AIR POLLUTION ABATEMENT AND ODOUR CONTROL

Sulphur oxides, nitrogen oxides, carbon monoxides, hydrogen sulphides, hydrocarbons and particulate matter are the major components of air pollution and are responsible for health and environmental hazards. Equally important are the substances that cause unpleasant/offensive odour. A range of malodorous substances, including phenol, styrene, trichloroethane (TCE), volatile organic compounds (VOCs), amines, hydrogen sulphide, mercaptans and ammonia, are present in gaseous effluents of various industries, treatment plants, animal rendering activities, etc. Deodourisation technology makes use of physical, chemical and biological means for odour removal. Biodeodourisation, though applied since 1923, is relatively less studied and less applied. Its importance and application is, however, increasing.

The Problem

Increase in environmental awareness has resulted in an increasing attention of the people to pollution problems. Offensive odour from pollution is easily sensed and regulations obviously target this aspect, leading to the development of deodourisation technology. A number of industries produce offensive waste gases. These include pesticide, petrochemical, explosive, mining, meat processing, paints and varnishes, textile, chemical, pharmaceutical, animal rendering, and fermentation industries.

The sources of malodorous substances are many and they may originate:
1. During the production process.
2. From storage areas.
3. From pumps and compressors which leak.
4. During transfer of material.
5. From open waste-water treatment plants and garbage composting plants.

Abatement of these odours is difficult because:
1. A number of different compounds are often involved.
2. Odour producing compounds may be present in very low concentrations — some mercaptans, for example, have an odour threshold of below 1 ppb.
3. Sources are often complex, multi-point and difficult to accurately trace.
4. Odours that escape from transfer, filtration and drying operations are more difficult to contain.

Deodourisation Processes

Preventive as well as corrective methods are useful for control of odour. In the first category, process modification and equipment modification can be included, while processes of corrective nature are roughly classified into physical, chemical and biological methods. Dispersion, water washing, adsorption, thermal incineration and catalytic incineration are amongst the dominant physical methods, while chemical methods include catalytic oxidation. Physical and chemical methods, in general, are not flexible for volume, concentrations, or composition of gas changes that may occur. This can be overcome by biological methods. Biological processes earlier required skilled control and large space, but recent developments have overcome some of these restrictions. These processes are now characterised by low running costs (one-third of other processes), easy operation/maintenance and control, and low energy intensity. There are five types of biological waste gas purification systems in operation—bioscrubbers, biofilters biobeds, trickling filters, biotrickling filters.

Applications of biological processes depend on physical and microbiological phenomena. While the former include: mass transfer between gas and liquid phase; mass transfer to micro-organisms; and average residence time of mobile phase; microbiological phenomena are dependent on rate of degradation; substrate/product inhibition; diauxy, etc. There is a lot in literature on laboratory experiments and successful field applications. Although biological deodourisation is considered to be an effective tool, applications are relatively limited. Biologically active materials, like peat, compost, humus, woody heather, bushwood carrying micro-organisms, activated sludge of effluent treatment units or mixture of organisms, or a single organism immobilised as a biofilm on an inert material or in suspended form, are used in biological oxidation of gases. In the biological deodourisation process, target molecules are decomposed by micro-organisms, but the deodourisation mechanism is not clear in many cases.

APPLICATION OF BIOTECHNOLOGY TO POLLUTION PROBLEMS OF VARIOUS INDUSTRIES

Biotechnology can work on five options (three corrective and two preventive) in tackling pollution. Let us see examples of the role played by biotechnology in pollution abatement in some select industries, as an indicator of the direction of efforts.

Food and Allied Industries

Wastes from this industry are characterised by high suspended solids, high BOD/COD, toxic matter in large volumes (although the quantity and load depends upon size of unit and variety of products). Effluents are generally rich in carbohydrates and may be relatively deficient in nitrogen. The question in the case of these manageable wastes is how economically we treat them. BOD/COD reduction is usually done along with generation of energy in the form of biogas and ethanol, or production of SCP/fodder yeast/biomass/aquaculture. Many new efficient biomethanisation reactors are available, and bioconversions of wastes to other chemical products is also possible. Solid wastes is a problem in fruit, vegetable, meat and poultry processing industries. Component separation and recovery of some useful products is common for meat and poultry processing industry which have slaughterhouses, while solid wastes of fruit and vegetable processing/canning industry are suitable for ethanol production or SCP/biomass production.

Dairy

Milk coagulating enzymes have the dominant market in this sector. India is one of the largest manufacturer of milk in the world. In recent years, microbial rennets and recombinant chymosin have very good market for cheese making from milk. The by-product of cheese manufacturing process (whey) can be enzymatically converted into glucose and galactose, which have an application as a sweetener.

Enzyme usage in starch, food and beverages industries

This can be classified into three areas:
1. Enzymes modifying carbohydrates like starch, sucrose, glucose, fructose, etc.
2. Enzymes modifying proteins, like rennets used in manufacturing of cheese and proteases used in the tenderisation of meat or production of protein hydrolysates.
3. Enzymes modifying lipids for hydrolysis of oils and fats to give monoglycerides, diglycerides and free fatty acids.

Starch conversion will be the major application of enzymes. The major enzymes used in this sector are alpha amylase, glucoamylase and glucose isomerase. Starch hydrolysis has gradually become a wholly enzymatic one in the last couple of years, and a range of starch degrading enzymes are now available in sufficient quantities at competitive rates.

Growth of this sector is related to the growth in demand for high-fructose corn syrup used by the soft drink industry. Soft drinks are yet very less explored in third world countries by major soft drink manufacturers and this can enhance the demand for enzymes.

Enzyme market for beverages is also likely to grow very fast. Pectinases are the most important enzymes used in juice and wine production. Beta glucanases find very large application in beer brewing. Increased demand for the production of low calorie beers will drive the growth of enzyme market in this segment. Growth of baked confectionary segment has been very high in India. Fungal alpha amylase, protease and hemicellulase enzymes are widely used in baking sector for bread and biscuit manufacturing. Enzymes are used in baked products and the result is high increase in market of food enzymes.

Animal feed

Enzymes are relatively a new class of animal feed additive. Their function is to act as catalysts to enable more efficient utilisation of animal feeds. They act as bio-catalysts and assist in digestion process for the total utilisation of nutrients and minerals, which are excreted unused. Solid state fermentation of agricultural wastes into vitamin rich products as well as conversion of such waste into pro-biotics or enriching poultry and animal feed are also picking up fast.

Paper Industry

Biotechnology has many contributions to offer to the modern pulp and paper industry and micro-organisms, enzymes, and newer technologies are being applied at various stages. Some major application areas are:

1. Biopulping (fungi used to degrade and reduce lignin contents of cellulose pulp).
2. Mechanical/chemical pulping.
3. Biobleaching (use of enzyme xylanase or fungi producing such enzymes to make pulp brighter) instead of chemical bleaching.
4. Ethanol production from sludge.
5. Growing yeasts or fungi on sulphite waste liquors.
6. Decolouration of pulp mill waste liquors with the help of fungal biomass, or degradation of chlorinated lignin derivatives by white rot fungus.
7. Biological deinking of paper (cellulases and hemicellulases to unhook ink from paper) and help its recycling.

Reduced use of chemicals, reduced pollution problems, higher yields, stronger/better quality paper are the advantages with biotechnological applications. Availability of enzymes at low cost has been responsible for increased applications.

There is a worldwide demand to replace chlorine used for bleaching in the pulp and paper industries due to environmental pressures. Enzymes like xylanase can be used in large quantities in this industry. Enzymes can also be used for lignin removal, deinking, paper coating, pitch removal and adhesive preparations. Only those Indian enzyme companies, which continuously innovate and bring out new products comparable to imported ones, both in quality and price, will be able to survive in longer run.

Chemical Industry

These industries will have the widest class of pollutants with each industry having its unique problems. However, xenobiotic ('stranger to life') substances, heavy metals, hazardous chemicals, toxic organics, extreme pH, high salt contents, high level of nitrogen or phosphorus etc. are the common characteristics. Fortunately, biotreatment techniques to tackle many of these problems are available. While an individual category of pollutant can be tackled successfully using these methods, together they pose a problem. Component separation and coordinated efforts using different technologies are required to get the desired results. For example waste-water from the petrochemical industry contains a large number of pollutants, with the composition depending upon the number, type and capacity of individual process plants, feedstock employed and products manufactured. Water miscible and immiscible liquids and water soluble gases are the major pollutants that need to be removed. Ethylene oxide, isopropanol, acetone, propylene oxide, acrylonitrile, cyclohexane, benzene, xylene, toluene, styrene, etc. are the various pollutants normally present. Biotreatment plants are operated on continuous basis and face limitation of carbon source. The usual BOD determination method does not give idea about real degradability since conditions for BOD test and that of biotreatment plant differ. Well-proven bioprocesses for treating pollutants like benzene, toluene, o-xylene, phthalic acid, kerosene, phenol, cresol, catechol, o-dichlorobenzene, resorcinol have been developed by the local industry.

Chemical synthesis

There are a number of chemicals. like acrylamide, vitamin C, monosodium glutamate, gluconic acid, lactic acid, and L-malic acid, produced by bio-transformation using enzymes. Even the conversion of glucose to fructose, penicillin-G to 6-APA, glycerides for emulsifiers is a major application of industrial enzymes. Also, enzymes are now being used in diagnostics, biosensors and medical research.

Tanning Industry

Biotechnology can play a significant role in tanning industry, both in preventing generation of wastes and also in effective treatment of wastes. Un-hairing and degreasing can be done with help of enzymes, avoiding chemicals like sulphides, alkylphenol ethoxylates, etc. The use of enzymes can cut down processes like bating and the hide structure will remain least disturbed. The use of fat-digesting enzymes for degreasing is beneficial as it can eliminate use of organic solvents and surfactants. This helps for recovery of proteins and fats from wastes as by-products. Fungi can be used for leaching out chromium from tannery effluents, and to remove toxic tannins present in tannery effluents.

India is today the fourth largest leather producer in the world and with environmental regulations becoming increasingly stringent, Indian tanneries will be forced to use more and more enzymes replacing currently used chemicals as far as possible. This is expected to result in a significant increase in the usage of enzymes in this sector.

Textile Industry

Textile industry is single largest foreign-exchange earner in India. Biotechnology can help to prevent pollution in this industry in processing of cotton, silk and wool. The industry can be considerably benefited by adopting such ecofriendly processes.

Enzyme usage in textile sector

This can be classified into five main areas, viz. desizing, bio-polishing, bio-washing, bleach clean-up and fibre modification. India's per capita consumption of denim is far behind most of the other countries in the world and this offers tremendous scope for growth.

Enzyme usage in detergent sector

Detergent and animal feed are two other major sectors where the usage of enzymes is expected to grow rapidly in India in the near future. Detergent enzymes have the largest market and accounts for nearly 33 per cent of the total market.

Household laundry, household dish washing and industrial laundry are the three segments of the market for enzyme-based detergents. However, in India, enzymes are mainly used for household laundries.

Proteases, amylases, lipases and cellulases are the main enzymes used here. Today, very few companies like Procter and Gamble (Ariel), Hindustan Lever Ltd. (Surf) and Henkel (Henko) are using detergent enzymes. Lots of other small detergent manufacturers in India have recently started using enzymes and the detergent enzyme market is expected to grow very soon.

Pesticide Industry

The tremendous diversity in chemistry of pesticides makes their detoxification process a difficult task. Manufacturing pesticides that are less persistent and more prone to biodegradation, manufacturing and using biopesticides which will have specificity of action and minimal environmental- or bio-hazard, and ultimately aiming for resistance within the crops by use of genetic engineering are all part of clean technology programmes. But till we get success in the prevention of pollution, treatment technologies to efficiently eliminate pollutants needs to be seriously examined. Pesticide industry waste-waters, residual pesticides in fields, and contaminated ground water are required to be decontaminated of pesticides and their intermediates. Micro-organisms possessing manipulated genes or enzymes with specific degradative capacity may be used for this purpose. There are various reports of use of enzymes like esterase, phosphatase, alkyl-sulphatase, oxygenase, hydrolase for detoxification of pesticides. Organisms like *Pseudomonas putida*, *Candida tropicalis*, *Fusarium flocciferum*, *Aureoasidium pullutants*, and *Aspergillus niger* can degrade herbicides of chlorobenzoate class. Though applications are limited today, the potential is proven.

Chapter 30

Instrumental Techniques for Environmental Pollution Monitoring

INTRODUCTION

The use of instrumentation is an exciting and fascinating part of chemical analysis that interacts with all the areas of chemistry and with many other fields of pure and applied sciences. Analytical instrumentation plays an important role in the production and evaluation of new products and in the protection of consumers and environment.

Scientists, engineers and technicians in a variety of fields often need information from chemical analysis in their work. Chemists themselves often require information provided by instrumental techniques outside their area of specialisation. Thus, instrumentation provides the lower detection limits required to assure safe foods, drugs, water and air.

For proper understanding of our environment, we must have a clear idea of the identities and quantities of pollutants and other chemical species in air, water, soil and biological samples, etc. These may be either in concentrated point source such as from factory smoke, stacks and sewage discharge or in diffused forms, such as from automobile's exhausts and runoff from agricultural land. These pollutants eventually endanger the life of both humans and animals. These pollutants can range from parts per billion (ppb) to hundreds of parts per million (ppm). In order to control the level of these pollutants in the environment, it is necessary to know their chemical or bio-chemical route of formation and degradation, the extent of their occurrence in the environment and their ecotoxicity. Analytical chemistry plays the most vital role in determining the extent of their occurrence in the environment while the degree of their ecotoxicity determines the priority of pollutants and the overall sensitivity required for analytical methods to be employed for their measurement.

To effectively monitor the environment as well as devise pollution abatement methods, it is necessary to know the quality of the environment-quantitatively. It is essential to be aware of the components responsible for environmental pollution and their concentration in waste-water and solid waste. In addition, the environment required for the growth of eco-friendly micro-organisms, which can be utilised to reduce the amount of pollutants in waste, is also required to be evaluated.

The air and water in our environment contain a wide assortment of toxic organic and inorganic pollutants. They enter the environment as emissions into the atmosphere or as discharges to water bodies. These may be either in concentrated point sources, such as from factory smoke, stacks and sewage discharges, or in diffuse forms, such as from automobiles' exhausts and run-off from agricultural land. These pollutants eventually endanger the life of both humans and animals. These pollutants can range from parts per billion (ppb) or below in rural areas to hundreds of parts per million (ppm) or higher in large industrial and urbanised areas.

In recent years, many chemicals previously considered only moderately toxic have been identified very toxic, e.g. potential carcinogens, and thus have been assigned lower threshold limit values (TLVs). In addition, the number of newly identified toxic substances are increasing everyday.

Hence, there is an increasing need for rapid screening and monitoring of toxic substances in air and water to meet the requirements of pollution control authorities.

In order to control the levels of these pollutants in the environment, it is necessary to know their chemical or biochemical route of formation and degradation, the extent of their occurrence in the environment and their ecotoxicity.

In the recent past, a great deal of advances have been made in analytical methodology and instrumentation. With the ever-increasing advancement of microprocessor technology, these analytical instruments have become more versatile. These are computer-aided instruments (so-called 'intelligent instruments) that offer highly improved detection limits (down to parts per billion (ppb) to parts per trillion (ppt) level), better precision, accuracy and increased specificity.

Thus environmental monitoring must be able, in many cases, to detect with accuracy and consistency contaminants present at very low levels. In the determination of the pollutants present, their fate, and their effect on the environment, biotechnology can be of considerable value, especially as molecular biology techniques are increasingly employed.

Using this information, the level of residual pollutants can be evaluated vis-a-vis the prescribed minimum. The results of such analysis are further used for mass and energy balance, design kinetics of biodegradation processes, design of bioreactors and fermenters, etc.

CHEMICAL ANALYSIS

For proper understanding of our environment, we must have a clear idea of the identities and quantities of pollutants and other chemical species in air, water, soil and biological samples, etc. These may be either in concentrated point source such as from factory smoke, stacks and sewage discharge or in diffused forms, such as from automobile's exhausts and runoff from agricultural land. These pollutants eventually endanger the life of both humans and animals. These pollutants can range from parts per billion (ppb) to hundreds of parts per million (ppm). In order to control the level of these pollutants in the environment, it is necessary to know their chemical or biochemical route of formation and degradation, the extent of their occurrence in the environment and their ecotoxicity. Analytical chemistry plays the most vital role in determining the extent of their occurrence in the environment while the degree of their ecotoxicity determines the priority of pollutants and the overall sensitivity required for analytical methods to be employed for their measurement.

TECHNIQUES FOR ANALYSIS

Spectrophotometric Methods

Absorption spectrophotometry

Absorption spectrophotometry of light-absorbing species in solution, historically called colorimetry when visible light is absorbed, is still used for the analysis of many water and some air pollutants. Basically, absorption spectrophotometry consists of measuring the percentage transmittance (%T) of monochromatic light passing through a light-absorbing solution as compared to the amount passing through a blank solution containing everything in the medium but the sought for constituent (100 per cent).

Atomic absorption and emission analyses

Atomic absorption analysis has become the method of choice for most metals analysed in environmental samples. This technique is based upon the absorption of monochromatic light by a cloud of atoms of the analyte metal. The monochromatic light is produced by a source composed of the same atoms as those being analysed. The source produces intense electromagnetic radiation, with a wavelength exactly the same as that absorbed by the atoms, resulting in extremely high selectivity. The basic components of an atomic absorption instrument are shown in Fig. 30.1.

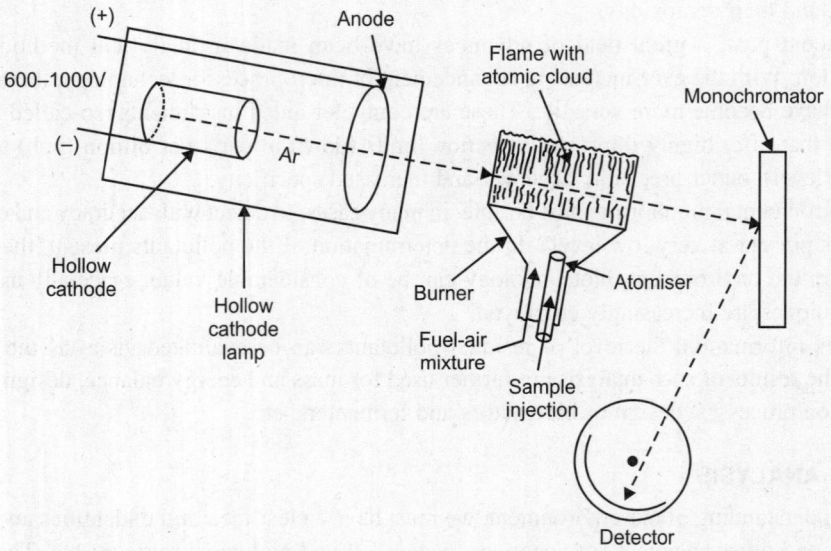

Fig. 30.1. The basic components of an atomic absorption spectrophotometer.

The key element is the hollow cathode lamp, in which atoms of the analyte metal are energised so that they become electronically excited and emit radiation, with a very narrow wavelength band characteristic of the metal. This radiation is guided by the appropriate optics through a flame into which the sample is aspirated. In the flame, most metallic compounds are decomposed and the metal is reduced to the elemental state, forming a cloud of atoms.

These atoms absorb a fraction of radiation in the flame. The fraction of radiation absorbed increases with the concentration of the sought-for element in the sample according to the Beer's law relationship. The attenuated light beam next goes to a monochromator to eliminate extraneous light resulting from the flame and then to a detector.

Atomisers other than a flame can be used. The most common of these is the graphite furnace, which consists of a hollow graphite cylinder placed so that the light beam passes through it. A small sample of up to 100–µl is inserted in the tube through a hole in the top. An electric current is passed through the tube to heat it–gently at first to dry the sample, then rapidly to vapourise and excite the metal analyte.

The absorption of metal atoms in the hollow portion of the tube is measured and recorded as a spikeshaped signal. A diagram of a simple graphite furnace with a typical output signal is shown in Fig. 30.2. The major advantage of the graphite furnace is that it gives detection limits up to 1000 times lower than those of conventional flame devices.

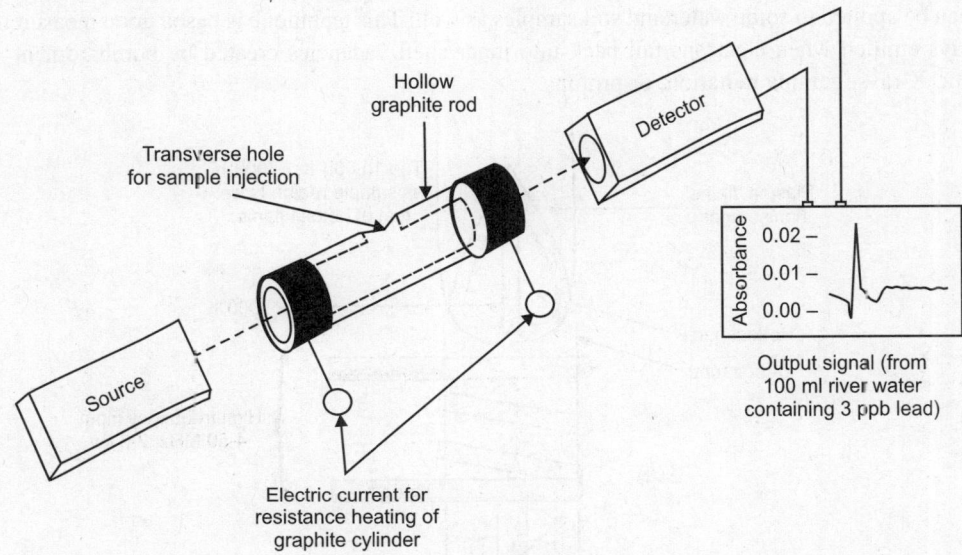

Fig. 30.2. Graphite furnace for atomic absorption analysis and typical output signal.

A special technique for the flameless atomic absorption analysis of mercury involves its room-temperature reduction to the elemental state by tin (II) chloride in solution, followed by sweeping it into an absorption cell with air. Nanogram (10^{-9}g) quantities of mercury can be determined by measuring mercury absorption at 253.7 nm.

Atomic emission techniques

Metals may be determined in water, atmospheric particulate matter, and biological samples very well by observing the spectral lines emitted when they are heated to a very high temperature. An especially useful atomic emission technique is inductively coupled plasma atomic emission spectroscopy. The 'flame' in which analyte atoms are excited in plasma emission consists of an incandescent plasma (ionised gas) of argon heated inductively by radiofrequency energy at 4–50 MHz and 2–5 kW (Fig. 30.3). The energy is transferred to a stream of argon through an induction coil, producing temperatures up to 10,000 K. The sample atoms are subjected to temperatures around 7000 K, twice those of the hottest conventional flames (for example, nitrous oxide-acetylene at 3200 K). Since emission of light increases exponentially with temperature, lower detection limits are obtained. Furthermore, the technique enables emission analysis of some of the environmentally important metalloids such as arsenic, boron and selenium. Interfering chemical reactions and interactions in the plasma are minimised as compared to flames. Of greatest significance, however, is the capability of analysing as many as 30 elements simultaneously, enabling a true multielement analysis technique. Thus, plasma atomisation combined with mass spectrometric measurement of analyte elements is a relatively new technique that is an especially powerful means for multielement analysis.

X-Ray fluorescence

X-ray fluorescence is another multielement analysis technique that can be applied to a wide variety of environmental samples. It is especially useful for the characterisation of atmospheric particulate matter, but it can be applied to some water and soil samples as well. This technique is based upon measurement of X-rays emitted when electrons fall back into inner shell vacancies created by bombardment with energetic X-rays, gamma radiation, or protons.

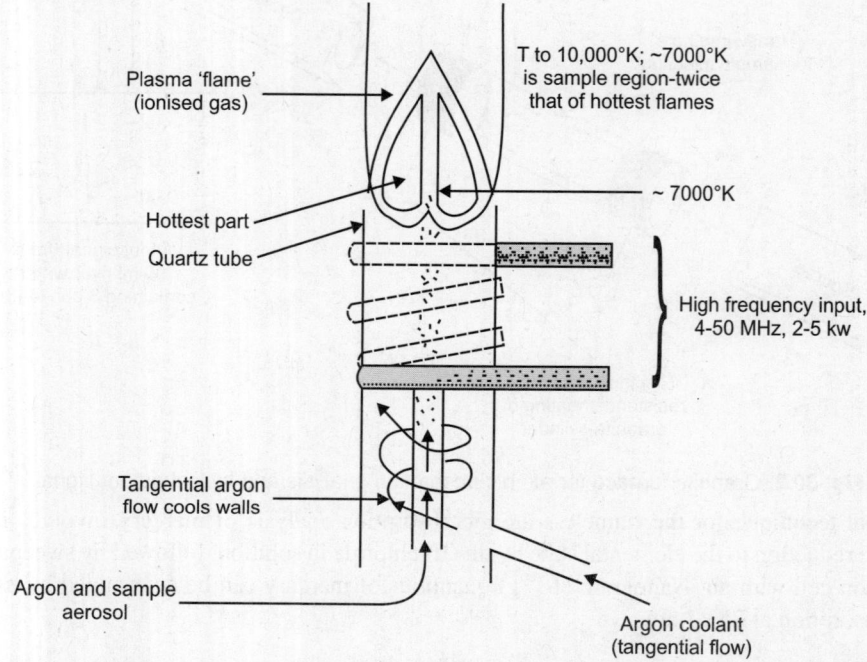

Fig. 30.3 Schematic diagram showing inductively coupled plasma used for optical emission spectroscopy.

The emitted X-rays have an energy characteristic of the parcticular atom. The wavelength (energy) of the emitted radiation yields a qualitative analysis of the elements and the intensity of radiation from a particular element provides a quantitative analysis. A schematic diagram of a wavelength-dispersive X-ray fluorescence spectrophotometer is shown in Fig. 30.4. An excitation source, normally an X-ray tube emitting 'white' X-rays (a continuum), produces a primary beam of energetic radiation which excites fluorescent X-rays in the sample. A radioactive source emitting gamma rays or protons from an accelerator may also be used for excitation. For best results, the sample should be mounted as a thin layer, which means that segments of air filters containing fine particulate matter make ideal samples.

The fluorescent X-rays are passed through a collimator to select a parallel secondary beam, which is dispersed according to wavelength by diffraction with a crystal monochromator. The monochromatic X-rays in the secondary beam are counted by a detector which rotates at a degree twice that of the crystal to scan the spectrum of emitted radiation.

Energy-selective detectors of the Si(Li) semi-conductor type enable measurement of fluorescent X-rays of different energies without the need for wavelength dispersion. Instead, the energies of a number of lines falling on a detector simultaneously are distinguished electronically. An energy-dispersive X-ray fluorescence spectrum from an atmospheric particulate sample is shown in Fig. 30.5. A significant advantage of X-ray fluorescence multielement analysis is that sensitivities and detection limits do not vary greatly across the periodic table as they do with methods such as neutron activation analysis or atomic absorption. Proton-excited X-ray emission is particularly sensitive.

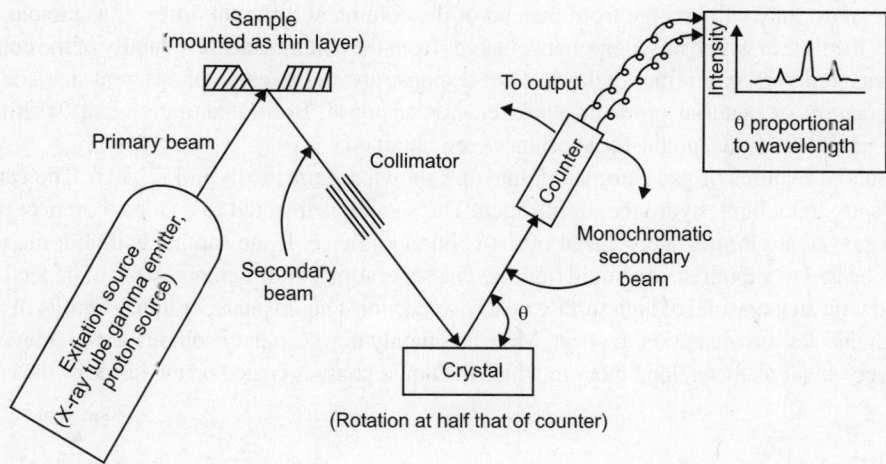

Fig. 30.4. Wavelength-dispersive X-ray fluorescence spectrophotometer.

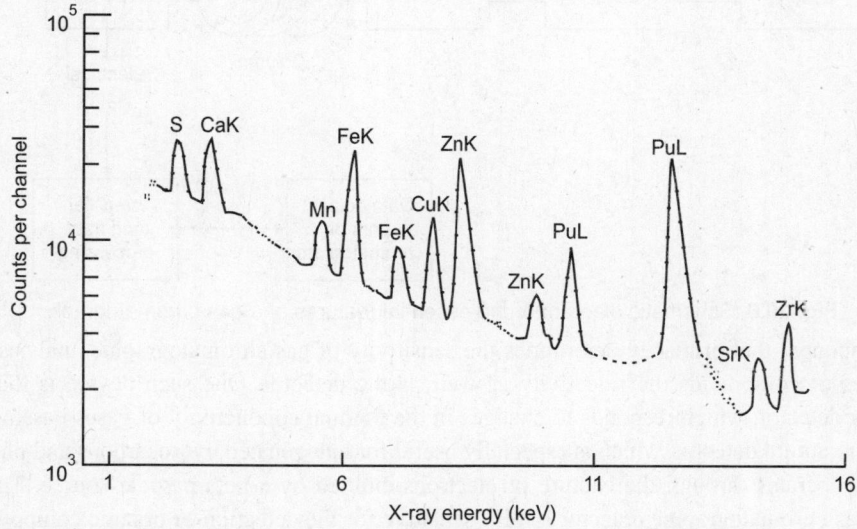

Fig. 30.5. Energy-dispersive X-ray fluorescence spectrum from an atmospheric particle sample.

Gas Chromatography

Gas chromatography has played an important role in the analysis of organic materials. Gas chromatography is both a qualitative and quantitative technique; for some analytical applications of environmental importance, it is remarkably sensitive and selective. Gas chromatography is based upon the principle that when a mixture of volatile materials transported by a carrier gas is passed through a column containing an adsorbent solid phase, or, more commonly, an absorbing liquid phase coated on a solid material, each volatile component will be partitioned between the carrier gas and solid or liquid. The length of time required for the volatile component to traverse the column is proportional to the degree to which it is retained by the non-gaseous phase. Since different components may be retained to different degrees, they will emerge from the end of the column at different times. If a suitable detector is available, the time at which the component emerges from the column and the quantity of the component are both measured. A recorder trace of the detector response appears as peaks of different sizes, depending upon the quantity of material producing the detector response. Both quantitative and (within limits) qualitative analyses of the sought-for substances are obtained.

The essential features of gas chromatography are shown schematically in Fig. 30.6. The carrier gas generally is argon, helium, hydrogen, or nitrogen. The sample is injected as a single compact plug into the carrier gas stream immediately ahead of the column entrance. If the sample is liquid, the injection chamber is heated to vapourise the liquid rapidly. The separation column may consist of a metal or glass tube packed with an inert solid of high surface area covered with a liquid phase, or it may consist of an active solid, which enables the separation to occur. More commonly now, capillary columns are employed which consist of very small diameter, long tubes in which the liquid phase is coated on the inside of the column.

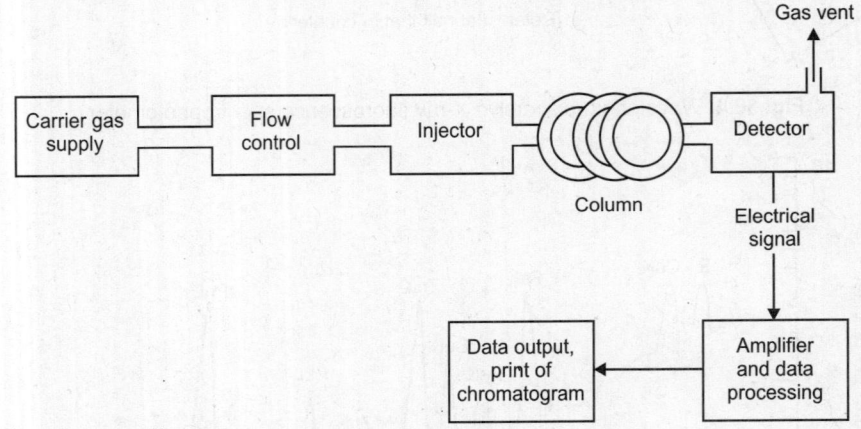

Fig. 30.6. Schematic diagram of the essential features of a gas chromatograph

The component that primarily determines the sensitivity of gas chromatographic analysis, and for some classes of compounds the selectivity as well, is the detector. One such device is the thermal conductivity detector, which responds to changes in the thermal conductivity of gases passing over it. The electron-capture detector, which is especially useful for halogenated hydrocarbons and phosphorus compounds, operates through the capture of electrons emitted by a beta-particle source. The flame-ionisation gas chromatographic detector is very sensitive for the detection of organic compounds. It is based upon the phenomenon by which organic compounds form highly conducting fragments, such as C^+, in a flame. Application of a potential gradient across the flame results in a small current that may be readily measured.

The mass spectrometer may be used as a detector for a gas chromatograph. A combined gas chromatograph/mass spectrometer (GC/MS) instrument is an especially powerful analytical tool for organic compounds. Gas chromatographic analysis requires that a compound exhibit at least a few mm of vapour pressure at the highest temperature at which it is stable. In many cases, organic compounds that cannot be chromatographed directly may be converted into derivatives that are amenable to gas chromatographic analysis. It is seldom possible to analyse organic compounds in water by direct injection of the water into the gas chromatograph; higher concentration is usually required.

Two techniques commonly employed to remove volatile compounds from water and concentrate them are extraction with solvents and purging volatile compounds with a gas, such as helium, concentrating the purged gases on a short column and driving them off by heat into the chromatograph.

High-performance liquid chromatography

A liquid mobile phase used with very small column-packing particles enables high-resolution chromatographic separation of materials in the liquid phase. Very high pressures up to several thousand psi are required to get a reasonable flow rate in such systems. Analysis using such devices is called high performance liquid chromatography (HPLC) and offers an enormous advantage, in that the materials analysed need not be changed to the vapour phase, a step that often requires preparation of a volatile derivative or results in decomposition of the sample. The basic features of a high-performance liquid chromatograph are the same as those of a gas chromatograph, shown in Fig. 30.6, except that a solvent reservoir and high-pressure pump are substituted for the carrier gas source and regulator. An HPLC chromatogram of some water pollutants is shown in Fig. 30.7.

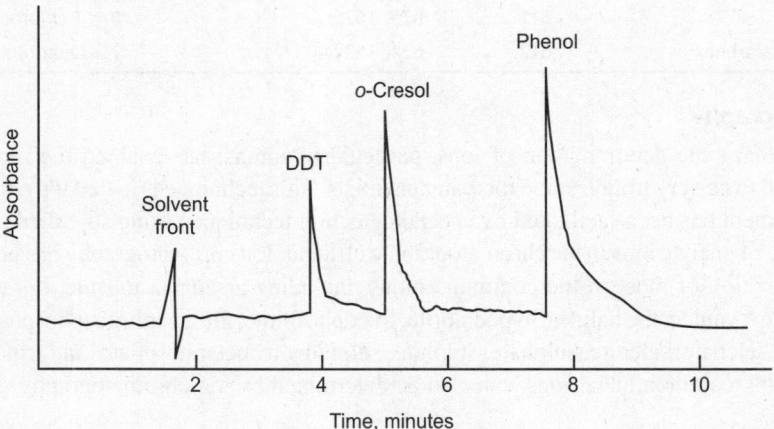

Fig. 30.7. HPLC chromatogram showing separation of some compounds extracted from water contaminated by coal gasification; ultraviolet absorption detection at 272 nm.

Refractive index and ultraviolet detectors are both used for the detection of peaks coming from the liquid chromatograph column. Fluorescence detection can be especially sensitive for some classes of compounds. Mass spectrometric detection of HPLC effluents has led to the development of LC/MS analysis. Somewhat difficult in practice, this technique can be a powerful tool for the determination of analytes that cannot be subjected to gas chromatography. High-performance liquid chromatography has emerged as a very useful technique for the analysis of a number of water pollutants.

Chromatographic analysis of water pollutants

A number of chromatography-based standard methods have also been developed for determining water pollutants. Some of these methods use the purge-and-trap technique, bubbling gas through a column of water to purge volatile organics from the water followed by trapping the organics on solid sorbents, whereas others use solvent extraction to isolate and concentrate the organics. These methods are summarised in Table 30.1.

Table 30.1. Chromatography methods for organic compounds in water.

Class of compounds	Method number			Example analytes
	GC	GC/MS	HPLC	
Purgeable halocarbons	601			Carbon tetrachloride
Purgeable aromatics	602	624, 1624	—	Toluene
Acrolein and acrylonitrile	603	604, 1624		Acrolein
Phenols	604	625, 1625		Phenol and chlorophenols
Phthalate esters	606	625, 1625		Bis(2-ethylhexylphthalate)
Nitrosamines	607	625, 1625		N-nitroso-N-dimethylamine
Organochlorine pesticides and PCBs	608	625		Heptachlor, PCB 1016
Nitroaromatics and isophorone	609	625, 1625		Nitrobenzene
Polycyclic aromatic hydrocarbons	610	625, 1625	610	Benzo[a]pyrene
Haloethers	611	625, 1625		Bis(2-chloroethyl) ether
Chlorinated hydrocarbons	612	624, 1624		1,3-Dichlorobenzene

Ion chromatography

Liquid chromatographic determination of ions, particularly anions, has enabled the measurement of species that used to be very troublesome for water chemists. This technique is called ion chromatography and its development has been facilitated by special detection techniques using so-called suppressors to enable detection of analyte ions in the chromatographic effluent. Ion chromatography has been developed for the determination of most of the common anions, including arsenate, arsenite, borate, carbonate, chlorate, chlorite, cyanide, the halides, hypochlorite, hypophosphite, nitrate, nitrite, phosphate, phosphite, pyrophosphate, selenate, selenite, sulphate, sulphite, sulphide, trimetaphosphate, and tripolyphosphate. Cations, including common metal ions, can also be determined by ion chromatography.

Mass Spectrometry

Mass spectrometry is particularly useful for the identification of specific organic pollutants. It depends upon the production of ions by an electrical discharge or chemical process, followed by separation based on the charge-to-mass ratio and measurement of the ions produced. The output of a mass spectrometer is a mass spectrum, such as the one shown in Fig. 30.8. A mass spectrum is characteristic of a compound and serves to identify it. Computerised data banks for mass spectra have been established and are stored in computers interfaced with mass spectrometers. Identification of a mass spectrum depends upon the purity of the compound from which the spectrum is taken. Prior separation by gas chromatography with continual sampling of the column effluent by a mass spectrometer, commonly called gas chromatography-mass spectrometry (GC/MS), is particularly effective in the analysis of organic pollutants.

Total organic carbon in water

Dissolved organic carbon exerts an oxygen demand in water; often this is in the form of toxic substances and is a general indicator of water pollution. Therefore, its measurement is quite important. The measurement of total organic carbon, TOC, is now recognised as the best means of assessing the organic content of a water sample. The measurement of this parameter has been facilitated by the development of methods which, for the most part, totally oxidise the dissolved organic material to produce carbon dioxide. The amount of carbon dioxide evolved is taken as a measure of TOC.

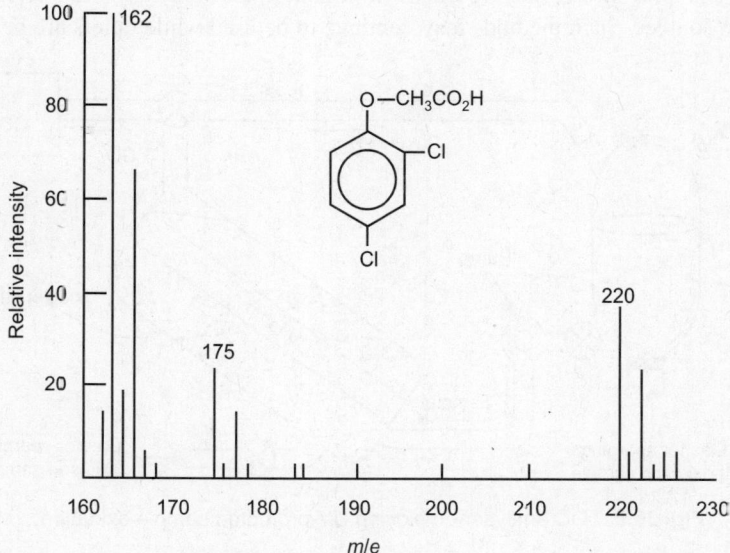

Fig. 30.8. Partial mass spectrum of the herbicide 2,4-dichlorophenoxyacetic acid (2,4-D), a common water pollutant.

TOC can be determined by a technique that uses a dissolved oxidising agent promoted by ultraviolet light. Potassium peroxydisulphate, $K_2S_2O_8$, is usually chosen as an oxidising agent to be added to the sample. Phosphoric acid is also added to the sample, which is sparged with air or nitrogen to drive off CO_2 formed from HCO_3^- and CO_3^{2-} in solution. After sparging, the sample is pumped to a chamber containing a lamp emitting ultraviolet radiation of 184.9 nm. This radiation produces reactive free radical species, such as the hydroxyl radical, HO. The active species bring about the rapid oxidation of dissolved organic compounds as shown in the following general reaction:

$$\text{Organics} + \text{HO}\bullet \longrightarrow CO_2 + H_2O$$

After oxidation is complete, the CO_2 is sparged from the system and measured with a gas chromatographic detector or by absorption in ultrapure water followed by a conductivity measurement. Fig. 30.9 is a schematic of a TOC analyser.

Atmospheric Monitoring

Good analytical methodology, particularly that applicable to automated analysis and continuous monitoring, is essential for the study and alleviation of air pollution. The atmosphere is a particularly difficult analytical system because of the very low levels of substances to be analysed; sharp variations in pollutant level with time and location; differences in temperature and humidity; and difficulties encountered in reaching desired sampling points, particularly those substantially above the Earth's surface. Furthermore, although improved techniques for the analysis of air pollutants are continually being developed, a need still exists for new analytical methodology and the improvement of existing methodology. Much of the earlier data on air pollutant levels (as well as much of the data currently being taken) were unreliable as a result of inadequate analysis and sampling methods. An atmospheric pollutant analysis method does not have to give the actual value to be useful. One which gives a relative value may still be helpful in establishing trends in pollutant levels, determining pollutant effects and locating pollution sources. Such methods may continue to be used while others are being developed.

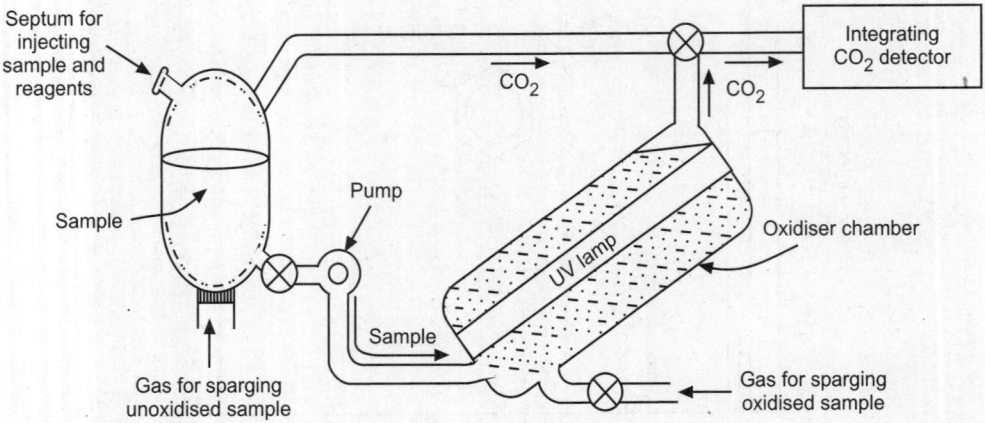

Fig. 30.9. TOC analyser employing UV-promoted sample oxidation.

Air pollutants measured

Air pollutants generally measured may be placed in several different categories. One such category contains materials for which ambient (surrounding atmosphere) standards have been set by the environmental protection agency. These are sulphur dioxide, carbon monoxide, nitrogen dioxide, nonmethane hydrocarbons and particulate matter. The standards are categorised as primary and secondary. Primary standards are those defining the level of air quality necessary to protect public health. Secondary standards are designed to provide protection against known or expected adverse effects of air pollutants, particularly upon materials, vegetation and animals. Another group of air pollutants to be measured consists of those known to be specifically hazardous to human health, such as asbestos, beryllium and mercury. A third category of air pollutants contains those regulated in new installations of selected stationary sources, such as coal-cleaning plants, cotton gins, lime plants and paper mills. Some pollutants in this category are visible emissions, acid (H_2SO_4) mist, particulate matter, nitrogen oxides and sulphur oxides. These substances often must be monitored in the stack to ensure that emission standards are being met. A fourth category consists of the emissions of mobile sources (motor vehicles)—hydrocarbons, CO, and NO_x. A fifth group consists of miscellaneous elements and compounds, such as certain heavy metals, fluoride, chlorine, phosphorus, polycyclic aromatic hydrocarbons (PAH), polychlorinated biphenyls, odourous compounds, reactive organic compounds and radio-nuclides.

For some species, an analytical method is well developed and reasonably satisfactory. For others, no really satisfactory method exists. The development of analytical techniques for air pollutants remains a fertile and challenging area for research.

Levels of air pollutants and other air-quality parameters are expressed in several different kinds of units. These are, for gases and vapours, $\mu g/m^3$ (alternatively, ppm by volume); for weight of particulate matter, $\mu g/m^3$; for particulate matter count, number per cubic metre; for visibility, kilometres; for instantaneous light transmission, percentage of light transmitted; for emission and sampling rates, m^3/min; for pressure, mm Hg; and for temperature, degrees Celsius. Air volumes should be converted to conditions of 10°C and 760 mm Hg (1 atm), assuming ideal gas behaviour.

Methods of Analysis

A very large number of different analytical techniques are used for atmospheric pollutant analysis. Some of these whose uses are not confined to atmospheric analysis were discussed earlier in this chapter. A summary of some of the main instrumental techniques for air monitoring is presented in Table 30.2.

Table 30.2. The main techniques used for air pollutant analysis.

Pollutant	Method	Potential interferences
SO_2 (total S)	Flame photometric (FPD)	H_2S, CO
SO_2	Gas chromatography (FPD)	H_2S, CO
SO_2	Spectophotometric (pararosaniline	H_2S, HCl, NH_3, NO_2, O_3 wet chemical)
SO_2	Electrochemical	H_2S, HCl, NH_3, NO, NO_2, O_3, C_2H_4
SO_2	Conductivity	HCl, NH_3, NO_2
SO_2	Gas-phase spectrophotometric	NO, NO_2, O_3,
O_3	Chemiluminescent	H_2S
O_3	Electrochemical	NH_3, NO_2, SO_2
O_3	Spectrophotometric (potassium iodide	NH_3, NO_2, NO, SO_2 reaction, wet chemical)
O_3	Gas-phase spectrophotometric	NO_2, NO, SO_2
CO	Infrared	CO_2 (at high levels)
CO	Gas chromatography	–
CO	Electrochemical	NO, C_2H_4
CO	Catalytic combustion-thermal detection	NH_3
CO	Infrared fluorescence	–
CO	Mercury replacement ultraviolet	C_2H_4 photometric
NO_2	Chemiluminescent	NH_3, NO, NO_2, SO_2
NO_2	Spectrophotometric (azo-dye reaction	NO, SO_2, NO_2, O_3 wet chemical)
NO_2	Electrochemical	HCl, NH_3, NO, NO_2, SO_2, O_3, CO
NO_2	Gas-phase spectrophotometric	NH_3, NO, NO_2, SO_2, CO
NO_2	Conductivity	HCl, NH_3, NO, NO_2, SO_2

Analysis of sulphur dioxide

The reference method for the analysis of sulphur dioxide is the spectrophotometric pararosaniline method first described by West and Gaeke, and subsequently optimised. It is applicable to the analysis of 0.005–5 ppm SO_2 in ambient air. Figure 30.10 illustrates the various components involved in a sampling train employed to sample the atmosphere for sulphur dioxide to be analysed by the West-Gaeke method.

The method makes use of a collecting solution of 0.04 M potassium tetrachloromercurate to collect sulphur dioxide according to the following reaction:

$$HgCl_4^{2-} + SO_2 + H_2O \longrightarrow HgCl_2SO_3^{2-} + 2H^+ + 2Cl^- \qquad \ldots (30.1)$$

Typically, this involves scrubbing 30 litres of air through 10 ml of scrubbing solution with a collection efficiency of around 95 per cent. Sulphur dioxide in the scrubbing medium is reacted with formaldehyde:

$$HCHO + SO_2 + 2H_2O \longrightarrow HOCH_2SO_3H \qquad \ldots (30.2)$$

The adduct formed is then reacted with uncoloured organic pararosaniline hydrochloride to produce a red-violet dye. Although NO_2 at levels above about 2 ppm interferes, the interference may be eliminated by reducing the NO_2 to N_2 gas with sulphamic acid, H_2NSO_3H.

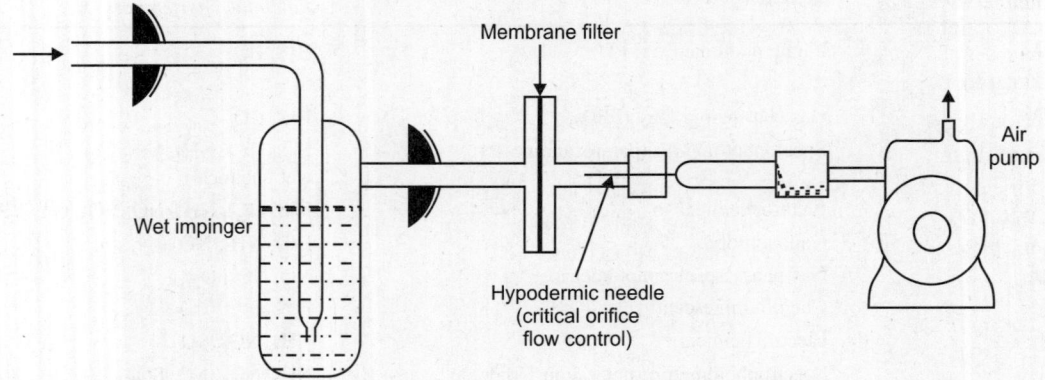

Fig. 30.10. Sampling train for collecting air samples for sulphur dioxide analysis with the West-Gaeke method.

Performed manually, the West-Gaeke method for sulphur dioxide analysis is cumbersome and complicated. However, the method has been refined to the point that it can be done automatically with continuous monitoring equipment. A block diagram of such an analyser is shown in Fig. 30.11.

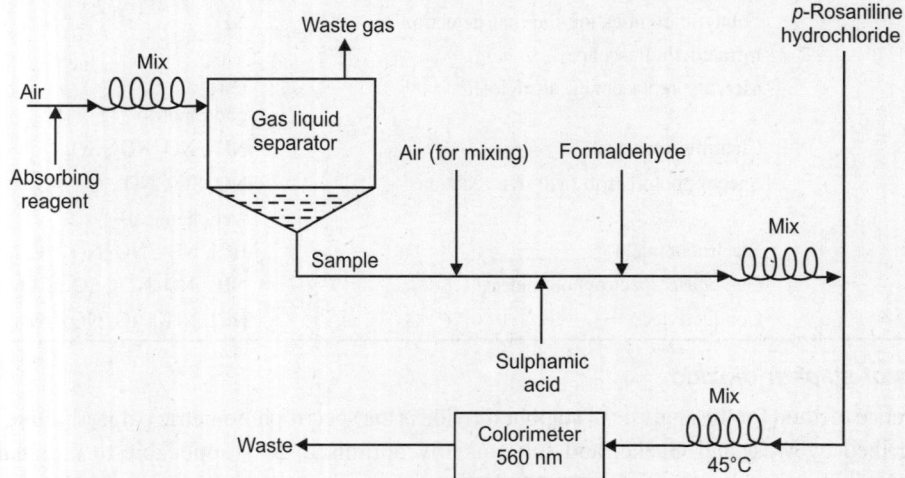

Fig. 30.11. Schematic diagram of an automatic analyser used for sulphur dioxide analysis by the West-Gaeke method.

Generally, sulphur dioxide is collected in a hydrogen peroxide solution and increased conductance of the sulphuric acid solution is measured. Several types of sulphur dioxide monitors are based on amperometry, in which an electrical current is measured that is proportional to the SO_2 in a collecting solution. Sulphur dioxide can be determined by ion chromatography, by bubbling SO_2 through hydrogen peroxide solution to produce SO_4^{2-}, followed by analysis of the sulphate by ion chromatography, a method that separates ions on a chromatography column and detects them very sensitively by conductivity measurement. Flame photometry, sometimes in combination with gas chromatography, is also used for the detection of sulphur dioxide and other gaseous sulphur compounds. The gas is burned in a hydrogen flame and the sulphur emission line at 394 nm is measured. Several direct spectrophotometric methods are used for sulphur dioxide measurement, including nondispersive infrared absorption, Fourier transform infrared analysis (FTIR), ultraviolet absorption, molecular resonance fluorescence and second-derivative spectrophotometry. The principles of these methods are the same for any gas measured.

Nitrogen oxides

Although as noted in Table 30.2, several methods have been used to determine nitrogen oxides, gas-phase chemiluminescence is the favoured method of NO_x analysis. It results from the emission of light from electronically excited species formed by a chemical reaction. In the case of NO, ozone is reacted with NO to produce NO_2, which loses energy and returns to the ground state through emission of light in the 600–3000 nm range. The emitted light is measured by a photomultiplier; its intensity is proportional to the concentration of NO. A schematic diagram of the device used is shown in Fig. 30.12.

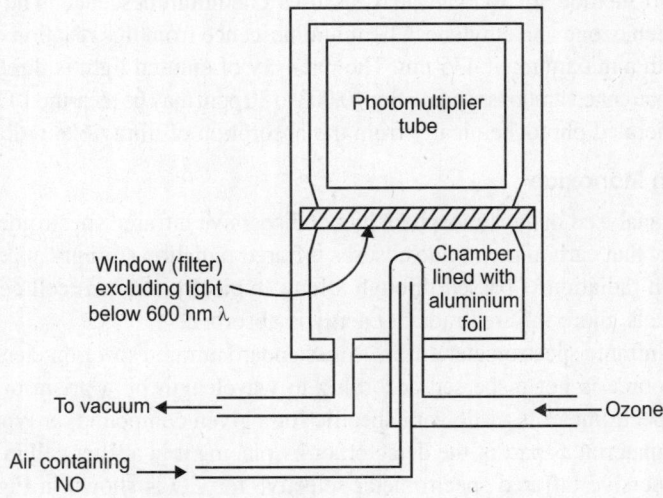

Fig. 30.12. Chemiluminescence detector for NO_x.

Since the chemiluminescence detector system depends upon the reaction of O_3 with NO, it is necessary to convert NO_2 to NO in the sample prior to analysis. This is accomplished by passing the air sample over a thermal converter. Analysis of such a sample gives NO_x, the sum of NO and NO_2. Chemiluminescence analysis of a sample that has not been passed over the thermal converter gives NO. The difference between these two results is NO_2.

Other nitrogen compounds besides NO and NO_2 undergo chemiluminescence by reacting with O_3, and these may interfere with the analysis if present in an excessive level. Particulate matter also causes interference which may be overcome by employing a membrane filter on the air inlet.

This analysis technique is illustrative of chemiluminescence analysis in general. Chemiluminescence is an inherently desirable technique for the analysis of atmospheric pollutants because it avoids wet chemistry, is basically simple and lends itself well to continuous monitoring and instrumental methods. Another chemiluminescence method, that is employed for the analysis of ozone, is described in the next section.

Analysis of Oxidants

Atmospheric oxidants that are commonly analysed include ozone, hydrogen peroxide, organic peroxides and chlorine. The classic manual method for the analysis of oxidants is based upon their oxidation of I^- ion followed by spectrophotometric measurement of the product. The sample is collected in 1 per cent KI buffered at pH 6.8. Oxidants react with iodide ion as shown by the following reaction of ozone:

$$O_3 + 2H^+ + 3I^- \longrightarrow I_3^- + O_2 + H_2O \qquad \ldots (30.3)$$

The absorbance of the coloured I_3^- product is measured spectrophotometrically at 352 nm. Generally, the level of oxidant is expressed in terms of ozone, although it should be noted that not all oxidants—PAN, for example, react with the same efficiency as O_3. Oxidation of I^- may be employed to determine oxidants in a concentration range of several hundreths of a part per million to approximately 10 ppm. Nitrogen dioxide gives a limited response to the method and reducing substances interfere seriously.

Now the favoured method for oxidant analysis uses chemiluminescence. The chemiluminescent reaction is that between ozone and ethylene. Chemiluminescence from this reaction occurs over a range of 300–6000 nm, with a maximum at 435 nm. The intensity of emitted light is directly proportional to the level of ozone. Ozone concentrations ranging from 0.003 to 30 ppm may be measured. Ozone for calibrating the instrument is generated photochemically from the absorption of ultraviolet radiation by oxygen.

Analysis of Carbon Monoxide

Carbon monoxide is analysed in the atmosphere by nondispersive infrared spectrometry. This technique depends upon the fact that carbon monoxide absorbs infrared radiation strongly at certain wavelengths. Therefore, when such radiation is passed through a long (typically 100 cm) cell containing a trace of carbon monoxide levels, more infrared radiant energy is absorbed.

A non-dispersive infrared spectrometer differs from standard infrared spectrometers in that the infrared radiation from the source is not dispersed according to wavelength by a prism or grating. The non-dispersive infrared spectrometer is made very specific for a given compound, or type of compound, by using the sought for material as part of the detector, or by placing it in a filter cell in the optical path. A diagram of a nondispersive infrared spectrometer selective for CO is shown in Fig. 30.13. Radiation from an infrared source is 'chopped' by a rotating device, so that it alternately passes through a sample cell and a reference cell. In this particular instrument, both beams of light fall on a detector which is filled with CO gas and separated into two compartments by a flexible diaphragm. The relative amounts of infrared radiation absorbed by the CO in the two sections depend upon level in the sample. The difference in the amount of infrared radiation absorbed in the two compartments causes slight differences in heating, so that the diaphragm bulges slightly toward one side. Very slight movement of the diaphragm can be detected and recorded. By means of this device, carbon monoxide can be measured from 0 to 150 ppm, with a relative accuracy of ±5 per cent in the optimum concentration range.

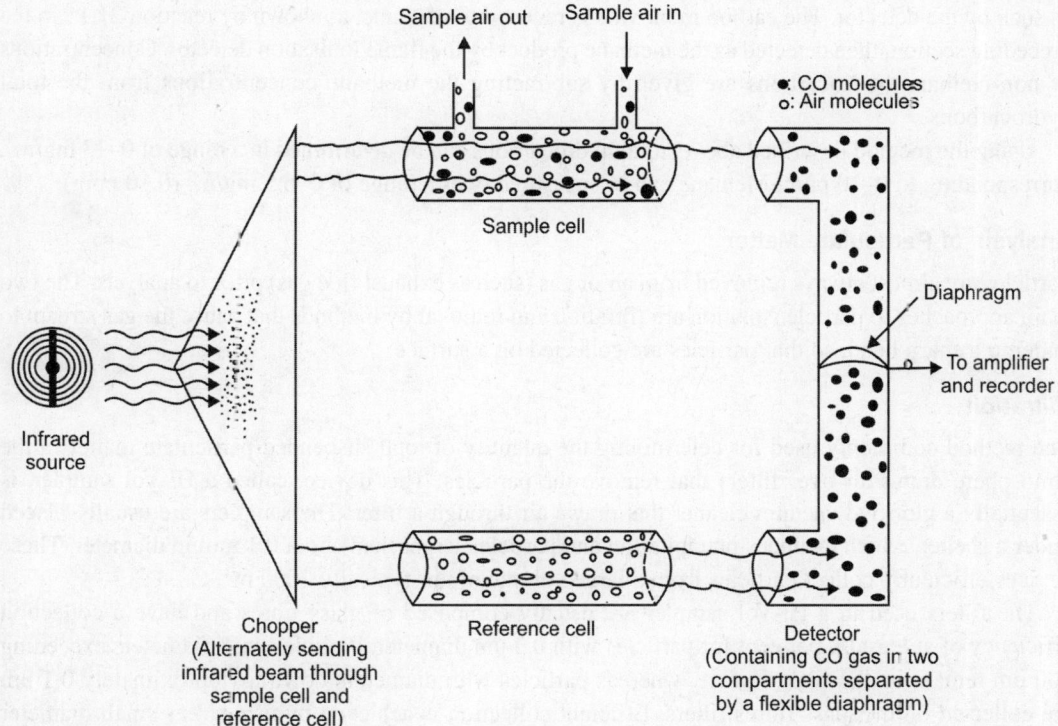

Fig 30.13. Nondispersive infrared spectrometer for the determination of carbon monoxide in the atmosphere.

Flame-ionisation gas chromatography detection can also be used for the analysis of carbon monoxide. It is selective for hydrocarbons and conversion of CO to methane in the sample is required.

This is accomplished by reaction with hydrogen over a nickel catalyst at 360°C:

$$CO + 3H_2 \longrightarrow CH_4 + H_2O \qquad \ldots (30.4)$$

A major advantage of this approach is that the same basic instrumentation may be used to measure hydrocarbons.

Carbon monoxide may also be analysed by measuring the heat produced by its catalytic oxidation to CO_2 over a catalyst consisting of a mixture of MnO_2 and CuO. Differences in temperature between a cell in which the oxidation is occurring and a reference cell through which part of the sample is flowing are measured by thermistors. A vanadium oxide catalyst can be used for the oxidation of hydrocarbons, enabling their simultaneous analysis.

Analysis of Hydrocarbons

Monitoring of hydrocarbons in atmospheric samples takes advantage of the very high sensitivity of the hydrogen flame ionisation detector to measure this class of compounds. Known quantities of air are run through the flame ionisation detector 4 to 12 times per hour to provide a measure of total hydrocarbon content. A separate portion of each sample goes into a stripper column to remove water, carbon dioxide, and non-methane hydrocarbons. Methane and carbon monoxide, which are not retained by the stripper column, are separated by a chromatographic column, passed through a catalytic reduction tube then to a flame ionisation detector. Eluting first, methane is not changed by the reduction tube, and is detected as such by the detector. The carbon monoxide is reduced to methane, as shown by reaction 21.15 in the preceding section, then detected as the methane product by the flame ionisation detector. Concentrations of non-methane hydrocarbons are given by subtracting the methane concentrations from the total hydrocarbons.

Using the method described above, total hydrocarbons can be determined in a range of 0–13 mg/m^3, corresponding to 0–10 ppm. Methane can be measured over a range of 0–6.5 mg/m^3 (0.10 ppm).

Analysis of Particulate Matter

Particles are almost always removed from air or gas (such as exhaust flue gas) prior to analysis. The two main approaches to particle isolation are filtration and removal by methods that cause the gas stream to undergo a sharp bend, so that particles are collected on a surface.

Filtration

The method commonly used for determining the quantity of total suspended particulate matter in the atmosphere draws air over filters that remove the particles. This device, called a Hi-Vol sampler, is essentially a glorified vacuum cleaner that draws air through a filter. The samplers are usually placed under a shelter, which excludes precipitation and particles larger than about 0.1 mm in diameter. These devices efficiently collect particles from a large volume of air, typically 2000 m^3.

The filters used in a Hi-Vol sampler are usually composed of glass fibres and have a collection efficiency of at least 99 per cent for particles with 0.3 μm diameter. Particles with diameters exceeding 100 μm remain on the filter surface, whereas particles with diameters down to approximately 0.1 μm are collected on the glass fibres filters. Efficient collection is achieved by using very small diameter fibres (less than 1 μm) for the filter material.

The technique described here is most useful for determining total levels of particulate matter. Prior to taking the sample, the filter is maintained at 15°–35°C at 50 per cent relative humidity for 24 hours, then weighed. After sampling for 24 hours, the filter is removed and equilibrated for 24 hours under the same conditions used prior to its installation on the sampler. The filter is then weighed and the quantity of particulate matter per unit volume of air is calculated.

The range over which particulate matter can be measured is approximately 2–750 μg/m^3, where volume is expressed at 25°C and 1 atm (760 mm Hg, 101 kPa) pressure. The lower limit is determined by limitations in measuring mass and the upper limit by limited flow rate when the filter becomes clogged.

Size separation of particles can be achieved by filtration through successively smaller filters in a stacked filter unit. Another approach uses the virtual impactor, a combination of an air filter and an impactor. In the virtual impactor, the gas stream being sampled is forced to make a sharp bend. Particles larger than about 2.5 μm do not make the bend and are collected on a filter. The remaining gas stream is then filtered to remove smaller particles. Results obtained by the analysis of particulate matter collected by the filters should be treated with some caution. A number of reactions may occur on the filter and during the process of removing the sample from the filter. This can cause serious misinterpretation of data.

For example, volatile particulate matter may be lost from the filter. Furthermore, because of chemical reactions on the filter, the material analysed may not be the material that was collected. Artifact particulate matter forms from the oxidation of acid gases on alkaline glass fibres. This phenomenon gives an exaggerated value of particulate matter concentration.

One of the major difficulties in particle analysis is the lack of suitable filter material. Different filter materials serve very well for specific application, but none is satisfactory for all applications. Fibre filters composed of polystyrene are very good for elemental analysis because of the low background levels of inorganic materials.

However, they are not useful for organic analysis. Glass-fibre filters have good weighing qualities and are therefore very useful for determining total particle concentration; however, metals, silicates, sulphates and other species are readily leached from fine glass fibres, introducing error into analysis for inorganic pollutant analysis.

Collection by impactors

Impactors cause a relatively high velocity gas stream to undergo a sharp bend, so that particles are collected on a surface impacted by the stream. The device may be called a dry or wet impactor, depending upon whether collecting surface is dry or wet; wet surfaces aid particle retention. Size segregation can be achieved with an impactor because larger particles are preferentially impacted and smaller particles continue in the gas stream. The cascade impactor, illustrated in Fig. 30.14, accomplishes size separation by directing the gas stream onto a series of collection slides through successively smaller orifices, which yield successively higher gas velocities. Particles may break up into smaller pieces from the impact of impingement; therefore, in some cases impingers yield erroneously high values for levels of smaller particles.

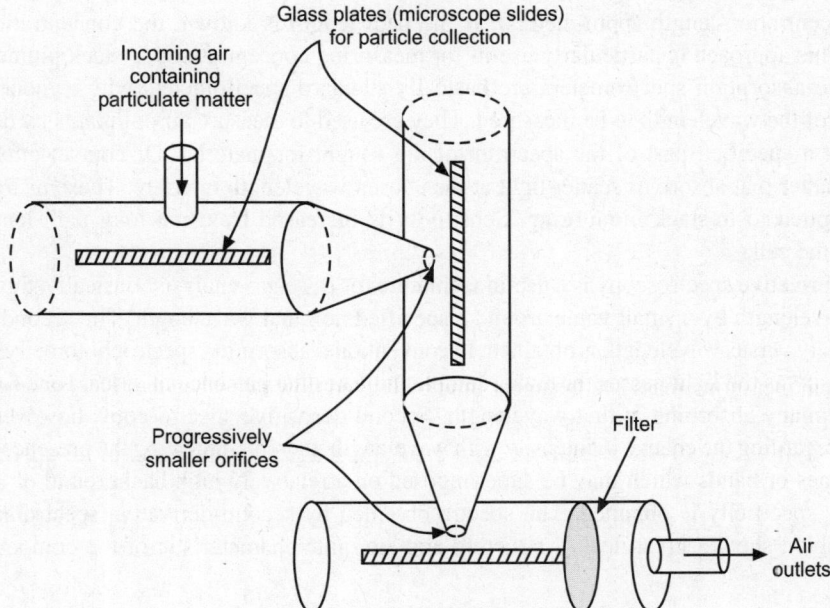

Fig. 30.14. Schematic representation of a cascade impactor for the collection of progressively smaller particles.

Particle analysis

A number of chemical analysis techniques can be used to characterise atmospheric pollutants. These include atomic absorption, inductively coupled plasma techniques, X-ray fluorescence, neutron activation analysis and ion-selective electrodes for fluoride analysis. Chemical microscopy is an extremely useful technique for the characterisation of atmospheric particles. Either visible or electron microscopy may be employed. Particle morphology and shape tell an experienced microscopist a great deal about the material being examined.

Reflection, refraction, microchemical tests and other techniques may be employed to further characterise the materials being examined. Microscopy may be used for determining the levels of specific kinds of particles and for determining their size.

Direct Spectrophotometric Analysis of Gaseous Air Pollutants

From the foregoing discussion, it is obvious that measurement techniques that depend upon the use of chemical reagents, particularly liquids, are cumbersome and complicated. It is a tribute to the ingenuity of instrument designers that such techniques are being applied successfully to atmospheric pollutant monitoring. Direct spectrophotometric techniques are more desirable when they are available and when they are capable of accurate analysis at the low levels required. One such technique, non-dispersive infrared spectrophotometry, has been described for the analysis of carbon monoxide. Three other direct spectrophotometric methods are Fourier transform infrared spectroscopy, tunable diode laser spectroscopy and differential optical absorption spectroscopy. These techniques may be used for point air monitoring, in which a sample is monitored at a given point, generally by measurement in a long absorption cell. In-stack monitoring may be performed to measure effluents. A final possibility is the collection of long-line data (sometimes using sunlight as a radiation source), an approach which yields concentrations in units of concentration-length (ppm-metres). If the path length is known, the concentration may be calculated. This approach is particularly useful for measuring concentrations in stack plumes.

Dispersive absorption spectrometers are basically standard spectrometers with a monochromator for selection of the wavelength to be measured. They are used to measure air pollutants by determining absorption at a specified part of the spectrum of the sought-for material. Of course, other gases or particulate matter that absorb or scatter light at the chosen wavelength interfere. These instruments are generally applied to in-stack monitoring. Sensitivity is increased by using long path lengths or by pressurising the cell.

Second-derivative spectroscopy is a useful technique for trace gas analysis. Basically, this technique varies the wavelength by a small value around a specified nominal wavelength. The second derivative of light intensity versus wavelength is obtained. In conventional absorption spectrophotometry, a decrease in light intensity as the light passes through a sample indicates the presence of at least one substance — and possibly many absorbing at that wavelength. Second-derivative spectroscopy, however, provides information regarding the change in intensity with wavelength, thereby indicating the presence of specific absorption lines or bands which may be superimposed on a relatively high background of absorption. Much higher specificity is obtained. The spectra obtained by second-derivative spectrometry in the ultraviolet region show a great deal of structure and are quite characteristic of the compounds being observed.

Chapter 31

Environmental Management in the Chemical Process Industries

INTRODUCTION

With increasing awareness of the effects of environmental degradation on account of green house gas emissions, ozone depletion, acid rain, entrophication, soil degradation, technological hazards, health and safety issues have become a major concern for the chemical process industry.

Industry operators are thus, challenged on the one hand to improve productivity, while on the other, to effectively tackle the environmental issues. Effective environmental management has, therefore, assumed prime importance.

This chapter reviews the current status of environment management measures followed by the chemical process units and describes the new efficient environmental protection and control measures that can be adopted to abate further environmental degradation.

Growth of industrial processing, guided mostly by the necessity of increasing productivity, has led to serious environmental degradation of water resources, soil and air around these plants. Worldwide the focus of pollution control in the industry has shifted from end-of-pipe treatment to source reduction, avoiding pollution, clean technology and sustainable development. Hence, it has become imperative that environmental considerations play a substantive role in the future development of the industry especially at a time when more and more industrial activities are being undertaken in the developing countries.

Several recent studies in different parts of the world focused on this issue with the objective of identifying key issues in environmental protection in the different industrial processes, assessing as to what extent the national and international norms or guidelines regarding pollution control and environmental management are implemented in the industries, understanding the problems encountered in environmental management and exploring the reasons for the noncompliance. Suggestions were also made on the basis of the above studies to develop guidelines of an environmental policy that will foster development without degrading the environment.

STATUS OF ENVIRONMENTAL MANAGEMENT

Today, most industrial processing units, be it in the developed or developing countries, have specific environmental policies and their emissions, effluents and waste disposal are guided by the stipulations of the state regulatory authorities. New plants are built with modern technologies where considerable technology integration has taken place at the inception stage itself to see that pollution prevention is a part of the process design itself. Already existing units are now operated with additionally built state-of-art pollution control facilities. The pollution and environment control departments attached to the plants usually exhibit meticulous care to see that the above objective is achieved. Thus, now-a-days effective control facilities exist in most of the processing units and they are operated with due diligence. The stipulations of the pollution control and environmental protection agencies are also within achievable limits of the available technology. Still excursions, at times, occur in the parameters on account of start up, shut down of plants or may be due to accidental situations. Existing facilities are capable of handling such situations also.

Several national as well as international standards covering a wide range of parameters have been developed to specify the emissions and effluents. These include pH, ammoniacal nitrogen, nitrates, fluorides, phosphates, total suspended solids, oils and fats and chemical and biological oxygen demands in the effluent streams, particulate matter, nitrogen oxides, sulphur oxides and carbon monoxide in exhaust streams. Radioactivity, toxicity, presence of heavy metals, organics, biological pollutants and pathogens, etc. are also monitored in specific cases.

CURRENT DEVELOPMENTS

Current developments in environmental chemistry and chemical engineering have helped industry operators reduce effluent generation at source and thus eliminate the need for treatment and disposal of effluents. Very often treatment of pollutants emanating from industrial operations is linked to the technologies adopted. Over the years the consumption of specific raw materials and energy for manufactured products has registered a continuous trend of improvement with the adoption of efficient technologies and best operating practices at the plant level. This invariably contributes to achieving better environmental standards through reduction in emissions, effluents and solid waste per ton of product manufactured.

Further improvements towards better environmental quality may require major design changes involving additional investment or going for a new proven and commercialised process. This is a costly option and hence efforts in this line are limited unless it brings about substantial economic incentive by way of increased productivity, lowering of energy consumption, etc.

In the case of products having high water intensity, there is an economic benefit in reusing treated effluents so that water conservation is achieved. The start up and shut down of plants are situations that may lead to an increased level of pollution of the environment compared to its normal operation. Hence most plants are equipped with specific provisions to take care of such situations.

Most of the pollution prevention methods implemented in the industries follow prescriptive approaches which follow standardised procedure built around questionnaires and check lists. In the new more descriptive approach, process operators are challenged to attack pollution problems and devise new and innovative ways for solving them.

Managements undertake substantial efforts to develop green belts and maintaining greenery around these plants to reduce the impact of green house gases. This is an important step in the direction of sustainable environmental control. Establishment of ISO 14000 environment management systems and a corporate environmental set up for regular monitoring and control is another major step in environmental protection. These systems are intended for continuous improvement of existing operations from the environmental angle. Certain industries have adopted zero effluent approach incorporating total recycle and reuse of effluents back to process, though it still remains more a concept than its effective implementation to a reasonable degree of reliability.

European process plant operators use the best available techniques (BAT) in their plants for environmental control. Both effluent specific standards and product specific standards are available.

In India and many other developing countries, systems are employed to the extent of controlling and reducing pollution from plants within the limits set by the statutory authorities, i.e. the pollution control boards (PCB). Operating units do not put in further efforts for reducing the pollution effects beyond the limits prescribed by the pollution control boards (PCB). This is primarily due to lack of incentives to encourage additional investment towards improved technology to go for better environmental quality.

Most operations emit large quantities of carbon dioxide (CO_2), which is a major green house gas to the atmosphere. There are no emission standards for carbon dioxide as prescribed by the statutory bodies. Attempt to reduce green house gas emissions all over the globe to tackle climate change will bring in specific limits for carbon dioxide emissions also in future. Every processing unit imposes certain environmental burden to the local environment and its impact categories are acidity, global warming, human health effects, ozone depletion, photochemical smog, aquatic oxygen demand, ecotoxicity to aquatic life, etc. A parametric assessment of the contribution of each of these components can be used to compare yearly performances of plants.

Extensive hazard and risk analysis using techniques such as hazard operability (HAZOP) Studies and quantitative risk assessment (QRA) are conducted based on which safe systems, work practices and risk reduction measures are adopted for processing facilities. Environment management plans of the production units are capable of mitigating the risk from most expected crisis situations barring those from nightmare incidents such as earthquakes, sabotage, etc.

Information to the public regarding the environmental consequences of these plants is very important. Communities associated with these units have a right to know the environmental risk they are subjected to. In most countries it is now mandatory that an environment impact assessment (EIA) be done prior to implementation of a project having large scale environmental consequences. A proper environment management plan (EMP) should be devised before a unit starts operating.

GLOBAL ENVIRONMENTAL CHALLENGES

Climate change across the world, depletion of ozone layer in the outer atmosphere, loss of biodiversity elements such as migratory species and important genetic resources, widespread degradation of land, urban air, forests and natural waters and marine ecosystems and accumulation of persistent organic pollutants in nature are major global environmental concerns. These issues have an impact that transcendent national boundaries and hence require global solutions. We have over 200 international legislations governing environmental issues. With currently available technology and adopting best practices mitigation of further degradation are possible. The far reaching measures to combat the effect of green house gases agreed upon in the Kyoto protocol is getting thwarted by many developed nations such as the US and Australia. In India, we have framed a comprehensive auto fuel policy that considers among other things, availability and security of supplies, vehicle technology, cost effective emission reduction, fiscal measures and institutional means to bring about progressive improvements by reducing vehicular emissions on ambient air. The fuel cell as a power source is becoming a viable alternative to the internal combustion engines with least environmental impacts. Thus concerted efforts are required, both at the national and international level, to stop further degradation and undo the damages already done.

STRESS ON ENVIRONMENTAL HEALTH

National environment policies shall foster efforts for sustaining environmental health of the people and shall call for a discrete assessment of pollutants entering the natural environment from human interventions in terms of their toxicity, persistence, mobility, bioaccumulation and methods available for source reduction and control mechanisms.

SOURCE REDUCTION

Reduction of pollution can be achieved through improvements in process chemistry, reaction kinetics, stoichiometry, conversion and yields. Similar approaches also include using different physical forms of catalysts, using water instead of volatile organic compounds (VOCs) in paints and coatings, using oxygen instead of air in oxidation reactions and thus preventing side reactions, using pigments and fluxes free of heavy metals and so on. Turpentine or citric acid based solvents replace chlorinated and inflammable solvents.

Engineering design modifications is another method for reduction of pollutants at source. Extreme temperature, pressure and concentration are reduced so as to render reactions to proceed in milder environments. Conversion of batch process to continuous process, whereby recycle of streams is possible, application of emulsion breakers for effective seperation, chemical synthesis from renewable sources rather than petrochemicals, use of different methods for handling reactants such as in the form of slurries, powders, etc. help to contain pollution at sources to a greater extent. Reduction of vents, spillages and emissions through improved instrumentation and better operating practices also result in reduced pollution loads to the environment. Management approaches with regard to inventories, quality, housekeeping and optimised operation are also important.

Efforts should be coordinated to reduce pollution at source. An example from agriculture will be able to illustrate this point. Mineral fertilisers are an essential input for improved agricultural productivity for which ammonia is an important raw material. Production of ammonia using natural gas as feedstock generates a lower volume of pollutants compared to other feedstock such as naphtha and fuel oil or coal and lignite.

During the cropping season plants absorb only part of the nutrient supplied through application of mineral fertilisers and the rest leaches out and finds its way into the environment. This invites twin problems financial loss to the farmer, as well as pollution. In order to reduce the ground water contamination and pollution arising out of excessive run off from applied agrochemicals and eutrophication of water bodies on account of limited nutrient absorption by plants, consumption of low analysis fertilisers (complex and blended) may be promoted in place of high nutrient-containing ones such as urea and diammonium phosphate.

PROMOTE INTEGRATED COMPLEXES

Refinery, fertiliser, power and petrochemicals are themselves major investment and high technology decisions and very often these units are put up by different agencies and function as independent companies. Technology brings in a lot of scope for exploiting the synergy within these units, which could play a major role in improving the bottom line of current operations of these units. Integration of refineries, fertiliser and petrochemical plants, and power generation units at the planning phase itself to develop integrated complexes will help to drastically reduce emission and other pollutants and ensure optimised operation.

REVIEW EXISTING CONTROL LIMITS FOR POLLUTANTS

The present standards for discharge of effluents from industrial units are just technological limits attainable through application of available technologies for abatement and control available at the time of specifying these norms and are not based on their longterm health effects. Revised standards based on the health impacts of each of the pollutants may be developed incorporating the advancement in this area. While doing so, care must be taken to see that the prescribed limits are achievable within the means and reach of the industry.

CALL FOR BEYOND COMPLIANCE OF STATUTORY STIPULATIONS

In industrial plants, systems are employed to the extent of controlling and reducing pollution from plants within the limits set by the statutory authorities. The units do not put in further efforts for reducing the pollution effects beyond the limits prescribed by the pollution control boards (PCB) in the interest of public health. This is primarily due to lack of incentives to encourage additional investment towards improved technology. Hence, industrial units should be encouraged to go beyond compliance and become more environment friendly.

HARNESS ENVIRONMENTAL BIOTECHNOLOGY

Environmental biotechnology employs living organisms, flora and fauna, engineered to exhibit specific traits in order to identify, control or prevent pollution. This technology has been applied to cleanup hazardous waste sites more efficiently than conventional methods, thereby reducing the need for incineration or extraction-based methods. Bioremediation has been applied for the cleanup of numerous varieties of pollutants, including heavy metals, persistent organic pollutants, explosives, sewage and industrial waste. Because of the prevalence of tropical climate, biological processes for pollution control have an edge over chemical processes and are more efficient. Modern developments such as recombinant and genetically engineered organisms find extensive application in biological processes for pollution control and bioremediation.

MUNICIPAL SOLID WASTE

Municipal solid waste management is a major area of environmental concern to all developing urban settlements. Still the socio-cultural response and the techno-economic considerations of the issue do not receive the required impetus in appropriate planning and implementation in our country. Lack of adequate civic sense, public awareness and participation, lukewarm approach of the local and state level governments have resulted in a situation in which most of the beautiful landscapes are slowly turn it or turning to litter zones. Thus, earnest efforts to thwart an impending disaster, as it happened in Surat a few years back are to be coordinated for reducing waste generation on the one hand and its effective disposal on the other involving primary collection, segregation at source, recycling to the extent possible and treatment through appropriate methods to reduce their harmful effects to a reasonably acceptable level.

Bio medical waste disposal has turned out to be a major problem to the civic bodies. As of now, disposal is done in certain locations only and that too in a haphazard manner. Incineration of the waste shall be done only in properly designed incinerators with burners so as to destroy the harmful products of the primary combustion of the waste. Identifying protected and specially designed landfill sites and use of engineered bacteria for biological degradation should be encouraged.

REDUCE EMISSION OF GREEN HOUSE GASES

There is sufficient scientific evidence to show that climate changes are caused due to increasing concentration of greenhouse gases (carbon dioxide, methane, nitrogen oxides, etc.) in atmosphere as a result of human activity. According to the recently published Stern review on the economics of climate change, the stock of greenhouse gases in terms of CO_2 equivalent has increased from 280 parts per million (ppm) before the industrial revolution to 430 ppm as of now and is expected to reach the 550 ppm level by 2050, if emissions did not increase beyond today's rate. But the fast growing economies with their huge investments in high carbon infrastructure in the energy, industry and transport sectors tend to accelerate the annual flow of emissions such that the 550 ppm level will be reached by 2035 leading to a global average temperature rise of over 2°C.

The consequences of warming up of the atmosphere are going to be disastrous both to human life as well as national economies. Hence, concerted efforts are needed to initially stabilise and then gradually bring down the concentration of greenhouse gases in the atmosphere. For this purpose, a multi-pronged approach of reducing demand for emission intensive goods and services, increasing efficiency of existing operations, avoiding deforestation and switching over to low carbon technologies for power, transport and heating needs is required

Encouragement through adequate financial incentives shall be made available to those intending for voluntary reduction of greenhouse gases and those resulting in climate change etc. The extension of natural gas pipelines, development of hydel and nuclear sources of energy should be encouraged.

ENSURE WATER AVAILABILITY

Availability of good quality water for the community and industry is going to be a major problem in the coming years in many countries. In order to address the issue of the non-availability of adequate drinking water, community and rural water supply schemes should be implemented on a priority basis over other development projects. Local expertise and public participation for protection of water sources, avoiding over-exploitation and promotion of water literacy could be encouraged in this matter. Conservation of water in all applications—domestic, commercial and industrial—shall be given utmost importance to ensure sustained availability within the limits of the existing resources.

Integrated water use and conservation, rain water harvesting and revival of traditional technologies are important in planning for future development of water resources. A participatory approach is needed in local level watershed development and management. Public consciousness will have to be mobilised through education, social interventions and mass movements for managing surface and ground waters in an equitable and sustainable manner.

DEVELOP POLLUTION INVENTORY DATABASE

In order to evaluate and improve environmental quality around industrial locations, a credible accounting of the sources and quantity of effluents and emissions is needed. This may be done through the development of an emission inventory for the overall location including adjacent metropolitan and rural areas. The idea is to develop an emission inventory and database system which may become the foundation for ongoing pollution inventory improvements. The environmental policy shall strive to further develop this effort to a national level pollution inventory database which could be used to ensure that pollutants are reduced over a period of time.

CHANGING ROLE OF THE REGULATOR

The general attributes of the current environmental regulatory system are to set national standards for health protection, ambient quality, promulgate rules and regulations; issue permits; inspect for compliance, monitor for environmental compliance and enforce as needed. Remediation of past releases and undertaking environmental research are also important. Environmental regulatory authorities in developing countries may be encouraged to become solution providers to the industry rather than being mere policing agents.

HAZARDOUS WASTE DISPOSAL

Management of hazardous waste materials generated in industries has become a major concern for plant operators from the environmental point of view. Hazardous waste may be a solid, semi-solid or non-aqueous liquid which because of its quantity, concentration or characteristics in terms of physical, chemical, infectious quality, is capable of significantly contributing to an increase in mortality or an irreversible damage. Left uncared or improperly treated, stored, transported and disposed, they are capable of posing a potential hazard to human health and neighbouring environment. Waste material is classified as hazardous if it exhibits whether alone or in contact with other wastes or substances, any of the characteristics such as corrosivity, reactivity, ignitability, toxicity, acute toxicity or infectious property. These substances either created as by-products of the industry or as residues of the process adopted, are highly toxic and are capable of causing irreversible damages to the environment.

A lot of hazardous waste is generated in countries as a result of several industrial operations and there are imports too for recovery of valuables, etc. Disposal of such waste is yet to gain the desired importance despite legislation in this regard for over a decade. Still many industrial areas are yet to identify disposal sites. Determined efforts from the part of local governments to put up facilities for treatment within a definite time frame have become necessary.

ENVIRONMENTAL HEALTH EDUCATION

Diseases caused to the public by environmental consequences are on the rise world over. The country's environmental policy should stipulate that environmental health education is mandatory to all segments of the society including students, households, workers and so on.

Science and technology should address issues relating to environmental health with a broader approach, with focus on infrastructural development in health care, strengthening institutions, practitioners and referral facilities and ensure speedy and effective delivery of primary health services, environmental sanitation, infection control, hospital waste management, nutrient deficiencies, occupational health problems, management of disaster and accident victims, trauma care, food and drug safety, prevention of non-communicable disease, etc. with a view to develop a national level holistic programme reaching all sections of the society.

EMERGENCY PLANNING FOR DISASTER MITIGATION

Emergency planning for disaster preparedness in case of natural calamities and man-made disasters is important. The programme may be coordinated on the lines of the awareness and preparedness for emergencies at local level (APELL), a project of the united nations environment programme (UNEP). It was developed in partnership with industry associations, communities and governments following some major industrial accidents that had serious impacts on health and the environment. This process creates awareness of hazards in communities close to industrial facilities, encourages risk reduction and mitigation, and develops preparedness for emergency response. Communication is often between the three main groups of stakeholders—company, community, and local authorities. Discussion on hazards usually leads to the identification of risk reduction measures, thus making the area safer than before.

The preparedness of the industry for management of abnormal situations, real time monitoring of systems and equipment and guard against human error should be well publicised. The social cause for the industries justifies such a risk level and the efforts to further sharpen the tools for process safety shall also continue.

KEY ISSUES IN ENVIRONMENTAL MANAGEMENT

Thus, the key issues in environmental management in the processing industries are identified as pollution from solid waste resulting in contamination of land space, liquid effluents endangering water streams and ground water resources and gaseous emissions degrading the quality of atmospheric air, risk to life from operational incidents to people and property in the industry and those in the neighbourhood of these units due to storage, handling, transport and use of large quantities of inflammable and hazardous chemicals and hydrocarbons, large scale depletion of natural resources, raw materials, energy resources and water and contribution to global warming due to emission of greenhouse gases.

Studies also reveal that the units have been successful in controlling pollution from their operations to the level prescribed by the statutory authorities and as required by the law. The best available technology for pollution control and environmental management are being used and it compares well with such practices being adopted internationally.

Generally, there is a good deal of compliance by all units to the standards prescribed for discharges of effluents. Often units are committed to attain the norms for various parameters as stipulated by the pollution control boards. Units even go to levels of pollution control beyond compliance if there are sufficient economic incentives for making the required additional investments. In other situations no attempt is made by units to achieve better control of pollution beyond the statutory limits.

ADDRESSING CHALLENGES: NATIONAL ENVIRONMENT POLICY

The important problems encountered in environmental management are lack of incentive for continuous improvement in the direction of pollution reduction beyond the compliance limits of the pollution control boards, integration of environmental concerns in to the core of the business strategy and lack of sufficient transparency with regard to environmental information. To effectively address the above problems and foster development, an apex level environmental policy with the following elements will be required.

Role of Top Management

The first and foremost guiding principle of an environmental policy facilitating growth of the industry is to ensure the unstinted commitment, involvement and action oriented approach of the top management of the organisation in achieving the set environmental goals.

Environmental Policy Statement

Top management need to codify their environmental commitment, values and perceptions in a documented policy. The policy should be relevant to its activities, products, and services and take into account its implications on different stakeholders. Attempts for improving energy efficiency, resource productivity and use of renewable source of energy and raw material need special mention in the policy.

Environment, Health, Safety (EHS) Vision Statement

Depending upon the nature and scale of its operation every unit should formulate an environment health and safety vision statement specifying its current thinking and aspirations of the future. The units should also adopt a national pollution prevention policy that encourages source reduction and environmentally sound recycling as a first option and also recognise safe treatment, storage and disposal practices as important components of an overall environment protection strategy.

Environmental Targets

The environmental targets, i.e. the qualitative and quantitative changes that are to be brought about to bring in more environment friendliness in the industry and acceptance to the community around are to be clearly set. Steps that are envisaged for minimising environmental impacts, reducing emissions of toxic gases and those causing global warming and improving the current levels of employee health, safety and pollution prevention are to be specified. The target must also address achieving zero accidents at work places, reducing incidents of work related diseases and overall reduction of the risk exposure to the employees as well as the community around. It should focus achieving sustainable development and eco-efficiency as a new business perspective for the industry through production and innovation, integrated environmental protection, responsible product stewardship and aim at total quality improvements.

Control Strategies

The policy shall provide for the use of legal, financial and social instruments, which influence the behaviour of companies, citizens, public bodies and authorities for achieving the objectives of the policy. Existing and innovative control mechanisms such as statutory provisions, stipulations of the various regulatory bodies may be used. Industry may be asked to go for the currently best available technology for pollution abatement. During the interim phase, strategy of monitoring comparison with set standards and penal action wherever required shall continue. Plants shall be operated to standards that will comply with the requirements of appropriate national and international legislation and codes of practice. The govt. should formulate country specific best available techniques (BATs) for the industry to facilitate continuous improvement in environmental management. Technically and economically feasible regulatory as well as non-regulatory measures are also suggested to improve environmental management in chemical processing operations. Fiscal incentives may be provided to encourage adoption of technologies that reduce pollution.

Risk Management

The management should ensure that potential health, safety, and environmental risks associated with the activities are assessed early to minimise and manage adverse effects and to identify opportunities for improvement. It is desirable to keep a workable disaster preparedness and emergency management plan (DPEMP) to mitigate any such situations in the unlikely event of its occurrence.

Staff Training

Necessary and state-of-the-art training may be given to the concerned people responsible for environmental management, keeping them abreast of the new developments, technologies and practical tools, accident investigation, environmental impact prediction, selecting appropriate protective equipment, implementing emergency response plans as and when necessary and so on. They may be trained to learn from previous incidents and similar experiences. They must be made conversant in the corporate environmental management systems and the proposed action plan for its implementation. In short necessary capabilities must be available in-house with all organisations to tackle probable emergency situations that are likely to arise.

Monitoring

A regular and meticulous environmental performance monitoring is necessary to keep track of the environmental burden imposed by the company and watch the direction of its progressing trends.

Quantitative as well as qualitative approaches may be used for this purpose. Issues such as emissions, waste streams, hazardous waste, disturbance, resource depletion, etc. should be addressed accordingly. Commitments towards targets for responsible care and social responsibility may also have to be assessed.

The current operations shall be regularly and systematically assessed for the purpose both of identifying and correcting any element which may put human beings, real property or the natural environment at risk of nuisance or damage and of establishing a basis of safety-related improvements of processes and products. Any new process and product as well as any new information of existing processes and products should be thoroughly analysed with regard of their health, safety and environmental implications.

The concerned authorities should be kept well informed of the operations and of their health, safety and environmental implications. Any incident entailing a risk of environmental disturbances or of conflict with existing regulations should be promptly reported to the proper authority.

Public Information

Necessary provision may be made in the policy for sharing information on environment safety and health aspects and reporting environmental compliance to the public. Involvement of the community and working with active environmental groups in the region in bettering the environmental situation should be encouraged, thereby enhancing public perception of the industry.

Annual Reports

The policy shall call for annual environmental status reports (AESR) along with the financial performance reports. Such reports are now available from many operators around the world. The feedback on these reports from the concerned stakeholders may be used for continued improvement of existing systems. The policy document shall be integrated with the national environmental plan of the country.

COST MANAGEMENT IN THE CHEMICAL PROCESS INDUSTRY

Cost management has taken a new meaning in the chemical industry, which earlier was busy in market development and product development. The industry is now looking for cost advantage in feedstocks, technology, research, procurement and utilities, in fact, the entire chain of operations in the chemical process industry.

Emphasis on optimising resource utilisation in water, energy and process inputs has begun to payoff, not just in terms of reduced costs but also in improving the environmental performance of the industry. New simulation tools and techniques are complementing software packages to better design and operate process plants.

Globally the chemical industry has never had it so difficult as now with margins squeezed, loss of consumer markets, and economic slowdown in Asia Pacific, Latin America, and South Asia—some of the large emerging markets.

Restructuring business operations has never been so widely carried out as now. Ranging from oil majors to bulk commodity to downstream speciality chemical firms-consolidation has become the watchword though routes are often different.

The world chemical industry is bracing for a challenging year ahead and economic revival in Asia will have much to do with its pace of recovery. In the meanwhile, the chemical industry is going through upheavals and revamps in all aspects of the business - research, production, business development and marketing.

While oil and petrochemical majors are busy downsizing and merging, life sciences, pharmaceuticals and crop protection chemicals businesses are cutting on research costs. Search for new chemical entities has become prohibitively costly for companies to go it alone. Keeping research costs down to low levels while keeping the molecular pipeline active is proving to be a major challenge for the industry.

Tackling Costs-Shifting from Products to Processes

In the early 90s, cost management did not receive the kind of attention that it deserved, as companies were all focusing on growth, new markets and new product development. But the events of the last two years within Asia and globally, has brought about radical changes in the way companies carry on with their operations. In the highly competitive and depressed markets of today, cost management has emerged as a strategic tool for the industry. Cost control now holds centrestage in the chemical industry and will continue to be so through much of the early part of the 21st century. This trend is likely to bring in its wake a totally new set of paradigms within the industry—from the way products are designed and developed to the way they are launched in the marketplace.

While in the past, cost control measures had to do with either reducing staff or salary freeze, today the practice is to have lean organisations where every function is being converted into a profit zone.

In the chemical industry the shift is from product oriented or result oriented focus to process oriented focus. This has brought in new cost measurement systems which measure specific activities. Activity based costing (ABC) has now emerged as a valuable tool within the industry. These measures often give the planner an idea of how activities in different departments of a chemical manufacturing plant influence costs. Industry has also begun to benchmark these costs against the best practices globally.

It was the Japanese corporations which first recognised and evolved the tools of cost management which enabled them to penetrate the highly competitive markets of USA and Europe in the early 70s. Producing high value at low cost became the central theme to many Japanese companies. They gave the world new tools such as Kaizen, Kanban, etc. which are being adopted by companies around the world today. Cost management within the chemical industry is also turning out to be a crucial tool to bridge the link between strategies being evolved and operational efficiencies, even as it serves to improve the competitive edge of the product as well as the company. In every segment of the industry companies are vying for cost leadership and focusing on how to serve the customers from a low cost base.

In key functions spanning R & D, production marketing, procurement and customer services, decisions are now centering on cost implications. Complementing quality management and change management tools such as TQM and business process reengineering (BPR), today total cost management (TCM) is forcing companies to seek across-the-board solutions for managing their costs better. In their endeavours to frame new systems for quality and cost management, companies are also realising that the process is a continuous one and not a one off process. The concept of TCM is reconstructed for different functions. While in R & D, it could focus on developing a low cost process, in production, the focus will be on optimising resource utilisation, and in marketing, it could be product launches. Key to TCM in chemical manufacturing is elimination of wastes and delays in operation as most operations are multi-step and multiutility oriented. Cutting in wastes is an issue which requires detailed focusing on waste management measures and tools adopted for waste management.

Combining Environmental Goals with Cost Optimisation

Cost optimisation and environmental goals have now emerged as key issues in the chemical industry where continuous search is on for new low cost options in technology and services.

International trade and environmental concerns will be increasingly viewed from a cost perspective. Innovations in technology are forcing shifts in raw material, energy usage and material balance within the industry. Technological edge is deciding the competitiveness and directions in most sectors of the chemical industry The shifts within the industry pose challenges as many alternative techniques and feedstock change bring with them additional cost. Conflicts between environmental goals and business bottom lines reflected in monetary terms will need to be resolved by radically different ways.

Discoveries in the laboratory are complementing modern tools on the shop floor and in the marketplace. Rational design of chemicals has become an intensive area of research, and analytical and research inputs needed for such activities will further push costs up. New approaches to technology development are being centred around the theme of maximising productivity through low, cost options,

Cost Control Through Waste Minimisation: Tools and Techniques

In process plants, the scope for cost control is enormous as process waste constitute direct and indirect cost ranging between 10 and 30 per cent. Cost of disposal, recycling and by-product processing, all keep the costs at prohibitively high levels.

Process integration has emerged as a key tool to develop new methodologies of waste minimisation that would enable cost savings. Process integration often acts complementary to the pollution prevention philosophy. There are three main areas of process integration:

1. Pinch analysis.
2. Knowledge-based approaches.
3. Numerical and graphical optimisation approaches.

Thermal pinch and water pinch analysis are emerging as important tools within the industry to optimise energy and water usage, where costs are as high as 20 per cent of the total manufacturing costs. Thermal pinch analysis is being used widely in process plants to get insights into how heat flows take place through the processes. The technique helps to optimise energy utilisation in process plants. In the context of environmental improvements, thermal pinch analysis is often useful in determining the scope for minimising energy consumption for the same amount of product manufactured. German majors such as BASF and Bayer have carried out detailed site pinch analysis to optimise their energy balance within their integrated sites and in the process, reducing emissions and energy consumption.

An approach which looks at systematic analysis of the overall process operation to seek reduction not only in waste water generation but also in fresh water use has been developed in recent years. Known as water pinch analysis, this method allows the plant engineer to look at various options to optimise water usage in plant operations.

New tools have been developed to design processes with low pollution load. Among such tools, 'process simulation tools' are gaining importance. Development of computer-aided chemical design software which incorporate environmental considerations in their design logic has advanced significantly in recent years.

Earlier software used for synthetic design did not cover environment, health and safety criteria. However, most software under current development incorporate existing chemical and functional group properties, toxicity data, structure/activity relationships and molecular composition, to generate retroactive and proactive analyses.

Commercial simulators like Aspen Plus, Pro II and Chemcad III are all effective in providing optimal solutions to pollution problems. Recently in USA, EPA and AIChE came up with key needs for preventing pollution in manufacturing processes. These relate to modules for pollution assessment and modules for pollution prevention to be used in combination with process stimulators, to handle pollution prevention aspects in many chemical manufacturing processes. Cost savings through such techniques have been enormous in process plants.

The syngen programme is a synthesis design programme which generates the shortest and lowest cost synthetic pathway for a given target molecule from a catalogue of commercially available organic feedstocks. The inherent benefit of this software is its ability to prioritise alternative chemical pathways in terms of both cost and environmental impacts.

'Aspen plus' estimates the molecular composition of reaction outputs (residuals, products and coproducts) when provided with the composition of process inputs and approximate knowledge of reaction conditions (temperature and pressure) and use of catalysts.

IT and ERP systems are also being increasingly used by the chemical industry enabling firms to rationalise their resource structures and cut on costs.

Cost Management

The chemical industry in India was used to a protected environment, and a cost-plus-pricing system did not have to manage cost structures so stringently in the past. With liberalisation of the economy, customers became the key determinant as quality goods and services were available from a large number of competing players. For companies not used to balancing cost structures in tune with lower sales, the future has become uncertain.

Amidst all the uncertainties, there are companies that are benchmarking their operations against the best in the world. One such case is of Gujarat Ambuja, which focused on TCM, to emerge as a low cost cement producer in the country today. By concentrating on coal and energy-the main cost drivers in cement manufacture—the company changed its fuel feed from coal to crushed sugarcane and also redesigned the operations to fit in with the new feed. The company also brought about technological changes in its mechanical transport systems and mining activities.

Normally, cost management frameworks rise out of structural and executional functions. On the executional front, factors such as capacity utilisation, optimal layout of plant and utilities, product slate mix and value addition along the chain, have become important features within the industry. In the chemical chain, optimal utilisation of capacity is determined by demand pushes from the market. Optimal use of market research, better shopfloor practices, rationalising inventories build-up, and enabling R & D to develop low cost options, have remained key issues.

In the pharmaceutical industry in India, companies such as Ranbaxy and Dr. Reddy's have exhibited ingenuity in process development and manufacturing process to emerge as low cost generics manufacturers. These companies are now moving into new molecules development and teaming up with global partners to put new products in the marketplace. Outsourcing also has become a key cost cutting trend in the industry with firms paying more attention on core areas and farming out the non core areas to outside parties. On the petrochemical front, companies like Reliance are backward integrating into upstream hydrocarbon and refining to enable cost optimisation. On the other hand, oil and gas majors such as Indian oil corporation and gas authority of India are busy looking at addition along the value chain by forward integrating into petrochemical and downstream products and also moving upstream into exploration and production. Whatever be the mode of integration, the stress is on achieving lowest cost per unit of product.

Other success stories of cost management is the backward integration of Nirma, Ltd. into linear alkyl benzene (LAB) and soda ash, leading to considerable savings in procurement costs that helped the firm to attain competitive positioning in the detergent industry in the late 80s. In the commodity chemical markets where the prime raw material is traded internationally, a key issue is, when to import and when to produce? The decision is always linked to prices in international markets. Here, utilising supply chain cost techniques can be of immense help in reducing costs. When the global prices are low, the companies should stop production and should import; when the prices pick up, they could start producing. To attain a balance between the import and export decision it is critical to have effective price monitoring and forecasting techniques.

Methodology for Cost Control

Managing the supply chain has now emerged as the key factor in the chemical industry's quest for cost control. In a multi-step industry such as the chemical industry there are far too many processes and structures which need to be looked at. In an industry where raw materials and production costs are the highest as compared to other areas of business, the scope for improvement is considerable. Most companies are actively looking at the links in the production chain to differentiate performing and non-performing areas.

Focus areas requiring cost control

1. Product identification.
2. Technology selection.
3. Evaluation of alternatives.
4. Product development.
5. Commercial scale-ups.
6. Optimising production processes.
7. Utility management.
8. Waste management.
9. Business development.
10. Customer services.

Steps for a typical cost management exercise

1. Benchmarking costs.
2. Process analysis through the value chain of the business.
3. Identifying critical cost influencers.
4. Differentiating between value creators and dead wood.
5. Setting targets for cost determinants.
6. Identifying methods for adoption.
7. Cross-functional teams.
8. Consensus management.

Globally integrated companies have successfully adopted techniques such as just-in-time, total quality management, statistical quality control, lean manufacturing, strategic outsourcing, supply chain management and integration in their operations to identify cost control niches.

In recent years the chemical industry globally has resorted to many functional areas to look for cost cutting initiatives. Manufacturing and purchasing were at the top of the chart. These were followed by recruitment and consulting. Many companies outsourced functions such as research, manufacturing, advertising and market research as well as data processing, to keep low costs. This was true in speciality and other high performance chemical sectors. The industry also looked at change management practices such as BPR and six sigma to enable cost control along its value chain. Six sigma has brought about radical changes in the way management of various companies have begun to look at their traditional practices. The statistical bias of the process enables a more quantitative assessment of the corrective steps to be taken. In chemical manufacturing, ABC has assumed prime importance.

On the manufacturing front, companies are looking at flexible manufacturing systems, integrated manufacturing planning systems and customer oriented systems.

Elements of supply chain management

1. Designing products based on vendor involvement.
2. Forecasting based on feedback from sales.
3. Strategic sourcing to identify common suppliers.
4. Vendor development to enable process modification.
5. Inventory management tools.

The sequential process helps to lower costs related to sourcing, inventories, supplies, procurement and carrying. The growth of the chemical industry and the need for just-in-time supplies across the world have led to chemical companies focusing on new initiatives in lowering costs of supplying a product to its customers.

In recent years the concept of regional hub distribution has come about enabling chemical companies to reach the product to its customers when needed. For global companies or regional companies seeking to reach their products across different terrain, keeping logistics costs low is critical. Lowering distribution hub costs by setting up management information control systems is emerging as a new trend.

In future, logistics will prove to be a major challenge for companies as they strive to reduce marketing costs and customer service costs. In a geographical spread like India, maintaining control over product movement and ensuring deliveries on time will prove to be a major issue for the chemical industry.

The chemical industry in the early 21st century is all set to usher in new business and market shifts in Asia; till a revival sets in, the chemical industry in the region will continue to seek new ways of keeping costs low while serving its customers and retaining its market share.

References

Adams, K.L., *Process Synthesis*, John Wiley & Sons, New York.
Andrew, Y.J., *Industrial Chemistry*, Marcel Dekker Inc., New York.
Austin, G.T., *Shereves Chemical Process Industries*, McGraw-Hill, New York.
Bradley, S.R., *Industrial Pollution*, Academic Press, London.
Bryan, C., *Air Quality Monitoring*, Applied Science Publishers, London.
Commoner, T., *Industrial Organic Chemistry*, John Wiley & Sons, New York.
Creighton, K.E., *Atmospheric Dispersion of Pollution*, Academic Press, London.
Dagley, R., *Industrial Health Engineering*, Plenum Publishing Corporation, London.
Evison, L., *Fundamentals of Air Pollution*, Reinhold Publishing Corporation, New York.
Franklin, B., *Atmosphere and Pollution*, Applied Science Publishers, London.
Horald, G., *Air Pollution and Its Control*, Prentice-Hall, London.
James, A., *Refinery Operations*, John Wiley & Sons, New York.
Jencks, W.P., *Chemical Analysis of Pollutants*, John Wiley & Sons, New York.
Kim, C.K., *Environmental Instrumentation and Analysis*, Marcel Dekker, New York.
Lemer, A., *Chemistry Toxicology and Pollutants*, Applied Science Publishers, London.
Lewis, R.L., *Air Quality Monitoring*, Chilton Book Co., USA.
Martin, B.H., *Pollution Prevention in Process Industries*, Academic Press, London.
Munn, P., *Industrial Health Engineering*, McGraw-Hill, New York.
Nyer, E., *Chemistry of Air Pollution*, Van Nostrand Reinhold, New York.
Odum, S.K., *Fundamentals of Ecology*, W.B. Saunders and Co., New York.
Perry, R.H., *Perry's Chemical Engineer's Handbook*, McGraw-Hill, New York.
Reid, G., *Industrial Pollution*, Reinhold Publishing Corporation, New York.
Smith, C.N., *Handbook of Air Pollution*, McGraw-Hill, Tokyo.
Stumm-Zollineger, E., *Air Pollution and Its Sources*, Wiley Interscience, New York.
Taylor, K.C., *Handbook of Environmental Engineering*, Longman Group Ltd., London.
Teal, J.M., *Air Pollution and Its Fundamentals*, Pergamon Press, New York.
Vollenweider, R.A., *Global Ecology and Environment*, Blackwell Scientific Publications, New York.
Whittaker, R.H., *Handbook of Environmental Chemistry*, John Wiley & Sons, New York.
Wolfrom, M., *Environmental Contaminants*, Reston Publishing Co., Reston, Virginia.

Index

A

Absorption spectrophotometry, 587
Absorption, 473
Acid wastes, 261
Acidic and alkaline effluents, 461
Acidification, 558
Acrylonitrile reactor risk-based input-output analysis of a reactor, 93
Activated carbon adsorption, 378, 399
Activated sludge method, 551
Activated sludge process, 559
Activated sludge, 575
Adsorption, 198, 212, 465, 474
Advanced scrubber technology solutions for pollution control, 181
Aerobic process, 564
Aerobic systems, 309
Aerobic treatment, 559, 574
Agricultural residues, 341
Air emission, 449
Air emissions characterisation, 293, 380, 468
Air emissions, 379, 429, 457
Air extraction, 438
Air pollution control measures, 271, 295, 380
Air pollution from big paper mills, 365
Air pollution from small paper mills, 365
Air pollution, 438, 546
Air quality issues, 12
Air stripping, 270
Air toxics, 17
Air-SO_3 sulphation/sulphonation, 373
Alkaline water, 264
Ammonium phosphate fertiliser and phosphoric acid plant, 424
Anaerobic filter and lagooning, 466
Anaerobic purification—septic tank method (digestion method), 552
Anaerobic systems, 309
Anaerobic treatment, 565, 572
Analysis of carbon monoxide, 600

Analysis of hydrocarbons, 601
Analysis of oxidants, 600
Analysis of particulate matter, 602
Analysis of sulphur dioxide, 597
Animal feed, 583
Anoxic biological treatment, 285
Antibiotics, 501
Antibiotics, vitamins and enzymes, 457
Application of biotechnology to pollution problems of various industries, 582
Approaches to better water management, 324
Aqueous liquid-phase reaction systems, 278
Areas for energy optimisation in chemical process industries (CPI), 169
Assessing dermal exposures, 51
Assessing opportunities for waste exchanges and by-product synergies, 160
Atmospheric monitoring, 595
Atom economy, 106
Atom-efficient processes, 119
Atomic absorption and emission analyses, 588
Atomic emission techniques, 589
Automation and controls in water and waste-water treatment plants, 202
Automation of tannery procedures, 332
Auxiliary operations, 499

B

Bagasse utilisation, 481
Bakery waste treatment 556
Bar screens, 268
Bar soaps, 372
Basic operations in dyeing, 446
Basic process steps in paper and board manufacture, 342
Batch and semi-continuous processes, 278
Batch kettle process, 370
Beverage alcohol, 500
Biofuels from biomass: a chemical industry's perspective, 249
Biogas production, 361

Biological inactivity, 505
Biological methods, 270
Biological oxidation or biofiltration, 177
Biological treatment, 196, 379, 559
Biomass platform molecules, 249
Biomethanation, 491
Biorefinery concept, 246
Bioremediation, 228
Biotechnology for air pollution abatement and odour control, 580
Bipyridinium herbicides, 389
Bituminous paint, 437
Black liquor segregation and disposal, 354
BOD of whole milk and by-products, 548
BOD reduction, 167
Boilers and steam usage, 169
Bottled water industry and beverage industry, 203
Breweries, 489
By-product recovery, 489

C

Calciners and bleachers, 526
Calciners, 525
Cancer and other toxic effects, 24
Carbamate insecticides, 389
Carbon adsorption, 270
Carbon monoxide (CO), 15
Carbon monoxide boiler, 275
Cascade engineered phase transfer catalysts (CEPTC), 110
Case study waste-water reclamation, 205
Catalysis, 107
Catalysts, 66, 165
Catalytic converters, 167
Catalytic oxidation, 179
Categorisation in phosphate production, 415
Caustic soda industry, 511
Centrifugation, 212
Centrifuges, 470
Characterisation of waste effluents, 448
Characteristics of textile waste-waters, 532
Characteristics of the composite waste-water, 303
Characteristics of the sectional waste-water, 302
Characteristics of waste-water management in small paper mills, 356
Characteristics of waste-water, 302, 317
Check list for saftey assessor, 34
Chemical coagulation, flocculation and sedimentation, 269
Chemical recovery operation, 342
Chemical synthesis, 584
Chemical treatment of wastes, 224
Chlorohydrin process, 119

Choice of chromium recovery and reuse system, 317
Choice of mass separating agent, 78
Choice of refrigerant for a low-temperature condenser, 5
Chromatographic analysis of water pollutants, 593
Classification of leather auxiliaries, 333
Classification of pesticides, 388
Classification of surfactants, 368
Clean hydrogen peroxide synthesis via a nanocatalyst process, 101
Clean technology for reduction of chromium usage, 314
Clean technology in the leather industry, 316
Cleaner suitable environmental technologies, 236
Cleaner technologies, 235
Cleaner technology for distillery industry, 491
Coagulation/Clarification, 536
Coagulation/flocculation/settling, 378
Coagulation-flocculation, 558, 566
Collection by impactors, 603
Combining environmental goals with cost optimisation, 615
Combustion controls, 276
Combustion turbines, 276
Comparative analysis of standards, 339
Composting, 233, 491
Comprehensive life-cycle management strategy needed to ensure sustainable future for Indian chemical industry, 150
Compressed air, 170
Condensate waters, 261
Condensation, 472
Conservation of water, 533
Control of air emissions in petrochemicals, 293
Control of air emissions in refinery, 271
Control of air emissions, 430
Control of fluorine emissions, 435
Control of gaseous pollutants, 331
Control of NO_x in the nitrophosphate plants, 434
Control of oxides of nitrogen, 433
Control of particulate matter in complex fertiliser plants, 432
Control of particulate matter, 432
Control of sulphur dioxide and acid mist emissions from sulphuric acid plants, 430
Control of sulphuric acid mist, 431
Control of water pollution, 306
Controlled chemical addition systems, 332
Controlled water addition systems, 332
Controlling ignition sources, 33
Conventional activated sludge, 466
Conventional synthetic chemistry based process, 102
Cooling towers, 171

Cooling water blow down recovery at Madras fertilisers Ltd., 206
Cost control through waste minimisation: tools and techniques, 616
Cost economics, 208
Cost management in the chemical process industry, 614
Cost management, 617
Cost of waste-water treatment, 468
Covering a liquid space, 273
Creation of a third liquid phase: L-L-L PTC, 109
Criteria air pollutants, 13
Crude oil producing and handling, 258
Crystallisers, 470
Cyanide destruction, 459
Cyclone, 275

D

Deliming and bating, 326
Deodourisation processes, 581
Dermal exposure to chemicals in the ambient environment, 54
Design for degradation, 107
Design for energy efficiency, 107
Designing safer chemicals, 57, 107
Detergent bars and cakes, 374
Detergent formulation and process wastes, 373
Detergent manufacture and waste streams, 372
Dialysis, 216
Diphenyl herbicides, 389
Direct discharge to POTWs, 535
Direct oxidation of propylene with oxygen, 121
Direct spectrophotometric analysis of gaseous air pollutants, 604
Disinfection, 228, 566
Dispersion models, 51
Disposal of spoiled products, 548
Disposal of the final treated effluent and monitoring, 414
Disposal of wastes from specific industries, 521
Dissolved air flotation (DAF), 462, 537, 559
Distillation, 214
Distillery effluents, 492
Distillery industry, 486
Distinction of SO_3 removal mechanism, 187
Dose-response curve, 27
Drivers for a sustainable industry, 238
Drum-dried detergent, 374
Dry detergent blending, 374
Dryers and predryers, 526
Dryers, predryers, and bleachers, 525
Dust and fibres, 31
Dust removal, 166
Dye and dye intermediates, 446

E

E.I. to finished leather, 299
Eco-efficient products, 241
Eco-friendliness of psychrometric evaporator, 252
Eco-friendly solvents, 116
Eco-industrial parks, 158
Ecology, 18
Economic considerations of seafood-processing and meat processing waste-water treatment, 568
Effect of chemical releases to surface waters on aquatic biota, 54
Effect of reactant concentration, 76
Effects of fertilisers in the environment, 394
Effects of pesticides in the environment, 390
Effects of the waste on receiving water, 485
Effects of wastes on receiving streams, 410
Effects on receiving stream, 506
Effluent control, 451
Effluent generation, 458
Effluent treatment, 195
Electrochemical techniques, 191
Electrocoagulation, 566
Electrodialysis, 216
Electronic information system on leather processing chemicals, 333
Electrostatic precipitators, 275
Elements of supply chain management, 619
Eletrodialysis (ED), 200
Emergence of green chemistry, 97
Emergency planning for disaster mitigation, 611
Emission control considerations, 472
Emission factors, 295
Emission sources, 468
Emissions reduction, 243
Emulsions, 260
Energy efficiency, 242
Engineered catalysts, 111
Engineering ethics, 42
Environment, health, safety (EHS) vision statement, 612
Environmental health education, 611
Environmental impact of dye and its intermediates, 452
Environmental performance of chemical processes, 152
Environmental targets, 613
Environmentally balanced industrial complexes, 427
Environmentally safe layout code for manufacturing units, 402
Enzymatic treatment, 364
Enzyme usage in detergent sector, 584
Enzyme usage in textile sector, 584
Epoxidation with *in situ* hydrogen peroxide, 122

Epoxidation with purchased hydrogen peroxide, 122
Equalisation for pulp wash waste-water, 349
Esterification, 118
Ethyl benzene hydroperoxidation process (SMPO), 120
Evaporation pond, 465
Exposure assessment for chemicals in the ambient environment, 53
Exposure assessment, 23, 28
Exposure pathways, 46
Exposure to toxic air pollutants, 53
Extended aeration, 466

F

Fans and blowers, 172
Fat splitting, 371
Fate of pesticides in the environment, 390
Fatty acid neutralisation, 371
Fermentation products, 496
Fermentation, 460
Fermentors, 469
Fertilisers, 389
Filters, 470
Filtration, 197, 212, 510
Flocculation, 509
Flotation or foam fractionation, 378
Flow equalisation and neutralisation, 557
Fluoride and phosphorus removal from a fertiliser complex waste-water, 429
Focus areas requiring cost control, 618
FOG separation, 558
Food and allied industries, 582
Forests, 341
Formulation of a paint solvent, 5
Formulation of an industrial cleaner, 4
Fugitive emission profiles, 88
Fumigants, 389
Fundamental principles for particulate removal, 186
Fungicides, 389
Furnace wastes from phosphorus manufacture, 424

G

Gas chromatography, 591
Gaseous emissions and air pollution, 365
Gaseous pollution, 330
Gases and vapours, 31
General overview of risk assessment concepts, 23
Global energy issues, 6
Global environmental challenges, 607
Global environmental issues, 6
Global warming, 7

Glycerine recovery process, 371
Good housekeeping and water conservation, 314
Good housekeeping, 441
Grades of paper and board and demand projection, 340
Granular carbon, 537
Gravity separators, 293
Green chemistry and the biorefinery, 245
Green chemistry based process, 102
Green chemistry to produce amines, 118
Greener routes to propylene oxide—an overview, 119
Greenhouse effect, 8
Grit chambers, 268
Ground water contamination, 55
Guidelines for quantum limits of pollutants in the treated waste-water, 315
Guidelines for selection of recycle schemes, 194

H

Hazard assessment for cancer, 25
Hazard assessment for non-cancer endpoints, 25
Hazard assessment, 23, 24
Hazardous waste disposal, 611
Heaters and boilers, 276
Heterocyclic herbicides, 389
Hides and skins, 297
High purity water systems, 204
High value applications for starch and cellulose, 248
High-performance liquid chromatography, 593
Hot condensate, 482
Hydrocarbon reuse, 273
Hydrolysis, 400
Hydro-oxidation of propylene, 121

I

Illustrations of implemented sustainable chemistry, 240
Impact of gaseous pollutants on environment, 330
Impact of liquid wastes on environment, 304
Impacts on waste-water treatment processes, 369
Important issues in green chemistry and examples of green processes, 98
Improving existing products, 146
In situ bioremediation, 233
Incineration and potash recovery, 492
Incineration, 219, 289
Industrial alcohol, 501
Industrial fermentations, 497
Industrial operation and waste-water, 370, 415
Inherently safer chemistry for accident prevention, 107
In-line sampler, 274

Inorganic chemical industry, 503
In-plant control and recycle, 375
Inplant control measures to minimise mercury loss in waste-water, 514
In-plant control measures, 306
In-plant control, 451, 533
In-plant control, recycle and process modification, 420
In-plant procedures and control, 507
Inspection and maintenance, 274
Integrating environment, energy, technology and business processes, 172
Integrating risk assessment with process design— a case study, 93
Internal-combustion engine, 276
Interpretation of life-cycle data and practical limits to life-cycle assessments, 140
Ion chromatography, 192, 594
Ion exchange and exclusion, 379
Ion exchange process, 201

K

Key issues in environmental management, 612

L

Lagooning, 291
Land treatment of textile waste-waters, 539
Landfill, 291
Leather processing chemicals menu, 336
Less hazardous chemical syntheses, 107
Life-cycle Assessment, 127
Life-cycle costing concept, 150
Life-cycle impact assessments, 136
Life-cycle inventories, 130
Lignin isolation, 361
Liming: Countercurrent approach, 326
Liquid effluents, 449
Liquid wastes, 301, 353
Low load activated sludge method—effluent lagoon, 552

M

Major air pollution episodes, 37
Manufacture of alcohol and generation of effluents, 492
Manufacture of resins and emulsions, 437
Mass balance model, 49
Mass spectrometry, 594
Mass transfer, 182
Material flows in chemical manufacturing, 156
Material selection, 161
Material use and selection for reactors, 65
Mechanical seals, 274

Membrane bio-reactor, 198
Membrane filter method, 193
Membrane separation, 216
Membranes, 167
Methodology of optimising energy use, 169
Methods for detection of ions and microbes, 190
Methods to reduce fugitive emissions, 89
Microbiological method, 192
Minimal national standards (MINAS) for oil refineries, 296
Minimal national standards for straight nitrogenous fertiliser industry, 414
Minimal national standards for straight phosphatic fertiliser (MINAS), 412
Modifications in pulping processes, 359
Molasses—the key raw material for distillery, 483
Multimedia filtration, 536
Multiple tube fermentation method, 192

N

Nanocatalysts in green chemistry, 100
Nanofiltration (NF), 199
Natural disasters, 21
Natural resources, 18
Neat soap manufacture and waste streams, 370
Need for chrome recovery and reuse, 317
Need for green chemistry, 96
Neutralisation of sulphuric acid esters and sulphonic acids, 373
New business models and new value chain for a sustainable, 244
New generation automatic demineralisation system, 203
New generation reverse osmosis-electro-deionisation (RO/EDI) system, 203
New generation waste-water recycling technology, 204
Newer routes for PO manufacture for a greener environment, 121
NO_x, hydrocarbons and VOCs—ground-level ozone, 14

O

Occupational exposures: recognition, evaluation and control, 44
Odours, 176, 471
Oil separation, 268
Oleum sulphonation/sulphation, 373
On-site treatment of textile waste-waters, 535
Open dumping, 291
Optimising energy use in chemical process industry, 168
Ore and product handling, 526
Organic synthesis, 457, 461
Organochlorine insecticides, 388

Overview of major environmental issues, 6
Oxidation of organic compounds, 8
Oxidation-reduction, 509
Oxidising agents like ozone, UV rays and hydrogen peroxide, 459
Ozonation, 538, 567
Ozone depletion in the stratosphere, 11

P

Paint manufacture, 436
Paper industry, 583
Particle analysis, 603
Particulate matter, 15
Particulates, 471
PEL (Permissible exposure limits), 44
Percolating filters, 549
Pesticide industry, 585
Pesticides/Fertilisers causing pollution, 388
Petrochemical and allied products, 276
Pharma industry, 203
Phenol removal, 166
Phosphate fertiliser manufacture, 420
Phosphoric acid and N-P-K fertiliser plant, 426
Phosphorus and phosphate compounds, 416
Photo chemical oxidation (PCO), 201
Photochemical smog, 8
Phthalic anhydride, 293
Pickle-tan closed loop system (no float, dry), 326
Pickling and tanning, 326
Pilot plant for chromium recovery, 318
Plate and frame RO system, 200
Pollution control and environmental protection in chemical industry, 166
Pollution control in single superphosphate (SSP) plant, 411
Pollution prevention applications for separative reactors, 84
Pollution prevention assessment integrated with HAZ-OP analysis, 91
Pollution prevention for chemical reactors, 63
Pollution prevention for separation devices, 77
Pollution prevention in material selection for unit operations, 61
Pollution prevention in storage tanks and fugitive sources, 87
Possible health hazards of pesticides residues, 394
Potential of energy, resource and water saving, 252
Potentiality of water hyacinth, 362
Preparation of the dye bath, 446
Preparation of the fibre, 446
Present trends in tannery automation, 332
Press mud—a source of energy, 481

Pretanning operation, 297
Pretreatment and tanning operations—their contribution to environmental problems, 322
Pretreatment of textile waste-water, 536
Primary clarifier and sludge drying, 350
Primary sedimentation, 268
Principles of green chemistry and green engineering, 239
Principles of green chemistry, 106
Process changes for reducing pollution load, 314
Process control in drum operations, 332
Process design and operation heuristics for separation technologies, 78
Process hazards, 111
Process modification, 451
Process modifications to reduce pollutional load, 306
Process patterns, 163
Process water, 277
Processes with process water contact as steam diluent or absorbent, 278
Processing facilities and wastes generated, 568
Psychrometric evaporation system, 250
Psychrometric evaporation: a novel zero discharge system, 250
Pyrolysis, 223, 290

Q

Qualitative techniques for inventories and characterisation, 142
Quick change blind/manifold design, 275

R

Raw hide in finished leather by chrome tanning, 299
Raw hide to chrome tanned semi-finished leather, 299
Raw hide to vegetable tanned semi-finished leather, 298
Reactors and process engineering, 115
Reactors, 469
Reagent kinetics, 183
Real-time analysis for pollution prevention, 107
Recovery and reuse, 314
Recovery and utilisation of by-products, 307
Recovery of wastes, 441
Recycle and reuse and management of waste-water in tanneries, 320
Recycle technologies, 195
Reducing emissions from fugitive sources, 88
Reducing the carbon rootprint, 150
Reducing toxicity, 58
Reduction of waste-water, 441, 451
Reduction of water usage in a tannery, 306
Refinery fuel gas, 273

Index

Refrigeration systems, 171
Regenerative catalytic oxidation of VOCs, 179
Removal of chlorine/hydrochloric acid vapours, 167
Removal of hydrogen sulphide, 167
Renewable resources biomass, 240
Requirements for acid gas removal, 182
Responsibilities for chemical process safety, 37
Responsibilities of chemical engineers for environmental protection, 40
Reverse osmosis (RO), 200, 218, 270
Risk characterisation of cancer endpoints, 29
Risk characterisation of non-cancer endpoints, 30
Risk characterisation, 23, 29
Risk management, 613
Rodenticides, 389
Role of automation in water and waste-water treatment industry, 202
Role of chemical processes and chemical products, 4

S

Safer solvents and auxiliaries, 107
Safety and environment protection, 168
Scrubbers, 276
Seafood-processing waste-water characterisation, 562
Secondary biological treatment, 308, 398, 487
Sedimentation, 211, 559, 563
Segregation of black liquor generating stream, 361
Segregation of salt laden effluents, 306
Semiconductor photocatalysis, 460
Separation of oil and grease, 563
Separation technologies for removal of organic and pesticidal chemicals from waste-water, 401
Separators with reactors for pollution prevention, 84
Slurry preparation, 379
SO_2, NO_x, and acid deposition, 16
SO_3 solvent and vacuum sulphonation, 373
Soaking, 325
Soap manufacture and processing, 370
Solid hazardous wastes, 449
Solid phase bioremediation, 232
Solid waste generation, 458
Solid wastes pollution in petrochemical industries, 286
Solid wastes, 353
Solidification and stabilisation, 223
Solvent vapour removal, 167
Solvents and effluents, 266
Sources and characteristics of waste-water, 279, 543
Sources of detergents in waters and waste-waters, 369
Sources of ignition in hazardous area and how to control them, 30
Sources of ignition, 33

Sources of pollution, 108
Sources of waste, 234, 547
Sources of waste-water and the characteristics of the wastes, 484
Special additives, 167
Spectrophotometric methods, 587
Spectrophotometry, 191
Spray drying, 380
Spray-dried detergents, 373
Standards to satisfy environmental requirements, 445
Status of environmental management, 605
Status of waste-water treatment in India, 443
Steam generation, 277
Steam stripping, 275
Steps for a typical cost management exercise, 618
Storage tank pollution prevention, 87
Straight nitrogenous fertilisers, 412
Stream requirements, 507
Streamlined data gathering for inventories and characterisation, 141
Streamlined life-cycle assessments, 141
Sugar mill wastes, 483
Sulphamic acid sulphation, 373
Sumitomo process, 120
Supercritical fluid extraction of secondary metabolites from plant surfaces, 246
Supercritical water, 99
Supramolecular chemistry, 99
Surfactant manufacture and waste streams, 373
Synthetic fibres, 531

T

Tackling costs-shifting from products to processes, 615
Tank vents and transportation, 296
Tanning industry, 584
Tanning process, 297
Technologies for recycling of effluents, 196
Technologies for recycling of waste-water, 193
Technology for removal of VOCs and odour, 177
Technology trends in pulp production, 343
Textile industry, 584
Thermal desorption, 222
Titration, 191
Total organic carbon in water, 594
Total water management approach to recycle of waste-water, 193
Towards a sustainable chemical industry, 238
Toxicity of chemical pesticides, 401
Treatability aspects of waste-water, 443
Treatment and disposal methods, 492
Treatment of air pollutants, 366

Treatment of distilling industry effluents, 487
Treatment of fertiliser waste-water, 411
Treatment of meat wastes, 568
Treatment of non-recoverable waste, 502
Treatment of phenolic waste-water, 282
Treatment of polymers, 293
Treatment of seafood processing wastes, 562
Treatment of textile waste-waters, 533
Treatment of the wastes, 485
Treatment of waste-water, 310
Treatment through non-conventional methods, 310
Treatment to reuse, 534
Trickling filter, 466
Trimming and sorting, 298
Type of hazards, 31
Types of emissions, 470
Types of tanning process and their unit operations, 298
Typical reactions when absorbing HCl, 184
Typical reactions when absorbing HF, 184
Typical reactions when absorbing SO_2, 183

U

Ultra filtration, 199
Ultrafiltration and hyperfiltration, 218
Ultrasound-promoted acylation of yara-yara in the presence of solid-acid catalysts, 118
Ultraviolet (UV) radiation, 568
Ultraviolet rays, 168
Use of computers in leather processing, 331
Use of renewable feedstocks, 107
Uses of life-cycle studies, 145
Using life-cycle concepts in early product design phases, 148
Utilisation of by-products, 548
Utilisation of distillery effluents, 492

V

Vaccination, microbial suspension antitoxin preparation, 461
Vacuum systems, 296
Value or risk assessment in the engineering profession, 22
Vanillin from lignin, 248
Vapour control systems, 271

Vibro-screen, 293
Vitrification, 215
Volatile organic compounds (VOCs), 176

W

Waste generation in petroleum refinery, 256
Waste heat source for psychrometric evaporation system, 250
Waste minimisation/management options, 358
Waste paper, 341
Waste reduction at source, 235
Waste reuse, 243
Wastes versus by-products, 497
Waste-water characterisation, 439
Waste-water characteristics and sources, 417
Waste-water characteristics from main sections, 440
Waste-water control and treatment, 375, 420
Waste-water from fertiliser plants, 410
Waste-water from refinery, 266
Waste-water from resin house, 440
Waste-water generation in pesticides industry, 395
Waste-water in leather industry, 320
Waste-water treatment methods, 376, 422
Waste-water treatment strategy, 361
Water pollution control in petrochemical industries, 281
Water pollution from boilers, 529
Water pollution from water-treatment plants, 529
Water pollution, 345
Water quality issues, 17
Water treatment process analysers, 204
Wet blue to finished leather, 299
White (industrial) biotechnology, 244
Widening the scope of green technologies through chemical engineering, 108
Wineries, 490

X

X-ray fluorescence, 589

Z

Zero discharge processes, 116
Zero discharge technologies, 162